The Canadian Atlas

OUR NATION, ENVIRONMENT AND PEOPLE

DOUGLAS & McINTYRE
VANCOUVER / TORONTO

ACKNOWLEDGMENTS

Reader's Digest Canada / Canadian Geographic
The Canadian Atlas

THE READER'S DIGEST ASSOCIATION (CANADA) LTD.	CANADIAN GEOGRAPHIC ENTERPRISES
PRESIDENT AND CEO: Pierre Dion	CEO AND PUBLISHER: John L. Thomson
VICE PRESIDENT, BOOKS & HOME ENTERTAINMENT: Deirdre Gilbert	EDITOR: Rick Boychuk
ART DIRECTOR: John McGuffie	ART DIRECTOR: Stephen Hanks, Design Matters

PROJECT TEAM

PROJECT EDITOR: Andrew R. Byers	MANAGING EDITOR: Eric Harris
DESIGNER: Andrée Payette	CARTOGRAPHER: Steven Fick
COPY EDITOR: Gilles Humbert	CONTRIBUTING EDITOR: Wendy Simpson-Lewis
PRODUCTION COORDINATOR: Susan Wong	
ADMINISTRATOR: Elizabeth Eastman	

THE CANADIAN ATLAS ADVISORY BOARD
Dr. Mark Graham; Bruce Amos; Jean-Marie Beaulieu;
Helen Kerfoot; Dr. Brian S. Osborne; Peter Puxley;
Dr. Frank Quinn; Dr. Joan Schwartz

CONSULTANTS: Ed Cowell, Richard Loreto
CONTRIBUTING EDITORS: Robert Ronald, Philomena Rutherford
RESEARCHER: Martha Plaine
PICTURE RESEARCHER: Rachel Irwin
COPY EDITOR: Judy Yelon
IMAGE TREATMENT: Yves Lachance

MAPS

MAPMEDIA Corp.
MAPS PRODUCED AND COPYRIGHT BY: MAPmedia
© MAPmedia Corporation, Toronto, Ontario

SATELLITE IMAGES

WorldSat
© WorldSat International Inc. 2004, All Rights Reserved

Published in Canada in 2004 by
THE READER'S DIGEST ASSOCIATION (CANADA) LTD., 1125 Stanley Street, Montréal, Quebec H3B 5H5
CANADIAN GEOGRAPHIC, 39 McArthur Ave., Ottawa, Ontario K1L 8L7
DOUGLAS & MCINTYRE LTD., Suite 201, 2323 Quebec Street, Vancouver, British Columbia V5T 4S7

For information about this and other products, please contact these 24-hour Customer Service hotlines:
Reader's Digest at 1-800-465-0780.
Canadian Geographic at 1-800-267-0824.
Douglas & McIntyre at 1-800-387-0117.

You can also visit these websites:
Reader's Digest at www.rd.ca.
Canadian Geographic at www.canadiangeographic.ca.
Douglas & McIntyre at www.douglas-mcintyre.com.

National Library of Canada Cataloguing in Publication Data
The Canadian atlas [cartographic material]: our nation, environment and people.

Includes index.
ISBN 0-88850-770-4 (Reader's Digest)   ISBN 1-55365-082-4 (Douglas & McIntyre)
1. Canada—Maps.   2. Canada—Gazetteers.
I. Reader's Digest Association (Canada).
G1115.C37   2004   912.71   C2004-901790-X

Douglas & McIntyre gratefully acknowledges the financial support of the Government of Canada
through the Book Publishing Industry Development Program (BPIDP) for its publishing activities.

Printed in Canada
04 05 06 07 / 5 4 3 2 1

FOREWORD

The Canadian Atlas encapsulates the complexity and majesty of our vast nation in the early years of the 21st century. This new and comprehensive visual reference work is the happy result of a collaborative effort between two of Canada's major publishers, Reader's Digest and Canadian Geographic. This atlas blends text, maps, images, charts, and tables in a way that provides informed understanding of all things Canadian. It offers a superb collection of up-to-date maps of virtually every part of the country, plus information-packed thematic spreads on issues of timely relevance and concern, with particular emphasis on sustainability. These illuminate aspects of many essential matters: the diversity of our natural regions, the abundance of our resource riches, our remarkable human past, the startling changes in our social values, and the dynamic drive of our ever-expanding urban centres. But *The Canadian Atlas* is more than just an inventory of national possessions and trends. It goes beyond the boundaries of the traditional atlas to explore many of the challenges that Canada faces today and tomorrow. True to this editorial vision, Reader's Digest and Canadian Geographic have created a truly unique reference work of matchless scope and insight, which they jointly present as a gift to Canada.

The Editors

Table of Contents

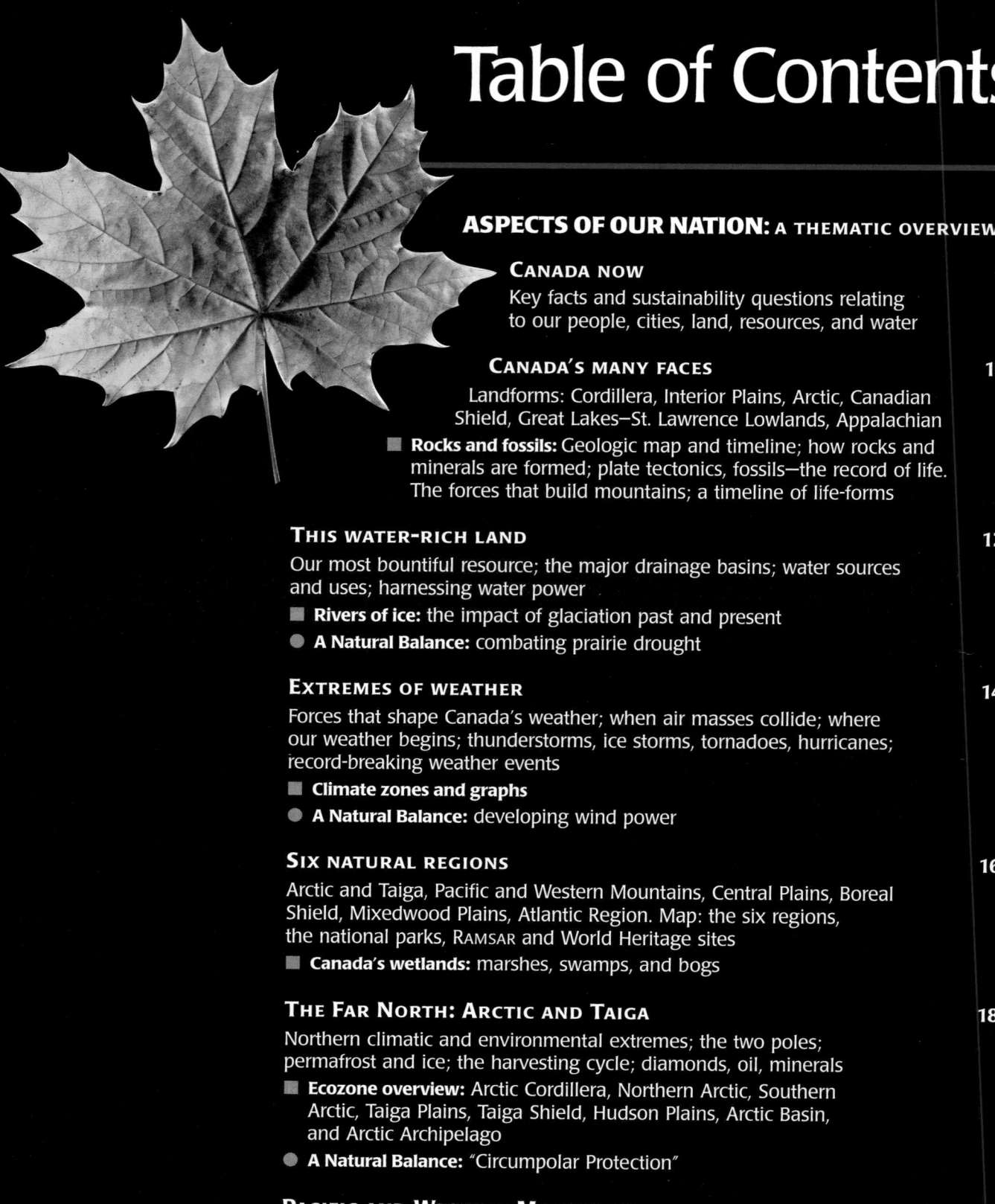

ASPECTS OF OUR NATION: A THEMATIC OVERVIEW

CANADA NOW   8

Key facts and sustainability questions relating
to our people, cities, land, resources, and water

CANADA'S MANY FACES   10

Landforms: Cordillera, Interior Plains, Arctic, Canadian
Shield, Great Lakes–St. Lawrence Lowlands, Appalachian
■ **Rocks and fossils:** Geologic map and timeline; how rocks and
minerals are formed; plate tectonics, fossils–the record of life.
The forces that build mountains; a timeline of life-forms

THIS WATER-RICH LAND   12

Our most bountiful resource; the major drainage basins; water sources
and uses; harnessing water power
■ **Rivers of ice:** the impact of glaciation past and present
● **A Natural Balance:** combating prairie drought

EXTREMES OF WEATHER   14

Forces that shape Canada's weather; when air masses collide; where
our weather begins; thunderstorms, ice storms, tornadoes, hurricanes;
record-breaking weather events
■ **Climate zones and graphs**
● **A Natural Balance:** developing wind power

SIX NATURAL REGIONS   16

Arctic and Taiga, Pacific and Western Mountains, Central Plains, Boreal
Shield, Mixedwood Plains, Atlantic Region. Map: the six regions,
the national parks, RAMSAR and World Heritage sites
■ **Canada's wetlands:** marshes, swamps, and bogs

THE FAR NORTH: ARCTIC AND TAIGA   18

Northern climatic and environmental extremes; the two poles;
permafrost and ice; the harvesting cycle; diamonds, oil, minerals
■ **Ecozone overview:** Arctic Cordillera, Northern Arctic, Southern
Arctic, Taiga Plains, Taiga Shield, Hudson Plains, Arctic Basin,
and Arctic Archipelago
● **A Natural Balance:** "Circumpolar Protection"

PACIFIC AND WESTERN MOUNTAINS   20

A scenic, resource-rich region; mountain zones; how the "rain shadow"
affects the weather; fisheries, farms, and forests of British Columbia
● **A Natural Balance:** "Green Getaways"–the ecotourism approach
■ **Ecozone overview:** Pacific Maritime, Montane Cordillera,
Boreal Cordillera, Taiga Cordillera, and Pacific Marine

CENTRAL PLAINS 22

A region of plains and boreal forests; rural depopulation; prairie crops; resource riches beneath the Prairies—potash, oil and natural gas production, the Alberta tar sands

● **A Natural Balance:** "The Pelican Project"

■ **Ecozone overview:** the Prairies—grassland and parkland; the Dry and Fertile Belts; Boreal Plains

BOREAL SHIELD 24

Canada's largest ecozone; mining and forest production; the impact of acid precipitation; clear-cutting concerns

● **A Natural Balance:** "The Greening of Sudbury"— reclaiming land devastated by industry and mining

■ **Ecozone overview:** natural hazards—fires and pests

MIXEDWOOD PLAINS 26

Canada's populous industrial powerhouse; the Great Lakes and the St. Lawrence Seaway; what the farms produce; where the people live; what the cities produce

● **A Natural Balance:** "Protecting the Oak Ridges Moraine"— safeguarding a significant natural area

■ **Ecozone overview**

ATLANTIC REGION 28

A region of fisheries, farms, and forests; the collapse of the cod fishery; the Hibernia oil project

● **A Natural Balance:** "A Model Forest"—ensuring the sustainability of Canada's forests

■ **Ecozone overview:** Atlantic Maritime, Atlantic and Northwest marine ecozones

FIRST PEOPLES 30

A diversity of aboriginal cultures—peoples of the Arctic, Subarctic, Pacific Northwest, Plateau, Plains, Eastern Woodlands; the arrival of the first peoples

■ **Aboriginals Today:** land treaties past and present; the growing population

▲ **Timeline:** 40,000 B.C.–A.D. 1400

SETTLING CANADA 32

Major explorers; the human tide—Vikings, Portuguese; French; Loyalists; Red River settlers; Irish; Asians, Eastern Europeans; interprovincial migration

■ **The New Mosaic:** how immigration is changing the fabric of Canadian society; visible minorities

▲ **Timeline:** 1487–1857

BUILDING CANADA 34

The nation's territorial evolution; the results of Canadian federal elections, 1867–2000; where bills become laws; governing Canada

■ **Provinces and Territories:** capitals, populations, areas

▲ **Timeline:** 1867–1999

CONNECTING CANADA 36

Communications: the Internet, satellites, cellular superlatives Transportation networks: highways and byways, waterways, railways, and airways

WHO WE ARE 38

A social profile of Canadians: births, languages, religions, education, marriage and divorce, family life, work and income, health, life expectancy

● **A Natural Balance:** "Caring for an Aging Population"

■ **Changing population patterns**

WHERE WE LIVE 40

Urban Canada; census metropolitan regions; city/regions; how to measure urban innovation; rural decline; suburban expansion; comparing amenities in five Canadian centres

● **A Natural Balance:** "Revitalizing Canada's Cities"

■ **Metro Canada:** populations, areas, densities

CANADA 2050 42

Future challenges: population, global warming, lifestyle and work; health; manufacturing, farming, and energy

MAPS OF CANADA 44

Relief map 46
Key map 48

CITY MAPS 134

MAP INDEX 150

Thematic subjects index 152

CREDITS 192

ASPECTS OF
THE NATION

Below: *A saltwater marsh where
the Fox River empties into Minas Channel,
just west of Parrsboro, New Brunswick. Visible
on the far shore is Cape Blomidon, Nova Scotia.*

Canada now

The challenge for Canada in the 21st century is no longer about wresting natural resources from a harsh and unpredictable environment. Today, Canada faces a different challenge—that of seeking "sustainability." This challenge embraces two main objectives: preserving resources while creating a buoyant long-lasting economy. With more area than any other country in the world except Russia, Canada is generously, if unevenly, endowed with raw materials. This resource base supports today's highly diversified Canadian economy, firmly centred in services and manufacturing, and the expanding urban scene where eight out of ten Canadians live and work. Canada is a country of complex interconnections, where the abuse of resources—the rampant exploitation of fisheries, forests, and mines, for example—may have serious widespread consequences. Achieving sustainability involves balancing our competing environmental, economic, and social concerns, some of which are surveyed on these pages. How such concerns will be resolved in the future is difficult to discern. When decisions are taken, however, the well-being of Canadians is of paramount importance. Canada ranks as one of the world's most desirable countries in which to live. Thinking of our concerns in the context of sustainability may ensure that it remains so.

KEY FACTS / SUSTAINABILITY QUESTIONS

Key facts about Canada appear to the right.
Sustainability questions direct readers to pages where *The Canadian Atlas* explores questions about present and future social and environmental issues

THE LAND AND THE PEOPLE

Key facts: At 9,984,670 km², Canada is the world's second largest country. But, with 31,752,842 people (in 2004), it has only 0.5 percent of the world's 6 billion. By world standards, Canada's growth rate—1.06 percent a year—is low. Moreover, its population is aging. According to the 2001 census, the average median age was 37.6 years—up from 35.3 years in 1996.
Sustainability questions: Will population growth continue to slow? Given Canada's aging population, can we sustain health care, pensions, and other benefits? Will immigration reverse this trend? See pages 30–31, 33, 42–43

CITIES

Key facts: Urban Canada is driving the country's affluent high-tech post-industrial economy. From 1996 to 2001, the nation's largest centres—Toronto, Montréal, Vancouver, and the Calgary-Edmonton corridor—grew 7.6 percent and seem set to expand indefinitely.
Sustainability questions: Can Canada's cities, currently underfunded, be revitalized? What steps have been taken to curb urban sprawl? What will urban Canada look like 50 years from now? See pages 40–41, 42–43

CLIMATE AND WEATHER

Key facts: Global warming's impact is already evident in Canada. In the Northwest Territories, warmer-than-usual weather is melting the permafrost and drier conditions are causing more forest fires. Other changes—for example, coastal flooding and droughts—are predicted. But experts are divided on global warming's eventual outcome.
Sustainability questions: How rapidly will global warming occur? Is it possible that this climate trend will be slower than expected? See pages 14–15, 42–43

ENERGY

Key facts: Canada abounds in energy sources—oil, natural gas, coal, hydroelectric, and nuclear. But it also uses more energy than most other countries, ranking sixth among the world's energy consumers. Demand has risen steadily since the 1960s and is expected to continue unabated, increasing pollution levels and heightening global warming.
Sustainability questions: Can Canada sustain a high level of energy consumption? In future, will we choose to use existing energy sources more efficiently, or will we turn to alternative energy sources? See pages 12–13, 22–23, 42–43

FARMS

Key facts: Seven percent of Canada's land area—roughly 680,000 km²—is used for agriculture. Dominating this sector are several trends: a shift toward more diverse crop outputs, a decline in the number of farms, and rural depopulation.
Sustainability questions: How have Canada's various agricultural regions responded to these trends? Will technology and new agricultural methods change the farms of the future? See pages 21, 22–23, 27, 29, 40, 42

FISHERIES

Key facts: Overfishing—the result of escalating world and domestic demand—reduced Atlantic and Pacific species. Canada's fisheries remain in constant flux—sometimes for the worse, sometimes for the better. From 1990 to 2001, the tonnage of fish caught in all fisheries fell 37 percent, while the value of the catch rose by 43 percent.

Sustainability questions: Will there be future declines in fishing stocks? How can such a scenario be avoided? See pages 21, 28, 42–43

FORESTS

Key facts: Canada's forests cover 4.2 million square kilometres—45 percent of the land area—and shelter two-thirds of the country's bird and animal species. In 2001, the timber-productive forests consisted of 2.45 million square kilometres. About a million hectares—only 0.4 percent of this area—is harvested annually. Since 1978, Canada's forest-product exports have increased almost fourfold, reaching $33.7 billion in 2002.

Sustainability questions: Will the profitable harvesting of Canada's productive forests destroy or seriously disrupt wildlife habitats? What steps are being taken to ensure Canada's forests remain sustainable? See pages 21, 25, 29

MINES

Key facts: Canada is a leading producer of potash, uranium, zinc, nickel, copper, platinum, cobalt, silver, and gold. Diamond production in the Far North is the latest Canadian mining success story. Output from new discoveries at Voiseys Bay in Labrador may some day rival those of the mineral-rich Sudbury region.

Sustainability questions: What have been the social and environmental consequences for mine-dependent communities when deposits run out or cease to be profitable? Will future technology disclose new extraction venues? See pages 16–17, 20–21, 23, 27

WATER

Key facts: Canada's freshwater is a renewable resource that most of the world would like to have. The average Canadian household, assured of a safe water supply, uses up 340 litres a day—an extravagant level when compared with other countries. In future, limiting water consumption may become imperative. Some experts predict that water will be "the oil of the 21st century." The export of Canada's water is already raising environmental concerns.

Sustainability questions: How well distributed is Canada's water supply? What is being done to save Canada's wetlands, a major natural asset? See pages 12–13, 16, 42–43

THE ECOZONE APPROACH

The Canadian Atlas adopts an "ecozone approach" to describe the complexity of the natural environment. The prefix "eco," from the Greek "oikos" for "home" or "household," is a reminder that ecozones are where we live—our home. Canada possesses 20 major ecozones: 15 terrestrial (land) and 5 marine (oceanic) ecozones, arranged under the headings of six natural regions (see pages 16 to 29). This approach provides a unique and timely perspective from which to explore some of the interconnections and relationships that exist between environmental and economic conditions. Sidebars give overviews of each ecozone's geographical and climatic characteristics, some of its plant and animal species, and human activities. "Natural Balance" features, identified by the globe (*below*), pinpoint regional sustainability success stories.

- ■ Precambrian
- ■ Paleozoic
- ■ Mesozoic
- ■ Cenozoic
- ■ Mixed Paleozoic, Mesozoic, and Cenozoic

Geologists break down the earth's long history into comprehensible intervals of time. The longest intervals—eons—are divided into eras (Paleozoic, Mesozoic, Cenozoic), which are themselves divided into periods such as the Quaternary (*see timeline below*). The periods are further subdivided into epochs such as the Holocene—our own time. The geologic map of Canada (*above*) locates the underlying rock formations dating from the four major eras; the timeline identifies present-day landform regions supported by these formations.

TIMELINE: GEOLOGY

Era	Time (millions of years ago)	Landform regions
Precambrian	4,600 to 600	Canadian Shield, Arctic (Baffin Island)
Paleozoic	600 to 250	Interior Plains (northern), Canadian Shield (Hudson Bay Lowlands), Great Lakes–St. Lawrence Lowlands, Appalachian, Arctic, and Innuitian
Mesozoic	250 to 65	Interior Plains (southern)
Cenozoic Quaternary*	65 to today	Cordillera Arctic (Mackenzie River delta), Cordillera (Fraser River delta)

* The Quaternary period includes the most recent epochs: the Pleistocene, or Ice Age, and Holocene (the present time)

ROCKS AND MINERALS

A jumble of ceaselessly recycled igneous, sedimentary, and metamorphic rocks makes up the earth's surface. Igneous rock (granite, gneiss) forms when molten rock cools. The Canadian Shield's igneous base abounds in copper, gold, iron, and nickel. Sedimentary rock (sandstone, limestone) consists of hardened layers of rock particles. Beneath the Prairies, sedimentary layers store oil, gas, and coal. Mineral-rich metamorphic rock, baked or compressed igneous or sedimentary rock, shows up in the Shield and Cordillera.

PLATE TECTONICS

Canada rides on one of 30 or so tectonic plates, or slabs, which underlie the earth's continents and oceans. The plates—each tens of kilometres thick—move in different directions, possibly in response to currents of molten material below the earth's crust. When plates collide, earthquakes and volcanic activity occur. In geological time, collisions uplift mountain chains. (See *Building Mountains,* facing page.) Canada's west coast is situated in one of the world's most active collision zones, where the westbound North American plate overrides the Pacific plate.

FOSSILS

The record of life is found in fossils, plant and animal remains preserved in rock. Yet some 90 percent of the record is missing. Geological forces have erased virtually all fossils from the Cryptozoic eon (Greek for "hidden life")—the earth's first 4 billion years. The Phanerozoic eon ("evident life") dawns with Paleozoic era, which left behind fossils revealing an explosion of life forms. One of the world's important Paleozoic finds is the Burgess Shale Site in Yoho National Park. (See *Timeline: Life Forms,* facing page.)

Canada's many faces

Viewed by satellite, the face of Canada reveals six clearly defined landform regions: Cordillera, Interior Plains, Canadian Shield, Great Lakes–St. Lawrence Lowlands, Appalachian, and Arctic Lands. All these regions occupy significant portions of Canada's vast expanse. Each possesses similar geologic structures, physical features, climatic conditions, soils, and vegetation. Considered as a whole, Canada's landforms encompass an unrivaled diversity of landscapes: spectacular mountain ranges, sweeping plains, rocky uplands, temperate lowlands, and frigid tundra. The forces of nature—our daily weather, for example—actively shape landforms. Some forces level landforms, others rebuild them. Over time, the impact of water, ice, and wind slowly and steadily reduce the mighty Rockies to rubble, while rivers bear away sediments to deltas and seabeds where new landforms wait to be born.

CORDILLERA

The towering peaks and plateaus of the Cordillera took shape millions of years ago, when the westward moving North American tectonic plate collided with the Pacific plate. The collision uplifted mountain chains in British Columbia and the Yukon. In the last 20,000 years, weathering, erosion, and glaciation have carved the sharp relief of Rocky Mountain peaks such as Rampart Mountain in Banff National Park (*left*). The westward movement of the North American plate continues unabated, exposing British Columbia's populous coastal lowlands to an ever-present threat of earthquakes.

CANADA'S LANDFORM REGIONS

1. Cordillera
2. Interior Plains
3. Canadian Shield
4. Great Lakes–St. Lawrence Lowlands
5. Appalachian
6. Arctic Lands

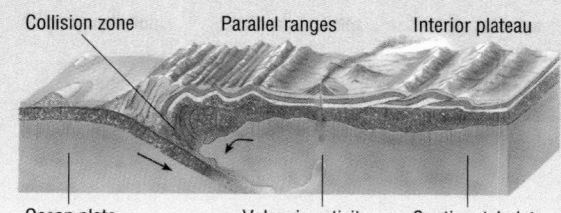

Collision zone　　Parallel ranges　　Interior plateau

Ocean plate　　Volcanic activity　　Continental plate

BUILDING MOUNTAINS

Mountains are often found along coastlines where continental and ocean plates meet. The collision sets off a process in which the ocean plate slides underneath the continental plate, buckling it and uplifting mountains. As the surface buckles, volcanoes force up fiery rock in fractures (faults) and wavy folds.

Mountain building is measured in millions of years. Some 250 million years ago, the collision of the North American plate with Europe and Africa built up the Appalachian and Laurentian mountains. Canada's east coast is relatively calm now, but its west coast lies on the Pacific's earthquake-prone "ring of fire." The North American plate began creeping westward 80 to 40 million years ago, eventually colliding with the Pacific plate. The collision compressed sedimentary rocks, thrusting up highly faulted and folded parallel chains from the Rockies to the Coast Mountains. It also created British Columbia's Interior Plateau, now the eroded remnant of another ancient range. In the Yukon's St. Elias Mountains, uplift continues, pushing up Canada's youngest, loftiest peaks, including its highest—5,959-m Mount Logan.

TIMELINE: LIFE FORMS

3,500 TO 600 MILLION YEARS AGO

PRECAMBRIAN The first life—algae, bacteria, and soft-bellied marine creatures—develops in the depths of primeval seas more than 3 billion years ago. Canada's oldest known evidence of life: 2.5-billion-year-old algae fossils, discovered at Steep Rock Mine, Ont.

600 TO 250 MILLION YEARS AGO

PALEOZOIC (Greek for "ancient life") The age of fishes begins about 400 million years ago. The first plants, descended from algae, have already taken root on land. As vegetation spreads across the bare rock, insects flourish in conifer forests. By the late Paleozoic era, amphibians and reptiles invade the land.

250 TO 65 MILLION YEARS AGO

MESOZOIC ("middle life") Reptiles, particularly dinosaurs, dominate this era. Broad-leaved trees and flowering plants replace conifers. Birds take wing, and small mammals appear. According to a prevailing view, dinosaurs die out in a global environmental disaster triggered by the impact of a huge asteroid.

65 MILLION YEARS AGO TO THE PRESENT

CENOZOIC ("recent life") The demise of the dinosaurs clears the way for a proliferation of mammals: horses, whales, and large carnivores. Grasslands develop. The Quaternary, starting about 2 million years ago, sees the arrival of the first known humans.

INTERIOR PLAINS

About 500 million years ago, shallow seas covered the Interior Plains. Rivers flowing into these waters deposited sediments, which were transformed into layer upon layer of sedimentary rock. In the southern part of the Interior Plains lie grasslands. The sedimentary materials here provide fertile soils for the patchwork of prairie farms. In the northern part, aspen parkland dwindles into sparsely treed taiga. Over time, weathering and erosion have cut deeply into the soft rock in parts of Alberta and Saskatchewan. At Horseshoe Canyon (left), near Drumheller, Alta., these processes have exposed multicolored layers from the Mesozoic era when dinosaurs roamed this region.

ARCTIC LANDS

Covering roughly 25 percent of Canada's landmass, the Arctic Lands embrace two distinct regions: Arctic and Innuitian. The land, permanently frozen to varying depths, supports only a thin surface mat of vegetation, which peters out completely in the high polar latitudes. The Arctic region includes the Yukon's narrow coastal plain, Banks and Victoria islands, and Baffin Island's lowlands. The Innuitian region runs along Baffin Island's rugged eastern upland, where sheer cliffs on Borden Peninsula overlook the ice-clogged waters of Lancaster Sound (left). This upland extends into Ellesmere Island, home of ice-capped mountain ranges, where elevations reach 1,500 m and higher. The island's loftiest pinnacle is 2,616-m-high Barbeau Peak.

CANADIAN SHIELD

Lion's Head (left), a wave-lashed arch of stone on the Lake Superior shores of Ontario's Sleeping Giant Provincial Park, exemplifies the Canadian Shield's rugged beauty. This region of rocks, lakes, and forests, also known as the Precambrian Shield, occupies more than half of Canada. Its enduring bedrock provides the geological foundation for adjacent regions, such as the Interior Plains. The capacious Shield includes the Hudson Bay Lowlands, one of the world's largest wetlands, the Torngat Mountains—Eastern Canada's highest peaks—and the uplands of central Baffin Island. Some 3 billion years ago, the Shield was a land of huge mountains and volcanoes. In the last ice age, glaciers stripped away the region's surface, exposing the world's oldest bedrock just east of Great Slave Lake.

GREAT LAKES–ST. LAWRENCE LOWLANDS

Stretching from Windsor to the city of Québec, this narrow plain is Canada's smallest landform region, but by far the most populous. East of Kingston, the Thousand Islands—an intrusion of the Canadian Shield—divides southern Ontario's lowlands from the St. Lawrence River valley. Around lakes Erie and Ontario, the bedrock is sedimentary, visible in the limestone strata of the Niagara Escarpment (left). An overlay of glacial debris, deposited during the last ice age, created southern Ontario's flat to rolling terrain. Along the St. Lawrence, the retreating ice-age Champlain Sea left behind a fertile riverine plain.

APPALACHIAN

The Appalachian Mountains, this region's dominant feature, extend from Newfoundland into the eastern United States. Once higher than the Rockies, these uplands formed 500 million years ago. Over time, weathering eroded the peaks. Except for Prince Edward Island's fertile fields, the interior is a land of rounded hills and narrow river valleys. Bold headlands rear up in the Gaspé Peninsula, shown left near Percé, Que. Along Nova Scotia's coast, the bays have long served as fishing ports and harbours.

5

4

This water-rich land

C anada's freshwater lakes and rivers cover 755,000 km², a bountiful 7.6 percent portion of the country. This abundance of water, flowing through an interconnecting network of waterways, has spurred Canada's development. Water provided the early routes for exploration, transportation, settlement, and trade. It is still put to myriad uses: hydroelectric power, shipping, irrigation, fishing, and recreation. Yet any undue future demand may overextend this resource. Many worry our water-rich land may one day export some of its supply to the United States. An overview of Canada's water already reveals a resource unevenly distributed and under threat. Whether a given region has an abundance or shortage of water is due in part to variations in precipitation. The rainy Pacific Coast enjoys a copious flow of water to the sea, while parched prairie drylands struggle with water deficits. These disparities aside, the pressure from industry and shipping along the Great Lakes and elsewhere pose ever-present pollution concerns.

Glaciers are flowing rivers of land ice, which originate with the accumulation and compression of snow. In the arctic climate of Ellesmere Island (*above*), glaciers overlay the land as they once did when they covered all of Canada. During the last 2 million years, ice swept over Canada four times. The ice retreated during warmer periods, including our present time. At the zenith of the last advance, 18,000 years ago, the Laurentide ice sheet extended beyond the Canadian Shield to depths of 3 km. In the West, the Cordilleran ice sheet stretched from the Rockies to the Pacific. The slow but relentless movement of ice scoured the face of Canada, leaving behind landmarks such as drumlins, eskers, and moraines (*below*). The only remnant from the distant icy past is the 100,000-year-old Barnes Ice Cap on Baffin Island. Today, three-quarters of Canada's glaciers cloak parts of the eastern arctic islands. The remaining glaciers cap the alpine summits of the Rockies and the West Coast ranges. In all, Canada's glaciers cover 200,000 km²—roughly half the area of Newfoundland and Labrador.

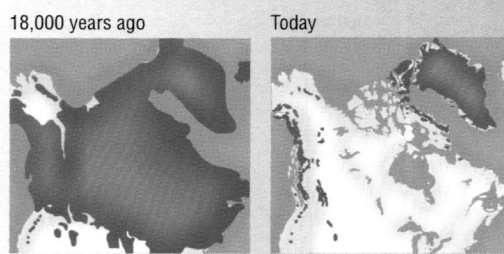

18,000 years ago   Today

DURING GLACIATION

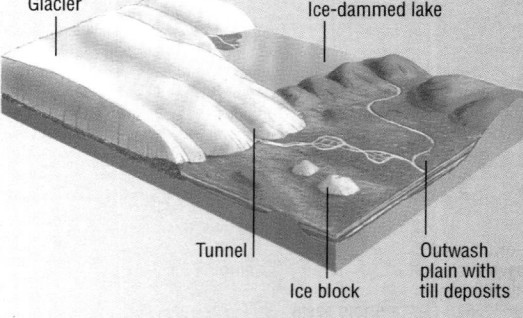

Glacier   Ice-dammed lake

Tunnel   Outwash plain with till deposits

Ice block

Some 12,000 years ago, meltwater of a glacier and an ice-dammed lake flowed from tunnels into streams, depositing mud, sand, and gravel. This debris, known as till, spread over the outwash plain where ice blocks fill steep-sided potholes.

AFTER GLACIATION

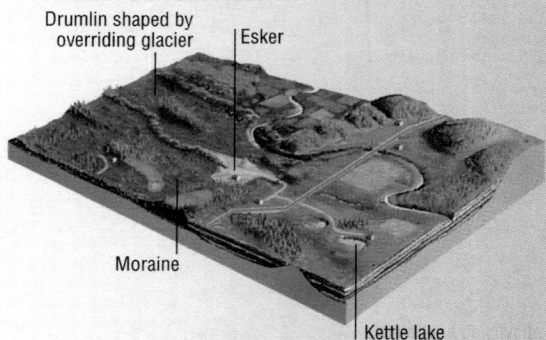

Drumlin shaped by overriding glacier   Esker

Moraine

Kettle lake

The glacial-lake bottom is now fertile land. Drumlins are hills composed of till; eskers, long or oval mounds that were the banks of the meltwater streams. Moraines mark the glacier's limit. Kettle lakes fill the potholes after the ice blocks melt.

Canada has five ocean drainage basins. The largest by far is the Hudson Bay basin; the smallest, situated in the southern Prairies, flows via the Missouri and Mississippi rivers into the Gulf of Mexico. Within each basin lie smaller subbasins. Drainage divides separate the basins. The best known of these is the Continental Divide of the Rockies, which separates rivers flowing westward to the Pacific from rivers flowing to the Arctic and Hudson Bay. The map at left locates Canada's five basins and main rivers, and lists figures for areas and streamflow to the sea.

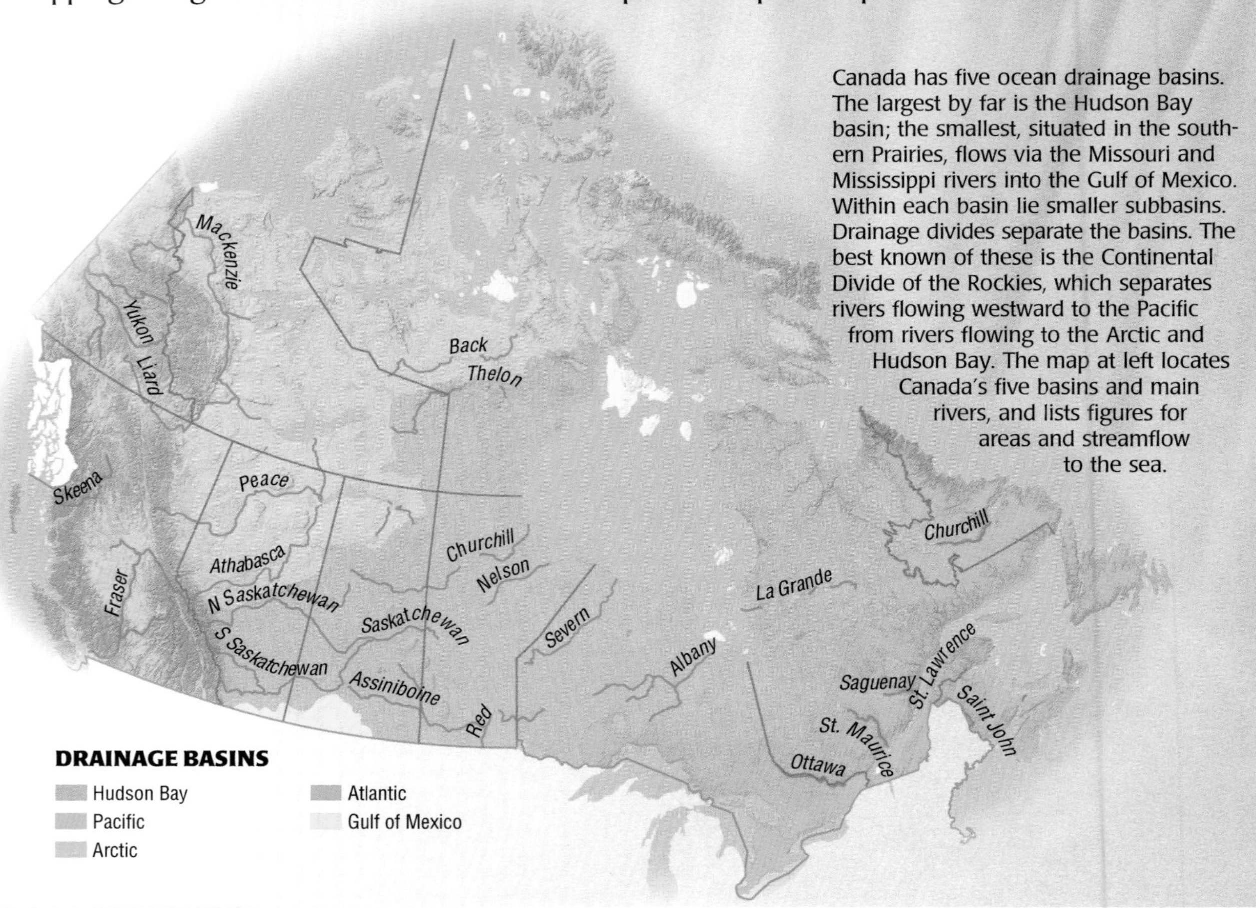

Mackenzie · Yukon · Liard · Back · Thelon · Skeena · Peace · Churchill · La Grande · Fraser · Athabasca · N Saskatchewan · S Saskatchewan · Saskatchewan · Nelson · Severn · Albany · Saguenay · St. Lawrence · Saint John · St. Maurice · Assiniboine · Red · Ottawa

DRAINAGE BASINS

- Hudson Bay
- Pacific
- Arctic
- Atlantic
- Gulf of Mexico

WATER SOURCES AND USES

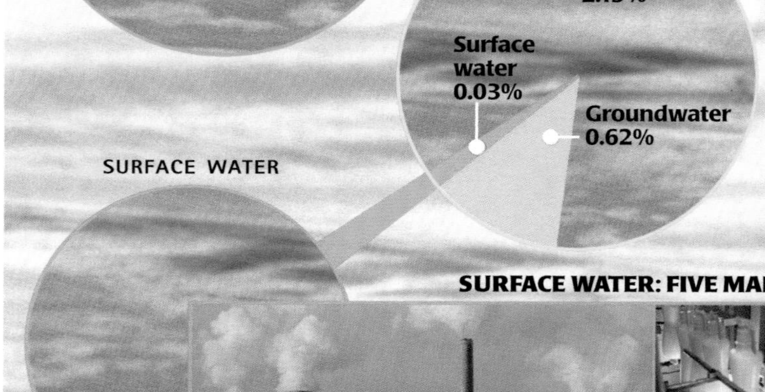

WORLD WATER SUPPLY

Salt water **97.2%**

Freshwater **2.8%**

SURFACE WATER

FRESHWATER

Ice **2.15%**

Surface water **0.03%**

Groundwater **0.62%**

Whereas instream, or natural, water uses in Canada include hydroelectric power generation, shipping, and fisheries, withdrawal uses involve less than 3 percent of our freshwater, used for domestic and industrial purposes and irrigation. Most of Canada draws its freshwater from rivers and lakes, but some smaller communities and farms depend on groundwater. The chart below identifies the major withdrawal users. Thermal and nuclear power plants, which use water to cool condensers and to drive generators, withdraw more than all the other users combined. Industry reuses the same water twice; mining, more than twice. Annual withdrawals involve 1.3 percent of Canada's water supply. The water returned to natural sources is less that the amount withdrawn, and it has usually been degraded through use. Agriculture returns the least water to natural sources.

SURFACE WATER: FIVE MAIN WITHDRAWAL USES

| Thermal power generation 64% | Manufacturing 14% | Municipal 12% | Agriculture 9% | Mining 1% |

A NATURAL BALANCE

AT-A-GLANCE FACTS: WATER FRESH AND FROZEN

● The world's water supply measures 1.36 billion cubic kilometres. Water, as liquid, ice, or vapour in the atmosphere, constantly circulates from the oceans to the land, and back again

● 97.2% of the world's water is found in oceans, but salty oceanic water is unsuitable for drinking or farming

● 2.8% of the world's water supply is fresh, but 2.15% is locked up, often for centuries, in ice sheets and glaciers

● 0.62% of the world's freshwater is groundwater; only 0.03% exists in lakes and rivers, and in the soil

● With 7% of the world's renewable freshwater, Canada ranks fourth among the league of water-rich nations, after Brazil (18%), China (9%), and the United States (8%)

● 755,000 km² of Canada is covered by freshwater lakes and rivers

● Roughly 60% of all Canada's rivers and streams flow northward

● One in three Canadians depend on the Great Lakes for water

● In glacial coverage, Canada ranks third in the world, after Antarctica and Greenland

● Canada has 25% of the world's wetlands, which occupy 16% of its area

● Two-thirds of Canada's irrigated farmland is found in Alberta

COMBATING DROUGHT

Most severe droughts occur in regions where farms depend on a limited water supply. One such region is the Prairie Dry Belt (*see map below*), where annual precipitation varies from less than 400 mm in southern Alberta to a high of 550 mm in southwestern Manitoba. Recent longer-than-usual summers and higher evaporation rates in this region have dried up soil moisture. It has revived fears of drought, dust storms, and crop failures that occurred in the 1930s. Some experts speculate the dry spell may be a foretaste of global warming; others see it as part of the recurrent cycle of wet and dry years. The dry years of the 1930s caused a drought that led to extensive topsoil erosion and drove 225,000 farm dwellers off the land. In response to this disaster, government drought-alleviation programs, such as the Prairie Farm Rehabilitation Administration (PFRA), were successfully introduced. Since the 1950s, PFRA-sponsored irrigation projects have revitalized the dry belt. Saskatchewan's Lake Diefenbaker irrigates 21,000 ha of farmland. Alberta's St. Mary and Waterton dams and reservoirs serve the 132,000-ha St. Mary Irrigation District, Canada's largest. The 466,000 ha of irrigated land in southern Alberta represents only 4 percent of the province's arable land. Yet this land supports 5,800 irrigation farms in a region where only 1,000 dryland wheat farms and ranches might normally thrive on existing water supplies. Moreover, these farms produce 18 percent of Alberta's agricultural output.

AB   SK   MB

Edmonton
Calgary   • Lloydminster
FERTILE BELT

Winnipeg
DRY BELT   • Regina
• Lethbridge

Niagara Falls is the world's greatest waterfall by volume. Canada's Horseshoe Falls (above), largest of the fall's two cataracts, is 670 m wide, 54 m high, with a flow of 155,000 liters per minute. Its power was first tapped in 1893 by an electric tramway company. Sir Adam Beck No. 1 and No. 2 power stations, opened in 1922 and 1954 respectively, were built downstream from the falls to fully exploit its potential. Today's torrent is spectacular but less so than it once was. A 1950 Canada–United States treaty reduced the visible flow to a minimum, setting aside the rest for the hydroelectric production.

HARNESSING THE WATERS

More than 60 percent of Canada's energy needs are supplied by water power. Several factors can influence a region's hydroelectric development: high precipitation, sloping landforms, and the proximity to markets. The Atlantic drainage basin was the site of the first hydroelectric power plants, built in the early 1900s at Niagara Falls, Trois-Rivières, and Shawinigan. But most hydroelectric production is found within the Canadian Shield area of the Hudson Bay basin. This area offers ideal conditions: Canada's highest streamflow, sudden drops in land elevation, and mighty rivers. The rivers of the Pacific basin, with the second highest streamflow (due to high precipitation along the coast), have also been extensively harnessed.

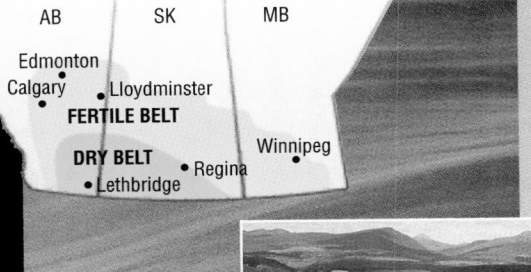

By the 1960s, virtually all the power sites close to markets had been developed. Since then, major hydroelectric power plants have been situated at remote northern sites such as Quebec's Manicouagan River (*left*). Canada's largest hydroelectric generating stations and their capacity include: LG-2 on La Grande Rivière, Que. (5,328 MW); Churchill Falls on the Churchill River, Nfld. (5,225 MW); and Gordon M. Shrum on the Peace River, B.C. (2,416 MW), the arctic drainage basin's only major hydroelectric development. To date, only 40 percent of Canada's hydroelectric potential has been realized.

Extremes of weather

A combination of the sun's heat and the earth's rotation powers the movement of winds carrying weather systems across the globe—and across the length and breadth of Canada. The temperature differences between the poles and the equator provide the energy that drives atmospheric circulation: low-pressure polar air sinks and migrates to the equator while high-pressure warm air from the tropics rises and moves to the poles. This global circulation system is the mechanism that drives Canada's daily weather. In low-pressure areas, warm air rises and cools, forming clouds, which bring rain, fog, snow, hail, and thunderstorms. In high-pressure areas, cold air descends and, as it falls, it is compressed and heats up, generally bringing clear, warm, and settled weather. Canada's vast size and varied landscape also influence day-to-day conditions, some of which include the record-breaking weather extremes described below.

Climate is a region's weather over a long period. Within its vast expanse, Canada embraces seven climatic zones (*see map above*). Each of the zones depends on conditions such as proximity to large bodies of water, altitude, and latitude. Southern Ontario enjoys warm, humid summers and short, cold winters, because of the moderating influence of the Great Lakes. The mountainous interior of British Columbia and the Yukon Territory support glaciers on the summits and semideserts in the valleys. Latitude—the distance north or south of the equator—influences whether a climate is cold or hot. The midlatitude Prairies experience continental extremes: cold winters and hot summers. The high-latitude Arctic endures intensely dry and frigid conditions. The largest zone by far is the Subarctic, which knows short, cool summers and long, cold winters, and low precipitation. All of southern Canada is classified as "temperate"— that is, it has four seasons. Winter touches all zones, save the Pacific, where warm winds promote a mild, rainy climate year-round. By contrast, the prevailing west-to-east winds moving across central Canada bring cool, humid summers and short, cool winters to Atlantic Canada.

Climate zones map legend:
- Pacific
- Cordillera
- Prairie
- Great Lakes–St. Lawrence Lowlands
- Atlantic
- Subarctic
- Arctic

CLIMATE GRAPHS

These show average temperatures and average snowfall and rainfall for each climatic zone. The growing season typically begins above 5°C.

- Average monthly temperature
- Average snowfall
- Average rainfall

RESOLUTE
Annual precipitation: 131 mm

WINNIPEG
Annual precipitation: 526 mm

DAWSON
Annual precipitation: 306 mm

TORONTO
Annual precipitation: 762 mm

VANCOUVER
Annual precipitation: 1,113 mm

QUÉBEC
Annual precipitation: 1,174 mm

MEDICINE HAT
Annual precipitation: 348 mm

HALIFAX
Annual precipitation: 1,282 mm

WHEN AIR MASSES COLLIDE

In a cold front, the leading edge of an advancing cold air mass meets less dense warm air and forces it up sharply like a blade, causing instability. Typically associated with low-pressure weather systems, cold fronts develop rapidly, often producing large cumulus and cumulonimbus clouds and triggering heavy rain and thunderstorms (*below*). Rainfall and winds are heaviest along the front.

In a warm front, the leading edge of a mass of warm air meets a stationary, cold air mass and gradually rises above it along a slope that can stretch for hundreds of kilometres. As the warm air rises, it cools, forming cirrus clouds. If higher clouds form, condensation will follow, causing widespread precipitation accompanied by strong winds.

COLD FRONT

WARM FRONT

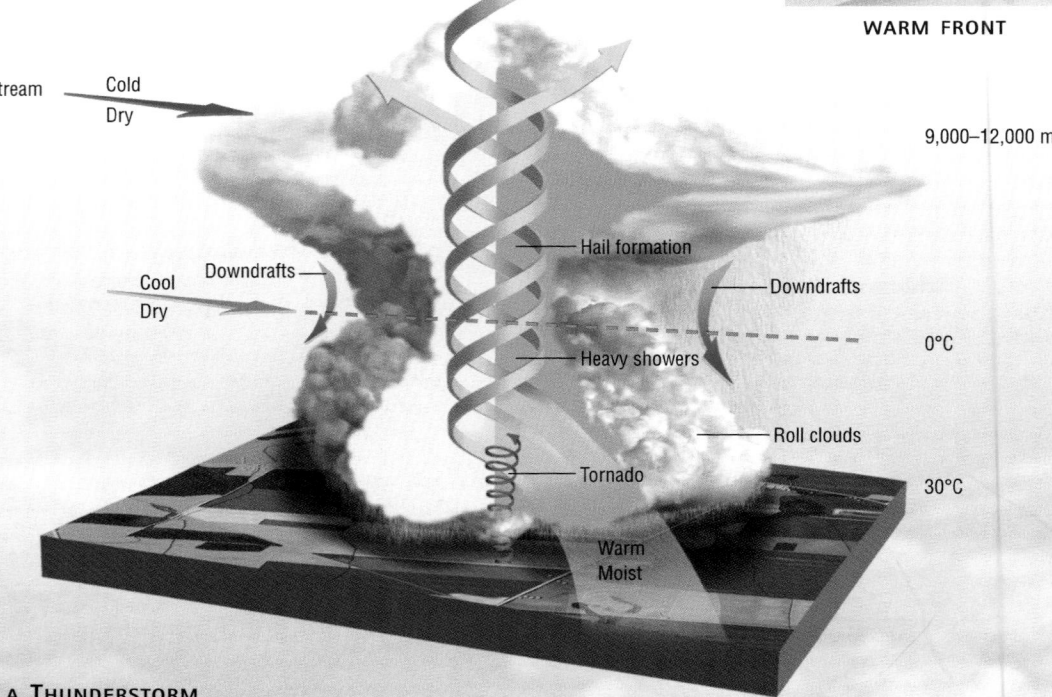

Jet stream · Cold Dry · Cool Dry · Downdrafts · Hail formation · Downdrafts · Heavy showers · Roll clouds · Tornado · Warm Moist · 9,000–12,000 m · 0°C · 30°C

ANATOMY OF A THUNDERSTORM

Thunderstorms begin when a parcel of warm, moist air begins to rise. As the air expands and cools, the water vapour within it condenses and forms a cloud. If there is sufficient atmospheric instability, the heat released by condensation will keep the air inside the cloud warmer than the air surrounding it, enabling it to grow larger and higher. The power of the rising air, or updraft, keeps millions of water droplets in suspension until they become so heavy they fall as rain. Above the freezing line (typically 12,000 to 15,000 m above the ground in summer), the droplets form supercooled ice crystals that can grow into hailstones. When the thundercloud reaches the cumulonimbus stage and hits the tropopause (where temperature stops decreasing with height, around 12,000 m in summer), the jet stream tugs the cloud into its famous "anvil" shape, and the rising air in the cloud falls back to earth in cool, dry currents of air surrounding the warm, moist core of the storm. These downdrafts can pool at the bottom of the thunderstorm and create microbursts—brief, violent gusts of wind and rain. When the static buildup between the clashing air masses in a thunderstorm (the downdrafts carry a positive charge; the updrafts a negative one) triggers an electrical discharge, lightning forks through the sky at 145,000 km/s. The lightning heats the surrounding air, which expands at supersonic speeds, creating the mighty crashes we recognize as thunder.

Highest temperature:	Lowest temperature:	Coldest month:	Greatest precipitation in one year:
Midale and Yellowgrass, Sask. **45°C, July 5, 1937**	**Snag, Y.T.** **−63°C, Feb. 3, 1947**	**Eureka, N.W.T.** **−47.9°C, Feb. 1979**	**Henderson Lake, B.C.** **9,479 mm, 1997**

WINTER AIR MASSES

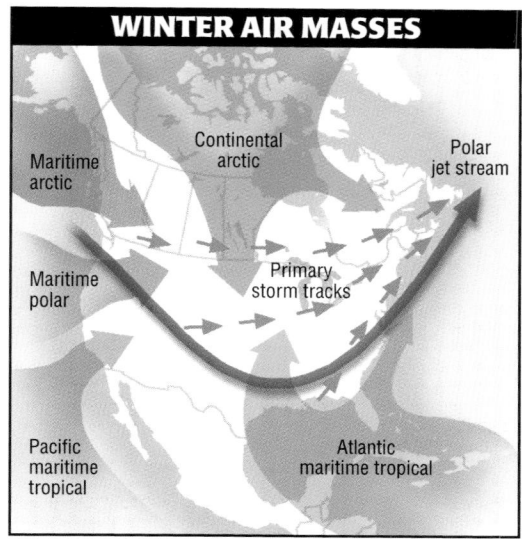

Maritime arctic • Continental arctic • Polar jet stream • Maritime polar • Primary storm tracks • Pacific maritime tropical • Atlantic maritime tropical

SUMMER AIR MASSES

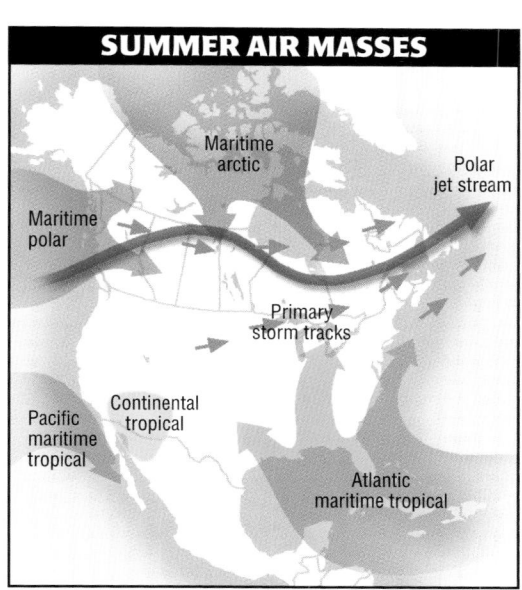

Maritime arctic • Polar jet stream • Maritime polar • Primary storm tracks • Pacific maritime tropical • Continental tropical • Atlantic maritime tropical

WHERE OUR WEATHER BEGINS

Five air masses affect Canada's weather: continental arctic, maritime arctic, maritime polar, maritime tropical, and continental tropical. Winds carry these great bodies of air across the country. Each air mass, extending hundreds or thousands of kilometres, has uniform temperature and moisture conditions, acquired from the underlying landmass or ocean where they developed. The very cold, dry continental arctic air mass, the source of Canada's bitter winters, originates over snow-covered barrens. In summer, its cool winds sweep south, bringing a welcome respite from heat waves. The maritime arctic air mass, traveling over large open bodies of water, is mild and moist. The maritime polar air masses of the Pacific and Atlantic soak coastal areas with rain, fog, and snow. The Atlantic maritime tropical air mass from the Gulf of Mexico scorches Eastern Canada with summer heat and humidity. By contrast, the Pacific maritime tropical has a cooling influence. The continental tropical air mass rarely reaches Canada because its hot, dry impact disappears as it moves north. The maps (*left*) show winter and summer air-mass movements. All weather changes are related to the interaction of these air masses along what are called fronts (*see opposite page*). The polar jet stream forms perhaps the biggest front, an ever-changing boundary where the cool winds from the north meet the warm winds from the south. In spring, the clashing air masses produce severe thunderstorms and tornadoes.

A NATURAL BALANCE

RIDING ON AIR

Wind power is the world's fastest growing power source. In 2000, the capacity of the world's wind-power facilities grew by 32 percent to 17,700 megawatts (MW); two years later, capacity exceeded 31,000 MW. In Canada, wind-power capacity increased by about 20 percent in 2002 to about 250 MW, and is expected to surpass 3,000 MW in 2005. Canada lags in wind-power development because of the lack of subsidies and research grants given to other energy utilities such as oil and natural gas. Yet, experts estimate Canada's wind-power potential is about 30,000 MW, sufficient for 15 percent of the country's electricity needs. Canada's winter winds are a plus factor. Winds are strongest in winter; in northern latitudes, they grow stronger, particularly during the day. So, Canada could expect wind power to meet demand at peak periods. Quebec, Alberta, and Saskatchewan are Canada's wind-power leaders; Ontario, Prince Edward Island, and the Yukon have entered the field with new plants. Canada's largest wind-power producer is the Le Nordais project in the Gaspé Peninsula. In 2003, a 95-m-high direct-drive wind turbine went into operation at Toronto's waterfront Exhibition Place. Its owners are the Toronto Renewable Energy Co-operative (TREC), a group of about 650 individuals, businesses, and organizations that sells the wind-powered output to Toronto Hydro Energy. Calgary Transit's "Ride the Wind" program uses coaches (*above*) run on power from wind farms at Pincher Creek, Alta. (*below*).

HORRIFYING HURRICANES

A hurricane is a cyclical storm of tropical origin thousands of square kilometres across with speeds between 65 and 240 knots (120–445 km/h). Canada's east coast is often visited by hurricanes between August and October, although most dissipate and are downgraded to gale force (34–47 knots or 65–90 km/h) storms by the time they reach our shores. Some hurricanes, however, have left their horrifying mark:
● On Aug. 24–25, 1927, a hurricane swept through Atlantic Canada, washing out roads, filling basements, and swamping boats. In Newfoundland, 56 people died at sea.
● On Oct. 15, 1954, Hurricane Hazel dumped an estimated 300 million tonnes of rain on Toronto, obliterating streets and washing out bridges. In all, 83 people died.
● On Sept. 11, 1995, the *QE2* ocean liner was struck by a 30-m wave during Hurricane Luis off the coast of Newfoundland, the largest measured wave height in the world.
● On Sept. 28, 2003, Hurricane Juan walloped Nova Scotia with winds in excess of 150 km/h (*below*). The storm killed 2 people, beached boats, uprooted hundreds of trees, and left thousands without power for days.

ICY AGONY

Freezing rain, precipitation that falls through a shallow layer of freezing temperatures before melting and then freezing upon impact, makes regular, transitory appearances during Canadian winters. But, on Jan. 5, 1998, freezing rain began to fall and continued for six days without letup, crippling eastern Ontario, southern Quebec, and the Maritimes. Trees snapped, roofs collapsed, and high-voltage towers crumpled under the weight of a 5-to-7.5-cm veneer of ice. The electrical system failed, leaving 4 million people in frozen darkness for at least 36 hours. Thousands took refuge in emergency shelters where many stayed for weeks. In the largest peacetime troop deployment in Canadian history, the army was called in to help. Storm-related claims totaled at least $2 billion. The ice storm caused 22 deaths in Quebec and 4 in Ontario.

TERRIFYING TWISTERS

All regions of Canada except the Arctic experience tornadoes, or twisters. Tornadoes occur from March to late October, peaking in early summer, with the most frequently affected areas being southern Ontario and the Prairie provinces. Forming in thunderstorms (*see opposite page*) rotated by high winds, a tornado occurs when a downward-spinning column of air inside the thunderstorm touches the ground, creating a funnel-shaped twister that can cut a path up to 1.6 km wide and 3.2 to 8 km long. Winds range from weak (65 km/h) to devastating (500 km/h). The 1912 Regina twister, Canada's deadliest, took 28 lives and destroyed the downtown area. Almost as destructive, the 1987 Edmonton tornado killed 27 people, injured over 200, and caused an estimated $250 million in damage.

Wind farm at Pincher Creek, Alta. Situated in foothills country, this prairie community captures energy from winds that reach more than 150 km/h between October and March.

Least annual precipitation:	**Greatest seasonal snowfall:**	**Greatest one-day snowfall:**	**Highest winds:**	**Heaviest hailstone:**
Arctic Bay, N.W.T. 12.7 mm, 1949	**Revelstoke/Mount Copeland, B.C.** 2,446.5 cm, 1971–72	**Tahtsa Lake, B.C.** 145 cm, Feb. 11, 1999	**Cape Hopes Advance (Quaqtaq), Que.** 201.1 km/h, Nov. 18, 1931	**Cedoux, Sask.** 290 g, Aug. 27, 1973

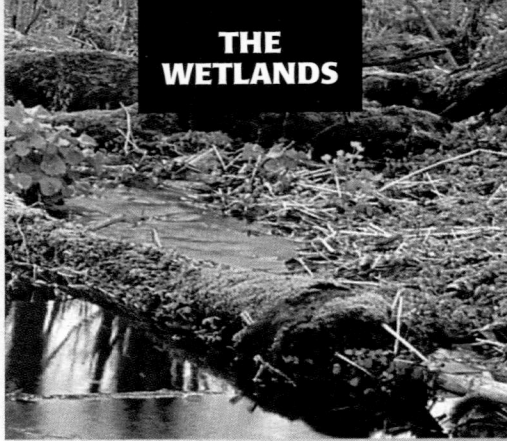

REALMS OF REFUGE AND RECREATION

Wetlands such as marshes, swamps, and bogs (*see below*) host animals ranging from amphibians and fish to large mammals. Birds, wetlands' most conspicuous creatures, use these refuges for nesting, permanent homes, and stopovers during migration.

Wetlands also aid human endeavour. They trap nutrients and purify water by filtering out pollutants. They store rainfall and runoff, preventing flooding and shoreline erosion. Wetlands provide food (wild rice, cranberries, fish, wildfowl), energy (peat, wood, charcoal), and building material (lumber). Finally, wetlands are valued for recreation.

Wetlands cover 6 percent of the world's surface. Canada holds a quarter of these wetlands, which occupy 1.27 million square kilometres, an area surpassed only by Russia's 1.5 million square kilometres. Wetlands exist in all provinces and territories, but the greatest number dot the Boreal Shield and Hudson Bay Lowlands. Here, bogs (also called muskeg) absorb carbon dioxide, storing it in thick layers of peat. This counters the impact of greenhouse gases produced by humans. An estimated 25 percent of the world's carbon may be locked up in the peat bogs of Canada's boreal forests.

MARSHES are the most biologically rich and diverse wetlands, characterized by mineral soils and by plant life dominated by grasses. Marshes occur at the mouths of rivers or in open shallows, where water flows sluggishly. Marsh beds have no distinct peat layer as organic materials decay rapidly and drain away to adjacent water bodies.

SWAMPS are low-lying areas of dense forest that are waterlogged or seasonally flooded. The boreal forest swamps host cedars and spruces; swamps in southern Canada have red and silver maples, ash, and yellow birch. Swamps are oxygen-rich and encourage plant growth. Because they also stimulate the rapid decomposition of organic matter, only thin peat layers form on swamp beds.

BOGS are poorly drained basins, blanketed by waterlogged, spongy sphagnum-moss mats. Bogs cannot break down organic materials. Instead, these materials collect in thick peat layers on bog floors, eventually building up and filling the space beneath the mats.

PROTECTING THE WETLANDS

In Canada, wetlands are in retreat. Most losses are due to agricultural expansion. Wetland areas hardest hit include the Fraser River delta (80 percent gone), the Prairies potholes region (71 percent), the lower Great Lakes and the St. Lawrence Valley (70 percent), and Atlantic coastal marshes (65 percent). In 1981, Canada entered into the Ramsar Convention, an international accord on wetlands protection, to stem further losses. Across Canada, 36 Ramsar sites protect 13 million hectares of wetlands.

ARCTIC AND TAIGA

Canada's northern ecozones cover all the arctic islands and offshore waters, and the taiga from the Mackenzie Delta to Labrador. Tuktut Nogait National Park (*above*), representative of the Southern Arctic ecozone, supports caribou, muskoxen (*right*), and Canada's rarest bird, the Eskimo curlew. Vegetation, confined by the frozen soil and the cold, dry climate, includes lichens, mosses, and dwarf shrubs, which burst forth during brief summers. Along the river gorges, a few stunted spruce mark the northern limit of tree growth.

PACIFIC AND WESTERN MOUNTAINS

Lofty ranges dominate this region's terrestrial ecozones: the Pacific Maritime, Montane Cordillera, Boreal Cordillera, and Taiga Cordillera. Landscapes range from Pacific Rim National Park's coastal greenery (*above*) and Mount Revelstoke National Park's forests and meadows to Yukon's Ivvavik National Park's barrens. The rugged mainland coast is home to the grizzly bear (*top*). Pacific Rim, a typical Pacific Maritime area, comprises surf-pounded beaches, seal-inhabited islands, and the wind-tangled woods of the West Coast Trail. Offshore lie the teeming waters of the Pacific Marine ecozone.

IVVAVIK
VUNTUT
KLUANE
YUKON
TUKTUT
NOGAIT
NAHANNI
N.W.T.
WOOD
BUFFALO
PACIFIC AND
WESTERN MOUNTAINS
GWAII
HAANAS
BRITISH
COLUMBIA
ALBERTA
JASPER
ELK ISLAND
Canadian Rocky
Mountain Parks
GLACIER
BANFF
PACIFIC RIM
MOUNT REVELSTOKE
YOHO
KOOTENAY
GULF ISLANDS
Alaksen National
Wildlife Area
Dinosaur
Provincial Park
WATERTON
LAKES

CENTRAL PLAINS

Riding Mountain National Park lies in a belt between this region's two ecozones: the cropland of the southern Prairies and the forests of the Boreal Plains. Rising 500 m above the surrounding expansive farmland, Riding Mountain contains prairie meadows (*above*) and a mix of conifers and hardwoods. No part of Canada has been altered as much as the Prairies. The largest remnant survives in Grasslands National Park. In the Boreal Plains ecozone, Wood Buffalo National Park shelters whooping cranes (*left*) and small herds of bison, once the monarchs of the grasslands.

BOREAL SHIELD

Largest of Canada's terrestrial ecozones, the Boreal Shield is one vast ecozone stretching from northern Alberta to Newfoundland's eastern coast. It superimposes a mantle of evergreens over much of the Canadian Shield, imperfectly hiding a rugged landscape still showing the scars of glacial assault and retreat. Pukaskwa National Park (*above*) on Lake Superior's northeastern shore sums up many of this region's essentials: rocky headlands, sand and cobble beaches, a dense forest of spruce and fir, rock-ringed lakes, and rolling rivers. Beaver and loons, ubiquitous denizens of the Shield, thrive here. Other typical wildlife include rare woodland caribou, moose, wolf, and black bear.

Six natural regions

Canada is a mosaic of natural regions, or ecozones, distinguished by their iconic features: the rain forest of the Pacific Coast, the flat-to-rolling horizon of the prairie, the evergreen wilderness of the Canadian Shield, and the polar barrens of the Arctic. A network of shared properties—geologic, vegetative, climatic, landforms and water, and human input—define the essence of each natural region. This network also encloses countless ecosystems—significant natural units such as wetlands. Altogether, Canada contains 15 terrestrial and 5 maritime ecozones, arranged here as follows: Arctic and Taiga, Pacific and Western Mountains, Central Plains, Boreal Shield, Mixedwood Plains, and Atlantic. The natural regions vary in size and shape. The Boreal Shield ecozone stretches across Canada; the Mixedwood Plains region lies along a narrow plain in the southern parts of Ontario and Quebec. The natural regions also vary in biodiversity. The spacious Central Plains are less abundant in plant and animal species than the Pacific ecozone, where rich layers of life ascend from ocean tidal pools to alpine mountain summits. The past and present impact of human endeavour, specifically agricultural development, resource exploitation, and urbanization, has radically altered the ecozones. Our national park system provides enclaves where their essence survives untouched. But a challenge for our time is learning to sustain our quality of life without straining their capacities further.

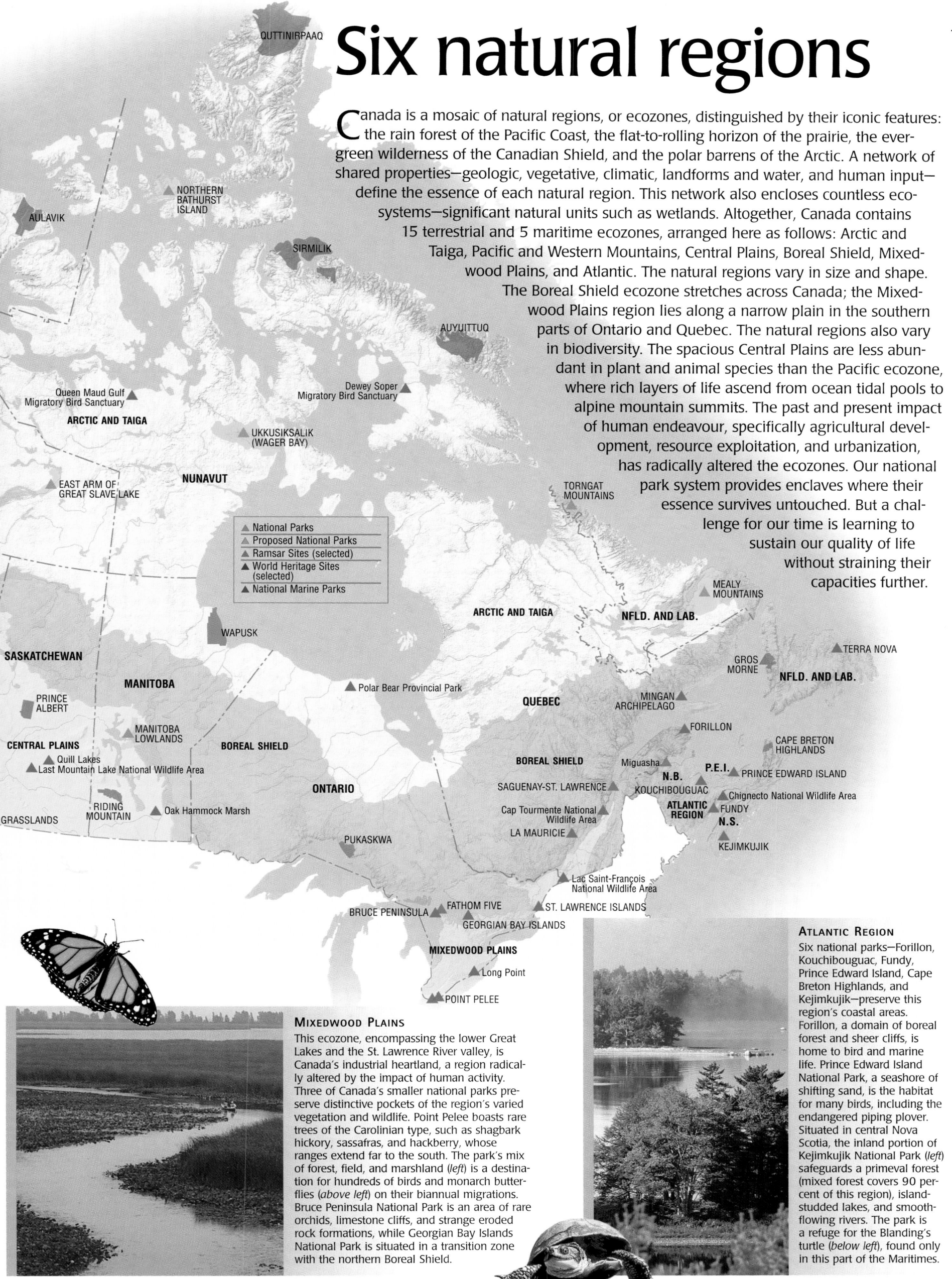

Map legend:
- ▲ National Parks
- ▲ Proposed National Parks
- ▲ Ramsar Sites (selected)
- ▲ World Heritage Sites (selected)
- ▲ National Marine Parks

Map labels: QUTTINIRPAAQ, AULAVIK, NORTHERN BATHURST ISLAND, SIRMILIK, AUYUITTUQ, Queen Maud Gulf Migratory Bird Sanctuary, Dewey Soper Migratory Bird Sanctuary, ARCTIC AND TAIGA, UKKUSIKSALIK (WAGER BAY), EAST ARM OF GREAT SLAVE LAKE, NUNAVUT, TORNGAT MOUNTAINS, MEALY MOUNTAINS, NFLD. AND LAB., ARCTIC AND TAIGA, WAPUSK, TERRA NOVA, GROS MORNE, NFLD. AND LAB., SASKATCHEWAN, MANITOBA, PRINCE ALBERT, Polar Bear Provincial Park, QUEBEC, MINGAN ARCHIPELAGO, MANITOBA LOWLANDS, BOREAL SHIELD, FORILLON, CAPE BRETON HIGHLANDS, CENTRAL PLAINS, Quill Lakes, Last Mountain Lake National Wildlife Area, BOREAL SHIELD, Miguasha, P.E.I., PRINCE EDWARD ISLAND, N.B., KOUCHIBOUGUAC, Chignecto National Wildlife Area, ONTARIO, SAGUENAY-ST. LAWRENCE, ATLANTIC REGION, FUNDY, RIDING MOUNTAIN, Oak Hammock Marsh, Cap Tourmente National Wildlife Area, N.S., GRASSLANDS, LA MAURICIE, KEJIMKUJIK, PUKASKWA, Lac Saint-François National Wildlife Area, FATHOM FIVE, ST. LAWRENCE ISLANDS, BRUCE PENINSULA, GEORGIAN BAY ISLANDS, MIXEDWOOD PLAINS, Long Point, POINT PELEE

MIXEDWOOD PLAINS

This ecozone, encompassing the lower Great Lakes and the St. Lawrence River valley, is Canada's industrial heartland, a region radically altered by the impact of human activity. Three of Canada's smaller national parks preserve distinctive pockets of the region's varied vegetation and wildlife. Point Pelee boasts rare trees of the Carolinian type, such as shagbark hickory, sassafras, and hackberry, whose ranges extend far to the south. The park's mix of forest, field, and marshland (*left*) is a destination for hundreds of birds and monarch butterflies (*above left*) on their biannual migrations. Bruce Peninsula National Park is an area of rare orchids, limestone cliffs, and strange eroded rock formations, while Georgian Bay Islands National Park is situated in a transition zone with the northern Boreal Shield.

ATLANTIC REGION

Six national parks—Forillon, Kouchibouguac, Fundy, Prince Edward Island, Cape Breton Highlands, and Kejimkujik—preserve this region's coastal areas. Forillon, a domain of boreal forest and sheer cliffs, is home to bird and marine life. Prince Edward Island National Park, a seashore of shifting sand, is the habitat for many birds, including the endangered piping plover. Situated in central Nova Scotia, the inland portion of Kejimkujik National Park (*left*) safeguards a primeval forest (mixed forest covers 90 percent of this region), island-studded lakes, and smooth-flowing rivers. The park is a refuge for the Blanding's turtle (*below left*), found only in this part of the Maritimes.

ECOZONE OVERVIEW

Harp seal

ARCTIC CORDILLERA

Landforms: Canada's most mountainous region outside the Rockies. Ice caps, glaciers, deep fjords

Climate: Very cold, dry, and windy

Vegetation: Arctic flowers and some ground-hugging shrubs flourish in the southern areas

Wildlife: Arctic hare, northern fulmar

Resources and industries: Hunting, tourism

NORTHERN ARCTIC

Landforms: Barren plains, some rocky outcrops

Climate: Cold and dry, with September-to-June snow cover

Vegetation: Herb-lichen tundra

Wildlife: Peary caribou, musk ox, red-throated loon, greater snow goose

Resources and industries: Hunting, mining

Arctic fox

SOUTHERN ARCTIC

Landforms: Plains, hills, and cold, clear lakes

Climate: Cold, dry; continuous permafrost

Vegetation: Shrublands, wet sedge meadows

Wildlife: Barren-ground caribou, wolf, arctic fox, arctic loon, snowy owl

Resources and industries: Hunting, trapping, tourism, mineral development

TAIGA PLAINS

Landforms: Plains bordering the Mackenzie River

Climate: Cold, semiarid to moist

Vegetation: Dense to open mixed forest

Wildlife: Moose, woodland caribou, wolf, black bear, red squirrel, northern shrike, spruce grouse

Resources and industries: Hunting, trapping, tourism, oil and gas development

TAIGA SHIELD

Landforms: Plains, some hills

Climate: Cold, moist to semiarid

Vegetation: Lichen-shrub tundra; evergreen-deciduous forest rests on the bedrock of the Canadian Shield

Wildlife: Barren-ground caribou, snowshoe hare, red-necked phalarope

Resources and industries: Tourism and recreation, mining, hunting, trapping

HUDSON PLAINS

Landforms: One of the world's largest wetlands

Climate: Cold to mild; discontinuous permafrost

Wildlife: Marten, arctic fox, Canada goose

Resources and industries: Hunting, trapping, recreation

ARCTIC BASIN

Landforms: Most of the northern polar waters. Water depths more than 2,000 m; year-round sea ice up to 2 m thick over 90 percent of the surface

Sealife: Virtually devoid of life

ARCTIC ARCHIPELAGO

Landforms: Arctic islands; Hudson and James bays

Sealife: In the northern parts, marine life occurs near polynyas and shoreleads; in the southern Arctic, a great diversity of bird and animal life

Resources and industries: Oil and gas, some fishing and hunting

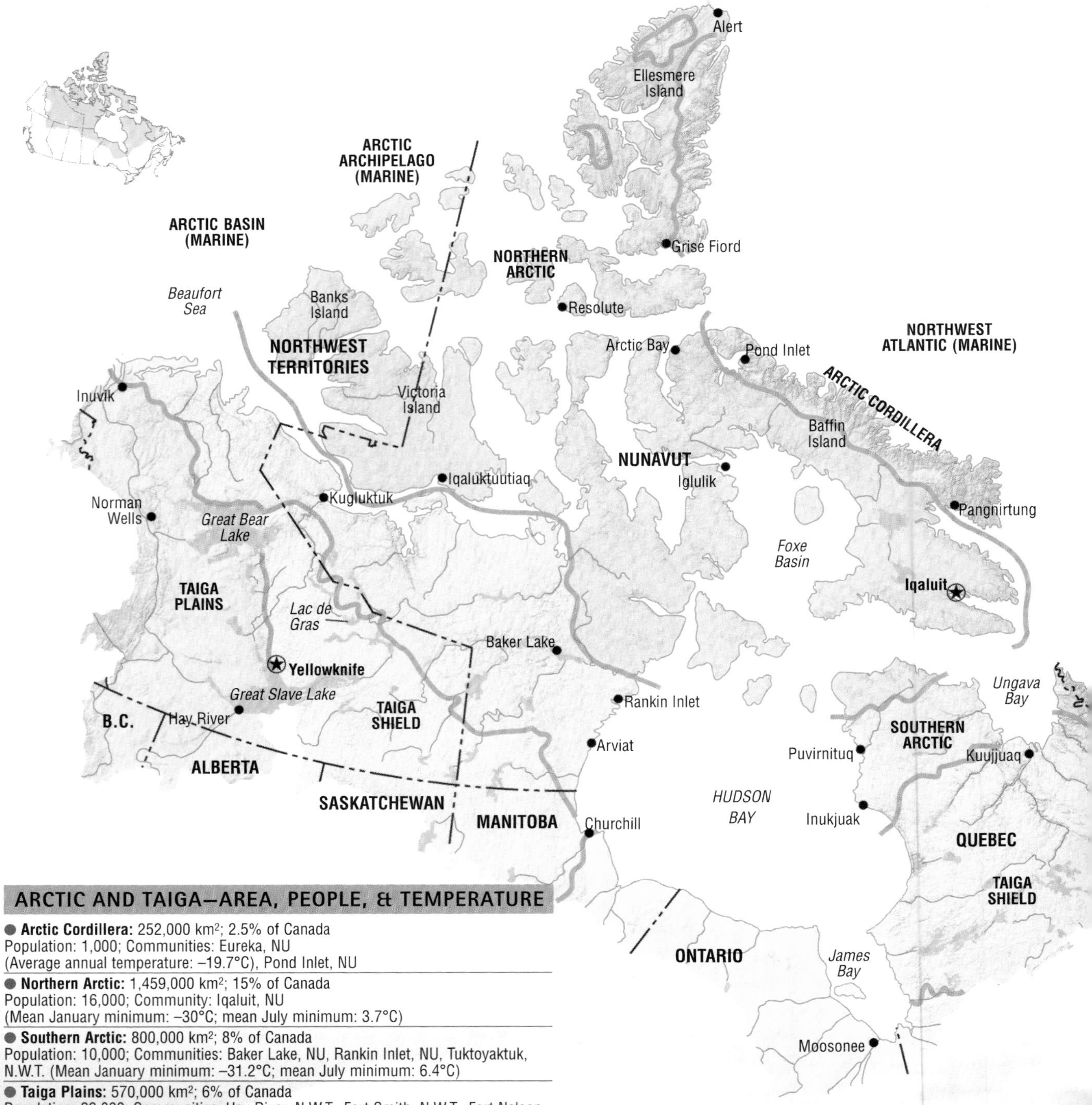

ARCTIC AND TAIGA—AREA, PEOPLE, & TEMPERATURE

● **Arctic Cordillera:** 252,000 km²; 2.5% of Canada
Population: 1,000; Communities: Eureka, NU
(Average annual temperature: −19.7°C), Pond Inlet, NU

● **Northern Arctic:** 1,459,000 km²; 15% of Canada
Population: 16,000; Community: Iqaluit, NU
(Mean January minimum: −30°C; mean July minimum: 3.7°C)

● **Southern Arctic:** 800,000 km²; 8% of Canada
Population: 10,000; Communities: Baker Lake, NU, Rankin Inlet, NU, Tuktoyaktuk, N.W.T. (Mean January minimum: −31.2°C; mean July minimum: 6.4°C)

● **Taiga Plains:** 570,000 km²; 6% of Canada
Population: 22,000; Communities: Hay River, N.W.T., Fort Smith, N.W.T., Fort Nelson, B.C. (Mean January minimum: −26.5°C; mean July minimum: 10.4°C)

● **Taiga Shield:** 1,302,000 km²; 14% of Canada
Population: 34,000; Communities: Yellowknife, N.W.T., Labrador City, Nfld.
(Mean January minimum: −32.2°C; mean July minimum: 12°C)

● **Hudson Plains:** 370,000 km²; 4% of Canada
Population: 10,000; Communities: Moosonee, Ont., Churchill, Man.
(Mean January minimum: −30.9°C; mean July minimum: 6.8°C)

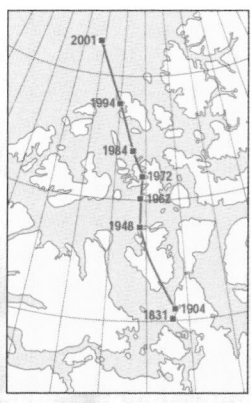

LAND OF THE TWO POLES

The Arctic boasts a geographic north pole and a north magnetic pole. The former, the earth's northernmost point, lies about 725 km north of Ellesmere Island in the Arctic Ocean. Year-round ice covers the north pole. Recently, open water was observed there, possibly a sign of global warming. The north magnetic pole is more than 1,600 km south of the geographic pole. Like its southern counterpart in Antarctica, this is the site where the earth's magnetic field reaches maximum intensity. It is the point toward which all compasses point, and it is constantly moving—at an average rate of 15 km a year. The map (*left*) traces its ever-shifting path. In the 1990s, the magnetic pole was located on Ellef Ringnes Island, Nunavut; by 2001, it had moved on to 80° North 110° West; and by 2004 it was located at 82.3° North 113.4° West.

THE POWER OF PERMAFROST

Permafrost is made up of two layers: a thin "active" upper layer that melts in summer, and a thick underlying base of frozen ground that never melts. The Arctic and the Taiga lie within zones of continuous, discontinuous, or sporadic permafrost. (Alpine permafrost, another type of permafrost, is found primarily in the Rockies and the Coast Mountains.) The permafrost varies in depth and extent from zone to zone. In the eastern Arctic, where 80 percent of the ground is frozen, permafrost is continuous and half a kilometre thick. In the northern Taiga, where frozen ground ranges from 30 to 80 percent, discontinuous permafrost occurs. In the southern Taiga, permafrost is sporadic, reaching depths of only a few centimetres below the active level (*see illustration*). Permafrost is formed when annual mean ground temperature—determined by air temperature, soil, drainage, and snow cover—remains below zero. The thawing of the "active" layer creates boggy conditions because the unyielding frozen layer hinders water drainage. Permafrost impedes mineral exploration and extraction. Thawing permafrost can undermine road surfaces, which require insulating layers of sand and gravel. Buildings must be steadied by supports capable of withstanding ground shifting. But thawing permafrost can be beneficial for it provides the mix of soil and water essential for the brief outburst of summer vegetation in the Canadian North.

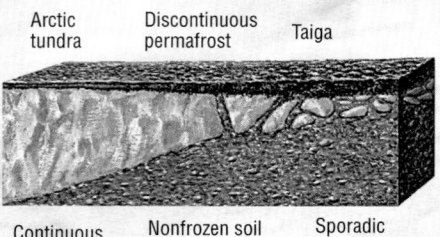

Arctic tundra / Discontinuous permafrost / Taiga / Continuous permafrost / Nonfrozen soil / Sporadic permafrost

The Far North: Arctic and Taiga

Half of Canada lies within the Arctic and Taiga ecozones, which manifest some of the nation's many climatic and environmental extremes. The Arctic is characterized by treeless tundra, carpeted with low-lying vegetation of mosses, lichens, herbs, and dwarf shrubs; and the Taiga, by sticklike forests of spruce and fir, interspersed with immense wetlands. The region's marine ecozones—the Arctic and the Arctic Archipelago, including Hudson and James bays—are as vast and complex as the terrestrial ecozones. Given the immensity of these regions, few generalizations can be made about them that are totally valid. Yet, one thing is sure: the environment of these ecozones is fragile, slow to change, and easy to disturb. In earlier times, Europeans hunted for whales, seals, and furs; today, developers search for minerals, gas, and oil in a rich—but vulnerable—environment. Politically, these regions encompass Canada's most sparsely populated regions: Nunavut, the Northwest Territories, the Yukon's coastal strip, and the northern parts of six provinces. The indigenous people who have long survived here are now reasserting control over their traditional domain.

Nain
Voiseys Bay
Davis Inlet
NEWFOUNDLAND AND LABRADOR

A NATURAL BALANCE

CIRCUMPOLAR PROTECTION

Developing policies that safeguard the Arctic and promote Inuit rights is a priority of the Inuit Circumpolar Conference (ICC). Working with the United Nations and other international forums, the organization represents some 155,000 Inuit living in Canada, Alaska, Greenland, Russia, and Scandinavia (see map above). In 2001, the ICC was instrumental in 151 countries signing the Stockholm Convention to reduce and eventually eliminate certain persistent organic pollutants (POPs), including chlordane, DDT, and other pesticides, industrial chemicals such as PCBs, and combustion by-products such as dioxins.

Although POPs have never been used or manufactured in the Arctic, tests show high levels of DDT, PCBs, dioxins, and other POPs in Inuit blood, lipid tissue, and breast milk samples. This happens because POPs travel long distances on air and in water, return to earth in precipitation, and enter the food chain. Accumulations are especially high in whales and seals, the traditional Inuit diet. Pregnant and nursing mothers in the Arctic are now advised to avoid marine mammals and fat, and to replace them with caribou meat and arctic char, which are low in POPs.

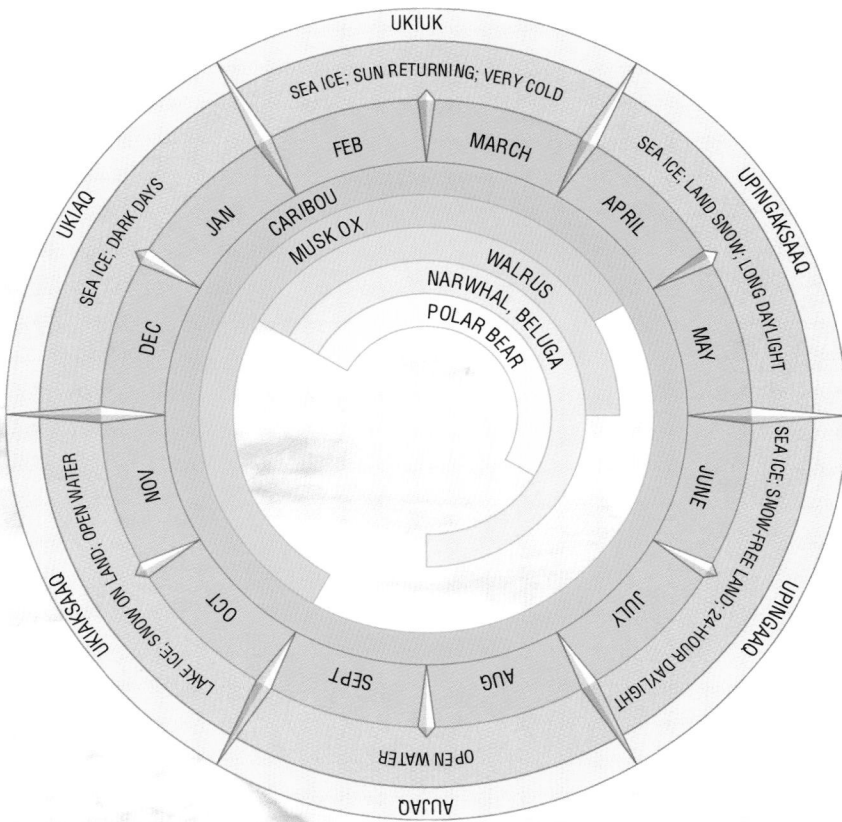

HARVESTING CYCLE

The aboriginal peoples of the Canadian North are strongly attached to "country food"—that is, meat and fish obtained by hunting. This food is more than just an alternative to store-bought foods. It is an intrinsic feature of aboriginal culture, where hunting lies at the heart of traditional community, family, and personal values. Country food is so important that the aboriginals have made wildlife a key issue in land-claim negotiations. The year-round harvesting cycle, above, depicts the changing conditions in weather and wildlife availability confronting the Inuit of Lancaster Sound in the Arctic.

ICEBOUND SEAS

In the Arctic, sea ice is land-like for part of the year; in many areas, for all of it. For the Inuit, the ice can be used to travel a region without roads. Even as ice melts or drifts away, it is replaced by other drifting ice, or by the new ice that forms in the autumn. In the southern Arctic, the sea freezes slowly; in the high Arctic, the sea ices over with startling rapidity, virtually overnight. In winter, ice envelops the arctic islands and coastlines, which remain ice-locked from October to June. Polynyas (open water areas) and shoreleads (long, linear cracks) appear in the midst of the sea ice. Created by strong winds or upwelling warm water, these openings serve as oases of sea life. Sea ice comes in many shapes. Initially, it is called brash, slush, or pancake. But as it grows stronger and thicker, ice collects in floes or packs. Icebergs, calved from Greenland's glaciers, float in Davis Strait, off Baffin Island. The south-flowing Labrador Current moves the icebergs toward Newfoundland and the Grand Banks. "Iceberg Alley," as these waters are called, is not ice-free until July.

RICHES OF THE NORTH

Major sites of mineral development in the Arctic and Taiga ecozone stretch from the oil and gasfields of Norman Wells on the Mackenzie River (above) to the newly discovered reserves of nickel, copper and cobalt at Voiseys Bay in Labrador. But it is the northern diamond mines that have stirred the greatest excitement. In 1991, diamonds with economic potential were discovered in diamond-bearing kimberlite rock at Lac de Gras, some 300 km northeast of Yellowknife. The discovery precipitated an exploratory rush unprecedented in Canadian mining history. The Ekati diamond mine at Lac de Gras began operating in 1998. Other sites may be mined in the future. The recovery of the diamonds from vast quantities of waste rock takes place in a processing plant. Cutting and polishing takes place at Yellowknife. In 2000, the Ekati mine reported total sales at $430 million worth of diamonds. Another mine—Diavik—is also operating in the Lac de Gras area. The money, jobs, and investment have boosted the northern economy. By 2003, Canada ranked third among the world's diamond producers.

Left: *The polar bear, the king of arctic mammals, spends its life at sea on the ice floes, searching for seals, which make up its winter diet. In summer, when the sea ice breaks up, the bears head for land where they feast on waterfowl, berries, and marine vegetation.*

PACIFIC MARITIME

Landscape: Mountainous coast, marine islands

Climate: More than 2,500 mm of precipitation a year along the coast. Average temperatures range from 4°C to 6°C (January) to 12°C to 18°C (July)

Vegetation: Coastal forests host western hemlock, western red cedar, Sitka spruce. In the rain-shadow areas, Douglas fir, mountain hemlock; on the Gulf Islands, rare Garry oak, arbutus

Cougar

Wildlife: Black and grizzly bears, cougar, Roosevelt deer (found only on Vancouver Island). Typical seabirds: black oystercatcher, tufted puffin

Resources and industries: Forestry, fish processing, agriculture, tourism. Construction in the lower mainland, the Gulf Islands, and Victoria, where three-quarters of all British Columbians live

MONTANE CORDILLERA

Landscape: Mountain chains, interior plateaus. Large, deep lakes, major river systems

Climate: Over 600 mm of precipitation a year at Prince George; well below 500 mm in the Okanagan Valley. Average temperatures at Kamloops range from −2.1°C (January) to 28.5°C (July)

Vegetation: Alpine shrubs, evergreen forests, sagebrush, grasslands

Wildlife: Mountain goat, Dall sheep inhabit the mountain heights; wapiti, woodland caribou, mule deer, bighorn sheep, and bobcat dwell in dense forests. Typical birds: black-billed magpie, blue grouse, golden eagle, Steller's jay

Resources and industries: Forestry, agriculture, mining, hydroelectricity, tourism

Mountain goat

BOREAL CORDILLERA

Landscape: Some of Canada's highest peaks, as well as plateaus and wide valleys

Climate: Over 1,000 mm of precipitation a year above the treeline; 500 mm at lower elevations. Average temperatures at Whitehorse: −14.4°C (January); 22.6°C (July)

Vegetation: White and black spruce, lodgepole pine at low levels; subalpine fir higher up

Ptarmigan

Wildlife: Moose, black and grizzly bears, Dall sheep, mountain goat, woodland caribou. Birds: ptarmigan, spruce grouse

Resources and industries: Forestry, mining, hunting and trapping, tourism

TAIGA CORDILLERA

Landscape: Some areas escaped ice-age glaciers. Flat-topped, ramplike mountains; broad valleys

Climate: At Old Crow, Y.T., the mean precipitation reaches 248 mm a year. Average temperatures: −36.7°C (January); 8.1°C (July)

Vegetation: In the south, forests host white and black spruce, dwarf birch, willow. In the north, only alpine plants thrive on partially frozen soil

Wildlife: Large migrating Porcupine caribou herd, Canada's most numerous wolverine population

Resources and industries: Trapping, hunting, mining, tourism and recreation, oil and gas

PACIFIC MARINE

Extent: Pacific Ocean basin and continental shelf

Climate: The Alaska Current creates the mild, moist coastal climate and drives west-to-east winds across southern Canada. Coastal sea ice is absent, except in sheltered bays and inlets

Sealife: Five species of salmon; clams, Dungeness crab. Marine animals: Steller sea lion, sea otter, northern fur seal, killer and gray whales

Industries: Commercial fishing, tourism

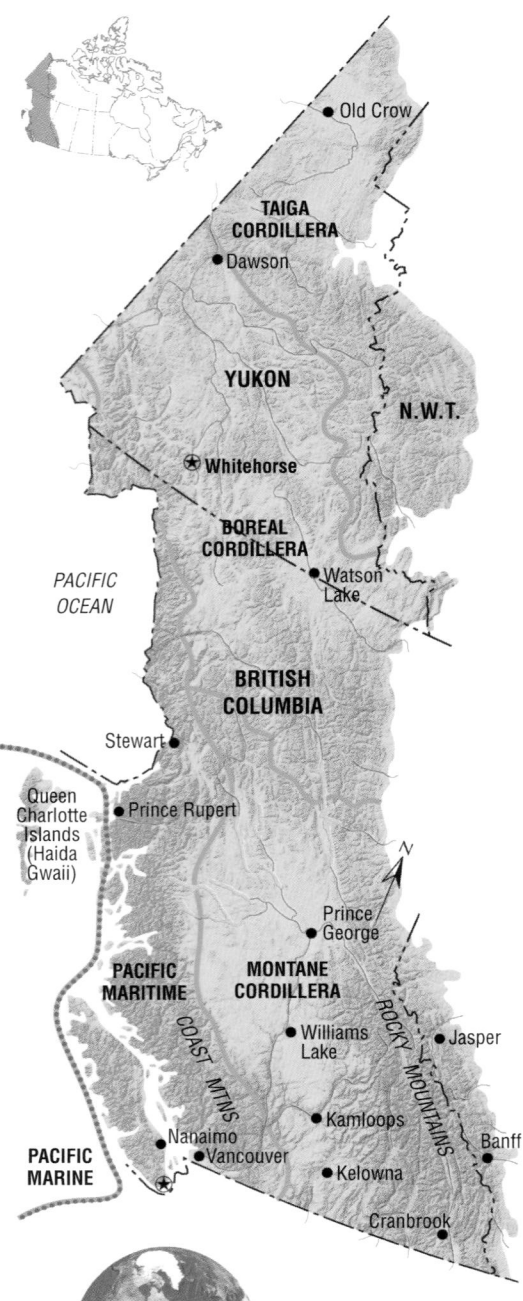

Pacific and Mountains

Resource-rich and breathtakingly beautiful, the Pacific and Western Mountains region is made up of four land ecozones: the Pacific Maritime coastal strip of mountain, rain forest, and fjord; and the three rugged Cordilleran zones: Montane Cordillera, Boreal Cordillera, and Taiga Cordillera. A fifth ecozone—the Pacific Marine—encompasses Canada's waters of the Pacific Ocean. The land zones comprise British Columbia and the Yukon Territory, but also include parts of Alberta and the Northwest Territories. Blessed with a diverse wildlife and vegetation, parts of this natural region—particularly the Pacific Maritime ecozone—are under pressure from rapid population growth, urban development, and resource exploitation.

Alpine zone. From the icy rock terrain on the mountaintop to the treeline: lichens, mosses, shrubs, sedges

Subalpine zone. From the treeline to the lower slopes: alpine fir, Engelmann spruce

Montane zone. On the densely forested lower slopes: lodgepole pine, Douglas fir, Douglas maple. At the mountain base lie the Parkland and Prairie zones

MOUNTAIN ZONES AND BOUNDARIES

Mountain slopes host a great variety of trees and plants according to elevation, as shown above. Zone elevations move downward in cold, northern locations. In southern B.C., the alpine-subalpine boundary, discernible at the treeline, is found at the 1,800-m level; in the Yukon, it lies near the foot of the mountain.

"GREEN GETAWAYS": THE NEW TOURISM

Outdoor recreation and tourism generate over $9 billion a year for British Columbia. But the annual tourist influx—22.6 million visitors in 2002—places heavy pressures on the province's much-visited natural attractions and affects local communities as well. In the 1980s, the adverse impact of large-scale tourism became a crucial issue for environmentalists, the tourism industry, and governments, both in Canada and abroad. In response to this issue, ecotourism has since evolved as a possible solution. Ecotourism promotes ideas of exploring, developing, and sustaining natural habitats in ways least likely to cause environmental damage or disruption. This "green getaways" approach may also produce beneficial spinoffs such as support for local environmental efforts and employment for guides or interpreters of local cultural traditions. Ecotours include wildlife and wilderness expeditions, whale-watching excursions (*above*), and aboriginal ecotours, which mix natural and cultural features.

A NATURAL BALANCE

**Wood products
(by value)**

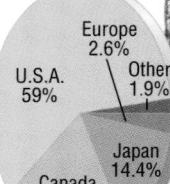

U.S.A. 59%
Europe 2.6%
Other 1.9%
Japan 14.4%
Canada 22.1%

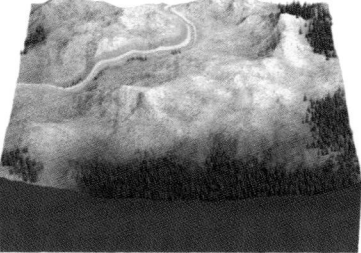

Coast Range   Interior B.C.   The Rockies

IN THE RAIN SHADOW: WHERE LUSH FORESTS GIVE WAY TO PARCHED LANDSCAPES

Parts of the Pacific Coast get 2,500 mm of rain and snow in a year; others, barely 500 mm. The Coast Range is responsible for these variations. When warm, moisture-laden ocean air hits the range and is forced upward, the rising air cools, losing its ability to retain moisture. This results in heavy rainfall that feeds lush forests on the western slopes. On the eastern slopes, the descending air becomes warmer and retains moisture. This effect, called the "rain shadow," creates the parched landscape of interior B.C. Moving eastward, the air recovers moisture, which falls as snow on the Rockies. If the air is moisture-starved for long periods, the result may be forest fires in the dry mountain valleys and drought on the Prairies.

**Pulp, paper,
paperboard
(by volume)**

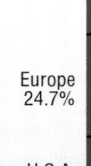

Asia 39.2%
Europe 24.7%
U.S.A. 24.5%
Rest of Canada 7.7%
Other 3.9%

FISHERIES: WILD VERSUS FARMED

British Columbia's commercial coastal and river fisheries harvest over 80 different kinds of fish and other species. Salmon is the most commercially valuable fish species, followed by herring, shellfish, and groundfish. The Pacific fishery has become more important since Atlantic groundfish catches declined around 1990. Pacific groundfish—hake, cod, rockfishes, halibut—represent the largest harvest by volume. Yet, the fishery faces declining catches. Excessive harvests have reduced fish stocks and, in turn, the number of fish caught. Other factors contributing to this decline include the growth in the size of the fishing fleets, and the use of radar and sonar equipment. Aquaculture, or fish farming, has expanded rapidly as catches declined. The fish farms, situated around Vancouver Island, keep fish (including nonindigenous Atlantic salmon) in net cages floating in seawater. Some experts claim aquaculture threatens the marine environment, citing the sea lice from fish farms that fatally infested wild pink salmon in the Broughton Archipelago. The debate over wild versus farmed fish intensified in September 2002, when B.C. lifted a seven-year moratorium on fish-farm expansion.

Halibut

FARM CASH RECEIPTS	
	in millions $
Floriculture & Nursery	394
Dairy	364
Cattle & Calves	348
Potatoes & Vegetables	335
Poultry & Eggs	331
Berries & Grapes	92
Tree Fruit	64
Grains & Oilseeds	52
Hogs	50
Ginseng	27
Other	155
Total	2,212

FARMING: PRODIGIOUS OUTPUT

Only 3 percent of British Columbia is suitable for farming. Most activity occurs in the lower Fraser River and Okanagan valleys, where much of the farmland, often ideal for fruit-growing, has been lost to urban sprawl. B.C. produces 5 percent of Canada's agricultural output by value. Some of this output, unique to this region, is prodigious. B.C. produces 18.4 kg of berries, kiwifruit, and grapes for every resident. The agricultural sector has introduced specialities such as ginseng and greenhouse peppers. The lower Fraser River valley has almost half of B.C.'s farms and generates over half of B.C.'s farm revenue of $1.8 billion. Dairy products, vegetables, berries, floriculture, and nursery products are its top commodities. The Okanagan Valley, second in output and revenue, is renowned as a top fruit-growing area. Ranching is centred in the Cariboo, Thompson-Nicola, and Kootenay regions.

FORESTRY: REGULATING LOGGING

British Columbia's forests cover 60 million hectares—64 percent of the province's land area. The coastal forests nourish 60-m-high Douglas firs and century-old western hemlocks. But interior B.C., with a diverse range of softwood trees, is the most timber-productive region. Coastal forests provide 32 percent of the timber harvest; the interior forest, 68 percent. In the 1980s, the B.C. forestry industry came under criticism for its clear-cutting of old-growth forest and its damaging logging practices. In response to the outcry of environmentalists, the B.C. government, owner of 95 percent of the forests, imposed regulations on logging. About 36 million hectares are now protected in parks or can never be touched; the remainder is open to logging. The B.C. forestry industry itself has improved its production, cutting, and reforestation practices. The annual cut of 190,000 ha represents a third of 1 percent of B.C.'s forest lands. B.C. produces most of Canada's plywood, half its softwood lumber, 15 percent of both its newsprint and its paper and paperboard. In 2000, the value of B.C. forest products abroad reached $16.8 billion.

**Newsprint
(by volume)**

U.S.A. 47.6%
Asia 28.6%
Rest of Canada 13.9%
Europe 3%
Other 6.9%

VALUE OF B.C. CATCHES

Farmed salmon $292 million
Wild and farmed shellfish $120 million
Wild salmon $52 million
Herring $45 million
Groundfish $133 million
Other $26 million

Male sockeye salmon

▼ *Waterton Lakes National Park*

In the Okanagan Valley (above), a hot, dry, sunny climate and fertile, well-irrigated soil create bountiful crops. The valley's northern sector produces dairy products and vegetables, while orchards in the dry southern sector grow more than 95 percent of B.C.'s fruit-tree output. Valley vineyards supply all the premium wine grapes in the province.

B.C. TIMBER HARVEST BY SPECIES

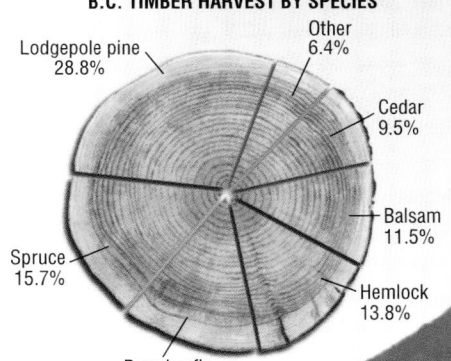

Lodgepole pine 28.8%
Other 6.4%
Cedar 9.5%
Balsam 11.5%
Hemlock 13.8%
Douglas fir 14.3%
Spruce 15.7%

Central Plains

THE PRAIRIES: GRASSLAND

Landforms: Flat or rolling plains, some hills

Climate: Long, cold winters; short, hot summers. The Rockies frequently block eastbound moisture-bearing clouds from reaching the Prairies. As a result, low precipitation, combined with a high evaporation rate, causes periodic droughts

Vegetation: Short-to-tall grasses

Wildlife: Prairie dog, pronghorn, coyote, Richardson's ground squirrel, northern pocket gopher. Birds: ferruginous hawk, greater prairie chicken, sharp-tailed grouse, burrowing owl (endangered)

Resources and industries: Grain farming, ranching operations by the foothills and on the dry plains. Oil and gas extraction and refining

Pronghorn

THE PRAIRIES: PARKLAND

Vegetation: This belt of land, separating grasslands from the boreal forest, is sometimes called "aspen parkland" for its most prevalent tree species. In western Manitoba and eastern Saskatchewan, the parkland supports bur oak. Mosses, willow shrubs, and tamarack thrive in the sloughs and marshes that dot the landscape

Coyote

Wildlife: The farmland perimeter is frequented by large mammal species: white-tailed deer, moose, black bear, coyote, wolf, red fox, caribou, beaver, elk. Myriad wetlands support one of the highest bird populations in North America, including ducks and geese

FARMING BELTS: THE DRY AND THE FERTILE

Soils, climate, and natural grassland vegetation define the two prairie agricultural zones. Most grassland soils are black, dark brown, and brown soils, all classified as chernozemic (Russian for "black earth"). The tall grasses that thrive on black or dark-brown humus-rich soils have fibrous roots that absorb water and nutrients from moist clays in the subsoil. The tallgrass zone extends from southwestern Manitoba to Edmonton and north to the prairie parkland. This zone—the Fertile Belt ❶—is ideal for crops and livestock. South of this belt, evaporation rates increase, transforming the land from tallgrass to short-grass prairie. Short grass is characteristic of this semiarid region—the Dry Belt ❷—that stretches across the southern Prairies from Estevan, Sask., to the Alberta foothills, and almost reaches Saskatoon. Here, brown, sandy, humus-poor soil is a half-metre deep. Beyond this point, roots and rainwater rarely penetrate, and the subsoil is permanently dry. But in recent years, irrigation has overcome this obstacle and made possible the cultivation of potatoes and lentils, crops new to this region.

BOREAL PLAINS

Landforms: Plains, foothills in western Alberta. Over 80 percent is forest; the rest is farmland. About 7 percent of the ecozone is covered by fresh water. The largest bodies of water are lakes Winnipeg and Winnipegosis

Climate: Harsh, cold winters; short, warm summers. High precipitation, plus low evaporation, creates moister conditions than exist on the plains

Vegetation: A mixed forest of conifers (white and black spruce, jack pine, tamarack, balsam fir) and deciduous species (poplar, trembling aspen, balsam poplar, white birch)

Wildlife: Woodland caribou, mule deer, moose, wapiti, black bear, beaver, muskrat. Small herds of bison, once dominant wildlife, are confined to park reserves. Birds: boreal and great horned owl, gray jay, white-tailed sparrow, rose-breasted and evening grosbeaks, Franklin's gull, brown-headed cowbird. Fish: walleye, northern pike, burbot

Bison

Resources and industries: Agriculture in Peace River country (Canada's most extensive northern agricultural area); forestry, pulp and paper, oil and gas development, hydroelectric production

Wheat fields, grain elevators, and remote farmsteads on the rolling prairie endure as persistent images of the Central Plains. But, on close inspection, this natural region presents a more varied and complex geographic and economic picture. The Central Plains consists of two distinct ecozones: the Prairies and the Boreal Plains, which cover much of the three Prairie provinces, as well as parts of British Columbia and the Northwest Territories. The Prairies, roughly triangular in shape, rolls westward from Winnipeg to the Rocky Mountain foothills, with Edmonton marking this ecozone's northern apex. The Boreal Plains to the north is an arc of boreal forest twice the size of the Prairies. Of all Canada's natural regions, the Prairies ecozone is the most greatly altered, largely through agriculture development. The long dominance of agriculture is now being challenged and transformed by resource extraction and industrial production that promise a more secure economic foundation for the Central Plains. Nowhere is this change more apparent than in Calgary, Edmonton, Saskatoon, Regina, and Winnipeg, which increasingly reflect more diversified economies and employment opportunities.

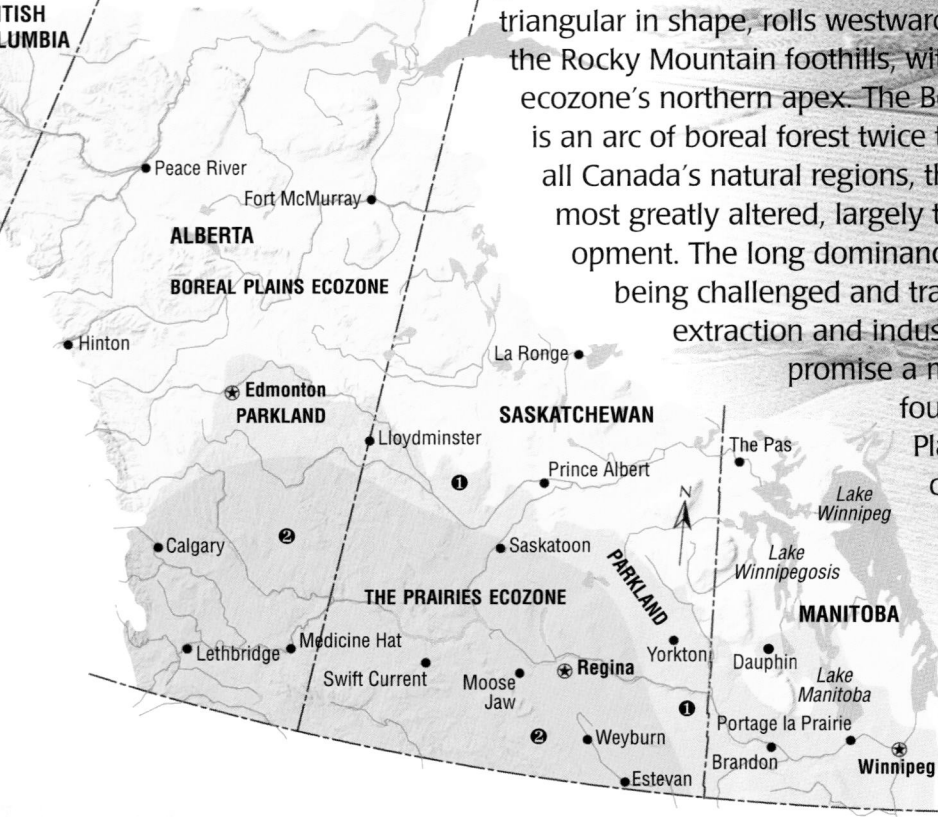

RESOURCE RICHES BENEATH THE PLAINS

Since the 1960s, development of resources such as oil, gas, and potash has moved the Central Plains in new economic directions. The 1990s saw fast-growing prairie cities move aggressively into high-tech and industrial enterprises. This trend offset the dampening effect of the crisis in agriculture. Although the agricultural economy is still vital, it is changing fast. The 1995 cancellation of the Crow Benefit, designed to allow farmers to ship their products by rail at a reduced rate, initiated an ongoing agricultural restructuring.

Europe 1.5%
United States 55%
Australia 3.5%
Latin America 11%
Asia 29%

SASKATCHEWAN—THE WORLD'S PREMIER POTASH PRODUCER

Saskatchewan boasts the world's largest and highest-quality potash deposit. More than 95 percent of Canada's potash comes from a belt stretching from the central to south-central parts of the province. The deposit, found about a kilometre below the surface of the prairie, lies in neatly arranged horizontal bands 2 to 3 m thick. Its thickest extent is around Saskatoon, where six of the province's nine potash mines are located. They consist of immense chambers and pillars, with corridors that stretch for hundreds of kilometres. Potash is used largely in fertilizers, while small amounts are used in soaps, glass, ceramics, dyes, drugs, water softeners, and explosives. In the late 1990s, Saskatchewan alone supplied 32 percent of the global output of potash and met 95 percent of the national demand. Canada exports 95 percent of its production to more than 30 countries. The United States is the major importer, buying 55 percent of Canada's exports.

Above left: *Top customers for Saskatchewan's potash include the United States, China, Brazil, Japan, Malaysia, Australia, and South Korea.* Above: *Machines of awesome proportions, such as this electrically powered trimmer, excavate vast underground chambers in order to extract potash.*

LEAVING PRAIRIE FARMS—RURAL OUTFLOW, URBAN GROWTH

Across Canada, the number of farms is falling rapidly. According to the 2001 census, farm numbers dropped 11 percent after the 1996 census—the biggest percentage decline since 1971. This means that 3 of every 10 farmers counted in 1996 had left farming in only five years. The steepest declines were recorded on the Prairies, where farms shrank in number, yet grew in size. Getting out or getting bigger is the choice facing prairie farmers who have been pressed by falling profits, rising production and transportation costs, and decade-long dry growing conditions. Farmers who stayed expanded their operations and produced more through the use of mechanization and technological innovation. As a result, fewer people are needed to work on farms. Rural migrants now head for Edmonton, Calgary, Winnipeg, and other burgeoning prairie cities, where jobs in service and high-technology industries are abundant. With the resulting depopulation, the small rural communities built to serve local needs struggle to survive.

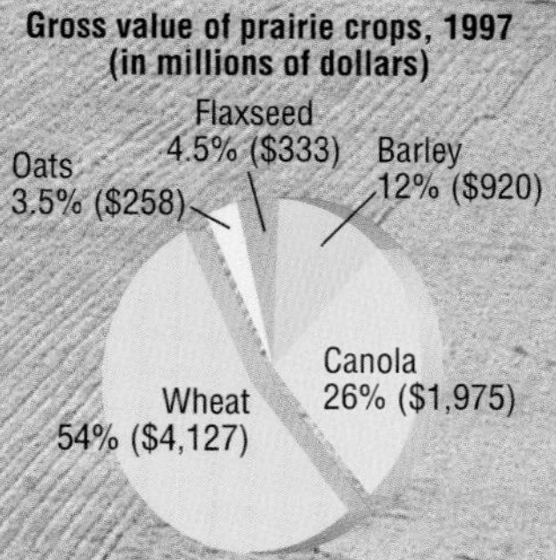

Gross value of prairie crops, 1997 (in millions of dollars)

- Flaxseed 4.5% ($333)
- Oats 3.5% ($258)
- Barley 12% ($920)
- Canola 26% ($1,975)
- Wheat 54% ($4,127)

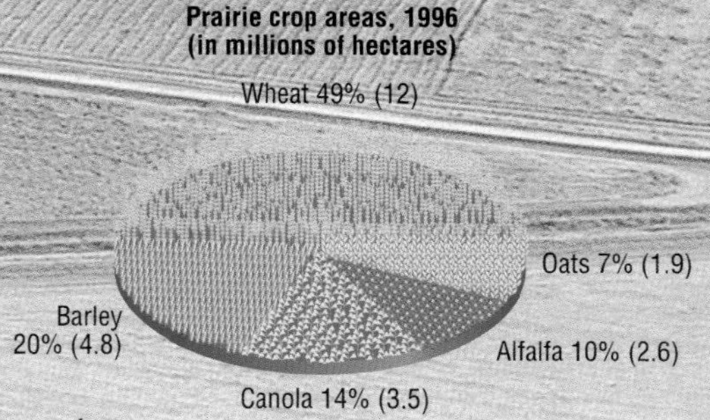

Prairie crop areas, 1996 (in millions of hectares)

- Wheat 49% (12)
- Oats 7% (1.9)
- Alfalfa 10% (2.6)
- Canola 14% (3.5)
- Barley 20% (4.8)

DIVERSIFYING PRAIRIE CROPS

Grains have long been the traditional crops on the Central Plains. The Prairie provinces produce most of Canada's grain output, which represents 5 percent of the world's supply of wheat, 9.9 percent of its barley, and 14 percent of its oats. But in recent decades, prairie farmers have diversified considerably, moving away from grains to more profitable oilseeds, dominated by canola, and specialty crops such as peas and lentils, mustard, and sunflower and canary seeds. The diversification has been prompted by low grain prices, rising costs of shipping grain by rail, the loss of the grain rail subsidy, and much higher grain subsidies in Europe and the United States. High-value canola, primarily used for salad oil and margarine, is now Canada's second major crop after wheat. From 1976 to 2001, the total area used for canola production increased more than five times to 3.5 million hectares, representing 10 percent of Canada's total agricultural land. By the late 1990s, Canada was responsible for 43 percent of the world's canola oil exports. Besides alternative crops, prairie farmers have turned to livestock and hog production. One of the most significant increases shown by the 2001 census was the 26-percent increase in hog numbers since 1996. International demand, from the United States, Japan, and new markets such as Mexico, helped to spur the increase.

THE PELICAN PROJECT

What started as a nonprofit project to protect a colony of 900 to 1,100 pairs of white pelicans in Saskatchewan's Redberry Lake Migratory Bird Sanctuary became a UNESCO Biosphere Reserve in 2001. Biosphere status confirmed that the project, with the support of the local community of Hafford, had committed to sustainability principles such as conserving biological diversity and creating opportunities for education and research. From the outset, the project has promoted sustainable viewing programs. The interpretative centre's programs and tours encourage knowledgeable appreciation of the area's varied wildlife. The 6,000-ha lake is a stopover for more than 200 migrating bird species, including threatened and endangered whooping cranes and peregrine falcons. The lake project uses a mobile satellite and computer telecommunications system to watch over its resident pelican colony. In recent years, Hafford has broadened its sustainability vision to deal with rural depopulation.

To stem outward migration, community plans call for enhancing the region's traditional agricultural mainstays of wheat and livestock and developing more sustainable farming methods, new markets, crops, and tourism-related activities.

Brightly painted, wooden grain elevators like these near Webb, Sask., have long dominated the prairie horizon. However, they are slowly disappearing and being replaced by large cylinder-shaped concrete terminals. Saskatchewan once possessed more than 3,000 traditional grain elevators. By 2001, fewer than 300 were left standing.

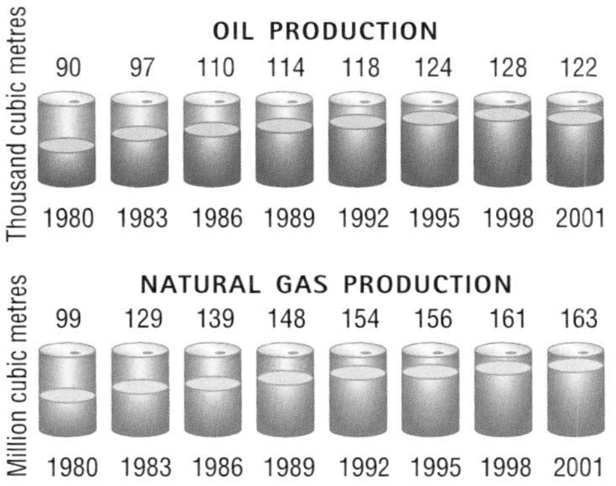

OIL PRODUCTION

Thousand cubic metres

1980	1983	1986	1989	1992	1995	1998	2001
90	97	110	114	118	124	128	122

Left: Canada's petroleum output, 1980–2001.

NATURAL GAS PRODUCTION

Million cubic metres

1980	1983	1986	1989	1992	1995	1998	2001
99	129	139	148	154	156	161	163

Below: The main oil and natural gas pipelines to markets in eastern Canada and the U.S.A.

Legend:
- Oil pipeline
- Gas pipeline
- Oil refinery
- Oil/gas well

Alberta
Saskatchewan
Manitoba
Edmonton
Calgary
Regina
Winnipeg

ALBERTA—TREASURE OF THE TAR SANDS

The most valuable resources on the Central Plains are its oil and natural gas deposits. The ecozone holds over 70 percent of Canada's oil reserves. Alberta sits on 65 percent of the oil—followed by Saskatchewan (12 percent)—and 82 percent of the natural gas. Pipelines carry western oil and gas to eastern Canada and the United States. In addition to these reserves, some 2.5 trillion barrels of oil, 300 billion of which is recoverable, are locked up in northern Alberta's tar sands. This oil-and-sand mixture, known as bitumen, lies in four different deposits over an area the size of New Brunswick. The tar sands put Canada after Saudi Arabia in proven oil reserves. At Fort McMurray, site of the largest deposit, bitumen is extracted from one of the world's largest open-pit mines (right). Once mined, bitumen is heated to separate oil from the sand. Although the tar-sands project has followed provincial environmental guidelines, its long-term effect is problematic. Already, Fort McMurray's two oil-sands plants are Canada's fourth largest source of carbon dioxide. In economic terms, the future is assured: by 2025, an estimated 70 percent of Canada's oil will come from the tar sands.

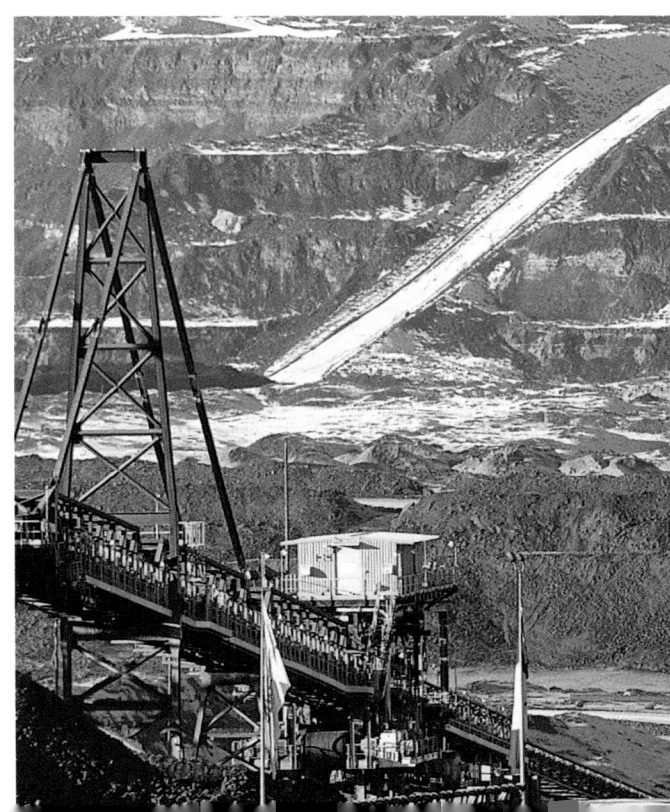

Boreal Shield

S ituated where an evergreen forest overlies the Canadian Shield, the Boreal Shield is the largest of Canada's terrestrial ecozones. This natural region—popularly known as "shield country"—extends 3,800 km from Saskatchewan to Newfoundland and Labrador. It covers 1.8 million square kilometres and encompasses almost 20 percent of Canada's landmass; its myriad rivers and lakes account for 22 percent of Canada's freshwater surface area. The Boreal Shield's rich supply of minerals and lumber plays a major role in fueling the economy of the "heartland" of southern Ontario and Quebec. Its bare rock, thin soils, and muskeg have restricted development to resource exploitation. Some major communities—Chicoutimi-Jonquière, Sudbury, Elliot Lake, and Thunder Bay—have successfully diversified their economic base, while other one-industry centres have experienced significant declines in population and primary-resource activity. Although less than 10 percent of Canadians reside here, the Boreal Shield's beauty attracts outdoor enthusiasts from urban centres in the south.

"SHIELD COUNTRY"

Otter

Landforms: Uplands and hills composed of Precambrian rock, exposed by repeated advances of ice-age glaciers. This rugged ecozone is strewn with lakes, rivers, streams, and wetlands, which form 20 percent of the Shield

Climate: Cold winters, warm to hot summers, with moderate precipitation in the heart of the ecozone. At Thunder Bay, the average monthly temperatures range from −8.9°C (January) to 24.4°C (July), while the average annual precipitation is 704 mm. At St. John's, temperatures are slightly lower, but precipitation much higher: 1,482 mm

Vegetation: White spruce, black spruce, and balsam fir create the uniform evergreen mass covering most of the Boreal Shield. Black spruce, the most northerly growing tree, fades out where the Boreal Shield borders the Hudson Bay Lowlands and the Taiga. Alder, birch, poplar, and willow are deciduous species found along river edges or in burnt-out clearings and cut-over areas. Tamarack grows in wetland areas; jack pine, in rocky clefts. Along its southeastern margins, evergreens blend with the deciduous species of the Mixedwood Plains ecozone. Here, maples enliven the landscape with brilliant splashes of colour in the fall

Lynx

Wildlife: The various natural settings of the boreal forest provide habitats for different animals. Woodland caribou seek out mature forests, which have an abundance of its favourite food, lichens. Ponds and swamps host creatures such as beaver, muskrat, and otter, as well as the mighty moose and predators such as the wolf and the lynx. Clear-cut areas attract the black bear, the porcupine, the striped skunk, and the white-tailed deer. Birds include the loon (a species adversely affected by the acidification of lakes), the boreal owl, the gray jay, and the woodpecker. Half a billion warblers, songsters of the forests, nest in this region from spring to fall. Boreal lakes host Brook char, lake trout, northern pike, perch, muskellunge, and walleye

Loon

Striped skunk

Resources and Industries:

Forestry, mining, pulp and paper production, tourism and recreation, hunting and trapping. Farming, limited by thin soil, is confined to a small number of fertile enclaves, including the Clay Belt near Cochrane, in north-central Ontario, and around Lac Saint-Jean in Quebec

NATURAL HAZARDS: FIRES AND PESTS

Across Canada, 9,000 to 12,000 forest fires occur annually, burning 2 to 7 million hectares. Wildfire is an inescapable aspect of the forest cycle, periodically removing overmature trees to clear the way for vigorous new growth. Many plants flourish in direct sunlight and the mineral-rich soil left by fire. In the Boreal Shield, fire benefits poplar and birch, which spread seeds widely after a conflagration. Although most fires result from human carelessness—an untended campfire or a smoldering cigarette—85 percent of the forest area burned annually is caused by lightning strikes. In built-up areas, fire suppression is essential to protect property and commercial timber stands. Firefighting may involve clearing fire lines and trenches to halt the fire, or using airplanes to douse a blaze with water.

Insect infestation is also a major forest menace. Spruce budworm, the dominant infestation of the Boreal Shield, takes hold in about 10 million hectares annually. According to some estimates, insects cause nine times more damage than fire.

Lake Athabasca

Wollaston Lake

Reindeer Lake

SASKATCHEWAN

Churchill River

La Ronge

Flin Flon

Nelson River

Thompson

MANITOBA

Norway House

Lake Winnipeg

Pine Falls

Kenora   Dryden

Fort Frances

ONTARIO

Lake Nipigon

Thunder Bay

Marathon

Wawa

Lake Superior

Sault Ste. Marie

■ Forestry-dependent communities
▲ Mining-dependent communities

Sudbury area before 1990

Sudbury area after 1990

MINING IN THE SHIELD

Ancient geological upheavals formed the bounty of minerals in the Boreal Shield. In the 1880s, the building of the CPR across northern Ontario first revealed the region's deposits of gold, silver, copper, nickel, cobalt, and zinc. The output of this region's mines (*see below*) ensures Canada's position as a leading mineral producer. Mining is the mainstay for more than 80 communities in the Shield. The map identifies some of the main mining centres. Although the mining industry is dominant in this region, its operations cover only 5,500 km² —or about 0.03 percent of this vast ecozone. Ontario and Quebec are the leaders in terms of the value of production. In the 1990s, for example, the annual value of Ontario's mineral production, based entirely in the Boreal Shield, was about $4 billion. The focus of uranium production, once dominated by Ontario, has moved to Rabbit Lake and other sites in northern Saskatchewan.

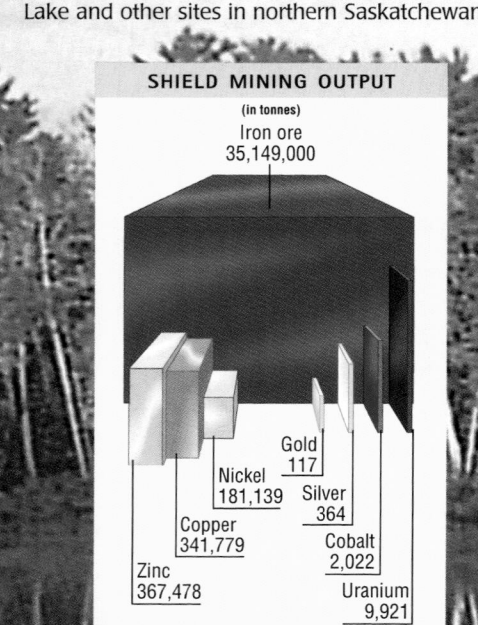

SHIELD MINING OUTPUT
(in tonnes)

Iron ore
35,149,000

Gold
117

Nickel
181,139

Silver
364

Copper
341,779

Cobalt
2,022

Zinc
367,478

Uranium
9,921

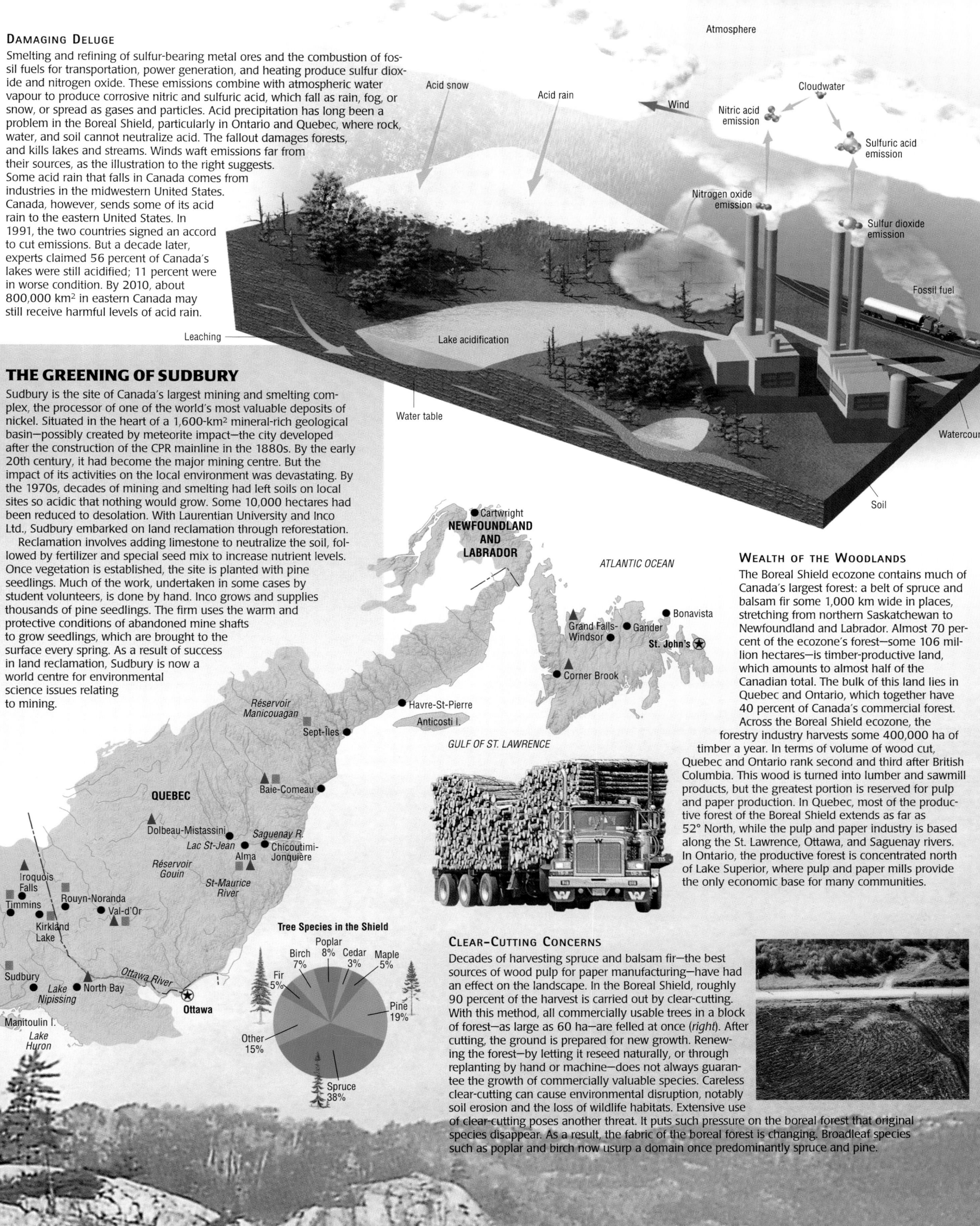

DAMAGING DELUGE

Smelting and refining of sulfur-bearing metal ores and the combustion of fossil fuels for transportation, power generation, and heating produce sulfur dioxide and nitrogen oxide. These emissions combine with atmospheric water vapour to produce corrosive nitric and sulfuric acid, which fall as rain, fog, or snow, or spread as gases and particles. Acid precipitation has long been a problem in the Boreal Shield, particularly in Ontario and Quebec, where rock, water, and soil cannot neutralize acid. The fallout damages forests, and kills lakes and streams. Winds waft emissions far from their sources, as the illustration to the right suggests. Some acid rain that falls in Canada comes from industries in the midwestern United States. Canada, however, sends some of its acid rain to the eastern United States. In 1991, the two countries signed an accord to cut emissions. But a decade later, experts claimed 56 percent of Canada's lakes were still acidified; 11 percent were in worse condition. By 2010, about 800,000 km² in eastern Canada may still receive harmful levels of acid rain.

Atmosphere

Acid snow

Acid rain

Wind

Cloudwater

Nitric acid emission

Sulfuric acid emission

Nitrogen oxide emission

Sulfur dioxide emission

Fossil fuel

Leaching

Lake acidification

Water table

Watercourse

Soil

THE GREENING OF SUDBURY

Sudbury is the site of Canada's largest mining and smelting complex, the processor of one of the world's most valuable deposits of nickel. Situated in the heart of a 1,600-km² mineral-rich geological basin—possibly created by meteorite impact—the city developed after the construction of the CPR mainline in the 1880s. By the early 20th century, it had become the major mining centre. But the impact of its activities on the local environment was devastating. By the 1970s, decades of mining and smelting had left soils on local sites so acidic that nothing would grow. Some 10,000 hectares had been reduced to desolation. With Laurentian University and Inco Ltd., Sudbury embarked on land reclamation through reforestation.

Reclamation involves adding limestone to neutralize the soil, followed by fertilizer and special seed mix to increase nutrient levels. Once vegetation is established, the site is planted with pine seedlings. Much of the work, undertaken in some cases by student volunteers, is done by hand. Inco grows and supplies thousands of pine seedlings. The firm uses the warm and protective conditions of abandoned mine shafts to grow seedlings, which are brought to the surface every spring. As a result of success in land reclamation, Sudbury is now a world centre for environmental science issues relating to mining.

Cartwright

NEWFOUNDLAND AND LABRADOR

ATLANTIC OCEAN

Grand Falls-Windsor • Gander • Bonavista

St. John's ⭐

Corner Brook

Havre-St-Pierre

Réservoir Manicouagan

Anticosti I.

Sept-Îles

GULF OF ST. LAWRENCE

Baie-Comeau

QUEBEC

Dolbeau-Mistassini

Lac St-Jean   Saguenay R.

Réservoir Gouin

Alma   Chicoutimi-Jonquière

St-Maurice River

Iroquois Falls

Timmins

Rouyn-Noranda

Val-d'Or

Kirkland Lake

Sudbury

Lake Nipissing   North Bay

Ottawa River

Ottawa ⭐

Manitoulin I.

Lake Huron

WEALTH OF THE WOODLANDS

The Boreal Shield ecozone contains much of Canada's largest forest: a belt of spruce and balsam fir some 1,000 km wide in places, stretching from northern Saskatchewan to Newfoundland and Labrador. Almost 70 percent of the ecozone's forest—some 106 million hectares—is timber-productive land, which amounts to almost half of the Canadian total. The bulk of this land lies in Quebec and Ontario, which together have 40 percent of Canada's commercial forest. Across the Boreal Shield ecozone, the forestry industry harvests some 400,000 ha of timber a year. In terms of volume of wood cut, Quebec and Ontario rank second and third after British Columbia. This wood is turned into lumber and sawmill products, but the greatest portion is reserved for pulp and paper production. In Quebec, most of the productive forest of the Boreal Shield extends as far as 52° North, while the pulp and paper industry is based along the St. Lawrence, Ottawa, and Saguenay rivers. In Ontario, the productive forest is concentrated north of Lake Superior, where pulp and paper mills provide the only economic base for many communities.

Tree Species in the Shield

Poplar 8%

Birch 7%

Cedar 3%

Maple 5%

Fir 5%

Pine 19%

Other 15%

Spruce 38%

CLEAR-CUTTING CONCERNS

Decades of harvesting spruce and balsam fir—the best sources of wood pulp for paper manufacturing—have had an effect on the landscape. In the Boreal Shield, roughly 90 percent of the harvest is carried out by clear-cutting. With this method, all commercially usable trees in a block of forest—as large as 60 ha—are felled at once (right). After cutting, the ground is prepared for new growth. Renewing the forest—by letting it reseed naturally, or through replanting by hand or machine—does not always guarantee the growth of commercially valuable species. Careless clear-cutting can cause environmental disruption, notably soil erosion and the loss of wildlife habitats. Extensive use of clear-cutting poses another threat. It puts such pressure on the boreal forest that original species disappear. As a result, the fabric of the boreal forest is changing. Broadleaf species such as poplar and birch now usurp a domain once predominantly spruce and pine.

Left: George Lake, Killarney Provincial Park, Ontario, embodies the distinctive essence of the Boreal Shield landscape

Mixedwood Plains

The Great Lakes–St. Lawrence Basin stretches from the western edge of Lake Superior to the Gulf of St. Lawrence. The southern portion, described here, is called the Mixedwood Plains. Canada's smallest ecozone, occupying 9 percent of the country, it consists of two sectors: Ontario's southward protruding peninsula, bounded by lakes Ontario, Erie, and Huron, and the long, narrow plains along the St. Lawrence in Quebec. The Thousands Islands is the point of division between the sectors. Well favored with navigable waterways, fertile soils, and a relatively mild climate, the Mixedwood Plains ecozone was the portal through which European explorers and settlers passed to enter the heart of Canada. Today, this region supports Canada's largest urban concentration, as well as a preponderant portion of its industrial and agricultural base.

Landforms: A relatively flat landscape with some hills. The thick and fertile claybed of eastern Ontario and southern Quebec is the legacy left by the Champlain Sea, which covered this area 10,000 years ago. In southern Ontario, moraines and other ice-age remnants lend a gently rolling aspect to the land. The outstanding physical features include the Oak Ridges Moraine (*see opposite page*), the Niagara Escarpment, and the Niagara Peninsula, famed for its fruit and wines. Another feature of interest is Point Pelee, the most southerly tip of the Canadian mainland. This 18-km-long peninsula is composed of a 70-m-thick deposit of sand, which sits on a submerged limestone ridge. Jutting into Lake Erie, Point Pelee lies at the same latitude as Northern California. In southern Quebec, the Monteregian Hills near Montréal are the most distinctive features in an otherwise flat-lying region

Climate: Cool to cold winters, warm to hot summers, moderate rainfall. Mean daily temperatures range from −3°C to −12°C in January to 18°C to 22°C in July; precipitation ranges from 720 mm to 1,000 mm annually. In southwestern Ontario, agriculture benefits from the moderating effect of the Great Lakes. Changeable weather throughout the region is caused by an alternating flow of air masses, such as the warm, humid air from the Gulf of Mexico and cold, dry air from the Arctic

Vegetation: The forests contain a mixture of deciduous and evergreen species. Little of the original woodland survives in this heavily urbanized and agricultural region. Deciduous species include beech and maple in mature forests; red and white oak, and walnut in younger woodlands. Conifers interspersed throughout the deciduous stands include white and red pine, and western hemlock. Southwestern Ontario is the northern limit of the Carolinian forest, where rich soil and a warm climate support a wide variety of colourful tree species, such as sassafras, hackberry, and shagbark hickory, as well as sycamore and tulip tree

Wildlife: Gray squirrel, groundhog, opossum, otter, raccoon, red fox, and woodchuck are typical woodland denizens. Large mammals such as the white-tailed deer and black bear, once found here, have largely disappeared due to human encroachment. Birds include blue jay, whippoorwill, Baltimore oriole, and red-headed woodpecker. More than 200 bird species pass over Point Pelee, where two major North American migratory routes—the Mississippi and Atlantic flyways—intersect. In the fall, the monarch butterfly also pauses here

Resources and Industries: Manufacturing and service industries, commerce and finance, construction, agriculture, tourism, and recreation

AT A GLANCE FACTS: THE GREAT LAKES

	Superior	Michigan	Huron	Erie	Ontario
Length in kilometres	563	494	332	388	311
Breadth in kilometres	257	190	295	92	85
Depth in metres	405	281	229	64	244
Volume of water in cubic kilometres	12,100	4,920	3,540	484	1,640

THE LAKES AND THE SEAWAY

The overall surface area of lakes Superior, Huron, Michigan, Erie, and Ontario is 246,000 km² —or 20 percent of the earth's surface freshwater—making the Great Lakes the world's largest lake system. The combined water volume of the lakes is a staggering 23,000 km³ and the combined drainage area is 766,000 km². The Great Lakes, which drop about 200 m in elevation from west to east, are linked to the Atlantic Ocean by the St. Lawrence River. The river flows 3,058 km northeastward from the eastern end of Lake Ontario to the Gulf of St. Lawrence. The river's drainage area is more than 1,344,000 km² and its mean discharge is 9,850 m³ per second—the largest of all rivers in Canada. Opened in 1959, the St. Lawrence Seaway—a system of canals, locks, dams, control structures, and harbours—allows vessels up to 222.5 m long, 23.2 m wide, and loaded to a maximum draft of 7.9 m, to sail 3,800 km from Montréal to the head of Lake Superior. There are six canals with a total length of 97 km on the Seaway. The canals have 19 locks, eight of which are on the Welland Canal, which bypasses the 99.5-m drop on the Niagara River—including 54-m-high Niagara Falls—which connects lakes Erie and Ontario. Overall, about 200 million tonnes of cargo move through the Seaway every year.

Seaway Cargoes

- Wheat 21%
- Other manufactured products 7%
- Iron and steel products 12%
- Other mine products 14%
- Iron ore 29%
- Other agriculture products 17%

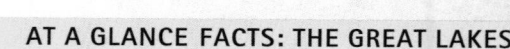

This view shows the St. Lawrence River as it flows northeastward from Lake Ontario to the Gulf of St. Lawrence. Along its 1,197-km course, the mighty river narrows or widens at different points (from bottom to top): ❶ the Thousand Islands; ❷ the junction of the Ottawa and St. Lawrence rivers in the lakes and rivers around Montréal; ❸ Lac Saint-Pierre, fed by the waters of the Richelieu River; ❹ the city of Québec and Île d'Orléans; and ❺ the estuary, where fresh and salt waters meet and the river is more than 50 km wide.

Land Cover in the Mixedwood Plains
Much of this densely populated region, as this chart indicates, is used for farming

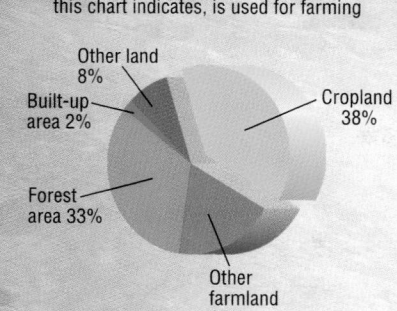

- Other land 8%
- Built-up area 2%
- Cropland 38%
- Forest area 33%
- Other farmland 19%

WHAT THE FARMS PRODUCE

The Mixedwood Plains' fertile soil, hot summers, plus an abundant water supply, provide ideal conditions for ample livestock and agricultural production. Roughly 37 percent of Canada's agricultural production comes from this ecozone.

■ Fodder crops, particularly hay and corn, support a livestock industry of cattle, sheep, pigs, and poultry in the St. Lawrence River valley.

■ The Niagara Peninsula—or the Fruit Belt—just east of Hamilton grows cherries, pears, peaches, and numerous other fruits in abundance. The area also grows grapes and is the centre of a vibrant wine industry.

■ Ontario's southwestern Essex and Kent counties—the province's Vegetable Belt—enjoys the warmest year-round climate in eastern Canada. The area produces the bulk of the nation's greenhouse cucumbers and tomatoes.

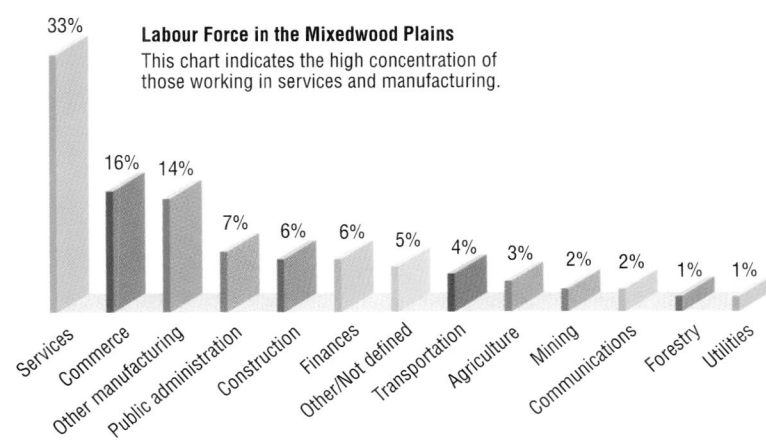

A NATURAL BALANCE

PROTECTING THE OAK RIDGES MORAINE

In 2002, the Ontario government issued a plan to ensure the sustainability of the Oak Ridges Moraine. The plan was intended to resolve a clash between developers and environmentalists over a proposed housing scheme on moraine land at Richmond Hill, just north of Toronto. The moraine is a conspicuous height of land—some 160 km long and up to 24 km wide—that extends from the Niagara Escarpment to Rice Lake. A popular recreational spot, the moraine supports a delicate environment of woods and wetlands. Beneath its surface lies Canada's largest aquifer, which feeds waterways such as the River Rouge (above). The plan divides the moraine into four zones. Natural Core Areas, occupying 38 percent of the moraine, get maximum environmental protection. The plan permits agriculture and recreation in Countryside Areas (30 percent) and more intensive development in Settlement Areas (8 percent). About a quarter of the moraine is set aside for a network of corridors, at least 2 km wide, which provide for the movement of wildlife. Although environmentalists debate the plan, some hail it as offering a significant degree of protection.

Industrial Profiles: Windsor to Québec

- ■ Nonmetallic minerals and miscellaneous
- ☐ Rubber, plastic, petroleum, coal, and chemicals
- ■ Primary metals and metal fabricating
- ☐ Machinery, transport equipment, and electrical products
- ■ Wood furniture, paper, publishing, and printing
- ■ Leather, textiles, knitting, and clothing
- ☐ Food and beverages

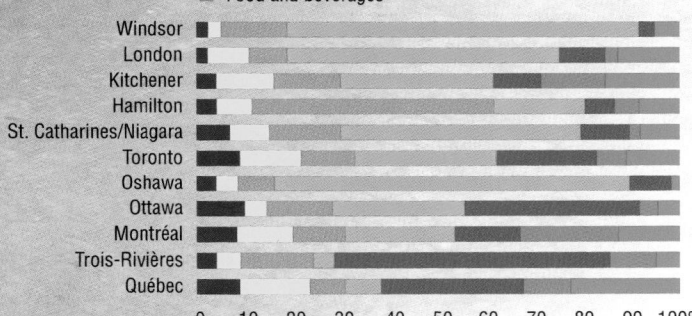

Windsor
London
Kitchener
Hamilton
St. Catharines/Niagara
Toronto
Oshawa
Ottawa
Montréal
Trois-Rivières
Québec

0 10 20 30 40 50 60 70 80 90 100%

WHERE THE PEOPLE LIVE

■ Canada's most highly urbanized concentration is located around the lower Great Lakes and along the St. Lawrence. According to the 2001 census, 12 of Canada's 27 census metropolitan areas (CMAs)—that is, urban areas with core centres of more than 100,000 people—are located here.

■ Population density is over 100 persons per square kilometre—10 times higher than anywhere else in Canada.

■ Ontario's population accounts for 38 percent of the Canadian total. More than 93 percent of the province's 12 million people live in the Mixedwood Plains.

■ Quebec has 24 percent of Canada's population. About 80 percent of the province's population is concentrated in three centres: Montréal, Québec, and Trois-Rivières, all of which lie in the Mixedwood Plains.

WHAT THE CITIES PRODUCE

The Mixedwood Plains is the site of Canada's industrial powerhouse. Ontario's part of this region produces over 40 percent of the country's goods; the Quebec portion, 22 percent.

■ No other region in Canada can match southern Ontario's advantages favourable to industry: a large labour force and consumer market, proximity to the United States, abundant supplies of raw materials and energy, and superior rail, road, and water transportation links. Toronto, with its many investment and insurance companies, and its bank and corporate head offices, is the city with the most diversified industrial base.

■ Around the western end of Lake Ontario, stretching from St. Catharines–Niagara to Oshawa, is the "Golden Horseshoe," an industrial concentration, which embraces St. Catharines–Niagara, Hamilton, Oakville, and Oshawa. This concentration extends westward from Toronto to Windsor and includes Brantford, Guelph, Kitchener, London, and Chatham-Kent.

■ Southern Quebec boasts a strong, diversified industrial base, with Montréal dominating high-technology fields such as aerospace, biotechnology, fiber optics, and computers. Quebec's industries also excel in clothing design and manufacturing, metal refining, printing, textiles, and transport equipment.

■ The chart (left) indicates the importance of different types of manufacturing in key industrial centres along the Windsor-Québec corridor.

Labour Force in the Mixedwood Plains
This chart indicates the high concentration of those working in services and manufacturing.

- Services 33%
- Commerce 16%
- Other manufacturing 14%
- Public administration 7%
- Construction 6%
- Finances 6%
- Other/Not defined 5%
- Transportation 4%
- Agriculture 3%
- Mining 2%
- Communications 2%
- Forestry 1%
- Utilities 1%

Atlantic Region

The Atlantic Maritime ecozone covers 2 percent of Canada's area. Within its embrace are Nova Scotia, New Brunswick, and Prince Edward Island, bound together by fisheries and forests. This ecozone extends to parts of Quebec: the Appalachian highlands and the Gaspé Peninsula. A diversity of physical features—wooded uplands, fertile lowlands, and an 11,200-km-long shoreline—endow this ecozone with incomparable beauty. Offshore lie the Atlantic and Northwest Atlantic Marine ecozones, which the Maritime provinces and Quebec share with Newfoundland. In the 20th century, the Atlantic provinces faltered with slow economic and population growth. Yet, as a new century dawns, offshore oil and gas development promises to quicken the economic pace.

ATLANTIC MARITIME

Landforms: Uplands, coastal lowlands

Climate: The Gulf Stream creates a temperate climate. Cool to cold winters, warm to hot summers, moderate to heavy precipitation

Vegetation: A mix of conifer (softwood) and deciduous (hardwood) species. Red and black spruce and balsam fir flourish along the coastlines and in the highlands of Cape Breton and the Gaspé

Wildlife: White-tailed deer, moose, black bear, raccoon, blue jay, eastern bluebird

Resources and industries: Forestry, agriculture, fish processing, tourism/recreation

Eastern bluebird

ATLANTIC ECOZONE

Extent: The chief feature of this marine ecozone is the continental shelf that spreads 500 km east from Newfoundland and 200 km southeast of Nova Scotia. Dozens of undersea plateaus, known as banks, lie across the shelf. Water depths over the banks are generally less than 100 m. They create an excellent environment for fish reproduction and growth. The largest, the Grand Banks of Newfoundland and the Georges Bank in the Gulf of Maine, have long been vital fishing areas

Climate: The Gulf Stream brings equatorial waters to the Maritimes before crossing the Atlantic to Europe. The waters off Nova Scotia and in the Bay of Fundy, also part of this ecozone, are temperate and ice-free

Fish: Capelin, cod, flounder, hake, haddock, herring, plaice, pollock, redfish, turbot; lobster and scallop; mackerel and Atlantic salmon

Seabirds: Great and double-crested cormorant, Atlantic puffin, common and thick-billed murre, black guillemot, razorbill

Marine mammals: Seals, whales, porpoises, dolphins

Resources and industries: Important commercial fisheries, oil and gas development

Humpback whale

NORTHWEST ATLANTIC

Extent: This marine ecozone stretches from the mouth of Lancaster Sound in the southern Arctic ecozone to the Grand Banks, where the frigid south-flowing Labrador Current merges with the Gulf Stream. It encompasses the southern Baffin Island Shelf, Hudson Strait, the Labrador Shelf, and the northern Newfoundland Shelf, and extends into the Gulf of St. Lawrence

Wildlife: More than 20 whale species, notably humpback, bluefin, and minke, are found in this ecozone. Harp seals thrive on an abundant marine food supply available along the Labrador coast

Resources and industries: Fishing, tourism

Razorbills

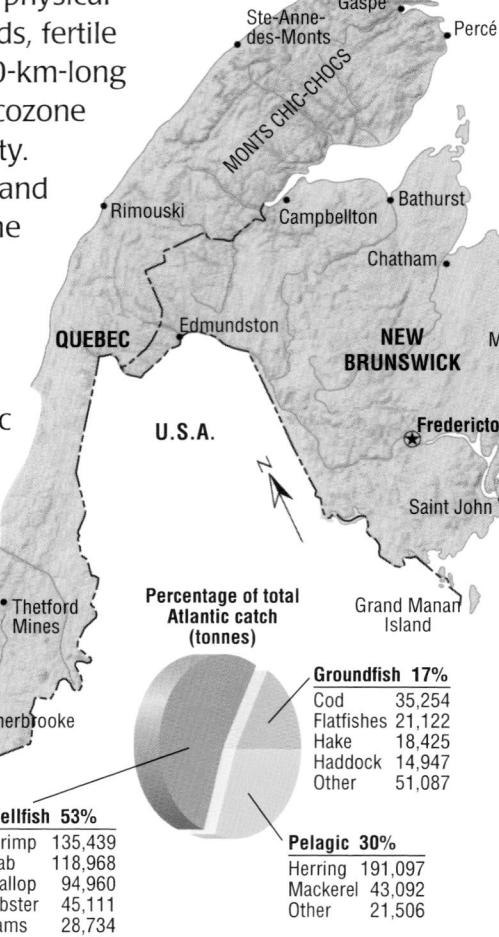

(Map labels:) Ste-Anne-des-Monts · Gaspé · Percé · Îles de la Madeleine · Glace Bay · GULF OF ST. LAWRENCE · Cape Breton Island · Sydney · Bras d'Or Lake · MONTS CHIC-CHOCS · Rimouski · Campbellton · Bathurst · PRINCE EDWARD ISLAND · Chatham · Summerside · Charlottetown · NORTHUMBERLAND STRAIT · Antigonish · Canso · QUEBEC · Edmundston · NEW BRUNSWICK · Moncton · New Glasgow · U.S.A. · Amherst · Truro · Fredericton · Minas Basin · NOVA SCOTIA · Saint John · Halifax · BAY OF FUNDY · ATLANTIC OCEAN · Thetford Mines · Grand Manan Island · Bridgewater · Digby · Liverpool · Sherbrooke · Yarmouth

Percentage of total Atlantic catch (tonnes)

Groundfish 17%
Cod	35,254
Flatfishes	21,122
Hake	18,425
Haddock	14,947
Other	51,087

Shellfish 53%
Shrimp	135,439
Crab	118,968
Scallop	94,960
Lobster	45,111
Clams	28,734
Other	24,116

Pelagic 30%
Herring	191,097
Mackerel	43,092
Other	21,506

Atlantic cod

ATLANTIC FISHERIES

The Maritime provinces, Newfoundland and Labrador, and Quebec account for roughly 75 percent of Canada's total fish catch. In the Atlantic Maritime, Nova Scotia and Prince Edward Island fishers ply the waters where they find a variety of fish, including cod, grey sole, flounder, redfish, and shellfish. Nova Scotia is the leading producer of fish (about 30 percent of total production), followed by Newfoundland and Labrador and the Pacific Coast fishery of British Columbia (each about 20 percent). In terms of value, the same rankings apply. Along the Atlantic coast, some 1,000 communities mainly depend on the fisheries and related industries such as fish processing plants and shipbuilding. Any change in the fishery industry exerts a powerful effect on the communities that support it.

Before the 1992 ban, Newfoundlanders depended mostly on cod. They still fish in the Grand Banks, where cod and other groundfish congregate in the shallow waters of the continental shelf. But focus on different fish and fishing techniques has superseded the onetime dependence on cod fishing, and has resulted in a rejuvenation of the Newfoundland and Labrador economy. Sales of snow crab and northern shrimp, whose catches require fishing farther from shore and in deeper waters, have increased dramatically since 1992. Increased production of crab and shrimp has opened new markets and has resulted in the construction and operation of new processing plants. Of the 30,000 fishers who lost their jobs in 1992, more than half are again employed.

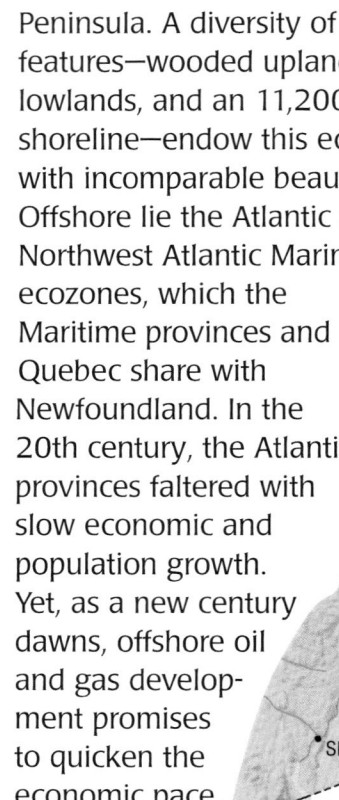

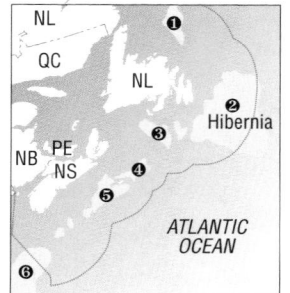

Newfoundland is part of the Boreal Shield, yet dependent on the fisheries of the Atlantic marine ecozone. This map locates the region's major fishing banks: Funk Island ❶; Grand Banks ❷; St. Pierre ❸; Banquereau ❹; Sable Island ❺; and Georges Banks ❻. Also shown, the Hibernia oil site. The White Rose deposit lies north of Hibernia; Terra Nova is situated to the south.

COD CALAMITY

For over 500 years, Atlantic Canada had one of the world's richest commercial fisheries. The Grand Banks were long a major source of cod, one of the world's leading food fishes. From the 1950s on, new fishing technologies—including dragnet fishing—and expanding markets for seafood caused overfishing. In 1977, Canada intervened to protect dwindling stocks and the Atlantic fishery by extending the offshore limit to 370 km (200 nautical miles). But by the late 1980s, groundfish stocks (fish living near the ocean floor, such as cod and halibut) were seriously depleted. In 1992, the steep decline in cod and other groundfish prompted Canada to curtail cod fishing off eastern Newfoundland and Labrador. What was originally a two-year government moratorium has since been extended indefinitely. Unfortunately, some foreign trawlers continue to overfish outside Canada's 200-mile protective limit.

Redfish

A MODEL FOREST

Model forests—there are 11 in Canada—enable individuals and groups to work together to maximize social and economic benefits from a forest region in ways that best sustain the forest. The Fundy Model Forest, for example, is a 34-member partnership, which occupies some 420,000 ha in south-eastern New Brunswick. It embraces Moncton, Fredericton, and Saint John, many communities, most of Fundy National Park (*above*), 3,500 privately owned woodlots, and major forestry operations.

The forest supports maple sugar products; woodcraft industries; Christmas trees—wreath production alone is worth more than $670,000 annually; firewood worth about $2 million a year to 40 percent of the region's households. Fishing brings in about $2.2 million a year; snowmobiling, cycling, downhill and cross-country skiing, and hunting (deer, moose, and bear) are all booming. Some 270,000 people visit Fundy National Park annually. Serving this tourist influx are 15 campgrounds and 31 motels, hotels, and bed-and-breakfasts.

To balance businesses' needs with forest survival, the model forest works with government, university, and industry researchers and scientists. As a result, loggers know that certain quantities of debris are needed to shelter small animals and provide nutrients for regrowth. Rather than working in isolation, woodlot owners jointly plan harvesting to make sure animal habitats and the watershed are preserved.

ATLANTIC FARMS

Farmland covers 9 percent of the Atlantic Maritime ecozone. Agricultural activity is concentrated in a few fertile areas: Prince Edward Island, New Brunswick's Saint John River valley, along the Northumberland Strait and the Bay of Fundy in Nova Scotia. Beyond these areas, the land is too rough and rugged for agricultural settlement. Favoured by a moist climate and silty, stone-free soils, farming is the leading industry on Prince Edward Island; potatoes are its best-known crop. The island also supports grain and dairy farms. Of all Canada's provinces, Prince Edward Island, called the "Garden of the Gulf," still uses almost half of its land area for agricultural purposes. In New Brunswick, the Saint John River valley produces potatoes and livestock, and there is mixed farming in the northwest of the province. Nova Scotia's best farmland is located along the Bay of Fundy and the Northumberland Strait, where the main agriculture activity revolves around dairy farming and poultry production. Nova Scotia's Annapolis Valley is renowned for its output of fruit—particularly apples. In the Atlantic Maritime region, as elsewhere across the country, farm size is increasing and the number of farms is decreasing.

Atlantic abundance
(crops per thousand hectares)

Tame hay and fodder	
Potatoes	
Barley	
Alfalfa	

0   50   100   150   200

The 16 teeth of Hibernia's platform can resist the impact of a 1-million-tonne iceberg

Hibernia's Topsides was placed on barges, then towed into position in the Atlantic

WEALTH FROM THE SHELF

Oil and gas developments hold the promise of revitalizing the economy of Atlantic Canada. Two oil projects in Newfoundland and Labrador have proved to be great successes: Hibernia and Terra Nova, in production since 1997 and 2002; a third major project, White Rose, is slated to begin production in 2005. Hibernia, located 315 km east of St. John's, Newfoundland, is a deposit of over 3 billion barrels of oil, lying underneath the Grand Banks—80 m below the water surface. The project required the construction of a specially designed 111-m-high storm- and iceberg-proof Gravity Based Structure (GBS). Atop this structure sits Topsides: two oil drill shafts, a riser shaft, and a utility shaft—each 17 m in diameter. By 2000, Hibernia accounted for 12 percent of Canada's oil output. Hibernia also holds a vast reserve of natural gas, which may be extracted in the future. The first offshore natural gas development in Canadian history was Nova Scotia's Sable project. Over 17 million cubic metres of gas are produced every day. In New Brunswick, oil and gas have been produced since the early 20th century.

ATLANTIC FORESTS

In the Atlantic Maritime ecozone, forests make up 90 percent of the total land cover. About half of this area is a mixed forest—a distinctive blend of deciduous (hardwood) and coniferous (softwood) species. Forestry has a long history here, longer than anywhere else in Canada. Old-growth hardwoods once covered much of the fertile land, but today only a few pockets of the true old-growth remain. In the early 1990s, it was estimated that more than 13 percent of the area harvested had been replanted or seeded several times. Unlike other regions, the Atlantic Maritime has a large proportion of privately owned forestland, totaling 90 percent in Prince Edward Island, 75 percent in Nova Scotia, and 50 percent in New Brunswick. In some rural areas, forestry may be the sole source of employment and the chief reason for a community's existence. The pulp and paper industry is the largest consumer of wood, using 65 percent of the annual harvest. About 24 percent is sawn into lumber. The Atlantic Maritime ecozone has two working-scale model forests—the Fundy and the Lower St. Lawrence—whose objective is the implementation of sustainable forest development. The Fundy Model Forest is described above.

Advocate Bay on Minas Channel, Nova Scotia, where the world's highest tides have been recorded

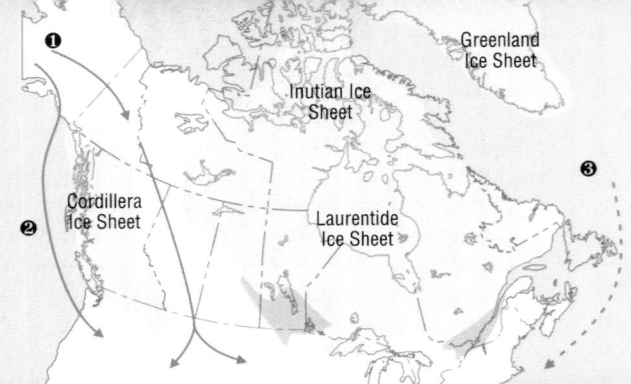

WHO WERE THE EARLIEST ARRIVALS?

The traditional theory holds that North America's first arrivals were Northeast Asian hunters of Mongoloid stock who followed mastodons across the Bering land bridge ❶ or sailed along the ice-free Pacific coast ❷. Some experts now think that ice-age Europeans known as Solutreans may have been the first arrivals. This new theory is based on the discovery of Solutrean-type spear points on the Atlantic coast. Extinct for 19,000 years, Solutreans came from France, Spain, and Portugal, and may have sailed to eastern North America along the ice-sheet edge ❸, subsisting on seals and seabirds. Skeletal remains from other archeological digs have given rise to another tantalizing hypothesis: that first arrivals were southern Asians, possibly Polynesians, who journeyed to North America via Australia, the Pacific, and South America.

ARCTIC PEOPLES

With no edible vegetation, and driftwood their only wood, the Arctic peoples—the Inuit—led a precarious life in the Far North. They hunted seals, whales, and walruses along the coastline, and followed the caribou inland. Birds, birds' eggs, and small mammals were other food sources. Blubber, fish, and meat were eaten raw. Dog teams pulled the snow sleds they made from driftwood, whalebone, and caribou antlers. Because of their keen sense of beauty, even everyday objects were fashioned with extraordinary care. All adornments had animal motifs. Groups of up to 100 wintered in snow house villages. Shamans and medicine men were highly regarded by these deeply spiritual people.

This 1901 photograph shows a long-abandoned Haida village on the Queen Charlotte Islands. Epidemics devastated the rich culture of the Haida as a result of contact with European traders in the late 1800s.

PACIFIC NORTHWEST

Salmon-swarming rivers, a bountiful sea, and the majestic rain forest gave power and wealth to the Haida and Coast Salish, the trading Tsimshian, the Nuu-chah-nulth (Nootka) whalers, the Tlingit, the Kwakwaka'wakw, and the Nuxalk (Bella Coolas). Food was varied and abundant. Towering red cedars yielded rot-resistant beams and framing for their fine homes, logs for their 22-m-long canoes, and rain-resistant bark for clothing and blankets. Renowned carvers of totems, masks, bowls, and helmets, they revered shamans for their links to the spirit world. The potlatch, a communal ritual of feasting, storytelling, dancing, and gift-giving, was all-important.

A DIVERSITY OF CULTURES

A network of independent native nations was spread across what is now Canada when the Europeans arrived. The nations spoke 55 languages and occupied vastly different environments. Their social and trading rituals, religious beliefs, and political structures were quite varied. Nevertheless, they could be classified into six major cultural groups (*see below*). They all shared a profound respect for nature, and all lacked immunity to smallpox, measles, and other imported diseases that would later all but wipe out the native population.

Right: An Inukshuk, "likeness of a person" in Inuktitut, is a cairn of stone used by the Inuit of the Arctic to show directions to travelers, or to mark a place of respect.

GWICH'IN (KUTCHIN)
INUVIALUIT
HARE
TUTCHONE
DOGRIB
TAHLTAN
DENE-THAH (SLAVEY)
ARCTIC
INUIT
HAIDA
TSIMSHIAN
NUXALK
DAKELH-NE (CARRIER)
SUBARCTIC
DUNNE-ZA (BEAVER)
CHIPEWYAN
NORTH-WEST COAST
NUU-CHAH-NULTH
SHUSWAP
SARCEE
CREE
PLATEAU
PLAINS CREE
CREE
KOOTENAY
PLAINS
BLACKFOOT
ASSINIBOINE
CREE
OJIBWA

PLATEAU

Fishing and foraging were mainstays of the Carrier, Lillooet, Okanagan, Shuswap, and other small tribes living between the Coast and Rocky mountains. They had great diversity of dress, religious beliefs, and language. They spoke dialects of four major language groups. Most wintered in semi-underground dwellings they entered through the roof; in summer they built bulrush-covered wooden lodges. The Columbia and Fraser rivers were their travel and trade routes and source of fish. Other foods were berries, wild vegetables, and game. The Plateau people fashioned canoes from the area's pine and cottonwood, and traded copper, jadeite, herbs, and oolichan oil to the coast Indians for otter pelts and decorative baskets.

c. 40,000–9000 B.C.
Waves of Asian hunters tracking mammoths cross over to North America via the Beringia land bridge

Some Asian migrants skirt the Pacific coast in small boats

c. 9000–6000 B.C.
Hunters' descendants, Paleo-Indians, occupy southern areas of what is now mainland Canada

c. 6000 B.C.
After mammoths become extinct, aboriginals hunt buffalo, caribou, small game, fish, and forage for fruit, nuts, roots, and berries

Ontario sites have yielded tools, axes, knives, and spear points from this era

c. 3000 B.C.
Paleo-Eskimos (the Denbigh people), coming from Siberia, via Alaska, settle in the Arctic. The Dorset People, their descendants, develop a technology suitable for the Arctic

Tools of this era have been found on Ellesmere Island

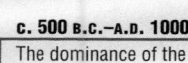

c. 500 B.C.–A.D. 1000
The dominance of the Dorset Culture peoples in the Arctic region

First Peoples

Aboriginal is the Canadian Constitution's collective term for our indigenous Indian, Inuit, and Métis peoples, now comprising more than 1.3 million (4.4 percent) of our population. Some 62 percent are native Indian, 30 percent are Métis, 5 percent are Inuit, and others belong to more than one aboriginal group. About one in five lives either in Ontario (22 percent) or British Columbia (19 percent). Less than half (47 percent) live on reserves, with the remainder in urban areas; half the aboriginal community is under 25 years of age. They may see their people's long struggle for land, resources, and self-government realized in their lifetimes because negotiations with the federal government and many provincial governments, more frequent and more fruitful in recent years, continue today. One of the most historic of hundreds of land settlements, ongoing since the 1970s, is the Inuit homeland, Nunavut. Ruling one-fifth of Canada's landmass, its legislature is the first in North America to be run by an aboriginal government.

SUBARCTIC

This culture encompassed mobile bands of Algonquian-speaking Cree and Innu east of Hudson Bay, and Athabascan-speaking Chipewyans, Dogrib, Hare, Dene-thah (Slavey), Dunne-za (Beaver), Gwich'in (Kutchin), Tutchone, Tahltan, and Dakelh-ne (Carrier) to the west. Skilled hunters, they occupied the taiga and boreal forests from Yukon to Newfoundland and Labrador. Caribou was their staple and this animal provided most of their necessities. Each summer, groups joined forces to socialize, pick berries, make canoes and snowshoes, and tan hides.

Right: *A shaman of the Western Subarctic once used this mask to enter the spirit world.*

TREATIES: PAST AND PRESENT

Historic Indian treaties encompass several agreements made between 1725 and 1923 by the Crown with various Indian tribes. Included were "peace and friendship" agreements in the East, and the "numbered" (1 to 11) treaties covering vast tracts in the West and North. Many tribes in Atlantic Canada, British Columbia, and Quebec were still without treaties when this phase ended.

In the 1970s, courts began ruling that aboriginal title rights exist in law. This prompted Ottawa to begin a flurry of modern treaties, the first being the 1975 James Bay and Northern Quebec Agreement with the Cree and Inuit. Another Inuit agreement in 1993 led to Nunavut's creation six years later.

Hundreds of modern treaties have now been made. Of 60 to 70 still in negotiation, some 50 are in British Columbia. An agreement with the Nisga'a in 2000 marked that province's first modern treaty. Negotiations elsewhere include Innu territory in Quebec, an Ottawa River valley claim by the Algonquin, and a Mississauga claim for compensation for some 102,000 ha on which Toronto and many of its suburbs sit.

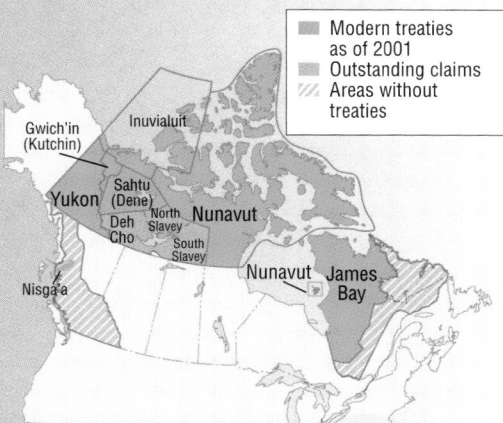

- ■ Modern treaties as of 2001
- ■ Outstanding claims
- ▨ Areas without treaties

Above: *Aboriginal land claims as of 2000. The map identifies aboriginal claims resolved by treaty, outstanding claims, and areas without treaties.*

A GROWING POPULATION

"We shall live again," proclaimed a ghost-dance song in the late 1800s, a time when many considered the aboriginal population a dying race. Data from the 2001 census affirms the song: at just over 1.3 million (4.4 percent of the total population, compared to 3.8 percent five years earlier), Canada's aboriginal population is on the rise. The upturn is most dramatic in the Métis community, which increased 43 percent in five years because of an increased fertility rate and better enumeration of community members.

The aboriginal community overall increased sevenfold in the second half of the 20th century, during which time Canada's general population merely doubled. This is all the more surprising since native birthrates have declined from 4 times to 1.5 times those of the general population.

Births, however, are only part of the picture. Lower infant death rates were a major factor, and about half the increase resulted from better reporting. More reserves were also better enumerated this time around (although some 30 were still not counted), and people were more willing than previously to acknowledge their aboriginal roots. Researchers attribute this to esteem-nurturing events such as the Royal Commission on Aboriginal People, court rulings on aboriginal rights, and the creation of Nunavut.

One-third of today's aboriginal population is 14 years or younger. In Nunavut, the median age is 19.1. Across the country, the median aboriginal age is 24.7—13 years younger than the mean age (37.7) for the general population.

EASTERN WOODLANDS

Two major language groups dominated this culture. Algonquian-speaking Ojibwa, Algonquins, Mi'kmaq, and Malaseet occupied land from Lake Superior to the Atlantic. The Iroquoian-speaking tribes included the Huron, located in southern Ontario, and the Iroquois, the Mohawk and others who lived in villages south of the Great Lakes and the St. Lawrence. Iroquoian speakers had a warring tradition. Men hunted and fished; women cultivated beans, maize, squash, and tobacco. When the soil was depleted in one place, they moved to new sites. The Algonquian speakers' lives were governed by the seasons: hunting in fall and winter; harvesting roots and berries in summer.

Left: *A 1965 recreation of a Plains Cree war bonnet. The eagle feathers on these elaborate headdresses traditionally represented acts of bravery in battle.*

PLAINS

The Plains Culture encompassed the nomadic Assiniboine, Blackfoot, Sarcee, and Plains Cree. Other than water and poles for their tepees, the buffalo met all their needs. Its meat was eaten at every meal. Hooves were boiled into glue; sinew became thread; stomachs served as pots; horns and bones were fashioned into tools and utensils; ribs became sled runners; hides made teepee covers, clothing, moccasins, and sleeping robes; buffalo hair made comfy cradle boards. Before horses, buffalo were hunted by herding them into enclosures or over cliffs. The arrival of horses in the early 1700s gave the hunters a distinct advantage and horses became a kind of currency on the Prairies. The Plains women played important roles in religious rituals.

A.D. 300–500
Eastern Woodland tribes cultivate beans, maize, squash. Villages develop

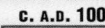
A Dorset mask (*right*), dating back 1,000 years, was likely used in rituals by shamans

C. A.D. 1000
Thule whalers from Alaska, ancestors of today's Inuit, displace the Dorset Culture population in most of the Arctic

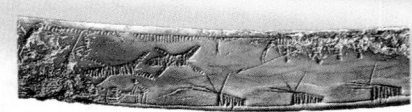
Ivory bow drill (*right*) with depictions of Thule life

C. A.D. 1142
Iroquois Confederacy frames its constitution Gayanashagowa, or Great Binding Law, which remains in force today

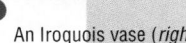
An Iroquois vase (*right*)

C. A.D. 1400
West Coast tribes establish vast trading network

Settling Canada

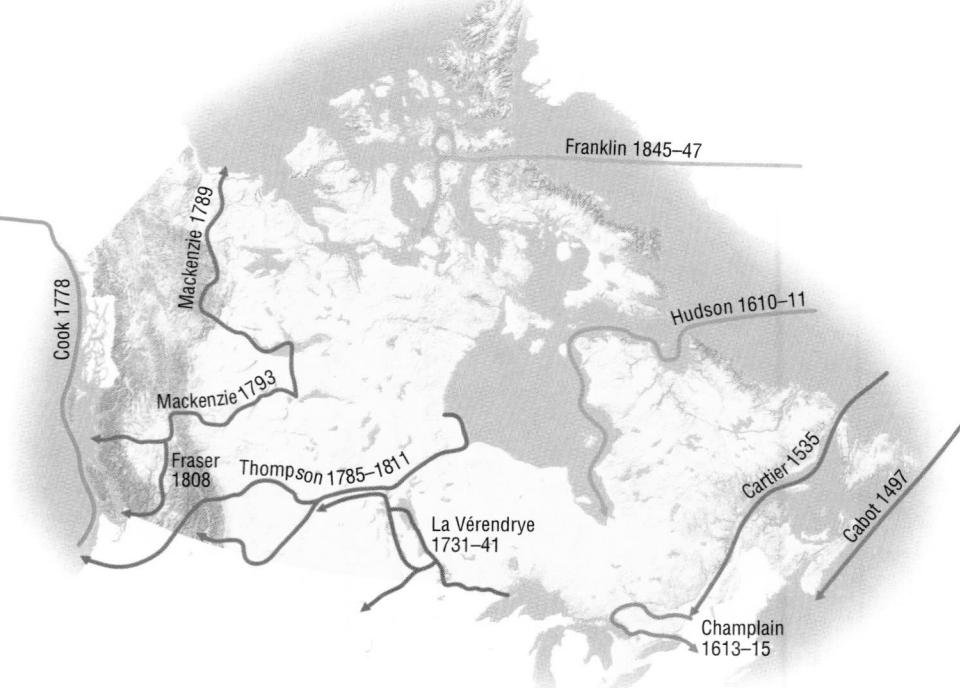

Aboriginal peoples were here for thousands of years before France planted a toehold on the Atlantic in 1605. When Britain acquired New France (1763), the toehold was a foothold on what are now the Atlantic provinces, Quebec, and Ontario. Loyalists arrived and a tradition began: people fleeing poverty, prejudice, and famine would be given refuge. In 1812, evicted Scottish farmers came to what is now Manitoba. Germans and Scandinavians, as well as persecuted Hutterites, Jews, and Mennonites, were among the million or so prairie pioneers who arrived between 1875 and 1914. To open the West, Canada bought Rupert's Land and made treaties with native peoples. In the 1880s, Chinese immigrants helped build a transcontinental railway.

Far in advance of settlers came myriad explorers whose expeditions into unknown territories and seas gradually unveiled the vastness of Canada. The map right outlines routes taken by major explorers of Canada between the 15th and 19th centuries.

■ John Cabot	1497	■ James Cook	1778
■ Jacques Cartier	1535	■ David Thompson	1785–1811
■ Henry Hudson	1610–11	■ Alexander Mackenzie	1789–93
■ Samuel de Champlain	1613–15	■ Simon Fraser	1808
■ Pierre de La Vérendrye	1731–41	■ John Franklin	1845–47

In the late 1890s, Canada opened the Prairies to eastern Europeans. Among the new arrivals, this Doukhobor family from Grandview, Manitoba, posed for their photograph in 1900.

To escape discrimination in British Columbia, the Chinese moved to the Prairies, then to Ontario and Quebec. This photograph of four Chinese in Calgary dates from 1910.

Dutch immigrants of the 1920s await departure from the Canadian Railways Terminal, Halifax. In the palmy twenties, many Dutch headed for Ontario, where jobs were plentiful.

■1000

VIKINGS
Africans, Britons, and Irish may have reached what is now Canada at various times between 500 B.C. and A.D. 500. The first confirmed European settlement, built about A.D. 1000, was the Viking enclave at L'Anse aux Meadows at the tip of Northern Peninsula in Newfoundland and Labrador. Discovered in 1960, the site has been reconstructed and declared a World Heritage Site. When occupied, it contained eight sod houses and an iron smelter. This might be the Vinland of Norse sagas, which describe encounters with an unknown people, possibly Dorset or Beothuk.

■1400

PORTUGUESE
Although Portuguese did not settle here in large numbers until the 1950s, place names such as Fundy (*Fundo*) and Labrador (*Lavrador*) are reminders of earlier visits. It is believed that they explored eastern Canada in 1452, 1470, and 1493, and Gaspar Corte-Real landed in Labrador in 1501. Portuguese fishermen hunted whales in the Arctic and fished the Grand Banks for cod for the next 200 years. Cod especially was in great demand in Europe where Catholics of the day abstained from meat on Fridays.

■1600

FRENCH
Jacques Cartier's voyages and discovery of the Gulf of St. Lawrence paved the way for settlement; first tiny colonies in present-day Nova Scotia (Port Royal, 1605) and present-day Québec (Champlain's fort, 1608), then other frail settlements on the Gulf. Jesuits preached to the Hurons, and *coureurs de bois* roamed inland in search of furs. Despite early hardships and Iroquois wars, a flourishing fur trade kept the colony alive. By 1688, New France had 10,000 inhabitants and extended from Hudson Bay to the Gulf of Mexico, and from Newfoundland to the Prairies.

■1783

LOYALISTS
British supporters in the American Revolution called themselves Loyalists. But in the fledgling United States many were beaten and lynched as traitors; 50,000 fled north between 1776 and 1783. Some settled in the St. Lawrence River valley and Great Lakes region. Most took refuge in what is now New Brunswick. Privation marked the early years but eventually they prospered, farming on land grants of 100 to 5,000 acres, lumbering, sawmilling, and trading with former enemies. The Loyalists were the first mass influx of English-speaking people into what is now Canada.

■1812

RED RIVER SETTLERS
The West's first European settlers were impoverished Highland Scots. They were brought there by Thomas Douglas, Earl of Selkirk, who owned a vast tract of the Red and Assiniboine river valleys. The first farmers arrived in 1812, occupying land the Métis considered theirs. Fearing disruption of the fur trade, the North West Company incited the Métis to resist. When 21 settlers were killed at Seven Oaks in 1816, the others abandoned the colony. Lord Selkirk hired a private army, rebuilt the settlement, and brought the settlers back. Ensuing court battles impoverished the Nor'Westers who were taken over by the Hudson's Bay Company in 1821.

1497 John Cabot lands in eastern Canada	**1534–36** Jacques Cartier charts the Gulf of St. Lawrence (1534); overwinters at what is now Québec (1535–36)	**1608** Samuel de Champlain builds a fort at present-day Québec	**1635** Champlain dies at Québec on Christmas Day	**1642** Paul de Chomedey, Sieur de Maisonneuve, founds Montréal	**1670** Newly founded Hudson's Bay Company granted all lands draining into Hudson Bay	**1713** Treaty of Utrecht cedes Hudson Bay, Newfoundland, and mainland Acadia to Britain	**1749–55** **1749:** Halifax founded by British **1755:** Britain deports Acadians	**1763** New France ceded to Britain by the Treaty of Paris

H.B.C.

CANADA'S CHANGING FACE

Canada's commitment to immigration is producing a rich ethnic mosaic. More immigrants came in the 1990s than in any decade in the last 100 years. The 2001 census shows 18.4 percent of the population was born outside Canada. Since the 1960s, there has been a major shift in where our immigrants come from. Unlike their predecessors, today's arrivals are choosing urban over rural lives, as the following charts show.

TOP 10 COUNTRIES OF BIRTH IN CANADA

1961		2001	
1. United Kingdom	24.3%	1. China	10.8%
2. Italy	16.5%	2. India	8.5%
3. Germany	10.8%	3. Philippines	6.7%
4. Netherlands	8.9%	4. Hong Kong	6.5%
5. Poland	5.0%	5. Sri Lanka	3.4%
6. United States	3.9%	6. Pakistan	3.2%
7. Hungary	3.1%	7. Taiwan	2.9%
8. Ukraine	2.4%	8. United States	2.8%
9. Greece	2.3%	9. Iran	2.6%
10. China	1.8%	10. Poland	2.4%

TOP 10 DESTINATIONS OF IMMIGRANTS IN 2001

1. Toronto, Ont.	111,564	6. Edmonton, Alta.	4,225
2. Montréal, Que.	32,998	7. Winnipeg, Man.	3,810
3. Vancouver, B.C.	29,922	8. Hamilton, Ont.	3,078
4. Calgary, Alta.	9,038	9. London, Ont.	1,709
5. Ottawa, Ont.	7,151	10. Québec, Que.	1,335

MINORITIES GAINING VISIBILITY

At Confederation, 61 percent of Canadians were of British, Irish, or Scottish origin; 31 percent were French; and the remaining 8 percent were aboriginals and other groups. Immigration patterns throughout the next century preserved this composition: some 90 percent of new Canadians were European-born, compared to about 3 percent from Asia.

Beginning in the 1960s, immigration patterns changed. Newcomers from southern Europe surpassed those from western and northern Europe. During the 1970s, Asians and Middle Easterners began arriving in larger numbers. Today, 58 percent of immigrants are Chinese, Indian, and other Asians; only 20 percent are European.

One result is a soaring number of visible minorities, Canada's term for people who are neither aboriginal or white. Because most new immigrants come from China, Canada's ethnic Chinese population now exceeds a million. One-third live in Vancouver. Other than Winnipeg, where they are outnumbered by a Filipino population, the Chinese are the largest visible minority in every city in western Canada. Minorities account for 16 percent of Calgary's population and 14 percent of Edmonton's.

Left: *Many immigrants to Canada crossed the Atlantic on liners in conditions less ideal than this welcoming poster suggests. Between 1928 and 1971, over a million of these newcomers first set foot in this country at Halifax's Pier 21, now a national historic site*

Today's immigrants head to major centres such as Toronto, Vancouver, and Montréal. In some cities, visible minorities are on their way to being the majority. Toronto, at 42.8 percent, has the largest visible minority, mostly Chinese, South Asians, and Blacks. Blacks are the largest visible minority in the Ottawa-Gatineau area, followed by Arabs, Southwest Asians, Chinese, and South Asians. Among provinces, British Columbia, at 18 percent, has the largest visible minority population, followed by Ontario (less than 16 percent), Alberta (10 percent), Manitoba (7 percent), and Quebec (7 percent).

INTERPROVINCIAL MIGRATION: UPROOTING AND RELOCATION

More than a million Canadians move every year, mostly within their immediate environs. Roughly one in five relocates to larger centres in other provinces, prompted by economic considerations. Migrants may leave remote or rural communities because they lack jobs or services. Between 1954 and 1972, Newfoundland uprooted 27,000 people from outports (*above*) in a bid to deliver better health care and education. In recent years, there has been an outflow from farms and resource-based communities. During the 1977–81 oil boom, Alberta was the top destination for Canadians from other provinces. When the oil boom dissipated in the1980s, Alberta was replaced as the choice location by Ontario and, then, by British Columbia in the early 1990s. The chart (*right*) shows the inward and outward flow of interprovincial migration from 1996 to 2001, with Ontario and Alberta again on top.

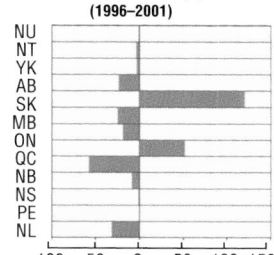

Net internal migrants by province (1996–2001)

(Bar chart with provinces listed: NU, NT, YK, AB, SK, MB, ON, QC, NB, NS, PE, NL; x-axis from -100 to 150)

Among the 40,000 war brides that arrived in Canada at the end of World War II was this happy group with their offspring.

After the early 1970s, most newcomers to Canada were of non-European ancestry, as reflected in this mixed group of new citizens, receiving language instruction in 2002.

■ 1848

IRISH

In the 1700s, Irish crewmen on British ships often stayed behind in Newfoundland. Also, Irish people comprised some 5 percent of New France's population (Riel—Reilly—is one of many French-Canadian corruptions of Irish names). Mass immigration, however, did not occur until the 1840s, when some 500,000 "Famine Irish" arrived—a pool of cheap labour that drove the economic expansion of the 1850s–60s. By 1871, Irish-Canadians accounted for more than 24 percent of Canada's population. Discriminated against because of their extreme poverty (much of what they earned was sent to their starving relatives), thousands migrated to the United States.

■ 1880

ASIANS

In the 1860s, the Chinese came to British Columbia to work on the Cariboo Road and in the goldfields. Many of the 15,000 who arrived between 1880 and 1885 worked on the railway, often assigned the most dangerous jobs and paying with their lives. Chinese- and Japanese-Canadians constantly faced discrimination. In Vancouver, mobs attacked their homes. They could not vote, practice law, or be elected to public office. With many subsisting on $20 monthly, an immigration head tax of up to $500 was implemented to restrict their numbers. To escape the hostility, many moved east to prairie towns, and to Ontario and Quebec.

■ 1900

EASTERN EUROPEANS

A "stalwart peasant in a sheepskin coat, born on the soil . . . with a stout wife and a half-dozen children is good quality," Minister of the Interior Sir Clifford Sifton told his critics. The majority of Canadians wanted only British and French homesteaders on the Prairies. Sifton disagreed, and paid recruiters $5 for each East European farmer, plus $2 for each family member, they signed up. Finns, Poles, Russian Doukhobors, Germans, and Ukrainians flocked to the West in 1885–1914. Place-names such as Valhalla, Gimli, and Steinbach signpost the communities where they settled. Ukrainians, in particular, became politically active and many, such as former Governor General Ray Hnatyshyn, have held prominent government posts.

1778

Capt. James Cook, anchors in Nootka Sound, on the west coast of Vancouver Island

1779–83
1779: North West Company established

1783: Canada-U.S. boundary fixed from the Atlantic to Lake of the Woods

1784
The first United Empire Loyalist refugees arrive at Saint John, N.B.

1792
David Thompson begins a 28-year career as a surveyor and mapmaker that sees him cover a distance of over 88,000 km

1793
York, present-day Toronto, is founded

1812
Britain and United States go to war

Red River Settlement is founded

Medal of 1813 proclaiming the successful defence of Upper Canada against United States

1818
49th parallel chosen as the boundary between Canada and United States, from Lake of the Woods to the Rocky Mountains

1821
Agreement to merge the North West Company and Hudson's Bay Company

1841
Upper Canada and Lower Canada united into Province of Canada

1857
Ottawa is named Canada's capital by Queen Victoria

Building Canada

From the time of Confederation in 1867 to the establishment of Nunavut more than 130 years later, Canada evolved from a scattering of provinces and communities to a transcontinental nation that has become a leader in the global economy. Yet the building of the nation was not without its struggles. The Canada of 1867 was far from complete. Prince Edward Island and Newfoundland refused to join Confederation. Vast, unsettled plains and mountain wilderness stood between the fledgling nation and British Columbia. It would take nearly 50 years for most of the provinces and territories to assume their present-day boundaries. The completion of the Canadian Pacific Railway in 1885 opened up the West. Over time, Hudson's Bay Company land was carved into new provinces and territories. Drafting boundaries was sometimes a straightforward matter of following lines of latitude and longitude; at other times, the source of dispute.

1867

1882

1895

1905

On July 1, 1867, New Brunswick, Nova Scotia, and the Province of Canada unite to form the Dominion of Canada. The Province of Canada is divided into Ontario and Quebec. In 1870, Canada acquires the North-West Territories from the Hudson's Bay Company, and Manitoba becomes Canada's fifth province.

In 1871, British Columbia joins Confederation as the sixth province, followed in 1873 by Prince Edward Island as the seventh. In 1876, the District of Keewatin is created from part of the North-West Territories. In 1880, British rights to the arctic islands pass to Canada. In 1881, the boundaries of Manitoba are extended eastward, an expansion that is contested by Ontario.

In 1895, the districts of Ungava, Mackenzie, Yukon, and Franklin join the existing districts in the North-West Territories. The creation of the Franklin District acknowledges the inclusion of the arctic islands in Canada. Three years later, the District of Yukon becomes a territory.

Alberta and Saskatchewan become the eighth and ninth provinces. The District of Keewatin is transferred back to the North-West Territories. The boundaries of the renamed Northwest Territories are redefined one year later.

FEDERAL ELECTION RESULTS, 1867 TO 2000

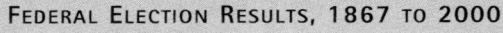

- Liberal Party
- Conservative Party
- New Democratic Party (CCF)
- Alliance Party (Reform)
- Bloc Québécois
- Social Credit
- Others

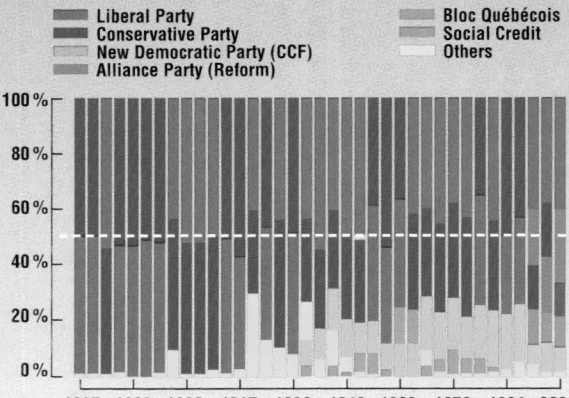

100 % · 80 % · 60 % · 40 % · 20 % · 0 %

1867 1882 1900 1917 1930 1949 1962 1972 1984 2000

POLITICAL PING-PONG

Throughout its history, Canada's laws, policies, and domestic and international affairs have been steered by one of two major political parties, the Conservatives or the Liberals. The chart above shows their fluctuating fortunes as reflected in the popular vote at every federal election since Confederation. Left-wing parties like the Prairie Co-operative Commonwealth Federation (the forerunner of today's New Democratic Party) have often run a distant third, yet have distinguished themselves by weaving the first strands of Canada's "social safety net," championing social services such as Medicare, family allowance, and unemployment insurance. In 1993, disaffected Quebecers and westerners voiced their frustration with the ruling Conservatives by voting in large numbers for two new regional parties, the Bloc Québécois and the western-based Reform Party, reducing the Tories to two seats in the House of Commons. After years of vote splitting, the Alliance (Reform) and Progressive Conservative parties united to form the Conservative Party of Canada in 2003.

WHERE BILLS BECOME LAW

Parliament Hill in Ottawa (*right*) is home to both Canada's House of Commons and the Senate, and is where bills become law. To be adopted into law, a bill must first gain a majority of votes from MPs in the House, then be approved by the Senate before receiving royal assent from the Governor General. The Gothic Revival structure was built between 1859 and 1866. A devastating fire in 1916 leveled the centre block, which was rebuilt by 1922 to a new design that included the 93-m Peace Tower.

1867
Confederation: Four of Canada's colonies are united as the Dominion of Canada on July 1 by the British North America Act

1870
Manitoba Act establishes province of Manitoba

1871
British Columbia joins Confederation

1873
Prince Edward Island joins Confederation

The North-West Mounted Police, forerunner of the RCMP, is established by an Act of Parliament

1878
A secret ballot is used for the first time in a federal election

1885
Last spike driven in the Canadian Pacific Railway at Craigellachie, B.C.

1898
Yukon Territory established

1905
Provinces of Alberta and Saskatchewan established

1910–1929
1914: Canada enters World War I
1919: Winnipeg General Strike

1921: A Canadian coat of arms is established

1929–1939
The Depression years
1931: Statute of Westminster gives Canada complete autonomy

1939: Canada declares war on Germany

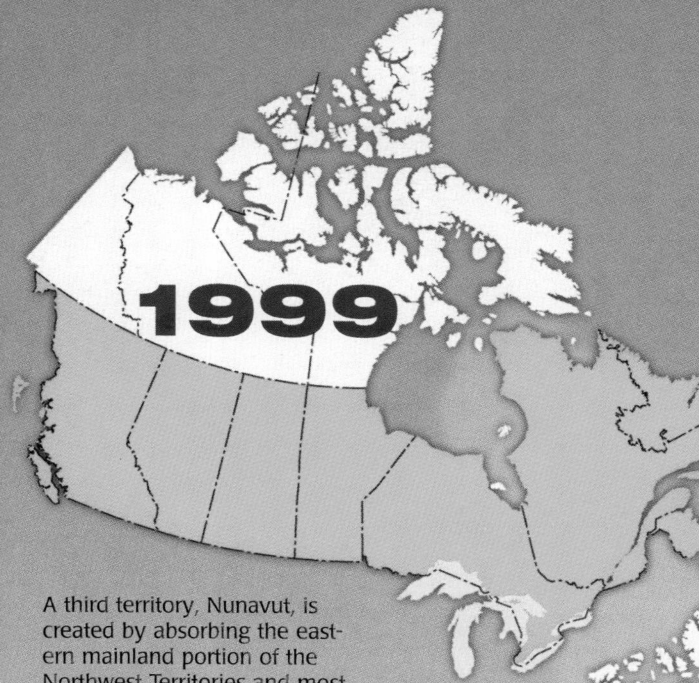

1999

A third territory, Nunavut, is created by absorbing the eastern mainland portion of the Northwest Territories and most of the northern arctic islands.

1949

After a series of closely contested referendums to decide its political future, Newfoundland enters Confederation as the tenth province of Canada.

1912

Ontario and Manitoba attain their present boundaries. Quebec's northern boundary is extended to Hudson Bay and Hudson Strait. In 1927, the boundary dispute between Quebec and Labrador is settled when Labrador is ceded to Newfoundland. The dispute began in 1902 when Newfoundland started lumber operations along the Churchill River.

GOVERNING CANADA

Canada is both a constitutional monarchy, with a governor general (the Queen of England's representative in Canada) as the head of state, and a self-governing democracy, represented by 301 Members of Parliament (MPs) who sit in the House of Commons and 104 appointed representatives who sit in the Senate. Citizens across the country elect an MP for their riding; the number of seats in the House per province is calculated by a formula that takes into account population distribution, with the more populous provinces retaining more seats (*right*). Senators representing the provinces are appointed by the Governor General on advice of the Prime Minister.

Number of Seats in the House of Commons

Ontario	103
Quebec	75
British Columbia	34
Alberta	26
Manitoba	14
Saskatchewan	14
Nova Scotia	11
New Brunswick	10
Newfoundland and Labrador	7
Prince Edward Island	4
Northwest Territories	1
Nunavut	1
Yukon	1

BRITISH COLUMBIA
Joined Confederation July 20, 1871
Capital: Victoria
Population (2002): 4,118,141
Area: 994,735 km² (9.46% of Canada)

ALBERTA
Joined Confederation September 1, 1905
Capital: Edmonton
Population (2002): 3,101,561
Area: 661,848 km² (6.63% of Canada)

SASKATCHEWAN
Joined Confederation September 1, 1905
Capital: Regina
Population (2002): 1,001,224
Area: 651,036 km² (6.52% of Canada)

MANITOBA
Joined Confederation July 15, 1870
Capital: Winnipeg
Population (2002): 1,150,038
Area: 647,797 km² (6.49% of Canada)

ONTARIO
Joined Confederation July 1, 1867
Capital: Toronto
Population (2002): 11,977,360
Area: 1,076,395 km² (10.78% of Canada)

QUEBEC
Joined Confederation July 1, 1867
Capital: Québec
Population (2002): 7,432,005
Area: 1,542,056 km² (15.44% of Canada)

NEW BRUNSWICK
Joined Confederation July 1, 1867
Capital: Fredericton
Population (2002): 756,939
Area: 72,908 km² (0.73% of Canada)

NOVA SCOTIA
Joined Confederation July 1, 1867
Capital: Halifax
Population (2002): 943,497
Area: 55,284 km² (0.55% of Canada)

PRINCE EDWARD ISLAND
Joined Confederation July 1, 1873
Capital: Charlottetown
Population (2002): 135,294
Area: 5,660 km² (0.06% of Canada)

NEWFOUNDLAND AND LABRADOR
Joined Confederation March 31, 1949
Capital: St. John's
Population (2002): 531,820
Area: 405,212 km² (4.06% of Canada)

NUNAVUT
Became Territory April 1, 1999
Capital: Iqaluit
Population (2002): 29,016
Area: 2,093,190 km² (20.96% of Canada)

NORTHWEST TERRITORIES
Became Territory July 15, 1870
Capital: Yellowknife
Population (2002): 40,071
Area: 1,346,106 km² (13.48% of Canada)

YUKON
Became Territory June 13, 1898
Capital: Whitehorse
Population (2002): 29,552
Area: 482,443 km² (4.83% of Canada)

1940–1950
1947: Canadian citizenship begins
1949: Newfoundland enters Canada as the tenth province
1959: Opening of the St. Lawrence Seaway

1965
Canada adopts the Maple Leaf flag
1970: FLQ kidnaps James Cross and Pierre Laporte, initiating the October Crisis

1980
Quebec votes no in a historic referendum on separation

1982
New Constitution and Charter of Rights and Freedoms comes into effect
1987: The "loonie" one-dollar coin is introduced as a cost-saving measure

1992
North American Free Trade Agreement (NAFTA) signed between Canada, United States, and Mexico

1995
Second Quebec referendum on sovereignty narrowly defeated

1999
Territory of Nunavut established

Connecting Canada

Canada's vastness has long posed a challenge for the exchange of materials and information. Yet, despite its expanse, Canada is solidly connected by ribbons of steel and asphalt, and invisibly crisscrossed by webs of airline routes and satellites. These interlocking networks form a system of telecommunications and transportation (*see facing page*) admired throughout the world. Canada's success in the competitive telecommunications field ensures the nation benefits from ongoing cutting-edge research and development. A paramount goal of Canada's communications system—from the establishment of the earliest telegraph line in 1846 to the launching of Anik F2 in 2004—is to provide a reliable and consistent flow of information to a population widely separated by vast distances and many time zones.

CANADA ONLINE

Vancouver author William Gibson coined the term "cyberspace" in 1984 to describe the netherworld inside a futuristic worldwide network of computers. Within a decade this science fiction fantasy would become a reality with the birth of the Internet.

Today the Internet is used daily by Canadian schools, libraries, governments, hospitals, police, and businesses large and small. According to Statistics Canada, 60 percent of Canadian households had at least one Internet user in 2001.

British Columbia and Alberta are the provinces with the highest proportion of households with at least one Internet user, at 65.3 percent. Ottawa remains Canada's most wired city, with 67 percent of households having at least one Internet user.

The different uses of the Internet are shown in the graph at right. Shopping on the Internet is becoming big business; in 2001, 2.2 million households spent almost $2 billion shopping online.

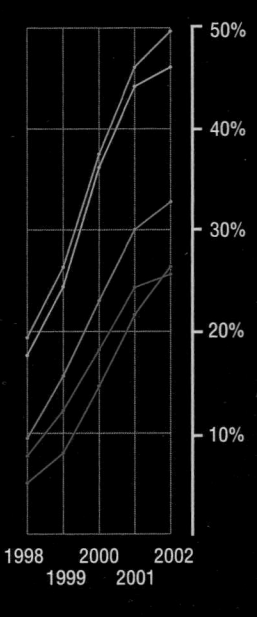

Graph y-axis: 50%, 40%, 30%, 20%, 10%
Graph x-axis: 1998, 1999, 2000, 2001, 2002

- E-mail
- General browsing
- Medical/health information
- Electronic banking
- Playing games

CELLULAR SUPERLATIVES
- Canadians send 800,000 text messages via their cell phones every day.
- The longest cellular corridor in the world is between Windsor and Québec.
- The wireless industry employs approximately 25,000 Canadians.
- Wireless revenues in Canada totaled $6.4 billion in 2001.

WIRELESS NATION
The introduction of the wireless or cellular phone in 1985 gave people a new life of mobility. Canadians were early adopters as cell phones became one of the fastest growing consumer products in history. As of 2003, over 12 million Canadians subscribed to a mobile phone service. By 2005, it is estimated that more than half of all Canadians will be cell phone users.

Scheduled to remain in a geosynchronous—or stationary—orbit 35,850 km above the Earth in 2004, Anik F2 will handle millions of cell phone calls and wireless Internet connections daily.

CANADA'S EYES IN THE SKY
With the launch of atmospheric studies satellites Alouette I in 1962 and Alouette II in 1965, Canada officially entered the space age. Our first telecommunications satellite, Anik A1, launched in 1972, made live television broadcasts possible. Today, six satellites orbiting the Earth allow Canadians to call, fax, e-mail, surf, and watch the information superhighway: Anik F1 (*pictured*) and Anik E2 carry all of Canada's television broadcasts; Nimiq 1 and Nimiq 2 carry hundreds of pay-per-view television channels; MSat, a communications satellite, is used by people in remote locations; and RADARSAT, Canada's first remote sensing satellite, was launched in 1995 to aid in scientific and geographical research.

YUKON
Whitehorse

NORTHWEST TERRITORIES

NUNAVUT
Iqaluit

Yellowknife

BRITISH COLUMBIA

ALBERTA
Edmonton
Calgary
SASKATCHEWAN

Churchill

MANITOBA

ONTARIO

QUEBEC

NEWFOUNDLAND AND LABRADOR
Gander   St. John's

Vancouver
1,148 TEUs*

Winnipeg
Thunder Bay

Québec

N.B.
P.E.I.

Halifax
501 TEUs*

N.S.

OTTAWA
Montréal
919 TEUs*

Toronto

Major highways
Major railways
Major airports
Major ports
* Volume of container cargo by port
(in twenty-foot equivalent units)

OUR SHRINKING COUNTRY

Technology has shrunk Canada's vast distances to the blink of an eye. The illustration above shows the ever-dwindling time it takes a letter to travel coast to coast.

- 1893: Steam train, 115 hours
- 1935: Steam train, 90 hours
- 1971: Diesel train, 60 hours
- 1930: Propeller-driven aircraft, 18 hours
- 1981: Jet aircraft, 5 hours
- ✳ 2004: Internet, 30 seconds

HIGHWAYS AND BYWAYS

Canada has more than 1.4 million kilometres of two-lane-equivalent roads serving some 24 million motor vehicles. Most are classed as urban streets or local roads, with freeways and primary and secondary highways accounting for only 217,000 km. Over two-thirds of Canada's road system lies west of the Atlantic

CANADA'S PUBLIC ROADS

	Two-lane-equivalent kilometres
Saskatchewan	250,000
Ontario	230,600
Quebec	228,300
Alberta	205,300
British Columbia	204,800
Manitoba	104,500
New Brunswick	76,600
Nova Scotia	48,700
Newfoundland and Labrador	27,100
Yukon	16,100
Northwest Territories & Nunavut	10,200
Prince Edward Island	6,500
Total	1,408,700

provinces (see above). Canada's longest road, the Trans-Canada Highway, is also the longest national highway in the world, spanning some 7,821 km and passing through 10 provinces. Canada's roads form a major backbone of our transportation network, with trucks moving 234 million tonnes of goods in 1998, everything from live animals to steel, consumer goods to cars, chemicals to forest products.

WATERWAYS

Goods transported along Canada's navigable waterways are generally bulk commodities of relatively low value per tonne, such as coal, ore, grain, and salt. Water transportation may be separated into three areas: ocean, inland water, and coastal transportation. Ocean transportation is vitally important to Canada, as about one-third of what Canada produces is exported. Much of this export traffic is carried overseas in containers by large, oceangoing vessels.

CONTAINER CARGO BY COMMODITY

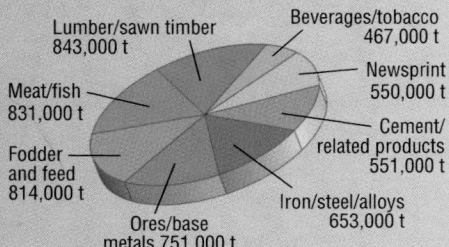

Lumber/sawn timber 843,000 t
Beverages/tobacco 467,000 t
Newsprint 550,000 t
Meat/fish 831,000 t
Cement/related products 551,000 t
Fodder and feed 814,000 t
Iron/steel/alloys 653,000 t
Ores/base metals 751,000 t

Almost one-third of all of these exports move through Vancouver, the largest port in Canada. (Tonnage handled by major Canadian ports is indicated on the chart above.) The Great Lakes and the St. Lawrence Seaway provide inland water transportation of grain, coal, and iron ore to and from the heartland of Canada on vessels called lakers. Coastal water transportation is important on the West Coast, where logs, lumber, chemicals, and other bulk commodities are moved by barge in British Columbia's coastal waters.

RAILWAYS

With 73,000 km of track, Canada's rail network is the world's third longest, after those of the U.S.A. and Russia. Canada's two major railways, Canadian National and Canadian Pacific, move millions of passengers and millions of tonnes of goods annually. The railways transport large quantities of bulk commodities over long distances at relatively low cost, so that our coal, potash, and lumber reach world markets efficiently and profitably.

RAIL CARGO BY COMMODITY

	Tonnes (000s)
Forest products (logs, pulpwood, newsprint)	42,121
Coal	39,522
Iron ore	39,063
Grain (barley, corn, oats, wheat)	30,619
Fertilizer materials (potash, sulfur)	21,928
Containers on flat car	18,075
Trailers on flat car	1,869
Other	91,566
Total	284,763

Movement by rail container is growing in importance. Most finished goods are now transported by this method. Until 2001, passenger traffic on Canadian railways declined because of airline deregulation and competition from other types of transport. In 1983, the number of rail passengers topped 7 million. In 1998, this figure fell to 3.9 million. Most of the passenger traffic was along the Windsor-Québec corridor. Since 2001, however, passenger traffic has increased by 1.4 percent.

AIRWAYS

Canada has 1,149 airports linked by 130,000 km of controlled airways. Of these, 26 major airports handle 94 percent of all airline passengers. Air transport also moves a wide variety of consumer goods and freight. In 2000, 853,110 tonnes of cargo were flown to and from Canada on domestic (61 percent), transborder (12 percent), and international routes (27 percent). Major airline routes are shown on the map above. Since deregulation in 1985, Canada's airways have been dominated by Air Canada. The airline industry has been struggling for survival since the terrorist attacks on the United States on September 11, 2001. In 1999, 85.3 million passengers boarded or deplaned at Canadian airports. This dropped to 80.1 million after 2001.

CANADA'S BUSIEST AIRPORTS

	Enplaned and deplaned passengers (000s)
Pearson International, Toronto	26,690
Vancouver International	15,137
Pierre Elliott Trudeau International, Montréal	8,188
Calgary International	8,102
Edmonton International	3,829
Macdonald-Cartier International, Ottawa	3,210
Halifax International	2,893
Winnipeg International	2,748
Mirabel International, Montréal	1,218
Total	72,015

Who we are

Canada is aging: our fastest-growing population group is 80 and over, and expected to number 1.3 million within a decade. Eighty percent of the population lives in large urban areas, half in the greater metropolitan areas of Vancouver, Edmonton-Calgary, Toronto, and Montréal. Our future growth will be determined by the influx of immigrants from 210 countries, increasingly Asian. Women are outpacing men in education and sometimes in the workplace; 66.6 percent of Alberta's workforce, for example, is female. Our 3.9 million 20-year-olds, 13 percent of the population, are the most educated generation ever. On the debit side, nearly 17 percent of Canadians live in poverty. The income gap between rich and poor is growing in every province, with the biggest gap in Alberta. And even though national crime rates are declining, 29 percent of women feel unsafe alone in their homes at night.

MARRIAGE AND DIVORCE

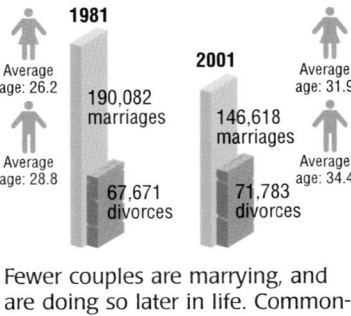

1981
Average age: 26.2
Average age: 28.8
190,082 marriages
67,671 divorces

2001
Average age: 31.9
Average age: 34.4
146,618 marriages
71,783 divorces

Fewer couples are marrying, and are doing so later in life. Common-law unions are on the increase. Such couples are twice as prevalent in Canada as in the United States. Of 1.1 million Canadian common-law couples, 44 percent live in the province of Quebec.
■ First unions increasingly are common-law, but most such couples will eventually marry.
■ On average, brides are aged 31.9 years; bridegrooms, 34.4.
■ People in their late 20s are the most likely to divorce.

■ The risk of divorce is greatest after five years of marriage. Based on current rates, one out of three married couples will divorce; two out of three marriages will endure.
■ Intermarriage between couples with different ethnic and cultural backgounds is on the rise. In Vancouver, 13 percent of all pairs in their 20s are mixed couples.

LANGUAGES

1. English
2. French
3. Chinese
4. Italian
5. German
6. Punjabi
7. Spanish
8. Portuguese
9. Polish
10. Arabic

Selected religions

Muslim 2%
Jewish 1%
Buddhist 1%
Hindu 1%
Sikh 1%
Others 1%

Other religions 7%
Other Christians 4%
No religion 16%
Protestant 29%
Catholic 44%

RELIGIONS

One in 3 British Columbians, 1 in 4 Albertans, and 4 in 10 Yukoners are among the 4.8 million Canadians (representing 16 percent of the population) belonging to no organized religion.
■ Seven out of every 10 Canadians are Roman Catholic or Protestant. Catholics make up roughly 44 percent of the population: Protestants, 29 percent. Both groups are declining.
■ Islam, Hinduism, Sikhism, and Buddhism are up substantially in Canada, a shift consistent with changing immigration patterns.
■ People of Muslim faith, the second most reported religion among 20-year-olds, have more than doubled—from 253,300 in 1991 to 579,600 in 2001. Of 1.8 million immigrants in the 1990s, Muslims accounted for 15 percent, Hindus, almost 7 percent, and Buddhists and Sikhs, 5 percent each.
■ Greek and Ukrainian Orthodox church memberships are declining, while Serbian and Russian Orthodox memberships have more than doubled in the last decade.
■ Some 1 percent of the population is of the Jewish faith.
■ Having lost much of their congregations, some Christian churches are increasingly being converted to day-care centres, food banks, and other uses.

EDUCATION

Levels of educational attainment
(population aged 25 and over)

College 16%
University 20%
Trades 12%
Less than high school 29%
High school 23%

Of Canadians 15 years and older, 51 percent have post-secondary qualifications. Among 25- to 34-year-olds, the figure is 61 percent. One-quarter of Ontarians, ages 25 to 64, have a university degree.

TOP FIELDS OF STUDY

1. Education
2. Engineering
3. Business
4. Finance
5. Psychology
6. Nursing
7. Computer science
8. Law
9. Economics
10. Medicine

Among 25- to 64-year-olds in the working population, 1.1 million have doctorates, master's degrees, and qualifications above the bachelor's degree level.
■ Fifty-one percent of all working-age university graduates, and 59 percent of all college graduates, are women.
■ Male university students are leaning to technology, females to education, and both to business and finance. At the college level, men choose computer science; women, finance and business.
■ Trades are dominated by men; 63 percent of all 2001 working-age trade graduates were male.
■ Building and construction lead in trades' studies, but data processing and computer science studies are increasing for men. Women are still drawn to beauty care and hairdressing.
■ More young people in every province are finishing high school.
■ More than 1 in 4 Canadians participate in adult education and training programs.

BIRTHS

Natural increase
(in thousands of persons)

[Line chart showing years 1976, 1981, 1986, 1991, 1996, 2001 on x-axis and values from -50 to 300 on y-axis]

Despite a slight upturn in 2001, the Canadian birthrate continues to slide downward, as shown in the chart above. Of the 333,744 babies born that year, Ontario couples welcomed nearly one in three births.
■ The fertility rate (the average number of children a woman will have in her lifetime) is 1.5, well below the 2.1 children per woman needed to maintain the population. If present trends continue, natural increase (births minus deaths) will be at zero in 2030, and immigration will be the only way the population can grow.
■ The number of children born per 1,000 women during their child-bearing years peaked in 1959 at 3,935 births. Today, the number is less than 1,500.
■ In remote regions and northern areas with large aboriginal populations, infant mortality rates are up to 3.1 times higher than the national rate (5.7 deaths for boys, 4.8 for girls, per 1,000 live births).
■ For aboriginal infants, the death rate from accidental injuries is four times the national rate.
■ Canada's current population growth is about three-quarters that of the United States. Eighteen percent of Canada's population was born outside the country.

One in 6 Canadians (and 4 in 10 Torontonians) are allophones whose mother tongue is neither English nor French. Allophones comprise 18 percent of the population; anglophones 59.1 percent, and francophones, 22.9 percent. The rate of English-French bilingualism increased 8.1 percent in the five years ending 2001.
■ Chinese is Canada's leading nonofficial language, as indicated in the top-ten list above. In the late 1990s, the use of Punjabi and Arabic increased 32.7 percent; Urdu, 50 percent; and Tagalog (the official language of the Philippines), 26.3 percent. Dravidian languages (spoken in India, Sri Lanka, and Pakistan) and Pashto (an Iranian language) are among the hundred other different mother tongues of new Canadians.
■ Only about 25 percent of aboriginal peoples can speak their own language, and many Native languages are endangered. Some aboriginal languages, such as Inuktitut, which is spoken by 9 out of 10 Inuit, is flourishing.

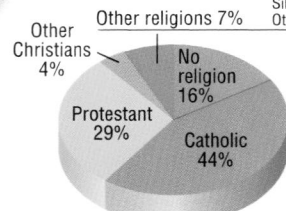

Household size declining

FAMILY LIFE

Over three decades, the size of the Canadian family has decreased from 3.7 to 3 persons.
■ Four times more children live with common-law parents than 20 years ago. About 19 percent of all children live with single parents, mostly mothers.
■ Step-families are on the rise. Half have only the female spouse's children; 1 in 10 has only the male's.
■ In what has become known as the "skip generation," nearly 1 percent of all grandparents, mostly women, are raising or living with children without the involvement of either of the child's parents.
■ Roughly 15 percent of female same-sex couples live with children, compared with only 3 percent of male same-sex couples.
■ Ottawa-Gatineau and Vancouver have the highest proportion of same-sex couples.
■ The number of childless couples and "empty nesters" is increasing.
■ More people, seniors especially, are living alone. In Quebec, Manitoba, Saskatchewan, British Columbia, and the Yukon, 3 out of 10 are one-person households. In Alberta, one-person households are increasing at double the national rate.

A NATURAL BALANCE

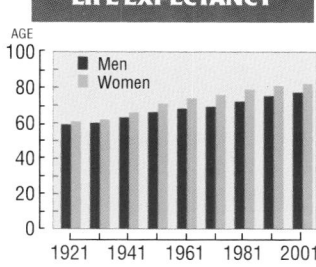

WORK AND INCOME

Seventy-four percent of Canada's labour force is in services; 15 percent, in manufacturing; and 6 percent, in farming, forestry, and other primary industries.
■ Three of the fastest growing jobs are service jobs such as call centre customer service workers, childcare workers, and financial planners.
■ Truck driving, the top job for men, increased 29 percent in the last decade. The demand for secretaries (excluding legal and medical) declined 35 percent over 10 years. Even so, it remains the second most common job for women after salesperson.
■ Half the country's 15.6 million workers have a postsecondary degree or diploma.
■ The median family income is $55,016. With average annual incomes of slightly more than $185,000, the top 10 percent of families earn 28 percent of total family income. Average income of the bottom tenth is $10,341.

CARING FOR AN AGING POPULATION

Thirteen percent of today's population is 65 or older. By 2020, seniors will number 6.7 million, about 19 percent of the population. Federal and provincial governments are already working on strategies to ensure this senior overload will not cripple either the health care system or the Canada/Quebec pension plans.

Seniors will need accessible public transit and safe, affordable places to live, so more funding for transportation and affordable housing may be in order. As older people are increasingly rejecting homes for the aged in favour of living alone, more home-care services are a must.

Serving the needs of retirees offers businesses a host of opportunities. Senior lifestyle communities, recreational clubs, and tour packages are just a few of the possible needs to fill. Some businesses may even find it advantageous to relocate to areas with large elderly populations.

With one-quarter of the population about to retire, and the ratio of younger to older workers declining, analysts predict shortages of doctors, nurses, professors, pipefitters, carpenters, bricklayers, plumbers, and electricians. Public and private sectors need to develop strategies to attract skilled workers to fill such jobs, possibly immigrants with young families. This would entail providing affordable housing, parks, schools, and day-care centres for the newcomers.

LIFE EXPECTANCY

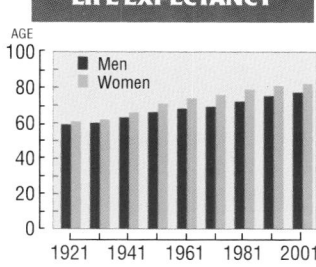

Canada ranks third in life expectancy, after Switzerland and Japan.
■ A woman's life expectancy is now 81.7 years; a man's, 76.3 years.
■ People in southern urban areas in and west of Ontario live longer. Mortality rates in these areas are 10 percent lower for all causes than the national average.
■ Life expectancy is four years less than the national average in remote parts of the country.
■ Life expectancy increases as the rate of unemployment decreases and the level of education increases.

Rate of disease onset (1994–97)

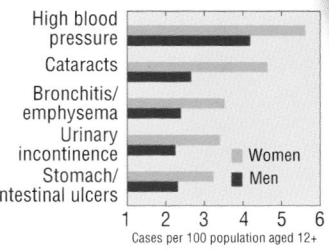

High blood pressure
Cataracts
Bronchitis/ emphysema
Urinary incontinence
Stomach/ intestinal ulcers

Women
Men

Cases per 100 population aged 12+

HEALTH

Compared to many countries, Canadians live longer and suffer fewer chronic illnesses and disabilities. People with higher incomes can expect to live longer and be healthier than those earning less. University educated people tend to have better health than those with little education. Single mothers often have poorer health than other groups.
■ Alcohol, smoking, unprotected sex, and inactivity reduce the quality of many lives. Obesity is on the rise for all age and sex groups—except women 20 to 34—and in all provinces, particularly Alberta.
■ Accidental injuries are a major cause of hospitalization and premature death. Cancer, heart disease, and stroke are the three top killers overall.

WHERE WE WORK

1. Sales and service
2. Business, finance, and administration
3. Trade and transport
4. Management
5. Manufacturing/ utilities
6. Natural and applied sciences
7. Social science, education, government
8. Health
9. Primary industries
10. Art, culture, recreation, sport

In the early years of the 21st century, Canada's population was 31,414,000, of which 25.9 percent was under 19 years old; 61.1 percent was 20 to 64 years; and 13 percent was 65 years and older. There were 75 senior men for every 100 senior women. As more and more young people flocked to large urban centres in search of jobs, small towns across Canada became communities of older people. This was particularly true in eastern Canada, but less so in western Canada, where, apart from British Columbia, the provinces and territories had the youngest populations. Of all the provinces, Alberta had the youngest in the country; Quebec had some of the oldest. Based on census statistics and demographic trends evident since the 1970s, statisticians projected how this population would most likely evolve.

Canada's age pyramid, 1900 Population: 5.3 million

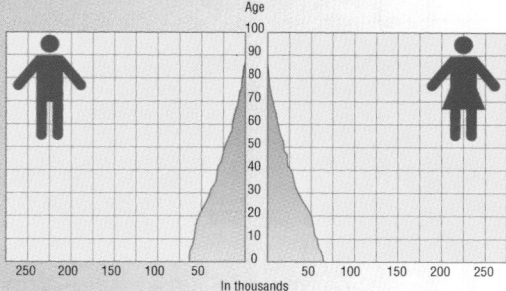

Canada's age pyramid, 2000 Population: 30 million

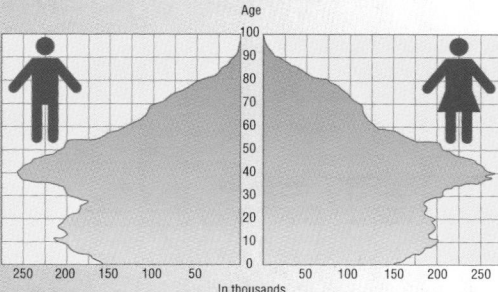

Above: *The shape of Canada's populations in 1900 and 2000. The middle-age bulge in the 2000 age pyramid represents the baby-boomer generation.*

FUTURE TRENDS

According to one scenario, Canada will continue growing for some 25 years, but at a decelerated rate. Due to a combination of an increased senior population and continued low fertility levels, by 2016, for the first time ever in Canadian history, people 65 years and over—10 million strong—will outnumber those 14 years and younger.

By 2026, some 8 million Canadians will be 65 or older; 1 in every 5 will be a senior (compared to 1 in 8 in 2000). With deaths exceeding births, the population will be in decline by 2046–50. By 2050, the median age in Canada will be 46.2; in 2000 it was 36.8. According to this scenario, the population decline in Canada could be as early as the 2030s.

All projections show the working population increasing up to 2016, then beginning to go into a decline. Such a decline would affect what is known as the "dependency ratio," the proportion of children and elderly to the working population. In 2003, there were some 100 people of working age per 46 children and seniors. Without the working-age people to replace future retirees, the children/seniors segment ratio will range from 55 to 60 as early as 2026.

Imbalance between old and young affects the "potential support ratio," the number of working-age people per senior. This will continue to spiral downward. In 2000, there were more than five working-age persons per senior; by 2026, it will be three per senior.

Immigration will be essential for growth and current immigration patterns offer some clues as to where this will occur. Up to 2026, some population gain is likely to occur in all provinces and territories, except Newfoundland and Labrador, New Brunswick, and Saskatchewan. Populations in Atlantic Canada, Quebec, Manitoba, and Saskatchewan will eventually decline. Alberta, British Columbia, and Ontario, already the fast-growing provinces, will continue to expand.

REVITALIZING CANADA'S CITIES

A survey of 205 cities worldwide ranks Vancouver (*above*) third for the overall quality of life. (Toronto, Ottawa, and Montréal also appear in the top 25.) Unfortunately, this quality of life is not available to all Canadian city dwellers.

Our cities' rapid growth is partly to blame. Affordable rental housing for low-income families is at an all-time low. Poor neighbourhoods are growing faster than cities overall, and the income gap between the affluent and the poor is widening. On a broader scale, the explosion in city populations has strained aging roads, bridges, and water and waste disposal systems. Compounding the gravity of the situation is the significant decline in federal and provincial funding for cities.

Revitalizing Canada's neglected cities has become the hot topic of the decade. Unlike in other countries, our cities are largely dependent on property taxes. Some municipal councils say they need user fees and additional tax tools to survive. Canada's big city mayors met in 2004 and sent a message to the federal government: funding at the national level is essential—even with provincial contribution. (Constitutionally, cities are a provincial responsibility.)

In early 2004, the federal Liberal Party set up a committee to explore urban issues. Under discussion was the possibility of giving municipalities a permanent share of gas and diesel-fuel taxes.

Bohemian Index
1 Vancouver
2 Toronto
3 Victoria

Mosaic Index
1 Toronto
2 Vancouver
3 Hamilton

Talent Index
1 Ottawa
2 Halifax
3 Toronto

TechPole
1 Montréal
2 Toronto
3 Ottawa-Hull

MEASURING INNOVATION

Cities recognize that innovation is as vital to the urban economy as raw materials. Measuring innovation, which involves surveying artistic creativity and social attitudes, has been used to predicate growth potential. A recent study of some Canadian cities, using the four innovation indexes: the Bohemian (the size of the creative community); the Mosaic (the numbers of foreign-born—an indicator of openness and diversity); the Talent (university-educated numbers); and the TechPole (level of high-tech employment). Top scorers in each index are shown at left.

Vancouver, with almost 10 per 1,000 population in the Bohemian category, led all others, followed by Toronto, Victoria, and Montréal. Toronto, where 42 percent of the population is foreign-born, dominated the Mosaic Index, followed by Vancouver, with just under 35 percent. The Ottawa area leads the Talent Index, with 23 percent of all adults educated at university, followed by Halifax and Toronto, each scoring 20 percent. Montréal dominates the TechPole Index. High-tech workers represent about 37 percent of all employment, largely due to the city's aerospace industry. Toronto follows at roughly 33, Ottawa-Hull at about 17, and Vancouver at about 12.

RURAL RETREAT

Statistics Canada defines "rural" in various ways: an area with 150 persons per square kilometre; an area with less than 1,000 population; or all territory outside urban areas. Another benchmark (and the basis for the figures that follow) is to define as rural any area, including small towns, outside an urban centre with a population of 10,000 or more. In 1996, such rural areas were home to 21.5 percent of Canadians. By 2001, this percentage had fallen to 20.3 percent or about 6.1 million people. Only Ontario, Manitoba, and Alberta bucked the trend. (Areas that grew generally had more than 1 in 3 residents commuting to larger centres.) Resource-rich communities of Northern Ontario (Greenstone, Kirkland Lake, and Elliott Lake) and British Columbia (Mackenzie, Prince Rupert, and Comox-Strathcona) were especially hard hit, with losses exceeding 12 percent of their populations.

Urban and rural population

Rural / Urban

100%, 80%, 60%, 40%, 20%, 0%

NORTH, B.C., ALTA., SASK., MAN., ONT., QUE., N.B., N.S., P.E.I., NFLD.

COMMUTING TO WORK

The journey to the job is on average a 62-minute daily chore for 11.4 million Canadians. Although 7 km is the average commute, 1 in 6 workers in population-dense Toronto, Montréal (*above*), and Vancouver travel 20 km or more. Driving is the preferred method everywhere except Nunavut. In 2001, 5.6 million men and 4.3 million women drove to work; 77 percent overall were alone in their cars. Drivers average 58 minutes on the road, compared to 100 minutes for bus/subway riders. And despite flexible work hours, traffic patterns still show 8 a.m. and 5 p.m. as the busiest times on the road.

Quebecers, Ontarians, and Manitobans led in public transit use, of which women and young adults are top users. A handful of workers walk to work (mostly women) or bike (mostly men).

Below: *In a time of rapid urban growth, Norton, New Brunswick, is a reminder of the enduring charm of many rural Canadian communities*

Where we live

Canadians, inhabitants of the world's second largest country, are moving by the thousands from rural areas into urban settings. In 2001, 79.4 percent—four out of five Canadians—lived in communities of 10,000 people or more. Some 64 percent resided in 27 regions called "census metropolitan areas"—or CMAs—where core populations are 100,000 or more. Two-thirds of Canadians live in city/regions—urban agglomerations of a million or more. Eight such urban entities exist: Ontario's Golden Horseshoe, centred around Toronto; Vancouver-Victoria; the Calgary-Edmonton corridor; Montréal and adjacent regions; Halifax-Dartmouth; Ottawa-Hull; Québec; and Winnipeg. The first four hold 51 percent of the Canadian population. All city/regions contain sizable proportions of provincial populations: almost 60 percent of Ontarians live around the Greater Toronto area and 72 percent of Albertans in the Calgary-Edmonton corridor. More than 70 percent of immigrants make their homes in Toronto, Montréal, and Vancouver. With immigration becoming the driving force behind Canada's population increase, the growth of these three areas seems certain.

EXPANSION AT THE EDGES

The "donut effect" is the term used to describe the sprawling suburbs and municipalities encircling the stagnant or declining cores of many CMAs. According to the 2001 census, Saskatoon and Regina exemplify this phenomenon. Between 1996 and 2001, Saskatoon's core population grew by 1.6 percent, while its surroundings grew by 14.6 percent. In this same period, Regina's core declined by 1.2 percent; its surroundings increased by 10 percent. A few exceptions are Abbotsford, B.C., and the Ottawa-Gatineau area, where the reverse applies: the core is outstripping surroundings.

People-scarce city cores have been part of the Canadian urban scene since the 1960s. Until then, many chose to live downtown to be near work. Immigrants settled in central districts where services for newcomers were available. Decades of suburban flight—accelerated by widespread car ownership and extensive expressway construction—have left many city cores empty and inhospitable. Many of the expanding outlying areas have severed connections with the core. These self-sufficient units are part of "the donut" that is just getting bigger. Between 1996 and 2001, immigrants to the Toronto city/region swelled the population of 4 of the 10 fastest-growing municipalities: Vaughan, a 37.3 percent increase; Barrie, 31 percent; and Richmond Hill, 29.8 percent. Interprovincial migration fueled the growth in communities along the Calgary-Edmonton corridor, the location of 5 of the 10 fastest-growing Canadian municipalities: Cochrane, a 58.9 percent increase; Sylvan Lake, 44.5 percent; Strathmore, 43.4 percent; Okotoks, 36.8 percent; and Rocky View No. 44, 31.6 percent.

Suburbs encircle Edmonton, expanding across the open prairie on the city's periphery.

Above: *Toronto's cluster of downtown towers symbolizes urban Canada's vitality*

URBAN AMENITIES: A COMPARISON OF FIVE CITIES

This chart ranks a handful of Canadian cities against amenities likely to influence where people choose to live and work. Affordability and safety are basic concerns. So is clean air. Most people appreciate convenient public transit, and ready access to cultural and recreational facilities. Such services, however, may mean higher taxes.

RANK IN CANADA	1	2	3	4	5
Availability of affordable housing	Regina	Saint John	Halifax-Dartmouth	Edmonton	Montréal
Green spaces	Edmonton	Vancouver	Winnipeg	Halifax	Regina
Cultural & recreational facilities	Toronto	Montréal	Vancouver	Ottawa	Calgary
Mass public transit network	Montréal	Toronto	Vancouver	Calgary	Ottawa
Air quality	Québec	St. John's	Winnipeg	Saint John	Regina
Low tax rate	Calgary	Halifax	Ottawa	Toronto	Montréal
Low crime rate	Québec	Toronto	Ottawa	Montréal	St. John's

CMA POPULATIONS—2003 ESTIMATES

1.	Toronto, Ont.	4,907,000
2.	Montréal, Que.	3,511,400
3.	Vancouver, B.C.	2,099,400
4.	Ottawa-Hull, Ont.-Que.	1,108,500
5.	Calgary, Alta.	969,600
6.	Edmonton, Alta.	954,100
7.	Québec, Que.	694,000
8.	Winnipeg, Man.	684,300
9.	Hamilton, Ont.	680,000
10.	Kitchener, Ont.	431,200
11.	London, Ont.	425,200
12.	St. Catharines–Niagara, Ont.	391,000
13.	Halifax, N.S.	359,100
14.	Victoria, B.C.	319,900
15.	Windsor, Ont.	313,700
16.	Oshawa, Ont.	304,600
17.	Saskatoon, Sask.	231,500
18.	Regina, Sask.	198,300
19.	St. John's, Nfld.	176,400
20.	Chicoutimi-Jonquière, Que.	158,800
21.	Sudbury, Ont.	157,900
22.	Sherbrooke, Que.	155,000
23.	Abbotsford, B.C.	147,370
24.	Kingston, Ont.	146,838
25.	Trois-Rivières, Que.	141,200
26.	Saint John, N.B.	127,300
27.	Thunder Bay, Ont.	125,700

METRO AREA (KM²)

1.	Edmonton, Alta.	9,419
2.	Toronto, Ont.	5,903
3.	Halifax, N.S.	5,496
4.	Ottawa-Hull, Ont.-Que.	5,318
5.	Saskatoon, Sask.	5,192
6.	Calgary, Alta.	5,083
7.	Winnipeg, Man.	4,151
8.	Montréal, Que.	4,047
9.	Sudbury, Ont.	3,536
10.	Regina, Sask.	3,408

POPULATION GROWTH—1996-2001

1.	Calgary, Alta.	15%
2.	Oshawa, Ont.	10%
3.	Vancouver, B.C.	8%
4.	Edmonton, Alta.	8%
5.	Kitchener, Ont.	8%
6.	Vancouver, B.C.	8%
7.	Abbotsford, B.C.	7%
8.	Windsor, Ont.	7%
9.	Hamilton, Ont.	6%
10.	Ottawa-Hull, Ont.-Que.	6%

DENSITY—POPULATION PER SQUARE KILOMETRE, 2002

1.	Montréal, Que.	847
2.	Toronto, Ont.	793
3.	Vancouver, B.C.	690
4.	Kitchener, Ont.	501
5.	Victoria, B.C.	499
6.	Hamilton, Ont.	438
7.	Oshawa, Ont.	328
8.	Windsor, Ont.	301
9.	St. Catharines–Niagara, Ont.	268
10.	Abbotsford, B.C.	235

Canada 2050

For Canadians, the journey to the world of 2050 holds many challenges. Some of these emerged in the last decades of the 20th century, when it became apparent that the land and its resources were neither limitless nor endlessly exploitable. Faced with stark realities as diverse as global warming and disappearing wetlands, a consensus developed around concepts of sustainability, which have been touched on throughout the preceding pages of *The Canadian Atlas.* These final pages speculate on how Canadians may live 50 years hence. Of all its resources, Canada's own people are its greatest asset. Whatever path they take, they are likely to ensure the country remains stable, dynamic, full of promise, and a welcoming destination for newcomers eager to experience the adventure of meeting Canada's future challenges.

HOW HOT WILL IT GET?

■ The world's climate will continue to get warmer. Industrial activity is responsible for the increase of greenhouse gases, such as carbon dioxide, in the atmosphere. But experts differ about the outcome of global warming. Some predict changes will come with savage suddenness within decades. Others believe that global warming will be slower than expected and that dramatic environmental changes will take hundreds of years. In 2050, "green methods" of manufacturing and greater use of alternative energy may have reduced greenhouse gas emissions. By that time, scientists may also have learned to purge air pollution, ending the threat of global warming.

HOW MANY OF US WILL THERE BE?

■ By 2050, Canada's population is expected to reach 40 million—give or take 5 million. Between now and then, population growth will have slowed due to continued low fertility levels, a measure of the number of children women have over their childbearing years. The population could level off at 35 million if the present low birthrate is not offset by a high rate of immigration.
■ At 1 senior to 8 working-age persons, the proportion of seniors in Canada was less than in many other industrialized countries as this century began. But in coming decades, the Canadian population will age more rapidly as baby boomers—those born in the two decades after World War II—reach the age of 65. By 2026, some 8 million Canadians, 1 person in every 5, will be 65 or older. By 2050, seniors may number 10 million, roughly 25 percent of the projected population.
■ Immigration from Asia, Africa, and Latin America will continue to drive Canada's population growth. Some analysts expect the newcomers will adopt the prevailing social values. They also believe Canada will move from a "mosaic" embracing a diversity of cultures and groups toward a homogenous society akin to the American "melting pot" model. Only aboriginals will resist assimilation and assert social and economic positions appropriate to their traditional values.

WHAT MAJOR CHANGES CAN WE EXPECT BY 2050?

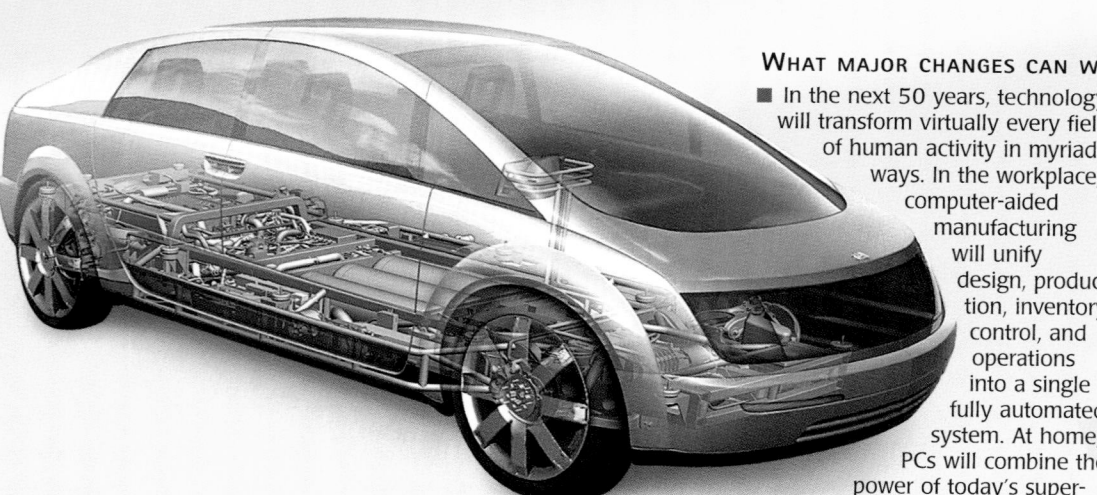

■ In the next 50 years, technology will transform virtually every field of human activity in myriad ways. In the workplace, computer-aided manufacturing will unify design, production, inventory control, and operations into a single fully automated system. At home, PCs will combine the power of today's supercomputers with some as yet undeveloped form of artificial intelligence. More and more, technology will be brought into play to ensure the sustainability of resources. Technology—combined with exploration—may reveal the whereabouts of mineral reserves in new extraction venues such as seabeds. It may also create new materials to replace scarce resources. "Green methods" of manufacturing will be used to recapture valuable by-products from industrial wastes.
■ By 2050, Canadian farmers will have incorporated alternative or organic farming techniques with traditional methods. The use of chemical fertilizers and pesticides will decline as they become more expensive and less effective. Computerized control of farming activities will be commonplace. With technological advance, farmers will produce crops that grow faster, resist disease, and are more robust. Some produce may have been engineered to provide medicinal properties. Improved aquaculture may provide much of our seafood if commercial fish species depleted by overfishing continue to decline.
■ By 2050, dependence on fossil fuels will decline as reserves run low or become more difficult to access. Canadians will call upon renewable sources—geothermal, hydroelectric, tidal, and wind—to meet our energy needs. Solar power may produce electricity at costs comparable to conventional sources.

Left: *Towers massed on a man-made island: a visionary model of a 21st-century high-density urban centre from famed Canadian architect Arthur Erickson*

How will we live and work?

■ In 2050, there will be fewer building starts, and more recycling of existing structures. Older buildings will be massively retrofitted with the latest technological innovations. New homes and workplaces will be smarter, built for energy efficiency and ease of maintenance. They will be constructed with self-repairing "intelligent materials," capable of responding to environmental changes. Home and business energy requirements will be supplied by wind and solar power, or by fuel cell or photovoltaic devices.

■ Urban sprawl, already infringing on Canada's first-rate farmland, will be halted by mid-century. A general awareness of the environmental impact and expense of building and maintaining highways and other infrastructures may diminish the appeal of the suburban, commuting lifestyle. By 2050, many people may have returned to the cities, where urban spaces will be extensively redeveloped for business and residential purposes. In city cores, moreover, there will be a high proportion of people living on their own, particularly young singles, widows, or widowers.

■ The city-bound flow will increase urban densities. Today more than 12 million Canadians live and work in Toronto, Montréal, Vancouver, and the Edmonton-Calgary corridor. Within 50 years, 8 to 10 million more will be concentrated in these centres.

■ By 2050, virtually all homes will have an entertainment centre, which combines interactive television, telephone, and computing capacity. Household robots will be commonplace. Garages will house small, quiet, nonpolluting battery-powered vehicles with ceramic engines and recycled plastic bodies. For the most part, products will be created from recycled materials. At least half of all products will be purchased through the use of the computer.

■ In 2050, Canadians will be better educated, constantly renewing skills to keep up with rapid changes in information and technology. Technological advances will enable most people to work at home. Retirement will be an outmoded concept. Many seniors will opt to work as long as their health, and their desire to do so, holds out.

■ The 2050 economy may be driven by the knowledge-intensive jobs (design work, for example), administration, education, social work, tourism, leisure, and the cultural field. Only a few will work in the primary sector (farms, fisheries, forests, and mines). Automation will mean small staffs in the manufacturing sector, which will produce high-quality goods that will be inexpensive. Some experts say most manufacturing may be sent "offshore"—a trend increasingly apparent in today's global economy. Whether some manufacturing is retained here or goes abroad will depend on comparative production costs.

How healthy will we be?

■ More Canadians in 2050 will be better fed, and live longer and healthier lives than ever before. This will be the result of improvements in lifestyle, nutrition, and medicine. By 2041, life expectancy at birth in Canada is expected to reach 81 years for men and 86 years for women.

■ In coming decades, aging baby boomers untouched by conditions such as Alzheimer's disease and cancer may find their quality of life challenged by some form of disability. A less active post-baby-boomer generation may, however, impose its burdens on the health-care system. According to some experts, the latter will experience a high incidence of diabetes and cardiovascular disease.

■ New and unpredictable diseases—caused by the cross-transmission of viruses from animals and birds to human—will mutate more rapidly as the world shrinks through air travel.

■ By 2050, a better understanding of life and living organisms will provide greater control over disease and disability. Treatments will involve more biochemical and engineered solutions. Computers and automated laboratory equipment will refine diagnostic capabilities and create new, purer pharmaceuticals. Such developments will reduce health-care costs and possibly overcome a decline in health-care personnel. By 2050, the number of doctors is projected to fall to two-thirds of the present level. Half of this reduction may occur as baby-boomer doctors reach retirement.

MAPS OF CANADA

Below: *A grizzly bear patrols the placid river shore in the Khutzeymateen Grizzly Bear Sanctuary, located north of Prince Rupert, British Columbia.*

Next page: *A satellite image of Canada*

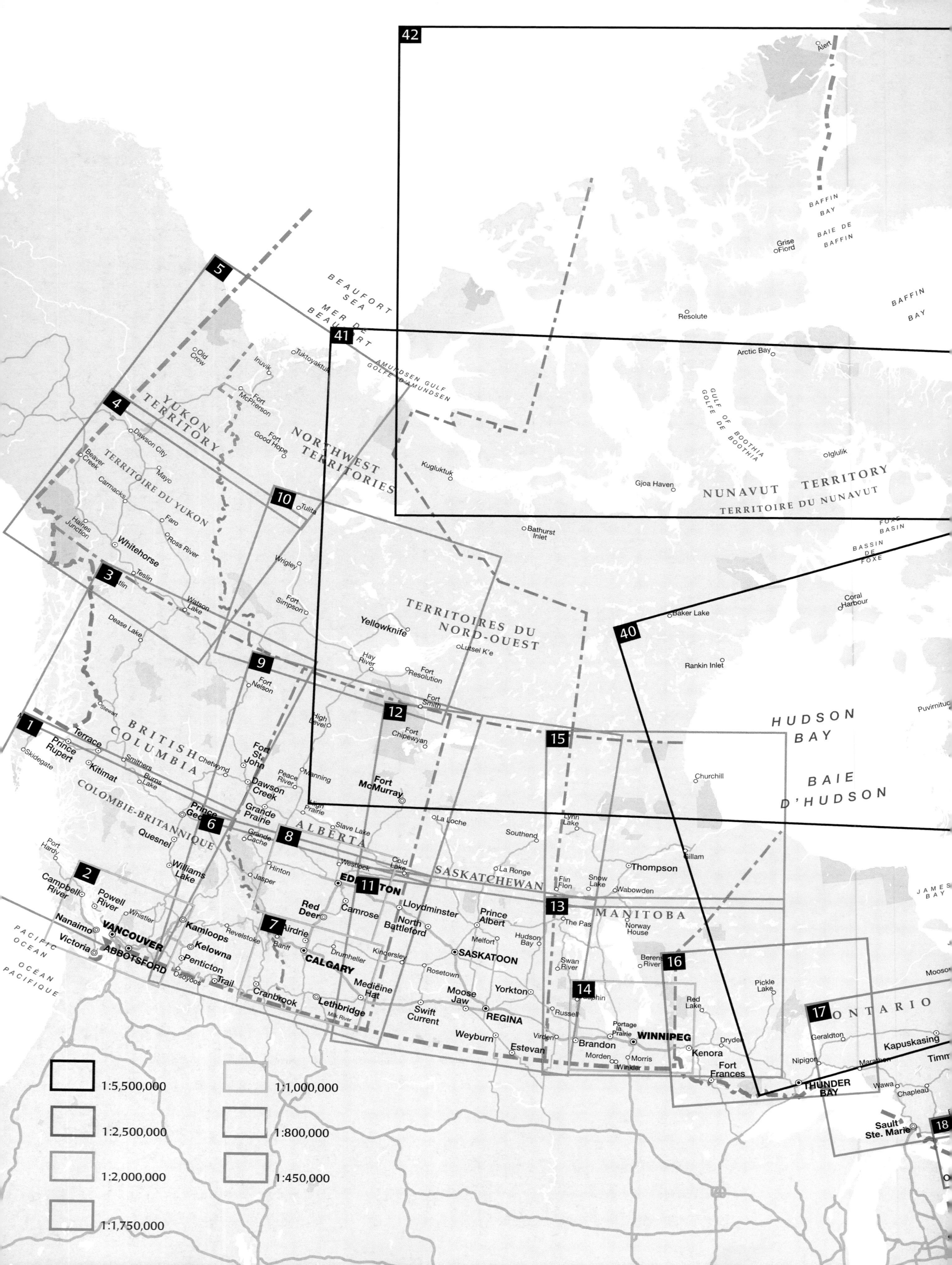

Key map

1. Western British Columbia
2. Southwestern British Columbia
3. Northern British Columbia
4. Southern Yukon
5. Yukon / Northwest Territories
6. Southeastern British Columbia / Southern Alberta
7. Southeastern British Columbia / Southern Alberta
8. Central Alberta
9. Northern Alberta
10. Northwest Territories
11. Southern Prairies
12. Northern Prairies
13. Southern Manitoba / Northwestern Ontario
14. Southern Manitoba
15. Northern Manitoba
16. Northwestern Ontario
17. Northeastern Ontario
18. Western Ontario
19. Southwestern Ontario
20. South Central Ontario
21. Southeastern Ontario
22. Northeastern Ontario / Northwestern Quebec
23. Northwestern Quebec
24. Southeastern Ontario / Southwestern Quebec
25. Central Quebec
26. Central Quebec
27. Southern Quebec
28. Lac Saint-Jean, Quebec
29. Gaspé Peninsula, Quebec / Northern New Brunswick
30. Gaspé Peninsula, Quebec
31. Southeastern New Brunswick
32. Southern New Brunswick / Central Nova Scotia
33. Western Nova Scotia
34. Eastern Nova Scotia
35. Prince Edward Island
36. Southwestern Newfoundland
37. Eastern Newfoundland
38. Northern Newfoundland
39. North Shore, Quebec / Newfoundland and Labrador
40. Northeastern Canada
41. North Central Canada
42. Arctic Islands

Map symbols

Provincial highway	Expressway service centre area
Ontario county highway	Interchange number
Yellowhead Highway	Kilometre distance
Crowsnest Highway	Cumulative kilometres distance
Trans-Canada Highway	Native community
Provincial tourism welcome centre	Dot indicative of population size
Point of interest	Hay Lakes, AB / Vanderhoof, BC / Flin Flon, MB / Kenora, ON / Shawinigan, QC / St. John's, NL
Spot height	
Mountain peak	Controlled access expressway (4 lane/divided)
Parks Canada facility	Controlled access expressway (2 lane/undivided)
Organized park – camping	Controlled access toll expressway (4 lane/divided)
Organized park – no camping	Expressway interchange
Unorganized park – camping	Primary 2-lane highway with interchange
Unorganized park – no camping	Multilane highway
Campground (inside large park)	Secondary highway
Interest area (inside large park)	Local road
Ontario conservation area (camping)	Winter road
Ontario conservation area (no camping)	Scenic parkway (limited access)
National wildlife facility	Scenic roadway
Provincial wildlife facility	Auto ferry
24-hour border crossing	Passenger ferry
Major ski centre	International boundary
Toll expressway full interchange	Provincial boundary
Toll expressway partial interchange	County/district/ MRC boundary
Expressway full interchange	Reorganized city boundary
Expressway partial interchange	Railway
Highway full interchange	Major walking trail
Highway partial interchange	Time zone boundary
Major airport	Urban built-up area
Minor airport	National Capital Commission facility
Remote access airport	Park
Expressway rest area	Wildlife area

GEOGRAPHICAL NAMES SOURCES

For bilingual forms of geographical names, the following sources were used:

Ontario: the Ontario Ministry of Natural Resources OnTerm GeoNames, Index (www.onterm.gov.on.ca/geo), and the Ontario Ministry of Tourism and Recreation *Guide touristique de l'Ontario*

New Brunswick: the official road map *New Brunswick Travel Map* and the New Brunswick Department of Tourism and Parks *Guide touristique du Nouveau-Brunswick*

Nunavut: Nunavut Tourism

Northwest Territories: the official *Explorers Map*

National Parks and World Heritage Sites: Parks Canada

National Wildlife Areas and National Migratory Bird Sanctuaries: Canadian Wildlife Service, Environment Canada

General source:

Geographical Names Board of Canada: This federal/provincial/ territorial committee, coordinated by Natural Resources Canada, is the primary source of Canadian geographical names. For further information, see http://geonames.nrcan.gc.ca.

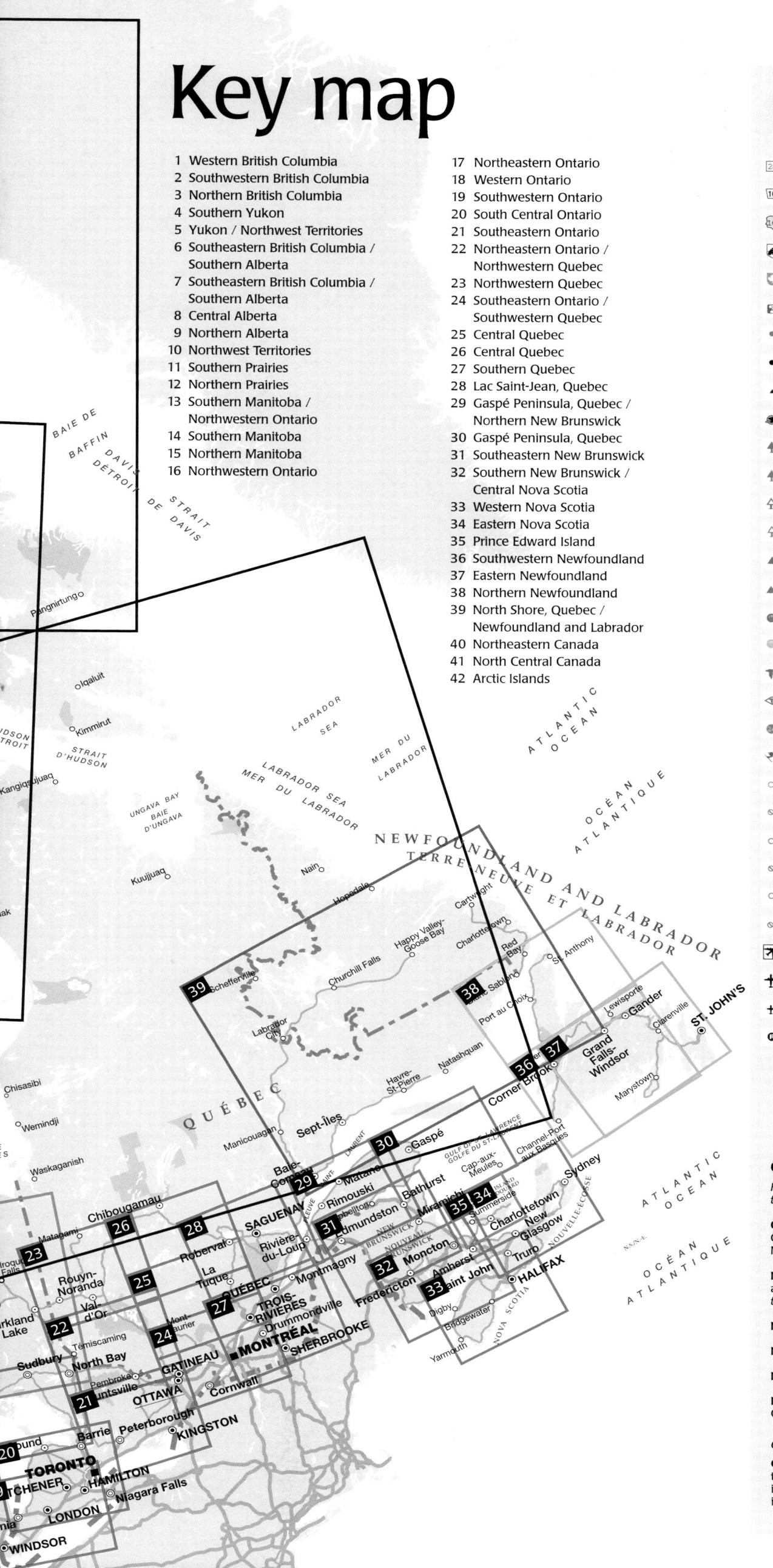

Dixon Entrance
Détroit de Dixon

Langara Point
Langara Island
Parry Passage
Cape Knox
Klashwun Point
Wiah Point
Beresford Bay
White Point
Morgan Point
Tian Head
Ingraham Bay
Port Louis
Hippa Island
ÎLE GRAHAM
Athlow Bay
Kano Inlet
Hunter Point
Kindakun Point
Cone Head
Cartwright Sound

Naden Harbour
Masset
Haida
Naikoon
Agate Beach
Rose Spit
Rose Point
Cape Ball
Tlell
Dead Tree Point
Misty Meadows
Port Clements
Juskatla
Pure Lake Provincial Park
GRAHAM ISLAND

SKEENA-QUEEN CHARLOTTE

Melville
Brown Passage
Stephens Island
Prescott
Porcher Island
Cape George
Goschen Island
Browning Entrance
McCauley Island
Baird Point
Cliff Point
Banks Island
Kelp Point

Museum of Northern British Columbia
Metlakatla
Prince Rupert
Port Edward
Diana Lake Prov. Park
Port Essington
Oona River
Kitkatla
Bonilla Island
Klewnuggit Inlet Provincial Marine Park
Lowe Inlet Provincial Marine Park
Hawkesbury Island
Hartley Bay
Farran

Exchamsiks River Provincial Park
Terrace
Kitsumkaylum
Lakelse Lake Provincial Park
Lakelse Lake
37
Gitnadoix River Provincial Recreation Area
Kitimat
Kitimat Centennial Museum
Kitamaat Village
Kitlope

KITIMAT RANGES

Eagle Peak 2093m
Andeste Peak 2379m
Powell Pk. 2012m
Mt. Dubose 2734m
Kemano

Barrett Lake
Houston
Topley
Babine Lake (Pendleton Bay) Prov. Park
Woyen
Tchesinkut La
François La
Ferryl
Noralee
Tatalrose
Takysie
Little Andrews Bay Provincial Marine Park
Wistaria Provincial Park
Ootsa Lake
North Tweedsmuir
Whitesail Lake
Whitesail Reach

QUEEN
CHARLOTTE
ÎLES DE
LA REINE-CHARLOTTE

Skidegate
Queen Charlotte City
Sandspit
Copper Bay
Cumshewa Head
Skedans Point
Moresby Camp
Sewell Inlet
Tasu Sound
MORESBY ISLAND
ÎLE MORESBY

HECATE STRAIT
DÉTROIT D'HÉCATE

Estevan Group
Campania Island
Aristazabal Island
Gil Island
Princess Royal Island
Surf Inlet
Green Inlet Provincial Marine Park
Fiordland Provincial Recreation Area

BRITISH COLUMBIA
Comet Mountain 2018m
Kimsquit
South Tweedsmuir
Tsitsutl Peak 2478m
Kalone Peak 2557m
Thunder Mountain 2681m
Bella Coola Historic Museum
Bella Coola
Hagensborg
Firvale
Mount Saugstad 2908m
Stuie

GWAII HAANAS NATIONAL PARK RESERVE AND HAIDA HERITAGE SITE
GWAII HAANAS NATIONAL MARINE RESERVE (Proposed)
RÉSERVE DE PARC NATIONAL ET SITE DU PATRIMOINE HAIDA GWAII HAANAS
RÉSERVE D'AIRE MARINE NATIONALE GWAII HAANAS (Projet)

Huxley Island
Burnaby Island
Ikeda Point
Benjamin Point
Houston Stewart Channel
Nagas Point
Nan Sdins National Historic Site/
Lieu historique national de Nan Sdins
SGaang Gwaii (Anthony Island)/
SGaang Gwaii (Île Anthony)
Kunghit Island
Cape St. James

Kitlope Heritage Conservancy Provincial Area

Klemtu
Swindle Island
Price Island
Roderick Island
Dowager Island
Ocean Falls
Denny Island
Shearwater
Waglisla-McLoughlin Bay
Bella Bella
Campbell Island
Codville Lagoon Provincial Marine Park
Hunter Island
Namu
CENTRAL COAST
Hakai Provincial Recreation Area
Calvert Island
Dawsons Landing
Good Hope
Rivers Inlet
Herbert Island
Cape Calvert
Penrose Island Provincial Marine Park
Smith Sound
Greaves
Cape Caution
Burnett Bay
Bramham Island
Seymour Inlet

King Island
Dean Channel
North Bentinck Arm
South Bentinck Arm
Monarch Icefield
Mona 3533r
COAST
CHAÎNE
Silverthrone Mountain 2896m

QUEEN CHARLOTTE
SOUND
Bassin de la
Reine-Charlotte

PACIFIC
OCEAN

OCÉAN
PACIFIQUE

Scott Islands
Lanz I.
Cox I.
Scott Islands Provincial Park
Cape Scott
Cape Scott Provincial Park
San Josef
Cape Palmerston
Raft Cove Provincial Park
Holberg
Coal Harbour
Winter Harbour
Cape Parkins
Quatsino Prov. Park
Lawn Point Lawn Point Provincial Park
Brooks Bay
Cape Cook
Brooks Peninsula
Brooks Peninsula Prov. Park
Clerke Point

MOUNT WADDINGTON
Sullivan Bay
God's Pocket Provincial Marine Park
Broughton I.
Echo Bay Provincial Marine Park
Kingcome Inlet
Wakeman Sound
Port Hardy
Port Hardy Historic Museum
Bear Cove
Kippase
Malcolm Island
Sointula
Cormorant Channel Prov. Marine Park
U'Mista Cultural Centre
Alert Bay
Beaver Cove
Telegraph Cove
Port McNeill
19
Quatsino Subdivision
Lower Nimpkish Prov. Park
Lower Tsitika River Prov. Park
Kokish
Minstrel Island
Port Neville
Cracroft I.
Hardwicke Island
Kelsey Bay
Sayward
Port Alice
Marble River Prov. Park
Nimpkish Lake Provincial Park
Woss
Mount Cain 1840m
Schoen Lake Provincial Park
White River Prov. Park
Morto Prov
Elk Falls Prov

VANCOUVER ISLAND
Chamiss Bay
Kyuquot
Fair Harbour
Artlish Caves Prov. Park
Dixie Cove PP.
Union I.
Tahsis Kwois Prov. Park
Zeballos
Tahsis
Esperanza
Ceepeecee
Hecate
Catala I. Prov. Marine Park
Big Bunsby Marine Park
Nootka Island
Nuchatlitz Provincial Park
Nuchatlitz Inlet
Bligh I. Prov. Park
Yuquot
Bajo Point
Santa Boca Provincial Park
Gold River
Mowachaht
Golden 2200m
Strathcona
White Ridge Prov. Park

ÎLE DE
VANCOUVER

Victoria Pk. 2163m
Can
Loveland Bay Prov. Park
STR

Hesquiat Peninsula Provincial Park
Hesquiat
Maquinna Provincial Park
Flores Island Provincial Park
Gibson Provincial Marine Park
Stewardson Inlet
Flores Island
Ahousat
Marktosis
Opitsat
Vargas Island
Tofino
Esowista
Clayoquot Sound
Dawley Passage Provincial Marine Park
Vargas Island Provincial Park
Sulphur Passage

PACIFIC RIM NATIONAL PARK RESERVE/
RÉSERVE DE PARC NATIONAL PACIFIC RIM

FJORDS AND MOUNTAINS

This view shows Quatsino Sound on northwestern Vancouver Island, Queen Charlotte Strait, and the mountainous, fjord-indented coastline of mainland British Columbia. Calvert, Hecate, and Hunter islands in Queen Charlotte Sound (left centre) lie at the entrance to Burke Channel. This spectacular fjord extends 130 km inland to the small saltwater port of Bella Coola. Beyond the coastal strip rise the Coast Mountains, some of which are more than 2,500 m high. Moist westerly Pacific winds, forced upward by this mountain chain, deluge the slopes and peaks with rain and snow.

Other points of interest

■ British Columbia's coastline extends 22,894 km, nearly three times as long as the Canada-United States border.

■ Canada's most powerful recorded earthquake, measuring 8.1 on the Richter Scale, occurred off the Queen Charlotte Islands [A1-C2] on Aug. 22, 1949.

■ Canada's wettest place is Ocean Falls [C4], where the average annual precipitation is 4,826 mm.

■ Glacier-fed Chilko Lake [D6], 80 km long and covering an area of 158 km², is North America's largest natural, high-elevation (1,171 m) freshwater lake.

■ At 948,597 km², British Columbia is nearly four times the size of Great Britain, 2.5 times larger than Japan, and larger than any American state except Alaska.

Scale 1:2,000,000

1 cm = 20 km

0   20   40   60   80 km

COAST MOUNTAINS CHAÎNE CÔTIÈRE

PACIFIC RANGES

COAST RANGES

MT WADDINGTON

SQUAMI

Good Hope Mountain 3240m

Ts'il-os Provincial Park

Monmouth Mountain 3194m

Gold Bridge

Kingcome Inlet

Homathko Estuary Provincial Park

Bishop River Provincial Park

Lillooet Glacier

Downton Lake

Mount Kennedy 2028m

Superb Mountain 2469m

Viscount

Comox-Strathcona

Upper Lillooet Provincial Park

Pemberton Meadow

Glendale Cove

Phillips Arm

Phillips Arm

Minstrel Island

Knight Inlet

Bute Inlet

Clendinning Provincial Park

Pemberton

19

POWELL RIVER

Pemberton Icefield

Hardwicke Island

West Thurlow I.

East Thurlow I.

Big Bay

Stuart Island

Walsh Cove Provincial Park

East Redonda I.

Homfray Channel

Princess Louisa Inlet

Princess Louisa Provincial Park

Callaghan Lake Provincial Park

Emerald Estates

99

Whistler

BRITISH COLUMBIA

Desolation Sound Provincial Marine Park

Powell Lake

Jervis Inlet

Whistler/Black

Brandywine Falls Provincial Park

Garibaldi

Garibaldi

VANCOUVER ISLAND

Campbell River

Quadra Island

Cortes Island

Refuge Cove

Lund

Okeover Arm Provincial Park

Sunshine Coast

Tantalus Provincial Park

Mount Garibaldi 2678m

Alice Lake Provincial Park

Garibaldi Heights

Gold River

Powell River

Paradise Valley

Saltery Bay Provincial Park

Skookumchuck Narrows Provincial Park

Squamish

Shannon Falls Provincial Park

Strathcona

Golden Hinde 2200m

Mt. Washington

101

Texada Island

Sechelt

Woodfibre

Britannia Beach

Courtenay

Comox

Royston

Union Bay

Denman Island

Hornby Island

Strait of Georgia Détroit de Georgie

Pender Harbour

Sechelt Peninsula

Halfmoon Bay

Howe Sound

Porteau Cove Provincial Park

Cumberland

COMOX-STRATHCONA

Qualicum Beach

Parksville

Gibsons

Langdale

Lions Bay

West Vancouver

North Vancouver

Port Moody

ALBERNI-CLAYOQUOT

Tofino

Port Alberni

Nanaimo

Gabriola

GREATER VANCOUVER

VANCOUVER

BURNABY

Ucluelet

PACIFIC RIM NATIONAL PARK RESERVE

RÉSERVE DE PARC NATIONAL PACIFIC RIM

Ladysmith

GULF ISLANDS NATIONAL PARK RESERVE
RÉSERVE DE PARC NATIONAL DES ÎLES-GULF

RICHMOND

Delta (Ladner)

White Rock

Bamfield

COWICHAN VALLEY

Lake Cowichan

North Cowichan

Chemainus

Duncan

Saltspring Island

Ganges

Sidney

North Saanich

PACIFIC OCEAN

ÎLE DE VANCOUVER

CAPITAL

Central Saanich

SAANICH

Victoria

Esquimalt

SAN JUAN

OCÉAN PACIFIQUE

Strait of Juan de Fuca Détroit de Juan de Fuca

CANADA U.S.A./É.-U.

Sooke

Colwood

Langford

Oak Bay

WASHINGTON

Friday Harbor

RIVER AND CANYON

The Thompson River (upper right) passes through British Columbia's arid interior plateau as it flows toward its junction with the Fraser River. South of the junction, near Lytton, the Fraser River is tightly constricted by the Coast Mountains to the west, and by the Cascade Mountains to the east. This 100-km stretch is the Fraser Canyon, where walls rise 1,000 m in some places. At Hell's Gate, the canyon narrows to a 30-m gap, through which the raging river rushes at 7 m a second. At its southern end, near Hope, the Fraser turns west abruptly and travels tamely through level farmland to the Pacific. Along this stretch, waters from the 65-km-long Harrison Lake (left centre) and other tributaries feed the river.

Other points of interest

■ Vancouver Island has more than 1,000 known caves. The Upana Caves, a network of caverns near Gold River [C1], have 15 entrances and a combined length of 450 m.

■ At 440 m, Della Falls is Canada's highest waterfall. It lies just south of Buttle Lake in Strathcona Provincial Park [C1-D1].

■ The Carmanah Giant, over 95 m high, is the world's largest Sitka spruce. More than 400 years old, this mighty tree may also be the world's oldest specimen. It is located in Carmanah Walbran Provincial Park [F3].

■ Volcanic formations are outstanding natural features of 1,958 km² Garibaldi Provincial Park [C5]. A notable example is 2,678-m Black Tusk Mountain, the eroded remains of an ancient volcanic core, which is located just north of Garibaldi Lake.

Scale 1:1,000,000
1 cm = 10 km

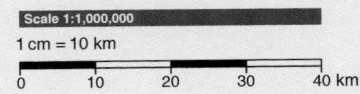

0 10 20 30 40 km

Map labels

THOMPSON-NICOLA

COLOMBIE-BRITANNIQUE

FRASER VALLEY

LILLOOET RANGE

CASCADE RANGE

OKANAGAN-SIMILKAMEEN

WASHINGTON

SKAGIT

WHATCOM

NORTH CASCADES NATIONAL PARK

MT. BAKER-SNOQUALMIE NATIONAL FOREST

ROSS LAKE NATIONAL RECREATION AREA

LAKE CHELAN NATIONAL RECREATION AREA

Kamloops, Merritt, Lytton, Hope, Cache Creek, Ashcroft, Clinton, Lillooet, Spences Bridge, Logan Lake, Princeton, Chilliwack, Abbotsford, Mission, Maple Ridge, Langley, Bellingham, Mount Vernon, Sedro Woolley, Burlington, Anacortes

CAMELSFOOT RANGE

H-LILLOOET

Stein Valley Nlaka'pamux Provincial Park

Skihist Mountain 2944m

Stein Mountain 2774m

Black Tusk

YUKON TERRITORY
TERRITOIRE DU

ROCKY MOUNTAINS

BRITISH COLUMBIA
COLOMBIE BRITANNIQUE

BOUNDARY RANGES

COAST CHAÎNE MOUNTAINS CÔTIÈRE

CASSIAR MOUNTAINS RANGES

STIKINE

OMINECA MOUNTAINS

SKEENA MOUNTAINS

KITIMAT-STIKINE

HOGEM RANGES

SWANNELL RANGES

ALASKA

Tongass National Forest

PRINCE OF WALES ISLAND

REVILLAGIGEDO ISLAND

CLEVELAND PENINSULA

MISTY FIORDS NATIONAL MONUMENT

PACIFIC OCEAN
OCÉAN PACIFIQUE

Dixon Entrance
Détroit de Dixon

U.S.A./É.-U.
CANADA

Place names and features:

Mount Bryde 1908m
Morley River Territorial Recreation Site
Rancheria Falls Recreation Site
Rancheria
Swift River
Watson Lake Campground
Watson Lake
Watson Lake Signpost Forest
Upper Liard
Lower Post
Liard Canyon Territorial Recreation Site
Good Hope Lake
Contact Creek
Smith River
Fireside
Coal River
Coal River Springs Territorial Recreation Site
Hyland River Provincial Park
Liard River Hotsprings Provincial Park

Simpson Peak 2173m
Atlin
Five Mile
Mount Canning 2212m
Atlin Provincial Recreation Area
Atlin Provincial Park

Juneau
Mendenhall Glacier Viewing Area
Wickersham State Hist. Park
Douglas I.

Mt. Nesselrode 2470m
Tongass National Forest
Bridget National Forest
Eagle Beach State Recreation Area
Ernest Gruening State Historic Park

Tuya Mountains Provincial Park
Jade City
Boya Lake Provincial Park

Muskwa-Kechika Management Area

Homeline Creek Provincial Park

Dease Lake
Denetiah Provincial Park
Gataga Mountain 2281m
Dune Za Keyih Provincial Park

Telegraph Creek
Glenora
Mount Buckley Lake
Grand Canyon of the Stikine
Forty Mile Flats
Stikine River Provincial Park
Iskut
Edziza Provincial Park
Mount Edziza 2787m
Mount Edziza Prov. Recreation Area
Tatogga
Mt. Cartmel 2175m
Spatsizi Plateau

Sheppard Peak 2515m
Tongass National Forest
Mt. Ratz 3136m

Glacial Mtn. 2306m
Ligne de partage des eaux

Mt. McNamara 2523m
Finlay-Russel Provincial Park

Devil's Thumb 2767m
Kinaskan Lake Provincial Park
Todagin South Slope Provincial Park
Iskut Hot Springs Provincial Park

Spatsizi Headwaters Provincial Park
Spatsizi Wilderness Park
Wilderness Park

Kupreanof
Kuiu
Kupreanof Island
Petersburg
Mitkof Island
Woewodski Island
Stikine
Choquette Hot Springs Provincial Park
Great Glacier Provincial Park

Bob Quinn Lake
Ningunsaw Provincial Park

Mt. Beirnes 2117m

COLOMBIE

Tattauti Provincial Park
Tatlatui Provincial Park

Zarembo Island
Wrangell
Petroglyph Beach State Historic Site
Bell II
Delta Peak 2298m
Mt. McEvoy 2125m
Dewar Peak 2240m
Fleet Peak 2326m

Point Baker
Kosciusko Island
Thorne
Etolin Island
Mt. Lewis Cass 6864 ft.
Border Lake Provincial Park
Lava Forks Provincial Park
Mt. Pattullo 2729m
Damdochax Provincial Protected Area

Cape Pole
Edna Bay
Coffman Cove
Deer Island
Mt. Willibert 6779 ft.
Meziadin Junction
Meziadin Lake Provincial Park
Motase Peak 2411m

Sustut Provincial Park

Bear Glacier Provincial Park
Hyder
Stewart
Swan Lake Kispiox River Provincial Park
Shelagyote Peak 2466m

Heceta
San Fernando Island
Klawock
Hollis
Craig
Kasaan
Yes Bay
Bell Island
Meyers Chuck
Settlers Cove State Historic Site
Ketchikan Gateway
Ketchikan
Saxman
Cranberry Junction
Kisgegas Peak 2347m

Waterfall
Suemez
Ward Cove
Street Historic Area
Black Sands Beach State Marine Park
Lavender Peak 2323m
Alice Arm
Nass Camp

Hydaburg
Metlakatla
Annette Island
Mt. Weber 2007m
Kitwancool Totem Poles
Kispiox
Mt. Thoen 2291m
Fort Babine
Nilkitkwa Lake PP
Rainbow Alley PP

Gitwinksihlkw
New Aiyansh
Nisga'a Memorial Lava Bed Provincial Park
Kitwanga Fort Kitwanga National Historic Site
Gitanmaax
'Ksan Indian Village
Hazelton
New Hazelton
Hagwilget
Netalzul Meadows PP
Smithers Landing
Babine Lake Marine Park

Laxgalts'ap (Greenville)
Gingolx (Kincolith)
Xhlawit (Alder Peak) 2220m
Oscar Peak 2304m
Kitwanga Mountain Provincial Park
South Hazelton
Seeley Lake PP
Kitseguecla (Gitsegukla)
Boulder Creek PP
Cedarvale
Blunt Mtn. 2286m
Ross Lake Prov. Park
Moricetown

Forrester Island
Cape Lookout
Cape Augustine
Dall Island
Lax Kw'alaams
Wales
Somerville Island
Khutzeymateen Valley Grizzly Bear Sanctuary
Seven Sisters Provincial Park
Weeskinisht Peaks 2755m
Rosswood
Driftwood Canyon Provincial Park
Smithers
Babine Mountains Provincial Park
Granisle
Red Bluff Provincial Park
Telkwa
Topley Landing

Cape Muzon
Point Marsh
Cape Fox
Zayas Island
Dundas Island
Gitwinksihlkw
Kitsumkalum Provincial Park
Mt. Kenney 2073m
Usk
Terrace
Kitsumkaylum
Kleanza Creek Provincial Park
Hudson Bay Mountain
Tyhee Lake Provincial Park
Fulton River
Topley Landing
Babine Lake (Pendleton Bay) Prov. Marine

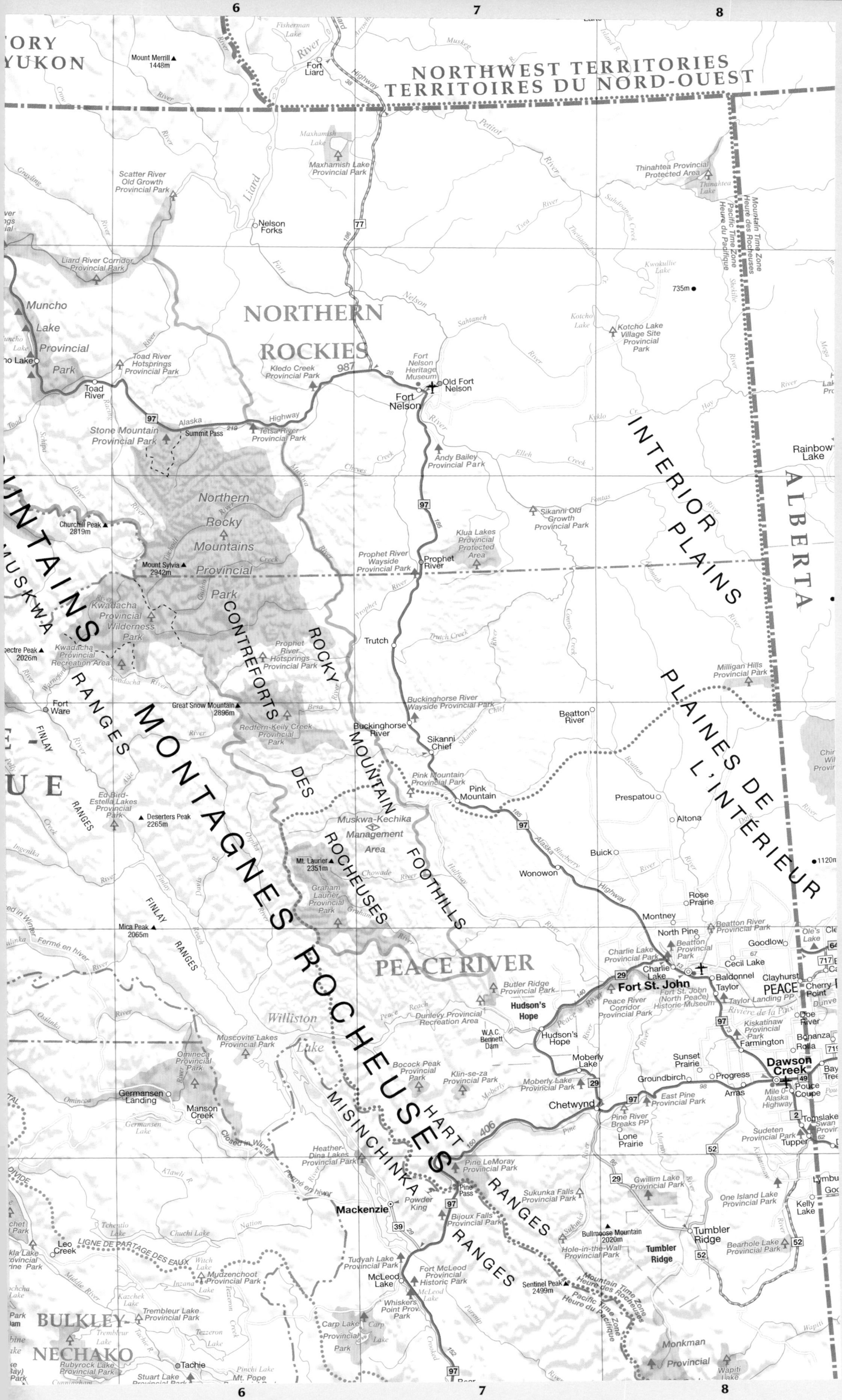

NORTHWEST TERRITORIES
TERRITOIRES DU NORD-OUEST

TORY
YUKON

Fisherman Lake

Mount Merrill ▲ 1448m

Fort Liard

Maxhamish Lake

Maxhamish Lake Provincial Park

Thinahtea Provincial Protected Area

Thinahtea Lake

A

Scatter River Old Growth Provincial Park

Nelson Forks

735m ●

NORTHERN

Liard River Corridor Provincial Park

Muncho Lake

Muncho Lake Provincial Park

Toad River Hotsprings Provincial Park

Kledo Creek Provincial Park

ROCKIES

Kotcho Lake Village Site Provincial Park

Kotcho Lake

Rainbow Lake

B

Toad River

97

Alaska Highway

Fort Nelson Heritage Museum

Fort Nelson

Old Fort Nelson

INTERIOR

Stone Mountain Provincial Park

Summit Pass

Tetsa River Provincial Park

Andy Bailey Provincial Park

ALBERTA

Churchill Peak 2819m

Northern Rocky Mountains

97

Sikanni Old Growth Provincial Park

PLAINS

C

Mount Sylvia 2942m

Provincial

Klua Lakes Provincial Protected Area

Kwadacha Provincial Wilderness Park

Park

Prophet River Wayside Provincial Park

Prophet River

PLAINES

Spectre Peak 2026m

Kwadacha Provincial Recreation Area

Trutch

Milligan Hills Provincial Park

Fort Ware

Great Snow Mountain 2896m

Prophet River Hotsprings Provincial Park

Buckinghorse River Wayside Provincial Park

Beatton River

DE

Redfern-Kelly Creek Provincial Park

Buckinghorse River

Sikanni Chief

FINLAY

Eon-Bird Estella Lakes Provincial Park

Deserters Peak 2265m

Pink Mountain Provincial Park

Pink Mountain

Prespatou

L'INTÉRIEUR

D

Muskwa-Kechika Management Area

Altona

Mica Peak 2065m

Mt. Laurier 2351m

Buick

97

Graham-Laurier Provincial Park

Wonowon

Blueberry

Rose Prairie

Montney

North Pine

Beatton River Provincial Park

Goodlow

Ole's

PEACE RIVER

Charlie Lake Provincial Park

Charlie Lake

Cecil Lake

717

Butler Ridge Provincial Park

Fort St. John

Baldonnel

Taylor

Clayhurst

Cherry Point

Muscovite Lakes Provincial Park

Williston

Dunlevy Provincial Recreation Area

Hudson's Hope

Peace River Corridor Provincial Park

PEACE

Taylor Landing PP

Kiskatinaw Provincial Park

Doe River

Omineca Provincial Park

Lake

W.A.C. Bennett Dam

Hudson's Hope

Bonanza

Farmington

Rolla

E

Germansen Landing

Bocock Peak Provincial Park

Moberly Lake

Sunset Prairie

Dawson Creek

Bay Tree

Manson Creek

Klin-se-za Provincial Park

Moberly Lake Provincial Park

Groundbirch

Progress

Mile 0 Alaska Highway

Pouce Coupe

Germansen Lake

Chetwynd

East Pine Provincial Park

Arras

Tomslake

Swan

Heather-Dina Lakes Provincial Park

Pine River Breaks PP

Lone Prairie

Sudeten Provincial Park

Tupper

52

Mackenzie

Pine LeMoray Provincial Park

Gwillim Lake Provincial Park

Lymbu Go

39

Powder King

Bijoux Falls Provincial Park

Sukunka Falls Provincial Park

One Island Lake Provincial Park

Kelly Lake

Leo Creek

Tudyah Lake Provincial Park

Bullmoose Mountain 2020m

Hole-in-the-Wall Provincial Park

Tumbler Ridge

Bearhole Lake Provincial Park

52

F

Tremblear Lake Provincial Park

McLeod Lake

Fort McLeod Provincial Historic Park

Whiskers Point Provincial Park

Sentinel Peak 2499m

Tumbler Ridge

BULKLEY-

Carp Lake Provincial Park

NECHAKO

Stuart Lake Provincial Park

Mt. Pope

Tachie

Monkman Provincial

MONTAGNES ROCHEUSES
ROCKY MOUNTAIN FOOTHILLS
CONTREFORTS DES ROCHEUSES
MUSKWA RANGES
FINLAY RANGES
MISINCHINKA RANGES
HART RANGES

VOLCANO COUNTRY

The Frank Mackie Glacier (left centre) in British Columbia's Coast Mountains is the source of the Unuk River (lower left). This remote river crosses the Canada-United States border and flows into Behm Canal, a coastal waterway of the Misty Fiords National Monument in the Alaska panhandle. Just north of the Unuk River, Lava Forks Provincial Park preserves the site of what is believed to be Canada's youngest volcano, which erupted in 1904.

Other points of interest

■ Dease Lake [B3] is one of the world's biggest producers of nephrite jade. This gemstone was first discovered here in 1965. Another high-quality deposit, roughly 50 km east of the community, was found in the 1990s.

■ Mount Edziza Provincial Park [C2-C3], covering some 2,300 km², preserves a spectacular region that has seen volcanic eruptions, probably within the last three hundred years. The park's dominant features are some 30 cinder cones, the largest of which is 2,787-m Mount Edziza, which towers over the surrounding 640-km² lava plain.

■ Within 2,170-km² Stikine River Provincial Park [C2-C5] lies one of Canada's most awesome sites, the 80-km-long Grand Canyon of the Stikine. Through the canyon flows the unruly 539-km Stikine River. The canyon has been carved by relentless river erosion. Its walls, up to 300 m high, are tinted gray, green, pink, and purple by various minerals. In the canyon depths, the width of the Stikine River ranges from 200 m across to a tiny 2-m gap through which the river tumbles.

Scale 1:2,000,000
1 cm = 20 km

0   20   40   60   80 km

Mount Warbelow 5553ft
Liberty
Clinton Creek
Fortymile
Harper 1874m
Tombstone Territorial Park
MOUNTAINS
BONNET PLUME RANGE

Jack Wade
Chicken
Taylor Mountain 5059ft
Boundary
Top of the World Highway
Forty-Mile, Fort Cudahy and Fort Constantine Historic Site
Yukon River Territorial Campground
Klondike Historic Complex National Historic Site / lieu historique national du Complexe-Historique-de-Dawson
Tombstone
Castle Mountain 2098m
NADALEEN RANGE
Rusty Mountain 1861m

Mount Fairplay 5541ft
Prindle Volcano 5125ft
Mount Hart 1621m
Dawson City
Bear Creek
Henderson Corner
Gold Dredge #4 National Historic Site / Lieu historique national de la Drague-Numéro-Quatre
Discovery Claim National Historic Site / Lieu historique national de la concession-Discovery
Klondike River Territorial Campground
Klondike Highway
Red Mountain 1762m
Steamboat Mountain 1120m
Mount Patterson 2088m
Keno
Keno City Mining Museum
Elsa
Mount Ortell 2063m

Junction
Reindeer Mountain 1578m
Stewart
Australia Mountain 1593m
McQuesten
Five Mile Lake Territorial Campground
Mayo Lake Rd.
Mt. Edwards 2088m
Mount Joy 2235m

Northway Junction
Northway
YUKON TERRITORY
Mount Stewart 1244m
Stewart Crossing
Silver Trail
Mayo
Ethel Lake Territorial Campground
Horseshoe Slough Territorial Habitat Protection Area
Mount Armstrong 2159m
RUSSEL RANGE

Tetlin National Wildlife Refuge
High Cache
Coffee Creek
Isaac Creek
Ddhaw Gro Habitat Protection Area
Big Kalzas Lake

Donjek
Selwyn
Fort Selkirk
Fort Selkirk Territorial Historic Site
Pelly Crossing
Uthsaw Wetland Territorial Habitat Protection Area
WILKINSON RANGE
Mt. McKenzie 1784m
Mount Selous 2176m

Wellesley Mountain 4966ft
Beaver Creek
Snag Junction Territorial Campground
Mount Baker
Snag
Mount Cockfield 1890m
Mount Pitts 1592m
Minto
McCabe Creek
Ta'Tla Mun Special Territorial Management Area
GLENLYON RANGE
Glenlyon Peak 2190m
ANVIL RANGE
1986m

Needle Peak 7586ft
Dry Creek
1974m
NUTZOTIN MOUNTAINS
Koidern
Lynx City
Mount Apex
Mt. Prospector 1976m
1435m
Tatchun Creek Territorial Campground
TATCHUN HILLS
Truitt Peak 2072m
Mt. Mye 2061m
Johnson Lake Territorial Campground

WRANGELL-ST. ELIAS NATIONAL PARK
Mount Natazhat 13435ft
Kluane Wildlife Sanctuary
2292m
Pickhandle Lake Territorial Recreation Site
Lake Creek Territorial Campground
NISLING RANGE
Mt. Klaza 1939m
Mt. Nansen
Mt. Nansen Rd.
Five Finger Rapids Territorial Recreation Site
Carmacks
Nunatuk Territorial Campground
Frenchman Lake Territorial Campground
Faro
Drury Creek Territorial Campground

TERRITOIRE DU YUKON
Mount Tittmann 9400ft
Mount Constantine
RUBY RANGE
1935m
Aishihik
Nordenskiold Wetland Territorial Protection Area
Mount Packers 1446m
Big Salmon
Mount Lokken 1835m
Fox Mount 2404m
Lapie Canyon Territorial Campground
Ross
Jackfish

CENTENNIAL RANGE
Mount Wood 4840m
Burwash Landing
Destruction Bay
Congdon Creek Territorial Campground
Kluane
Sheep Mountain Visitors Centre
Aishihik Road
Aishihik Lake Territorial Campground
Otter Falls Territorial Recreation Site
Hootalinqua
Fox Lake Territorial Campground
Kingston Mountain
Mount Anticline 1362m
1848m
Mount Caribou
Pass Peak 2162m
Mount St.Cyr 2050m

KLUANE NATIONAL PARK AND RESERVE
Mount Lucania 5226m
ST. ELIAS RANGES
Kluane Wildlife Sanctuary
Pine Lake Territorial Campground
SIFTON RANGE
Pilot Mount 2054m
Lake Laberge
KIMMERS RANGE
Mount Black 2158m
Quiet Lake Territorial Recreation Site

King Peak
Mt. Queen Mary 3886m
PARC NATIONAL ET RÉSERVE KLUANE
Mt. Cairnes 2789m
Mount Archibald
Haines Junction Visitors Centre
Canyon Creek
Ibex Valley
Takhini Hotsprings
Takhini
Crestview
Mount Cap 1801m
Mount Bynd
Mt. Murphy

Mount Logan 5959m Highest Point in Canada / Point le plus élevé du Canada
Snowshoe Peak
Spruce Beetle Trail
Champagne
162
Alaska Highway
1876m
Porter Creek
Takhini River Territorial Campground
Whitehorse
S.S. Klondike National Historic Site / Lieu historique national du S.S. Klondike
1898m
Nisutlin River National Wildlife Reserve / Réserve nationale du Delta-de-la-Rivière-Nis...

Table Mount 9360ft
Mount St. Elias 5489m
Mount Augusta 4289m
Mount Vancouver 4785m
MC Hubbard 4577m
Kathleen Lake
Mount Bratnober
Beloud Post
Mount Granger 2035m
MacRae
Wolf Creek Territorial Campground
Kusawa Lake Territorial Recreation Site & Campground
Mt. Arkell
Beringia
Robinson
Mt. Lorne
Marsh Lake Territorial Campground
Marsh Lake
Squanga Lake
Mount Streak
Johnsons Crossing

Mount Cook 4194m
Seward Glacier
Chaix Hills 3787ft
Goatherd Mountain
Kluane Wildlife Sanctuary
Dezadeash Lake Territorial Campground
Klukshu
2259m
Mount Lorne
Mount Skukum 2382m
Carcross Desert
Carcross
Tagish Bridge Terr. Rec. Site
Tagish
Hayes Peak
Jake's Corner
Brooks Brook
Teslin Lake Territorial Campground
1791m

WRANGELL-ST. ELIAS NATIONAL PRESERVE
Malaspina Glacier
Mount Hendrickson 4590ft
Knight I.
Point Manby
Mount Seattle 3072m
Dalton Post
Dalton Post Territorial Historic Site
Million Dollar Falls Territorial Campground
2219m
Windy Arm
Snafu Lake Territorial Campground
Tarfu Lake Territorial Campground
Mount Bryde 1908m
Teslin

Khantaak I.
Yakutat
Ocean Cape
Situk
Mount Wade 7960ft
Mount Armour 8770ft
Mount Ruhamah 5620ft
ALSEK RANGES
256
Tatshenshini-Alsek Provincial Wilderness Park
Highway Subject to Periodic Winter Closings / Route parfois fermée en hiver
Canada U.S.A.
Chilkoot Pass National Historic Site / Lieu historique national de la Piste-Chilkoot
COAST MOUNTAINS CHAÎNE CÔTIÈRE
BRITISH COLUMBIA
Atlin

PACIFIC OCEAN / OCÉAN PACIFIQUE
Mount Reaburn
Tongass National Forest
Crescent Mountain 4770ft
GLACIER BAY NATIONAL PRESERVE
Mosquito Lake State Park
Chilkat State Park
Klukwan
Pleasant Camp
Eagle Council Grounds
Chilkat Bald Eagle Preserve
Chilkat Pass
Chilkoot Gold Rush National Historical Park
Chilkoot Pass
White Pass
Fraser
Skagway
Dyea
Atlin Provincial Recreation Area
Five Mile

ALASKA
Mount Root 12860ft
GLACIER BAY NATIONAL PARK
Mount Canning 2112m
Mt. Foster 2173m
Chilkat Inlet
Haines
Chilkat State Park
Portage Cove State Park
Chilkat Islands
COLOMBIE
Atlin Provincial Park

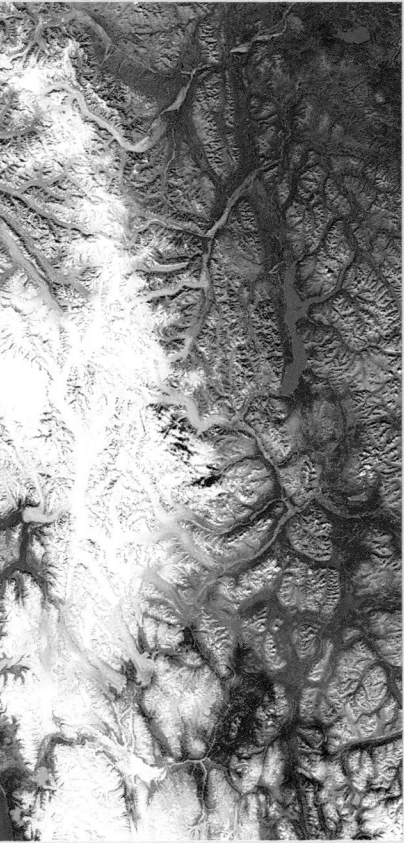

PEAKS AND ICEFIELDS

At 22,015 km², Kluane National Park and Reserve is Canada's second largest park and the site of the country's highest and largest mountains, the St. Elias. The mountains are split into two ranges—the Icefield (left) and the Kluane (centre). The Icefield ranges have more than 20 peaks higher than 4,200 m, including 5,959-m Mount Logan and 5,226-m Mount Lucania (upper right). The park contains the world's largest nonpolar icefields, which, as this view shows, spread glaciers from the heights into the valleys. The Kluane ranges, about 2,500 m on average, are visible from the Alaska Highway, which runs between the park's eastern boundary and Kluane Lake (upper right). Kluane is a World Heritage Site.

Other points of interest

■ The 3,185-km-long Yukon River system [A1-E4] rises in Tagish Lake on the British Columbia border and empties into the Bering Sea. It flows north and northwest across 1,149 km of rugged terrain in Canada, draining water from 65 percent of the Yukon Territory. The river continues 2,036 km through Alaska. The Yukon—from the Loucheux Indian word for "big river"—is North America's fifth longest river.

■ Gold may have been discovered near Dawson City [A2] in 1896, but today's mining activity in this area centers on asbestos, copper, lead, silver, and zinc.

■ Beaver Creek [C1] is the westernmost community in Canada.

■ Carcross [E4] boasts that it is the site of the "World's Smallest Desert," which consists of 260 ha of sand dunes, all that remains of an ancient glacial lake.

Scale 1:2,000,000

1 cm = 20 km

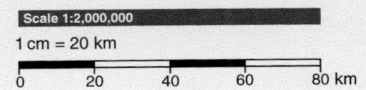

0   20   40   60   80 km

Map labels

MACKENZIE
BACKBONE
SELWYN
HESS MOUNTAINS
CANYON
NORTHWEST TERRITORIES
RANGES
TERRITOIRES DU NORD-OUEST
RANGES
MOUNTAINS
MACKENZIE MOUNTAINS

Horn Peak 2515m
Macmillan Pass
Dall Mtn. 2034m
Christie Pass
Mt. Wilson 2276m
Mt. Sheldon 2114m
Mt. Pike 2097m
Mt. Tidd
Mt. Sir James MacBrien 2762m
Tungsten
Rabbitkettle Lake Warden Station
NAHANNI NATIONAL PARK RESERVE
RÉSERVE DE PARC NATIONAL NAHANNI
Nahanni
LOGAN MOUNTAINS

YUKON
McEvoy Lake
Finlayson Lake Territorial Historic Site
Mount Hogg 2065m
2353m
Frances Lake
Frances Lake Territorial Campground & Historic Site
Mount Billings 2106m
Nahanni Range Territorial Campground
Norah Willis Michener Game Preserve
Mountain Time Zone Heure des Rocheuses
Pacific Time Zone Heure du Pacifique

CASSIAR MOUNTAINS
CONTINENTAL DIVIDE
ENGLISHMANS RANGE
1687m
2130m
Tuchitua
Mount Murray 2163m
Simpson Lake Campground Territorial Park
2079m
Morley River Territorial Recreation Site
Swift River
Rancheria Falls Territorial Recreation Site
Rancheria
Simpson Peak 2173m
Big Creek Territorial Recreation Site
Watson Lake Territorial Campground
Watson Lake
Watson Lake Signpost Forest
Upper Liard
Lower Post
Liard Canyon Territorial Recreation Site
Contact Creek
Hyland River Provincial Park
Coal River Springs Territorial Park
Smith River
Coal River
Fireside
Liard River Hotsprings Provincial Park
Liard River
Muncho Lake

COLUMBIA
BRITANNIQUE
Jade City
Good Hope Lake
Boya Lake Provincial Park
Muskwa-Kechika Management Area
Ligne de partage des eaux

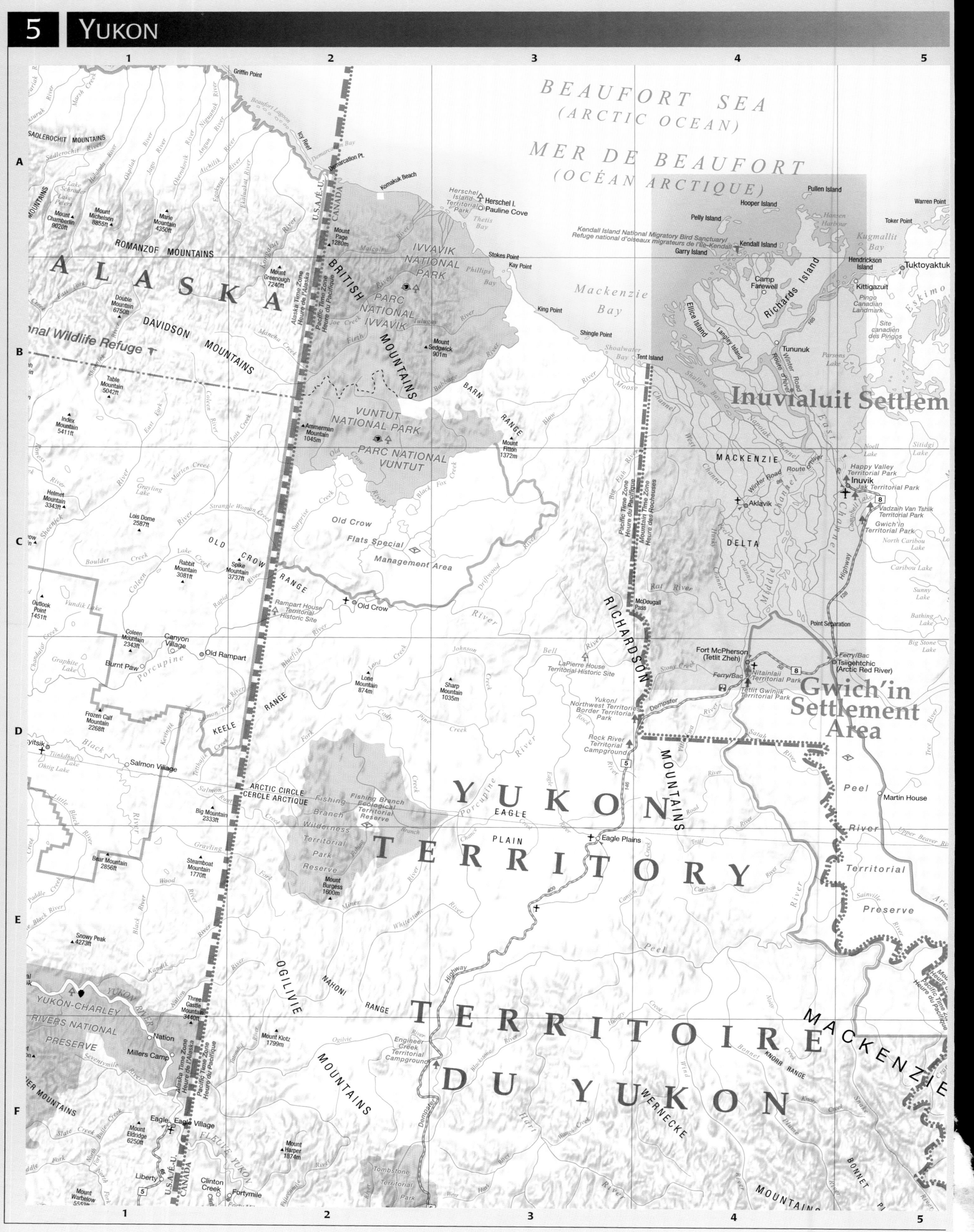

Map labels (north to south, grid A–F, columns 1–5)

BEAUFORT SEA
(ARCTIC OCEAN)

MER DE BEAUFORT
(OCÉAN ARCTIQUE)

Griffin Point

SADLEROCHIT MOUNTAINS

Beaufort Lagoon

Icy Reef

Demarcation Pt.

U.S.A./É.-U.
CANADA

Komakuk Beach

Herschel Island Territorial Park
Herschel I.
Pauline Cove

Thetis Bay

Pullen Island

Warren Point

Hooper Island

Pelly Island

Hansen Harbour

Toker Point

Kendall Island National Migratory Bird Sanctuary/
Refuge national d'oiseaux migrateurs de l'île-Kendall

Kugmallit Bay

Kendall Island

Hendrickson Island

Garry Island

Eskimo

Mount Chamberlin 9020ft

Mount Michelson 8855ft

Marie Mountain 4350ft

ROMANZOF MOUNTAINS

Mount Page 1280m

BRITISH

IVVAVIK NATIONAL PARK

Stokes Point

Kay Point

Camp Farewell

Richards Island

Tuktoyaktuk

Kittigazuit

Pingo Canadian Landmark / Site canadien des Pingos

A L A S K A

Double Mountain 6750ft

Mount Greenough 7240ft

PARC NATIONAL IVVAVIK

King Point

Mackenzie Bay

al Wildlife Refuge

DAVIDSON MOUNTAINS

Shingle Point

Ellice Island

Tununuk

Inuvialuit Settlem

Table Mountain 5042ft

COLUMBIAS

Mount Sedgwick 901m

Langley Island

Tent Island

Shoalwater Bay

Parsons Lake

Index Mountain 5411ft

VUNTUT NATIONAL PARK

BARN RANGE

Mount Fitton 1372m

MACKENZIE

Sitidgi Lake

Helmet Mountain 3343ft

Ammerman Mountain 1045m

PARC NATIONAL VUNTUT

Winter Road

Happy Valley Territorial Park

Inuvik

Jak Territorial Park

Lois Dome 2587ft

OLD

Old Crow

Aklavik

Vadzaih Van Tshik Territorial Park

Gwich'in Territorial Park

Rabbit Mountain 3081ft

CROW

Spike Mountain 3737ft

Old Crow Flats Special Management Area

DELTA

North Caribou Lake

Outlook Point 1451ft

Coleen Mountain 2343ft

Canyon Village

Rampart House Territorial Historic Site

McDougall Pass

Point Separation

Sunny Lake

Bathing Lake

Burnt Paw

Old Rampart

Big Stone Lake

Graphite Lake

Porcupine

Bell

LaPierre House Territorial Historic Site

Fort McPherson (Tetlit Zheh)

Ferry/Bac

Tsiigehtchic (Arctic Red River)

Frozen Calf Mountain 2268ft

KEELE

Lone Mountain 874m

Sharp Mountain 1035m

Ferry/Bac

Nitainlaii Territorial Park

Gwich'in Settlement Area

vitsik

Salmon Village

RANGE

RICHARDSON

Yukon/ Northwest Territories Border Territorial

Tetlit Gwinjik Territorial Park

Ohtig Lake

ARCTIC CIRCLE
CERCLE ARCTIQUE

Rock River Territorial Campground

Dempster

MOUNTAINS

Peel

Martin House

Big Mountain 2333ft

Fishing Branch Ecological Territorial Reserve

Y U K O N

Eagle Plains

River

Bear Mountain 2856ft

Fishing Branch Wilderness Territorial Park Reserve

EAGLE

Territorial

Steamboat Mountain 1770ft

PLAIN

T E R R I T O R Y

Preserve

Snowy Peak 4273ft

Mount Burgess 1600m

OGILVIE

Whitestone

Peel

YUKON-CHARLEY RIVERS NATIONAL PRESERVE

Three Castle Mountain 3440ft

NAHONI RANGE

T E R R I T O I R E

M A C K E N Z I E

Nation

Mount Klotz 1799m

Engineer Creek Territorial Campground

D U Y U K O N

WERNECKE

KNORR RANGE

Millers Camp

OGILVIE MOUNTAINS

Eagle

Eagle Village

Mount Eldridge 6250ft

Mount Harper 1874m

Liberty

Clinton Creek

Tombstone Territorial Park

BONNET PL

Mount Warbelow 5550ft

Fortymile

MOUNTAINS

U.S.A./É.-U.
CANADA

FLEUVE YUKON

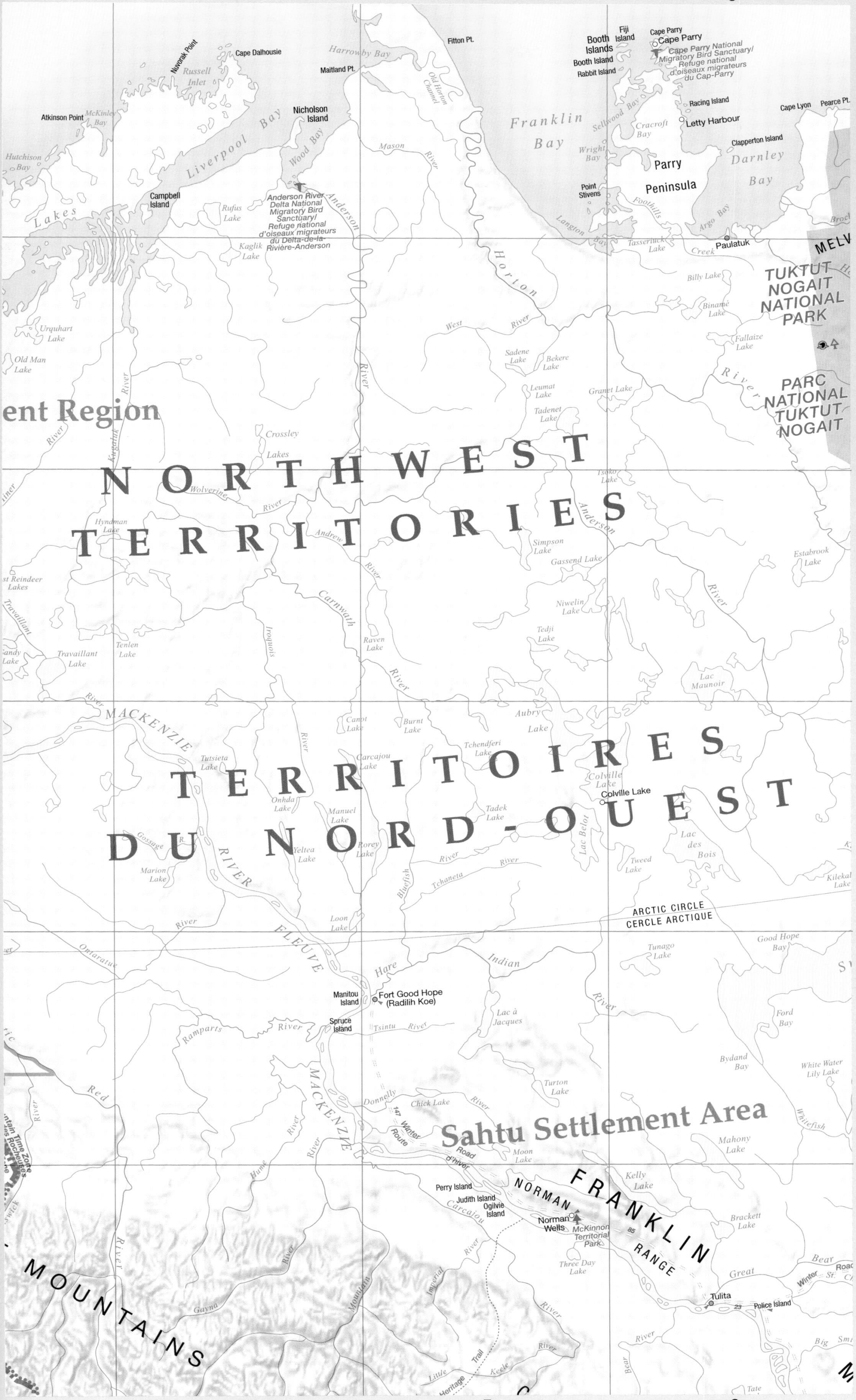

Map labels (grid references 6, 7, 8 across top; A–F down right side):

Nuvorak Point · Cape Dalhousie · Harrowby Bay · Fitton Pt. · Booth Islands · Fiji Island · Cape Parry · Cape Parry

Maitland Pt. · Cape Parry National Migratory Bird Sanctuary/ Refuge national d'oiseaux migrateurs du Cap-Parry

Booth Island · Rabbit Island

Atkinson Point · McKinley Bay · Russell Inlet · Nicholson Island · Sellwood Bay · Racing Island · Cape Lyon · Pearce Pt.

Franklin Bay · Cracroft Bay · Letty Harbour · Clapperton Island

Hutchison Bay · Liverpool Bay · Wood Bay · Mason · River · Wright Bay · Darnley Bay

Campbell Island · Rufus Lake · Anderson River Delta National Bird Sanctuary/ Refuge national d'oiseaux migrateurs du Delta-de-la-Rivière-Anderson

Point Stivens · Parry Peninsula · Argo Bay

Kaglik Lake · Horton · River · Langton Bay · Foothills · Tasseriuck Lake · Brock

Urquhart Lake · Paulatuk · MELV

Old Man Lake · West · River · Billy Lake · TUKTUT NOGAIT NATIONAL PARK

Sadene Lake · Bekere Lake · Biname Lake · Fallaize Lake

ent Region · Crossley Lakes · Leumat Lake · Tadenet Lake · Granet Lake · River · PARC NATIONAL TUKTUT NOGAIT

Wolverine · River

NORTHWEST TERRITORIES

Hyndman Lake · Andrew River · Simpson Lake · Anderson · Estabrook Lake

st Reindeer Lakes · Carnwath · River · Gassend Lake

Travaillant Lake · Iroquois · Niwelin Lake · River

Tenlen Lake · Raven Lake · Tedji Lake

Sandy Lake · Travaillant Lake

TERRITOIRES DU NORD-OUEST

MACKENZIE · River · Canot Lake · Burnt Lake · Aubry Lake · Lac Maunoir

Tutsieta Lake · Carcajou · River · Tchendferi Lake

Onhda Lake · Manuel Lake · Tadek Lake · Colville Lake · Colville Lake

Gossage · Yeltea Lake · Rorey Lake · River · River · Lac Belot · Lac des Bois

Marion Lake · RIVER · Tchaneta · Tweed Lake · Kilekal Lake

Loon Lake · ARCTIC CIRCLE CERCLE ARCTIQUE

River · Hare · Indian · Tunago Lake · Good Hope Bay

FLEUVE · Bluefish · River · River

Ontaratue · Manitou Island · Fort Good Hope (Radilih Koe) · Lac à Jacques · Ford Bay

Red · Spruce Island · Tsintu · River · Ramparts · Bydand Bay · White Water Lily Lake

MACKENZIE · Donnelly · Chick Lake · River · Turton Lake

Sahtu Settlement Area

Winter Route · Moon Lake · Mahony Lake · Kelly Lake

ic · RIVER · Perry Island · Judith Island · Ogilvie Island · NORMAN · Brackett Lake · FRANKLIN

Carcajou · Norman Wells · McKinnon Territorial Park · 85 · RANGE

Three Day Lake · Great · Bear · Road

tain Time Zone Rockies · Guynn · River · Winter Rd. Sr.

Tulita · MOUNTAINS · Little · Heritage · Trail · Keele · Police Island · 23 · Big Sm

River · River · Tate · M

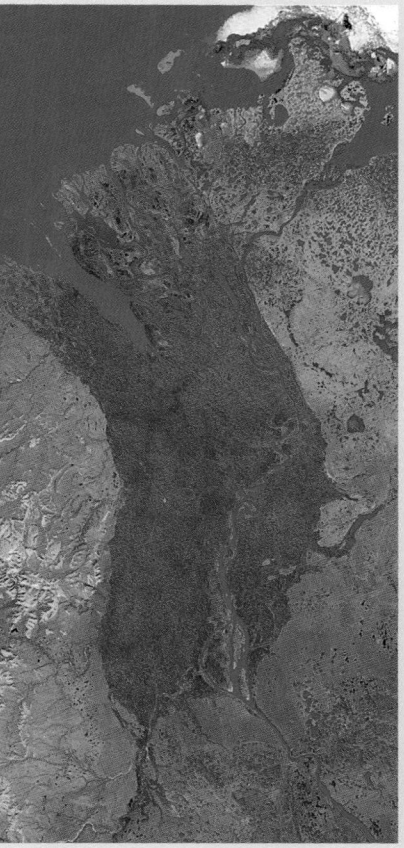

AN ARCTIC DELTA

The Mackenzie River, about one to two kilometres wide, flows majestically and placidly until it meets the Peel River (lower right). At this point, the broad stream dissolves into the labyrinthine Mackenzie Delta (centre), which extends 210 km to Mackenzie Bay and the Beaufort Sea in the Arctic Ocean (top). At 12,272 km², the Mackenzie Delta is the largest coastal delta in Canada. It contains some 20,000 lakes and ponds, and an intricate network of channels thousands of kilometres long. The Middle Channel, visible above, is the largest of three main waterways that flow through the delta and are used for navigation purposes. The natural curiosities of this region are pingoes—cone-shaped, ice-hills that rise up to 50 m. The delta has roughly 1,450 pingoes.

Other points of interest

■ In Ivvavik National Park [A2-B3] lies part of a region that escaped the impact of ice-age glaciation. More than 90 percent of this 10,168-km² park is dominated by the British Mountains, where some of the peaks reach 1,800 m. This is Canada's only non-glaciated range. South of Ivvavik, Vuntut National Park [B2-C3] is the site of an important waterfowl habitat.

■ Opened in 1979, the Dempster Highway [F2-C4] is the northernmost public highway in Canada, and the only one to traverse the Arctic Circle. Its construction took 20 years in often inhospitable conditions. It stretches from Dawson City to Inuvik, a distance of 720 km. Most of the highway had to be elevated on gravel "berms" to insulate the road from permafrost and prevent excess thawing.

Scale 1:2,000,000
1 cm = 20 km
0 · 20 · 40 · 60 · 80 km

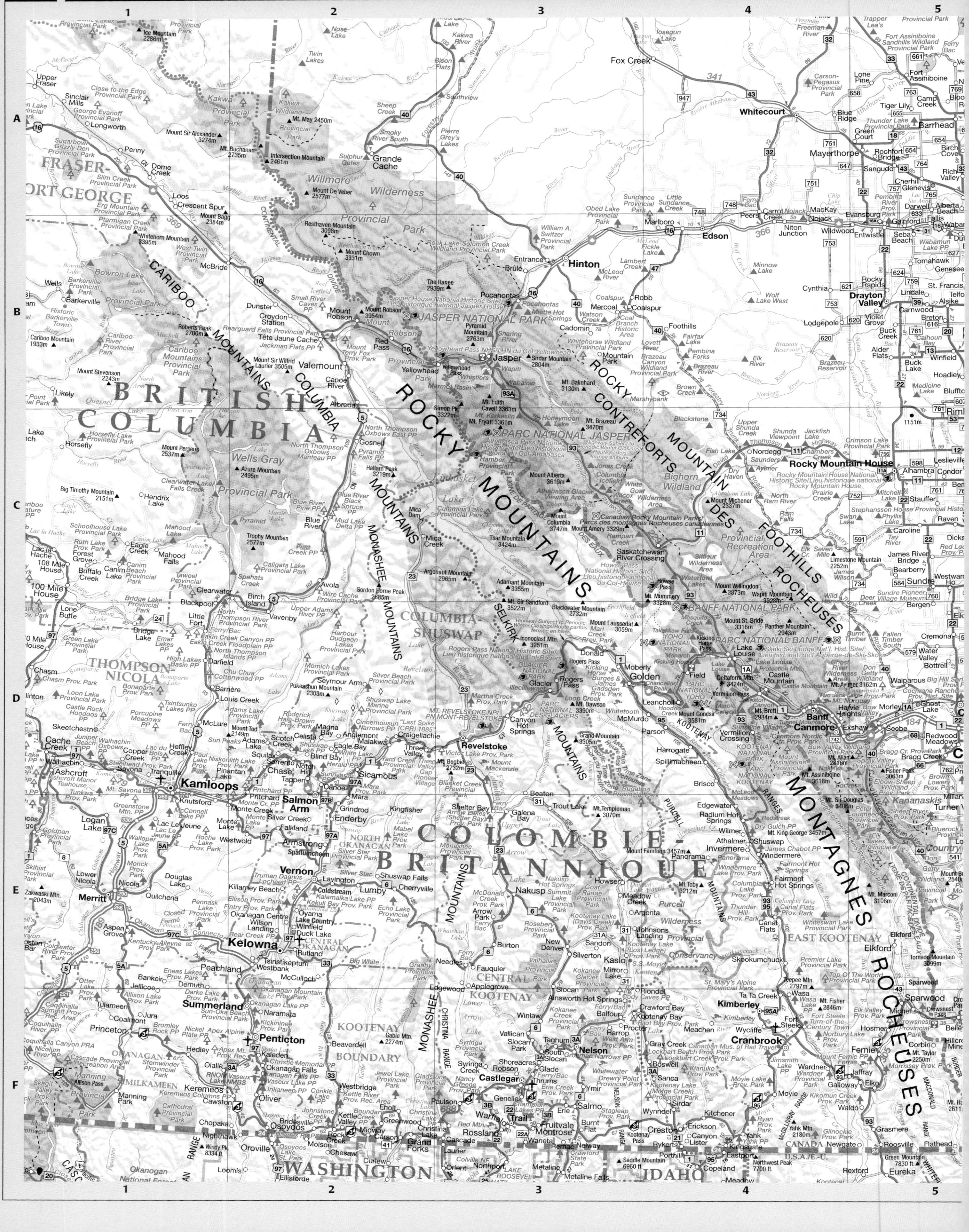

1 2 3 4 5

BRITISH COLUMBIA

ROCKY MOUNTAINS

COLOMBIE-BRITANNIQUE

MONTAGNES ROCHEUSES

FRASER-FORT GEORGE

CARIBOO MOUNTAINS

MONASHEE MOUNTAINS

COLUMBIA-SHUSWAP

SELKIRK MOUNTAINS

PURCELL MOUNTAINS

THOMPSON-NICOLA

NORTH OKANAGAN

CENTRAL OKANAGAN

CENTRAL KOOTENAY

EAST KOOTENAY

KOOTENAY BOUNDARY

WASHINGTON

IDAHO

Whitecourt · Edson · Hinton · Jasper · Rocky Mountain House · Nordegg · Banff · Canmore · Lake Louise · Golden · Field · Revelstoke · Kamloops · Salmon Arm · Vernon · Kelowna · Penticton · Summerland · Peachland · Westbank · Merritt · Princeton · Nelson · Castlegar · Trail · Rossland · Grand Forks · Osoyoos · Oliver · Cranbrook · Kimberley · Fernie · Sparwood · Elkford · Creston · Nakusp · Kaslo · New Denver · Silverton · Valemount · McBride · Dunster · Blue River · Clearwater · 100 Mile House · Cache Creek · Ashcroft · Radium Hot Springs · Invermere · Fairmont Hot Springs · Fort Steele

Wells Gray Provincial Park · Bowron Lake Provincial Park · Cariboo Mountains Provincial Park · Jasper National Park · Banff National Park · Yoho National Park · Glacier National Park · Mount Revelstoke National Park · Kootenay National Park · Manning Provincial Park

A B C D E F

LAKES IN THE TRENCH

The Rocky Mountain Trench is a 1,600-km-long valley, which extends through British Columbia from just below the Canada-United States border into the Yukon. The damming of rivers to produce hydroelectric power covered large areas of the valley with long, linear lakes, such as Kinbasket Lake (shown above). Williston Lake, farther north, is also a trench lake. Kinbasket Lake lies behind Mica Dam, which confines the flow of Columbia River (lower left). Towering over Kinbasket Lake are the Columbia (left) and Rocky Mountains. The Rocky Mountain Trench marks the Rockies' western boundary.

Other points of interest

■ Banff, Jasper, Kootenay, and Yoho are the four national parks [B2-E4] that make up the Canadian Rocky Mountain Parks, a United Nations World Heritage Site. Also part of this site are three adjacent B.C. provincial parks: Hamber [C3], Mount Assiniboine [D4], and Mount Robson [B2].

■ The Columbia Icefield [C3], covering an area of 230 km² at a depth of up to 365 m, is the largest accumulation of glacial ice in the Rocky Mountains. It is called "the mother of rivers" because its meltwaters feed the North Saskatchewan, Columbia, Athabasca, and Fraser river systems. It contains some 30 glaciers. The Athabasca Glacier [C3], adjacent to the Icefields Parkway in Jasper National Park, is one of the largest.

■ In Yoho National Park [D4], the Burgess Shale Site preserves the fossils of some 120 species of marine creatures, dating back 515 million years. The site was discovered in 1909 by American paleontologist Charles Walcott.

Scale 1:2,000,000

1 cm = 20 km

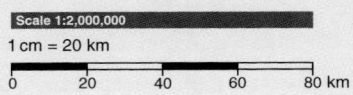

0   20   40   60   80 km

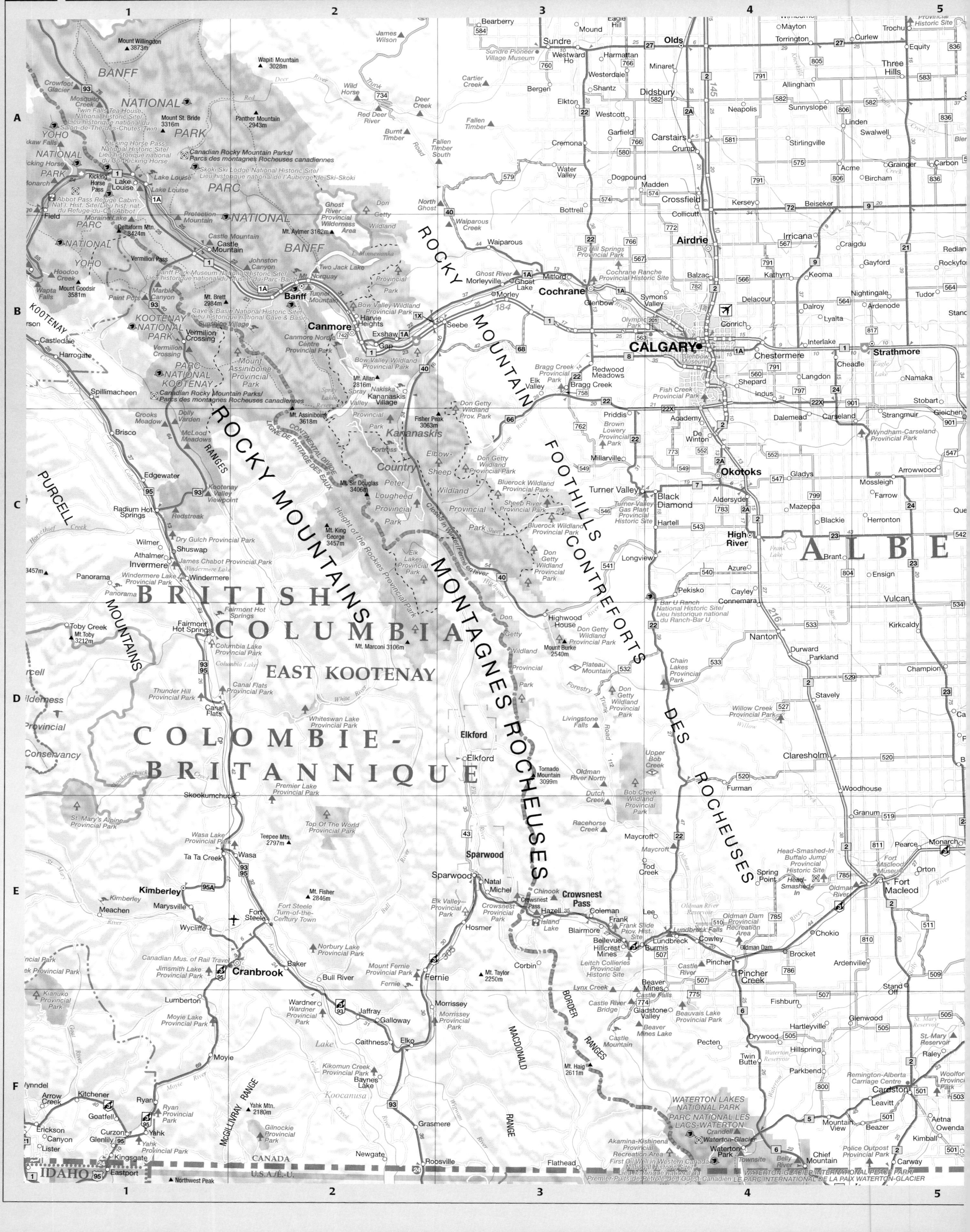

BANFF NATIONAL PARK

YOHO NATIONAL PARK

KOOTENAY NATIONAL PARK
PARC NATIONAL KOOTENAY

PURCELL MOUNTAINS

ROCKY MOUNTAINS
MONTAGNES ROCHEUSES

ROCKY MOUNTAIN FOOTHILLS
CONTREFORTS DES ROCHEUSES

BRITISH COLUMBIA
COLOMBIE-BRITANNIQUE

EAST KOOTENAY

ALBE[RTA]

BORDER RANGES

MCGILLIVRAY RANGE

MACDONALD RANGE

WATERTON LAKES NATIONAL PARK
PARC NATIONAL LES LACS-WATERTON

IDAHO

Major places:
Lake Louise, Field, Banff, Canmore, Radium Hot Springs, Edgewater, Invermere, Windermere, Fairmont Hot Springs, Canal Flats, Kimberley, Marysville, Wasa, Ta Ta Creek, Fort Steele, Cranbrook, Wycliffe, Moyie, Yahk, Kitchener, Curzon, Eastport

Sparwood, Elkford, Natal, Michel, Hosmer, Fernie, Hazell, Crowsnest Pass, Coleman, Blairmore, Bellevue, Frank, Burmis, Cowley, Pincher, Pincher Creek, Lundbreck, Brocket, Fort Macleod

CALGARY, Cochrane, Airdrie, Olds, Didsbury, Carstairs, Crossfield, Okotoks, Black Diamond, Turner Valley, High River, Longview, Nanton, Claresholm, Stavely, Granum, Cardston, Waterton Park

Kananaskis, Kananaskis Village, Nakiska, Kananaskis Country, Peter Lougheed Provincial Park, Elbow-Sheep Wildland, Don Getty Wildland Provincial Park, Sheep River Provincial Park, Bluerock Wildland Provincial Park, Highwood House, Plateau Mountain, Livingstone Falls

Head-Smashed-In Buffalo Jump National Historic Site
Bar U Ranch National Historic Site
Frank Slide Prov. Hist. Site
Leitch Collieries Provincial Historic Site
Oldman Dam
Oldman River Reservoir

Mountains: Mount Willingdon 3873m, Wapiti Mountain 3028m, Panther Mountain 2943m, Mount St. Bride 3316m, Deltaform Mtn. 3424m, Mount Goodsir 3581m, Mt. Brett 2984m, Mt. Aylmer 3162m, Mt. Assiniboine 3618m, Mt. Allan 2816m, Fisher Peak 3063m, Mt. Sir Douglas 3406m, Mt. King George 3457m, Mt. Marconi 3106m, Mt. Burke 2540m, Mt. Toby 3212m, Teepee Mtn. 2797m, Mt. Fisher 2846m, Tornado Mountain 3099m, Oldman River North, Mt. Taylor 2250m, Mt. Haig 2611m, Yahk Mtn. 2180m, Northwest Peak

CANADA
U.S.A./É.-U.

IN THE BADLANDS

For some 300 km, the windswept, weathered, and eroded terrain of the Alberta Badlands flanks the Red Deer River. In this view, the river snakes through a section of the arid, barren valley just west of Drumheller. Little Fish Lake Provincial Park (upper right) and the Finnegan Field Wetlands (lower left) brighten the scene with patches of vegetation. Once the domain of dinosaurs, this Badlands region abounds with the remains of these ancient creatures. Drumheller's famed Royal Tyrrell Museum of Palaeontology reveals fine examples, unearthed locally.

Other points of interest

■ Waterton Lakes National Park [F4], with an area of 525 km², embraces different natural environments ranging from rolling grasslands to snowcapped peaks, almost 3,000 m high. Its features include the three Waterton lakes. Upper Waterton Lake, at a depth of more than 150 m, is the deepest in the Rockies.

■ Big Rock, 7 km west of Okotoks [C4], is the world's largest glacial erratic. This 18,000-tonne quartzite boulder, measuring 40 x 18 x 9 m, was thought to have been carried from Jasper by advancing glaciers during the last ice age.

■ More than 500 skeletal remains of 50 dinosaur species have been discovered in Dinosaur Provincial Park [C7]. The park's most ancient fossil find dates back 77 million years.

■ Once described by Rudyard Kipling as the "City With All Hell for a Basement," Medicine Hat [D8] sits atop a natural gas field occupying 390 km². This field produces over 5 million cubic meters of natural gas daily.

Scale 1:1,000,000

1 cm = 10 km

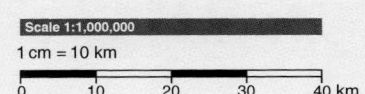

0   10   20   30   40 km

MONTANA

CANADA
U.S.A./É.-U.

1   **2**   **3**   **4**   **5**

A

SWAN HILLS

Goose Mountain

43

Iosegun Lake

Iosegun Lake

Smoke Lake

Fox Creek

Kaybob

Freeman Lake

Swan Hills

33

32

Trapper Lea's

Timeu

661

Doris

Fort Assiniboine Sandhills Wildland Provincial Park

Fort Assiniboine

658

33

Camp Creek

763

Tiger Lily

Mystery Lake

655

Tiger Lake

Campsie

18

341

43

947

32

Carson-Pegasus Provincial Park

Lone Pine

B

Benbow Junction

Benbow

Whitecourt

Lonira

43

Blue Ridge

658

35

Connor Creek

Peavine

Thunder Lake Provincial Park

763

Ba

Glenister

Moss
Lav

Stev

65

Mahaska

32

751

Green Court

18

16

Mayerthorpe

Rochfort Bridge

Cosmo

654

Meadowvie

Ballantine

Haddock

Highway

Anselmo

647

Sangudo

43

764

Lisburn

Cherhill

C

Pioneer

748

McLeod Valley

Ferry/Bac

Hattonford

751

Rangeton

757

Stanger

Park Court

765

Glenevis

Wes
Cove

Darwell

Obed

16

Hargwen

Pedley

Obed Lake Provincial Park

75

Marlboro

Galloway

Rosevear

Peers

Bickerdike

Yates

Wolf Creek

Edson

Carrot Creek

16

Niton

MacKay

Chip Lake

Ravine

Pembina River Provincial Park

633

Magnolia

Gainford

16

Niton Junction

Nojack

Wildwood

Evansburg

Entwistle

Falls
Wabam

40

Rock Lake

Rock Lake-Solomon Creek Wildland Provincial Park

Entrance

Hinton

Little Sundance Creek

Sundance Provincial Park

McLeod River

Fickle Lake

Erith

Hill Creek

Minnow Lake

753

A L B E R T A

22

Seba Beach

Water's Edge

759

Waba

31

Horen

627

Highwa

D

Brûlé

Pocahontas

Jasper House Nat'l Hist. Site/ Lieu hist. nat'l Jasper House

18

Pocahontas

Miette Hot Springs

40

Luscar

Cadomin

78

Watson Creek

Fidler

Lambert Creek

Weald

McLeod River

47

Robb

Coalspur

Mercoal

Coalspur

Coal Branch Historic Area

40

Foothills

Lovett River

Fairfax Lake

Wolf Lake West

Moon Lake

624

Round Valley

Northleigh

Burtonsville

Cree

Cynthia

621

Rocky Rapids

Drayton Valley

39

Carnwood

Poplar Ridge

22

Violet Grove

620

22

Henry House National Historic Site/ Lieu historique national Henry House

Snaring River

Jasper

Wapiti

Medicine Lake

Sirdar Mountain 2804m

Whitehorse Wildland Provincial Park

Mountain Park

Pembina Forks

Brazeau River

Brazeau Reservoir

Brazeau Dam

Brazeau Reservoir

753

Lodgepole

Pembina

Buck Creek

616

761

Buck Lake

Calhoun Bay

Norbu

Buck Lake

761

34

Pendryl

Alder Flats

13

E

Whistlers

Basin

Wabasso

93

93A

Mt. Edith Cavell 3363m

Mt. Fryatt 3361m

JASPER NATIONAL PARK

Maligne Lake

Mt. Kerkeslin

PARC NATIONAL JASPER

Honeymoon Lake

Sunwapta Falls

Mt. Balinhard 3130m

Canadian Rocky Mountain Parks/ Parcs des montagnes Rocheuses canadiennes

Mt. Brazeau 3470m

R O C K Y

C O N T R E F O R T S

Brazeau Canyon Wildland Provincial Park

Marshybank

Brown Creek

M O U N T A I N

D E S

Blackstone

734

Nordegg

22

Medicine Lake

Willesden Green

1151m

Crimson Lake Provincial Park

761

Leed

Carlos

R O C K Y   M O U N T A I N S

M O N T A G N E S   R O C H E U S E S

93

Athabasca Pass National Historic Site/ Lieu historique national du col Athabasca

Hamber Provincial Park

Maligne Lake

Highway Subject to Periodic Winter Closings

Jonas Creek

Stanley Falls

Mount Alberta 3619m

Upper Shunda Creek

Thompson

Fish Lake

Harlech

Beaverdam

Shunda Viewpoint

Jackfish Lake

Highway

11

Chambers Creek

Saunders

Crimson Lake

Frisco

12

Bingley

761

F

Cummins Lake Provincial Park

Clemenceau Icefield

BRITISH COLUMBIA

COLOMBIE-BRITANNIQUE

Kinbasket Lake

Tsar Mountain 3424m

Mount Columbia 3747m

Columbia Icefield

Athabasca Glacier Viewing Area

Columbia Icefield

Mount Amery 3329m

93

Rampart Creek

229

Saskatchewan River Crossing

Saskatchewan River Crossing

Mistaya Canyon

Waterfowl Lakes

White Goat Provincial Wilderness Area

Thompson Creek

Siffleur Provincial Wilderness Area

11

Kootenay Plains

Kootenay Plains

Crescent Falls

David

Dry Haven

11

Goldeye Lake

Aylmer

Mount Michener 2337m

Abraham Lake

734

North Ram River

Ram Falls

752

Kootenay Plains

Peppers Lake

Elk Creek

Seven Mile

Limestone Mountain 2252m

734

F O O T H I L L S

Horburg

Horburg

Rocky Mountain House

11A

Garth

Ferrier

11

Rocky Mountain House National Historic Site/ Lieu historique national Rocky Mountain House

752

Alhambra

11

Prairie Creek

Congresbury

22

Dovercourt

South Fork

Mitchell Lake

Cheddarville

Phyllis Lake

Caroline

591

Tay River

22

Crammond

James Riv
Bridge

Ravc

1   **2**   **3**   **4**   **5**

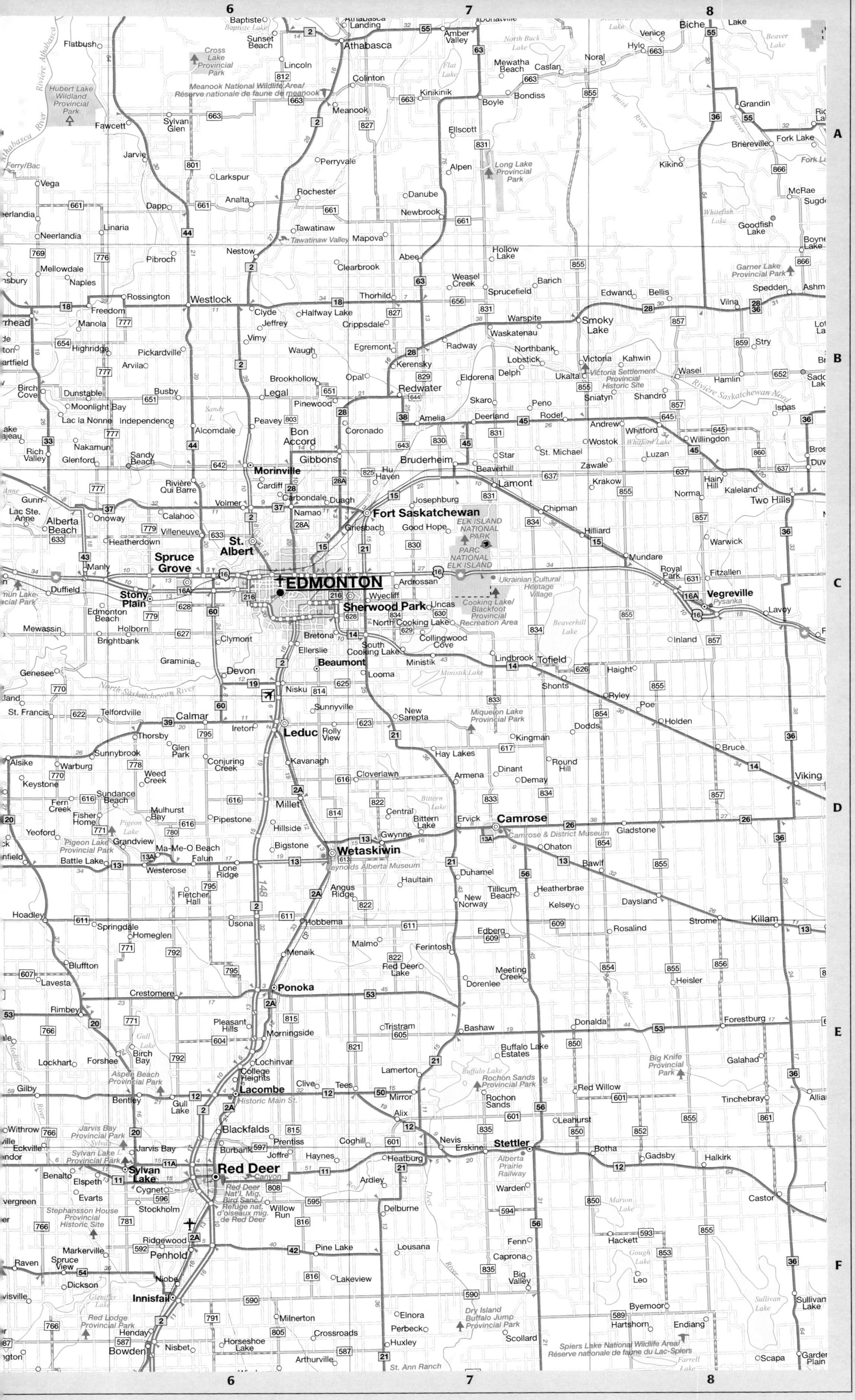

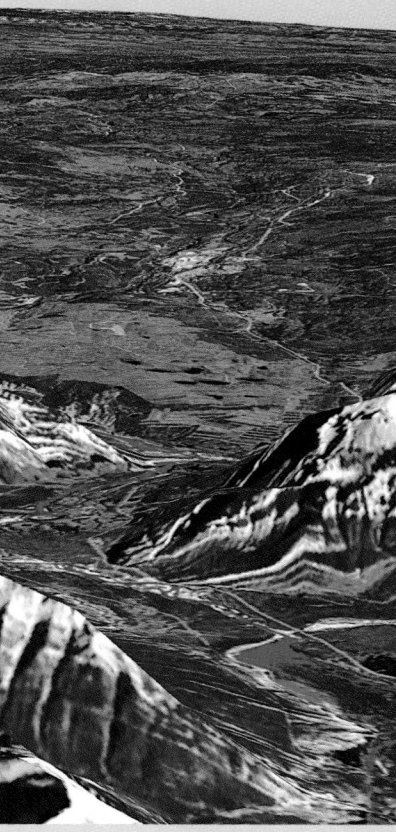

FROM PARK TO PRAIRIE

This view of Jasper National Park shows the wide scenic valley just north of the park townsite. The Trans-Canada Yellowhead Highway (Highway 16) runs between Jasper and Talbot lakes (lower right) and, farther on, along the banks of the Athabasca River. At the gap in the Rockies, the Yellowhead exits the park, passes Brulé Lake (left), and heads toward the horizon on the Prairies. The bold shape of Roche Miette (right centre) hides Fiddle Valley, where a road leads to one of Jasper's popular attractions, the Miette Hot Springs.

Other points of interest

■ Elk Island National Park [C7] is a 194-km² wildlife refuge. It is the only national park completely surrounded by a fence. The fenced boundary prevents park animals—elk, bison, moose, and deer—from straying, and stops predators—wolves and bears—from entering. Because the elk population increases at a rate of 20 percent per year, the management of the herd involves relocating excess numbers to other parts of Canada and the United States. Elk Island also preserves a remnant of Alberta's Beaver Hills.

■ On Feb. 13, 1947, "black gold" gushed from the Leduc No. 1 Oilwell site, located several kilometres north of Leduc [D6]. The Imperial Oil crew drilled 133 dry wells before it struck oil at a depth of 1,771 m, finally tapping the 300-million-barrel Leduc oil field. A plaque at the site commemorates the event. The oil derricks are long since gone as oilmen seek new sources underground or in Alberta's oil-rich tar sands.

Scale 1:1,000,000
1 cm = 10 km

0   10   20   30   40 km

NORTHWEST TERRITORIES · TERRITOIRES DU NOR[D]

BRITISH COLUMBIA

COLOMBIE-BRITANNIQUE

INTERIOR PLAINS · PLAINES DE L'INTÉRIEUR

ALBER[TA]

Mackenzie

CARIBOU MOUNTAINS

Caribou Mountains Wildland Provincial Park

BUFFALO HEAD HILLS

CLEAR HILLS

SWAN HILLS

Selected places and features:

Fort Nelson, Old Fort Nelson, Fort Nelson Heritage Museum, Andy Bailey Provincial Park, Kotcho Lake Village Site Provincial Park, Prophet River, Klua Lakes Provincial Protected Area, Sikanni Old Growth Provincial Park, Horse River Provincial Park, Beatton River, Pink Mountain, Prespatou, Altona, Wonowon, Buick, Montney, North Pine, Rose Prairie, Charlie Lake, Fort St. John, Fort St. John North Peace Historic Museum, Baldonnel, Taylor, Cecil Lake, Clayhurst, Hudson's Hope, Moberly Lake, Moberly Lake Provincial Park, Chetwynd, Groundbirch, Progress, Arras, Dawson Creek, Pouce Coupe, Sunset Prairie, Farmington, Rolla, Bonanza, Doe River, Bay Tree, Gordondale, Demmitt, Tupper, Tomslake, Swan Lake Provincial Park, Lymburn, Goodfare, Beaverlodge, Hythe, La Glace, Valhalla Centre, Buffalo Lake, Huallen, Wembley, Elmworth, Grovedale, Grande Prairie, Grande Prairie Pioneer Museum, Sexsmith, Clairmont, Bezanson, DeBolt, Goodwin Corner, Ridgevalley, Sturgeon Heights, Williamson Provincial Park, Calais, Valleyview, Little Smoky, Fox Creek, Waskahigan River, Wallace Mountain 1259m, Goose Mountain, House Mtn 1067m, Swan Hills, Freeman River, Fort Assiniboine Sandhills Wildland Provincial Park

Thinahtea Provincial Protected Area, Zama City, Habay, Hay-Zama Lakes Wildland Provincial Park, Chateh (Assumption), Rainbow Lake, Meander River, Tugate, Hutch Lake, High Level, MacKenzie Crossroads Museum, Rocky Lane, Boyer, Fort Vermilion, North Vermilion, Machesis Lake, La Crete, Buffalo Head Prairie, Paddle Prairie, Keg River, Carcajou, Kemp River, Twin Lakes, Notikewin Provincial Park, Hotchkiss, Notikewin, Manning, North Star, Deadwood, Dixonville, Chinook Valley, Peace River, St. Isidore, Grimshaw, Shaftesbury, Mane-Reine, Nampa, Reno, Berwyn, Whitelaw, Bluesky, Brownvale, Fairview, Dunvegan, Historic Fort Dunvegan, Queen Elizabeth Provincial Park, Hines Creek, Eureka River, Worsley, Cleardale, Silver Valley, Bear Canyon, Cherry Point, Silver Valley, Blueberry Mountain, Moonshine Lake Provincial Park, Rycroft, Spirit River, Woking, Peoria, Wanham, Eaglesham, Tangent, Girouxville, Falher, Donnelly, McLennan, Guy, Kathleen, Jean Côté, Watino, Big Prairie, Grouard, Grouard Mission Native Culture Arts Museum, Hilliard's Bay Provincial Park, Enilda, Joussard, Driftpile, Kinuso, Faust, High Prairie, New Fish Creek, Winagami Wildland Provincial Park, Winagami Lake Provincial Park, Heart River, Salt Prairie, Sunset House, Lesser Slave Lake, Lesser Slave Lake Wildland Provincial Park, Slave Lake, Widewater, Canyon Creek, Marten Beach, Grizzly Ridge Wildland Provincial Park

Steen River, Lutose, Slavey Creek, Indian Cabins, 60th Parallel Territorial Park, Cameron Hills, Wentzel Lake, Fox Lake, John D'or Prairie, Peerless Lake, Loon Lake, Cadotte Lake, Red Earth Creek, Little Buffalo, Trout Lake, Gift Lake, Atikameg, Atikamisis Lake

Tumbler Ridge, Monkman Provincial Park, Wapiti Lake Provincial Park, Arctic Pacific Lakes Provincial Park, Ice Mountain 2266m, Sentinel Peak 2499m, Bullmoose Mountain 2020m, Hole-in-the-Wall Provincial Park, Bearhole Lake Provincial Park, Gwillim Lake Provincial Park, Sukunka Falls Provincial Park, Pine River Breaks PP, Lone Prairie, East Pine Provincial Park, One Island Lake Provincial Park, Sudeten Provincial Park, Kelly Lake, Mountain Time Zone / Heure des Rocheuses, Pacific Time Zone / Heure du Pacifique

Highways: 97, 29, 43, 52, 59, 2, 49, 40, 33, 32, 35, 58, 88, 697, 695, 692, 691, 690, 689, 688, 685, 986, 743, 741, 740, 739, 737, 735, 733, 732, 731, 730, 729, 727, 726, 725, 724, 723, 722, 721, 720, 719, 717, 681, 682, 683, 680, 671, 672, 667, 668, 670, 665, 666, 669, 676, 677, 679, 744, 745, 747, 749, 750, 751, 661, 64, 64A, 2A, 625

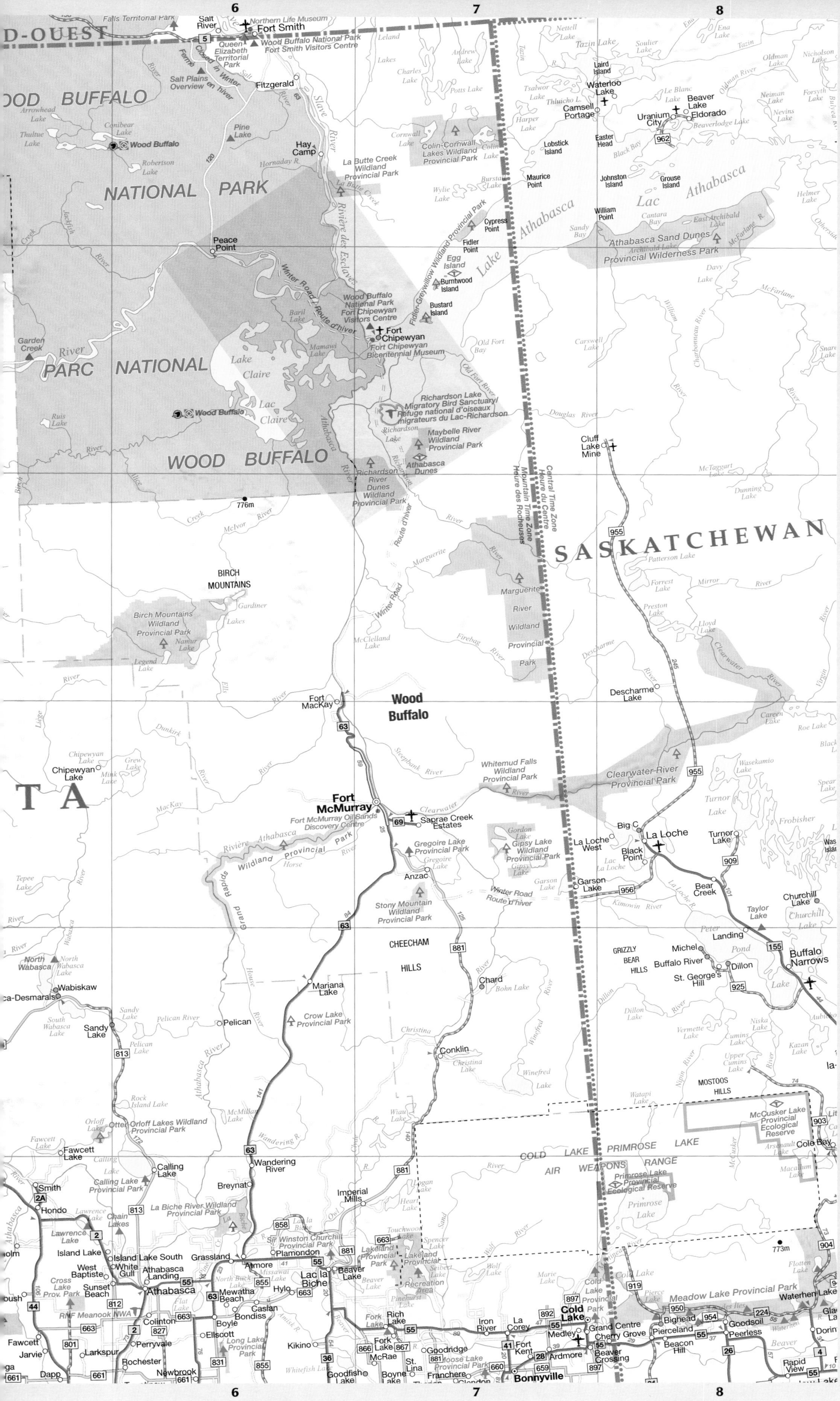

AN INLAND DELTA

The Peace-Athabasca Delta occupies an area of 3,200 km², which contains meadows, marshes, and lakes. This sprawling wetland is the world's largest inland delta. Much of it lies within the southeastern corner of Wood Buffalo, Canada's largest national park, which covers 44,802 km². This view shows parts of the region's largest lakes, Claire (left) and Athabasca (right). The Peace-Athabasca Delta really consists of three deltas—those of the Peace, Athabasca, and Birch rivers. The Peace River delta is the river's junction with the Slave River (top). The largest delta, the Athabasca, has filled the west end of Lake Athabasca. Before reaching the delta, the Athabasca River winds past dunes (lower left) and Richardson Lake. The delta of the Birch River [B6] expands into Lake Claire, which, in turn, empties into Lake Athabasca. The Peace-Athabasca Delta is a key intersection of North America's four migratory bird flyways.

Other points of interest

■ Athabasca Dunes [A7-B8] and Richardson River Dunes Wildland Provincial Park [B7-C7] contain the largest area of actively migrating sand in Alberta, and are part of Canada's largest dune field. The dunes, said to be 8,000 years old, are 7 km long and 1.5 km wide. Some of the dunes reach heights of 35 m. The dunes are moving in a southeast direction at a rate of 1.5 m per year.

■ Near Fort McMurray [D7], the Athabasca Oil Sands is one of the world's great oil deposits. The oil sands cover some 77,000 km², equivalent to the area of New Brunswick. Reserves are estimated to be about 300 billion barrels.

Scale 1:2,000,000

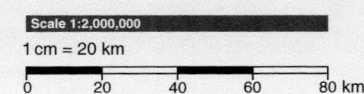

1 cm = 20 km

0   20   40   60   80 km

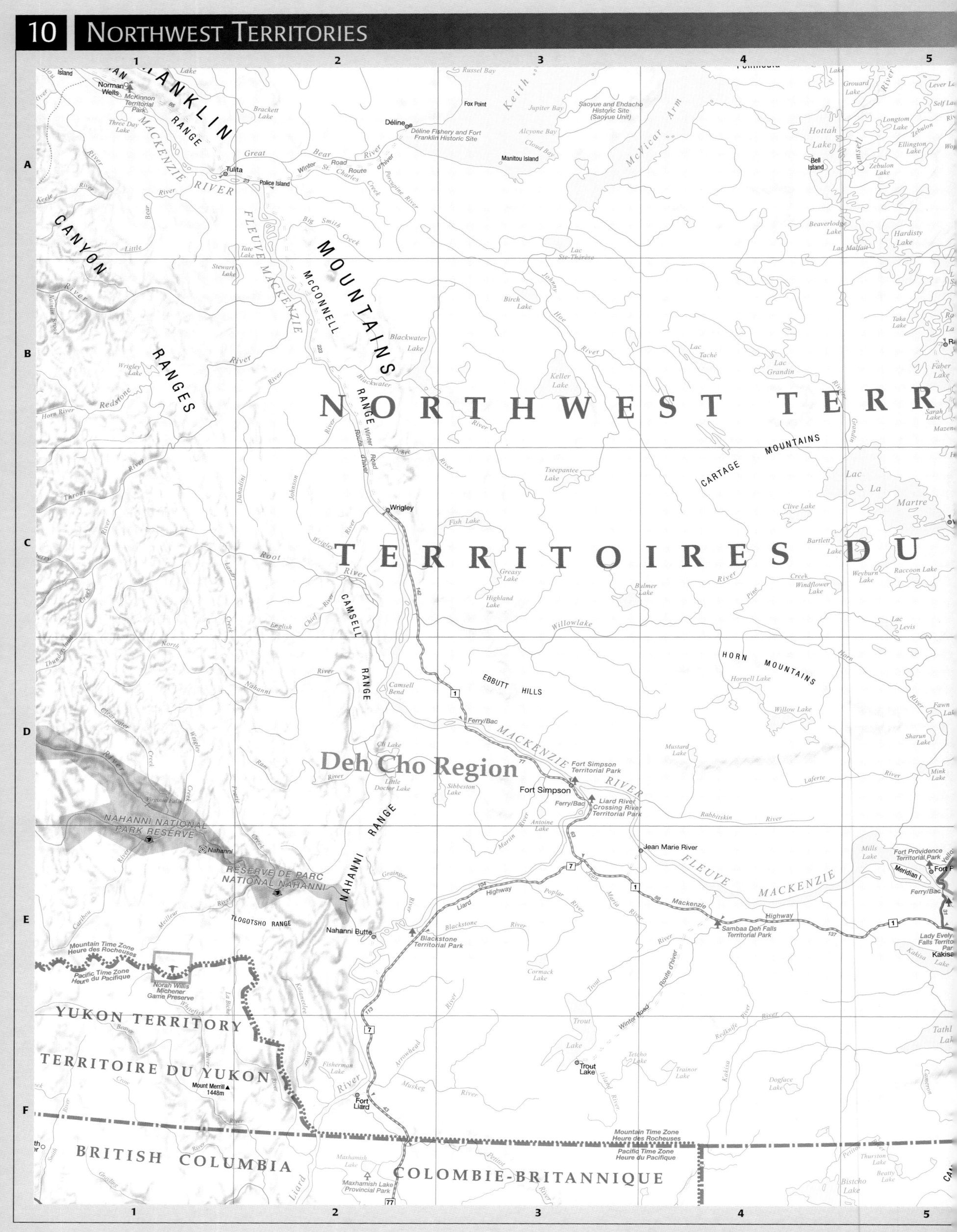

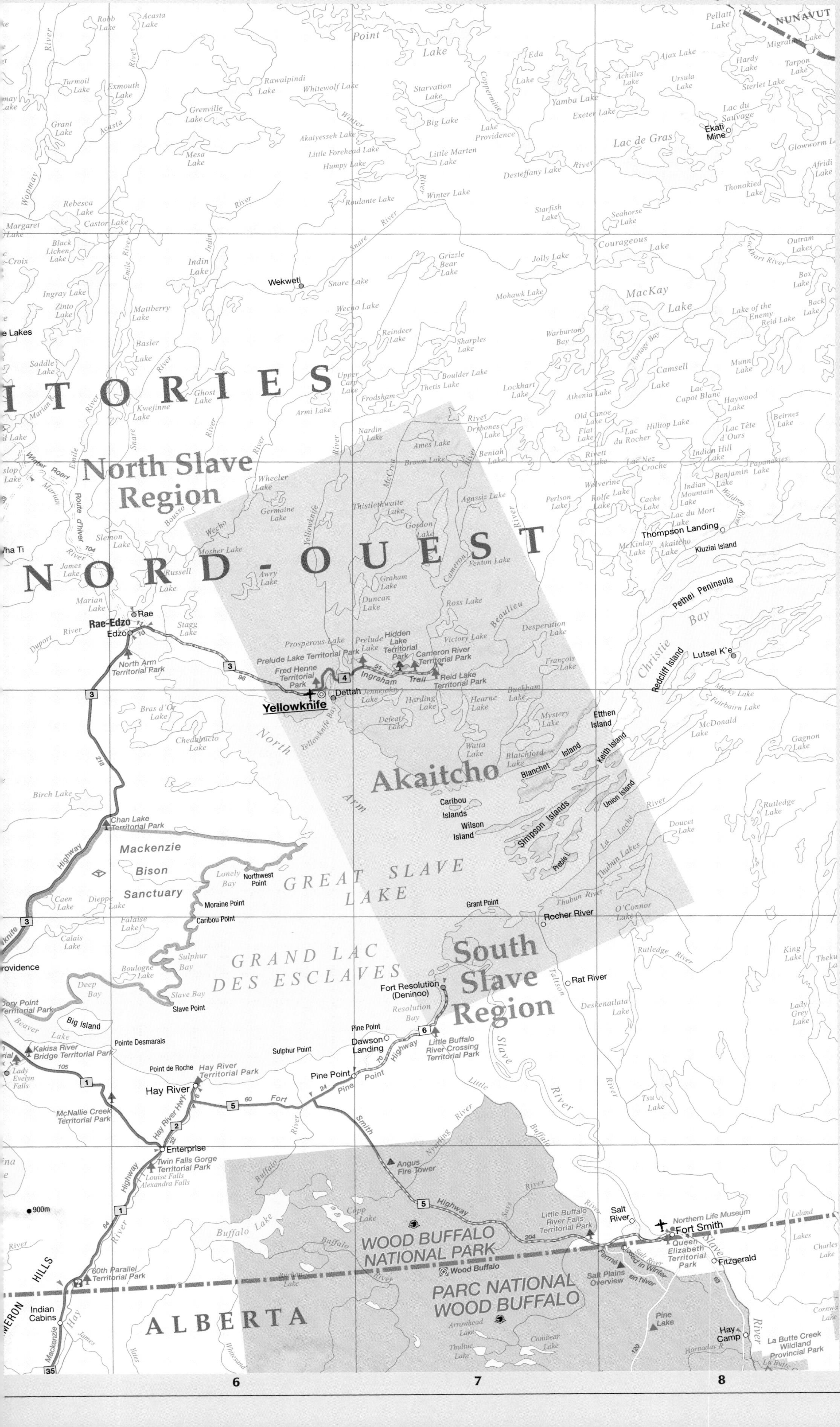

NUNAVUT

ITORIES

North Slave Region

NORD-OUEST

Rae-Edzo
Rae
Edzo

North Arm
Territorial Park

Prelude Lake Territorial Park
Fred Henne
Territorial Park
Yellowknife
Dettah

Ingraham
Trail

Prosperous Lake
Prelude Lake
Hidden Lake Territorial Park
Cameron River Territorial Park
Reid Lake Territorial Park

Akaitcho

Chan Lake
Territorial Park

Mackenzie
Bison
Sanctuary

Northwest Point
Lonely Bay
Moraine Point
Caribou Point
Falaise Lake
Calais Lake

GREAT SLAVE LAKE

Caribou Islands
Wilson Island
Simpson Islands

Grant Point

Rocher River

GRAND LAC DES ESCLAVES

South Slave Region

Rat River

Fort Resolution (Deninoo)
Slave Point
Resolution Bay

Pine Point

Big Island

Kakisa River Bridge Territorial Park

Pointe Desmarais
Point de Roche
Sulphur Point
Dawson Landing

Hay River Territorial Park
Pine Point

McNallie Creek Territorial Park

Hay River

Fort

Little Buffalo River Crossing Territorial Park

Enterprise
Twin Falls Gorge Territorial Park
Louise Falls
Alexandra Falls

Angus Fire Tower

Little Buffalo River Falls Territorial Park

Salt River
Northern Life Museum
Fort Smith

900m

WOOD BUFFALO NATIONAL PARK
Wood Buffalo

Queen Elizabeth Territorial Park
Fitzgerald

Salt Plains Overview

PARC NATIONAL WOOD BUFFALO

Arrowhead
Thultue Lake
Conibear Lake
Pine Lake

Hay Camp

La Butte Creek Wildland Provincial Park

HILLS
Indian Cabins
60th Parallel Territorial Park

ALBERTA

Ekati Mine
Lac de Gras
Thompson Landing
Kluziai Island

Pethei Peninsula
Christie Bay
Redcliff Island
Lutsel K'e

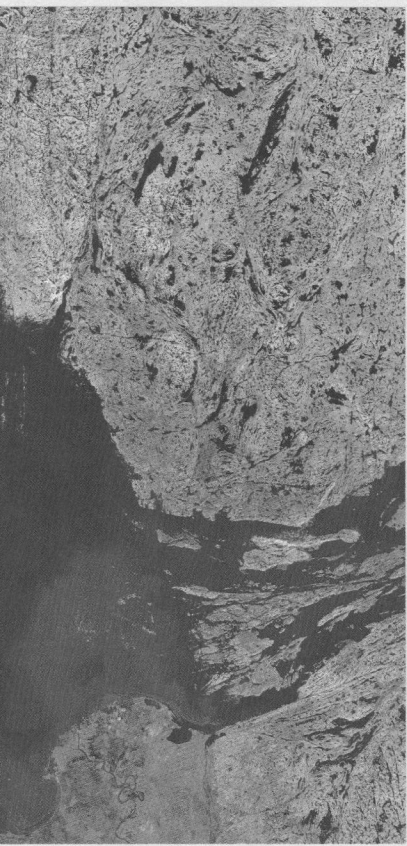

THE DEEPEST LAKE

At 28,568 km², Great Slave Lake is Canada's fourth largest and the world's eleventh largest lake. It is also Canada's deepest lake, reaching a maximum depth of 614 m. The forests of the Boreal Shield extend as far as the southern shore of the lake. But most of the region around the lake is a transition zone where the Shield gives way to the sparser vegetation of the taiga. This view shows island-congested Christie Bay at the entrance to the East Arm (right), which has been proposed as the site of a national park. Yellowknife, the capital of the Northwest Territories and its largest city, is visible at the head of Yellowknife Bay on the North Arm (upper left). In the jumble of lakes and woods northwest of Yellowknife lie the world's most ancient rocks—granites that formed more than 3.96 billion years ago.

Other points of interest

■ The 320-km-long South Nahanni River runs through 4,700-km² Nahanni National Park Reserve [D1-E2]. The turbulent river rushes through three huge canyons more than 1,000 m deep and plunges over 90-m Virginia Falls. Other natural features of this remote park include hot springs and icy caves.

■ In 1985, geologists found kimberlites—ancient volcanic formations containing diamonds—at Lac de Gras [A8] in the treeless tundra of the Northwest Territories. Six years later, Ekati Mine, the first Canadian diamond mine, announced the discovery of substantial deposits in this region. The news started one of the largest staking rushes in Canadian history. Ekati Mine now produces between 3 million and 5 million carats of diamonds per year.

Scale 1:2,000,000
1 cm = 20 km
0   20   40   60   80 km

ALBERTA
SASKATCHEWAN
MONTANA
CANADA
U.S.A./É.-U.
TOOLE
LIBERTY
Meadow Lake Provincial Park
PRINCE ALBERT NATIONAL PARK
PARC NATIONAL DE PRINCE ALBERT
GRASSLANDS NATIONAL PARK/ PARC NATIONAL DES PRAIRIES
Cypress Hills Interprovincial Park
Fort Walsh National Historic Site/ Lieu historique national du Fort-Walsh
THE MIDDLE SAND HILLS
Canadian Forces Base Suffield
THE GREAT SAND HILLS
CYPRESS HILLS
COTEAU HILLS
BAD HILLS
Mountain Time Zone / Heure des Rocheuses
Central Time Zone / Heure du Centre
Lloydminster
Vegreville
Vermilion
Wainwright
St. Paul
Bonnyville
Elk Point
Two Hills
Killam
Viking
Coronation
Castor
Hanna
Oyen
Provost
Consort
Brooks
Medicine Hat
Redcliff
Bow Island
Foremost
Aden
Port of Whitlash
Port of Wild Horse
Wild Horse
Willow Creek
Port of Willow Creek
Maple Creek
Elkwater
Eastend
Shaunavon
Swift Current
Gull Lake
Leader
Kindersley
Rosetown
Outlook
Biggar
Unity
Macklin
Wilkie
Kerrobert
North Battleford
Battleford
Meadow Lake
Big River
Spiritwood
Shellbrook
Prince Albert
Rosthern
Warman
Martensville
SASKATOON
Humboldt
Watrous
Davidson
Elbow
Moose Jaw
Gravelbourg
Assiniboia
Coronach
Val Marie
Mankota
Kyle
Eston
Cabri
Herbert
Morse
Chaplin
Central Butte
Lucky Lake
Lake Diefenbaker
Duck Lake
Batoche
Waskesiu Lake
Lavallee Lake
Montreal Lake
Grey Owl's Cabin

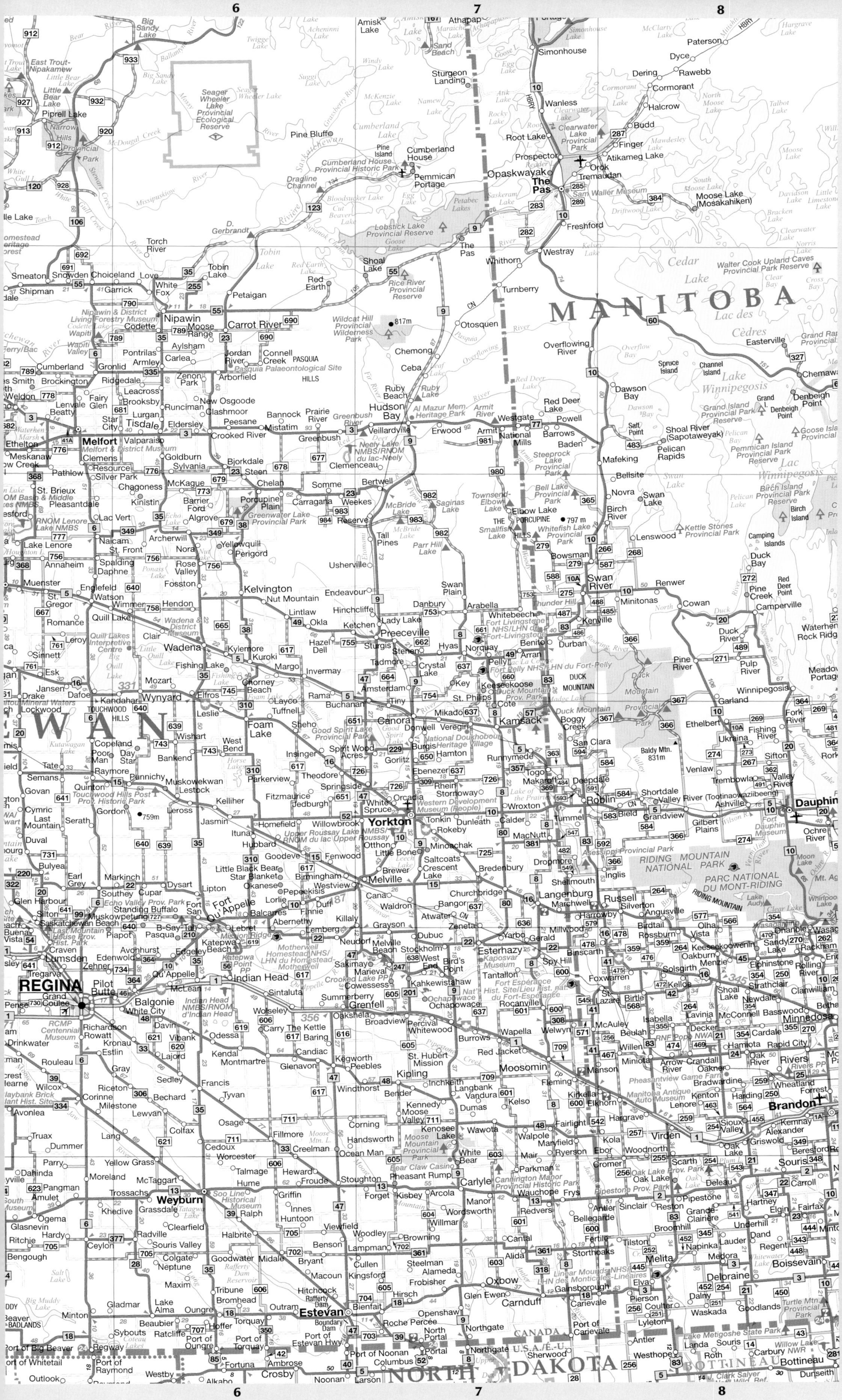

A PRAIRIE RESERVOIR

The serpentine shape of Lake Diefenbaker, some 220 km long but never more than 2 to 3 km wide, stands out indelibly in the patchwork prairie landscape of south-central Saskatchewan. This man-made reservoir, the fifth largest in Canada, stores 94 million cubic metres of water, collected from the east-flowing South Saskatchewan River. The earth-fill Gardiner Dam, 5 km wide and 64 m high, is clearly visible at the top of Lake Diefenbaker's north arm. The dam generates hydroelectric power, supplies 45 percent of Saskatchewan's drinking water, and irrigates some 43,000 ha of the farmland in this semiarid region. North of the Gardiner Dam, the South Saskatchewan River resumes a narrow course northward toward Saskatoon.

Other points of interest

■ Lloydminster [B2] is the only city in Canada that is divided between two provinces: the 110th meridian of longitude—which forms the boundary between Alberta and Saskatchewan—runs down the middle of Main Street.

■ Little Manitou Lake [D5], the remnant of an ancient glacial lake, has a salinity level of 12 percent, making it 3.5 times saltier than sea water. It also has a specific gravity of 1.06. This statistic indicates that Little Manitou Lake is denser than the Dead Sea in the Middle East.

■ Estevan [F6], the "Sunshine Capital of Canada," averages 2,500 hours of sunshine per year, more than any other city in the country. Estevan has also the highest annual number of hours per year with clear skies: 2,979 hours.

Scale 1:2,000,000

1 cm = 20 km

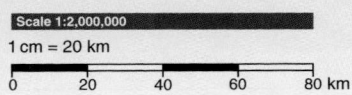

0   20   40   60   80 km

NORTHWEST TERRITORIES

Mountain Time Zone / Heure des Rocheuses
Central Time Zone / Heure du Centre

WOOD BUFFALO NATIONAL PARK

PARC NATIONAL

WOOD BUFFALO

Salt Plains Overview
Arrowhead Lake
Thulnai Lake
Conibear Lake
Wood Buffalo
Robertson Lake
Pine Lake
Hay Camp
Peace Point
Sweetgrass Landing
Baril Lake
Wood Buffalo
Garden Creek
Ruis Lake
776m
Claire Lake
Mamawi Lake

Fort Smith Visitors Centre
Fitzgerald
Wood Buffalo National Park
La Butte Creek Wildland Provincial Park
Charles Lake
Andrew Lake
Colin-Cornwall Lakes Wildland Provincial Park
Cornwall Lake
Wylie Lake
Cypress Point
Fidler Point
Egg Island
Burntwood Island
Bustard Island
Fort Chipewyan
Fort Chipewyan Bicentennial Museum
Fort Chipewyan Fort Chipewyan Visitors Centre
Richardson Lake National Migratory Bird Sanctuary / Refuge national d'oiseaux migrateurs du Lac-Richardson
Richardson Lake
Maybelle River Wildland Provincial Park
Athabasca Dunes
Richardson River Dunes Wildland Provincial Park

Slave River
Rivière des Esclaves
Winter Road d'hiver

Lake Athabasca
Lac Athabasca
Sandy Bay
Old Fort Bay
Old Fort River

Tazin Lake
Laird Island
Waterloo Lake
Carnsell Portage
Lobstick Island
Easter Head
Uranium City
Eldorado
Beaverlodge Lake
962
Johnston Island
Grouse Island
Maurice Point
William Point
Athabasca Sand Dunes Wilderness Provincial Park
Helmer Lake
Archibald Lake

590m
Fond-du-Lac
Stony Rapids
Stony Lake
964
966
Black Lake
Fir Island
Giles Lake
Athabasca Winter Road/Route

Cluff Lake Mine
955

BIRCH MOUNTAINS
Birch Mountains Wildland Provincial Park
Namur Lake
Legend Lake
Gardiner Lakes

Winter Road
Marguerite River Wildland Provincial Park
Marguerite River

ALBERTA
Fort MacKay
63
Wood Buffalo
Fort McMurray
Fort McMurray Oil Sands Discovery Centre
69
Saprae Creek Estates
Gregoire Lake Provincial Park
Gipsy Lake Wildland Provincial Park
Anzac
Stony Mountain Wildland Provincial Park
63
CHEECHAM HILLS
Mariana Lake
881
Pelican
Crow Lake Provincial Park
Chard
Bohn Lake
Christina Lake
Conklin
Winefred Lake
63
Wandering River
McMillan Lake
Breynat
881
Grassland
Imperial Mills
858
Plamondon
881
Lac La Biche
Sir Winston Churchill Provincial Park
Beaver Lake
Lakeland Provincial Park
Hylo
Caslan
Bondiss
Mewatha Beach
663
Boyle
Ellscott
Long Lake Provincial Park
831
Newbrook
855
Kikino
Goodfish Lake
Smoky Lake
Warspite
Waskatenau
Bellis
Vilna
St. Lina
Iron River
La Corey
Fort Kent
Cold Lake
892
Medley
Ardmore
Bonnyville
Glendon
Hoselaw
St. Vincent
897

SASKATCHEWAN
Clearwater River Provincial Park
Descharme Lake
955
Big C
La Loche West
La Loche
Black Point
956
Garson Lake
Bear Creek
909
Turnor Lake
Michel
Buffalo River
Dillon
925
St. George's Hill
Buffalo Narrows
Churchill Lake
155
Peter Pond Lake
Taylor Lake
Patuanak
Primeau Lake
Knee Lake
Elak Dase
Dipper Lake
918
Île-à-la-Crosse
908
Gordon Lake
Pinehouse Lake
914
Île-à-la-Crosse South Bay
Fort Black
La Plonge
Canoe Narrows
965
Beauval
Canoe Lake
Jans Bay
Cole Bay
903
McCusker Lake Provincial Ecological Reserve
MOSTOOS HILLS
GRIZZLY BEAR HILLS
Primrose Lake
COLD LAKE AIR WEAPONS RANGE / PRIMROSE LAKE
Primrose Lake Provincial Ecological Reserve
919
904
773m
Cold Lake
Cold Lake Provincial Park
Grand Centre
Cherry Grove
Crane Lake
Ethel Lake
Wolf Lake
950
Meadow Lake Provincial Park
Bighead
Goodsoil
954
Pierceland
224
Waterhen Lake
Glaslyn
903
941
951
Dore Lake
917
929
THUNDER HILLS
Michel Point
Beatty Lake
155
Waterhen Lake
Green Lake
Dorintosh
Barnes Crossing
Sled Lake
924
922
923
Beacon Hill
Flat Valley

Key Lake Mine
914
637m
Cree Lake
Grey Island
MacKenzie Bay
Highrock Lake
Costigan Lake
Nameless Lake
Carswell Lake
165
910
935
Sucker River
Morin Lake
Little Hills
La Ronge
Kitsakie
Napatak
Big Island
Doré Lake
Solenite Point Provincial Ecological Reserve
938
921
939
Molanosa
916
969
Weyakwin
2
Montreal River
Nipekamew Sand Dunes Provincial Protected Area

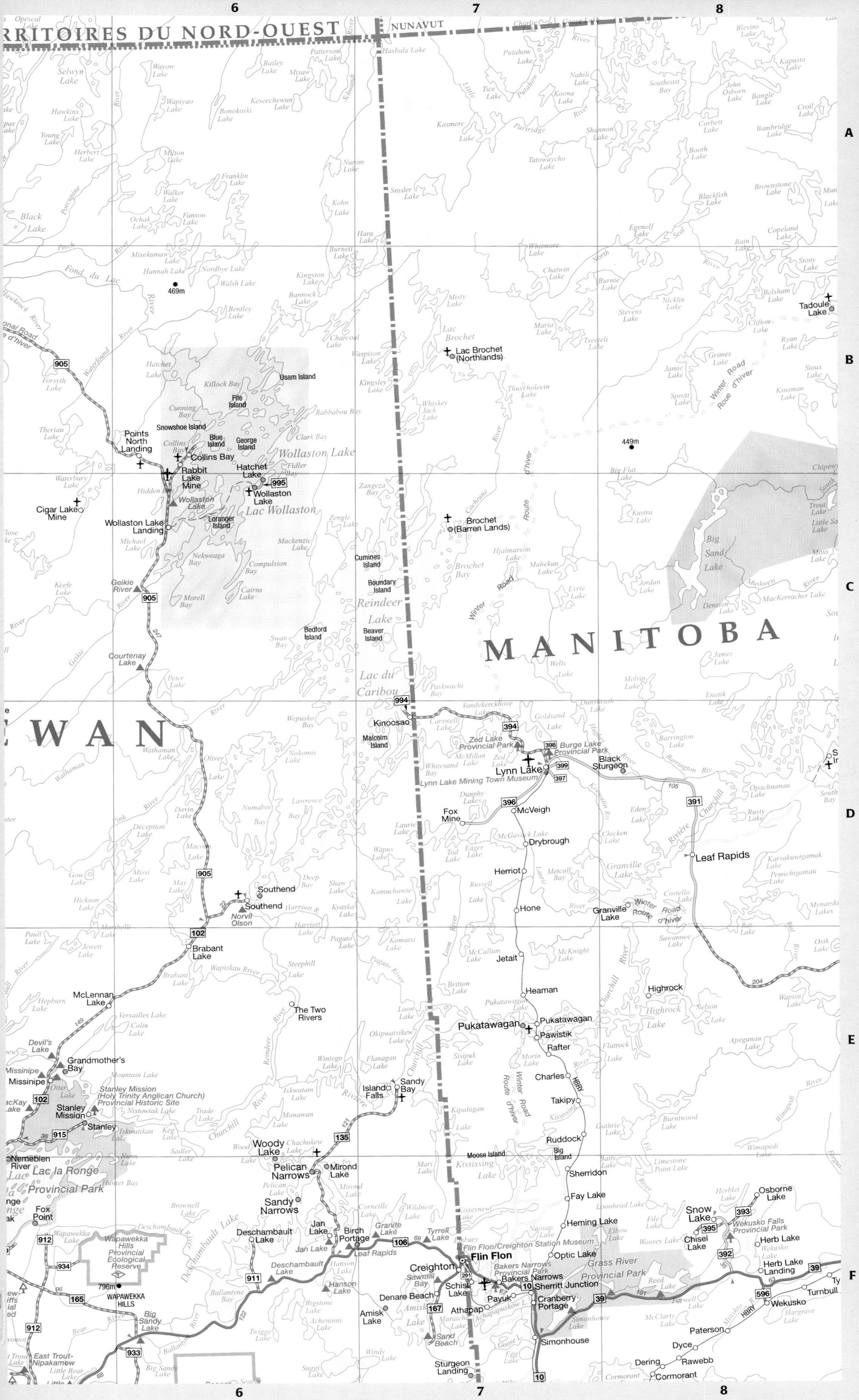

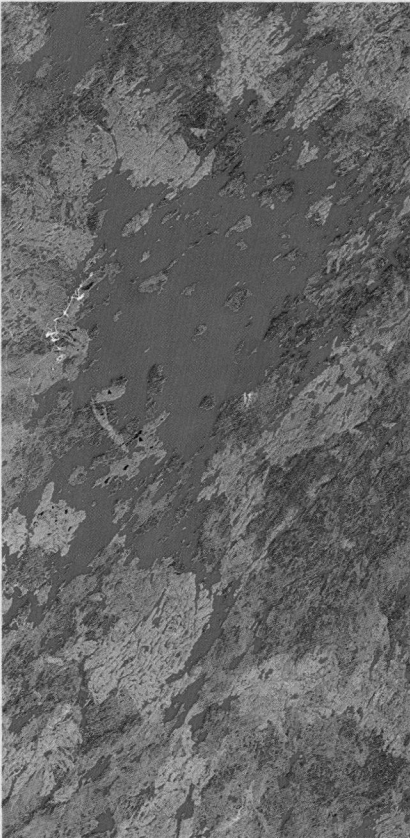

A LAKE THAT FLOWS TWO WAYS

At 2,681 km², Wollaston Lake is Canada's 19th largest lake, where waters drain in two directions. The lake flows northwest through Hatchet Lake (upper left) and the Fond du Lac River into Lake Athabasca and the Mackenzie River; and northeast, past Usam Island (upper right) via the Cochrane River into Reindeer Lake and Churchill River. With a 4,300-km shoreline and myriad islands, the lake is a sports fishing destination, renowned for its trophy-size catches of northern pike, lake trout, walleye, and arctic grayling. The road visible on Wollaston Lake's west side (left centre) leads to Rabbit Lake Mine, one of northern Saskatchewan's major uranium operations.

Other points of interest

■ Clearwater Wilderness Provincial Park [C3-D3] encompasses much of the 280-km Clearwater River, which flows west to join the Athabasca River at Fort McMurray [Plate 9, D7]. Along the river lie canyons and waterfalls. Once a fur-trade route, the Clearwater has been designated a Canadian Heritage River.

■ In 1,600-km² Meadow Lake Provincial Park [F2-F3] lie forests, meadows, and marshes. A significant park feature is a chain of 25 lakes linking Cold Lake [F2] with Waterhen Lake [F3].

■ A cairn at Montréal River [F5] marks the geographical centre of Saskatchewan. To the north lies 1,414-km² Lac La Ronge and Lac La Ronge Provincial Park [E5-F5]. At 3,345 km², the park is a popular year-round recreational destination. Farther north is Saskatchewan's oldest building, the Holy Trinity Anglican Church (built in the 1850s) situated at Stanley Mission.

Scale 1:2,000,000

1 cm = 20 km

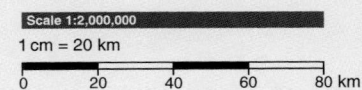

0   20   40   60   80 km

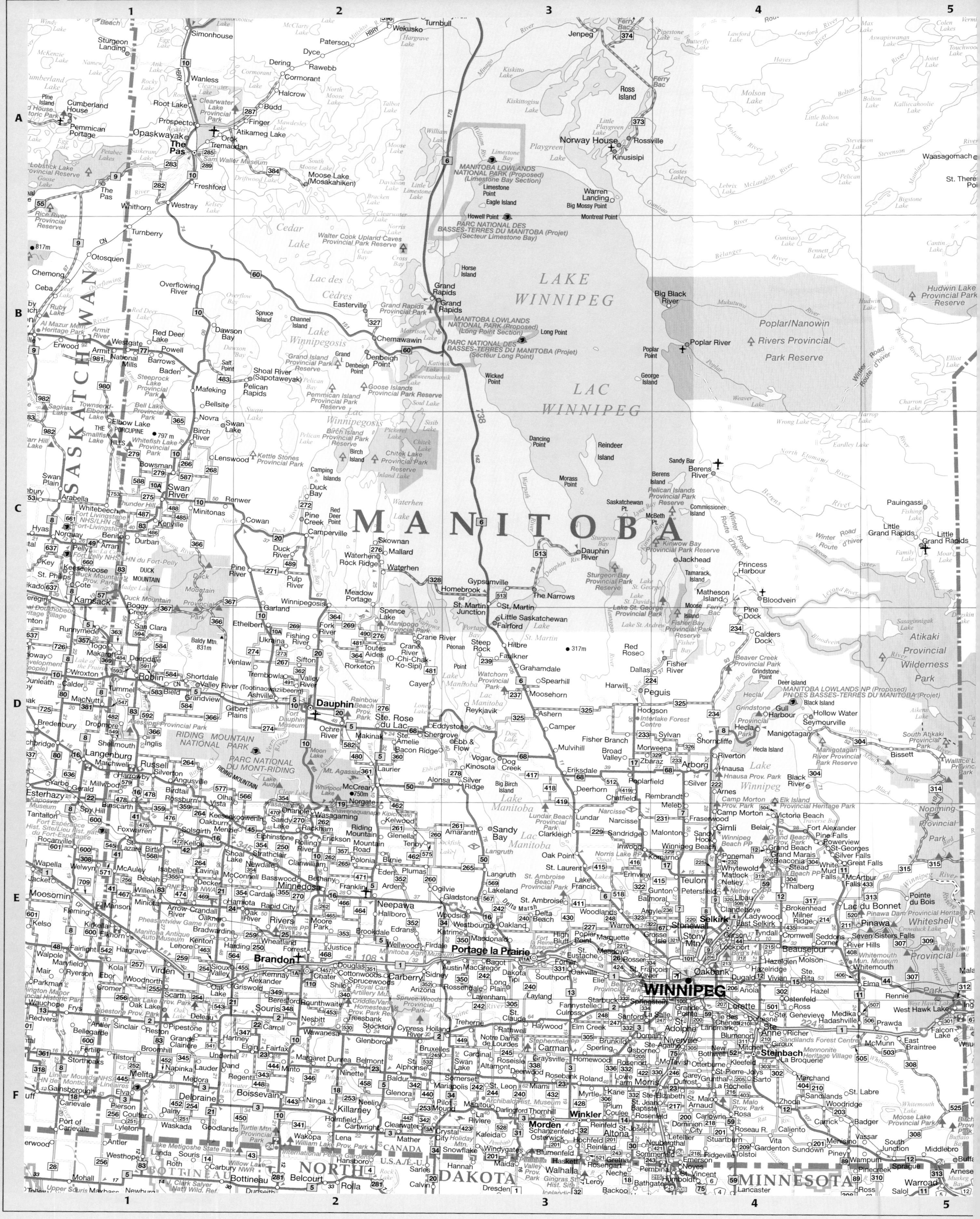

MANITOBA
SASKATCHEWAN
LAKE WINNIPEG
LAC WINNIPEG
Lake Winnipegosis
Lac Winnipegosis
Cedar Lake
Lac des Cèdres
Lake Manitoba
Lac Manitoba
RIDING MOUNTAIN NATIONAL PARK
PARC NATIONAL DU MONT-RIDING
MANITOBA LOWLANDS NATIONAL PARK (Proposed) (Limestone Bay Section)
PARC NATIONAL DES BASSES-TERRES DU MANITOBA (Projet) (Secteur Limestone Bay)
MANITOBA LOWLANDS NATIONAL PARK (Proposed) (Long Point Section)
PARC NATIONAL DES BASSES-TERRES DU MANITOBA (Projet) (Secteur Long Point)
Poplar/Nanowin Rivers Provincial Park Reserve
Atikaki Provincial Wilderness Park
Whiteshell Provincial Park
Nopiming Provincial Park
WINNIPEG
Brandon
Portage la Prairie
Dauphin
Selkirk
Steinbach
Morden
Winkler
The Pas
Opaskwayak
Flin Flon
Swan River
Grand Rapids
Norway House
Gimli
Virden
Neepawa
Minnedosa
Roblin
Russell
Killarney
Boissevain
Melita
Souris
Carman
Altona
CANADA
U.S.A./É.-U.
NORTH DAKOTA
MINNESOTA

CONTRASTING SHORES

Lake Winnipeg, with lakes Manitoba and Winnipegosis, dominates central Manitoba. At 24,400 km², Lake Winnipeg is Canada's sixth largest, but it is relatively shallow, with a maximum depth of 18 m. The view above shows its northern expanse, from Sturgeon Bay (lower right) to Big Mossy Point (upper right), where the lake flows northward through Playgreen Lake into the Nelson River and, eventually, Hudson Bay. Lake Winnipeg presents contrasting shores: the eastern shore has the granite outcrops of the Canadian Shield; and the western shore, the level landscape of the Manitoba Lowlands, where a proposed national park would include Limestone Stone Bay (upper left) and Long Point Peninsula (left centre).

Other points of interest

■ The southeastern corner of Duck Mountain Provincial Park [C1] is the site of Manitoba's highest point, 831-m Baldy Mountain [D1].

■ Lake of the Woods [F5-F6], shared by Ontario, Manitoba, and Minnesota, has 104,000 km of shoreline and contains 14,000 islands.

■ Manitoba's 100,000 lakes hold most of the 900 trillion litres of surface water covering 16 percent of the province.

Scale 1:2,000,000
1 cm = 20 km

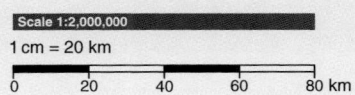

0   20   40   60   80 km

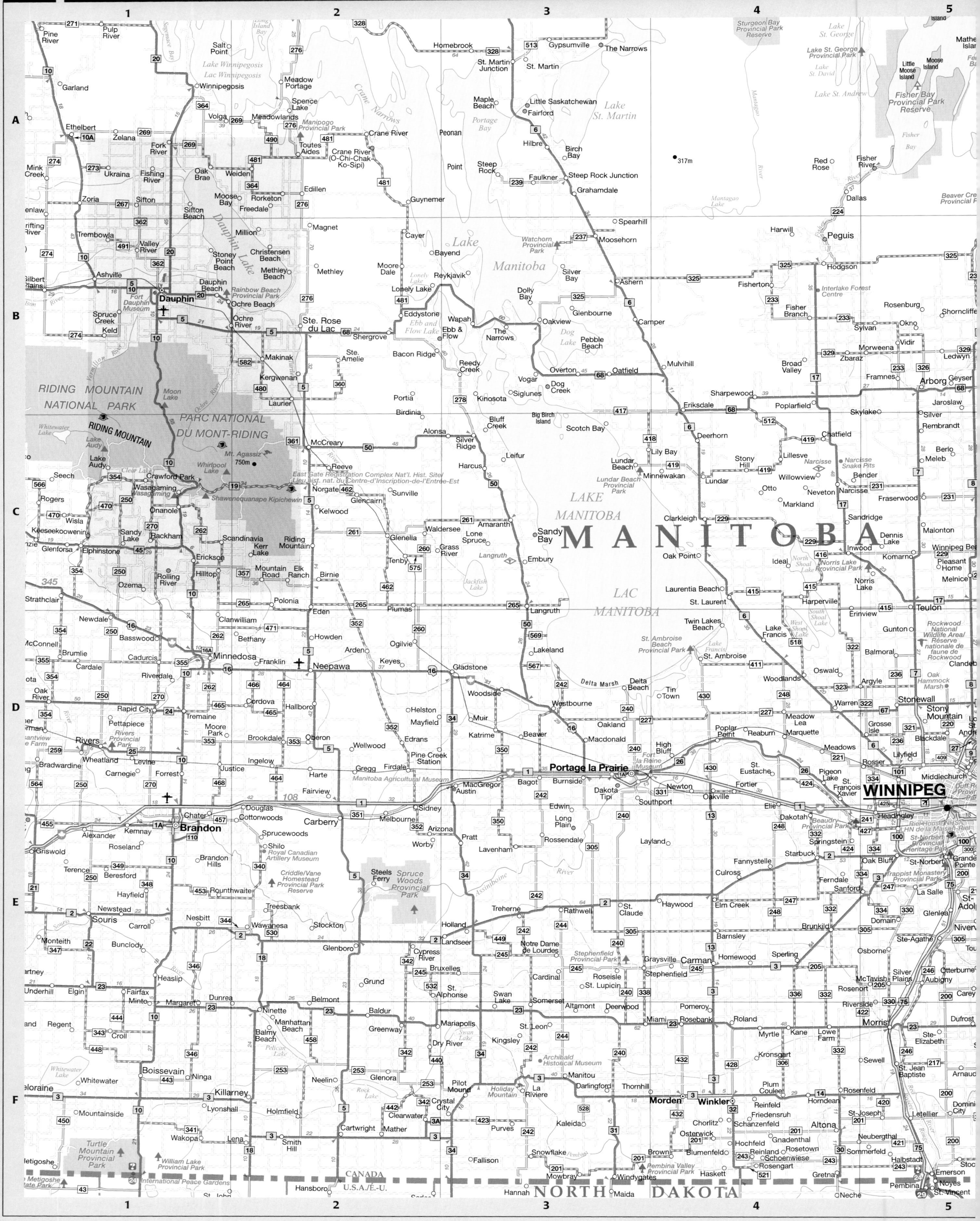

MANITOBA

LAC MANITOBA

Lake Manitoba

Lake Winnipegosis / Lac Winnipegosis

RIDING MOUNTAIN NATIONAL PARK

PARC NATIONAL DU MONT-RIDING

WINNIPEG

Brandon

Dauphin

Portage la Prairie

Morden • **Winkler**

Turtle Mountain Provincial Park

Spruce Woods Provincial Park

CANADA / U.S.A./É.-U.

NORTH DAKOTA

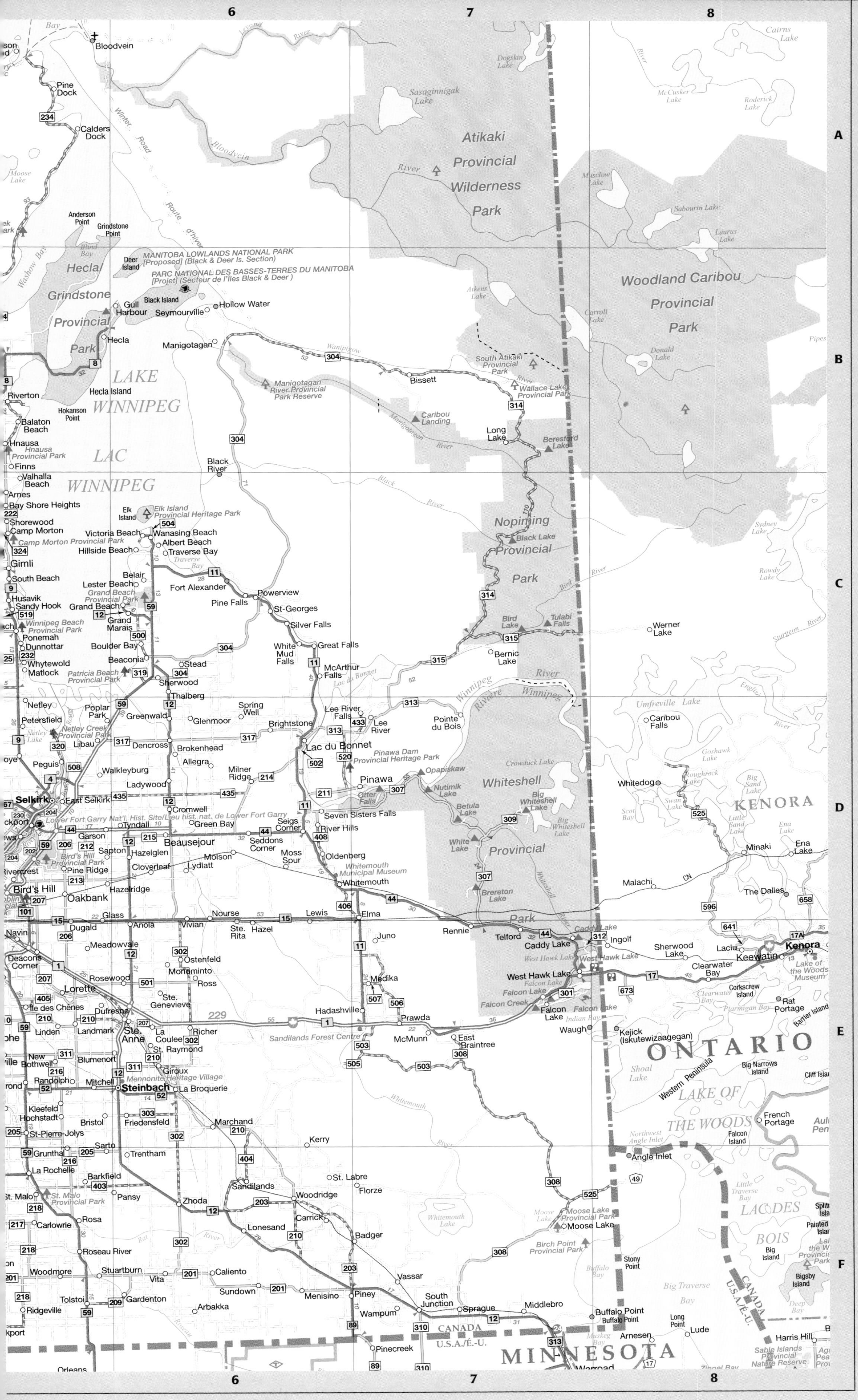

A PRAIRIE ESCARPMENT

This view shows the dark green plateau of Riding Mountain National Park (centre) and the bright, blue waters of Dauphin Lake (upper centre). Riding Mountain National Park is situated on the Manitoba Escarpment—a series of uplands that angle northwest through Manitoba and Saskatchewan. The park's northern boundary, sharply defined above, rises 450 m above a landscape checkered with farm fields. Within its 3,000 km², the park contains boreal and deciduous forest, aspen parkland, open grassland, and meadows. Also visible above is Clear Lake at Riding Mountain's southern entrance. Dauphin Lake, like larger Winnipeg, Winnipegosis, and Manitoba lakes, is a remnant of Lake Agassiz, which covered much of southern Manitoba 12,000 years ago.

Other points of interest

■ South of Carberry, Spruce Woods Provincial Park [E2] preserves an untamed desert formed thousands of years ago when this region was once the delta of an ancient river. Spruce Woods has two natural curiosities: Spirit Sands, a 5-km tract of shifting dunes towering 30 m above the prairie, and The Devil's Punchbowl, a sunken pit created by currents in an underground stream.

■ The 877-km-long Red River [C5-F5] may flood when its headwaters in the United States thaw before the river in Manitoba is ice-free. This has caused disastrous floods twice: in 1950 and 1997. During the last flood, more than 8,600 members of the Canadian Forces called in to control the rising waters. It was Canada's largest military operation since the Korean War.

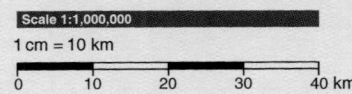

Scale 1:1,000,000

1 cm = 10 km

0   10   20   30   40 km

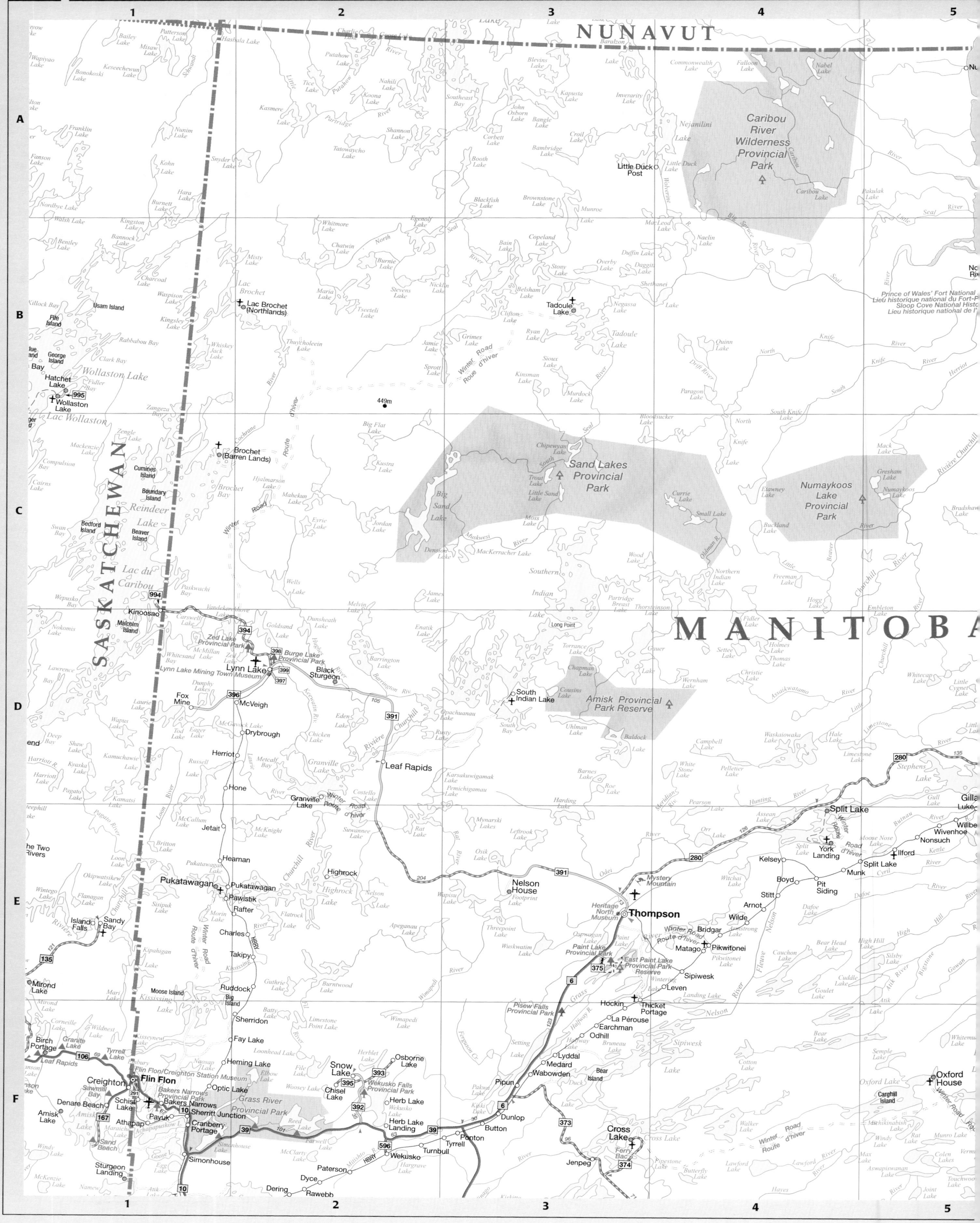
1
2
3
4
5
A
B
C
D
E
F
NUNAVUT
SASKATCHEWAN
MANITOBA
Caribou River Wilderness Provincial Park
Little Duck Post
Nejanilini Lake
Lac Brochet (Northlands)
Tadoule Lake
Wollaston Lake
Lac Wollaston
Hatchet Lake
Brochet (Barren Lands)
Reindeer Lake
Sand Lakes Provincial Park
Big Sand Lake
Numaykoos Lake Provincial Park
Prince of Wales' Fort National
Lieu historique national du Fort-P
Sloop Cove National Histo
Lieu historique national de l'
Southern Indian Lake
Lac du Caribou
Kinoosao
Zed Lake Provincial Park
Burge Lake Provincial Park
Lynn Lake
Lynn Lake Mining Town Museum
Black Sturgeon
Fox Mine
McVeigh
Drybrough
South Indian Lake
Amisk Provincial Park Reserve
Leaf Rapids
Herriot
Hone
Granville Lake
Jetait
Heaman
Pukatawagan
Pawistik
Rafter
Charles
Takipy
Ruddock
Big Island
Sherridon
Fay Lake
Heming Lake
Highrock
Highrock Lake
Nelson House
Thompson
Heritage North Museum
Mystery Mountain
Split Lake
York Landing
Split Lake
Ilford
Kelsey
Boyd
Pit Siding
Munk
Stitt
Arnot
Wilde
Bridgar
Matago
Pikwitonei
Paint Lake Provincial Park
East Paint Lake Provincial Park Reserve
Sipiwesk
Leven
Hockin
Thicket Portage
La Pérouse
Earchman
Odhill
Lyddal
Medard
Wabowden
Pisew Falls Provincial Park
Pipun
Dunlop
Button
Ponton
Tyrrell
Turnbull
Wekusko
Paterson
Dyce
Dering
Rawebb
Herb Lake
Herb Lake Landing
Wekusko Falls Provincial Park
Snow Lake
Osborne Lake
Chisel Lake
Flin Flon
Creighton
Flin Flon/Creighton Station Museum
Bakers Narrows Provincial Park
Bakers Narrows
Payuk
Schist Lake
Sherritt Junction
Cranberry Portage
Grass River Provincial Park
Optic Lake
Simonhouse
Athapap
Denare Beach
Amisk Lake
Sand Beach
Sturgeon Landing
Birch Portage
Granite Lake
Tyrrell Lake
Leaf Rapids
Mirond Lake
Island Falls
Sandy Bay
Cross Lake
Jenpeg
Oxford House
Carghill Island
Gillam
Wivenhoe
Willbeck
Nonsuch
Winter Road
Route d'hiver
449m
995
994
394
396
399
397
391
280
391
6
375
373
374
39
596
392
393
395
10
167
291
106
135
Churchill River
Nelson River
Seal River
Knife
North Knife
South Knife
Cochrane
Burntwood
Grass
Hayes

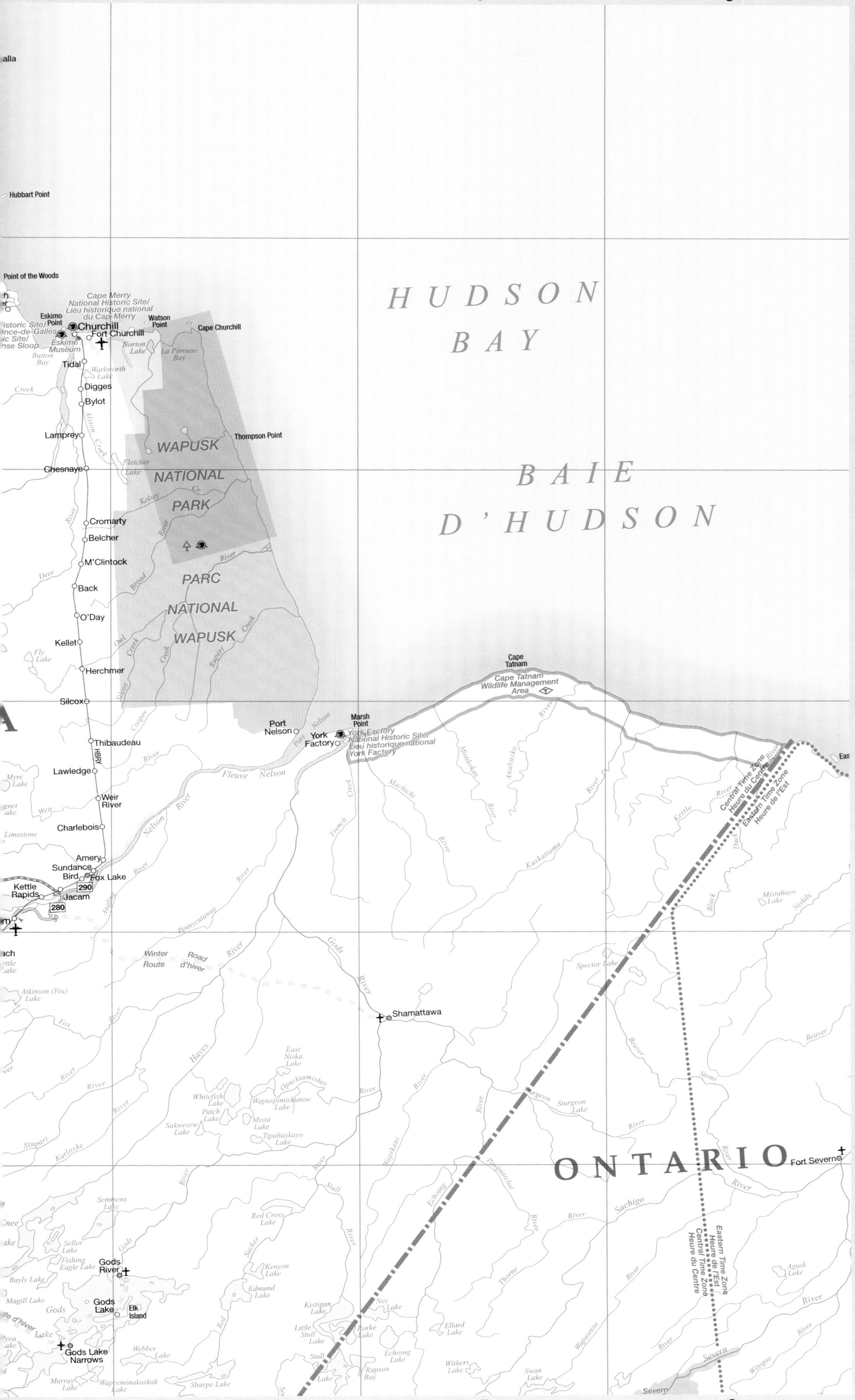

6   7   8

A

HUDSON

BAY

BAIE

D'HUDSON

B

C

D

E

F

Hubbart Point

Point of the Woods

Cape Merry
National Historic Site/
Lieu historique national
du Cap-Merry

Eskimo
Point
Churchill
Fort Churchill
Eskimo
Museum

Historic Site/
rince-de-Galles
ric Site/
nse Sloop

Watson
Point
Cape Churchill

Button
Bay
Tidal

Warkworth
Lake
La Pérouse
Bay

Norton
Lake

Digges

Bylot

Lamprey

Thompson Point

WAPUSK

Chesnaye

Fletcher
Lake

NATIONAL

Kelsey
Lake

PARK

Cromarty

River

Belcher

M'Clintock

PARC

Back

NATIONAL

O'Day

WAPUSK

Kellet

Fly
Lake

Oxel

Creek

Herchmer

Cape
Tatnam

Silcox

Cape Tatnam
Wildlife Management
Area

Cooper

Port
Nelson
Marsh
Point
York
Factory
York Factory
National Historic Site/
Lieu historique national
York Factory

Thibaudeau

HBRY

River

Fleuve Nelson

Lawledge

Myre
Lake

Weir
River

Nelson River

Central Time Zone
Heure du Centre

Eastern Time Zone
Heure de l'Est

East

Charlebois

Limestone

Amery
Sundance
Bird
Fox Lake
290
Kettle
Rapids
Jacam
280

Kettle

Mistahayo
Lake

Neskhi

Winter
Route
Road
d'hiver

Gods

River

Spector Lake

Beaver

Atkinson (Fox)
Lake

Fox

Shamattawa

East
Niska
Lake

Opikiamirches

Sturgeon
Lake

Shone

ONTARIO

Whitefish
Lake
Patch
Lake

Mistu
Lake
Tipahaskayo
Lake

Wapusimiskanow
Lake

Sakweswa
Lake

River

Stupart

Karloike

Fort Severne

Semmens
Lake

Gods

Thorne
Lake

Agnsk
Lake

Red Cross
Lake

Kenyon
Lake

Seller
Lake
Fishing
Eagle Lake

Gods
River

Edmund
Lake

Eastern Time Zone
Heure de l'Est
Central Time Zone
Heure du Centre

Magill Lake

Gods
Lake
Elk
Island

Kistigan
Lake

re d'hiver
Gods Lake
Narrows

Webber
Lake

Little
Stull
Lake

Porke
Lake

Red

Ellard
Lake

Murray
Lake
Wapueminakusskak
Lake

Stull
Lake

Echoing
Lake

Withers
Lake

Swan
Lake

Rapson
Bay
Sharpe Lake

Severn

Severn

6   7   8

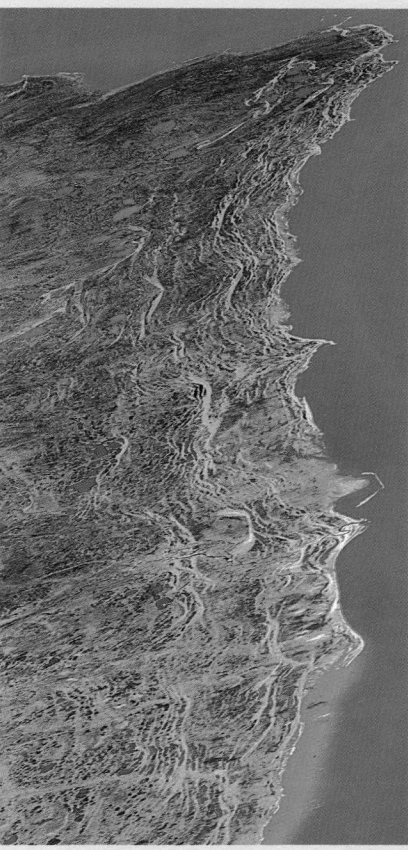

A NORTHERN WETLAND

Hudson Bay's cold waves crash on the beaches of Wapusk National Park, where Cape Churchill (top) is the northernmost promontory. Canada's eighth largest park, Wapusk protects an 11,475-km² portion of the Hudson Bay Lowlands, which stretch along Hudson Bay shoreline from the Manitoba-Nunavut boundary to James Bay. The lowlands—Canada's largest wetland region and the world's third largest wetland—occupy an area of 300,000 km². In Wapusk National Park, about half the surface is covered with lakes, bogs, fens and streams, and rivers. Along the coast, the park is a treeless tundra, carpeted with willows, sedges, and other low arctic vegetation. The beach ridges—the shorelines of earlier times—indicate the slow coastal rebound (at about a metre per century) from the last ice age's immense glacial weight.

Other points of interest

■ Churchill [B5], Canada's only seaport on the Arctic Ocean, is open only three months a year for navigation.

■ Hudson Bay is an immense inland sea, covering 822,324 km² and penetrating deeply into northern Canada. Its maximum length is 1,500 km; its greatest width is 830 km. The Hudson Bay Lowlands form the southwestern shore, but the remaining coast is mainly part of the Canadian Shield. Many rivers enter the bay, among them the Thelon, Churchill, Nelson, Severn, La Grande, and Eastmain. The total area of the Hudson Bay drainage is 3.8 million square kilometres. The bay is shallow, mostly between 100 and 200 m deep. The bay's floor is slowly rising, exposing more and more of the coastline.

Scale 1:2,000,000

1 cm = 20 km

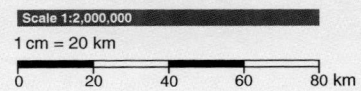

0   20   40   60   80 km

1 2 3 4 5

A

Lewis Lake
Carr-Harris Lake
Black Birch L.
Cherrington L.
Deer Lake
Winter Road
Deer Lake
North Caribou Lake
Pauingassi
Little Grand Rapids
Little Grand Rapids
Poplar Hill
North Spirit Lake
North Spirit Lake
MacDowell Lake
Windigo Lake
Upper Windigo Lake
Nango Lake
Menako Lake
Horseshoe L.
Pineimuta

B

MANITOBA
Atikaki Provincial Wilderness Park
Pikangikum
Pikangikum Lake
Berens Lake
Cat Lake
Cat Lake
Pickle Lake
Central Patricia
Parc provincial
Woodland Caribou
Bissett 304

ONTARIO

C

KENORA
Red Lake
Cochenour
Balmertown
Madsen
Starratt-Olsen
105
Snake Falls
Bruce Lake
Ear Falls 804
Goldpines
Mishkeegogamang (New Osnaburgh)
599
Savant Lake
Allan Water
Lac Seul
516

D

West Hawk Lake
Kenora
Keewatin
Minaki
Redditt
Jones
Red Lake Road
Vermillion Bay
Minnitaki
Dryden
Wabigoon
Sioux Lookout
Hudson
Sam Lake
Superior Junction
Ghost River
O'Briens Landing
Silver Dollar
Umfreville
Dinorwic
Dyment
Borups Corners
Ignace
Martin
English River
Graham
Upsala

E

Middlebro
Buffalo Point
Warroad
Rainy River
Baudette
Stratton
Emo
Nestor Falls
Crow Lake
Caliper Lake
Sioux Narrows
Whitefish Bay
RAINY RIVER
Atikokan
Sapawe
Kawene
Kashabowie
Burchell Lake
Shebandowan

F

MINNESOTA
Fort Frances
International Falls
Kabetogama
VOYAGEURS NATIONAL PARK
Ash Lake
Crane Lake
Lac la Croix
Hunter Island
Quetico
Provincial Park
Saganaga Lake
Pigeon River
CANADA
U.S.A.-É.-U.

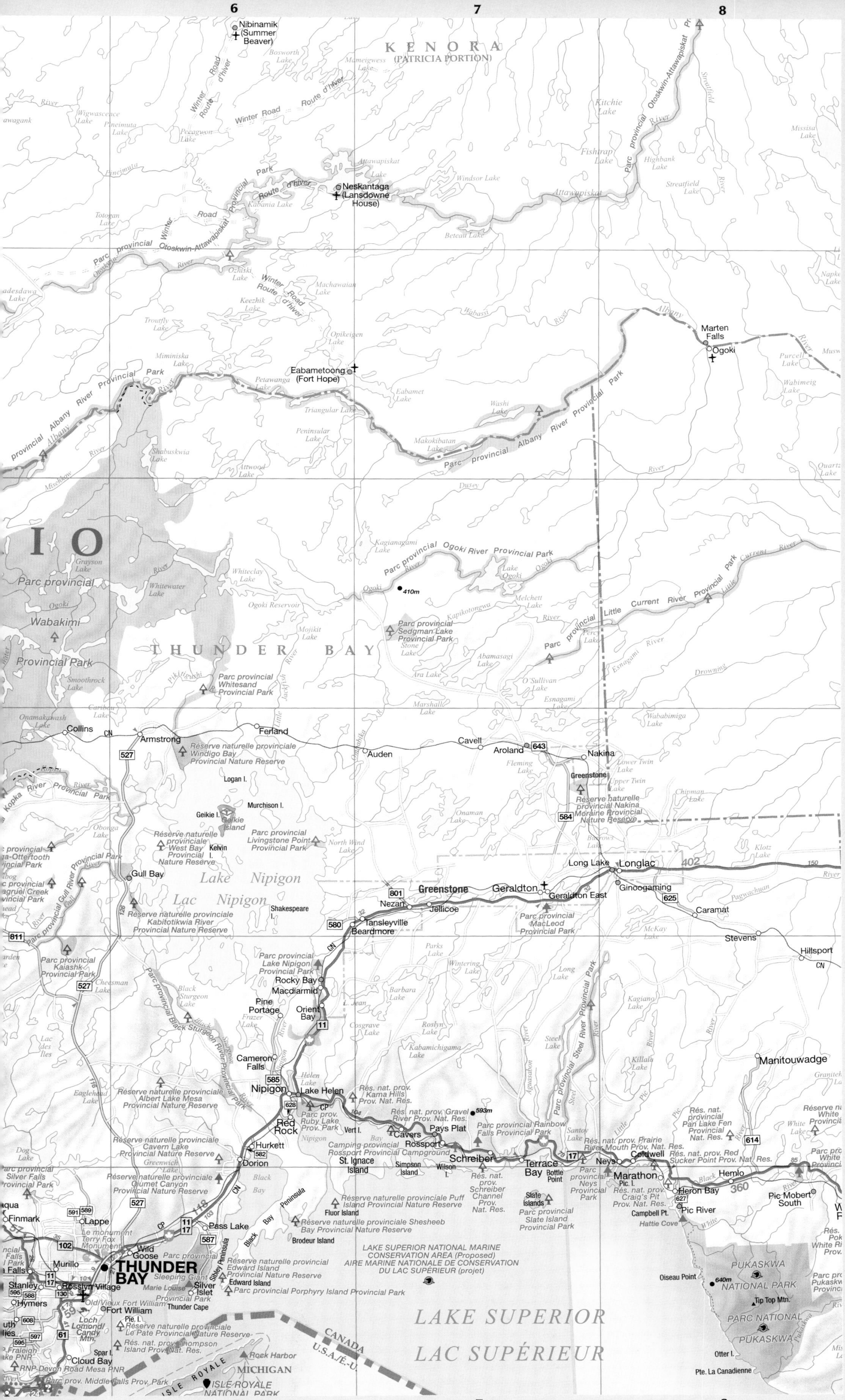

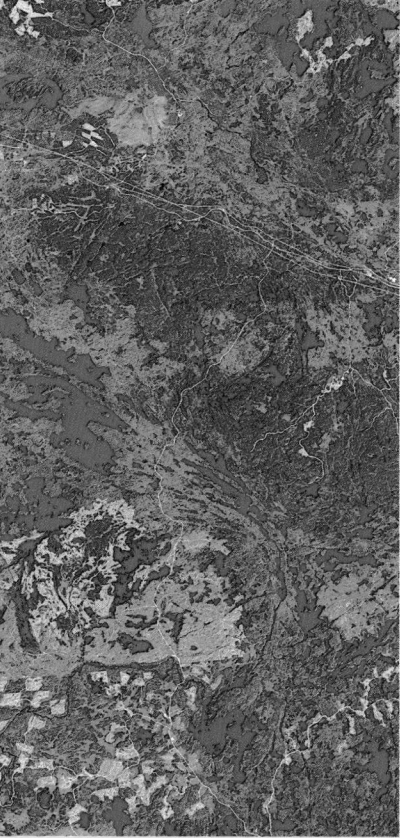

IN SHIELD COUNTRY

In northwestern Ontario, the Trans-Canada Highway has two sections: Highway 17 (Kenora–Thunder Bay) and Highway 11 (Fort Frances–Thunder Bay). The road at the top of the view is a section of Highway 17, just east of Dyment, Ontario. The road south is Provincial Highway 622, a secondary route linking up with Highway 11 at Atikokan. In this view, the 622 threads through rugged Shield country [E3], where blocks of clear-cut forest are visible.

Other points of interest

■ Sleeping Giant Provincial Park [F6] covers most of Sibley Peninsula. At the southwest corner of the 243-km² park lies the Sleeping Giant, a cluster of mesas rising 300 m high. From Thunder Bay, some 25 km away, the mesas resemble a huge recumbent figure.

■ Ouimet Canyon Provincial Nature Reserve [F6] preserves a 5-km-long, 150-m-wide gorge where sheer cliffs plunge 100 m to the canyon floor.

■ Lake Nipigon [D6], the fourth largest in Ontario, drains south into Lake Superior through the Nipigon River. The lake receives water from the Albany River [B5–B7], via the Ogoki River and Reservoir [B8–C6]. This diverted water increases the lake's capacity and allows hydroelectric generation at three plants along the Nipigon River.

■ In 1981, Hemlo, Ontario [F8], was the site of a major gold discovery. Within 10 years, Hemlo had become the largest source of gold in Canada. Its development revived the importance of gold to Canada's economy. Hemlo's three mines produce 25 percent of Canada's supply.

Scale 1:1,725,000

1 cm = 17.25 km

0 10 20 30 40 50 60 70 km

(Map of Northwestern Ontario)

KENORA (PATRICIA PORTION)

THUNDER BAY

LAKE SUPERIOR
LAC SUPÉRIEUR

THUNDER BAY

PUKASKWA NATIONAL PARK / PARC NATIONAL PUKASKWA

ISLE ROYALE NATIONAL PARK

MICHIGAN

CANADA / U.S.A./É.-U.

LAKE SUPERIOR NATIONAL MARINE CONSERVATION AREA (Proposed)
AIRE MARINE NATIONALE DE CONSERVATION DU LAC SUPÉRIEUR (projet)

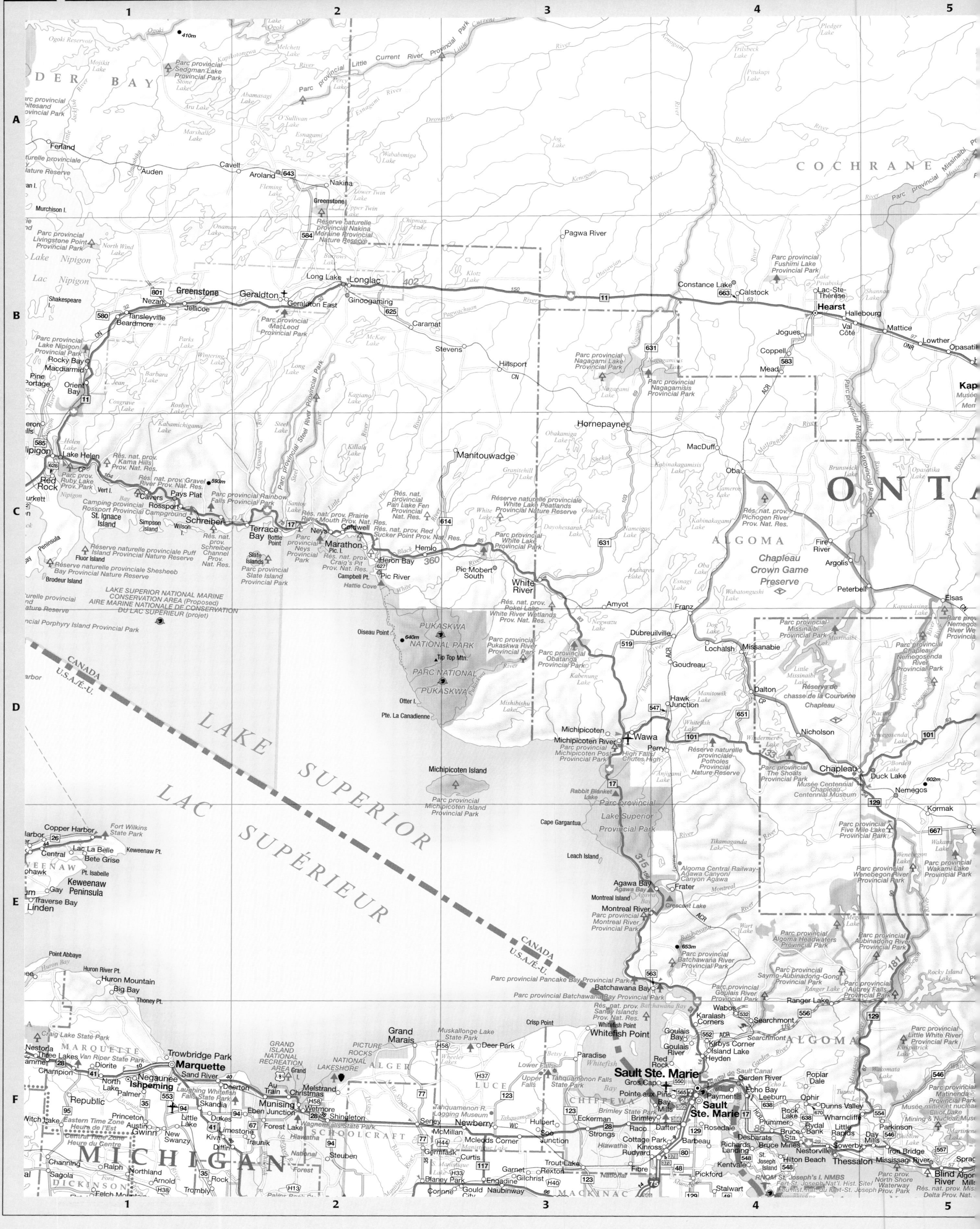

THUNDER BAY

COCHRANE

Ferland
Auden
Cavell
Aroland 643
Nakina
Greenstone
584
Pagwa River

Murchison I.
North Wind Lake

Lac Nipigon
Lake Nipigon
Shakespeare

801 Greenstone
Geraldton
Long Lake
Longlac
402
11
Constance Lake 663 Calstock
Hearst
Hallebourg
Val Côté
Mattice
Lowther

580 Nezah
Jellicoe
Geraldton East
Ginoogaming
625
11
631
Jogues
Coppell
583 Mead
ONR
Kap

Tansleyville
Beardmore
MacLeod Provincial Park
Caramat
Stevens
Hillsport
CN

Macdiarmid
Orient Bay
11
McKay Lake
Nagagami Provincial Park
Nagagamisis Provincial Park

585 Nipigon
Lake Helen
626 Red Rock
Schreiber
Terrace Bay
17
Marathon
Hemlo
614
Manitouwadge
Hornepayne
MacDuff
Oba
631
ONTARIO
ALGOMA

Rossport
St. Ignace Island
Simpson Island
Wilson
Neys Provincial Park
Pic River
Heron Bay
Pic Mobert South
White Lake Provincial Park
Fire River
Argolis
Peterbell

Slate Islands Provincial Park
Campbell Pt.
Hattie Cove
White River
Amyot
Franz
Dubreuilville
519
Goudreau
Lochalsh
Missanabie
Dalton
Chapleau Crown Game Preserve

LAKE SUPERIOR
LAC SUPÉRIEUR

CANADA
U.S.A./É.-U.

PUKASKWA NATIONAL PARK
PARC NATIONAL PUKASKWA
Tip Top Mtn.
Otter I.
Pte. La Canadienne

Hawk Junction
547
Michipicoten
Michipicoten River
Wawa
Perry
101
651
Nicholson
Chapleau
Duck Lake
Nemegos
129
Kormak

Michipicoten Island
Michipicoten Island Provincial Park

Cape Gargantua
Leach Island
Lake Superior Provincial Park
Kormak
667

Copper Harbor
Fort Wilkins State Park
Lac La Belle
Bete Grise
Keweenaw Pt.
Pt. Isabelle
Keweenaw Peninsula
Gay
Traverse Bay
Linden

Agawa Bay
Montreal Island
Montreal River
Frater
Algoma Central Railway
Agawa Canyon

Huron Bay
Point Abbaye
Huron River Pt.
Huron Mountain
Big Bay
Thoney Pt.

Batchawana Bay
563
Pancake Bay Provincial Park
Batchawana Bay Provincial Park
Ranger Lake
556
129

Grand Marais
Muskallonge Lake State Park
Deer Park
Crisp Point
Whitefish Point
Goulais
Searchmont
532
Heyden
Goulais Bay
Kirbys Corner
Island Lake

PICTURE ROCKS NATIONAL LAKESHORE
GRAND ISLAND NATIONAL RECREATION AREA

Trowbridge Park
MARQUETTE
Marquette
Sand River
Au Train
Christmas
Munising
H58
Paradise
Red Rock
Sault Ste. Marie
Gros Cap
Garden River
Echo Bay
Leeburn
Poplar Dale
546

Negaunee
Ishpeming
553
Palmer
Skandia
Deerton
Wetmore
Shingleton
Seney
Newberry
WC
Hulbert
Eckerman
Brimley State Park
Pointe aux Pins
Bay Mills
Raco
Dafter
550
Sault Ste. Marie
17
638
Leeburn
Wharncliffe
554

Republic
Witch Lake
94
95
North Lake
41
Princeton
Gwinn
Little Lake
41
Limestone
Forest Lake
McMillan
H44
Soo Junction
Strongs
Cottage Park
Kinross
Rudyard
80
Barbeau
Rosedale
129
638
Red Lake
Ophir
Dunns Valley
670
Bruce Station
Rydal Bank
548
Sowerby

MICHIGAN
DICKINSON
Channing
Sagola
Felch Mountain
Northland
Ralph
Arnold
Tromby
Sagola
Steuben
Blaney Park
Gould City
Naubinway
Curtis
Germfask
Trout Lake
Rexton
Engadine
Corinne
Gilchrist
Garnet
Fibre
Pickford
Stalwart
Kentvale
St. Joseph
Hilton Beach
Thessalon
Mississagi River
Blind River
557
DELTA
LUCE
SCHOOLCRAFT
MACKINAC
CHIPPEWA
ALGOMA

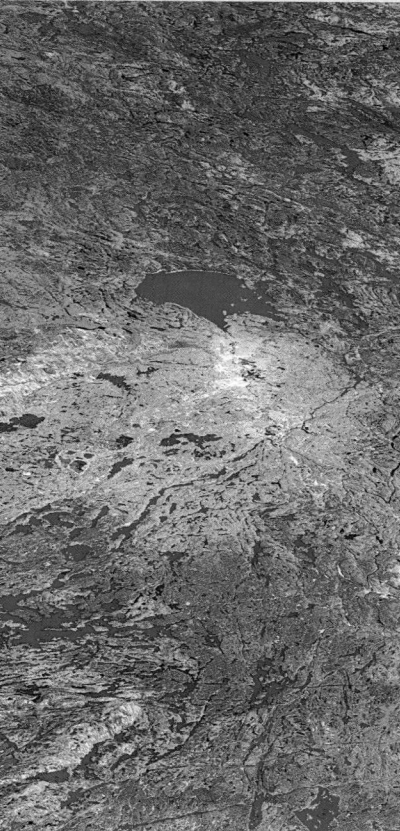

A BASIN OF MINERALS

Wanapitei Lake (centre) lies northeast of Greater Sudbury, the leading regional centre in northeastern Ontario. Nickel Centre is the site of the Falconbridge smelter. It is situated in the brown and yellow circular area below the lake. The colour here and elsewhere in this view shows where the forest cover is thin as a result of pollution. The city of Sudbury itself is visible just below Nickel Centre. Greater Sudbury is situated in a 60 x 27-km oval-shaped geological basin, possibly created either by the impact of a meteor or volcanic eruption. Whatever the cause, it endowed the region with a wealth of minerals: gold, silver, cobalt, platinum, and the world's largest known deposit of nickel. Since the 1980s, the reclamation of the local landscape, ravaged by a century of mining and industrial pollution, has been remarkably successful.

Other points of interest

■ Wawa [D3] supplies iron ore for the steel mills at Sault Ste. Marie [F4], Canada's second largest steel-producing center (after Hamilton). South of Wawa is High Falls, a 25-m-high, 40-m-wide waterfall on the Magpie River.

■ Pukaskwa National Park [C2-D3] occupies 1,878 km² of land on Lake Superior. Its natural features include sheer cliffs along the Lake Superior shoreline and 640-m Tip-Top Mountain inland. Lake Superior Provincial Park [D3-E4] is the site of towering granite headlands such as Cape Gargantua.

■ In 1903, the world's richest silver vein was discovered at Cobalt [E8]. Since then, Cobalt has produced more than 500 million ounces of silver.

Scale 1:1,725,000

1 cm = 17.25 km

0 10 20 30 40 50 60 70 km

Map labels (selected): QUÉBEC, NORD-DU-QUÉBEC, Baie-James, ONTARIO, Timmins, Cochrane, Iroquois Falls, Kirkland Lake, Rouyn-Noranda, New Liskeard, North Bay, Sudbury, Espanola, Elliot Lake, Temagami, Cobalt, Haileybury, Sturgeon Falls, Chelmsford, Garson, Capreol, Kapuskasing

1 2 3 4 5

A

ALGOMA

Dunns Valley
Wharncliffe
Kynoch
Cummings Lake
Club Lake
546
639
Keelor Lake
Dunlop Lake
Ten Mile Lake
Big Moon Lake
Quirke Lake
Whisky Lake
Wiggle Lake
Dowling
Phelans
Larchwood
Chelm
129
Little Rapids
Sowerby
Day Mills
Parkinson
Patton
554
546
Blind Lake
Elliot Lake
Esten Lake
Horseshoe Lake
Musée minier et nucléaire d' Elliot Lake / Elliot Lake Mining & Nuclear Museum
Elliot Lake
557
108
Marshland Lake
553
Emerald Lake
Granary Lake
High Falls
Worthington
4
144
Vermilion Lake
69
Whitefish
Nau
17
Thessalon Point
Thessalon
Iron Bridge
Dean Lake
15
Mississagi River
Algoma Mills
538
Spragge
17
Pronto East
Serpent River
Cutler
Walford
Spanish
Massey
17
Cecil Lake
Parc provincial Chutes Provincial Park
Webbwood
McKerrow
Espanola
Nairn Centre
17
Turbine
Whitefish
10
Round Lake

B

NORTH CHANNEL
CHENAL NORTH

Drummond Island
MICHIGAN
West Grant Island
East Grant Island
Hennepin Island
Réserve naturelle provinciale Mississagi Delta Provincial Nature Reserve
Sandford Island
Turnbull Island
John Island
Aird Island
McBean Harbour
Sagamok
Eagle Island
Fox I.
La Cloche Lake
Raven Lake
Moose Lake
6
Parc provincial Chiblow Provincial Park
La Cloche
Whitefish Falls
Willisville
Parc provincial Fairbank Provincial Park
Parc provincial Killarney Lakelands
SUDBU
Cockburn Island
Tolsmaville
Meldrum Bay
Sheshegwaning
Barrie Island
Gore Bay
540A
540B
Kagawong
540
Honora
Little Current
Northeastern Manitoulin and the Islands
Sucker Creek
Strawberry
Badgeley Island
Killarney
Philip
Edward Island
6
Parc provincial La Cloche Provincial Park
Bay of Islands
Great La Cloche Island
Birch Island
Parc provincial Killarney Coast & Islands Provincial Park
McGregor Bay
South La Cloche
Killarney
637
Mountains
Parc provincial Killarney Provincial Park

C

Meldrum Bay
540
The Queen Elizabeth The Queen Mother M'Nidoo M'Nissing Provincial Park
Silver Water
Burnt Island
Greene Island
MANITOULIN
Tobacco Lake
542
Long Bay
West Bay
540
Sheguiandah
Big Burnt I.
Squaw Island
Western Duck Island
Réserve naturelle provinciale Misery Bay Provincial Nature Reserve
Elizabeth Bay
Evansville
Perivale
Poplar
Britainville
Spring Bay
542
551
Mindemoya
542
Manitowaning
Wikwemikong
GEOR
Great Duck Island
Outer Duck Island
Grimsthorpe
551
Providence Bay
Michael's Bay
Tehkummah
542A
Sandfield
6
Buzwah
Musée Assiginack Museum
Wikwemikonsing
Kaboni
Rabbit Island
Ba
Blue Jay Creek Provincial Park
South Baymouth
Wall Island
Club Island
Lonely Island
Owen Channel
Fitzwilliam Island

D

LAKE HURON

Thompson's Harbour State Park
North Point
Old Presque Isle Lighthouse
Presque Isle
23
PRESQUE ISLE
Besser
Middle Island
Parc marin national Fathom Five National Marine Park
Cove Island
Flowerpot Island
Bears Rump Island
Res. nat. prov. Little Cove Prov. Nat. Res.
BRUCE PENINSULA NATIONAL PARK / PARC NATIONAL DE LA PÉNINSULE-BRUCE
Cabot Head
Réserve naturelle provinciale Cabot Head Provincial Nature Reserve
BRUCE PENINSULE P
ONT
Russel Island
Tobermory
Cape Hurd
Dyer's Bay
ALPENA
Alpena Lighthouse
Alpena
32
Thunder Bay Island
North Point
Singing Sands
Eagle Point
St. Edmunds
Parc provincial Johnston Harbour/ Pine Tree Point Provincial Park
Miller Lake
6
Northern Bruce Peninsula
Cape Chin
Réserve naturelle provinciale Smokey Head-White Bluff Provincial Nature Reserve

E

MICHIGAN
LAC HURON

Thunder Bay
Ossineke
Michigan Islands National Wildlife Refuge
South Point
Negwegon State Park
Spruce
Black River
23
Huron National Forest
Alcona
F41
9
Pine Tree Point
Parc provincial Ira Lake Provincial Park
Clarke's Corners
Stokes Bay
Parc provincial Black Creek Provincial Park
Greenough Point
Réserve naturelle provinciale Stokes McMaster Island Provincial Nature Reserve
Lyal Island
Lion's Head
Réserve naturelle Provincial Nature
Barrow Bay
9
Hope Bay
ALCONA
Sturgeon Point
Lincoln
72
17
Harrisville
Harrisville State Park
F41
Pike Bay
Adamsville
Howdenvale
Red Bay
9
Mar
9
18

F

U.S.A./É.-U.
CANADA

F30
Mikado
Greenbush
Oscoda
Au Sable
IOSCO
Tuttle Marsh
Au Sable Point
Tawas Lake
East
23
F41
Oliphant
Chiefs Point
Sauble Falls
Parc provincial Sauble Falls Provincial Park
Sauble Beach North
Sauble Beach
Sauble Beach South
8
Hepworth
Park Head
14
Shal
Scotch Settlement
Chippawa Hill
Southampton
Chantry Island National Migratory Bird Sanctuary/ Refuge national d'oiseaux migrateurs de l'île-Chantry
Port Elgin
Saugeen Shores
Parc provincial MacGregor Point Provincial Park
Brucedale
North Bruce
Allenford
Cruikshank
13
Wiarton
South Bruce Peninsula
Clavering
Oxe
Spirit Ro
6
Tara
Arkwright
17
Invermay
16
Arran-Elderslie
Dobbinton
Saugeen
21
40

1 2 3 4 5

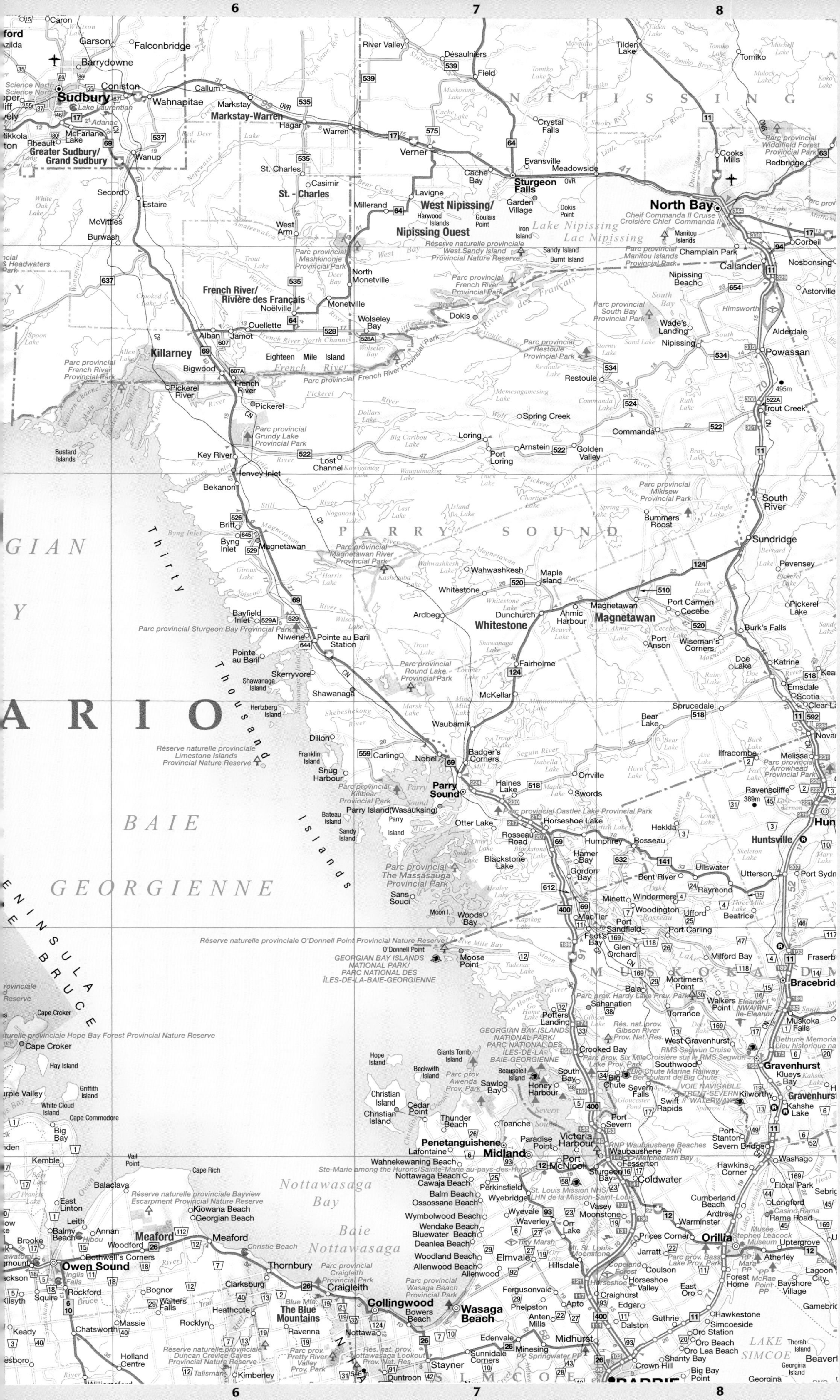

PENINSULA AND ISLAND

In this view, the Bruce Peninsula points northwest across Main and Fitzwilliam channels toward Manitoulin Island (top). The peninsula's northern tip is the site of Bruce Peninsula National Park (still under development), as well as Fathom Five National Marine Park (Canada's first marine park), which protects Cove, Flowerpot, and other islands visible offshore in Lake Huron (left) and Georgian Bay (right). "The Bruce" is part of Ontario's 725-km-long Niagara Escarpment. This height of land runs from the Niagara Peninsula to Manitoulin Island and beyond. At 2,700 km², Manitoulin is the world's largest island in a freshwater lake. Beyond Manitoulin on the far shore of the North Channel lies La Cloche Provincial Park.

Other points of interest

■ Located on the southern edge of the Canadian Shield, 345-km² Killarney Provincial Park [B4-B5] contains rock formations more than 2 billion years old. Its rugged beauty inspired the Group of Seven painter A. Y. Jackson. It was through his efforts that this area became a park.

■ The 110-km-long French River [B6-B7] connects Lake Nipissing to Georgian Bay. Fur traders once plied its maze of channels and secluded bays, and portaged its rapids and falls. It has been designated as one of Canada's most historic waterways.

■ At 832 km², Lake Nipissing is the fifth largest lake in Ontario—excluding the Great Lakes. The lake connects the Ottawa and French rivers, part of the historic route of the early explorers and fur traders heading westward.

Scale 1:800,000

1 cm = 8 km

0   10   20   30 km

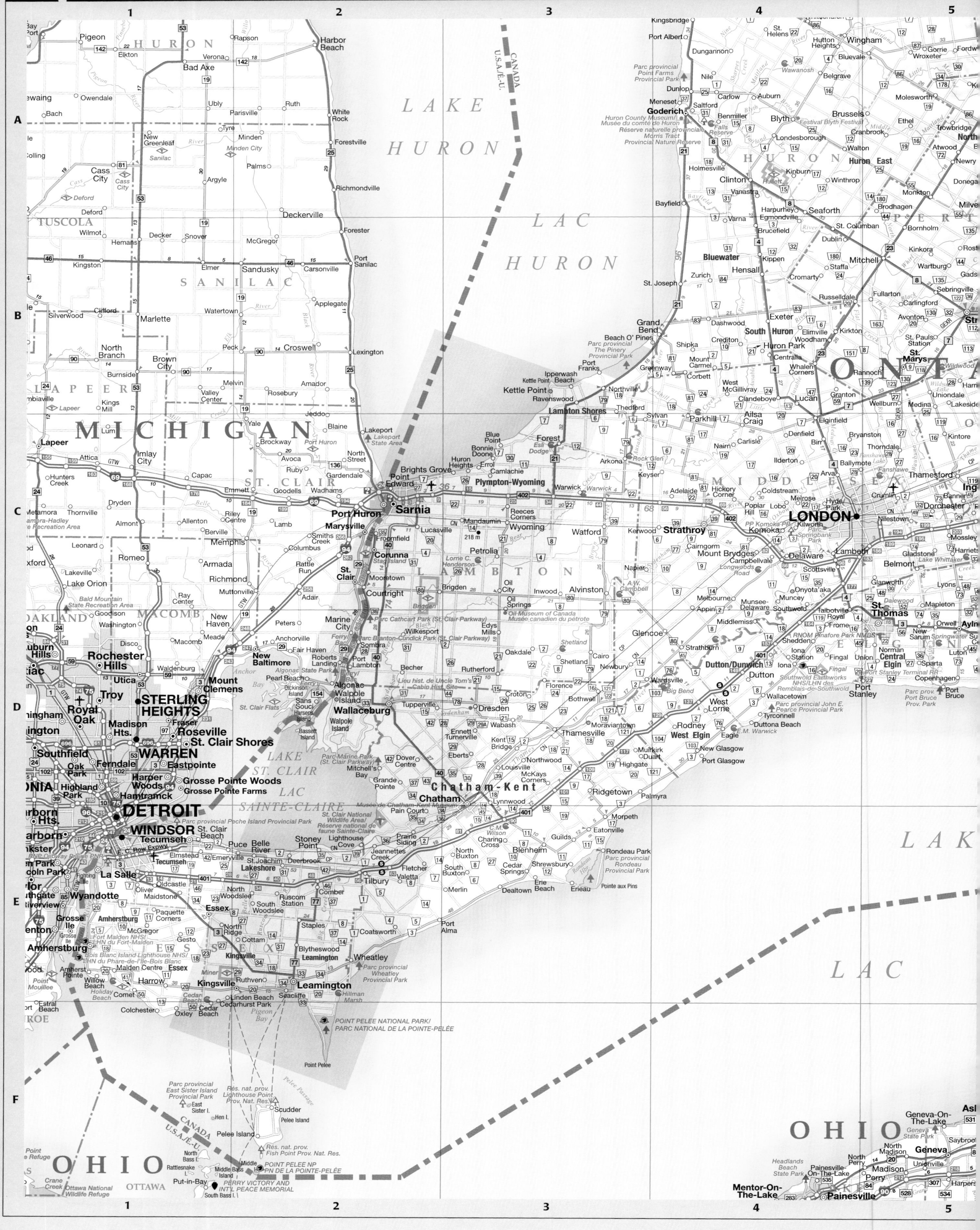
LAKE HURON
LAC HURON
MICHIGAN
LAKE ST. CLAIR
LAC SAINTE-CLAIRE
OHIO
CANADA
U.S.A./É.-U.
HURON
SANILAC
TUSCOLA
LAPEER
ST. CLAIR
MACOMB
OAKLAND
MONROE
OTTAWA
LAMBTON
MIDDLESEX
ELGIN
ESSEX
PERTH
ONTARIO
Chatham-Kent
DETROIT
WINDSOR
LONDON
Sarnia
Port Huron
STERLING HEIGHTS
WARREN
Goderich
Strathroy
St. Thomas
Leamington
Kingsville
Chatham
Wallaceburg
Petrolia
Tilbury
Amherstburg
Essex
Lakeshore
La Salle
Tecumseh
Rochester Hills
Troy
Royal Oak
Roseville
St. Clair Shores
Mount Clemens
Port Stanley
Dutton/Dunwich
West Elgin
Lambton Shores
Plympton-Wyoming
Bluewater
Mitchell
St. Marys
Painesville
Mentor-On-The-Lake
Geneva
Pelee Island
Point Pelee
POINT PELEE NATIONAL PARK/ PARC NATIONAL DE LA POINTE-PELÉE
LAKE ERIE
LAC ÉRIÉ
1
2
3
4
5
A
B
C
D
E
F

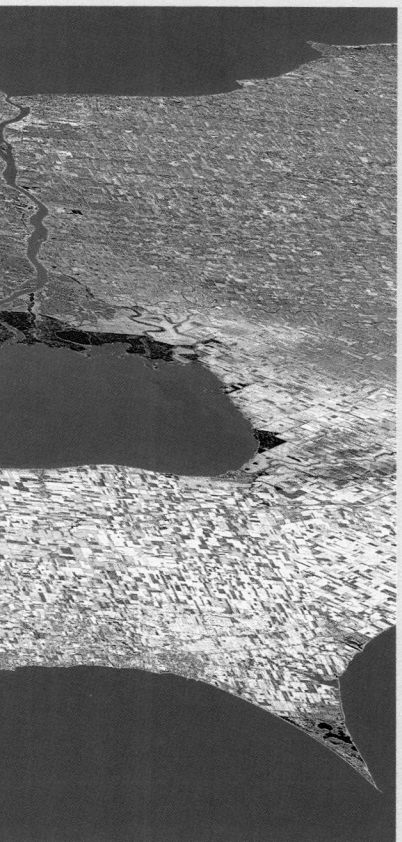

LAKES AND FARMLANDS

Southwestern Ontario is bounded by three lakes: Huron (top), St. Clair (centre), and Erie (bottom). At 1,114 km², Lake St. Clair is "the smallest of the Great Lakes." It is connected to Lake Huron to the north by the St. Clair River and empties into Lake Erie to the south by the Detroit River. The delta of the St. Clair River is visible on the lake's north shore. The Thames River, which enters at the lake's south end, flows through some of Canada's best farmland. This view shows the highly productive fields in Ontario's Essex and Chatham-Kent counties. The 18-km-long triangular spit jutting into Lake Erie is Point Pelee, the site of the second smallest of Canada's national parks, after St. Lawrence Islands, and mainland Canada's southernmost point.

Other points of interest

■ At 59,600 km², Lake Huron is the second largest of the Great Lakes, and the world's fifth largest. The Canadian portion occupies 36,000 km².

■ Lake Erie is the shallowest of the Great Lakes, with a maximum depth of 64 m, and also the smallest by volume (484 km³). The Canadian portion occupies slightly less than half of the lake's 25,700 km². Erie's waters fall 100 m to Lake Ontario; roughly half of the drop occurs at Niagara Falls.

■ Uninhabited Middle Island [F2] is Canada's southernmost bit of territory.

■ Oil Springs, Ont. [C3], is the site of North America's first commercial oil well, which James Miller Williams put into production in 1857. A museum near the site traces the history of oil production in this region. In 1999, southern Ontario still produced crude oil valued at $44 million.

Scale 1:800,000
1 cm = 8 km
0   10   20   30 km

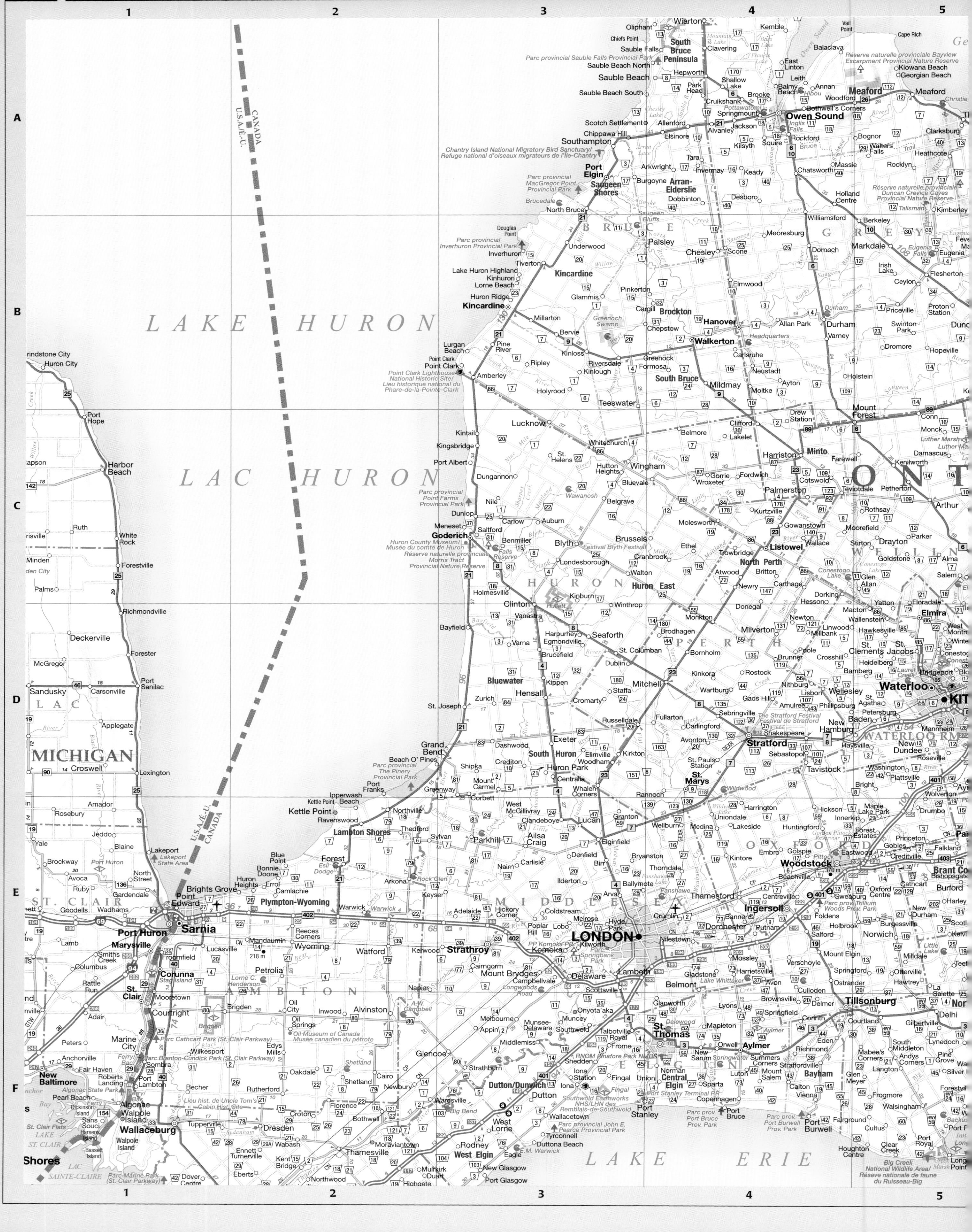

LAKE HURON

LAC HURON

MICHIGAN

LAKE ERIE

LAC SAINTE-CLAIRE

CANADA
U.S.A./É.-U.

Major places:

- South Bruce Peninsula
- Owen Sound
- Meaford
- Port Elgin
- Saugeen Shores
- Southampton
- Kincardine
- Hanover
- Walkerton
- South Bruce
- Arran-Elderslie
- Mount Forest
- Minto
- Listowel
- North Perth
- Waterloo
- KIT (Kitchener)
- Stratford
- St. Marys
- Goderich
- Huron East
- South Huron
- Huron Park
- Bluewater
- Woodstock
- Ingersoll
- Sarnia
- Port Huron
- Marysville
- Corunna
- St. Clair
- Plympton-Wyoming
- Strathroy
- LONDON
- Thamesford
- St. Thomas
- Central Elgin
- Aylmer
- Bayham
- Tillsonburg
- Dutton/Dunwich
- West Elgin
- Wallaceburg
- New Baltimore
- Walpole Island
- Lambton Shores
- Forest

Counties/Regions:
- BRUCE
- GREY
- HURON
- PERTH
- WELLINGTON
- MIDDLESEX
- LAMBTON
- OXFORD
- ELGIN
- ST. CLAIR
- WATERLOO RM
- BRANT CO

ONTARIO'S GOLDEN HORSESHOE

The Golden Horseshoe wraps around the western end of Lake Ontario from St. Catharines to Toronto and beyond. With almost 7 million inhabitants, the region has Canada's densest population cluster. The area takes its name from the lake's horseshoe-like curve, as well as the good fortune bestowed by its industrial and financial activities. Some of its features visible above include: Port Weller Harbour, the gateway to the Welland Canal east of St. Catharines (lower right); central Hamilton and the Skyway Bridge (lower left); and the expanse of Metro Toronto, lying just north of its two lakeshore harbours. Pickering—the site of a nuclear power station—sits on the promontory at the far right. At the top of this view is Cook's Bay on Lake Simcoe.

Other points of interest

■ Long Point National Wildlife Area [F5-F6] is a 32-km-long spit of sand dunes and marshland. Some 237 bird species—roughly 75 percent of all species recorded in Ontario—have been spotted at Long Point. It is an important stopover for migrating waterbirds, bats, and butterflies.

■ Niagara Falls [E8] is the world's greatest by volume (2,832 m³). The Canadian, or Horseshoe, Falls is 54 m high, 675 m wide, and flows at a rate of 155 million litres per minute. Some 12,500 years ago, as the last ice age was ending, Niagara Falls was born when Lake Erie plunged over the edge of Niagara Escarpment. At that time, the cataract began retreating at the rate of 1.2 m a year. It is now more than 11 km from its birthplace at Queenston.

Scale 1:800,000

1 cm = 8 km

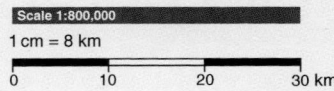

0   10   20   30 km

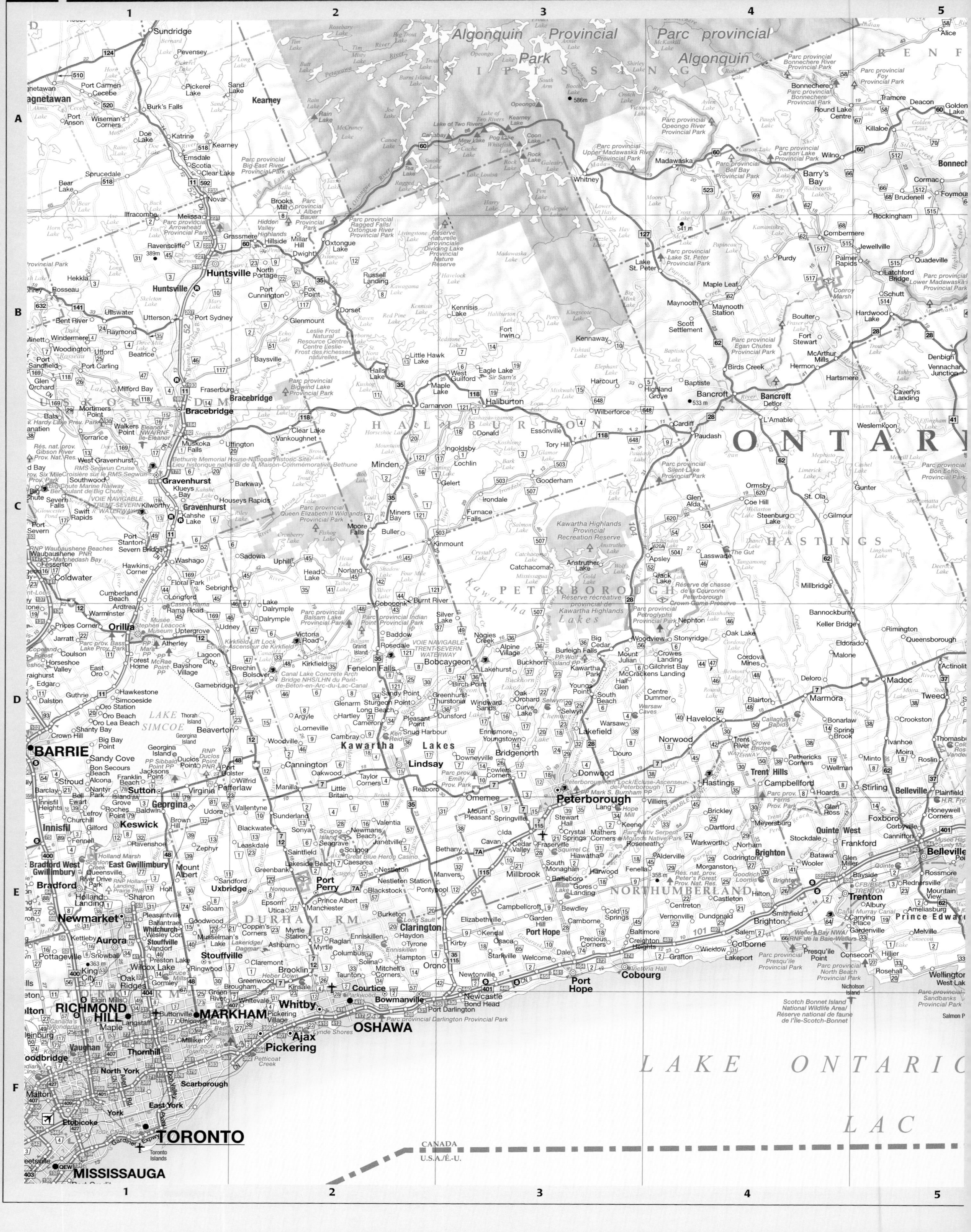

Algonquin Provincial Park / Parc provincial Algonquin

ONTARIO

HALIBURTON

HASTINGS

PETERBOROUGH

KAWARTHA LAKES

NORTHUMBERLAND

DURHAM RM.

LAKE SIMCOE

LAKE ONTARIO / LAC ONTARIO

TORONTO

MISSISSAUGA

BARRIE

Peterborough

Lindsay

Huntsville

Bracebridge

Gravenhurst

Orillia

Bancroft

Newmarket

Aurora

Richmond Hill

MARKHAM

OSHAWA

Whitby

Ajax

Pickering

Scarborough

North York

Etobicoke

Bowmanville

Courtice

Port Hope

Cobourg

Trenton

Belleville

Prince Edward

Barry's Bay

Madawaska

Haliburton

Minden

Bobcaygeon

Fenelon Falls

Port Perry

Uxbridge

Stouffville

Bradford

Sutton

Keswick

Georgina

Innisfil

CANADA / U.S.A./É.-U.

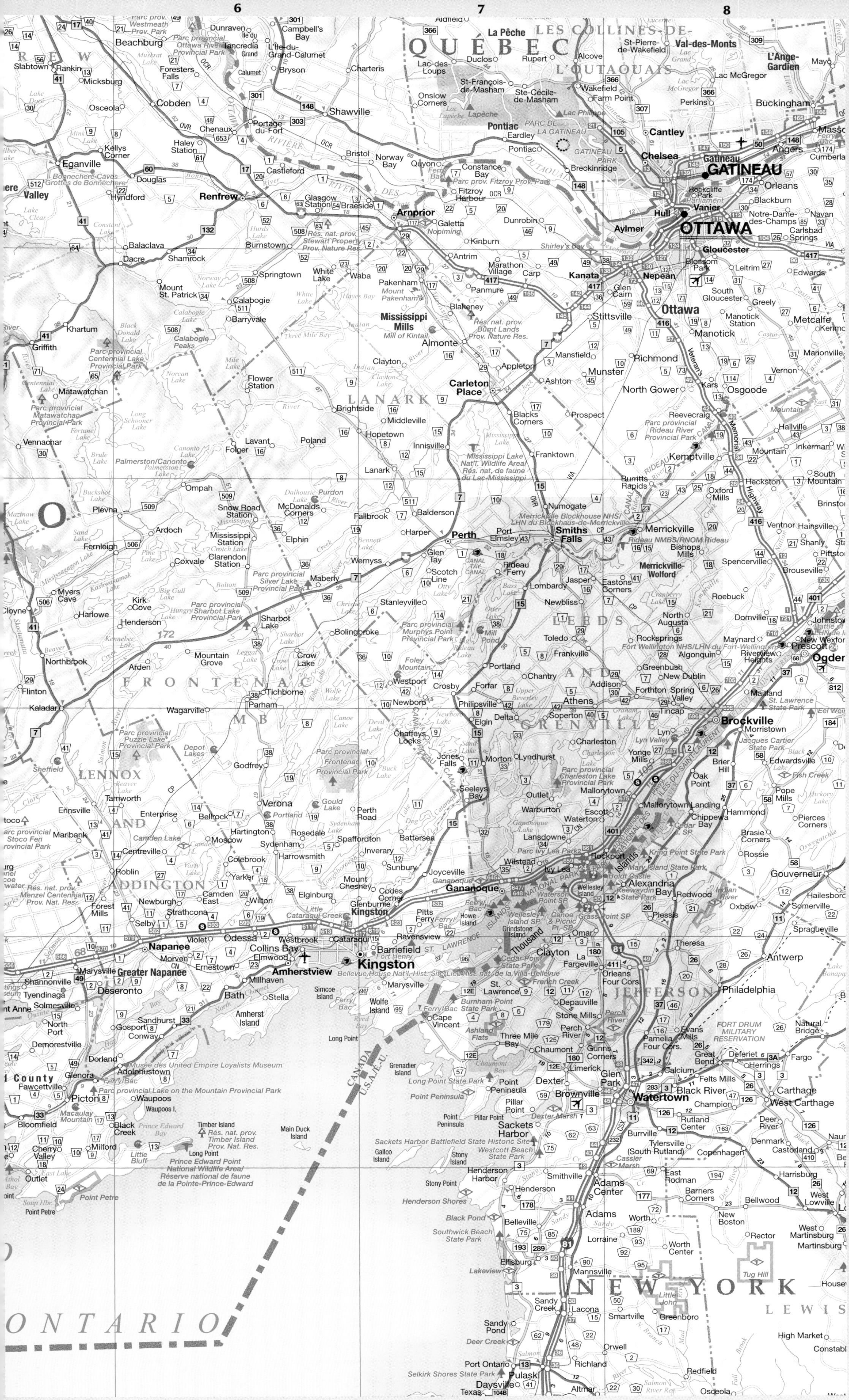

ISLANDS IN THE RIVER

This view of the Thousand Islands on the St. Lawrence River shows the section from Gananoque to just beyond Mallorytown Landing. The Thousand Islands—summits of ancient hills now submerged by the St. Lawrence—are formed by the Frontenac Axis, an 80-km-wide belt of granite that connects the Canadian Shield with the Adirondack Mountains in New York State. Visible (left to right) are the tips of Howe and Wolfe islands, and Grenadier Island on Canada's side of the river; Grindstone and Wellesley islands, on the American side. North of the St. Lawrence lies Shield country, where a string of lakes (top) links the 202-km-long Rideau Canal that connects Kingston and Ottawa.

Other points of interest

■ Kawartha Lakes [D2-D4], a chain of 14 lakes, forms a boundary between southern Ontario's rolling farmland and the Canadian Shield. The Kawarthas— "bright waters and happy lands" in the Huron language—are connected by the canals and locks of the Trent-Severn waterway. The 388-km-long waterway begins at Trenton [E4] and uses Rice Lake, the Kawarthas, and Lake Simcoe to reach Port Severn on Georgian Bay. A major link in the waterway is the Lift Lock at Peterborough [D3].

■ At Sandbanks Provincial Park [E5-F5], dunes up to 25 m high stretch 10 km along the shores of Lake Ontario.

■ Lake on the Mountain Provincial Park [E5], situated 62 m above the Bay of Quinte, has a turquoise-coloured lake. Its source was long a mystery until divers discovered that it is fed by underground streams flowing through subterranean layers of limestone.

Scale 1:800,000

1 cm = 8 km

0   10   20   30 km

1   2   3   4   5

A

B

C

D

E

F

11
65
Témiscamingue
Notre-Dame-du-Nord
Dymond
New Liskeard
558
Haileybury
North Cobalt
11B   Cobalt
Latchford
567
St-Bruno-de-Guigues
101
St-Eugène-de-Guigues
Angliers
Moffet
Laforce
Winneway
Lac Simard
R0815
Lac des Fourches
Lac Gaonnaga

Parc. prov. W.J.B. Greenwood Prov. Park
Lake Timiskaming
Témiscamingue
382
391
Laverlochère
Fugèreville
Latulipe
382
Belleterre
Lac aux Sables
Lac Soufflot
Lac Bay
Lac Winawiash
Grand-Lac-Victor
Lac Gaonnaga
R0816

382
Lorrainville
Ville-Marie
391
Béarn
Lieu historique national du Fort-Témiscamingue/ Fort Témiscamingue National Historic Site
Lac des Bois
R0813
Lac Devlin
Lac à la Truite
Lac Moore
Lac la Perche
Victoria
Lac Cartier
R0817

Fabre
Lac Guay
R0814
R0812
Lac Lescon
Lac Bobinet
Lac des Baies
R0817

TÉMISCAMINGUE
ZEC   Kipawa
Laniel
R0829
Ostaboningue
Lac Saveginaga
Lac Ogascanane
Lac Samleau
Lac de l'Original
R0704

Hunter's Point
Lac Kipawa
R0812
Lac Pammeroy
Lac des Loups
Lac aux Foins
Lac aux Écorces
R0710
Lac Larive

Réserve de Chasse de la Couronne Nipissing Crown Game Reserve
Wickstell Lake
Lac Kikiwissi
Lac du Marécage
Lac Planse
Lac Gardner

Marten River
Réserve nat. prov. Kenny Forest Prov. Nat. Reserve
101
Marua
Lac McKillop
Lac McLachlin
Lac Florio
Lac Marion
Lac Dix
Milles

64
Parc prov. Marten River Prov. Park
Kebaowek
Kipawa
Lac Tee
Lac Smith
Lac Plantin
Lac Sairs
R0819
R0819
ZEC
Lac Ramé

Tilden Lake
Thorne
Témiscaming
Tee Lake
Lac Grindstone
Lac Booth
Lac du Goéland
ZEC
Restigo
Lac Bruce

NIPISSING
Eldee
R0852
Lac Beauchêne
Lac Bleu
Lac Des Jardins
Lac du Pin Blanc
Lac Sept Milles

Tomiko Lake
Lac Duncan
Lac Russell
Lac Hamilton
QUÉB

11
63
Lac Marin
Lac du Fils
Lacs Aumont

Balsam Creek
533
Lac Maganasipi
ZEC
Maganasipi
Lac Moulin
R0706

17
CFB North Bay
Redbridge
656
Lac McCracken
ZEC
R0834
Dumoine
ZEC
ZEC de
R0706

Beaucage Point
Jocko Point
North Bay
11B
Mattawa
526m
Rapides-des-Joachims

Manitou Islands
Parc prov. Manitou Islands Prov. Park
94
Corbeil
17
Rutherglen
Parc provincial Samuel de Champlain Provincial Park
Bac/Ferry
208
17
Deux Rivières
Bissett Creek
Parc provincial Driftwood Provincial Park
Rapides-des-Joachims (Da Swisha)
ZEC St-Patrice

Lake Nipissing
Cross Point
Deepwater Point
Callander
531
Bonfield
Nosbonsing
Eau Claire
Greenbough
Waterloo Lake
Parc provincial Bissett Creek Provincial Park
Parc provincial Grant's Creek Provincial Park
Mackey
Rolphton
635
Point Alexander

Lac Nipissing
654
Astorville
Lake Nosbonsing
ONTARIO
Deep River
17

Sand Lake
Hintsworth
Parc prov. South Bay Prov. Park
630
Parc provincial Amable du Fond Provincial Park
Papineau Creek
Thompson Lake
Brent
Radiant Lake
Carl Wilson Lake
Cedar Lake
Lake Travers
CFB/BFC PETAWAWA

PP Restoule PP
Restoule
534
Nipissing
Kiosk
Kioshkokwa Lake
Manitou Lake
Gilmour Lake
Erables Lake
Maple Lake
Lane
Cartier Lake
Fo
Petawawa Po

Restoule Lake
524
534
Powassan
Three Mile Lake
North Tea Lake
Biggar Lake
Catfish Lake
Parc provincial Barron River Provincial Park
Black Bay

McQuaby Lake
Commanda
522
Trout Creek
522B
Kawawaymog Lake
Nipissing River
Algonquin Provincial Park
Barron
Pe

Pickerel Lake
PARRY SOUND
Spring Lake
Parc prov. Mikisew Prov. Park
South River
Deer Lake
Eagle Lake
Nipissing
Barnroot Lake
Hogan Lake
Lake Luvieille
Clemow Lake
White Partridge Lake
Barron Lake
Grand Lake
Chalk River

Sundridge
Lake De Muir
NIPISSING
Merchant Lake
Big Crow Lake
Redrock Lake
Dickson Lake
Sec Lake
McKay
28
Pen

510
124
11
Magnetawan
Port Carmen
Cecebe
Bernard Lake
Pickerel Lake
Lake Tim
Big Trout Lake
Happy Isle Lake
Produt Lake
Clover Lake
Alice
58

520
Burk's Falls
Seguin Lake
Spence Lake
Rainy Lake
Pickerel Lake
Three Mile Lake
Butt Lake
McIntosh Lake
Misty Lake
Trout Lake
Round Island Lake
McKaskill Lake
Shirley Lake
Robitaille Lake
Parc provincial Bonnechere River Provincial Park
RENF

Katrine
518
Kearney
Scotia
Parc provincial Opeongo River Provincial Park
South Arm
Booth Lake
Abseyer Lake
Wilkins Lake
Bonnechere
Parc prov. Bonnechere Prov. Park
Parc prov. Foy Prov. Park

Bear Lake
518
Ernsdale
592
Smoke Lake
Rock Lake
Lake Louisa
Victoria Lake
Aylen Lake
Paugh
Round Lake Centre
Tramore
Deacon
60
58

Sprucedale
Novar
60
Bella Lake
Lake Opeongo
60
Whitney
Madawaska
Parc provincial Carson Lake Provincial Park
Wilno
512

Horn Lake
Ravenscliffe
45
Melissa
Millar Hill
8
Parc provincial Arrowhead Provincial Park
Parc provincial J. Albert Bauer Provincial Park
Parc provincial Oxtongue River-Ragged Falls Provincial Park
Rugged Falls
Pen Lake
Harry Lake
Clydegale Lake
Lower Hay
Parc provincial Bell Bay Provincial Park
523
62
Barry's Bay
Brudenell
Cormac

11
17
ONR
CN
122
75
40
60

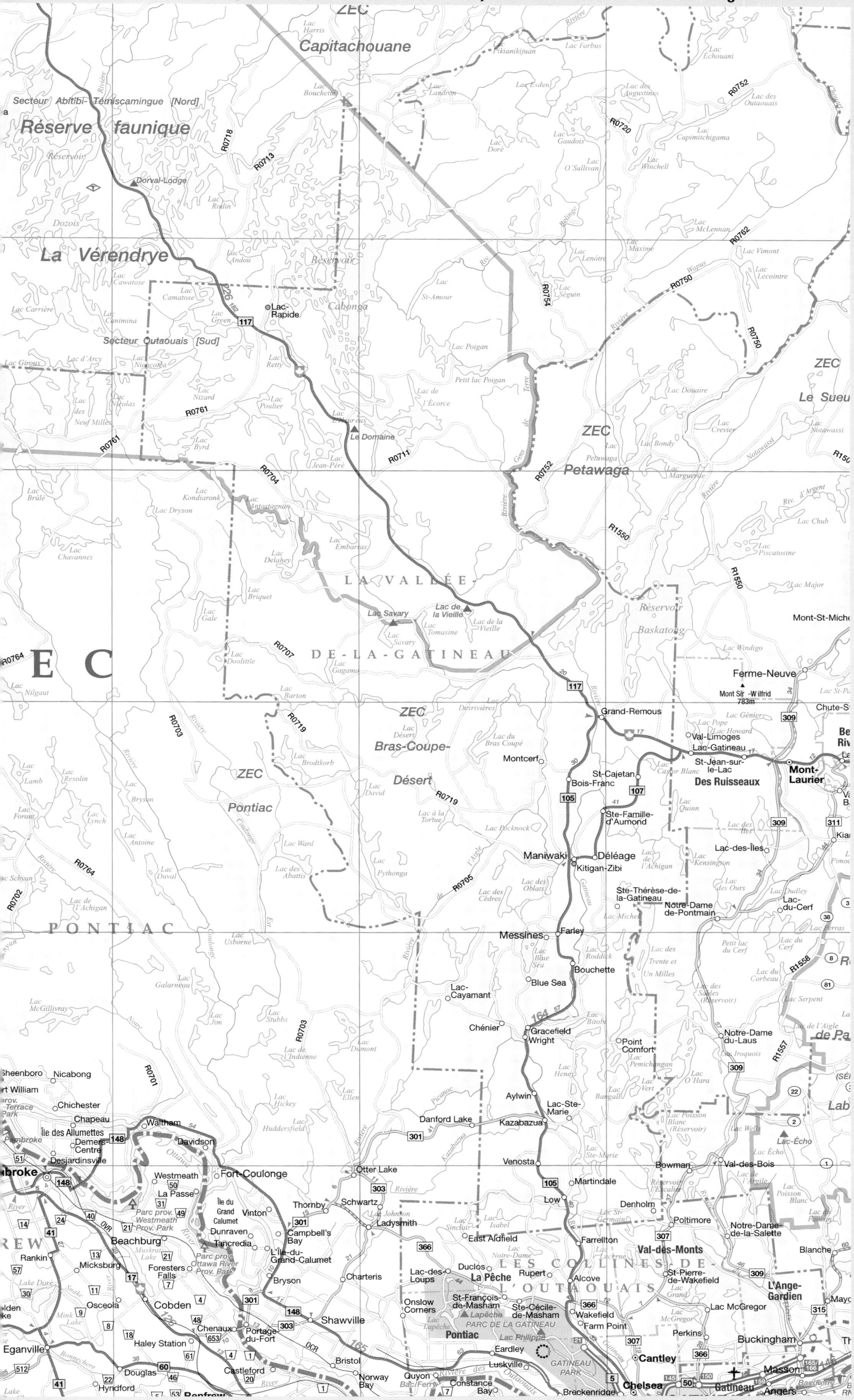

UPSTREAM ON THE OTTAWA

The 1,271-km-long Ottawa River origi-nates in a chain of Quebec lakes and reser-voirs: Réservoir Dozois, Grand Lac Victoria, Lac Granet, Réservoir Decelles, Lac Simard, and Lake Timiskaming (see also Plate 23, F5, F4, F3, and F1). South of Lake Timis-kaming, the Ottawa—which stretches diag-onally across this view—becomes a wide, powerful river, and a natural boundary between Quebec (right) and Ontario. Highway 101, visible on the Quebec side, runs parallel to the river until it reaches the pulp-and-paper town of Témiscaming. Here, travelers cross the river to reach Highway 63 in Ontario. The X-shaped expanse of water on the Quebec side is Lac Kipawa; just below it is Lac Beauchêne.

Other points of interest

■ Algonquin Provincial Park [E2-E4, F2-F4], covering 7,723 km² of Shield country contains some 2,500 lakes. Wildlife includes deer, moose, and wolves, as well as 240 species of birds. Established in 1893, it is Canada's oldest and still one of the largest of provincial parks.

■ The 13,615-km² Réserve Faunique La Vérendrye [A4-A7, B4-B7, C6-C7] has more than 4,000 lakes and rivers, including Réservoir Dozois, a source of the Ottawa River. This wildlife reserve, once the domain of the Algonquin Indians, fur trappers, and lumberjacks, was created in 1939. In 1950, it was renamed to commemorate the 200th anniversary of the death of French explorer La Vérendrye.

Scale 1:800,000
1 cm = 8 km
0   10   20   30 km

ONTARIO

QUÉB (Quebec)

COCHRANE

TIMISKAMING

ABITIBI-OUEST

ABITIBI

Rouyn-Noranda

Matagami
Joutel
Val-Paradis
Villebois
Beaucanton
Normétal
St-Lambert
Val-St-Gilles
Clermont
St-Eugène-de-Chazel
La Reine
Chazel
Dupuy
La Sarre
Clerval
Ste-Hélène-de-Mancebourg
Colombourg
Macamic
Authier
Languedoc
Guyenne
St-Gérard-de-Berry
St-Dominique-du-Rosaire
St-Nazaire-de-Berry
St-Félix-de-Dalquier
L'Île-Nepawa
Palmarolle
Poulaires
Authier-Nord
Taschereau
St-Maurice-de-Dalquier
Lac-Castagnier
Roquemaure
Gallichan
Ste-Germaine-Boulé
Launay
Villemontel
Pikogan
La Morandière
Amos
Landrienne
Rapide-Danseur
Laferté
Duparquet
Reneault
Destor
Manneville
Ste-Gertrude-Manneville
St-Marc
Barville
D'Alembert
Cléricy
Mont-Brun
Preissac
Barraute
Lac-Dufault
La Corne
La Motte
Évain
Rouyn-Noranda
Maison Dumulon
Arntfield
Granada
McWatters
Cadillac
Rivière-Héva
Vassan
St-Edmond
Kirkland Lake
King Kirkland
Chaput Hughes
Virginiatown
Kearns
Beaudry
Ste-Agnès-de-Bellecombe
Centre éducatif forestier du Lac-Joannès
Malartic
L'Île-Siscoe
Val-Senneville
Larder Lake
Montbeillard
Cloutier
St-Roch
Rapide-Deux
Dubuisson
Val-d'Or
Louvicou
Boston Creek
Tarzwell
Englehart
Charlton
Tomstown
Rollet
Rapide-Sept
Earlton
Hilliardton
Roulier
Rémigny
Nédélec
Guérin
Kenabeek
Thornloe
Belle Vallée
Témiscamingue
Notre-Dame-du-Nord
New Liskeard
Dymond
Angliers
Moffet
Winneway
Haileybury
North Cobalt
St-Bruno-de-Guigues
St-Eugène-de-Guigues
Laverlochère
Laforce
Grand-Lac-Victor

Parc provincial Abitibi-de-Troyes Provincial Park
Lake Abitibi
Parc (Québec) d'Aiguebelle
Réserve faunique d'Aiguebelle
Réserve naturelle provinciale Thackeray Provincial Nature Reserve
Parc provincial Esker Lakes Provincial Park
Réserve naturelle provinciale Pushkin Hills Provincial Nature Reserve
Parc prov. Kap-Kig-Iwan Prov Park
Parc provincial Larder River Provincial Park
Musée régional des mines
Village minier de Bourlamaque

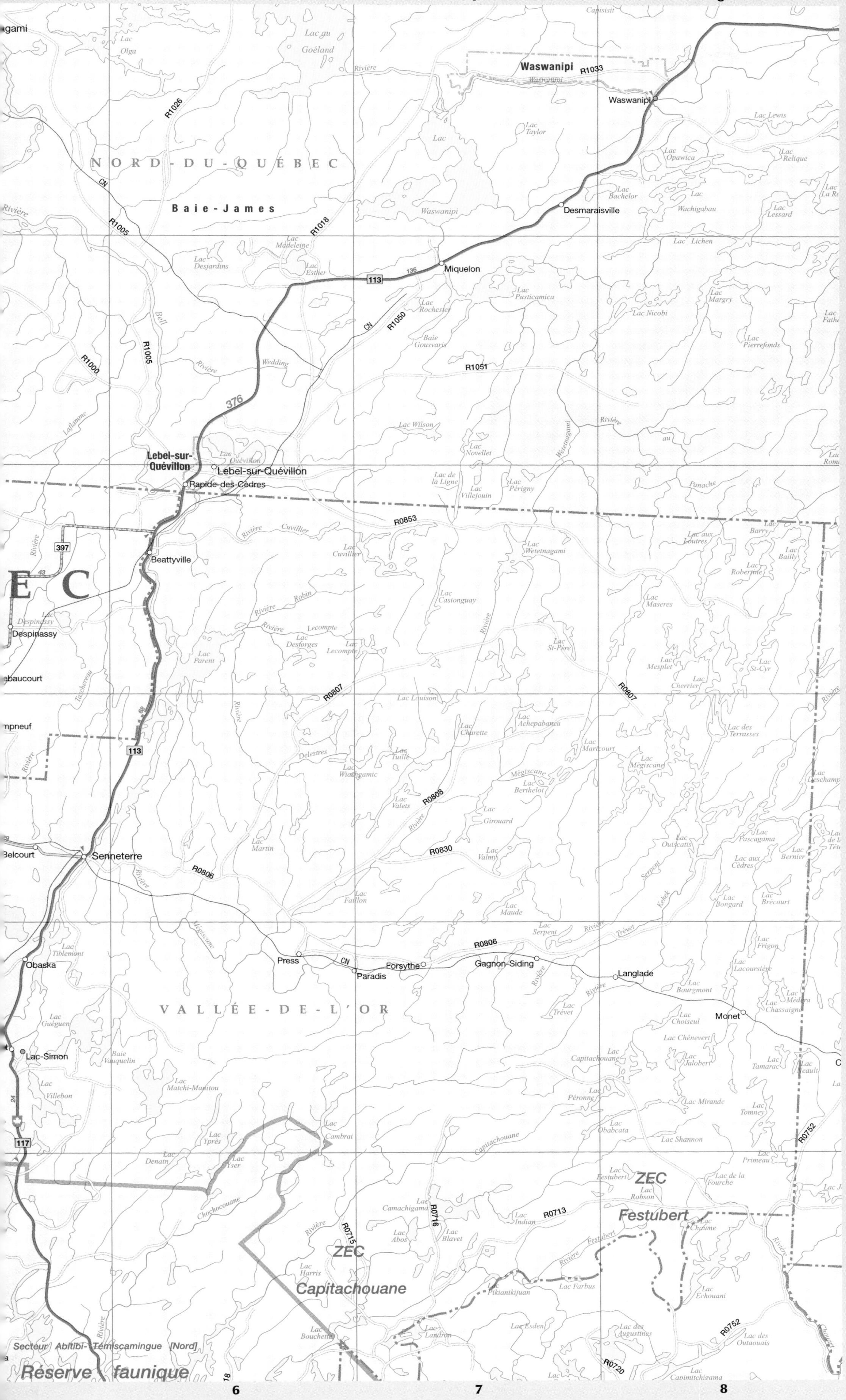

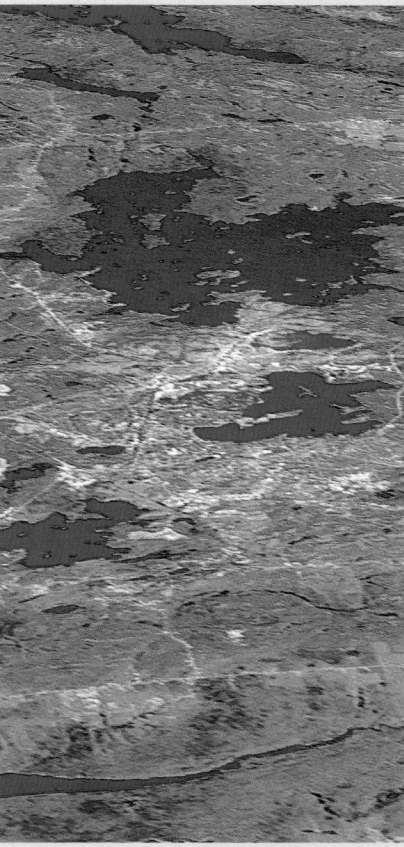

THE ROUYN-NORANDA REGION

The view looks northeast across the mining region of Rouyn-Noranda toward Lac Dufault (centre) and Lac Dufresnoy (top). Rich copper and gold deposits were discovered in this part of the Canadian Shield in the mid-1920s. The prosperity of Rouyn-Noranda (the latter name combines the words "Nor[th]" and "[C]an[a]da") still rests on smelting and processing ore from the region's many mines. Rouyn-Noranda is located on Lac Osisko (far right, just below Lac Dufault). The junction of the region's two major provincial highways, 117 and 101, is visible just east of the city.

Other points of interest

■ In 1912, the prospector Harry Oakes discovered the gold that sparked the growth of Kirkland Lake [D1]. Although none of the original mines are still active, the opening of new mines in the vicinity, and the processing of old mine tailings, have helped revive the city's fortunes.

■ Abitibi-de-Troyes Provincial Park [C1], located on the southern shore of Lake Abitibi, is a mixture of forest and swamp. One of the park's natural features is a 12-km-long peninsula jutting into the lake.

Scale 1:800,000
1 cm = 8 km

0   10   20   30 km

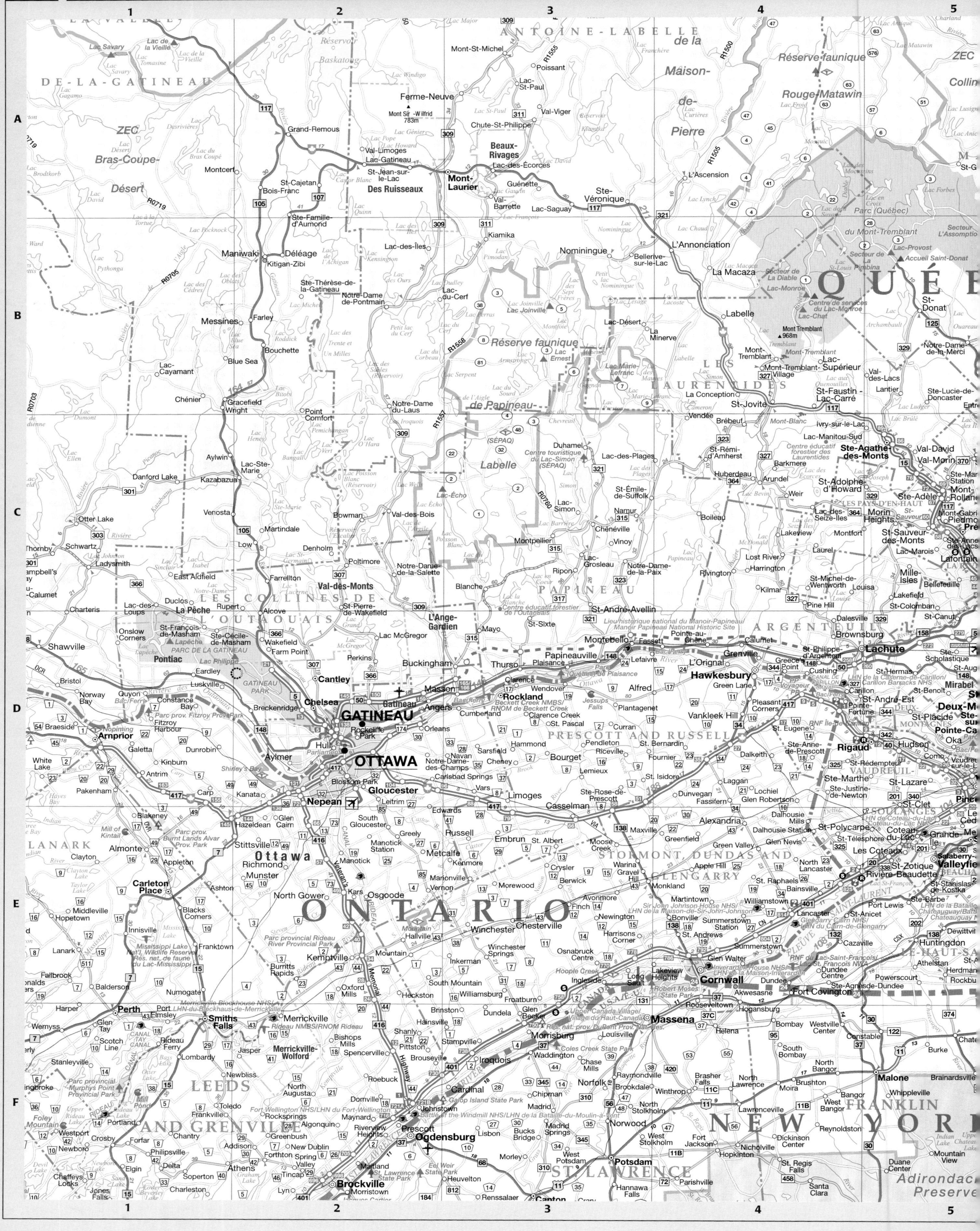

QUÉBEC

ONTARIO

NEW YORK

Major places: Mont-Laurier, Maniwaki, Gatineau, Hull, OTTAWA, Nepean, Gloucester, Arnprior, Carleton Place, Perth, Smiths Falls, Merrickville-Wolford, Brockville, Ogdensburg, Cornwall, Massena, Potsdam, Malone, Hawkesbury, Lachute, Mont-Tremblant, Ste-Agathe-des-Monts, Rigaud, Valleyfield

Regions: DE-LA-GATINEAU, ANTOINE-LABELLE, LES LAURENTIDES, ARGENTEUIL, PAPINEAU, LES COLLINES-DE-L'OUTAOUAIS, PRESCOTT AND RUSSELL, STORMONT, DUNDAS AND GLENGARRY, LANARK, LEEDS AND GRENVILLE, FRANKLIN, ST. LAWRENCE, ADIRONDACK Preserve

Réserve faunique de la Maison-de-Pierre, Réserve faunique Rouge-Matawin, Réserve faunique, PARC DE LA GATINEAU / GATINEAU PARK

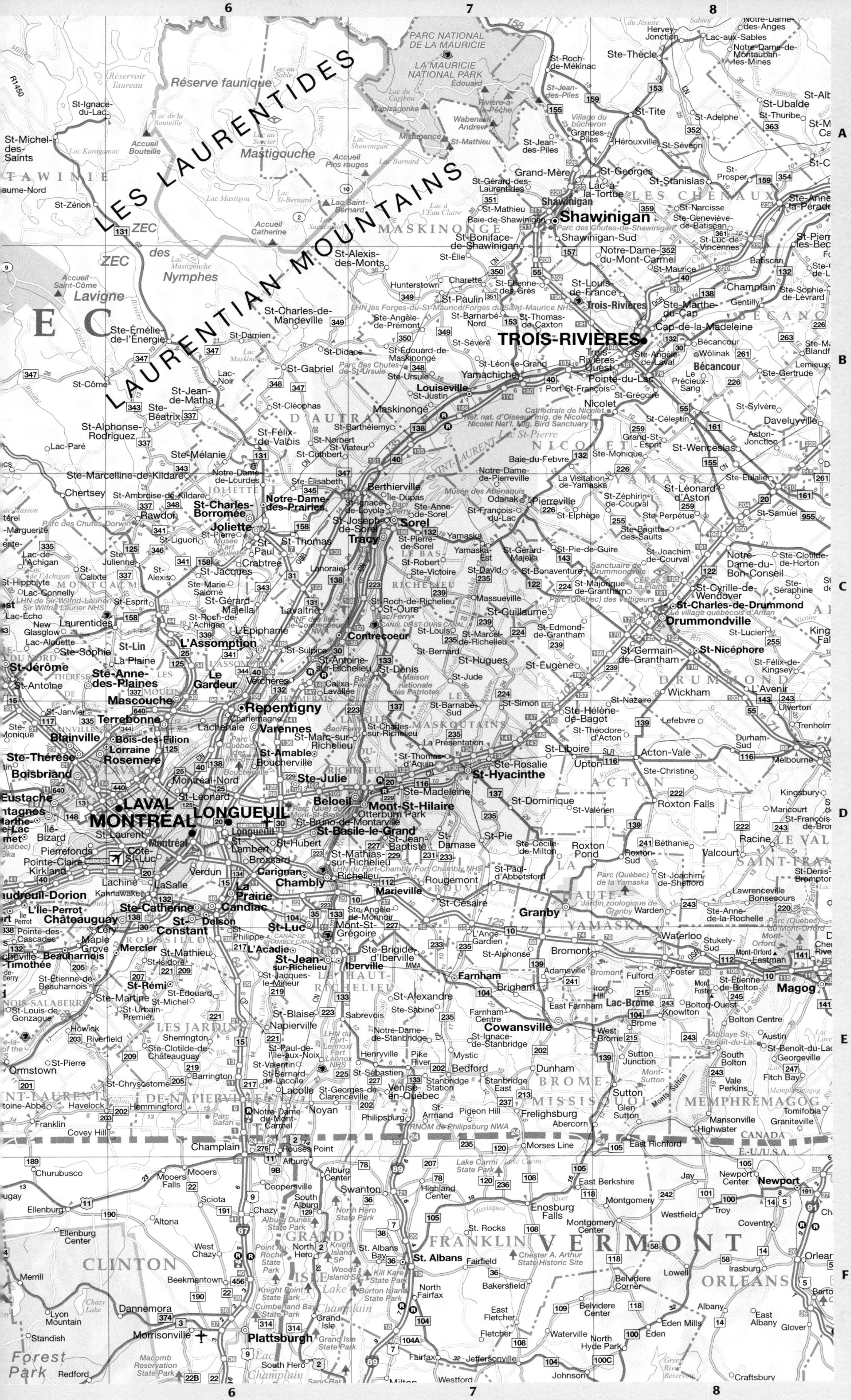

A LAKE IN THE ST. LAWRENCE

In New France, seigneurs parceled out farmland in long, thin strips to ensure their tenants had river frontage. The persistence of this land pattern is clearly evident in this view of the southern Quebec region where the Richelieu, Yamaska, and Saint-François rivers (left to right) flow north into the St. Lawrence River. At the mouth of the Richelieu lies the industrial city and seaport of Sorel. At this point, the St. Lawrence River widens to form Lac Saint-Pierre, a shallow, sluggish reach of water some 10 km across at its widest point. Île Dupas, Île aux Ours, and other low, mud-banked islands clog the entrance to the lake.

Other points of interest

■ Gatineau Park [D1-D2] encompasses 35,600 ha of forest, swamp, peat bogs, and lakes. Rich in flora and fauna, the park is home to 1,000 plants and 40 species of trees. The park supports 54 mammal species, including a population of 2,000 white-tailed deer, as well as 230 bird species.

■ The Laurentian Mountains form a 100- to 200-km-wide belt across the southern edge of the Canadian Shield in Quebec. This highland region stretches from the Gatineau Park in the west to the Saguenay River in the east—a distance of roughly 550 km. Geologically, the Laurentians were formed more than a billion years ago. Originally rising to much greater heights, they have been worn down into rounded hills.

Scale 1:800,000

1 cm = 8 km

0   10   20   30 km

Grid columns: 1 2 3 4 5
Grid rows: A B C D E F

Region labels:
QUÉBEC · LE HAUT-SAI · E-L'OR · LA VALLÉE · DE-LA-GATINEAU · ANTOINE-LABELLE · de la Maison- · de- · Pierre

ZEC areas:
ZEC Festubert · ZEC Capitachouane · ZEC Petawaga · ZEC Le Sueur · ZEC Bras-Coupé-Désert · ZEC Mitchinamécus · ZEC Normandie · ZEC Mazana · ZEC Boulé · ZEC Collin

Towns / places:
Paradis · Forsythe · Gagnon-Siding · Langlade · Monet · Clova · Oskélanéo · Parent · Casey · Manawan · Le Domaine · Ste-Anne-du-Lac · Mont-St-Michel · Poissant · Lac-St-Paul · Ferme-Neuve · Val-Viger · Chute-St-Philippe · Beaux-Rivages · Lac-des-Écorces · L'Ascension · Grand-Remous · Val-Limoges · Lac-Gatineau · St-Jean-sur-le-Lac · Mont-Laurier · Des Ruisseaux · Guénette · Ste-Véronique · Montcerf · St-Cajetan · Bois-Franc · Cèdre Blanc · Val-Barrette · Lac-Saguay · Ste-Famille-d'Aumond · Kiamika · Maniwaki · Déléage · Kitigan-Zibi · Lac-des-Iles · Nominingue · Bellerive-sur-le-Lac · L'Annonciation · La Macaza · Ste-Thérèse-de-la-Gatineau · Notre-Dame-de-Pontmain · Labelle · Messines · Farley · Bouchette · Blue Sea · Minerve · Mont-Tremblant · St-Donat · Accueil Saint-Donat · Notre-Dame-de-la-Merci

Réserves:
Réserve faunique Rouge-Matawin · Réserve faunique · Parc (Québec) du Mont-Tremblant · Secteur de La Diable · Secteur de la Pimbina · Centre de services du Lac-Monroe · Secteur de L'Assomption

Highways:
117 · 107 · 105 · 309 · 311 · 321 · 125 · 329 · 47 · 63 · 57 · 51 · 6 · 45 · 4 · 41 · 42 · 22 · 28 · 3

Peaks:
Mont Sir-Wilfrid 783m · Mont Tremblant 968m

Rivers / road numbers:
R0830 · R0806 · R0713 · R0715 · R0716 · R0711 · R0705 · R0719 · R0720 · R0750 · R0752 · R0754 · R0762 · R1500 · R1501 · R1502 · R1503 · R1505 · R1550 · R1555 · R1556 · R1450 · R1556 · R0400 · R0405 · R0457 · CN

Lakes (selection):
Lac Faillon · Lac Maude · Lac Valmy · Lac Bongard · Lac Brécourt · Lac des Cèdres · Lac Bernier · Lac des Cinq Milles · Lac Vantes · Lac Leblanc · Lac Peter · Lac Choiseul · Lac Chénevert · Lac Buies · Lac Oskélanéo · Lac Carmen · Lac Tassé · Lac du Gros Ours · Lac Festubert · Lac Robson · Lac Indian · Lac de la Fourche · Lac Gosselin · Lac Thaumur · Lac Choquette · Lac Dandurand · Lac Niverville · Lac Capintit · Lac Harris · Lac Abos · Lac Blavet · Lac Farbus · Lac Échouani · Lac Long · Lac Pierre · Lac Liliber · Lac Bouchette · Lac Landron · Lac Esten · Lac des Augustins · Lac des Outaouais · Lac Capimitchigama · Lac Doré · Lac Gaudois · Lac O'Sullivan · Lac Winchell · Lac Girène · Lac Nasigon · Lac Leleau · Lac Bassinet · Lac Adonis · Lac Némiscachingue · Kempt · Lac McLennan · Lac Maxime · Lac Vimont · Lac Lecointre · Lac Lajoue · Lac De La Bidière · Lac Duchastel · Lac Turnbull · Lac à la Croix · Lac Manou · Lac Cabonga · St-Amour · Lac Poigan · Petit lac Poigan · Lac Douaire · Lac Notawassi · Lac Mitchinamécus · Lac Duplessis · Lac Sproule · Lac Waterloo · Lac Badajoz · Lac Tourbis · Lac du Repos · Lac Lieu · Lac de l'Écorce · Lac Crevier · Lac Dieppe · Lac Bressan · Lac Mazana · Lac Castelveau · Lac Jean-Péré · Lac Bondy · Lac Marguerite · Lac Chopin · Lac Gorman · Lac Lengain · Lac Rupert · Lac Embarras · Lac Chub · Lac d'Argent · Lac des Polonais · Lac Beauregard · Lac Revelstoke · Lac Sunset · Lac Rouge · Lac Charland · Lac Piscatosine · Lac Tapani · Lac Major · Lac Windigo · Lac de la Hache · Lac Franchère · Lac de la Maison de Pierre · Lac Antique · Lac Matawin · Lac Lusigny · Lac Lynch · Lac Forbes · Lac Gagamo · Lac Desrivières · Lac du Bras Coupé · Lac Howard · Lac Génier · Lac Pope · Lac St-Paul · Réservoir Kiamika · Lac David · Lac Nominingue · Lac Chaud · Lac de la Savane · Lac Provost · Lac-Chat · Lac Savary · Lac de la Vieille · Lac Tomasine · Lac Quinn · Lac Gagnon · Lac François · Lac Saguay · Lac Brodkorb · Lac David · Lac à la Tortue · Lac de l'Achigan · Lac des Iles · Lac Kensington · Lac Pimodan · Lac Petit Nominingue · Lac Pythonga · Lac des Oblats · Lac des Ours · Dudley · Lac-du-Cerf · Lac des Sept Frères · Lac Joinville · Lac Macaza · Lac des Cèdres · Lac Michel · Lac des Trente et Un Milles · Petit lac du Cerf · Lac du Cerf · Lac Gatineau · Lac Roddick · Blue Sea · Lac Désert · Lac Armstrong · Lac Ernest · Lac Monroe · Lac-Tremblant · Lac Archambault · Lac Ouareau · Lac en Croix · Réservoir Baskatong · Réservoir · Reservoir

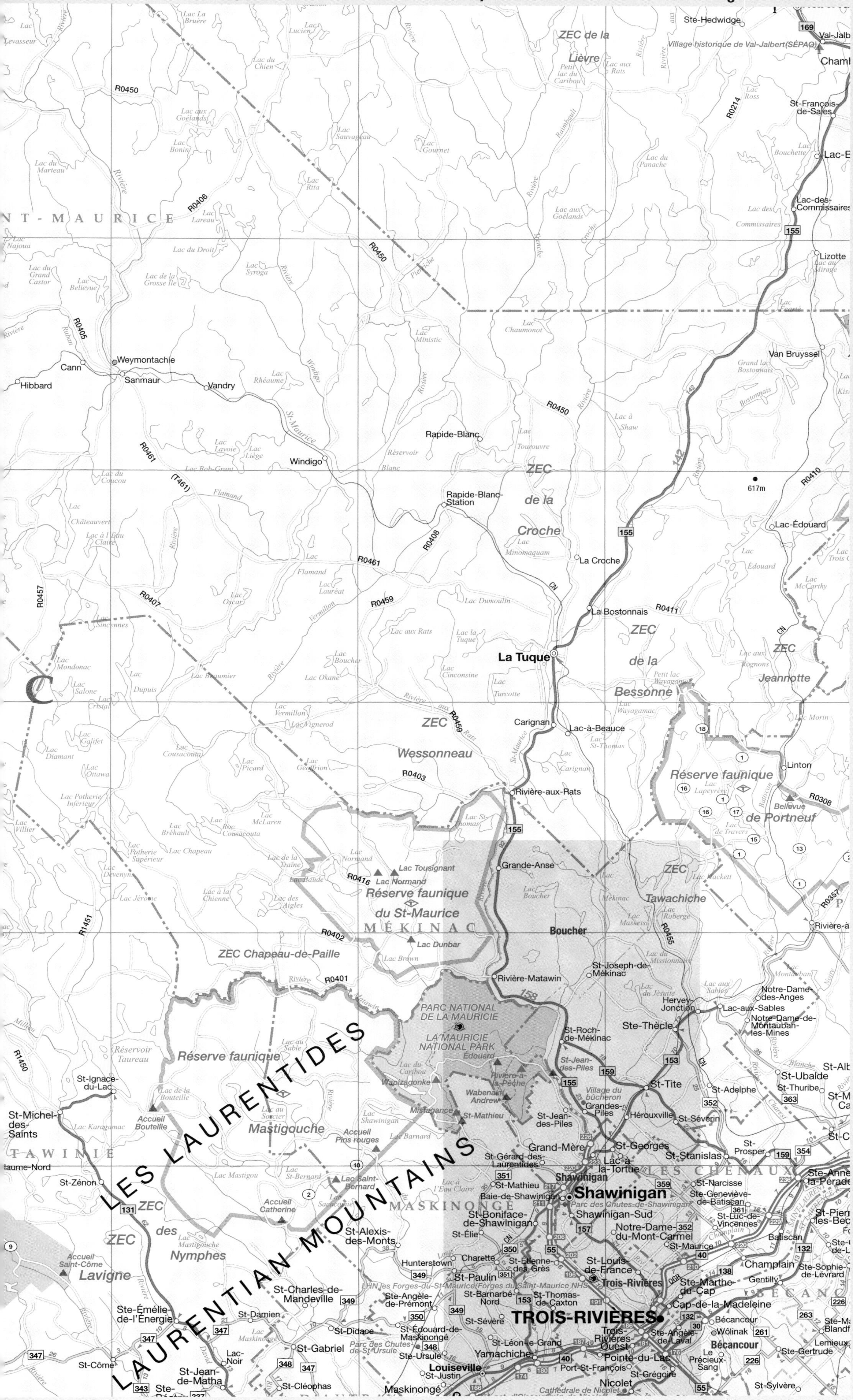

LA MAURICIE

Downstream from Lac Saint-Pierre, the St. Lawrence River resumes a fairly regular channel some two and a third kilometres wide. Trois-Rivières (lower right), where the 563-km-long St. Maurice River flows into the St. Lawrence, is halfway between Montréal and Québec. Founded in 1634, Trois-Rivières is Canada's second oldest city after Québec. This region, known as La Mauricie, provides logs and hydroelectric power for saw- and pulp-and-paper mills. La Mauricie National Park is located at the upper right of this view. Shawinigan, situated just before the bend in the river (left centre), processes metals and minerals, and produces chemicals.

Other points of interest

■ Parc du Mont-Tremblant [F4-F5] covers 1,500 km² of rugged Laurentian wilderness, encompassing some 500 lakes, seven rivers, and many streams and waterfalls. Mont Tremblant—at 986 m, the park's highest peak—is a favourite with skiers. The park's wildlife includes moose, white-tailed deer, black bear, and fox. Originally established as a forest reserve in 1894, Mont-Tremblant is Quebec's oldest provincial park and one of Canada's most popular.

■ The 549-km² La Mauricie National Park [E7] is divided into sectors. The southern sector has been developed for campers; the northern sector is still a rugged wilderness. Lac Wapizagonke is a favourite with canoeists and sailing enthusiasts.

Scale 1:800,000

1 cm = 8 km

0 10 20 30 km

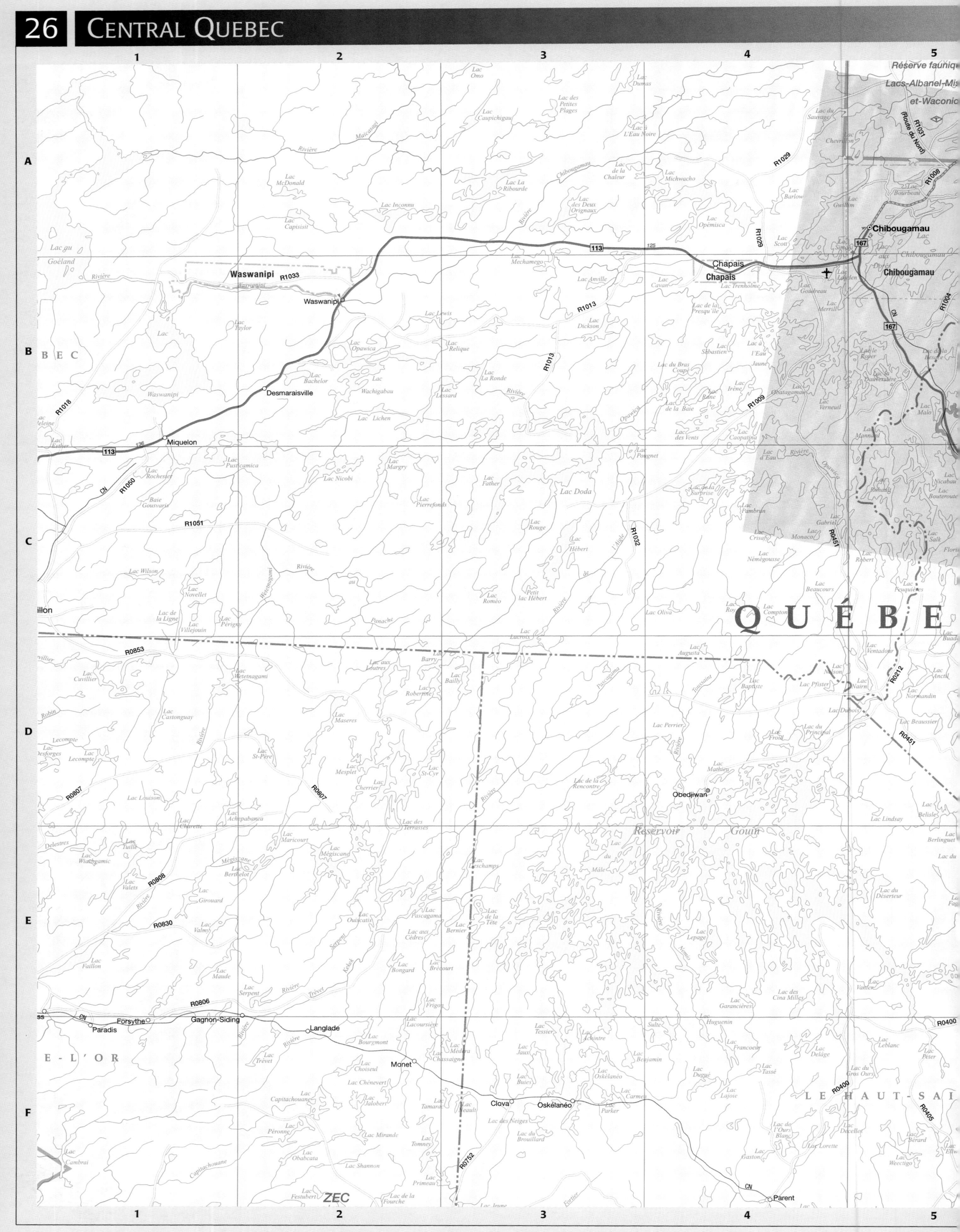

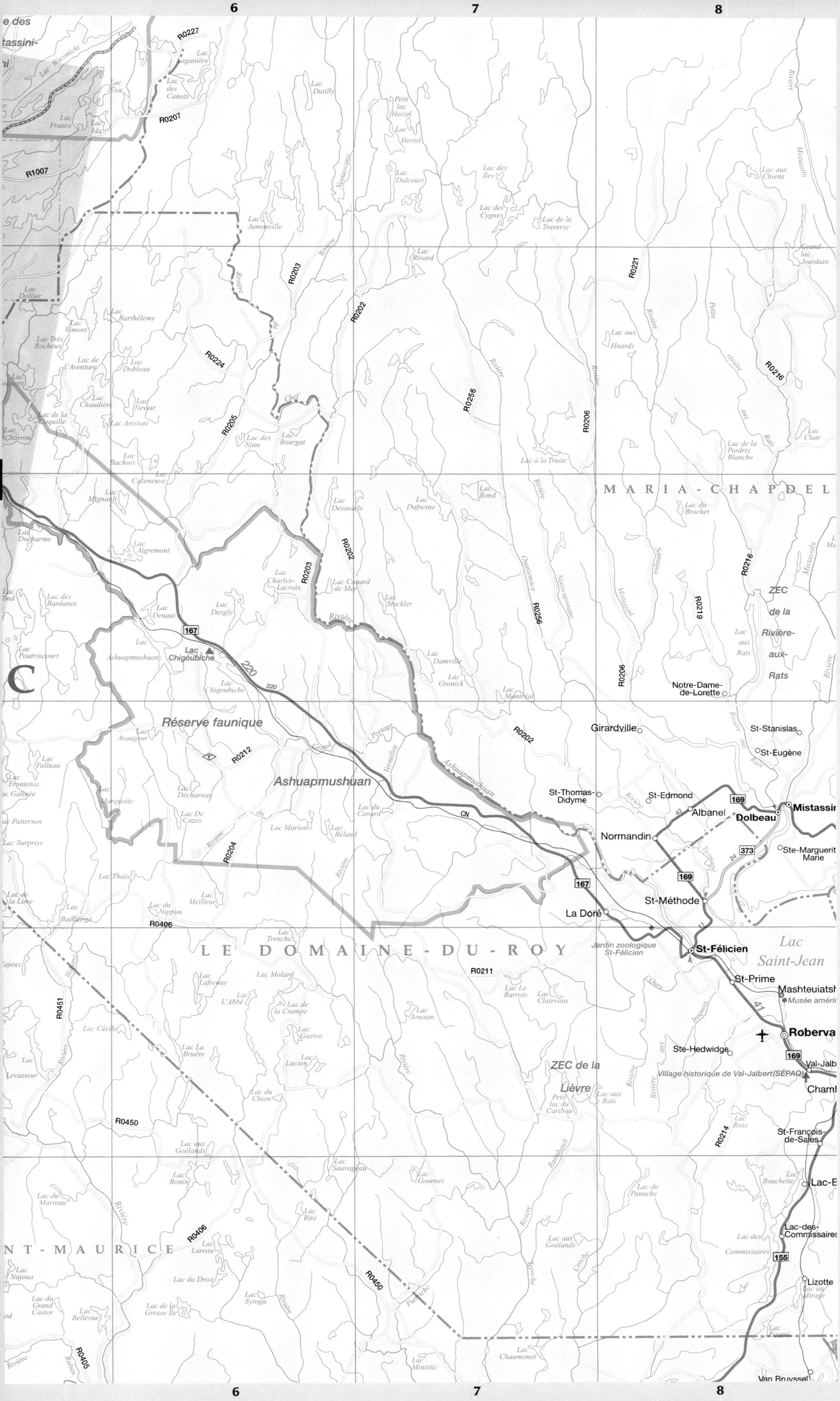

AROUND LAC CHIBOUGAMAU

In this view, the large expanse of water is Lac Chibougamau (top right). Just to the south of the lake, Highway 167 leads to the isolated mining town of Chibougamau, situated west of the lake. The townspeople work in the many copper and zinc mines located in this forested region. North of Lac Chibougamau, the Réserve Faunique des Lacs-Albanel-Mistassini-et-Waconichi has more than 5,000 lakes within its area of 28,285 km².

Other points of interest

■ The 4,500-km² Réserve Faunique Ashuapmushuan [C6, D6-D7] takes its name from the Montagnais "the place where the moose are seen." Black bear, fox, wolf, and hare inhabit this wildlife park. Brook trout, walleye, and northern pike abound in the park's 1,200 lakes. The park has two forest zones: firs and hardwoods dominate the southern sector; black spruce and jack pine, the northern sector.

■ Réservoir Gouin [D4, E3-E5] is a 1,570-km² artificial lake, accessible to visitors only by water and air. It was completed in 1918 as part of a hydroelectric project to control the flow of water in the Upper Mauricie River.

Scale 1:800,000

1 cm = 8 km

0   10   20   30 km

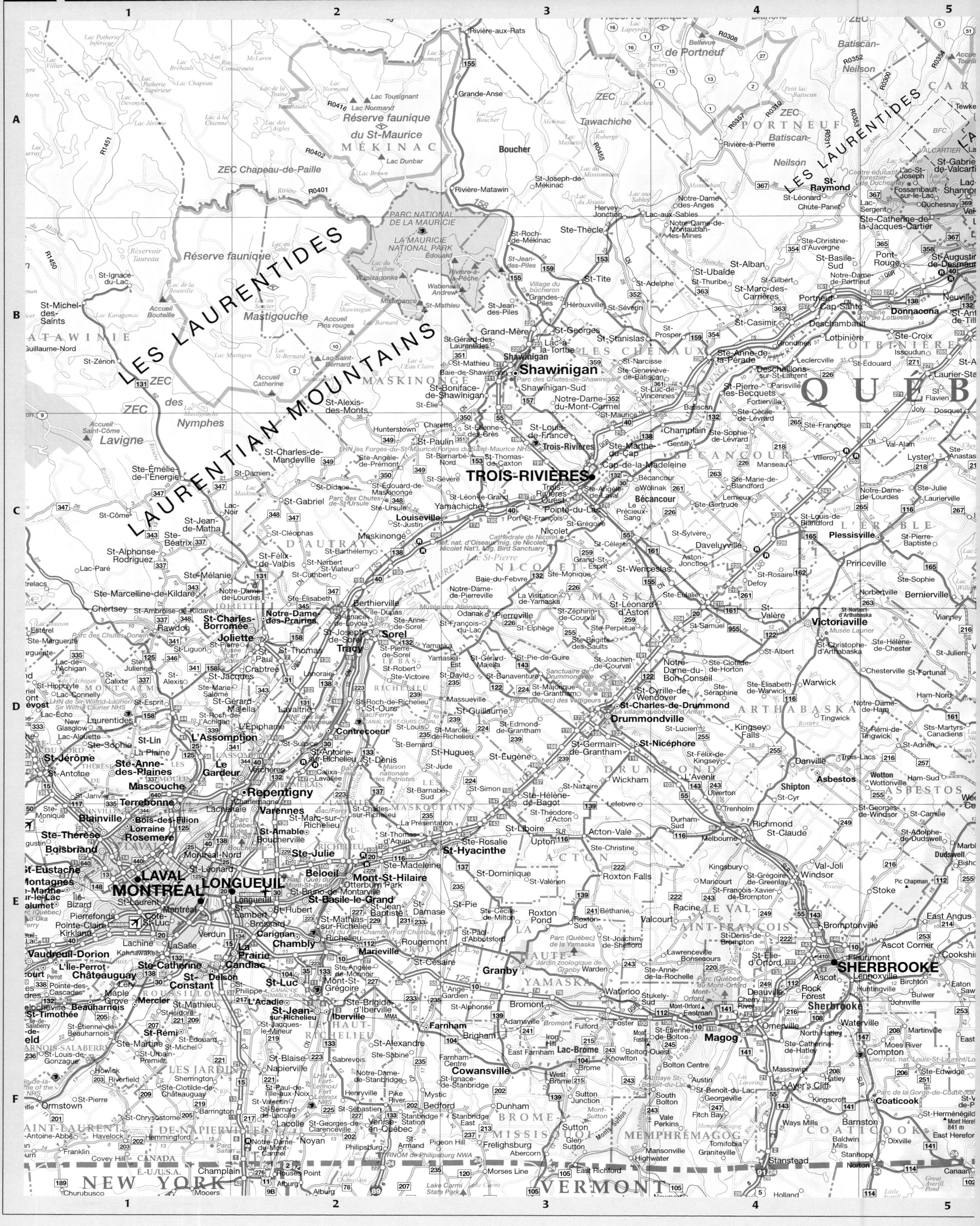

LES LAURENTIDES

LAURENTIAN MOUNTAINS

QUÉBEC

Grid columns: 1 2 3 4 5
Grid rows: A B C D E F

Major cities and towns

TROIS-RIVIÈRES

Shawinigan

MONTRÉAL

LAVAL

LONGUEUIL

SHERBROOKE

Sherbrooke

DRUMMONDVILLE

Victoriaville

Joliette

Sorel — **Tracy**

Magog

Granby

St-Hyacinthe

St-Jérôme

Terrebonne

Plessisville

Asbestos

Donnacona

St-Raymond

Selected place names

Rivière-aux-Rats, Grande-Anse, Boucher, Rivière-Matawin, Réserve faunique du St-Maurice, MÉKINAC, ZEC Chapeau-de-Paille, Réserve faunique, PARC NATIONAL DE LA MAURICIE, LA MAURICIE NATIONAL PARK, St-Michel-des-Saints, St-Zénon, Guillaume-Nord, St-Côme, St-Gabriel, Maskinongé, Louiseville, Berthierville, Nicolet, Pierreville, Odanak, Yamaska, St-Gérard-Majella, Massueville, St-Guillaume, St-Germain-de-Grantham, St-Nicéphore, St-Charles-de-Drummond, Wickham, L'Avenir, Shipton, Danville, Windsor, Richmond, Melbourne, Acton-Vale, St-Rosalie, Upton, Béloeil, Mont-St-Hilaire, Marieville, Rougemont, Granby, Waterloo, Bromont, Farnham, Cowansville, Sutton, Knowlton, Lac-Brome, Magog, Mont-Orford, Eastman, Waterville, Coaticook, Compton, Lennoxville, Ascot, Fleurimont, Bromptonville, East Angus, Windsor, Valcourt, Racine, Stoke, Deauville, Omerville, North Hatley, Ayer's Cliff, Stanstead

St-Jean-sur-Richelieu, Iberville, Chambly, Carignan, St-Hubert, Brossard, La Prairie, Candiac, Ste-Catherine, Châteauguay, Mercier, Beauharnois, St-Rémi, Napierville, Lacolle, Noyan, Philipsburg, Bedford, Stanbridge

Repentigny, Varennes, Verchères, Contrecoeur, Lavaltrie, Lanoraie, St-Sulpice, L'Assomption, Le Gardeur, Mascouche, Blainville, Ste-Thérèse, Boisbriand, Rosemère, Lorraine, St-Eustache, Deux-Montagnes, Pointe-Claire, Kirkland, Lachine, LaSalle, Verdun, Dorval, Pierrefonds

St-Tite, St-Georges, St-Stanislas, St-Casimir, Portneuf, Cap-Santé, Deschambault, St-Marc-des-Carrières, Grondines, Lotbinière, Ste-Croix, Issoudun, Bécancour, Gentilly, Champlain, Batiscan, Ste-Anne-de-la-Pérade, St-Luc-de-Vincennes, Cap-de-la-Madeleine, Pointe-du-Lac, Yamachiche, Charette, St-Étienne-des-Grès, Grand-Mère, St-Gérard-des-Laurentides, Lac-à-la-Tortue, St-Boniface-de-Shawinigan, Shawinigan-Sud

NEW YORK, NEW YORK (É.U./U.S.A.), VERMONT, CANADA, Champlain, Rouses Point, Mooers, Alburg, Lake Champlain, East Richford, Highwater, Stanstead

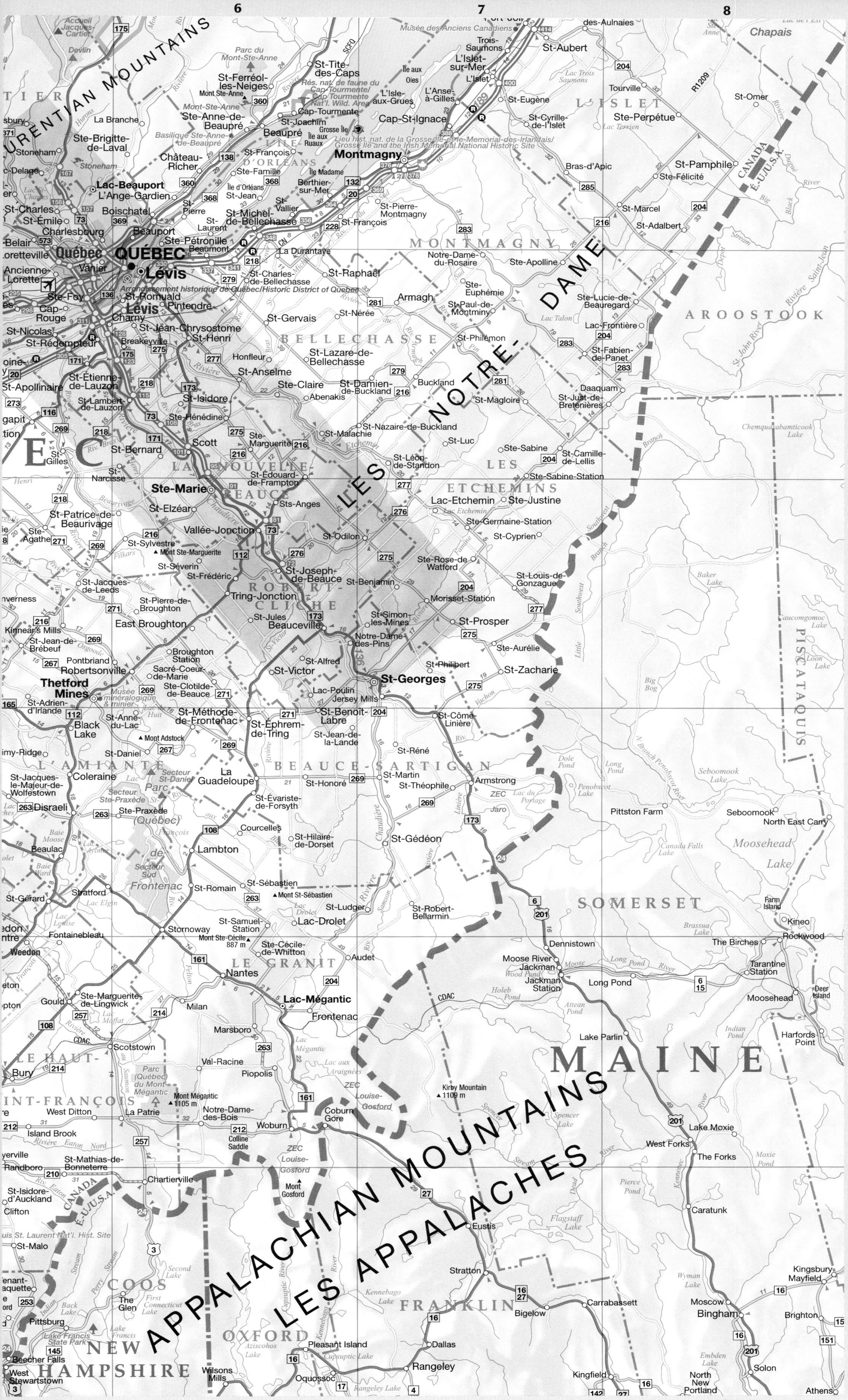

WHERE A MIGHTY RIVER NARROWS

This view looks northward toward Québec, visible on 98-m-high Cap Diamant above the St. Lawrence. At Québec, high cliffs narrow the river to a width of a kilometre. The bridge downstream from Québec (right centre) connects the north shore of the St. Lawrence with the picturesque 193-km² Île d'Orléans, the second largest island in the St. Lawrence after the island of Montréal. The island marks the tidal zone of the St. Lawrence where the river's fresh-water currents meet and mix with salt-water tides. Other north-shore sites: Lac Saint-Charles (top left) and the Rivière Montmorency (centre right), east of the bridge leading to the Île d'Orléans. South of the St. Lawrence, the Rivière Chaudière flows north through the farmland of the serene Beauce region.

Other points of interest

■ The 200-m-high Mont Royal in the heart of Montréal [E1] is one of 10 Monteregian Hills, which rise abruptly from agricultural lowlands of southern Quebec. Geologists believe the hills were formed when igneous rock thrust into layers of sedimentary rock below the earth's surface. Millions of years of erosion wore down the sedimentary rock, eventually exposing the hills.

■ The 400-m-high Mont Saint-Hilaire [E2], the highest of Monteregian Hills, is the site of a McGill University conservation area, which is a well-known bird sanctuary and Canada's first UNESCO biosphere reserve.

Scale 1:800,000

1 cm = 8 km

0   10   20   30 km

Grid columns: 1 2 3 4 5
Grid rows: A B C D E F

Row A
Lac Ducharme · Lac Aigremont · Lac Cavard de Mer · Lac Stackler · Lac Malfait · Lac Connelly · Lac Doucet · 720m
Lac des Bardanes · Lac Denau · Lac Desgly · Rivière · Lac Charles-Lacroix · Lac Damville · Lac Bellemare
Lac Poutrincourt · 167 · Lac Chigoubiche · Lac Gronick · Lac Montréal · ZEC de la Rivière-aux-Rats · Lac de l'Ouest · ZEC des Grandes Pointes
Lac Ashuapmushuan · 220 · Lac Chigoubiche · Notre-Dame-de-Lorette · Lac des Aigles · des Passes · Lac Bernabé
Réserve faunique · R0212 · R0202 · Girardville · St-Stanislas · Lac Noir

Row B
Ashuapmushuan · R0204 · St-Eugène · Ste-Élisabeth-de-Proulx · R0250
LE DOMAINE-DU-ROY · Lac Marion · Lac Béland · St-Thomas-Didyme · St-Edmond · 169 · Ste-Jeanne-d'Arc · St-Ludger-de-Milot
Lac Palluau · Lac Meilleur · Albanel · Dolbeau · Mistassini · Notre-Dame-du-Rosa
Lac Frontenac / Lac Galinée · Lac du Nippon · Normandin · 373 · Ste-Marguerite-Marie · St-Augustin · Lame
Lac Thaïs · 169 · St-Méthode · Pénbonka · Ste-Monique
Lac de la Lionne · R0406 · La Doré · Parc (Québec) de la Pointe-Taillon · St-Henri-de-Taillon · L'Ascension · Labrec
Lac Baillarge · Jardin zoologique St-Félicien · St-Félicien · 169 · Delisle · St-Lé

Row C
Lac Cécile · R0451 · Lac Molard · Lac Le Barnois · Lac Clairvaux · Lac Saint-Jean · Lac de la Grande Décha · St-Nazi
Lac La Bruère · L'Abbé · Lac de la Crampe · St-Prime · Mashteuiatsh · Musée amérindien · 169 · Alma
Lac Levasseur · Lac Gaston · Lac Jenssen · Roberval · Ste-Hedwidge · Val-Jalbert · St-Gédéon · 170 · St-Bruno
ZEC de la Lièvre · Lac Lucien · Lac aux Rats · Village historique de Val-Jalbert (SÉPAQ) · Métabetchouan · Hébertville-Station · Larouche
Petit lac du Caribou · Lac du Chien · Chambord · Desbiens · Lac-à-la-Croix · Hébertville
R0450 · Lac Ross · St-François-de-Sales · LAC-ST-JEAN-EST
Lac aux Goélands · Lac Sauvageau · Lac Bonin · St-André-du-Lac-St-Jean · Lac-Bouchette · Belle-Rivière · Accueil des Écorces

Row D
SAINT-MAURICE · R0406 · R0450 · R0214 · Lac-des-Commissaires · Lac St-Paul · QUÉBEC · 169
Lac du Marteau · Lac Lareau · Lac aux Goélands · 155 · Lizotte · Lac au Mirage · 36
Lac Najoua · Lac du Grand Castor · Lac Bellevue · Lac Rita · Lac Chaumonot · Lac Métabetchouane · Mont
Lac Syroga · Lac de la Grosse Île · Lac Ministic · Van Bruyssel · R0213 · 32
R0405 · Weymontachie · Cann · Sanmaur · Vandry · Grand lac Bostonnais · 361 · Réserve faunique
Hibbard · Lac Rhéaume · Lac Lescarbot · Lac Métasconac · 184 · LES

Row E
R0461 (T461) · Lac Lavoie · Lac Liège · Rapide-Blanc · Lac Tourouvre · Lac à Shaw · 366 · des
Lac du Coucou · Lac Bob-Grant · Windigo · Réservoir Blanc · ZEC de la Croche · 617m · R0410 · Lac Ventadour · 36
Lac Châteauvert · Lac à l'Eau Claire · Rapide-Blanc-Station · Lac Minomaquam · 155 · Lac-Édouard · Lac des Trois Caribous · 167
R0457 · R0407 · R0461 · R0408 · La Croche · Lac Édouard · Lac McCarthy · LAUR
Vermillon · R0459 · La Bostonnais · R0411 · ZEC de la Bessonne · Lac Brûlé

Row F
Lac Mondonac · Lac Beaumier · Lac Dumoulin · La Tuque · Jeannotte · Lac aux Rognons · ZEC · Petit lac Jacques-Cartier
Lac Salone · Lac Dupuis · Lac Okane · Lac Cinconsine · ZEC de la Rivière-Blanche · Lac Batiscan · 5
Lac Crystal · Lac Vermillon · Carignan · Lac-à-Beauce · Lac Wayuagamac · 18 · Lac Morin · R0302
Lac Giglfet · Lac Vignerod · ZEC Wessonneau · Linton · R0357 · Batiscan-Neilson · 51
Lac Diamant · R0403 · Rivière-aux-Rats · Réserve faunique de Portneuf · R0308 · 5 · R0352 · Accueil Touri
Lac Ottawa · Lac Potherie Inférieur · 155 · Grande-Anse · R0354 · CAR

A LAURENTIAN "KINGDOM"

Prosperous farmland and rugged Laurentian Highlands encircle the 1,002-km² Lac Saint-Jean. The many rivers that feed the lake originate in the surrounding highlands, including the Ashuapmushuan and Mistassini (top left and right), which flow into the lake from the west, and the Péribonka (right), which enters from the north. Lac Saint-Jean is drained by the Saguenay River (right), which flows east through a majestic 150-km-long fjord to the St. Lawrence. The overall length of the Saguenay to its source at the head of the Péribonka in the Laurentian Highlands is 698 km. The populous but remote Lac-Saint-Jean–Saguenay region is often called the "Kingdom of the Saguenay."

Other points of interest

■ The 1,138-km² Saguenay–St. Lawrence Marine Park [C6-D8], a federal-provincial park, protects the waters of the Saguenay and St. Lawrence, which support seal and porpoises, as well as minke, beluga, and blue whales.

■ The 228-km² Parc du Saguenay [C7-D8], a provincial park, occupies strips of land on either side of the Saguenay River. It overlooks the Saguenay fjord, carved from the Precambrian rock by glaciers during the last ice age. This majestic inlet, 2 km wide and 250 m deep, is flanked by cliffs up to 500 m high. The southernmost of Canada's fjords, the Saguenay is unique in that it flows into a larger river, not into the ocean as is usual elsewhere.

Scale 1:800,000

1 cm = 8 km

0   10   20   30 km

1 **2** **3** **4** **5**

A

de LA HAUTE-
Labrieville
Lac Isidore
Lac Leman
385
R0954
Labrieville-Sud
CÔTE-NORD
LES LAURENTIDES
ZEC de Forestville
Lac Lessard
LAURENTIAN MOUNTAINS
LAURENTIDES
MANICOUAGAN
Barrage Manic-Deux
R0901
Lac-au-Loup Marin
Lac Nipi
Secteur Mingan (Hauterive)
Secteur Marquette
Barrage Manic-Un
Baie-Comeau
389
Franquelin
Pointe St-Pancrace
Pointe à la Croix
Godbout
Pointe-des-Monts
Chute-aux-Outardes
Péninsule de Manicouagan
Les Buissons
Pointe-Lebel
Pointe de Manicouagan
Ragueneau
Ruisseau-Vert
Pointe-aux-Outardes
138
Baie aux Outardes
Parc régional de Pointe-aux-Outardes
Papinachois

B

R0952
Betsiamites
Rivière-Bersimis
Pointe de Betsiamites
FLEUVE SAINT-LAURENT
ST. LAWRENCE RIVER
Grosses-Roche
L'Anse-à-la-Croix
Ste-Félicité
Petit-Matane
385
Les Îlets-Jérémie
Colombier
Cap Colombier
St-Marc-de-Latour
193
R0912
R0952
Forestville
Matane
132
St-Adelme
St-Luc
St-Ulric
31
195
Baie-des-Sables
St-Léandre
ZEC Rivière des Cèdres
D'Iberville
Ste-Anne-de-Portneuf
Les Boules
Métis-sur-Mer
64
Réserve faunique de la Rivière-Matane
St-René-de-Mat
Nordique
Pointe-à-Boisvert
St-Paul-du-Nord
Jardins de Métis
Grand-Métis
Centre d'interprétation du saumon Atlantique
Ste-Flavie
Ste-Paule
St-Damase
297

C

Sault-au-Mouton
Baie-des-Bacon
Mont-Joli
Ste-Luce
St-Joseph-de-Lepage
St-Octave-de-Métis
St-Noël
234
St-Moïse
132
Val-Brillant
St-Tharcisius
195
Pointe-au-Père
Pointe-au-Père Lighthouse National Historic Site
Lieu historique national du Phare-de-Pointe-au-Père
Réserve nationale de faune de Pointe-au-Père/Pointe-au-Père National Wildlife Area
Luceville
Sayabec
73
St-Donat
Ste-Angèle-de-Mérici
Ste-Jeanne-d'Arc
St-Cléophas
25
Les Escoumins
Essipit
Rimouski-Est
St-Barnabé
Musée régional de Rimouski
St-Anaclet
298
QUEBEC
Amqui
Rimouski
Île du Bic
132
20
505
Ste-Odile
Neigette
St-Gabriel
Les Hauteurs
La Rédemption
Ste-Irène
Lac-au-Saumon
132
St-Alexandre-des-Lacs
Cap à l'Original
Parc (Québec) du Bic
Le Bic
St-Valérien
Ste-Blandine
234
298
LA MITIS
Lac Inférieur
Sacré-Cœur
Petites-Bergeronnes
172
138
Cap de Bon-Désir
Grandes-Bergeronnes
Pointe Sauvage
St-Fabien-sur-Mer
St-Fabien
Mont-Lebel
St-Marcellin
St-Charles-Garnier
Lac-Humqui
St-Léon-le-Grand
195
Albertville

D

Parc (Québec) du Saguenay
Centre d'interprétation des mammifères marin
PARC MARIN DU SAGUENAY-SAINT-LAURENT
Tadoussac
Pointe-Noire
Baie-Ste-Catherine
St-Simon
61
St-Mathieu
132
Lac St-Mathieu
Île aux Basques
St-Narcisse-de-Rimouski
St-Eugène-de-Ladrière
232
Lac Taché
R0114
ZEC
RIMOUSKI-NEIGETTE
BAS-
NOTRE-DAME
Lac Ferré
Lac Supérieur
St-
Accueil Sifroi
LES
Ste-Florence
Réserve Rivière
R0104
R0115
138
Réserve nationale de faune de la Baie-de-l'Isle-Verte/Baie de l'Isle-Verte National Wildlife Area
SAGUENAY-ST. LAWRENCE MARINE PARK
Île Verte
Trois-Pistoles
Rivière-Trois-Pistoles
Ste-Françoise
296
LES BASQUES
Ste-Cécile
La Trinité-des-Monts
Esprit-Saint
46
1
15
14
Réserve faunique de Rimouski
124
A2
A2
142
Laurent
Réserve faunique de la Rivière-Patapédia
Lieu hist
Bath
Réserve national de faune des Îles-de-l'Estuaire/Îles-de-l'Estuaire National Wildlife Area
Notre-Dame-des-Sept-Douleurs
St-Éloi
293
Ste-Rita
Ste-Françoise
47
L'Isle-Verte
St-Médard
St-Jean-de-Dieu
St-Guy
296
232
Lac-des-Aigles
Lac des Aigles
295
Accueil Biencourt
Lac-Rimouski
31
211
A2
St-Mistigougèche
St-François-d'Assise
St-Jean-de-Matapédia
L'Ascension-de-Patapédia
R0115

E

St-Georges-de-Cacouna
Cacouna
507
Île aux Lièvres
St-Arsène
291
St-Épiphane
St-Paul-de-la-Croix
St-Clément
La Société
Biencourt
South
296
Patapédia River
Réserve faunique de Kedgwick
Rivière-du-Loup
Musée du Bas-St-Laurent
St-Modeste
St-François-Xavier-de-Viger
293
St-Cyprien
Squatec
232
R0108
Kedgwick River
Kedgwick Game Refuge
Menneval
Grog
St-Patrice
Notre-Dame-du-Portage
132
185
RIVIÈRE-DU-LOUP
St-Hubert
Lamy
St-Pierre-de-Lamy
ZEC Owen
Third Lac
Second Lac
Green
Kedgwick
St-Jean-de-Rest
Les Pèlerins
St-Antonin
Whitworth
Petit-Témis
291
St-Honoré
232
Fort Ingall
Cabano
185
Auclair
Lejeune
Grand lac Squatec
Lots-Renversés
Owen
Réserve faunique de Kedgwick
265
Kedgwick
260
St-André
488
20
St-Alexandre
289
230
St-Joseph-de-Kamouraska
St-Louis-du-Ha! Ha!
St-Juste-du-Lac
Lac Touladi
First Lac
Kedgwick River
Whites Brook
St-Martin-de-Restigouche
Five Fingers
180

F

KAMOURASKA
Ste-Hélène
CN
Sanctuaire du Parke
St-Elzéar
Notre-Dame-du-Lac
Dégelis
NEW BRUNSWICK
17
Pohénégamook
Lac Pohénégamook
St-Eusèbe
232
TÉMISCOUATA
120
185
Moulin-Morneault
St-Joseph-de-Madawaska
St-Quentin
80
St-Éleuthère
289
Estcourt
Sully
Packington
Pied-du-Lac
Lac Meruenticook
Parc provincial Jardins de la République/Provincial Park
MADAWASKA
Notre-Dame-de-Lourdes
VICTORIA
287
76
G-109
ZEC Lac de l'Est
Chapais
R0111
Rivière-Bleue
St-Jean-de-la-Lande
St-Jacques
Musée des automobiles d'autrefois/Antique Automobile Museum
St-Basile
St-David
Rivière-Verte
Quisibis
46
Ste-Anne-de-Madawaska
17
St-Marc-du-Lac-Long
289
Lac Baker
Edmundston
144
Verret
Madawaska
St-Hilaire
2
Grand Isle
Cleveland
2
St-Armand
Belfleur
255
AROOSTOOK
Lac-Baker
215
120
Baker Brook
Caron Brook
St-François-de-Madawaska
Clair
Frenchville
Upper Frenchville
St. Agatha
Notre Dame
Siegas
144
St-Léonard Parent
Riley Brook
CANADA
U.S.A.
St-François
161
205
Wheelock
Fort Kent Mills
161
Daigle
162
Sinclair
Long Lake
Keegan
1A
St-Léonard
Dickey
Bradbury
Fort Kent St. Historic Site
Fort Kent
St. John
10
St. Agatha
Van Buren
St-André
MAINE
Allagash
161
15
Soldier Pond
Plaisted
Wallagrass
Cross Lake
Guerette
Ouellette
Eagle Lake State Reserve
Nic

1 **2** **3** **4** **5**

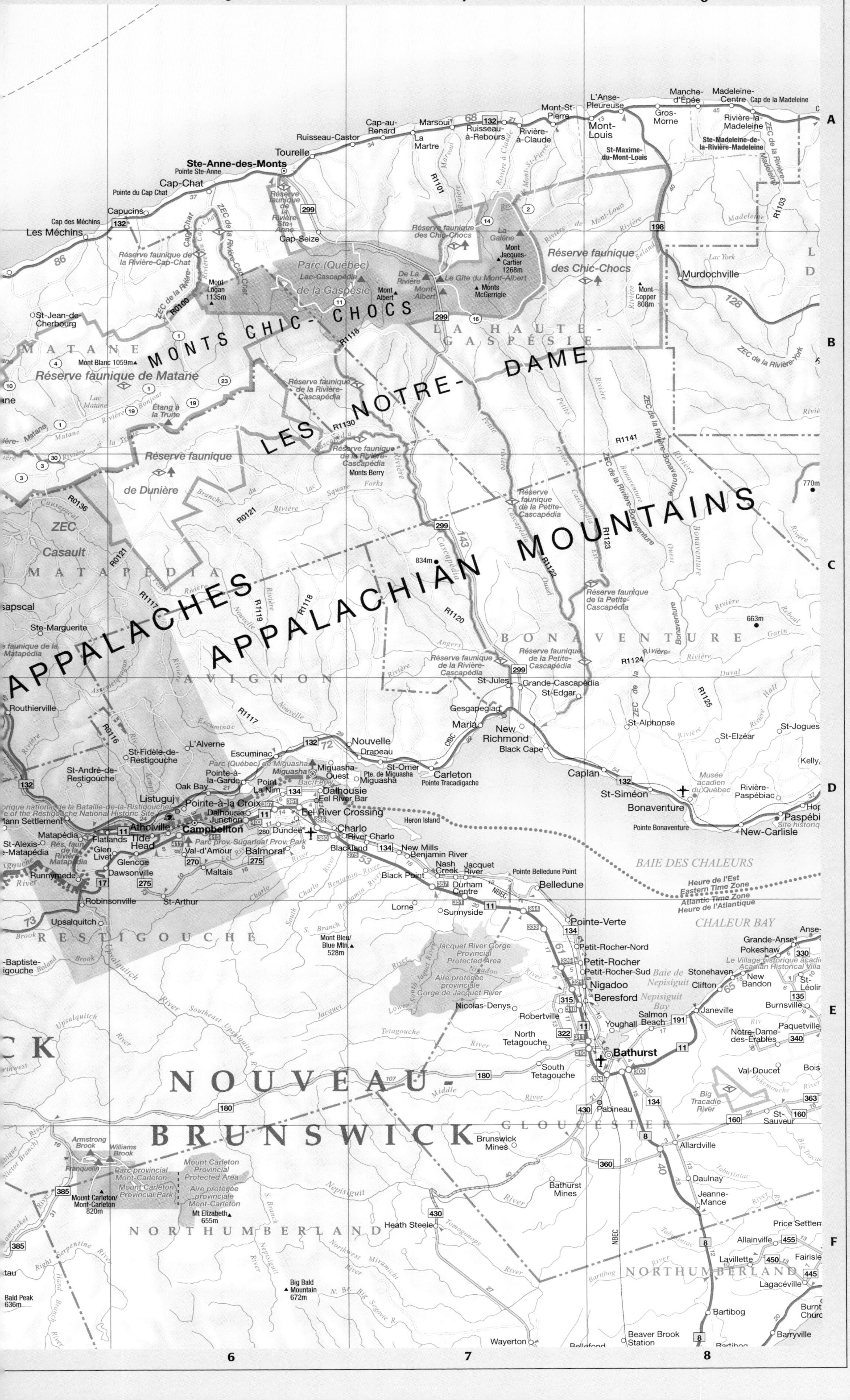

Map grid columns: 6   7   8

MONTS CHIC-CHOCS

LA HAUTE-GASPÉSIE

LES NOTRE-DAME

APPALACHES

APPALACHIAN MOUNTAINS

BONAVENTURE

AVIGNON

MATAPÉDIA

NOUVEAU-BRUNSWICK

GLOUCESTER

NORTHUMBERLAND

RESTIGOUCHE

BAIE DES CHALEURS

CHALEUR BAY

Place names (selection):
Ste-Anne-des-Monts, Pointe Ste-Anne, Cap-Chat, Capucins, Cap des Méchins, Les Méchins, St-Jean-de-Cherbourg, Mont Logan 1135m, Réserve faunique de la Rivière-Cap-Chat, Cap-Seize, Marsoui, La Martre, Tourelle, Cap-au-Renard, Ruisseau-Castor, Ruisseau-à-Rebours, Rivière-à-Claude, Mont-St-Pierre, L'Anse-Pleureuse, Mont-Louis, St-Maxime-du-Mont-Louis, Gros-Morne, Manche-d'Épée, Madeleine-Centre, Cap de la Madeleine, Rivière-la-Madeleine, Ste-Madeleine-de-la-Rivière-Madeleine, Murdochville, Mont Copper 808m, Réserve faunique des Chic-Chocs, Mont Jacques-Cartier 1268m, Monts McGerrigle, La Galène, De La Rivière, Mont-Albert, Le Gîte du Mont-Albert, Lac-Cascapédia, Parc (Québec) de la Gaspésie, Réserve faunique des Chic-Chocs, Mont Blanc 1059m, Réserve faunique de Matane, Lac Matane, Étang à la Truite, Réserve faunique de Dunière, ZEC Casault, Ste-Marguerite, Routhierville, Réserve faunique de la Rivière-Cascapédia, Monts Berry, Réserve faunique de la Petite-Cascapédia, Gesgapegiag, Maria, New Richmond, Black Cape, Caplan, St-Siméon, Bonaventure, St-Alphonse, St-Elzéar, St-Jogues, Kelly, Musée acadien du Québec, Rivière-Paspébiac, Paspébiac, New-Carlisle, Pointe Bonaventure, L'Alverne, Escuminac, Nouvelle, Drapeau, St-Omer, Carleton, Pointe Tracadigache, Grande-Cascapédia, St-Jules, St-Edgar, Miguasha, Miguasha-Ouest, Pte. de Miguasha, Oak Bay, Pointe-à-la-Garde, Point-à-la-Croix, La Nim, Dalhousie, Eel River Bar, Dalhousie Junction, St-Fidèle-de-Restigouche, St-André-de-Restigouche, Listuguj, Pointe-à-la-Croix, Eel River Crossing, Charlo, River Charlo, Heron Island, Blackland, New Mills, Benjamin River, Nash Creek, Jacquet River, Pointe Belledune Point, Belledune, Campbellton, Atholville, Tide Head, Val-d'Amour, Balmoral, Glen Livet, Glencoe, Dawsonville, Maltais, Sugarloaf Prov. Park, Dundee, Matapédia, Flatlands, St-Alexis-de-Matapédia, Runnymede, Robinsonville, St-Arthur, Upsalquitch, Baptiste-igouche, Mont Bleu/Blue Mtn. 528m, Jacquet River Gorge Protected Area, Black Point, Durham Centre, Lorne, Sunnyside, Pointe-Verte, Petit-Rocher-Nord, Petit-Rocher, Petit-Rocher-Sud, Nigadoo, Beresford, Nicolas-Denys, Robertville, North Tetagouche, South Tetagouche, Bathurst, Pabineau, Brunswick Mines, Bathurst Mines, Allardville, Daulnay, Jeanne-Mance, Heath Steele, Big Bald Mountain 672m, Mount Carleton Provincial Park, Mount Carleton 820m, Mt Elizabeth 655m, Bald Peak 636m, Armstrong Brook, Williams Brook, Franquelin, Stonehaven, Clifton, New Bandon, Burnsville, Janeville, Salmon Beach, Youghall, Notre-Dame-des-Érables, Val-Doucet, Paquetville, St-Léolin, Pokeshaw, Grande-Anse, Le Village historique acadien/Acadian Historical Village, Baie de Nepisiguit, Nepisiguit Bay, Big Tracadie River, St-Sauveur, Allainville, Lavillette, Fairisle, Lagacéville, Bartibog, Beaver Brook Station, Wayerton, Bellefond, Barryville, Price Settlement, Bellefond

Heure de l'Est / Eastern Time Zone
Heure de l'Atlantique / Atlantic Time Zone

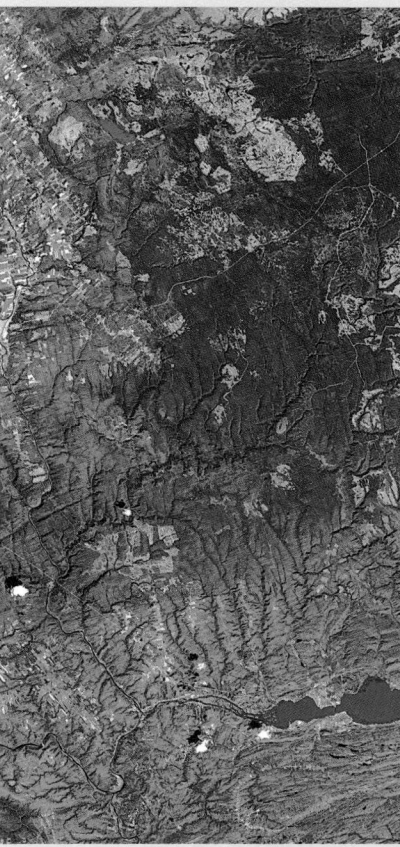

RESTIGOUCHE AND MATAPÉDIA

This view shows the wide estuary of the 200-km-long Restigouche River, which is slightly pinched where it empties into the waters of Chaleur Bay (right bottom). The Restigouche, with the Patapédia [Plate 31, A3], forms part of the boundary between Quebec and New Brunswick. One of its tributaries is the Rivière Matapédia, whose confluence with the Restigouche is clearly visible above (left bottom). The Matapédia flows south from the Notre Dame Mountains and through the Appalachians. Both are part of the Appalachian mountain system that extends from the eastern United States into this region.

Other points of interest

■ Réserve Faunique de Matane [B6-C5] is a 1,300-km² wilderness park with one of the highest concentrations of moose in Quebec. The terrain changes from rolling hills in the south to 1,000-m-high peaks of the Chic-Chocs in the north. Popular with hikers, the reserve contains a 72-km stretch of the International Appalachian Trail. Wildlife includes white-tailed deer, black bear, fox, and coyote, as well as 150 species of birds.

■ The 174-km² Mount Carleton Provincial Park [F5-F6] is named for its most outstanding physical feature, Mount Carleton—at 820 m, the highest mountain peak in the Maritimes. The remote park is renowned for its hiking trails and salmon spawning grounds.

Scale 1:800,000
1 cm = 8 km
0   10   20   30 km

Réserve faunique SÉPAQ Anticosti

Cap Ottawa

ÎLE

Détroit d'Honguedo
Strait of Honguedo

Baie des Sables

Rivière-Chicotte

Atlantic Time Zone
Heure de l'Atlantique
Heure de l'Est
Eastern Time Zone

A

Manche-d'Épée
Madeleine-Centre
Cap de la Madeleine
Gros-Morne
Rivière-la-Madeleine
Cap Barré
Petite-Vallée
Pointe-à-la-Frégate
Grande-Vallée
Ste-Madeleine-de-la-Rivière-Madeleine
Petite-Anse
Cloridorme
St-Yvon
Maxime-Mont-Louis
R1103
196

Pointe à la Renommée
L'Anse-à-Valleau
Pointe-Jaune
St-Maurice-de-l'Échouerie
Petit-Cap
Petit-Rivière-au-Renard
Rivière-au-Renard
L'Anse-à-Fugère
Centre d'accueil
Pointe Nord-Ouest

B

Lac York
Mont Copper 808m
Murdochville
LA CÔTE-DE-GASPÉ
R1140
128
R1126
Morris
St-Majorique
Gaspé
Wakeham
York Centre
ZEC de la Rivière-York
ZEC York-Baillargeon
R1102

PARC NATIONAL FORILLON
FORILLON NATIONAL PARK
Centre d'accueil
Penouille
L'Anse-au-Griffon
Jersey Cove
Cap des Rosiers
Cap-des-Rosiers
Des-Rosiers
Cap-aux-Os
Cap-Bon-Ami
Musée de la Gaspésie
Sandy Beach
Petit-Gaspé
Grande-Grave
Haldimand
Douglastown
Cap de Gaspé
Baie de Gaspé

C

QUÉBEC
770m
R1151
ROCHER PERCÉ
R1150
Réserve faunique de la Rivière-St-Jean
R1102
Réserve faunique de la Rivière-St-Jean
Cap du Bois Brûlé
Fort-Prével
St-Georges-de-Malbaie
Belle-Anse
Pointe-St-Pierre
Barachois
Pointe Verte
Bridgeville
Coin-du-Banc
Percé
Cannes-de-Roches
Rameau
Percé
Rocher Percé
Val-d'Espoir
L'Anse-à-Beaufils
St-Isidore
Cap Blanc
Parc (Québec) de l'Île-Bonaventure-et-du-Rocher-Percé
Île Bonaventure
Parc de la Baie-de-Percé
La Malbaie

D

R1141
R1124
663m
R1128
ZEC des Anses
Grande-Rivière
Petite-Rivière-Ouest
Ste-Thérèse-de-Gaspé
Cap-d'Espoir
Cap d'Espoir
Petit Pabos
Pabos
Chandler
Pabos Mills
Pointe de Newport
Newport
132
Réserve faunique de Port-Daniel
R1125
St-Alphonse
St-Joques
Clemville
St-Elzéar
Port-Daniel
Kelly
Marcil
Shigawake
St-Godefroi
170
Pointe au Maquereau
L'Anse-aux-Gascons
Baie de Port-Daniel
Pointe de l'Ouest

GOLFE DU
SAINT-

E (upper)

32
Siméon
Bonaventure
Musée acadien du Québec
Rivière-Paspébiac
Hope Town
Paspébiac
Pointe Bonaventure
New-Carlisle
Site historique du Banc-de-Paspébiac

BAIE DES CHALEURS
Heure de l'Est
Eastern Time Zone
Atlantic Time Zone
Heure de l'Atlantique
CHALEUR BAY

Île Miscou
Miscou Island
113
Miscou Centre
Pte. Sandy Pt.
Miscou Harbour

Petite-Rivière-de-l'Île
Petit-Shippagan
313
Pigeon Hill
Ste-Cécile
Île Lamèque
Lamèque Island
113
310
Petite-Lamèque
305
Lamèque
Ste-Marie-St-Raphaël

E

er-Nord
Grande-Anse
Pokeshaw
330
Le Village historique acadien/
Acadian Historical Village
Anse-Bleue
320
Maisonnette
303
Caraquet
11
145
Bas-Caraquet
St-Léolin
Bertrand
335
Shippagan
Aquarium and Marine Centre
L'Aquarium et Centre Marin
St-Simon
113
Le Goulet
Haut-Pokemouche
345
Inkerman
Pokemouche
New Bandon
Clifton
Stonehaven
Baie de Nepisiguit
Nepisiguit Bay
Janeville
Burnsville
325
191
Salmon Beach
135
Paquetville
350
Rang-St-Georges
355
Notre-Dame-des-Érables
340
Val-Doucet
Bois-Blanc
11
Bathurst
11
135
Pont-Landry
160
St-Isidore
365
Losier Settlement
Tracadie

F

134
8
Big Tracadie River
NEW BRUNSWICK
160
St-Sauveur
160
St-Irénée
363
370
Sheila
Tracadie-Sheila
Parc provincial Val Comeau Provincial Park
Val-Comeau
Allardville
Daunay
Jeanne-Mance
NOUVEAU-BRUNSWICK
Pont-Lafrance
Rivière-du-Portage
Brantville
460
Wishart Point
Price Settlement
Tabusintac
Tabusintac Lagoon
Allainville
455
445
11
Lavillette
450
Fairisle
NORTHUMBERLAND
445
Neguac
Plage Neguac Beach
Lagacéville

Map labels

6

Parc (Québec) d'Anticosti

Rivière
Rivière aux Saumons
Anse Harvey
Pointe Joseph

D'ANTICOSTI

Baie Prinsta
Cap de la Table

QUÉBEC

Baie du Renard
Baie-du-Renard

ANTICOSTI ISLAND

Lac du Renard

Rivière de la Chaloupe
Rivière Bell
Rivière Schmitt
Rivière Prinsta

Réserve faunique SÉPAQ Anticosti

Pointe de l'Est

Rivière-de-la-Chaloupe

Baie du Naufrage

Pointe du Sud
Pointe au Cormoran
Pointe Heath

7

8

A

B

FE DU

LAURENT

C

GULF OF

T. LAWRENCE

D

Refuge national d'oiseaux migrateurs des Rochers-aux-Oiseaux/ Bird Rocks National Migratory Bird Sanctuary

Rochers aux Oiseaux

Refuge écologique de l'Île-Brion
Île Brion

E

Grosse-Île-Nord
La Grosse-Île
Grosse-Île
199
Île de l'Est
Réserve nationale de faune de la Pointe-de-l'Est/ Pointe de l'Est National Wildlife Area

Old Harry
Île de la Grande Entrée

Pointe-aux-Loups
Grande-Entrée

Les Îles-de-la-Madeleine
Île Shag
ÎLES DE LA MADELEINE
QUÉBEC

Île du Havre aux Maisons

Fatima
Havre-aux-Maisons

Les Caps
L'Étang-du-Nord
La Vernière
Cap-aux-Meules
Île du Cap aux Meules
Gros-Cap

Île-d'Entrée

199
Baie de Plaisance
L'Île-d'Entrée

Île du Havre Aubert
Étang-des-Caps
Cap du Sud-Ouest
Havre-Aubert
L'Anse-à-la-Cabane
Bassin

F

6   **7**   **8**

HEADLANDS OF THE GASPÉ

The 3,200-km-long Appalachian mountain system reaches a terminus at the end of the Gaspé Peninsula. This view shows the prominent bays and high headlands (top to bottom): Cap de Gaspé (the site of Forillon National Park), Baie de Gaspé, Pointe Verte, La Malbaie, Cap Blanc, and Cap d'Espoir. Just beside Cap Blanc, the shape of famed Percé Rock can be glimpsed. Some 3.5 km farther offshore lies Île Bonaventure, home of one of the world's largest colonies of seabirds.

Other points of interest

■ At the northern tip of the Gaspé Peninsula, 240-km² Forillon National Park [B2-B3] embraces limestone cliffs, pebble beaches, sandy coves, and a boreal inland forest.

■ At the entrance of Chaleur Bay lies 64-km² Miscou Island [D2-E2], New Brunswick's "Land's End," a picturesque spot with white sand beaches and saltwater lagoons.

■ At 8,000 km², Anticosti Island [A4-A7] is larger than Prince Edward Island, but its population is only 300 people. Its physical features include sheer limestone cliffs, deep river canyons, and offshore reefs. In 1895, the wealthy French chocolate manufacturer Henri Menier purchased Anticosti for use as his own sports preserve. The island's 100,000 white-tailed deer are the descendants of the 220 that Menier introduced to the island.

■ Îles de la Madeleine [E8-F7], an archipelago of 16 islands, islets, and reefs in the Gulf of St. Lawrence, is some 100 km long, and has more than 300 km of sandy beaches.

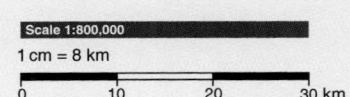

Scale 1:800,000
1 cm = 8 km
0   10   20   30 km

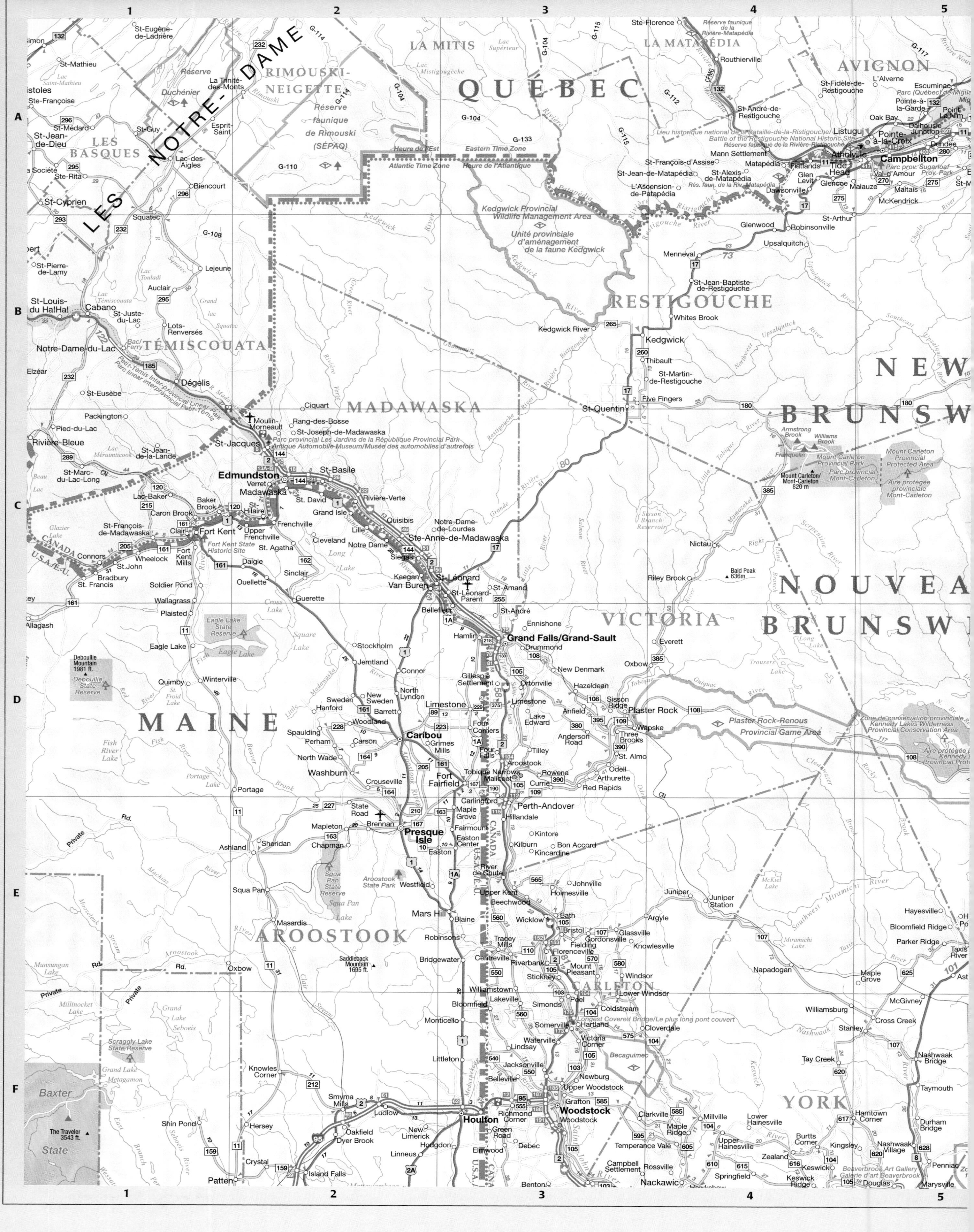

QUÉBEC

LA MITIS

LA MATAPÉDIA

AVIGNON

RESTIGOUCHE

NEW BRUNSWICK

NOUVEAU BRUNSWICK

MADAWASKA

VICTORIA

MAINE

AROOSTOOK

CARLETON

YORK

LES NOTRE-DAME

LES BASQUES

RIMOUSKI-NEIGETTE

TÉMISCOUATA

CANADA
U.S.A./É.-U.

Major places:
Campbellton, Atholville, Listuguj, Dalhousie Junction, Edmundston, Madawaska, St-Basile, Grand Falls/Grand-Sault, St-Léonard, St-Quentin, Kedgwick, Caribou, Presque Isle, Fort Fairfield, Perth-Andover, Plaster Rock, Woodstock, Houlton, Hartland, Mars Hill, Fort Kent, Notre-Dame-du-Lac, Cabano, Dégelis, Mount Carleton Provincial Park / Parc provincial Mont-Carleton 820 m

Longest Covered Bridge/Le plus long pont couvert — Hartland

Aire protégée provinciale Mont-Carleton
Mount Carleton Provincial Protected Area

Kennedy Lakes Wilderness Provincial Conservation Area
Zone de conservation provinciale

Plaster Rock–Renous Provincial Game Area

Baxter State
The Traveler 3543 ft.

Atlantic Time Zone / Heure de l'Atlantique
Eastern Time Zone / Heure de l'Est

Eagle Lake State Reserve
Deboullie Mountain 1981 ft. / Deboullie State Reserve
Scraggly Lake State Reserve
Aroostook State Park
Squa Pan State Reserve

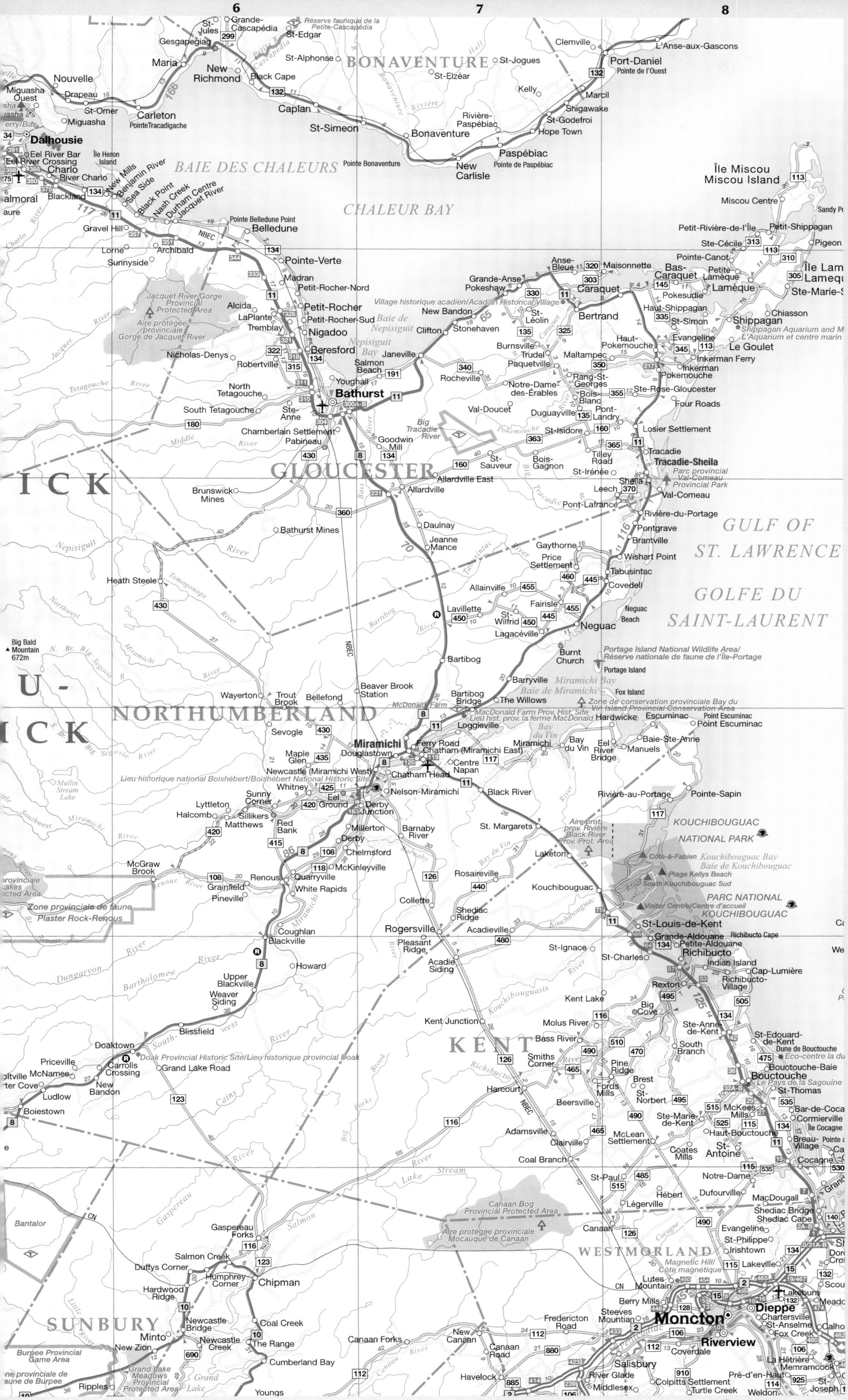

THE NORTHUMBERLAND SHORE

The shallow, warm waters of Northumberland Strait lap the gently sloping shoreline of eastern New Brunswick. The main sites on the 60-km-long stretch of the Northumberland shore shown above include (top to bottom): the Richibucto River estuary, Richibucto Cape, Baie de Buctouche, Cocagne Harbour, and Shediac Bay. Just north of Richibucto River lies Kouchibouguac National Park, which encloses the region's characteristic natural features: mixed forests inland, and sand dunes, lagoons, and beaches along the coast.

Other points of interest

■ The Saint John River rises in the forests of northern Maine and runs along the southern edge of New Brunswick's Madawaska County [C1], where it forms the boundary between Canada and the United States. At Edmundston [C2], the 673-km river turns southeast. From Grand Falls [D3], the Saint John flows through prosperous farming country and widens as it approaches the Bay of Fundy. The 23-m-high Grand Falls [D3] has been tamed for hydroelectric power. Hartland [F3] is the site of the world's longest covered bridge.

■ The Miramichi River flows 217 km from Juniper [E4] across central New Brunswick to the Gulf of St. Lawrence [D7]. The river is the offspring of the Southwest and Northwest Miramichi branches, which connect near Newcastle [D6]. The Miramichi, long one of Canada's major salmon rivers, has suffered through overfishing and pollution.

Scale 1:800 000
1 cm = 8 km
0   10   20   30 km

BONAVENTURE

BAIE DES CHALEURS

CHALEUR BAY

GLOUCESTER

NORTHUMBERLAND

GULF OF ST. LAWRENCE

GOLFE DU SAINT-LAURENT

KENT

WESTMORLAND

SUNBURY

KOUCHIBOUGUAC NATIONAL PARK

PARC NATIONAL KOUCHIBOUGUAC

Bathurst

Miramichi

Moncton

Dieppe

Riverview

Dalhousie

Caraquet

Tracadie-Sheila

Bouctouche

Richibucto

Rogersville

Column/row grid: 1 2 3 4 5 / A B C D E F

Major region labels

NEW BRUNSWICK / NOUVEAU BRUNSWICK

CARLETON · YORK · SUNBURY · QUEENS · KINGS · ST. JOHN · CHARLOTTE · WASHINGTON · MAINE

THE FUNDY ISLES / LES ÎLES DE FUNDY

GRAND MANAN ISLAND / ÎLE DU GRAND MANAN

BAY OF FUNDY / BAIE DE FUNDY

GULF OF MAINE / GOLFE DU MAINE

CANADA / U.S.A. – É.-U.

Selected place names

Perth-Andover, Carlingford, Maple Grove, Fairmount, Easton Center, Kilburn, Kintore, Bon Accord, Kincardine, Red Rapids, Juniper, Juniper Station, Argyle, Knowlesville, Johnville, Holmesville, Upper Kent, Beechwood, Bath, Wicklow, Bristol, Gordonsville, Glassville, Fielding, Mount Pleasant, Windsor, Lower Windsor, Coldstream, Williamsburg, Hayesville, Holtville, McNamee, Priceville, Bloomfield Ridge, Porter Cove, Ludlow, New Bandon, Carrolls Crossing, Grand Lake Road, Doaktown, Blissfield, Doak Provincial Historic Site / Lieu historique provincial Doak

Coughlan, Blackville, Rogersville, Pleasant Ridge, Acadie Siding, Upper Blackville, Weaver Siding, Howard, Kent Junction

Woodstock, Upper Woodstock, Houlton, Richmond Corner, Green Road, Grafton, Clarkville, Millville, Lower Hainesville, Upper Hainesville, Temperance Vale, Zealand, Keswick, Keswick Ridge, Douglas, Marysville, FREDERICTON, St. Marys, New Maryland, Lincoln, Oromocto, Hanwell, Waasis, Rusagonis, Waterville, Beaver Dam, French Lake, Geary, Gagetown, CFB/BFC Gagetown

Astle, McGivney, Cross Creek, Stanley, Nashwaak Bridge, Taymouth, Durham Bridge, Hamtown Corner, Kingsley, Nashwaak Village, Penniac, Napadogan, Maple Grove

Salmon Creek, Duffys Corner, Humphrey Corner, Hardwood Ridge, Chipman, Minto, Newcastle Bridge, Coal Creek, Newcastle Creek, The Range, New Zion, Canaan Forks, Cumberland Bay, Youngs Cove, Princess Park, Douglas Harbour, Scotchtown, Waterborough, Cambridge-Narrows

Kings Landing Historical Settlement / Village historique de Kings Landing, Prince William, Kingsclear, Longs Creek, Lake George, Magundy, Newmarket, Nasonworth, Acton, Cork, Tracy, Fredericton Junction, Central Blissville, Blissville, Hoyt, Patterson, Enniskillen, Wirral, Clarendon

Harvey, York Mills, Manners Sutton, Thomaston Corner, Coburn, Tweedside, Upper Brockway, Brockway, Lawrence Station, McAdam, St. Croix, Vanceboro, Lambert Lake, Basswood Ridge, Oak Hill, Moores Mills, Waweig, Oak Bay, Little Ridge, Scotch Ridge, Princeton, West Princeton, St. Stephen, Calais, Milltown, Baring, Woodland, Upper Mills, Red Beach, Bayside, Chamcook, St. Andrews, Robbinston, Bocabec, Digdeguash, St. George, Pennfield, Pocologan, Bonny River, Rollingdam, Elmsville, Utopia, New River Beach, Lepreau, Musquash, Maces Bay, Dipper Harbour, Chance Harbour, Black Beach, Lorneville

Saint John, Quispamsis, Rothesay, Gondola Point, Nauwigewauk, Hampton, Norton, Sussex, Apohaqui, Bloomfield, Hammondvale, Hampstead, Wickham, Evandale, Oak Point, Browns Flat, Welsford, Greenwich Hill, Moss Glen, Long Reach, Kingston, Lakeside, Upham, Barnesville, Hanford Brook, Loch Lomond, West Quaco, St. Martins, Fairfield, Quaco Head, Cape Spencer, Mispec, Red Head, Gardner Creek, Black River

Grand Bay, Westfield, Nerepis, Morrisdale, Hardings Point, Bayswater, Summerville, Kingshurst, East Riverside, Renforth, Fairvale, Baxters Corner, Shanklin

St-Croix Island Int'l Historic Site / Lieu hist. int. de l'Île Sainte-Croix, St. Andrews Blockhouse NHS / LHN du Blockhaus-de-St-Andrews, Moosehorn National Wildlife Refuge, Crawford, Meddybemps, North Perry, Charlotte, Ayers, Grove, Cooper, Perry, West Pembroke, Pembroke, Dennysville, Marion, Eastport, Welshpool, Campobello Island, Wilsons Beach, North Lubec, West Lubec, Lubec, South Lubec, Whiting, Trescott, South Trescott, Cutler, Machias, Machiasport, Jonesboro, East Machias, Marshfield, Whitneyville, Jacksonville, Northfield, Columbia Falls, Addison, Jonesport, Beals, Roque Bluffs, Indian River, West Jonesport, South Addison, Starboard, Bucks Harbor

Deer Island, Lords Cove, Leonardville, Fairhaven, Lambertville, Richardson, Back Bay, Blacks Harbour, Letang, L'Etete, The Wolves, Ross Island, Seal Cove, White Head Island, North Head, Castalia, Grand Harbour, Woodwards Cove, Dark Harbour

Parc provincial Herring Cove Provincial Park, Roosevelt-Campobello International Park / Parc international Roosevelt-Campobello, Parc provincial Castalia Provincial Park, Grand Manan National Migratory Bird Sanctuary / Refuge national d'oiseaux migrateurs Grand Manan, Machias Seal Island National Migratory Bird Sanctuary / Refuge nationale d'oiseaux migrateurs de l'Île-Machias-Seal, Point Lepreau Nuclear Generating Station / Station nucléaire Point Lepreau

Digby, Annapolis, Port Royal, Upper Clements, Granville Ferry, Granville Centre, Deep Brook, Bear River, Smiths Cove, Cornwallis, Culloden, Gulliver's Cove, Waterford, Rossway, Centreville, Brighton, Hillgrove, Marshalltown, Trout Cove, Acaciaville, Gilbert Cove, Plympton, Morganville, Barton, Sandy Cove, Lake Midway, Clementsvale, Bloomfield, Hainsfield, Greenland

Fort Anne & Scots Fort National Historic Site / Lieu historique national du Fort-Anne-& du Fort-Scots, Melanson Settlement National Historic Site / Lieu historique national de l'Établissement Melanson, Port-Royal National Historic Site / Lieu historique national de Port-Royal

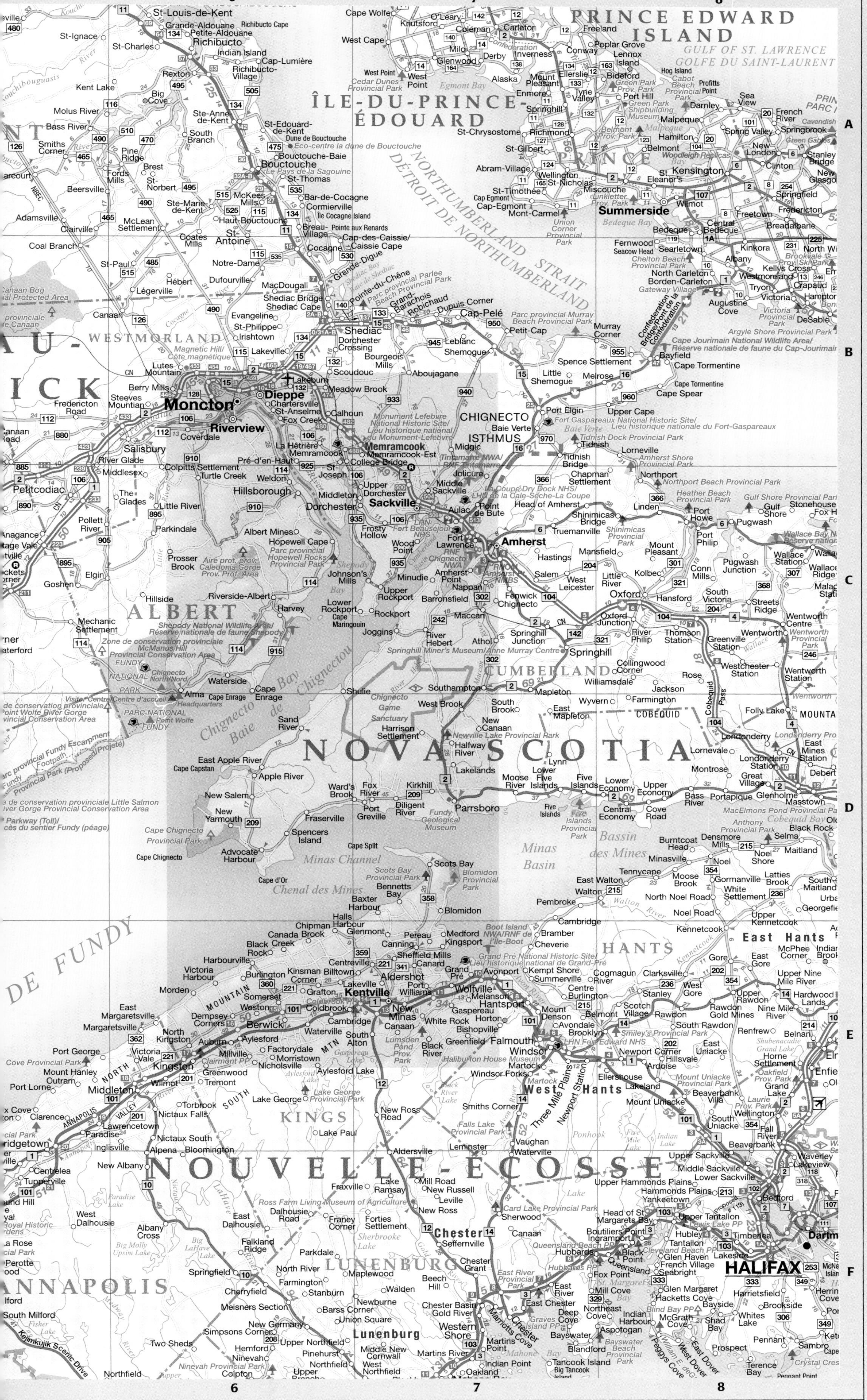

BAYS AND CHANNELS

The eastern end of the deep and narrow Bay of Fundy is indented with smaller bays and channels (clockwise, from top to bottom): Chignecto Bay and its two arms, Shepody Bay and Cumberland Basin on the New Brunswick side, and Minas Channel and Minas Basin on the Nova Scotia side. Minas Basin is the site of the world's highest tides, which can reach as high as 16 m. Moncton is located at the bend in the Petitcodiac River (top); Fundy National Park lies just beyond Cape Enrage (centre left). The upland that rims the Bay of Fundy's southern shore provides a protective shield for the apple orchards of Nova Scotia's Annapolis Valley (bottom).

Other points of interest

■ Grand Manan Island is one of three major islands at the entrance to the Bay of Fundy. The 24-km-long and 10-km-wide island is frequented by more than 400 species of migratory birds. Offshore waters are visited by humpback, minke, and other whales.

■ On Campobello Island [E2-F2], the centrepiece of the Roosevelt-Campobello International Park is the 34-room mansion, where U.S. President Franklin D. Roosevelt spent his summers until 1921, when he was stricken with polio. Off Deer Island [E2] lies Old Sow, the world's second largest whirlpool (after Norway's Maelstrom).

■ Fundy National Park [C6-D6] contains 206 km² of wooded hills, deep valleys, and 13 km of craggy Fundy shoreline. The spruce-fir forest on the coast and the birch-maple forests of the interior of the park support beaver, bobcat, coyote, and moose, as well as peregrine falcons and migrating shorebirds.

Scale 1:800,000

1 cm = 8 km

0   10   20   30 km

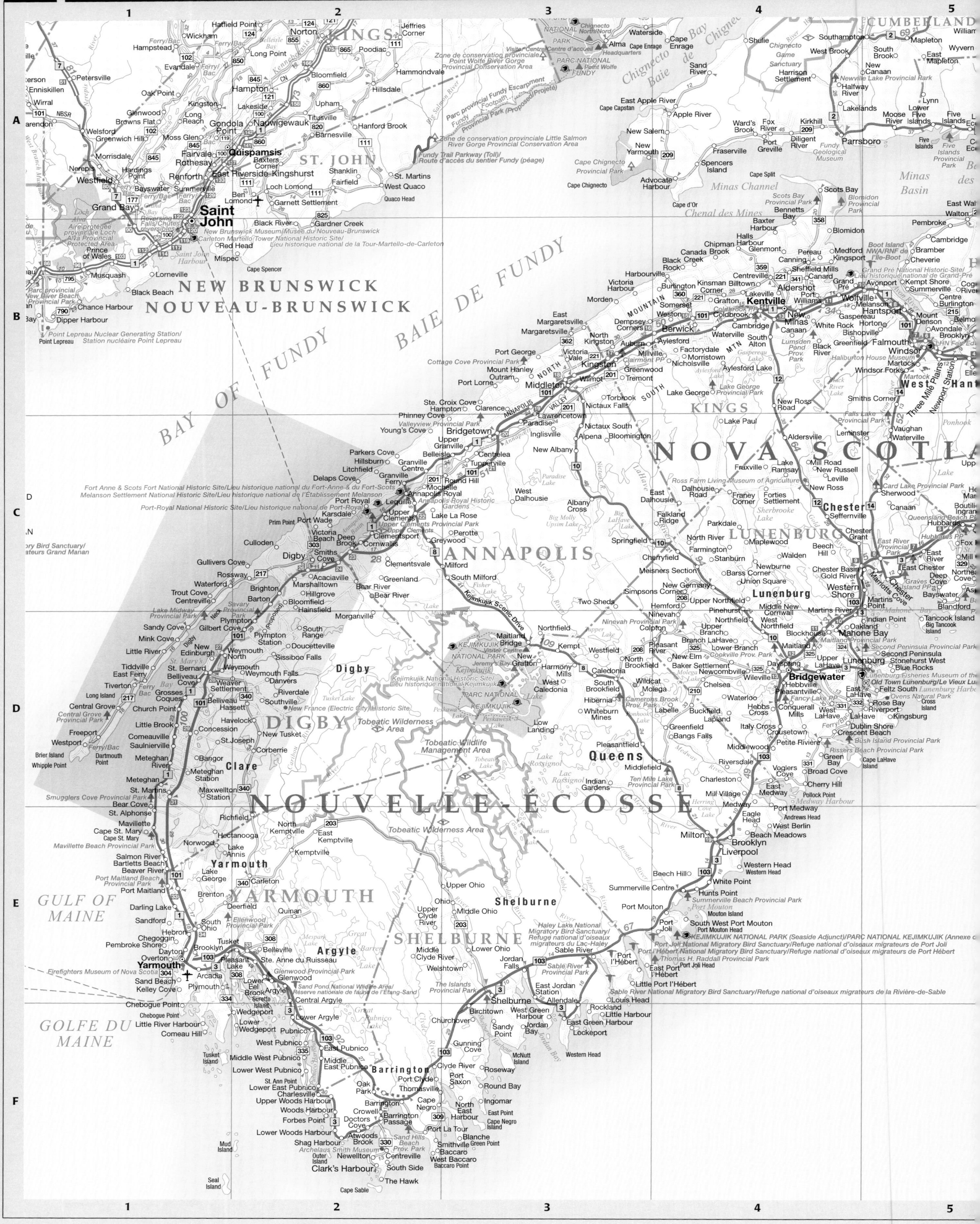

Grid columns: 1 2 3 4 5

Grid rows: A B C D E F

Major Labels

NEW BRUNSWICK / NOUVEAU-BRUNSWICK

NOVA SCOTIA / NOUVELLE-ÉCOSSE

BAY OF FUNDY / BAIE DE FUNDY

GULF OF MAINE / GOLFE DU MAINE

CUMBERLAND

KINGS

ANNAPOLIS

DIGBY

CLARE

YARMOUTH

ARGYLE

SHELBURNE

QUEENS

LUNENBURG

ST. JOHN

Minas Channel

Minas Basin

Chignecto Bay / Baie de Chignecto

Chenal des Mines

Selected Place Names

Saint John, Quispamsis, Rothesay, Grand Bay, Westfield, Hampton, Kingston, Norton, Sussex, St. Martins, West Quaco, Quaco Head, Black River, Gardner Creek, Musquash, Dipper Harbour, Chance Harbour, Lorneville, Cape Spencer

Parrsboro, Advocate Harbour, Cape Chignecto, Cape d'Or, Scots Bay, Blomidon, Cape Split, Apple River, New Salem, Spencers Island

Kentville, Wolfville, Windsor, Berwick, Kingston, Middleton, Bridgetown, Annapolis Royal, Digby, New Minas, Canning, Falmouth, Hantsport, Avonport, Grand Pré

Annapolis Royal, Port Royal, Granville Ferry, Bear River, Weymouth, Church Point, Meteghan, Saulnierville, Comeauville, Tiverton, Freeport, Westport, Brier Island, Long Island

Yarmouth, Arcadia, Tusket, Wedgeport, Pubnico, West Pubnico, East Pubnico, Argyle, Barrington, Clark's Harbour, Cape Sable, The Hawk, Shag Harbour, Woods Harbour

Shelburne, Lockeport, Liverpool, Brooklyn, Milton, Port Mouton, Port Joli, Port l'Hébert, Jordan Falls, Clyde River

Lunenburg, Bridgewater, Mahone Bay, Chester, LaHave, Rose Bay, Blue Rocks, Western Shore, New Germany, Chelsea, Petite Rivière

Kejimkujik National Park / Parc National Kejimkujik

Tobeatic Wilderness Area / Tobeatic Wildlife Management Area

National Historic Sites
Fort Anne & Scots Fort National Historic Site / Lieu historique national du Fort-Anne & du Fort-Scots
Melanson Settlement National Historic Site / Lieu historique national de l'Établissement Melanson
Port-Royal National Historic Site / Lieu historique national de Port-Royal
New Brunswick Museum / Musée du Nouveau-Brunswick
Carleton Martello Tower National Historic Site / Lieu historique national de la Tour-Martello-de-Carleton
Grand Pré National Historic Site / Lieu historique national de Grand-Pré
Point Lepreau Nuclear Generating Station / Station nucléaire Point Lepreau

DIGBY NECK & THE FRENCH SHORE

At the western end of the Bay of Fundy, Brier and Long islands and the 45-km-long Digby Neck shelter St. Mary's Bay. Across the bay lies the "French Shore," where picturesque Acadian villages line Highway 101 along the Gulf of Maine. Away from the coast are the mixed forests typical of inland Nova Scotia. The fishing port of Digby—a popular tourist destination—is situated on the west side of Annapolis Basin (top). A ferry service connects Digby with Saint John, N.B., 50 km across the Bay of Fundy.

Other points of interest

■ Lake Rossignol [D3], 26 km by 16 km, is Nova Scotia's largest fresh-water lake. It drains into the Atlantic by way of the Mersey River.

■ Cape Sable Island [F2], Nova Scotia's most southerly point, is a haven for the white ibis and other exotic birds. A 1,200-m causeway connects the island with Barrington Passage. Clark's Harbour, the island's largest community, is the 1907 birthplace of the 12-m-long Cape Island boat, still used (now with inboard motors) by inshore fishermen.

■ Kejimkujik National Park consists of two inland and seashore sectors. The 380-km² interior [D3] contains island-dotted lakes, low-lying hills, and one of Nova Scotia's finest forests. More than 500 species of plants, including many different types of ferns, orchids, and aquatic plants, are found here. The park's Seaside Adjunct [E4] serves as a migratory bird sanctuary.

■ Shubenacadie Provincial Wildlife Park [B6] shelters North American wildlife, such as lynx, moose, and reindeer.

Atlantic [The Bank Fishery-The Age of Sail/
L'exposition de la pêche sur les bancs, à l'ère de la voile]

Scale 1:800,000
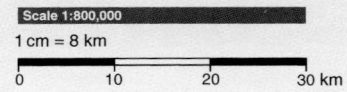
1 cm = 8 km

0   10   20   30 km

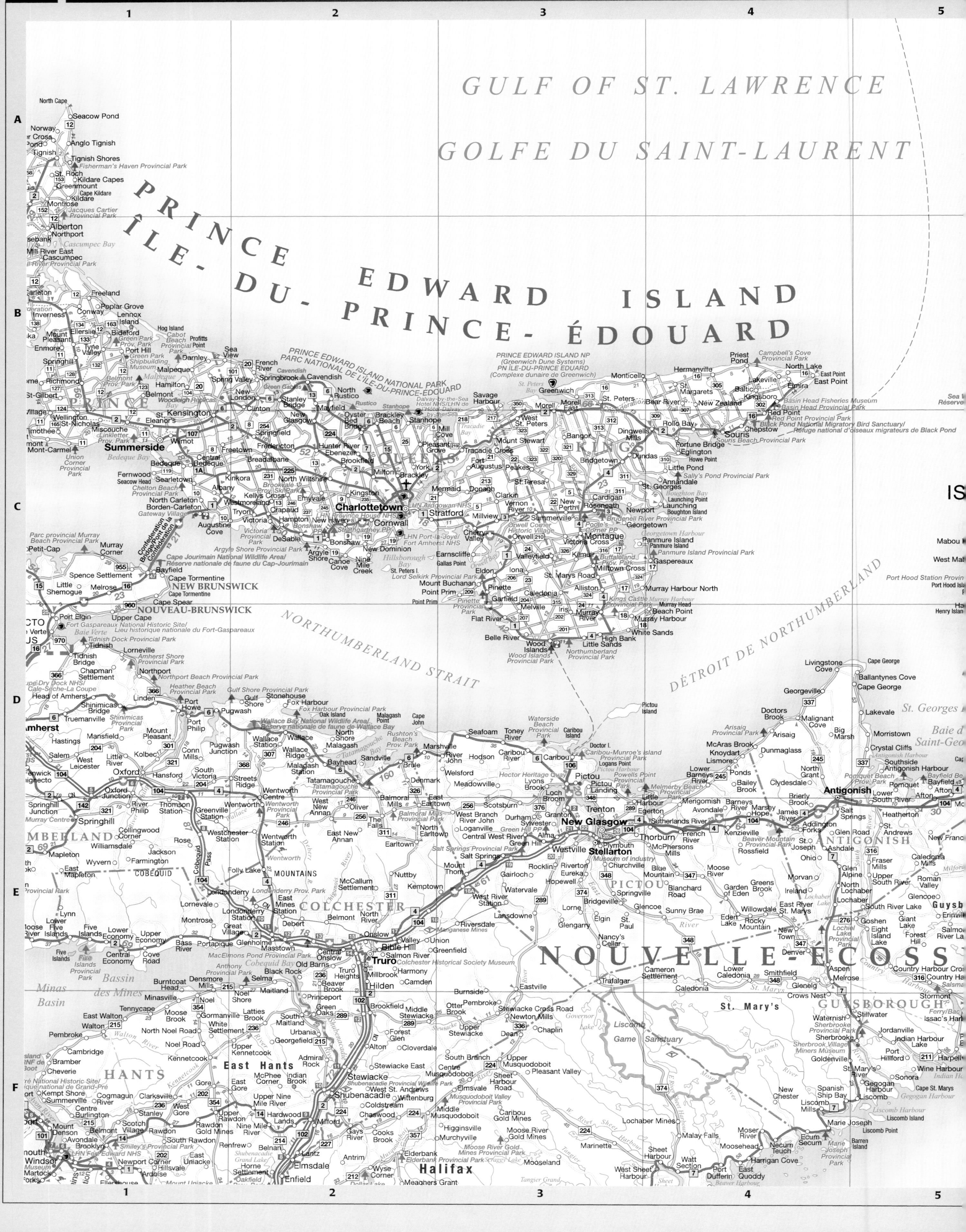

GULF OF ST. LAWRENCE

GOLFE DU SAINT-LAURENT

PRINCE EDWARD ISLAND
ÎLE-DU-PRINCE-ÉDOUARD

PRINCE EDWARD ISLAND NATIONAL PARK
PARC NATIONAL DE L'ÎLE-DU-PRINCE-ÉDOUARD

PRINCE EDWARD ISLAND NP
(Greenwich Dune Systems)
PN ÎLE-DU-PRINCE ÉDUARD
(Complexe dunaire de Greenwich)

Charlottetown
Summerside
Stratford
Cornwall
Montague
Georgetown
Souris
Kingsboro
East Point
Basin Head Provincial Park
Black Pond National Migratory Bird Sanctuary /
Refuge national d'oiseaux migrateurs de Black Pond

NEW BRUNSWICK
NOUVEAU-BRUNSWICK

NORTHUMBERLAND STRAIT

DÉTROIT DE NORTHUMBERLAND

Confederation Bridge / Pont de la Confédération

Wood Islands
Northumberland Provincial Park

Cape George
Ballantynes Cove
Georgeville
Cape George

Amherst
Oxford
Springhill
Pugwash
Wallace
Tatamagouche
River John
Pictou
New Glasgow
Stellarton
Westville
Trenton
Antigonish

COBEQUID MOUNTAINS

Truro
Bible Hill
Onslow
Colchester
Shubenacadie
Stewiacke
Elmsdale
Enfield

Minas Basin
Bassin des Mines

HANTS
East Hants
Windsor

COLCHESTER

PICTOU

ANTIGONISH

GUYSBOROUGH

NOUVELLE-ÉCOSSE

St. Mary's

Liscomb Game Sanctuary

Sherbrooke
Sheet Harbour

Halifax

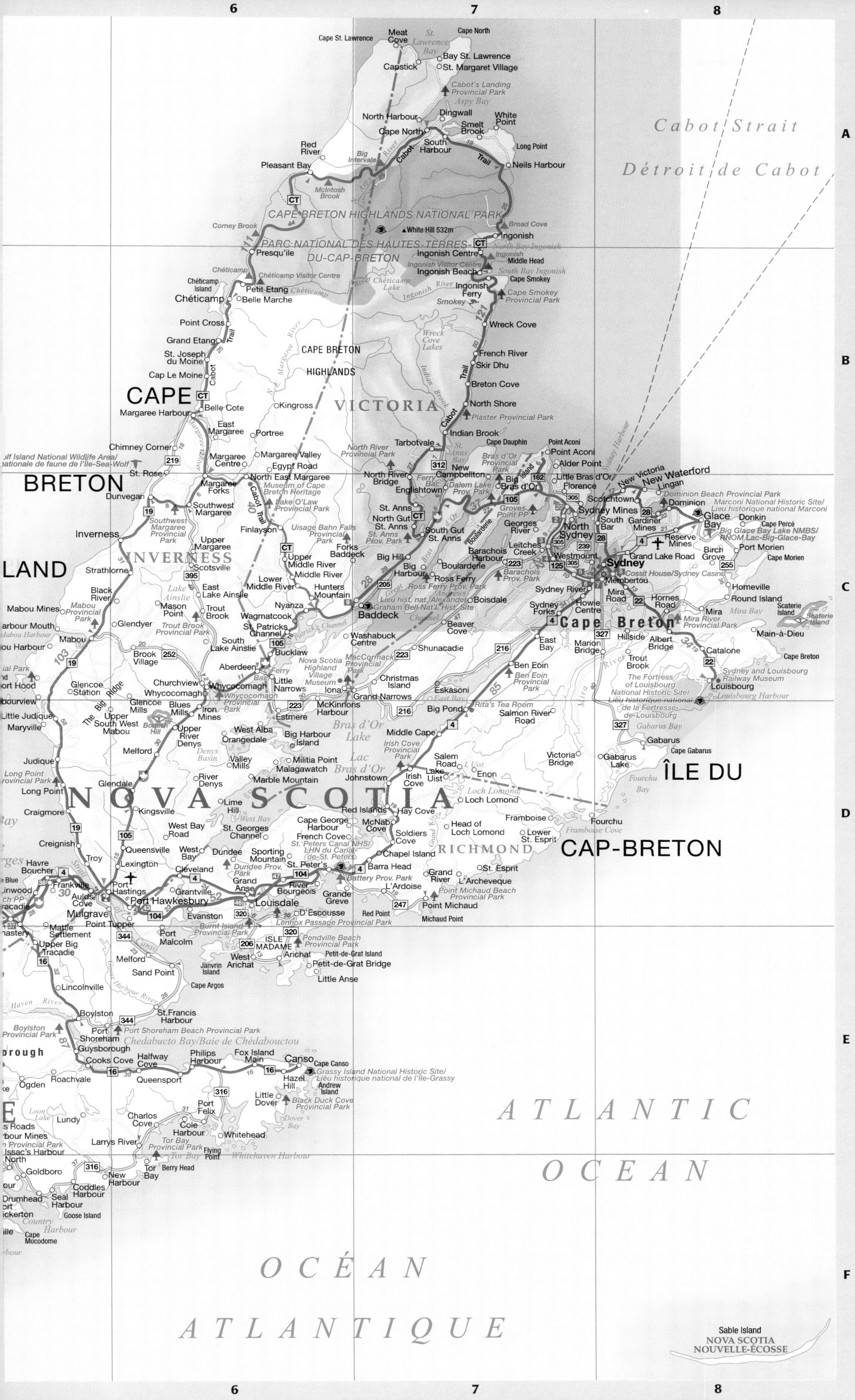

THE HIGHLANDS OF CAPE BRETON

This view sweeps across northeastern Cape Breton Island. Along the curve of the coast facing Cabot Strait lie these sites (counterclockwise, top to bottom): Cape North, the island's uppermost tip; Aspy Bay, just north of the site of Cape Breton Highlands National Park; North and South Ingonish bays (separated by a small peninsula); St. Anns Bay; Great Bras d'Or and St. Andrews channels, which lie north and south of Boularderie Island. The industrial city of Sydney (bottom right) is visible at the east side of Sydney Harbour.

Other points of interest

■ Bras d'Or Lake is an arm of the Atlantic Ocean, covering 1,098 km² and almost dividing Cape Breton Island in two. At its north end, Great Bras d'Or and St. Andrews are natural channels linking "the lake" to the Atlantic Ocean. At the south end, the St. Peter's Canal is another—man-made—channel that cuts through a kilometre-wide isthmus. Built between 1854 and 1869, the ship continues to provide a shortcut through Cape Breton Island that is safer than the route around the island.

■ Sable Island [F8], 35 km long and 1.6 km wide, is an exposed tip of the eastern continental shelf. Wind and water constantly change the shape of its shifting sandy terrain. The island is famed for its bands of wild horses, whose numbers vary from 175 to 450. It was thought the horses were shipwreck survivors, but recent evidence suggests they may have arrived during an ill-advised attempt to create an island settlement. Lying 300 km east of Halifax, the island is now the focus of offshore natural gas exploration.

Scale 1:800,000
1 cm = 8 km
0   10   20   30 km

1   2   3   4   5

A

North Cape
North Cape Interpretive Centre/ Aquarium
Elephant Rock
Seacow Pond
Nail Pond
Norway 161
Christopher Cross
Skinners Pond 14
Ascension 160 159
Anglo Tignish
Waterford 156 158
Judes Point
Peterville
Leóville 157
Tignish Shore
St. Felix
Cape Gage 14
Palmer Road 155 153
St. Roch
Miminegash 152 St. Louis
Greenmount
Kildare Capes
Kildare Capes Red Sandstone Cliffs
St. Edward 152 162
Cape Kildare
Mimingash Pond 154 Kildare
St. Lawrence 151 Alma Montrose
Jacques Cartier Provincial Park

B

Roseville Huntley 152
Elmsdale 150 152
Campbellton Brockton Alberton
Alberton Museum & Genealogy Centre
Piusville 148 149 Alberton Harbour
Burton 14 Mill River East Rosebank 152
Bloomfield 145 Northport (Alberton South)
Bloomfield Corner 146 172 Cascumpec Sand Hills
Seal Point 143 Mill River Provincial Park 136 Cascumpec
Cape Wolfe 146 Bloomfield Provincial Park 142 Cascumpec Bay
147 144 Howlan 143 Foxley Bay
O'Leary Museum
West Cape 14 176 Knutsford 142 O'Leary Unionvale 12
Springfield West 140 Carleton 137 Roxbury
Milburn 148 Coleman 168 Foxley River
141 139 Glenwood Milo Freeland 174
West Point Lighthouse Museum 14 176 164 Derby 175 Conway 173
West Point 170 Brae Harbour 138 West Devon 12 Poplar Grove
Indian Point Sand Hills North Enmore 134 McNeills Mills

PRINCE

C

Cedar Dunes Provincial Park
Alaska
Egmont Bay
Grande Digue Point
Enmore 133 Bideford
Ellerslie Lennox Island
Mi'kmaq Cultural Centre
Green Park Provincial Park
Shipbuilding Museum & Yeo House
Mossy Point Mount Pleasant 169 Tyne Valley 12 Port Hill
Saint-Chrysostome 135 Springhill Northam 131 Red Point
Victoria West 130 132 Birch Hill
Harmony 167 Arlington
Saint-Philippe 127 Bayside Malpeque Bay
Baie- 126 Richmond 131 South West Lot 16
Egmont 125 Saint-Gilbert Provincial Park
Abram-Village 124 179 Central Lot 16
Urbainville 165 Wellington Belmont
Maximeville 11 Saint-Raphaël Acadian Museum/ Le musée acadien
Saint-Timothée 177 Saint-Nicholas Bentick Cove
Cap-Egmont 11 Miscouche St. Eleanors
Mont-Carmel 165 Sherbrooke
Acadian Pioneer Village Linkletter 121 New Annan
Cap-Egmont Mont-Carmel Provincial Park Linkletter Provincial Park
Sunbury Point Wilmot Valley
Union Corner Provincial Park Summerside
Union Corner MacCallums Point

D

Cap-de-Cocagne
Cap de Cocagne
Cap-des-Caissie/ Caissie Cape
Bourgeois
de-Digue
Pointe de Grande-Digue Point
de Shediac Island
NEW BRUNSWICK
Baie de Shediac Bay
Parc provincial Parlee Beach Provincial Park
Cap-Brûlé
Boudreau
Barachois
Robichaud 133
Ohio-du- Barachois 15
Gallant Settlement
Cap-Pelé Cape
Bourgeois Mills
Cormier- Village
St-André- de-Shediac
Leblanc 15
Pointe Fagan Point
Cap-Pelé
Trois- Ruisseaux
Botsford
Petit- Cap
NOUVEAU- BRUNSWICK
Haute- Aboujagane Basse- Aboujagane
Shemogue 940
Petit- Cap
Cap Shemogue Head
Pointe Cadman Point
Havre de Little Shemogue Harbour
Anderson Settlement
Mates Corner
Chapmans Corner 955
Cadman Corner
Parc provincial Murray Beach Provincial Park
Murray Corner
Murray River
Île Jourimain Island
Confederation Bridge/ Pont de la Confédération
Île-du-Prince-Édouard

North Enmore
Salutation Cove
Graham Head
North Bedeque
Lower Bedeque
Fernwood 119 Bedeque
Seacow Head
Chelton
Chelton Beach Provincial Park

WESTMORLAND

Little Shemogue
Woodside
Spence Settlement 16
Bayfield
Cape Tormentine
Melrose 960 Cape/Cap Tormentine
Cape Spear
Cape/Cap Spear

E

Bourgeois Mills
Haute-Aboujagane
Cookville
Centre Village
Timber River 970 Hardy
Port Elgin 960 Upper Cape
Coburg 16 Bayside
Brooklyn Road
Baie Verte Road 970
Fort Gaspareaux National Historic Site/ Lieu historique national du Fort-Gaspareaux
Baie Verte Tidnish Head
Tidnish Dock Provincial Park
Carleton Borden
Cape Traverse Amherst Cove
North Tryon Augustine Cove
Tryon Crapaud
Victoria Richard Point
Tryon Head
Passage Abegweit Passage
Passage de la Pointe
Charlottetown
Cornwall
Stratford

F

Woodhurst
Mount View 940
Midgic
Ward
Fairfield
Cherry Burton
Upper Sackville
Middle Sackville
Mount/ Allison University
Ogden Mill
Frosty Hollow
Sackville
Westcock
Aulac 935
Mount Whatley
Fort Beauséjour National Historic Site/ Lieu historique national du Fort-Beauséjour
Jolicure
Upper Point de Bute
Point de Bute
La Coupe Dry Dock National Historic Site
Lieu historique national de La Coupe-Sèche-De-La-Coupe
CUMBERLAND
Brooklyn
Baie Verte
Parc linéaire Tantramar Rail Trail
Tintamarre National Wildlife Area/ Réserve national de faune de Tintamarre
Rafts Hill
Tidnish Bridge
Beecham Head
Tidnish 366 Seagrove
Lorneville
Amherst Shore Amherst Shore Provincial Park
Chapman Settlement
Coldspring Head
NOVA SCOTIA
Lower Shinimicas
Amherst Head
Shinimicas Bridge
Linden
West Linden
East Linden 366
Northport
Northport Beach Provincial Park
Birch Head
NOUVELLE-ÉCOSSE
Heather Beach Provincial Park
Gulf Shore Provincial Park
Upper Gulf Shore
Lower Gulf Shore
Pugwash Harbour

GULF OF ST. L...
GOLFE DU SAIN...

PRINCE ÎLE- DU- EDWAR... PRINC...

PRINCE EDWARD ISLAND NATIONAL PARK
PARC NATIONAL DE L'ÎLE-DU-PRINCE-ÉDOUARD

Springbrook
Cavendish
Green Gables House/ Maison Green Gables
North Rustico
New London
Stanley Bridge
L.M. Montgomery Heritage Museum
French River
Long River
Margate
Clinton
Kensington
New Glasgow
Stanhope
Brackley Beach
Covehead
Oyster Bed Bridge
Wheatley River
Harrington
Brookfield
Winsloe
Marshfield
Charlottetown
Cornwall
Stratford
Rocky Point
Port-la-Joye/ Fort Amherst National Historic Site
Lieu historique national de Port-la-Joye/Fort-Amherst
New Dominion
Nine Mile Creek
Rice Point
Canoe Cove
Argyle Shore Provincial Park
Hillsborough Bay
DÉTROIT DE NORTHUMBERLAND

NORTHUMBERLAND STRAIT
Bedeque Bay

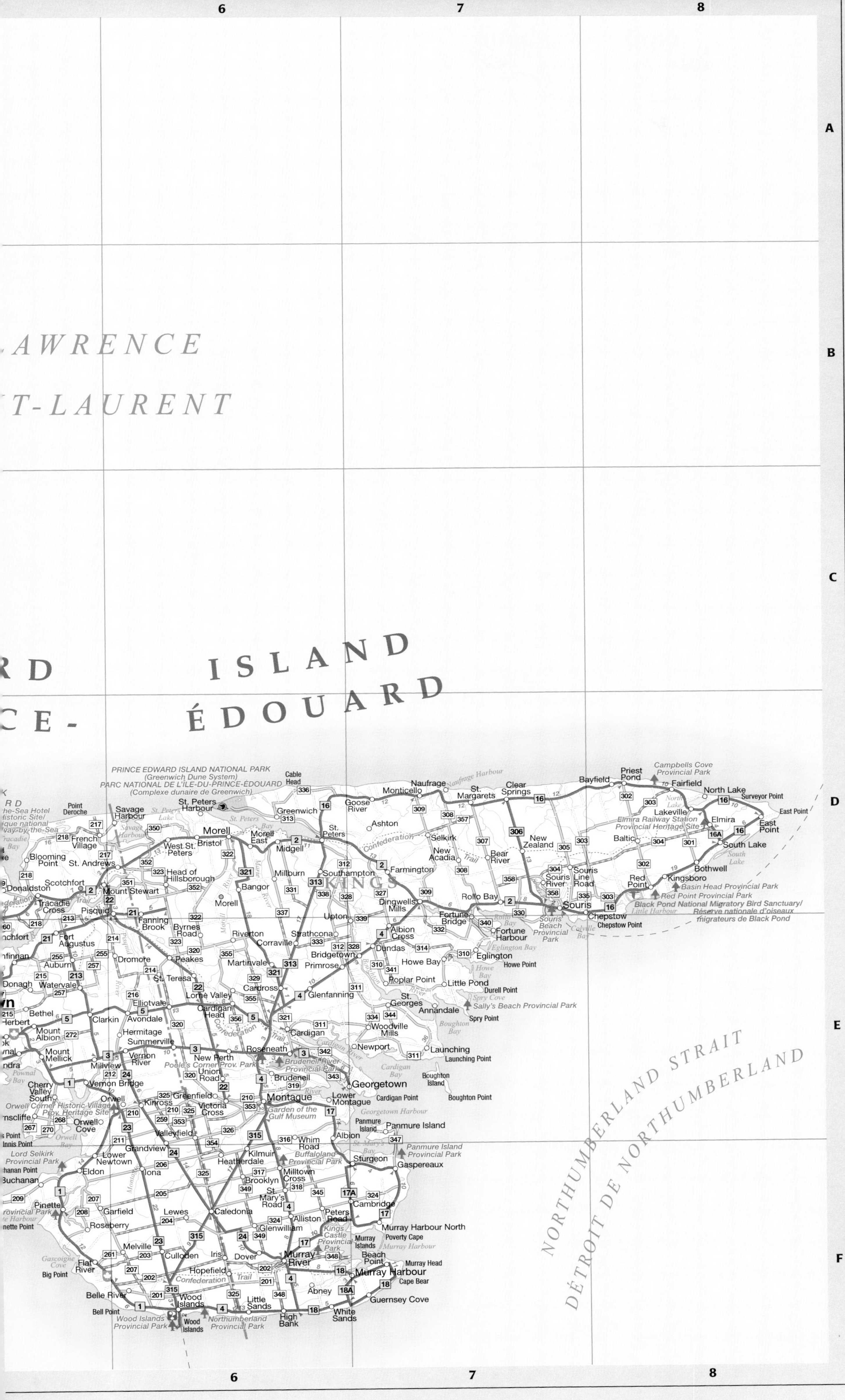

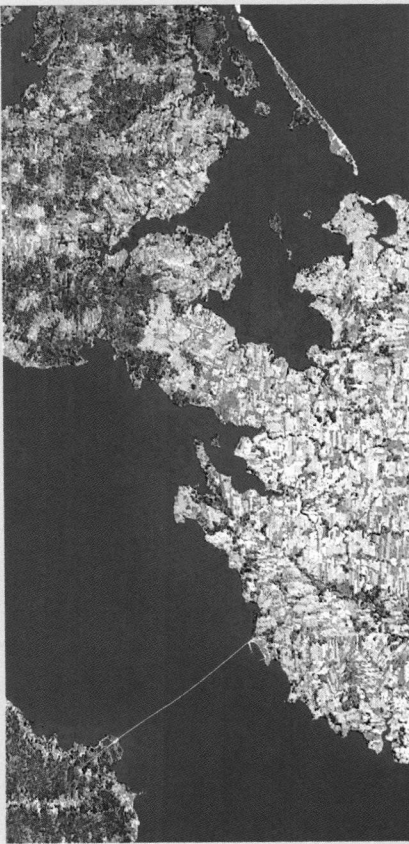

THE BAYS AND THE BRIDGE

The capes and coves of Malpeque Bay lie north of the isthmus connecting Prince (left) and Queens (right) counties on Prince Edward Island. At the entrance to the bay, Hog Island acts as a slender barrier against the shivery waters of the Gulf of St. Lawrence. On the south side of the isthmus is Summerside, P.E.I.'s second largest community, which is visible on Bedeque Bay. The patchwork of farms is a reminder that P.E.I. is Canada's smallest, most densely populated province, with the highest proportion of the people still living on farms. What looks like a slender thread stretched across Northumberland Strait is the 13-km-long multispan Confederation Bridge linking P.E.I. and New Brunswick.

Other points of interest

■ Prince Edward Island National Park [D4-D6], a mere 18 km², preserves a 40-km coastal strip of sand, bluffs, and wetlands along the province's northern shore. The notable features include red sandstone cliffs—rising up to 30 m—and some of the finest beaches in North America. The deep, stringy roots of marram grass (*Ammophila arenaria*, meaning "sand loving") anchor the fragile dune system against wind and salt spray. Great white heron and piping plover frequent the saltwater marshes and freshwater ponds.

■ Basin Head [D8] is the site of a fisheries museum that recounts the history of inshore commercial fishery in Prince Edward Island.

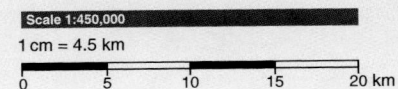

Scale 1:450,000
1 cm = 4.5 km
0   5   10   15   20 km

Map labels:

LAWRENCE
T-LAURENT

ISLAND
CE-   ÉDOUARD

KINGS

NORTHUMBERLAND STRAIT
DÉTROIT DE NORTHUMBERLAND

PRINCE EDWARD ISLAND NATIONAL PARK
(Greenwich Dune System)
PARC NATIONAL DE L'ÎLE-DU-PRINCE-ÉDOUARD
(Complexe dunaire de Greenwich)

Cable Head, Campbells Cove Provincial Park, Priest Pond, Fairfield, Bayfield, North Lake, Surveyor Point, Naufrage, Monticello, St. Margarets, Clear Springs, Red Point, Lakeville, Elmira, East Point, South Lake, Bothwell, Kingsboro, Basin Head Provincial Park, Black Pond National Migratory Bird Sanctuary/Réserve nationale d'oiseaux migrateurs de Black Pond, Souris, Chepstow, Rollo Bay, Fortune Bridge, Albion Cross, Dundas, Howe Bay, Eglington, Little Pond, Durell Point, St. Georges, Annandale, Boughton Bay, Spry Point, Sally's Beach Provincial Park, Woodville Mills, Launching, Cardigan, Newport, Brudenell River Provincial Park, Georgetown, Boughton Island, Panmure Island Provincial Park, Gaspereaux, Sturgeon, Cambridge, Murray Harbour North, Poverty Cape, Murray Harbour, Cape Bear, Guernsey Cove, White Sands, High Bank, Wood Islands Provincial Park, Belle River, Bell Point, Big Point, Flat River, Culloden, Hopefield, Abney, Murray River, Little Sands, Dover, Iris, Melville, Garfield, Rosaberry, Caledonia, Lewes, Glenwilliam, St. Mary's Road, Alliston, Peters Road, Brooklyn, Milltown Cross, Heatherdale, Kilmuir, Iona, Eldon, Lower Newtown, Grandview, Valleyfield, Albion, Whim Road, Victoria Cross, Montague, Lower Montague, Cardigan Point, Greenfield, Kinross, Orwell, Vernon River, Vernon Bridge, New Perth, Roseneath, Rosebank, Brudenell, Union Road, Millview, Summerville, Hermitage, Avondale, Clarkin, Elliotvale, Cardigan Head, Lorne Valley, Glenfanning, Cardross, St. Teresa, Peakes, Dromore, Auburn, Watervale, Donagh, Fort Augustus, Byrnes Road, Corraville, Riverton, Strathcona, Bridgetown, Primrose, Martinvale, Poplar Point, Dingwells Mills, Upton, Morell, Bangor, Head of Hillsborough, West St. Peters, Bristol, Morell East, Midgell, St. Peters, Greenwich, Goose River, Ashton, Selkirk, New Acadia, Bear River, Farmington, Southampton, Millburn, New Zealand, Baltic, Souris Line Road, Savage Harbour, Pisquid, Fanning Brook, Mount Stewart, Tracadie Cross, Scotchfort, St. Andrews, Donaldston, Blooming Point

GULF OF
ST. LAWRENCE

GOLFE DU
SAINT-LAURENT

NEWFOU
AND LA
TERRE-
ET- LA

St. George's Bay

LONG RANGE MOUNTAINS

ANGUILLE MOUNTAINS

ATLANTIC

OCEAN

OCÉAN

ATLANTIQUE

Cabot Strait
Détroit de Cabot

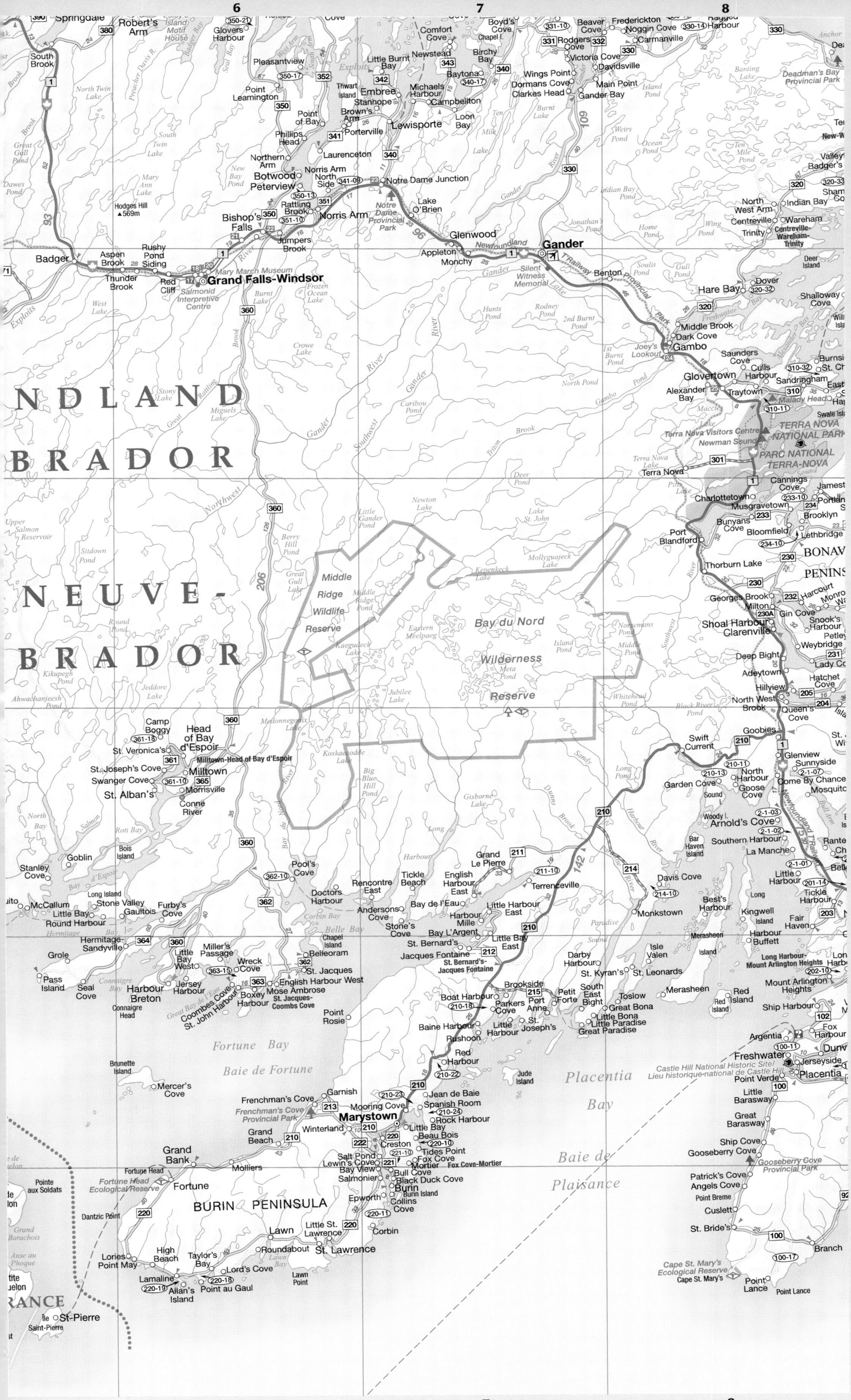

A GATEWAY TO THE ISLAND

Southwestern Newfoundland is a windswept land of coastal barrens, bold headlands, placid wilderness lakes, and densely forested mountains. Channel-Port aux Basques (bottom) connects the island to the rest of Canada by ferry (via North Sydney, Nova Scotia) and by highway. The Trans-Canada Highway starts here and continues to St. John's some 850 km away. The Long Range Mountains (right bottom), part of the Appalachian mountain system (see also Plates 29, 30, and 38), run the full length of western Newfoundland. Cape Anguille (bottom left) is the westernmost point on the island of Newfoundland.

Other points of interest

■ At 344 km², Grand Lake [B3-A4] is the largest lake on the island of Newfoundland. It is accessible from Deer Lake [A3] by way of T'Railway Provincial Park, Newfoundland's portion of the Trans Canada Trail.

■ At 2,895 km², Bay du Nord Wilderness Reserve has ample roaming room for a herd of 15,000 caribou, the largest on the island of Newfoundland. The reserve encompasses a variety of environments: bogs, fens, and barrens, and forests of black spruce, balsam fir, and trembling aspen. It can be accessed by challenging canoe routes and hiking trails. (A provincial travel permit required.) The Bay du Nord River, flowing south into Fortune Bay, is a Canadian Heritage River, whose rapids and lower reaches require expert skill.

■ Fortune [F6] provides a ferry service to the French-controlled islands of Saint-Pierre and Miquelon [E5-F5].

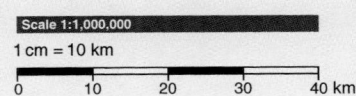

Scale 1:1,000,000
1 cm = 10 km

0   10   20   30   40 km

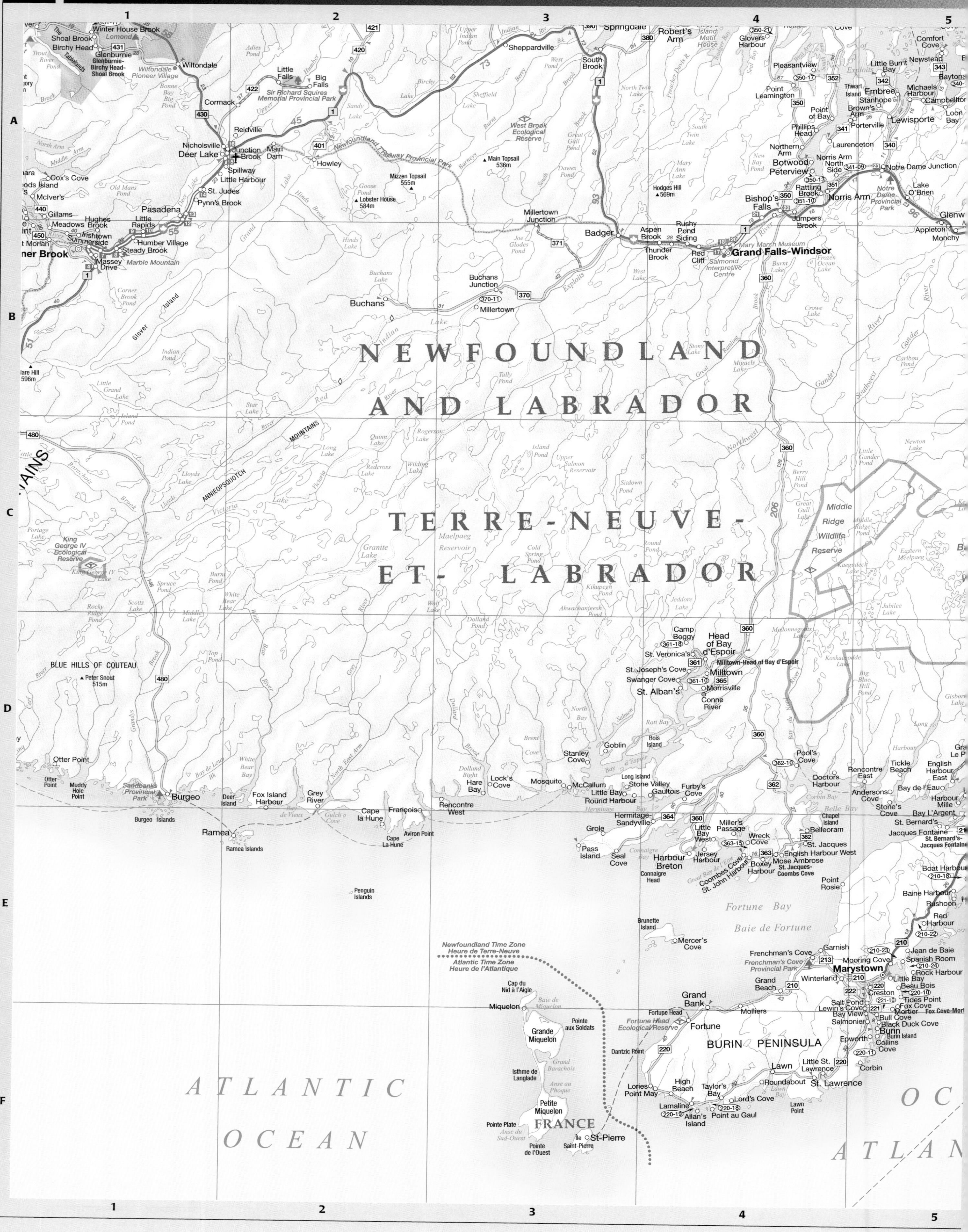

NEWFOUNDLAND AND LABRADOR

TERRE-NEUVE- ET- LABRADOR

ATLANTIC OCEAN

FRANCE

BURIN PENINSULA

Fortune Bay
Baie de Fortune

Newfoundland Time Zone
Heure de Terre-Neuve
Atlantic Time Zone
Heure de l'Atlantique

Grande Miquelon

Petite Miquelon

St-Pierre

Deer Lake
Corner Brook
Pasadena
Grand Falls-Windsor
Lewisporte
Botwood
Bishop's Falls
Buchans
Millertown
Marystown
Grand Bank
Fortune
Burgeo
Ramea
Harbour Breton
St. Alban's
Head of Bay d'Espoir

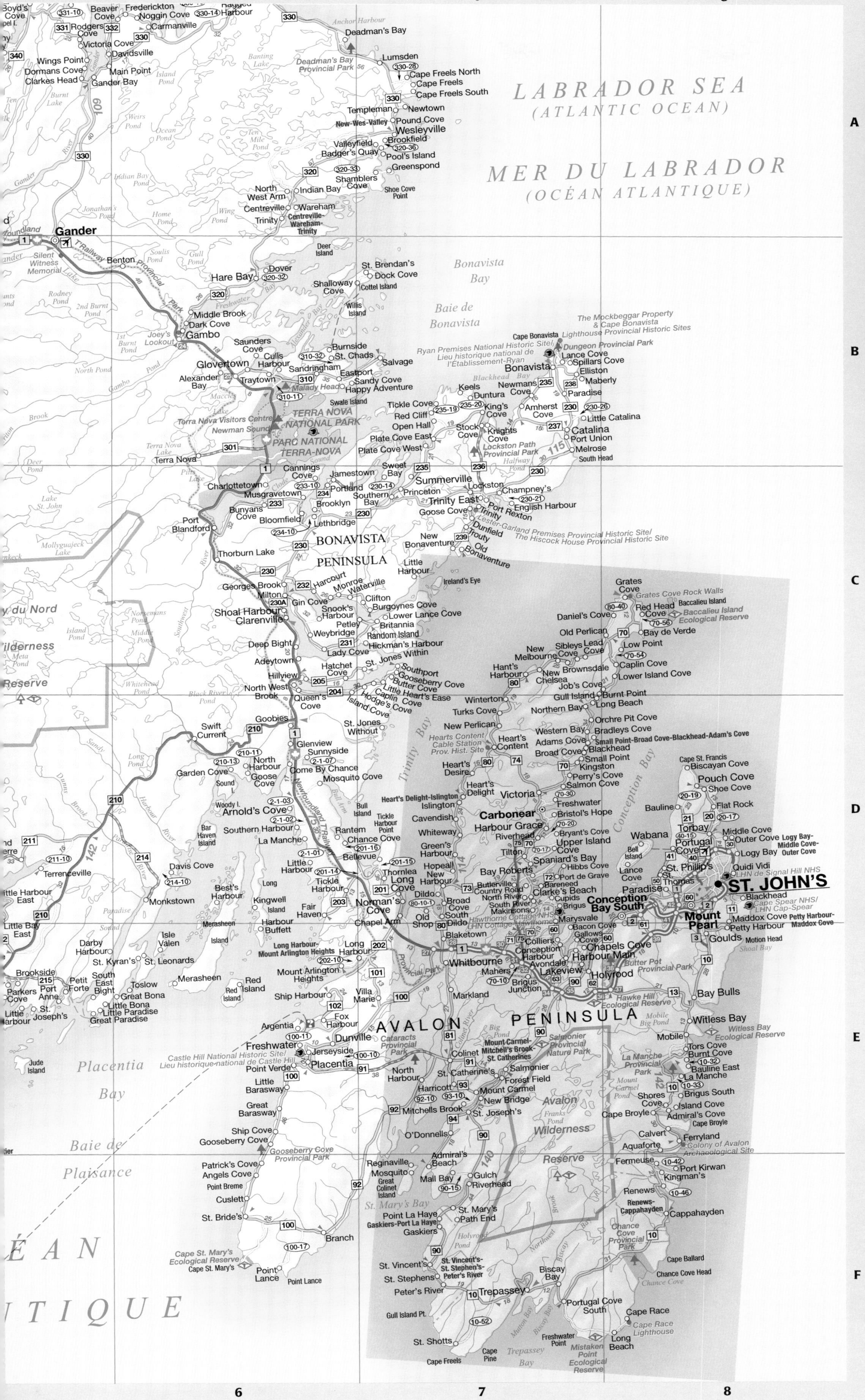

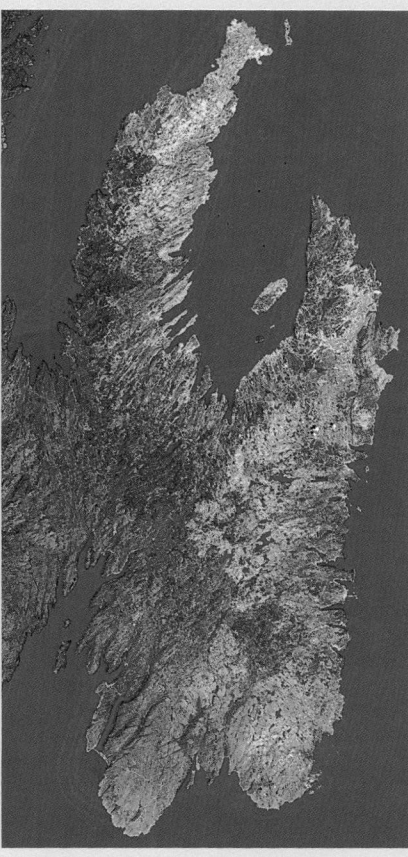

THE AVALON PENINSULA

The bulky coastal cliffs of Newfoundland's 10,360-km² Avalon Peninsula confront the cold, turbulent waters of the North Atlantic. Four capacious bays (clockwise, top to bottom)—Trinity, Conception, Trepassey, and St. Mary's—shelter myriad tiny fishing ports from the unruly ocean. The narrow, almost landlocked, harbour of St. John's can be detected just above the bulge of Cape Spear, Canada's easternmost point (upper right).

Other points of interest

■ The natural features of 296-km² Terra Nova National Park [B6-C6] include towering headland and tranquil beaches along the coast of Bonavista Bay. Inland are gently rolling hills and forests with ponds, streams, and bogs. The landscape of Terra Nova—Canada's most easterly national park—was created by ice-age glacial activity. Until midsummer, icebergs borne along by the icy Labrador Current drift offshore in the Atlantic's "Iceberg Alley."

■ Cape St. Mary's Ecological Reserve [F6], at the southwest tip of the Avalon Peninsula, is accessible by a 16-km road from Route 100. The outstanding natural attraction is 76-m-high Bird Rock, which is linked to the shore by a land bridge. More than 50,000 seabirds—gannets, murres, and kittiwakes—gather on the mighty rock and the coastal cliffs.

Scale 1:1,000,000

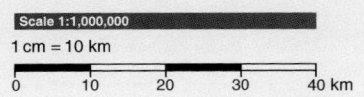

1 cm = 10 km

0   10   20   30   40 km

LABRADOR SEA
(ATLANTIC OCEAN)

MER DU LABRADOR
(OCÉAN ATLANTIQUE)

Bonavista Bay

Baie de Bonavista

TERRA NOVA NATIONAL PARK
PARC NATIONAL TERRA-NOVA

BONAVISTA PENINSULA

AVALON PENINSULA

Placentia Bay
Baie de Plaisance

Trinity Bay

Conception Bay

Trepassey Bay

St. Mary's Bay

Gander

St. John's
Mount Pearl
Conception Bay South

Bonavista

Carbonear
Harbour Grace

Placentia

Trepassey

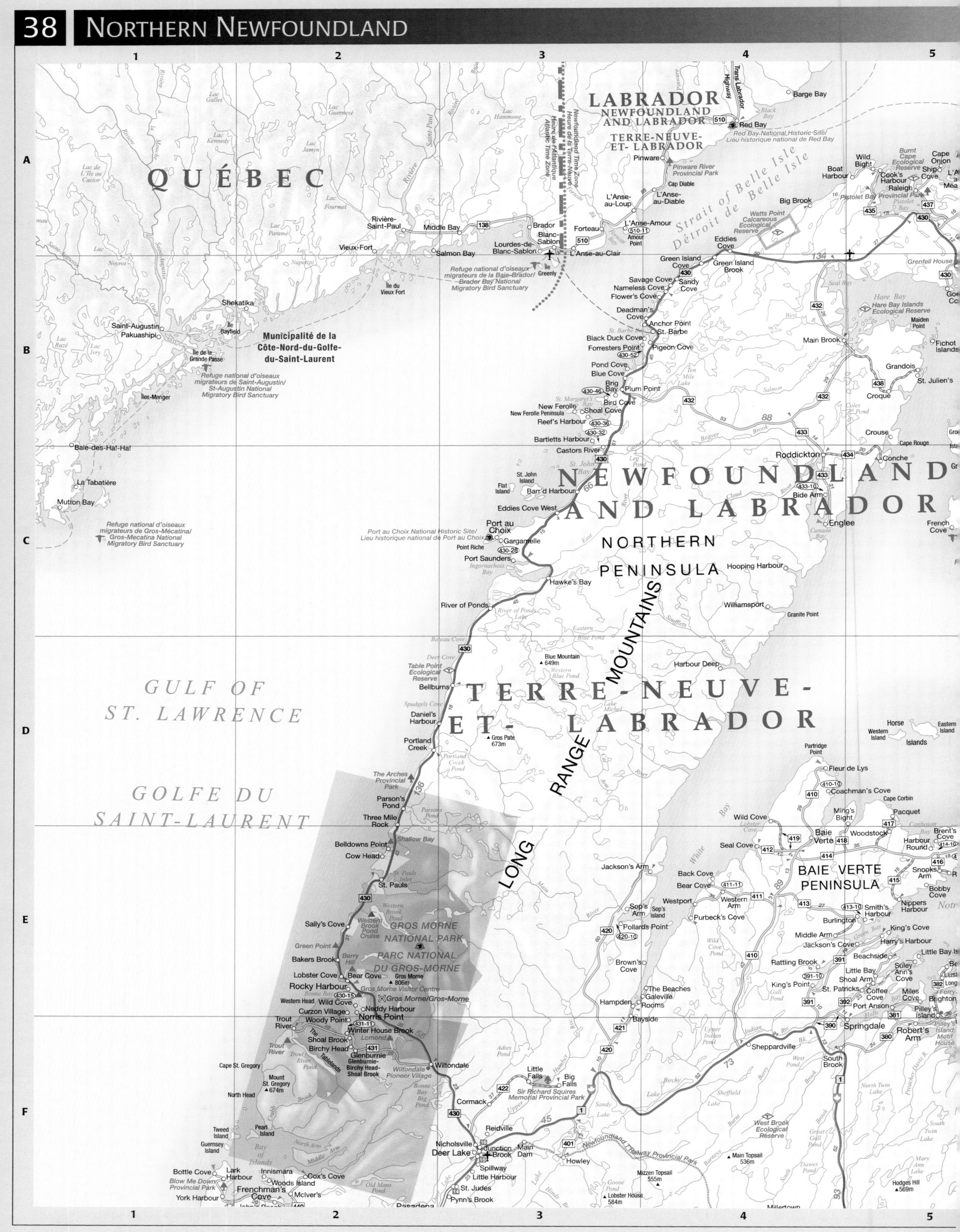

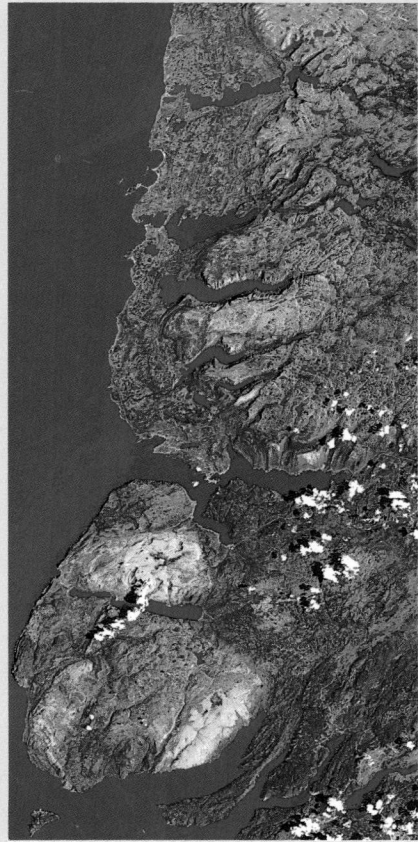

6 **7** **8**

A

L'Anse
aux
Meadows
Quirpon
Island
L'Anse aux Meadows National Historic Site/
Lieu historique national de L'Anse aux Meadows
Straitview
Quirpon
436-11
Gunners Cove
Griquet
St. Lunaire-Griquet
St. Lunaire
436
Little Brehat
Great Brehat
330-75
St. Anthony Bight
330-76
St. Carols
St. Anthony

B

LABRADOR SEA
(ATLANTIC OCEAN)

Islands

MER DU LABRADOR
(OCEAN ATLANTIQUE)

Bell
Island
Shepherd Island National Migratory Bird Sanctuary/
Refuge national d'oiseaux migrateurs de l'Île-Shepherd
Île aux Canes National Migratory Bird Sanctuary/
Refuge national d'oiseaux migrateurs de l'Île-aux-Canes

C

D

Cape St. John
La Scie
Manful
Bight
414-13
Shoe Cove
14-12
Tilt Cove
ound Harbour

e Dame
Bay

Baie de
Notre-
Dame

Little
Fogo
Islands

Funk Island
Ecological
Reserve
Funk Island

Barr'd
Islands
Joe Batt's Arm-Barr'd Islands-Shoal Bay
Fogo Head
Joe Batt's Arm
Sandy Cove
Fogo
334
Tilting
Crow Head
N. Twillingate I.
Twillingate I.
Change
Islands
Hare
Bay
Deep
Bay
Shoal
Bay
Cape Fogo
Wild Cove
Back Harbour
Merritt's
Harbour
Pikes Arm
335-14
Fogo
Island
Cape Cove
Twillingate
340
Kettle
Cove
Togood Arm
340-3
330-18
Seldom
Fogo Island Region
Bayview
345
346
Newville
Cobbs Arm
Island
Harbour
330-21
Little Seldom
Tizzard's Harbour
New World
333
Valley Pond
Fairbank
Island
Moreton's Harbour
Hillgrade
Wadham Islands
Bridgeport
Chanceport
Carter's Cove
Farewell
Stag
Harbour
Exploits
344
Virgin Arm
Port Albert
Hamilton
Triton
380-16
Jim's
Cove-
Card's Harbour
Luke's Arm
Cottle's
Island
Summerford
335
Sound
Ladle
Cove
Card's Harbour
Fleury
Bight
Fortune
Harbour
Boyd's
Cove
Stoneville
340
Aspen
Cove
Musgrave Harbour
Jim's Cove
Leading
Tickles
Islands
Chapel I.
331-10
Frederickton
Doting Cove
Glovers
Harbour
Comfort
Cove
Rodgers
Cove
Beaver
Cove
Noggin Cove
Carmanville
330-14
Ragged
Harbour
330
Newstead
331
332
Pleasantview
Little Burrit
343
Birchy
Bay
Victoria Cove
Deadman's Bay
350-17
352
Thwart
Island
342
Baytona
340-17
Davidsville
Anchor Harbour
Point
Leamington
350
Michaels
Harbour
Stanhope
340
Wings Point
Dormans Cove
Main Point
Deadman's Bay
Provincial Park 56
Lumsden
330
Embree
Campbellton
Clarkes Head
Gander Bay
Island
Pond
330-28
Cape Freels North
Brown's
Arm
26
Loon
Bay
Banting
Lake
Cape Freels
Phillips
Head
341
Porterville
Lewisporte
Ten
Mile
Lake
Burnt
Lake
Ocean
Pond
330
Cape Freels South
Templeman
Newtown
340
New-Wes-Valley
Pound Cove
Wesleyville
Laurenceton
Weirs
Pond
Brookfield
Northern
Arm
Botwood
Norris Arm
North
Side
341-09
Notre Dame Junction
River
330
Ten
Mile
Pond
Valleyfield
Badger's Quay
320-36
Pool's Island
Peterview
351
Indian Bay
Pond
Greenspond
Rattling
Brook
350-13
Gander
320
Shamblers
Cove
320-33
Bishop's
Falls
350
351-10
Norris Arm
Notre
Dame
Provincial
Park
Lake
O'Brien
Jonathan's
Pond
Home
Pond
North
West Arm
Centreville
Shoe Cove
Point
New
Bay
Pond
Glenwood
Wing
Pond
Indian Bay
Cove
Wareham
Centreville-
Wareham-
Trinity

E

F

6 **7** **8**

A WORLD HERITAGE SITE

Along western Newfoundland's Northern
Peninsula, the Long Range Mountains rise
dramatically above a low coastal plain.
The coastline is deeply indented with fjords
and bays (top to bottom): Parsons Pond,
St. Pauls Inlet, Western Brook Pond, and
Bonne Bay. Gros Morne National Park,
covering 1,943 km², encircles Bonne Bay.
Its terrain was formed by continental collision 450 million years ago, but its present
landscape was shaped by ice-age glaciers.
Among its sites are Western Brook Pond
(centre)—a fjordlike lake—and 806-m-high
Gros Morne Mountain, visible above the
point where Bonne Bay splits into two
arms. Because of its intriguing geological
history and formations, the park has been
recognized as a World Heritage Site.

Other points of interest

■ L'Anse aux Meadows National
Historic Site [A5] preserves the remains
of a Viking settlement, established here
about A.D. 1000. The site contains eight
structures discovered by a team of
Norwegian archeologists in 1960.
L'Anse aux Meadows was declared
a World Heritage Site in 1978.

■ At St. Anthony [B5], the Grenfell
Hospital Mission serves the remote
ports of northern Newfoundland and
southern Labrador. The Grenfell
House Museum recounts the work
of Dr. Wilfred Grenfell, who founded
the mission in 1893.

■ Port au Choix Historic Site [C3] was
unearthed in the late 1960s. The site
is a reminder of the "red paint people"
who lived in this vicinity some 5,000
years ago. Relics belonging to these
aboriginal peoples are displayed here.

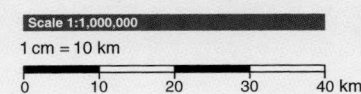

Scale 1:1,000,000
1 cm = 10 km

0   10   20   30   40 km

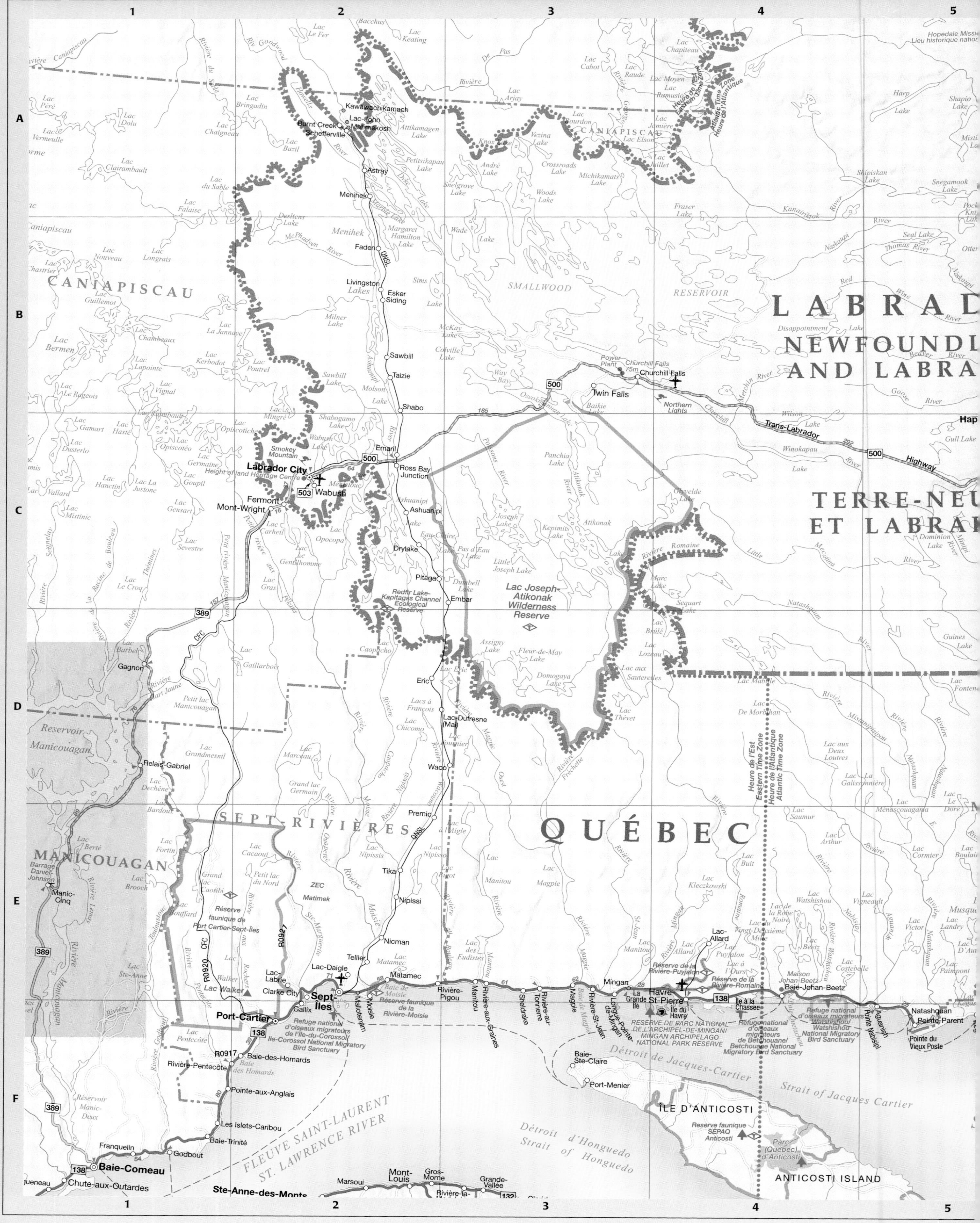

LABRADOR

NEWFOUNDLAND AND LABRADOR

TERRE-NEUVE ET LABRADOR

QUÉBEC

CANIAPISCAU

SEPT-RIVIÈRES

MANICOUAGAN

ÎLE D'ANTICOSTI

ANTICOSTI ISLAND

Kawawachikamach, Lac-John, Matimekosh, Burnt Creek, Schefferville, Astray, Menihek, Faden, Livingston Lakes, Esker Siding, Sawbill, Taizie, Shabo, Molson Lake, Emarii, Labrador City, Height of Land Heritage Centre, Fermont, Mont-Wright, Wabush, Ross Bay Junction, Ashuanipi, Drylake, Pitaga, Embar, Eric, Lac-Dufresne (Mai), Waco, Premio, Tika, Nipissi, Nicman, Tellier, Lac-Daigle, Lac-Labrie, Clarke City, Sept-Îles, Gallix, Port-Cartier, Rivière-Pentecôte, Baie-des-Homards, Pointe-aux-Anglais, Les Islets-Caribou, Baie-Trinité, Franquelin, Godbout, Baie-Comeau, Chute-aux-Outardes, Matamec, Rivière-Pigou, Manitou, Sheldrake, Rivière-au-Tonnerre, Magpie, Longue-Pointe-de-Mingan, Mingan, Havre-St-Pierre, La Grande Île, Île du Havre, Baie-Johan-Beetz, Natashquan, Pointe-Parent, Pointe du Vieux Poste, Baie-Ste-Claire, Port-Menier

Churchill Falls, Twin Falls, Northern Lights, Trans-Labrador Highway, Power Plant

Smokey Mountain, Redfir Lake-Kapitagas Channel Ecological Reserve, Lac Joseph-Atikonak Wilderness Reserve

Réserve faunique de Port Cartier-Sept-Îles, Réserve faunique de la Rivière-Moisie, Réserve de la Rivière-Puyjalon, Réserve faunique SEPAQ Anticosti, Parc (Québec) d'Anticosti

Réserve de Parc National de l'Archipel-de-Mingan / Mingan Archipelago National Park Reserve

Refuge national d'oiseaux migrateurs de l'Île-du-Corossol / Île-du-Corossol Island Migratory Bird Sanctuary

Refuge national d'oiseaux migrateurs de Betchouane / Betchouane National Migratory Bird Sanctuary

Refuge national d'oiseaux migrateurs de Watshishou / Watshishou National Migratory Bird Sanctuary

FLEUVE SAINT-LAURENT / ST. LAWRENCE RIVER

Détroit de Jacques-Cartier / Strait of Jacques Cartier

Détroit d'Honguedo / Strait of Honguedo

RESERVOIR Manicouagan, Réservoir Manic-Deux, Barrage Daniel-Johnson, Manic-Cinq, Manic-Trois

Ste-Anne-des-Monts, Marsoui, Mont-Louis, Gros-Morne, Grande-Vallée

Heure de l'Est / Eastern Time Zone — Heure de l'Atlantique / Atlantic Time Zone

Routes: 138, 389, 500, 503

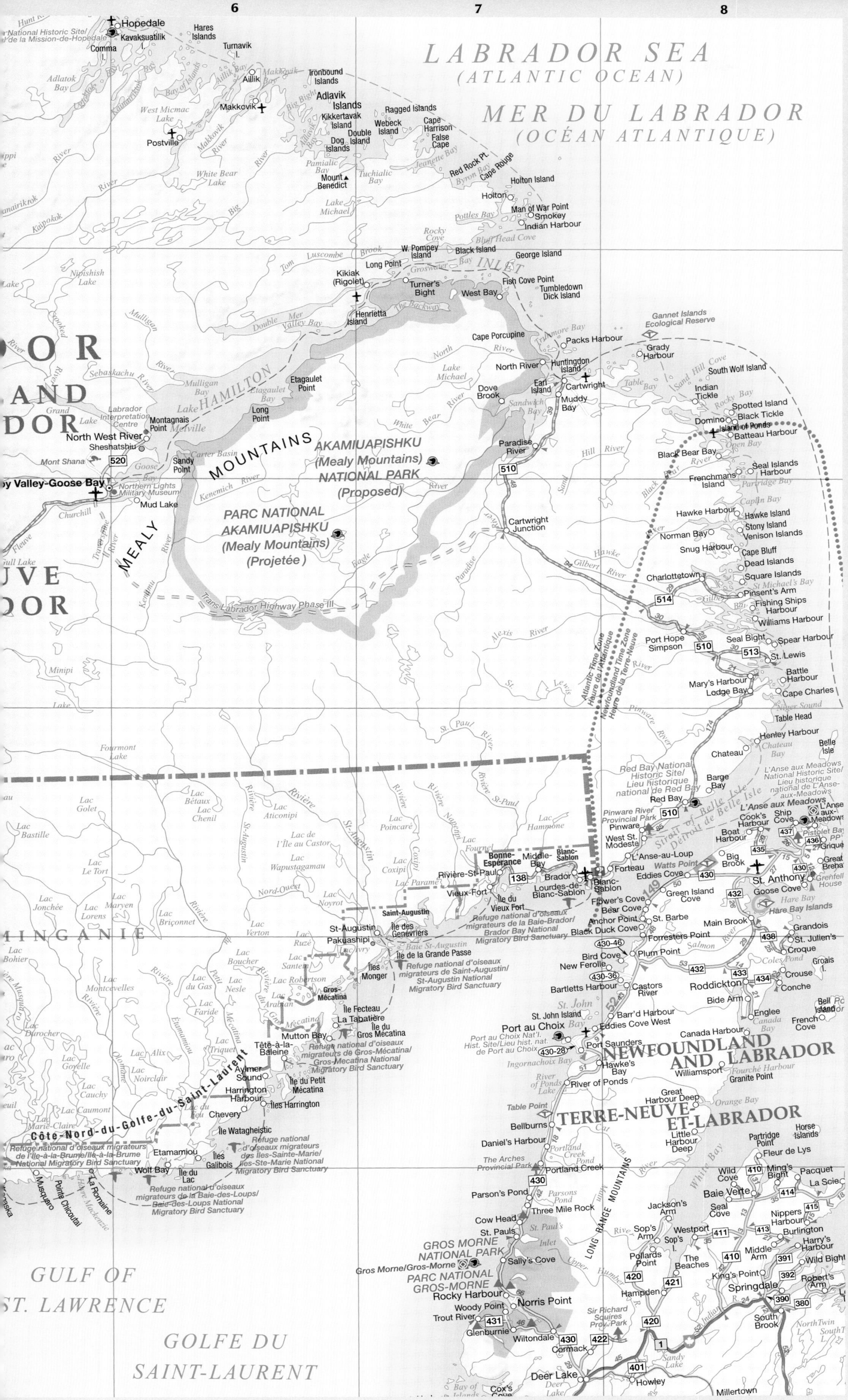

LABRADOR SEA
(ATLANTIC OCEAN)

MER DU LABRADOR
(OCÉAN ATLANTIQUE)

GULF OF
ST. LAWRENCE

GOLFE DU
SAINT-LAURENT

AKAMIUAPISHKU
(Mealy Mountains)
NATIONAL PARK
(Proposed)

PARC NATIONAL
AKAMIUAPISHKU
(Mealy Mountains)
(Projetée)

NEWFOUNDLAND
AND LABRADOR

TERRE-NEUVE
ET-LABRADOR

GROS MORNE
NATIONAL PARK
PARC NATIONAL
GROS-MORNE

A LAKE OF METEORITE ORIGIN

Réservoir Manicouagan, Quebec's second largest natural lake after Lac Mistassini (Plate 40, F5), was created by the impact of a meteorite millions of years ago. At the centre of the lake lies Île René-Levasseur, capped by 952-m-high Mont de Babel. The waters of the lake flow south to the St. Lawrence by way of Rivière Manicouagan (centre left). The hydroelectric development of the river followed the 1971 opening of the Daniel Johnson Dam (Manic 5). The dam, one of the world's largest, is located some 40 km downstream from the lake.

Other points of interest

■ With an area of 6,423 km², Réserve faunique de Port-Cartier–Sept-Îles [E1-E2], with over 1,000 lakes and 15 rivers, is prized for Quebec red and brook trout, and Atlantic salmon.

■ Labrador's Mealy Mountains [B6-C7] are the site of the proposed Akamiuapishku National Park, which embraces several environments—tundra, boreal forest, and wetland. The region is a haven for caribou, moose, osprey, and harlequin duck.

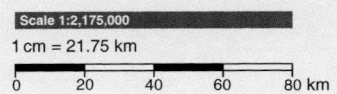

Scale 1:2,175,000
1 cm = 21.75 km

0   20   40   60   80 km

NUNAVUT

HUDSON BAY

BAIE D'HUDSON

SOUTHAMPTON ISLAND
ÎLE SOUTHAMPTON

Cape Comfort
Cape Dorchester
Foxe Peninsula

Coral Harbour (Salliq)

Kinngait (Cape Dorset)

Chesterfield Inlet (Igluligaarjuk)

Rankin Inlet (Kangiqiniq)

Ijiralia

Fall Caribou Crossing Historic Site

Whale Cove (Tikirarjuaq)

Arviat
Arvia'juaq Historic Site

Cape Fullerton
Cape Kendall
Bay of Gods Mercy
Cape Low

Beli Peninsula
Seahorse Point
Mill Island
Mallikjuaq

Leyson Point
Nottingham Island
Salisbury Island

Coats Island
Cape Pembroke

Cape Southampton

Fisher Strait
Evans Strait

Fair Ness

HUDSON DÉTROIT

Charles Island NUNAVUT
Cap de Nouvelle-France

Digges Islands NUNAVUT
Ivujivik
Salluit
Déception
Purtuniq
Wales

Mansel Island

Kangiqsujuaq

Smith I. NUNAVUT
Akulivik
Pingualuit
Cratère du Nouveau-Québec (Chubb Crater) 657 m

PÉNINSULE

D'UNGAVA

Puvirnituq
Povungnituk Bay

Kovik Bay

Ottawa NUNAVUT Islands

Farmer Island NUNAVUT
Hopewell NUNAVUT Islands
Inukjuak

Sleeper Islands NUNAVUT

King George Islands NUNAVUT
Nastapoka NUNAVUT Islands

Umiujaq
Lac Guillaume-Delisle
Lacs-Guillaume-Delisle-et-à-l'Eau-Claire
Lac à l'Eau Claire
Lac d'Iberville

Belcher Islands NUNAVUT
Sanikiluaq

MANITOBA

PRINCE OF WALES FORT
Churchill
Cape Churchill
WAPUSK

Sundance
York Factory
Gillam
280

Shamattawa
Fort Severn
Cape Tatnam

Wabuk Point
Peawanuck
Polar Bear
Cape Henrietta Maria

Kuujjuarapik
Whapmagoostui NUNAVUT

Long Island NUNAVUT
Pointe Louis-XIV

QU

ONTARIO

Big Trout Lake
Sandy Lake
en Hill
Opasquia

Weagamow Lake
North Caribou Lake

Winisk Lake
Lansdowne House

Pickle Lake
599

Armstrong
Sioux Lookout
72

Attawapiskat
Akimiski Island NUNAVUT
Fort Albany

JAMES BAY

BAIE JAMES

Chisasibi
La Grande Rivière
Radisson
Keyano
Sakami

North Twin Island NUNAVUT
South Twin Island NUNAVUT

Wemindji
Eastmain
Charlton Island NUNAVUT

Nemiscau
Waskaganish

QU

Moosonee
Moose Factory
Hannah Bay

Baie de Rupert

Lacs-Albanel-Mistassini-et-Waconichi

Nakina
Geraldton
Longlac
584
506
11

Fraserdale
Matagami
Chapais

Hearst
Kapuskasing
634
652
Smooth Rock Falls
Greenwater
167

Mistissini
Lacs-Albanel-Mistassini-et-Waconichi

Chibougamau
Desmaraisville

Atikokan
Nipigon
Manitouwadge
Terrace Bay
Hornepayne
St. Ignace
527
614
631
Kakabeka Falls
102
17
11
Quetico

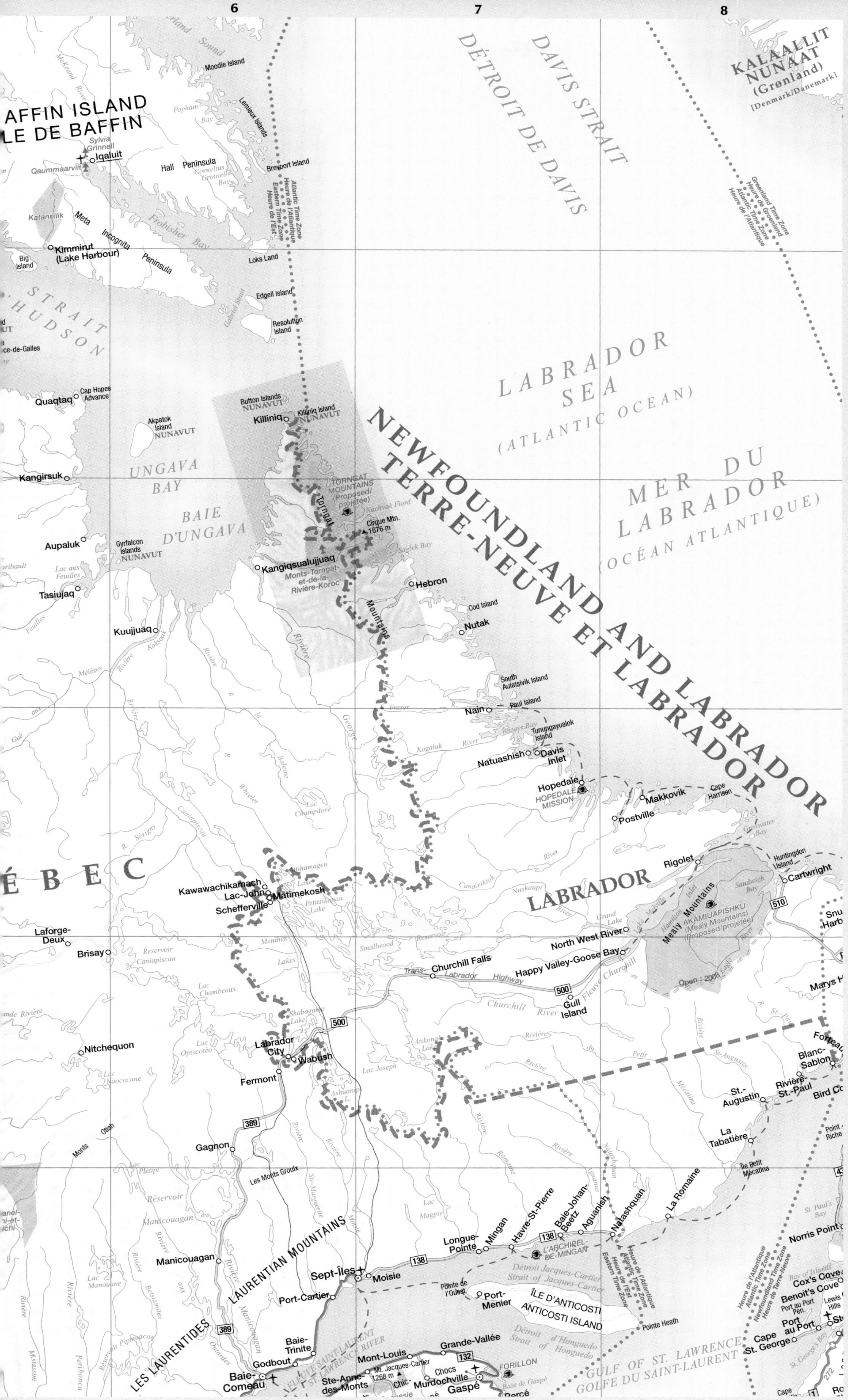

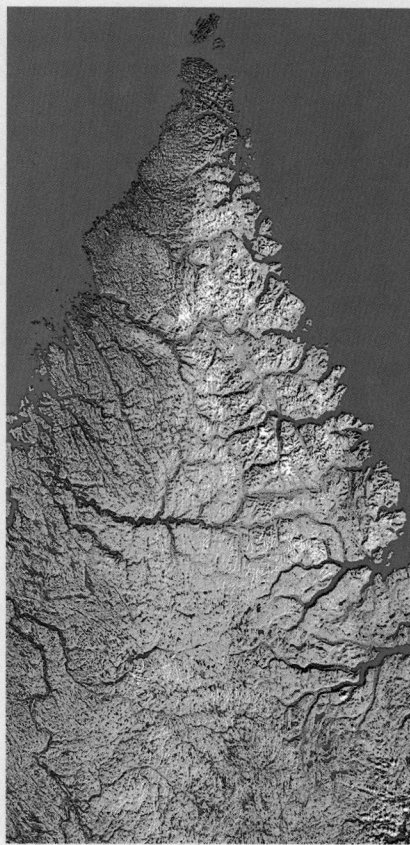

PEAK AND PLATEAU

The Labrador Peninsula has two distinct regions: the George Plateau in Quebec (left) and the Torngat Mountains, largely in Labrador (right). The George Plateau is a flat, rocky plain, cut deeply by rivers flowing into Ungava Bay. The Torngats, the highest peaks east of the Rockies, rise up 900 m from the frigid Labrador Sea. Along this coast, fjords extend some 30 to 80 km inland. Visible in this view (right centre) are Nachvak Fiord, Saglek Bay, and the inner reaches of Hebron Fiord. At 1,676 m, Cirque Mountain is the region's highest. Its snowy summit is visible just south of the Nachvak Fiord. At the very tip of the peninsula are Killiniq and the Button Islands of Nunavut. A national park has been proposed for this spectacular region.

POINTS OF INTEREST

■ James Bay [E3–E4] is 160 km wide between Cape Henrietta Maria and Pointe Louis-XIV. The heart of Quebec's James Bay Project is Réservoir Robert-Bourassa [E4], which receives water diverted from the Eastmain and other rivers. Nunavut administers Akimiski and all the other islands in the bay.

■ At 6,527 km², Smallwood Reservoir [D7–E7] is Canada's tenth largest freshwater body, created by the damming of the Churchill River for hydroelectric power production. Smallwood Reservoir is one of the world's largest man-made lakes.

■ Mingan Archipelago National Park Reserve [F7] comprises 40 islands, myriad islets, and reefs that extend more than 150 km. Spectacular rock formations—the products of erosion by winds and waves—are among the outstanding features of this park.

Scale 1:5,600,000

1 cm = 56 km

0 50 100 150 200 km

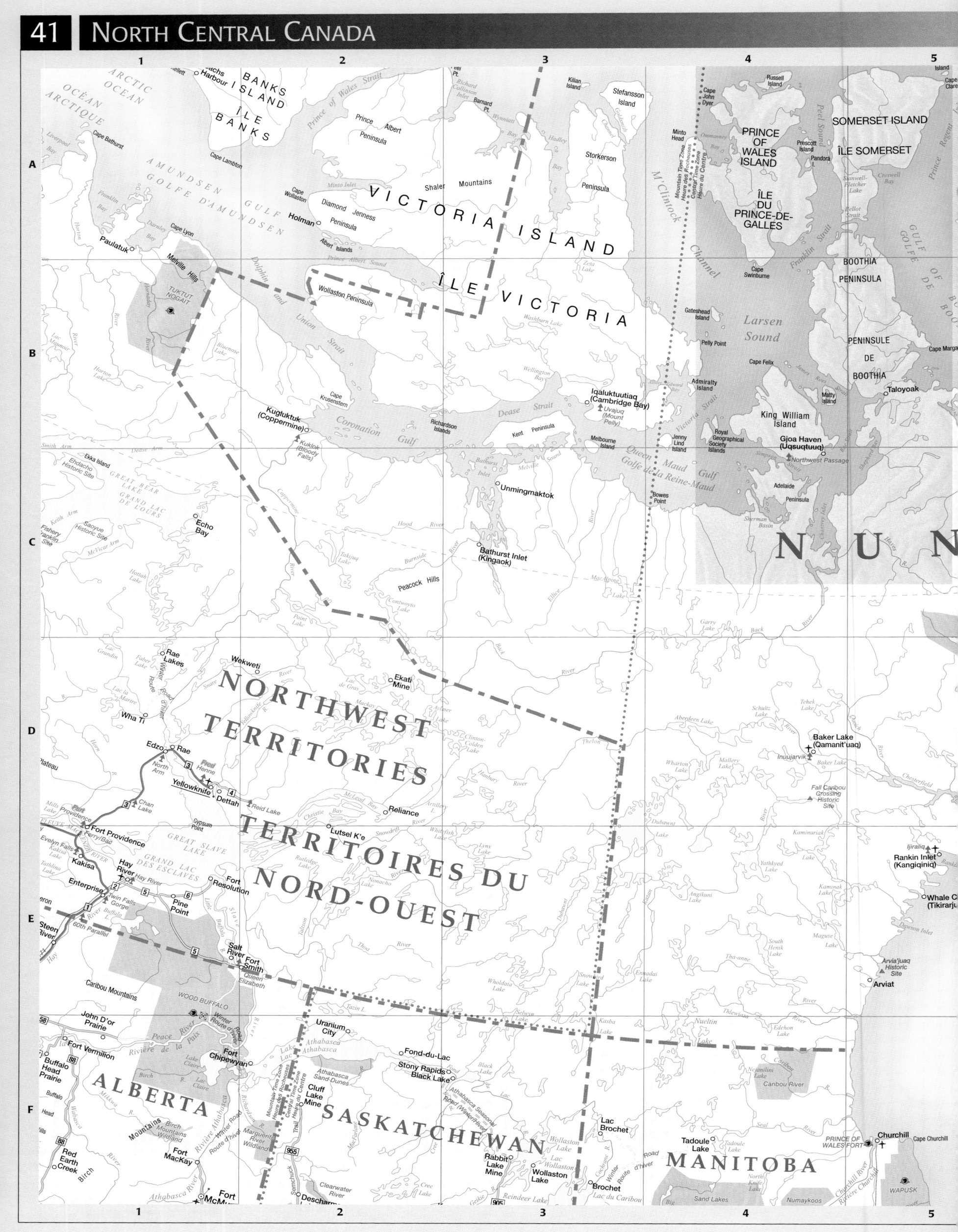

6 **7** **8**

Cape Crauford
Nanisivik
Arctic Bay
(Ikpiarjuk)
BORDEN
PENINSULA
BRODEUR
PENINSULA
SIRMILIK
Bylot
Island
Cape Graham Moore
Pond Inlet
Qilalugat
Pond Inlet
(Mittimatalik)
Cape Macculloch
Cape Jameson

BAFFIN
BAY
BAIE DE
BAFFIN

Atlantic Time Zone
Heure de l'Atlantique
Eastern Time Zone
Heure de l'Est

A

Cape Adair
Clyde River
(Kangiqtugaapik)
Cape Raper
Auliiting Island
Cape Henry
Kater

B A F F I N   I S L A N D
Î L E   D E   B A F F I N

Bernier Bay
Berlinguet Inlet
Eastern Time Zone
Heure de l'Est
Central Time Zone
Heure du Centre

Steensby
Inlet

Crown Prince
Frederick Island
Fury and Hecla Strait
Cape
Engelfield
Jens
Munk
Island
Rowley
Island
Bray
Island
Baird
Peninsula
Foley
Island

AUYUITTUQ
CUMBE

B

...rrison
...ands
Cape Chapman
Igloolik
(Igloolik)
Hall Beach
(Sanirajak)

Koch
Island

Air
Force I.

ARCTIC CIRCLE
CERCLE ARCTIQUE

Mt. Odin
2147 m

Simpson
Peninsula
Committee
Bay
MELVILLE
Hall
Lake
Purry
Bay
Prince
Charles
Island

Pangnirtun
(Pangniqtu

A   V   U   T
PENINSULA
Pelly Bay
(Arviliqjuat)
Wales
Island

FOXE
BASIN

BASSIN
DE
FOXE

Cape Penrhyn
Cape Dominion

Great Plain
of the
Koukdjuak

Koukdjuak R.

Nettilling
Lake

Nettilling
Fiord

Kek

C

...THIA

...Bay

Repulse Bay
(Naujaat)
White
Island
Winter Island
Vansittart Island
Cape
Dorchester

Bowman
Bay

Amadjuak
Lake

McKeand River

Tessik
Lake

Chorkbak
Inlet

Mingo
Lake

Qaummaarviit

UKKUSIKSALIK
(WAGER BAY)
Wager Bay

Foxe   Peninsula

Markham
Bay

Katannilik

...ve
...q)
Cape Comfort
Mallikjuaq
Kinngait
(Cape Dorset)
Fair Ness
Big
Island

HUDSON
DÉTROIT
STRAIT

D

SOUTHAMPTON
ISLAND
Coral Harbour
(Salliq)
ÎLE
SOUTHAMPTON
Ross Welcome Sound

Seahorse
Point
Bell   Peninsula
Mill
Island
Salisbury
Island

HUDSON
Charles Island
NUNAVUT
Cap de
Nouvelle-France
Wales Island
NUNAVUT
Cap du
Prince-de-G...

Cape
Fullerton
Cape Kendall
Bay of
Gods Mercy
Native
Bay
Leyson Point
Nottingham
Island

Déception
Purtuniq
Joy Bay

Chesterfield Inlet
(Igluligaarjuk)
Cape Low
Evans Strait
Cape
Pembroke
Digges
Islands
NUNAVUT
Ivujivik
Salluit
Kangiqsujuaq

Fisher Strait

D

...Inlet

Marble
Island

...nlet

Coats
Island

Mansel
Island

Cape
Southampton

Povungnituk
Pingualuit
Cratère du Nouveau-Québec
(Chubb Crater) 657 m

Lac Nantais

P É N I N S U L E
D ' U N G A V A

E

Smith I.
Akulivik
NUNAVUT
Mosquito Bay

Lac
Couture

Rivière

Kovik Bay

Lac Payne

H U D S O N   B A Y

Puvirnituq
Povungnituq Bay

Q U É B E C

Kogaluk Bay

Riviere Kogaluk

E

Ottawa
NUNAVUT
Islands

B A I E   D ' H U D S O N

Hopewell
NUNAVUT
Islands
Inukjuak

Farmer
Island
NUNAVUT

Lac Minto

F

Eastern Time Zone
Heure de l'Est
Central Time Zone
Heure du Centre

Sleeper
Islands
NUNAVUT

King George
Islands
NUNAVUT

Nastapoka

Riv. Nastapoca

Umiujaq
Lac Guillaume-
Delisle

6 **7** **8**

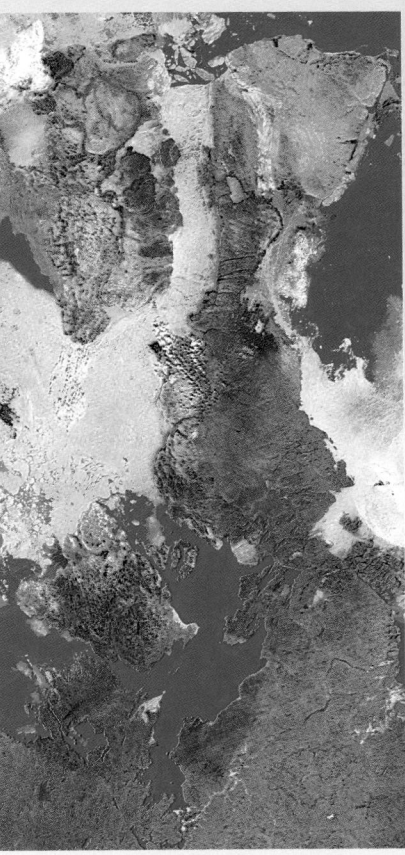

AT THE END OF THE CONTINENT

Icebound Boothia Peninsula (centre right) is North America's northernmost point. It is connected to the continent by a 40-km isthmus and separated from Somerset Island (top right) by Bellot Strait, only 2 km wide but clearly visible above. In 1831, James C. Ross discovered the magnetic north pole on the west side of the peninsula. (Since then, the pole has migrated farther north.) At 24,786 km², Somerset Island is ninth largest in the Arctic Archipelago; Prince of Wales Island (top left), at 33,339 km², ranks eighth. King William Island (bottom left) is associated with Sir John Franklin, whose search for the Northwest Passage ended in disaster and mystery. Franklin is known to have died on June 11, 1847.

POINTS OF INTEREST

■ Arviat [E5], Nunavut, is the geographical centre of Canada.

■ At 22,200 km², Sirmilik National Park [A6] encompasses three distinct environments: Bylot Island, which has a landscape of mountains, glaciers, and coastal lowlands; Borden Peninsula, which is cut deeply by broad river valleys; and a narrow fjord named Oliver Sound. Bylot Island supports a vast seabird colony.

■ When 21,469-km² Auyuittuq National Park [B8] was established in 1976, it was the world's first park above the Arctic Circle. In 1988, it was joined by Sirmilik (see above); in 1992, by 12,200-km² Aulavik National Park on Banks Island [Plate 42, D1]; and 37,775-km² Quttinirpaaq National Park, on Ellesmere Island [Plate 42, A5].

Scale 1:5,600,000

1 cm = 56 km

0   50   100   150   200 km

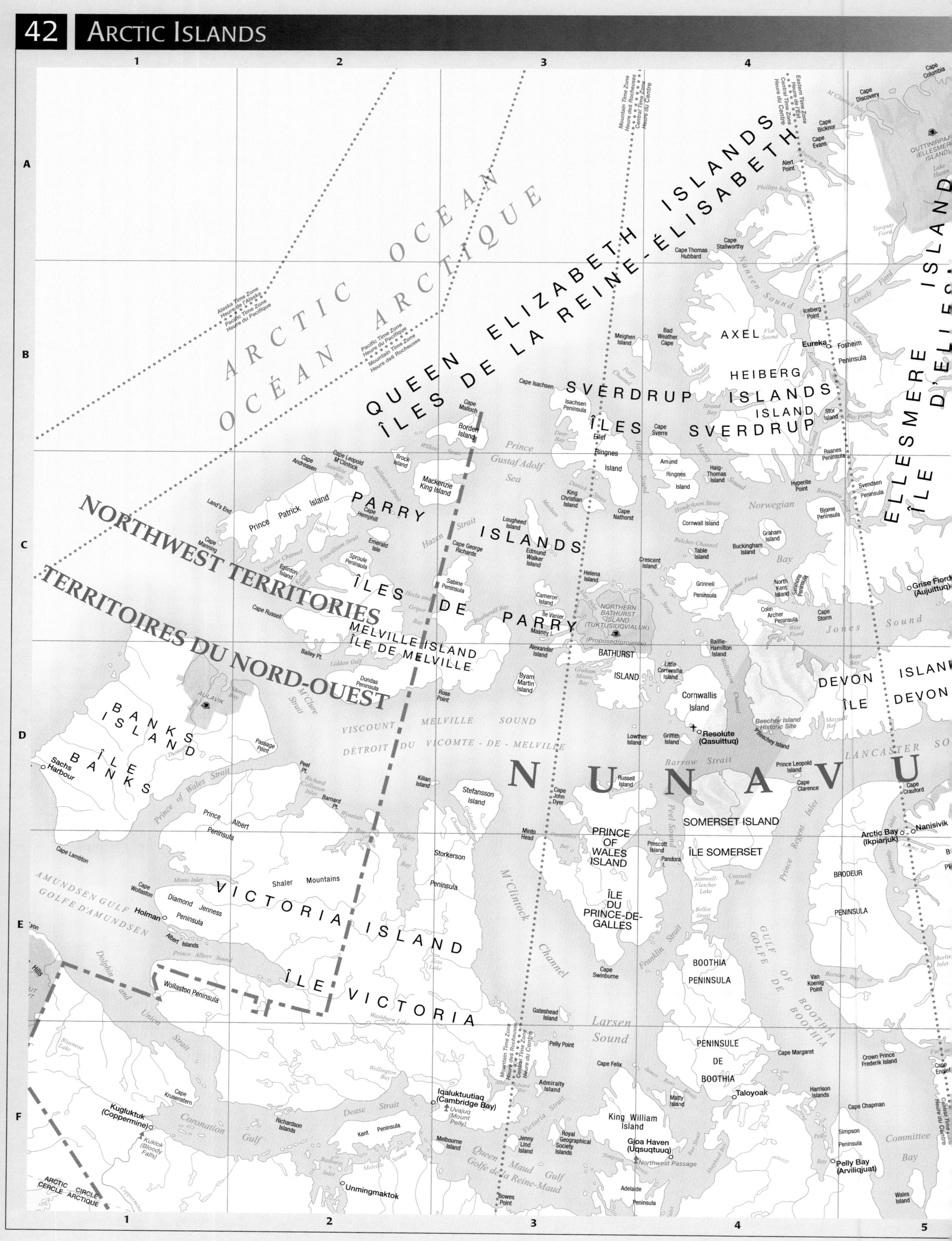

ARCTIC OCEAN
OCÉAN ARCTIQUE

QUEEN ELIZABETH ISLANDS
ÎLES DE LA REINE-ÉLISABETH

NORTHWEST TERRITORIES
TERRITOIRES DU NORD-OUEST

PARRY ISLANDS
ÎLES DE PARRY

SVERDRUP ISLANDS
ÎLES SVERDRUP

AXEL HEIBERG ISLAND

ELLESMERE ISLAND
ÎLE D'ELLESMERE

BANKS ISLAND
ÎLE DE BANKS

MELVILLE ISLAND
ÎLE DE MELVILLE

VICTORIA ISLAND
ÎLE VICTORIA

DEVON ISLAND
ÎLE DE DEVON

NUNAVUT

SOMERSET ISLAND
ÎLE SOMERSET

PRINCE OF WALES ISLAND
ÎLE DU PRINCE-DE-GALLES

BOOTHIA PENINSULA
PÉNINSULE DE BOOTHIA

VISCOUNT MELVILLE SOUND
DÉTROIT DU VICOMTE-DE-MELVILLE

AMUNDSEN GULF
GOLFE D'AMUNDSEN

Larsen Sound

Coronation Gulf

Queen Maud Gulf
Golfe de la Reine-Maud

Lancaster Sound

Jones Sound

Northern Bathurst Island
(TUKTUSIUJUALUK)
(Proposed/projeté)

BATHURST ISLAND

Cornwallis Island

Resolute (Qasuittuq)

Grise Fiord (Aujuittuq)

Arctic Bay (Ikpiarjuk)

Nanisivik

Taloyoak

King William Island

Gjoa Haven (Uqsuqtuuq)

Iqaluktuutiaq (Cambridge Bay)

Kugluktuk (Coppermine)

Kuklok (Bloody Falls)

Holman

Sachs Harbour

Aulavik

Pelly Bay (Arviliqjuat)

BRODEUR PENINSULA

GULF OF BOOTHIA
GOLFE DE BOOTHIA

Committee Bay

ARCTIC CIRCLE
CERCLE ARCTIQUE

Mountain Time Zone
Heure des Rocheuses
Central Time Zone
Heure du Centre

Eastern Time Zone
Heure de l'Est

Alaska Time Zone
Heure de l'Alaska
Pacific Time Zone
Heure du Pacifique

QUTTINIRPAAQ
(ELLESMERE ISLAND)

Eureka

Alert Point

Cape Columbia

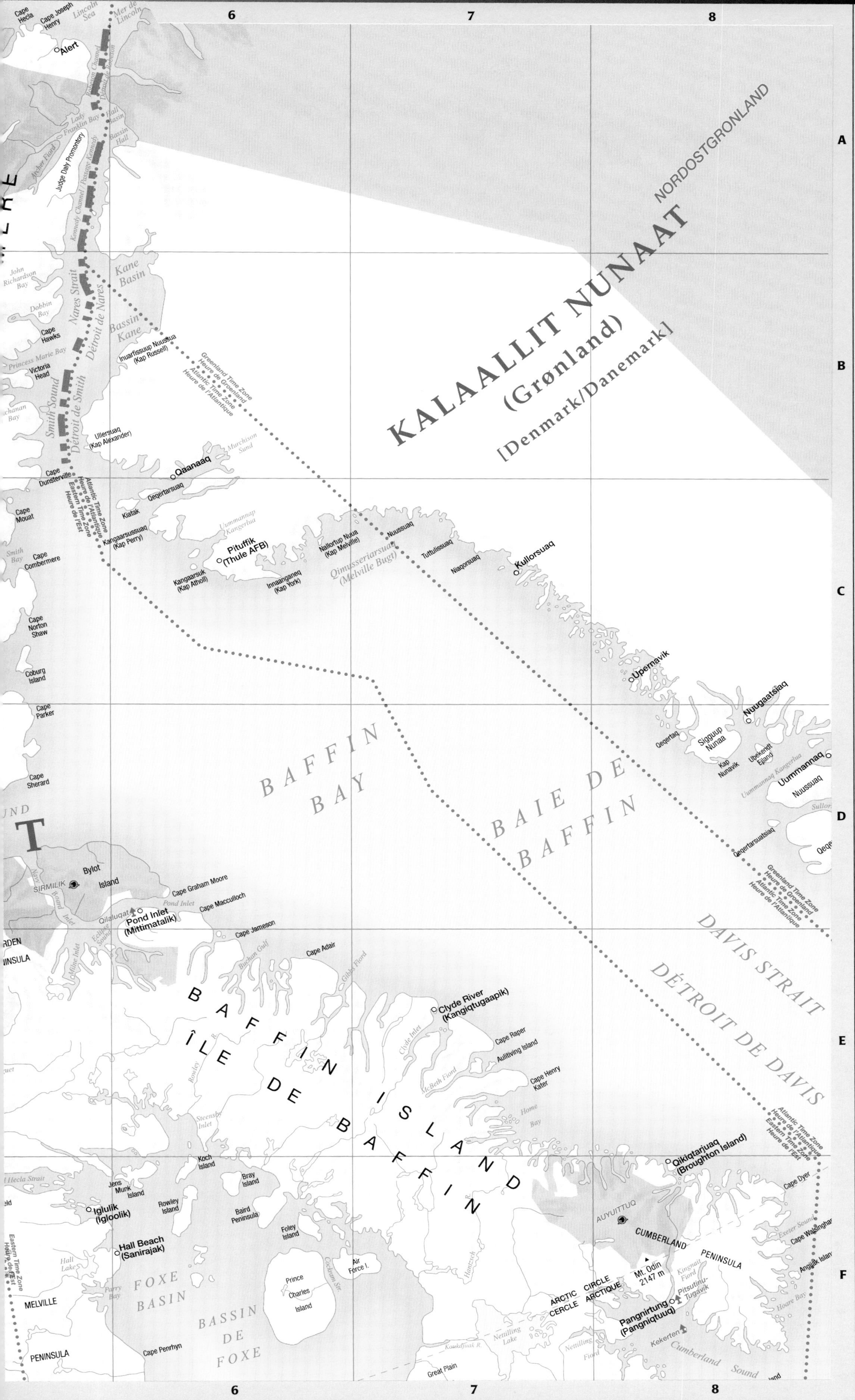

Map labels

6 7 8

Cape Hecla
Cape Joseph Henry
Lincoln Sea
Mer de Lincoln
Alert
Cape Columbia

A

NORDOSTGRONLAND

John Richardson Bay
Kennedy Channel
Kane Basin
Lady Franklin Bay
Judge Daly Promontory

Dobbin Bay
Cape Hawks

Princess Marie Bay
Victoria Head
Buchanan Bay
Naves Strait
Détroit de Smith
Smith Sound

KALAALLIT NUNAAT
(Grønland)
[Denmark/Danemark]

B

Inuarfissuaq Nuussua (Kap Russell)
Ullersuaq (Kap Alexander)
Greenland Time Zone
Heure de l'Atlantique

Cape Dunsterville
Détroit de Smith
Qaanaaq
Murchison Sound

Cape Mouat
Qeqertarsuaq
Kiatak
Kangaarsussuaq (Kap Perry)
Pituffik (Thule AFB)
Kangarsuk (Kap Atholl)
Innaangaveq (Kap York)
Qimusseriarsuaq (Melville Bugt)

Nallortup Nuua (Kap Melville)
Nuussuaq
Tuttulissuaq
Naajorsuu

Kullorsuaq

C

Smith Bay
Cape Combermere

Cape Norton Shaw

Upernavik

Coburg Island

Nutaatsiaq

Cape Parker

Qeqertaq
Siggiup Nunaa
Kap Nunarvik
Uummannap Kangerlua
Uummannaq
Nuussuaq

Cape Sherard

BAFFIN BAY

BAIE DE BAFFIN

D

T
SIRMILIK
Bylot Island
Cape Graham Moore
Cape Maculloch
Pond Inlet (Mittimatalik)
Qilaluqat
Pond Inlet
Cape Jameson

Qeqertarsuatsiaq
Qeq

Greenland Time Zone
Heure de Groenland
Atlantic Time Zone
Heure de l'Atlantique

DAVIS STRAIT

DÉTROIT DE DAVIS

Cape Adair

E

Clyde Fiord
Clyde River (Kangiqtugaapik)
Cape Raper
Auditing Island
McBeth Fiord
Cape Henry Kater
Home Bay

BAFFIN
ÎLE DE BAFFIN
ISLAND

Sceensby Inlet

Koch Island
Bray Island
Baird Peninsula

Qikiqtarjuaq (Broughton Island)
Cape Dyer

Jens Munk Island
Rowley Island

F

Iglulik (Igloolik)
Foley Island

AUYUITTUQ
CUMBERLAND PENINSULA
Cape Wragsinghar
Avajalik Islan

Hall Beach (Sanirajak)
Prince Charles Island
Air Force I.

MELVILLE
Hall Lake
Eastern Time Zone
Heure de l'Est

FOXE BASIN
BASSIN DE FOXE

Great Plain

ARCTIC CIRCLE
CERCLE ARCTIQUE
Mt. Odin 2147 m
Pangnirtung (Pangniqtuuq)
Pitsulini-Tugavik
Kekerten

Kingnait Fiord

PENINSULA
Cape Penrhyn
Nettilling Lake
Cumberland Sound

6 7 8

A MARS LANDSCAPE IN THE ARCTIC

The three Arctic islands shown above are Devon (top right), Cornwallis (centre left) and the northern shore of Somerset Island (bottom). Devon Island is the site of the Haughton Crater—the light-blue circular shape visible top right. More than 20 km in diameter, the crater was formed by the impact of an asteroid or a comet some 23 million years ago. In 1997, NASA, the United States National Research Council, and the Geological Survey of Canada chose the crater as a simulation site for the study of missions to Mars. Experts say the crater ressembles conditions on Mars: the landscape is a cold, dry, and rocky desert, virtually unvegetated and drenched in ultraviolet light in summer. Tiny Beechey Island is visible in Lancaster Sound just south of Devon Island. The island served as a base for the ill-fated Franklin expedition and, later, for others who came in search of Franklin after his disappearance.

Other points of interest

■ The remote Canadian outpost of Alert [A5] is the world's northernmost community, closer to Moscow than it is to Toronto. A population of less than 100 inhabitants works at the weather station or the Canadian military base.

■ In 1985, paleontologists from the Geographical Survey of Canada discovered the fossils of a 40-million-year-old forest at the Geodetic Hills on Axel Heiberg Island [B4]. The fossils include swamp cypress, dawn redwood, and other semitropical trees, some of which reached as high as 35 m.

Scale 1:5,600,000
1 cm = 56 km
0 50 100 150 200 km

CITY MAPS

Below: *Edmonton's downtown towers add their glow to a winter evening.*

VANCOUVER BC

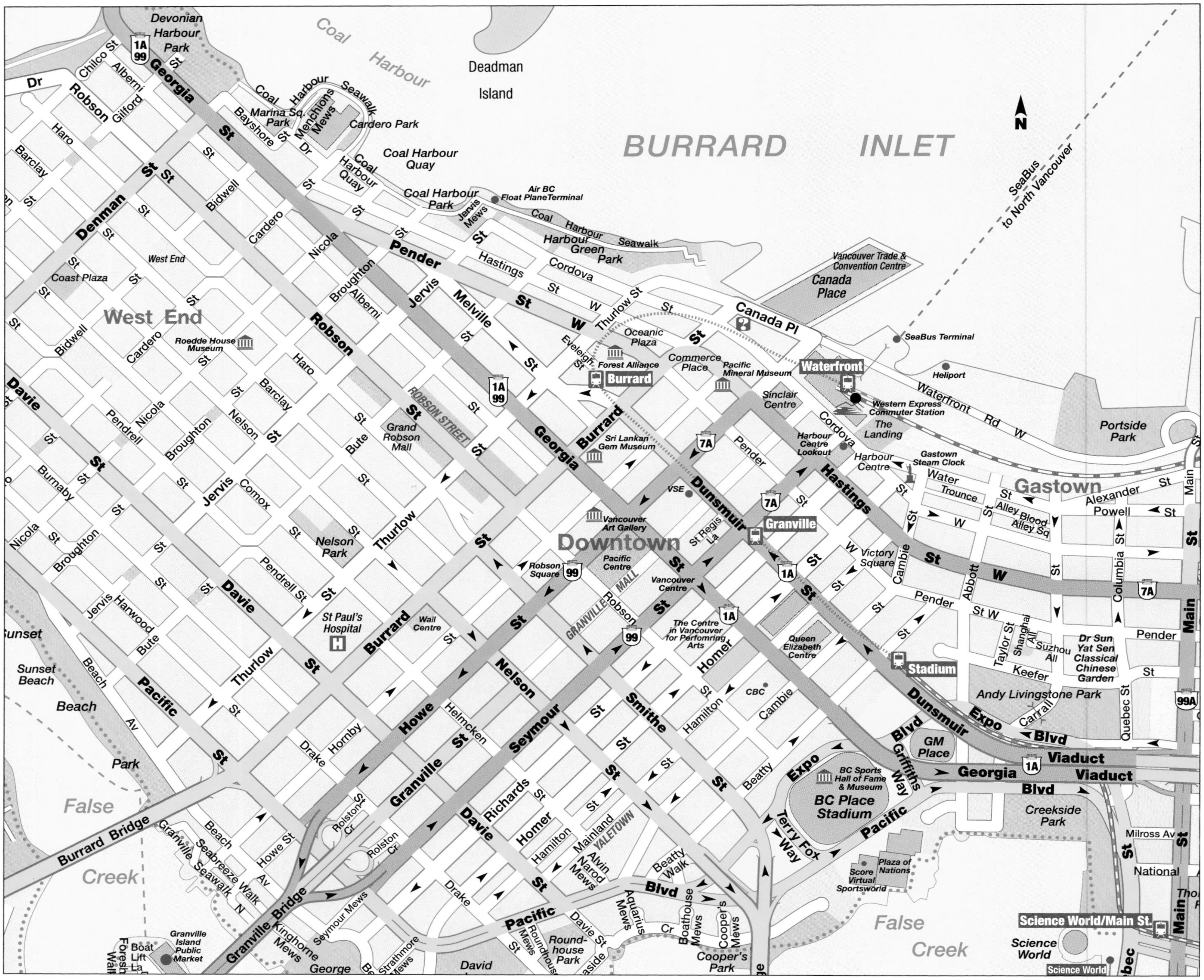

CALGARY AB

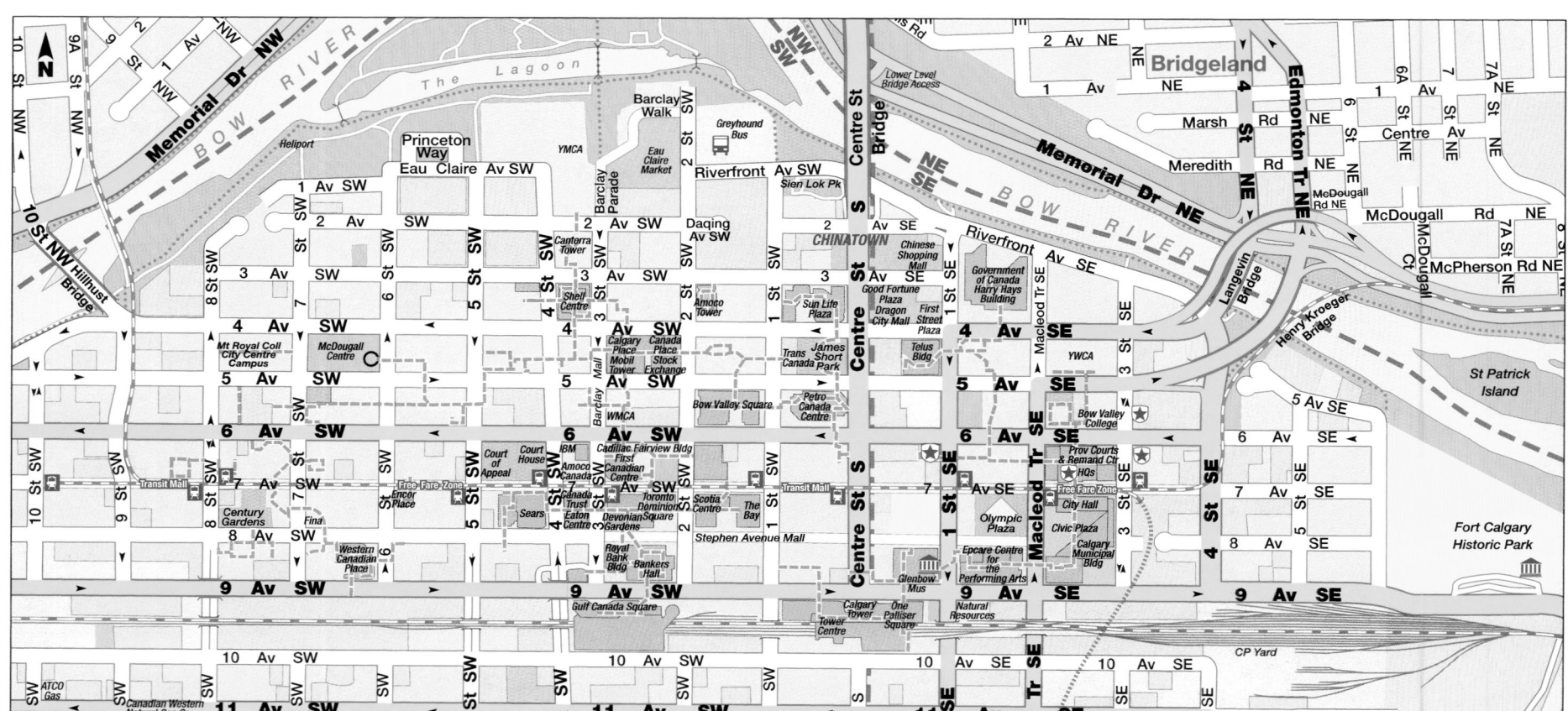

YELLOWKNIFE NT

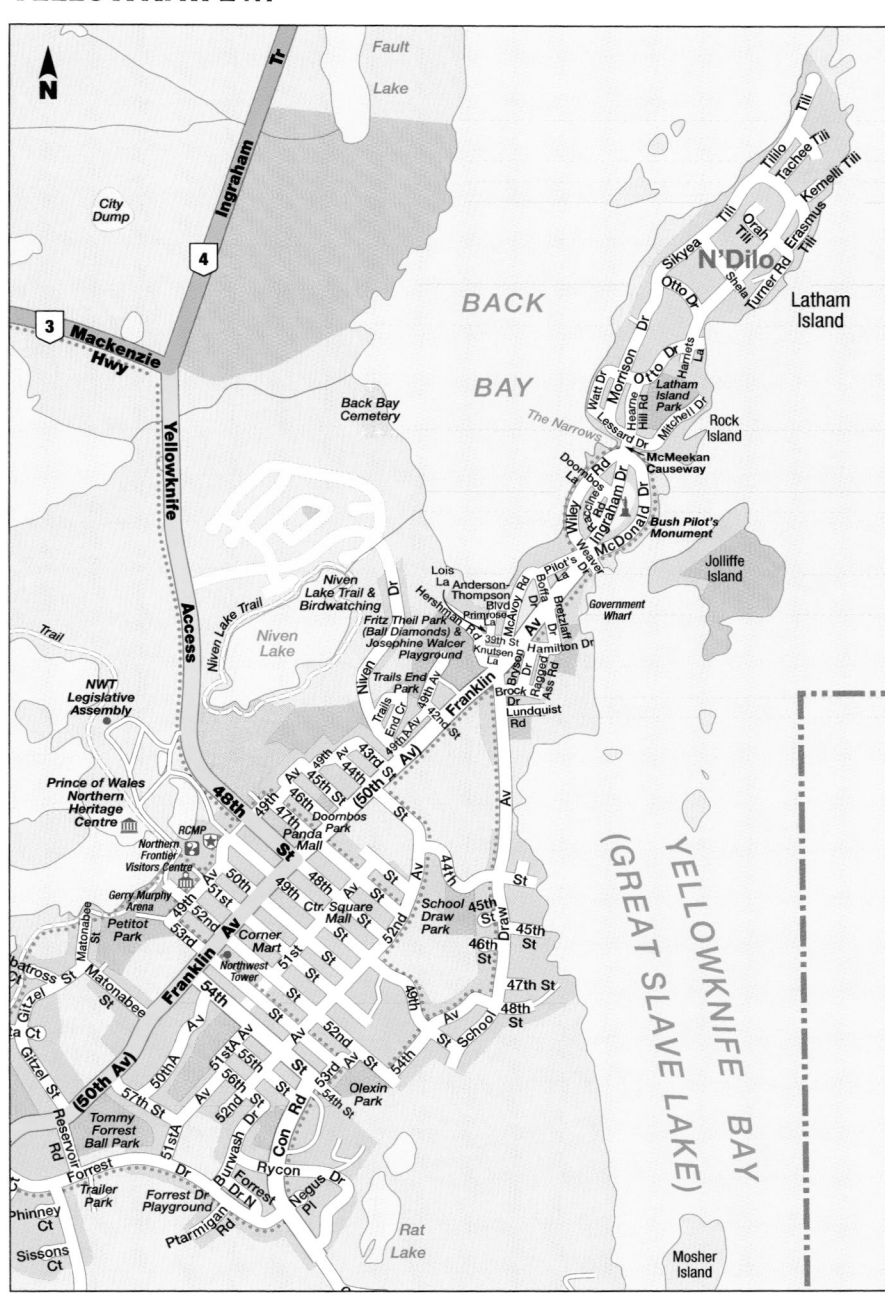

WHITEHORSE YT

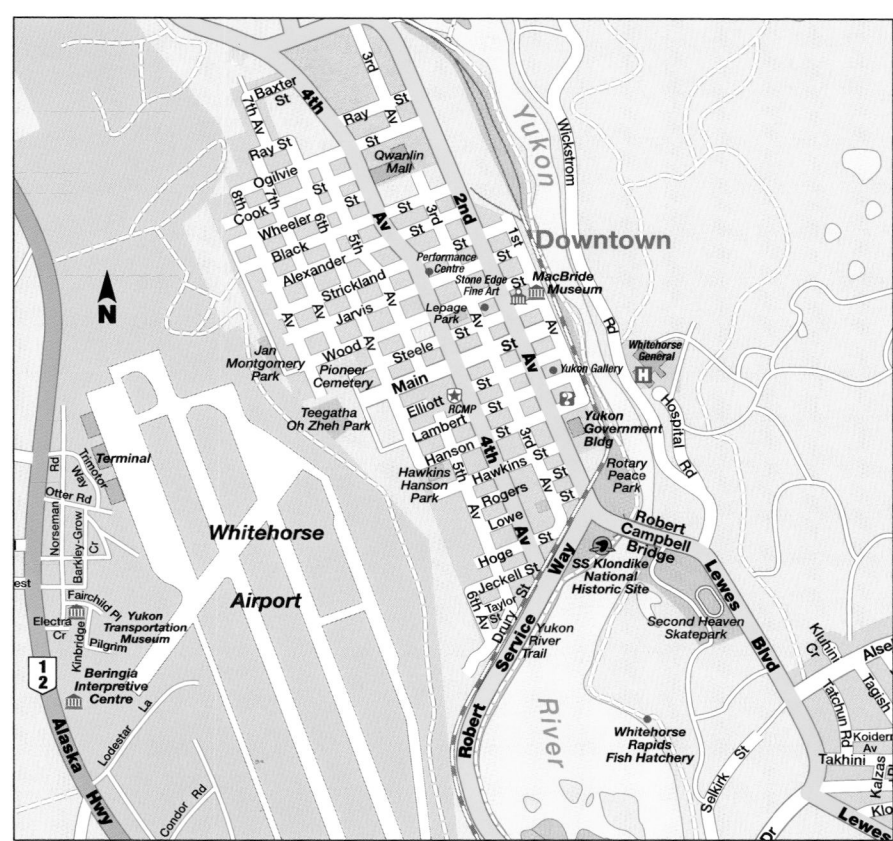

EDMONTON AB

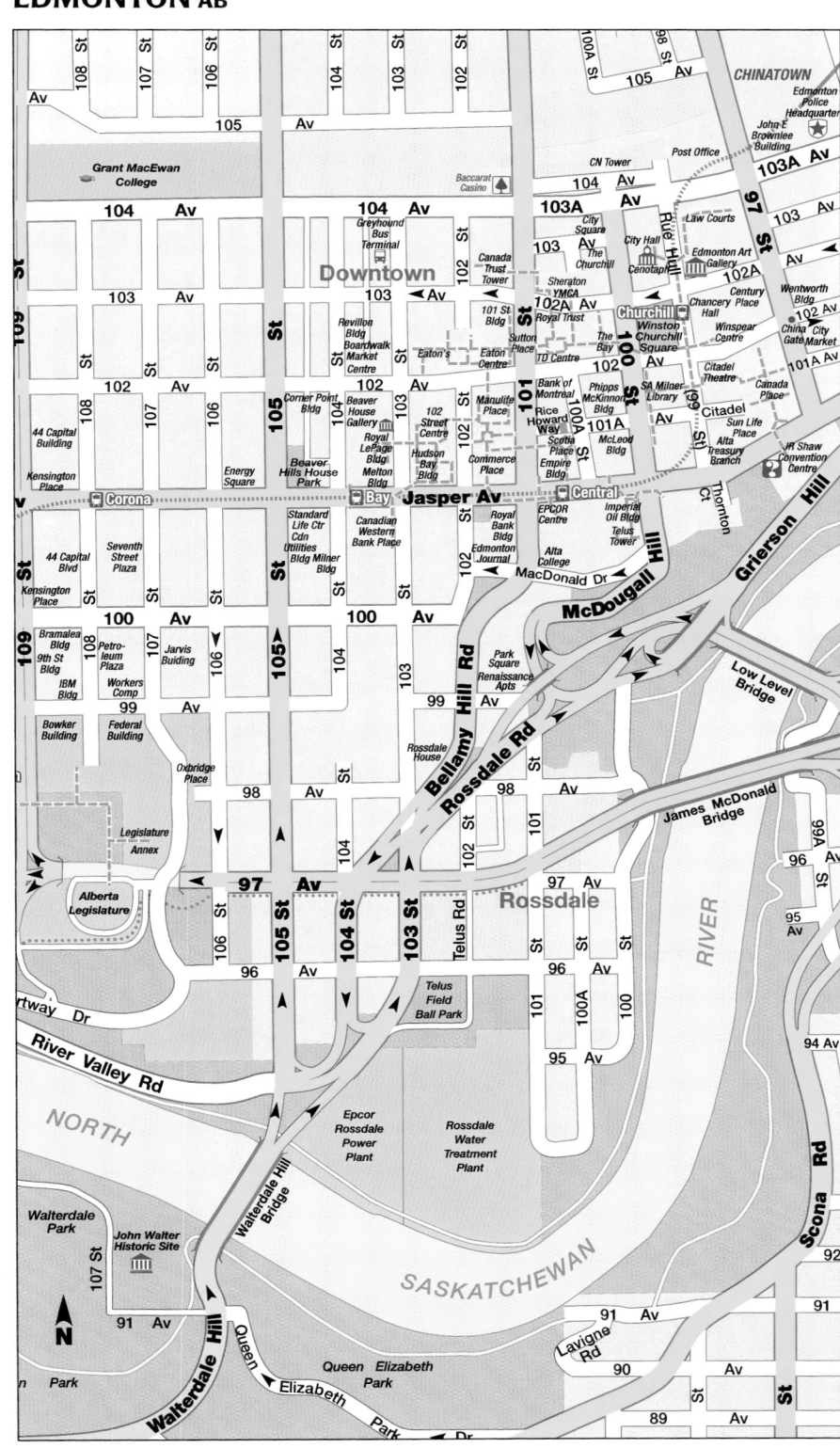

VICTORIA BC

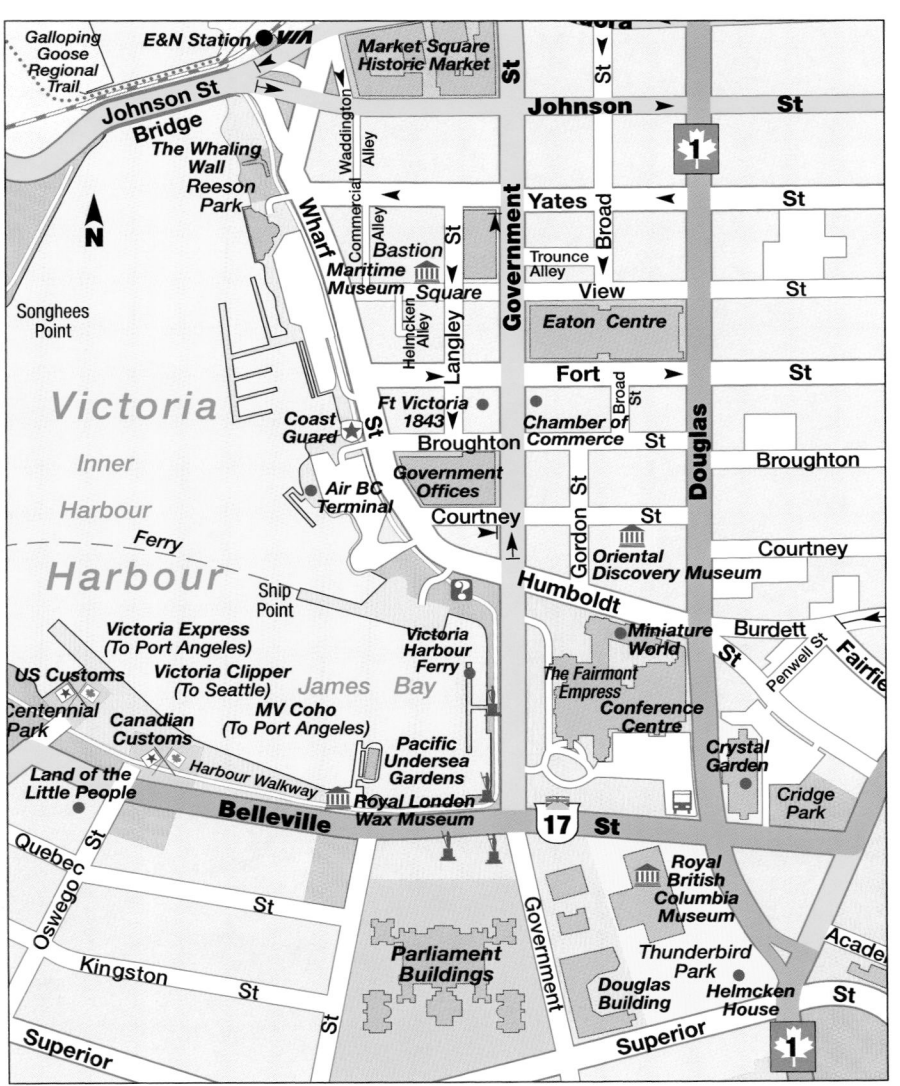

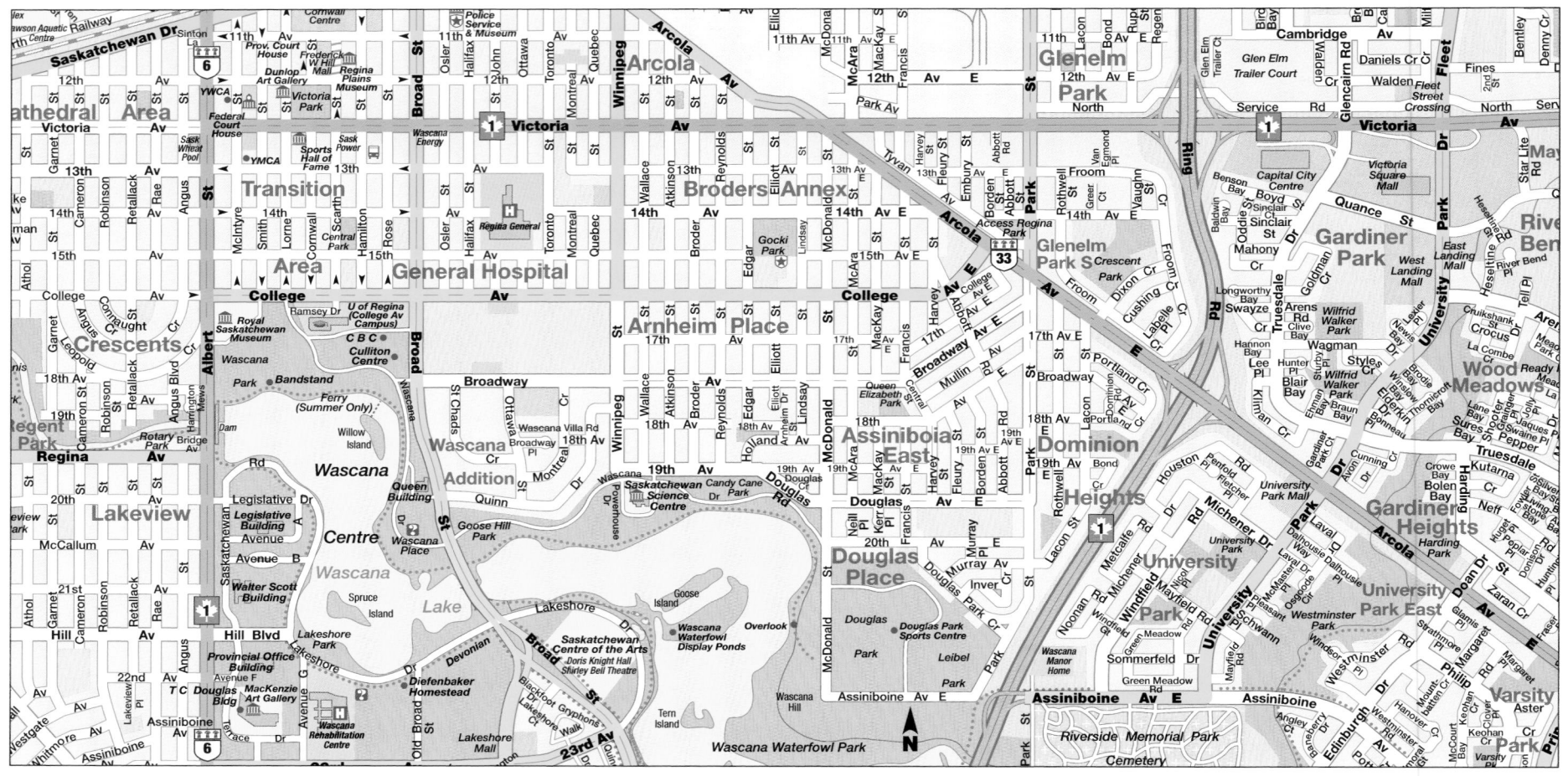

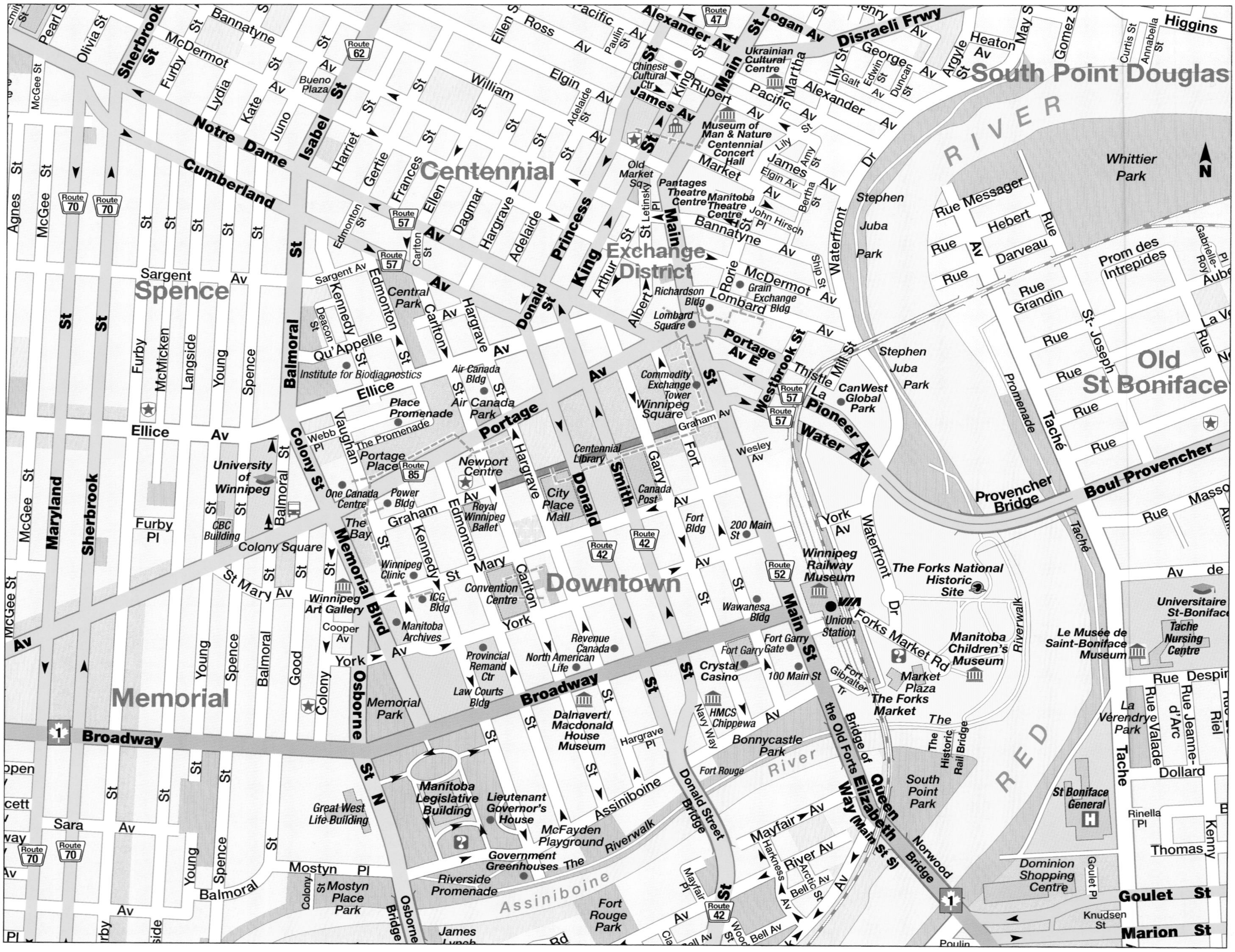

SASKATOON SK

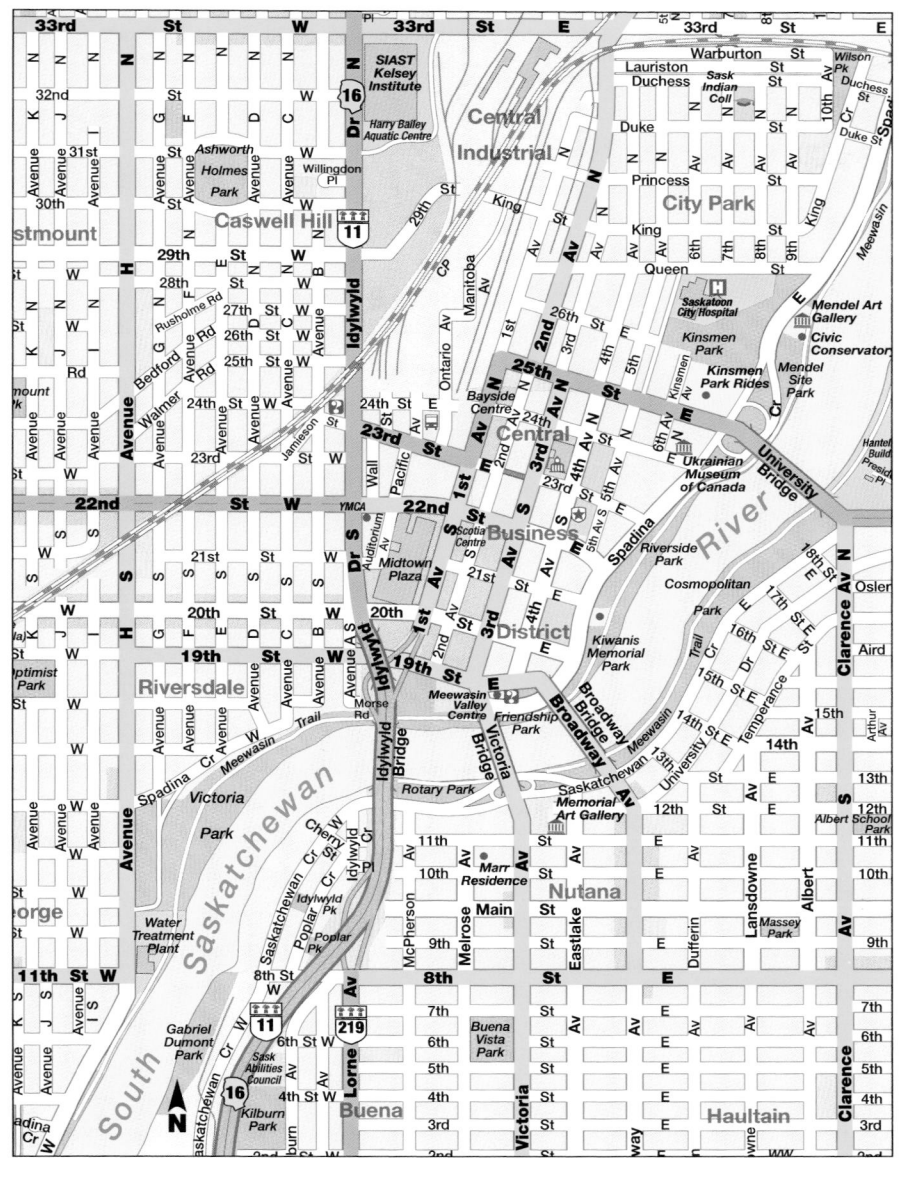

THUNDER BAY ON

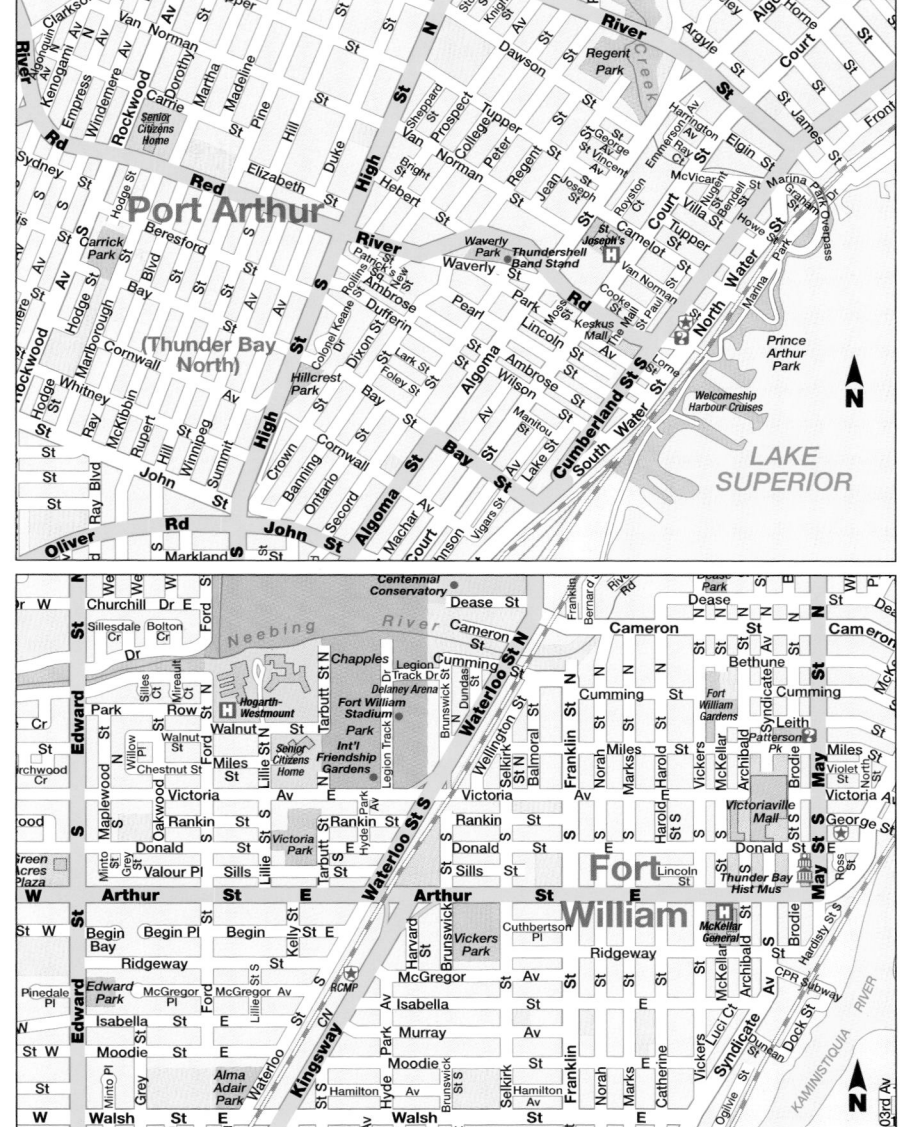

SUDBURY ON

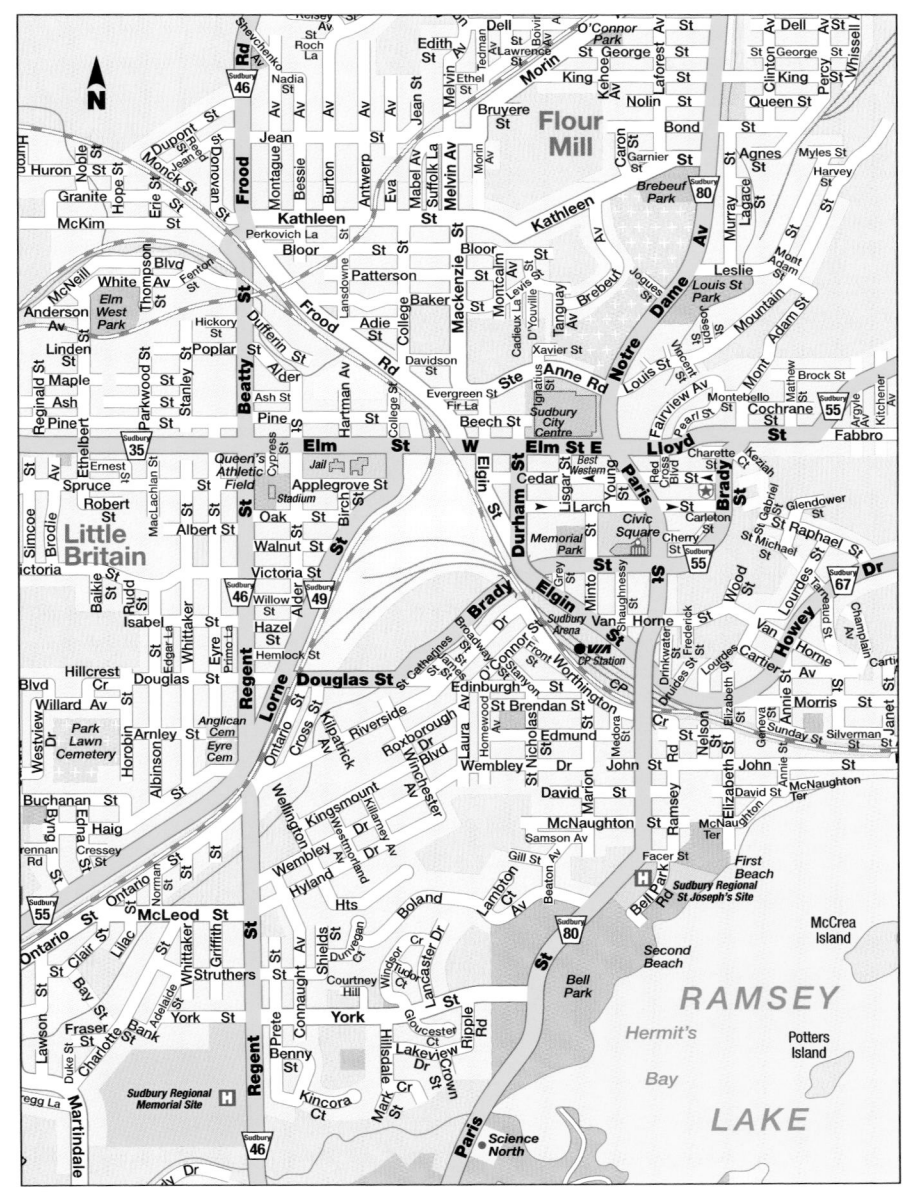

WINDSOR ON

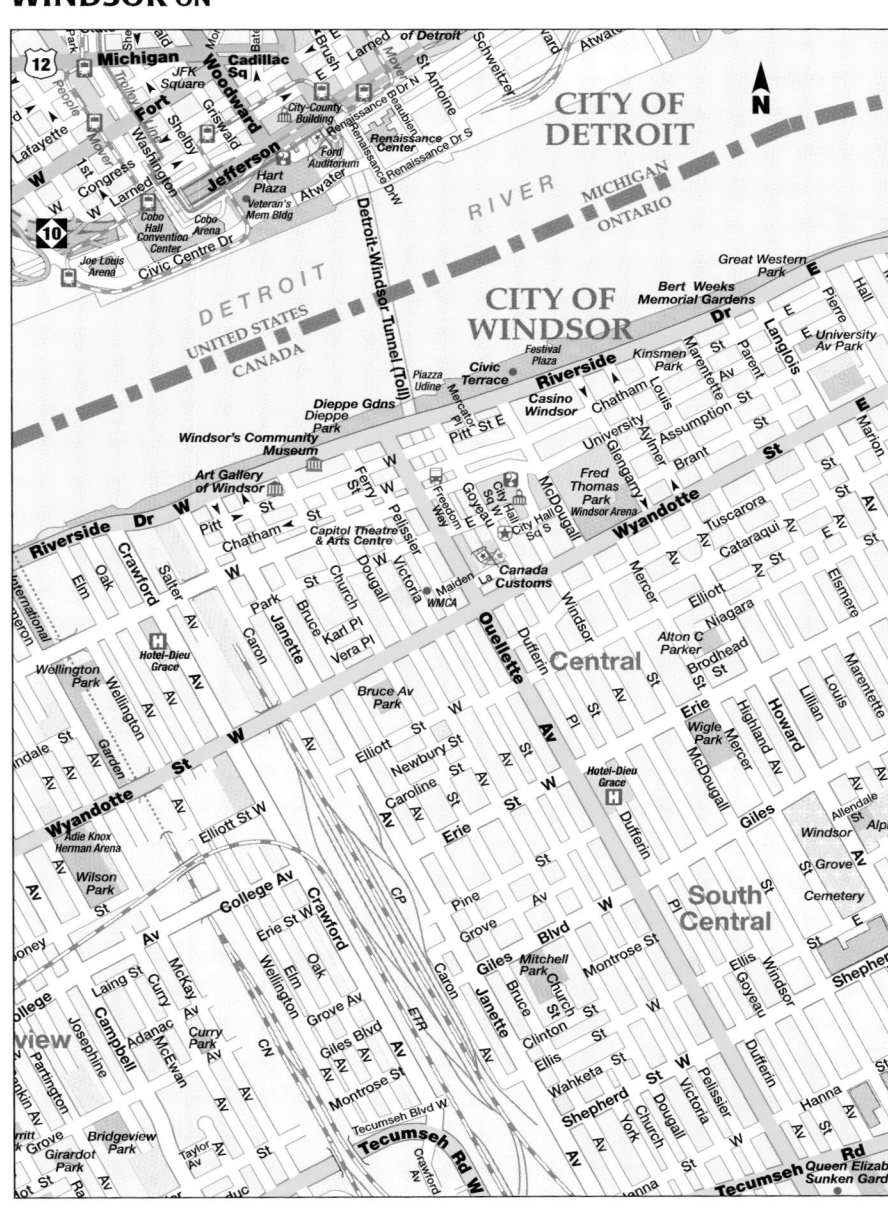

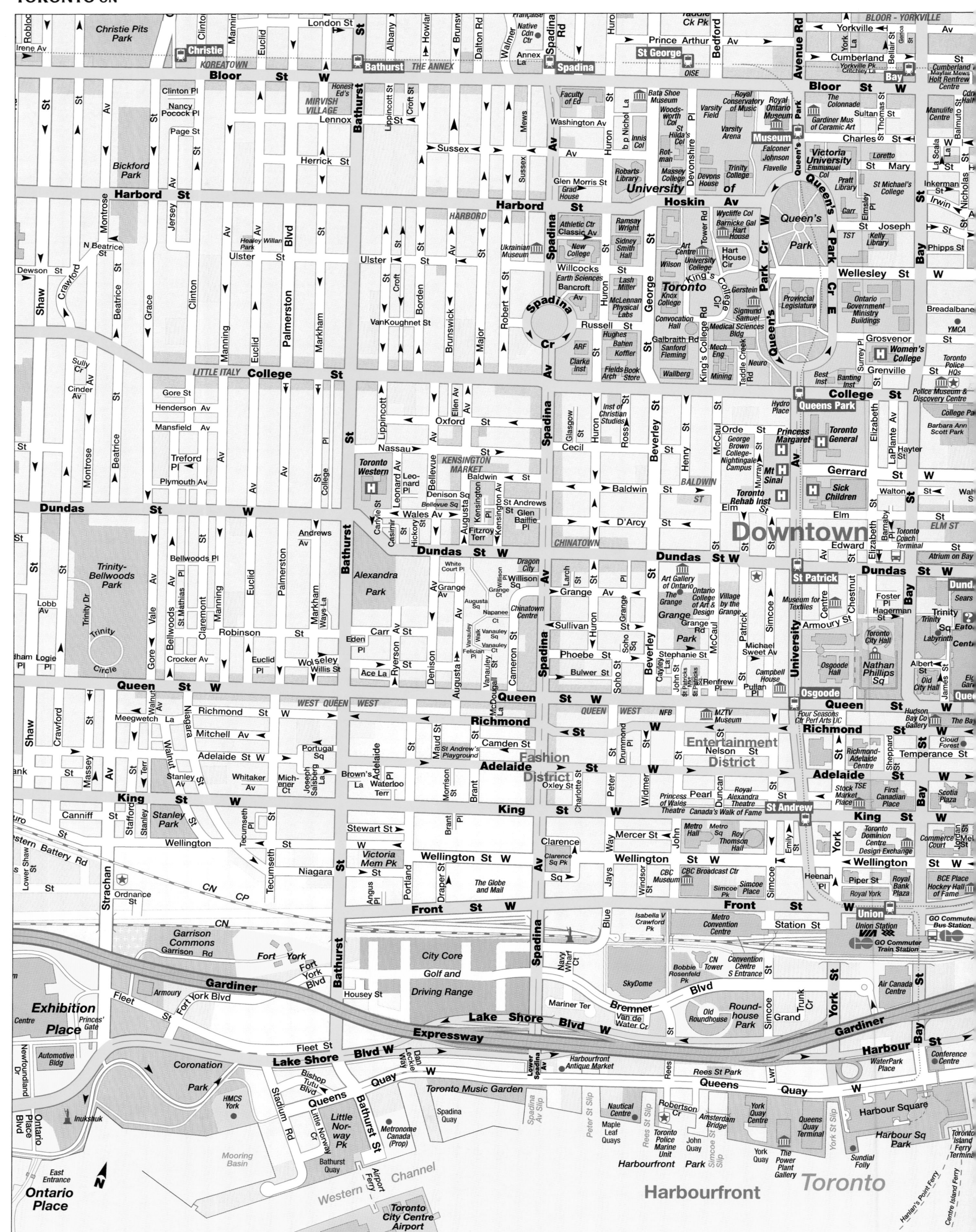

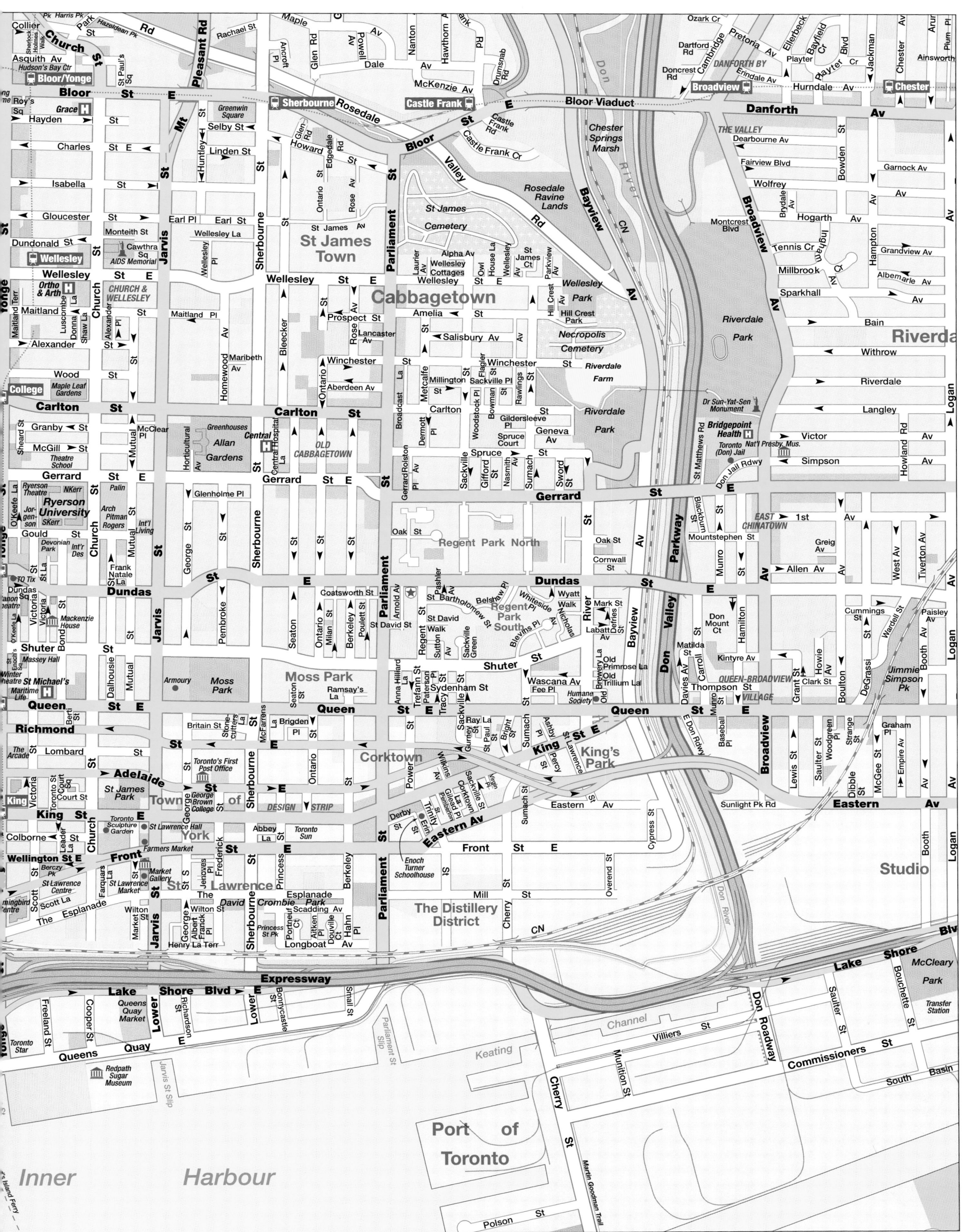

LONDON ON

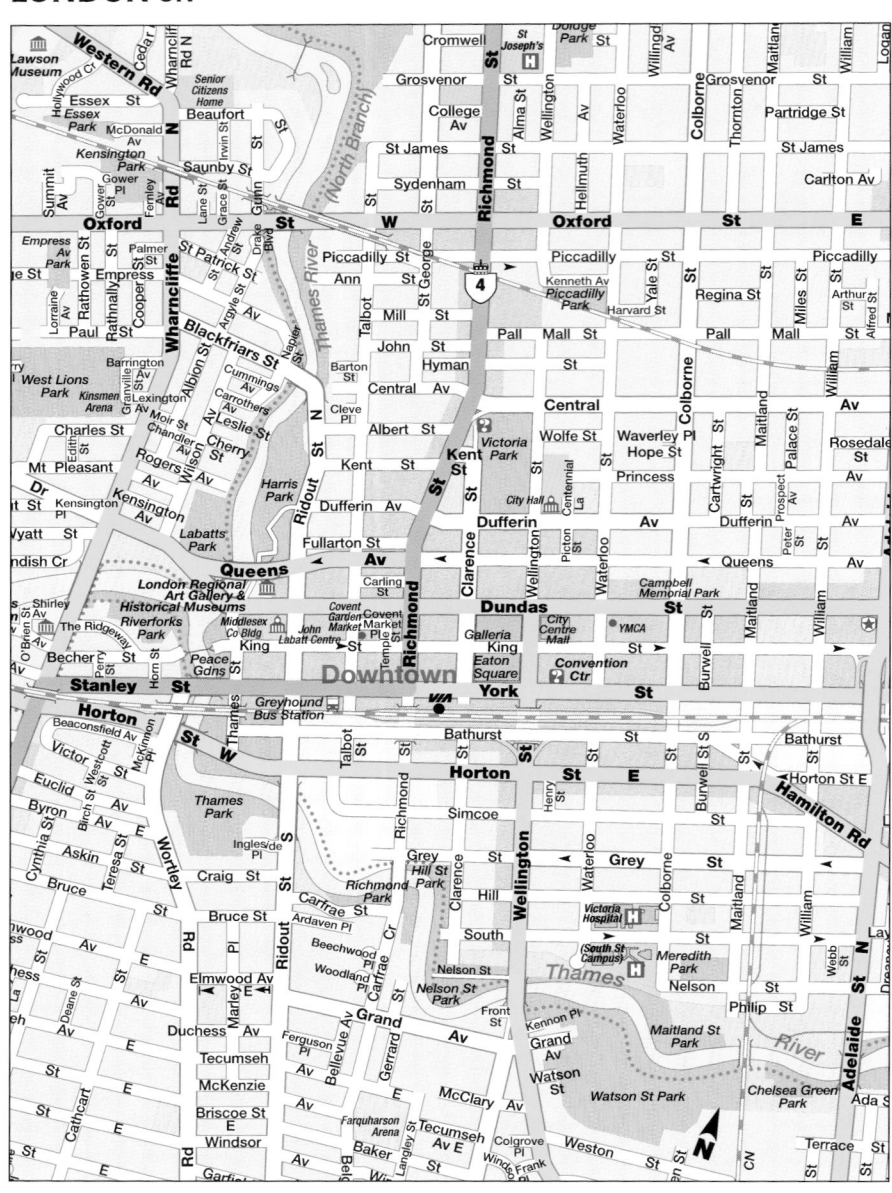

KITCHENER/WATERLOO ON

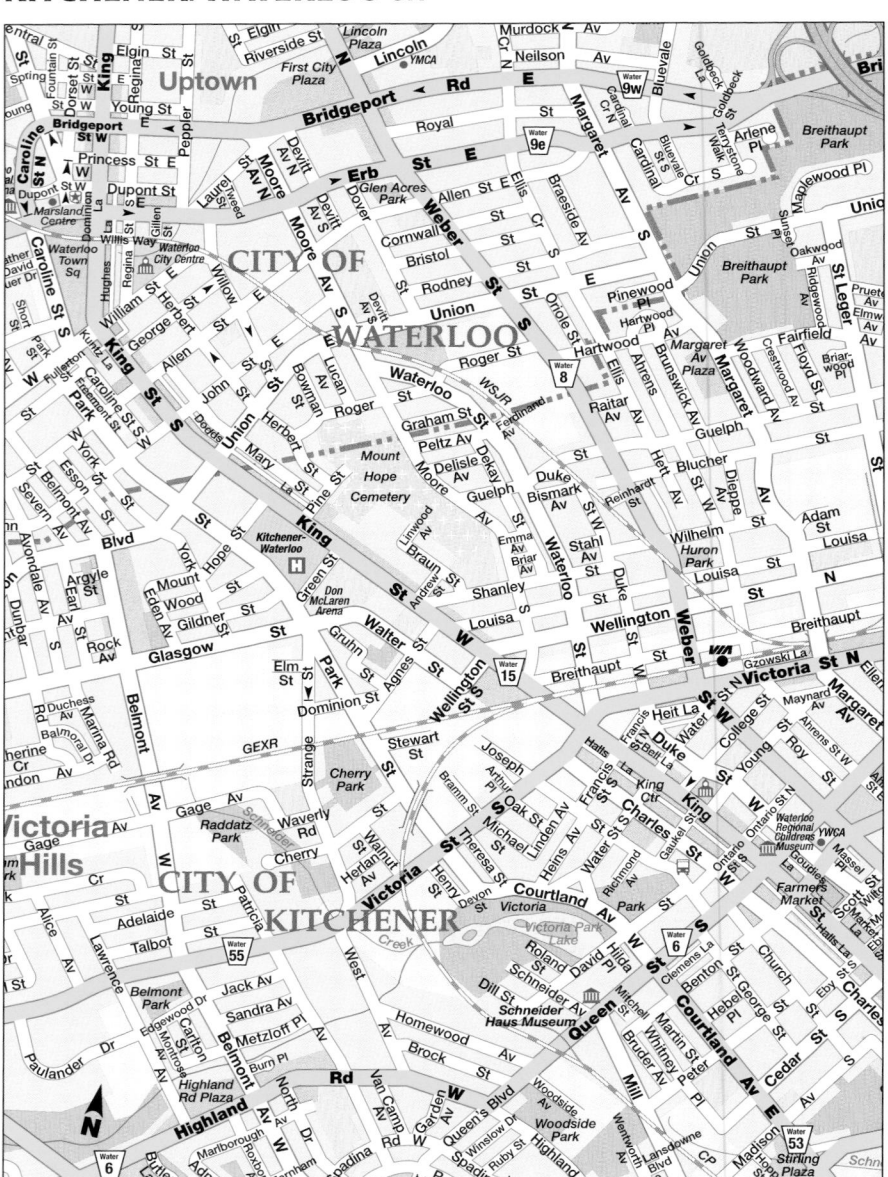

ST. CATHARINES ON

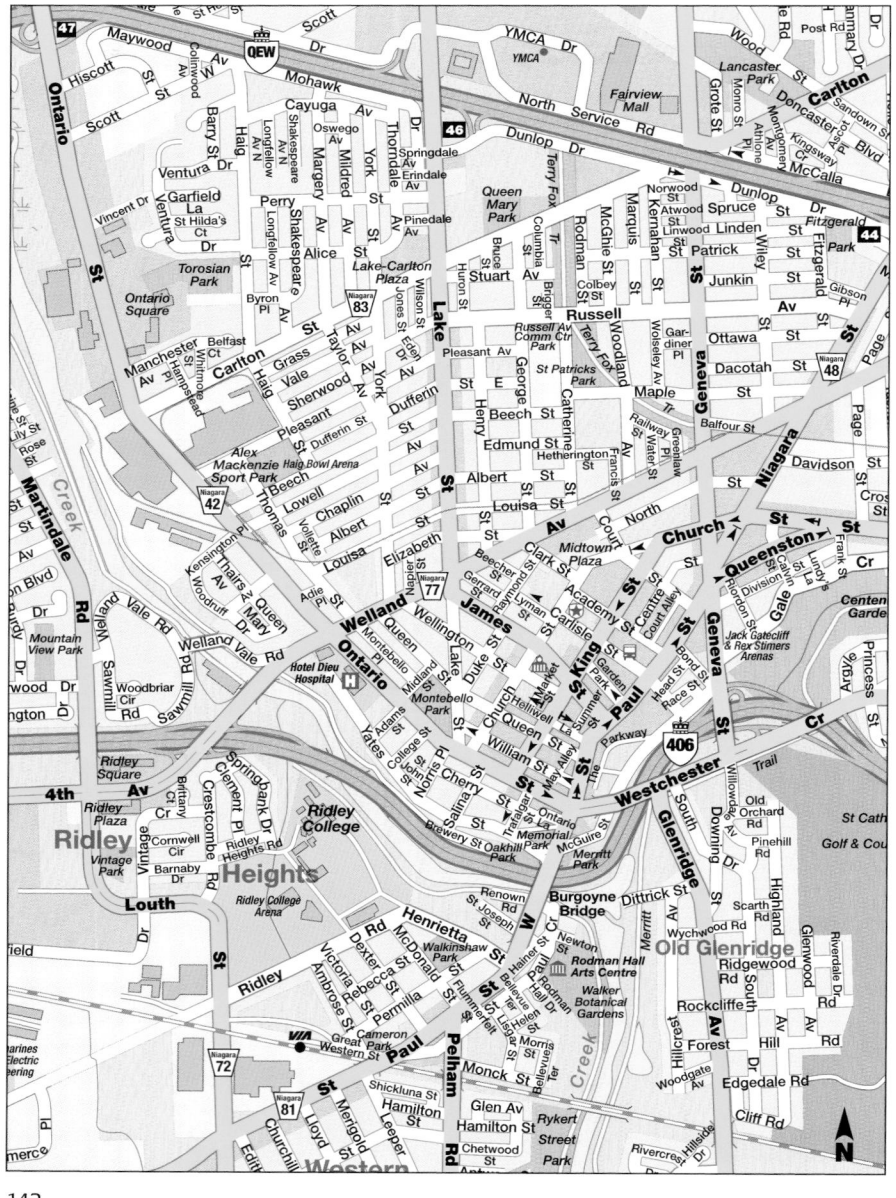

HAMILTON ON

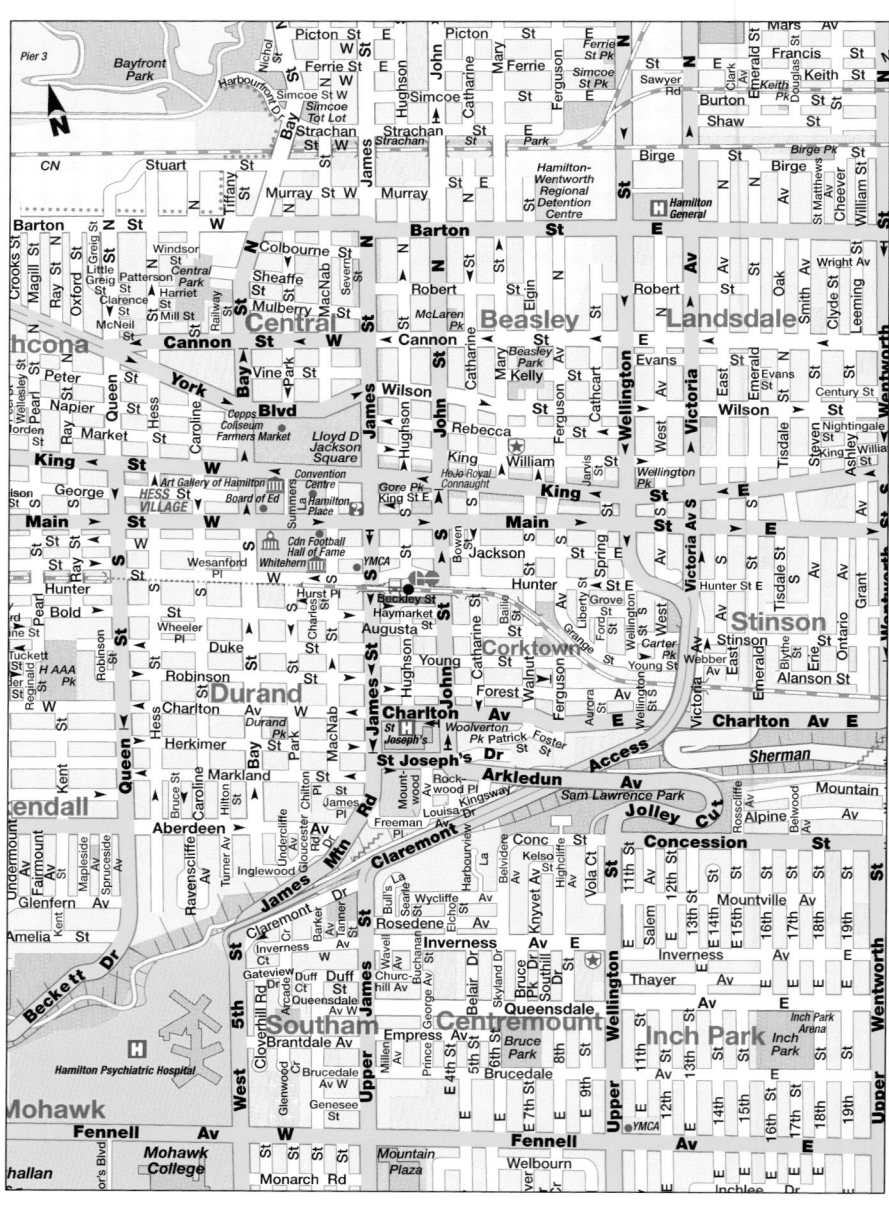

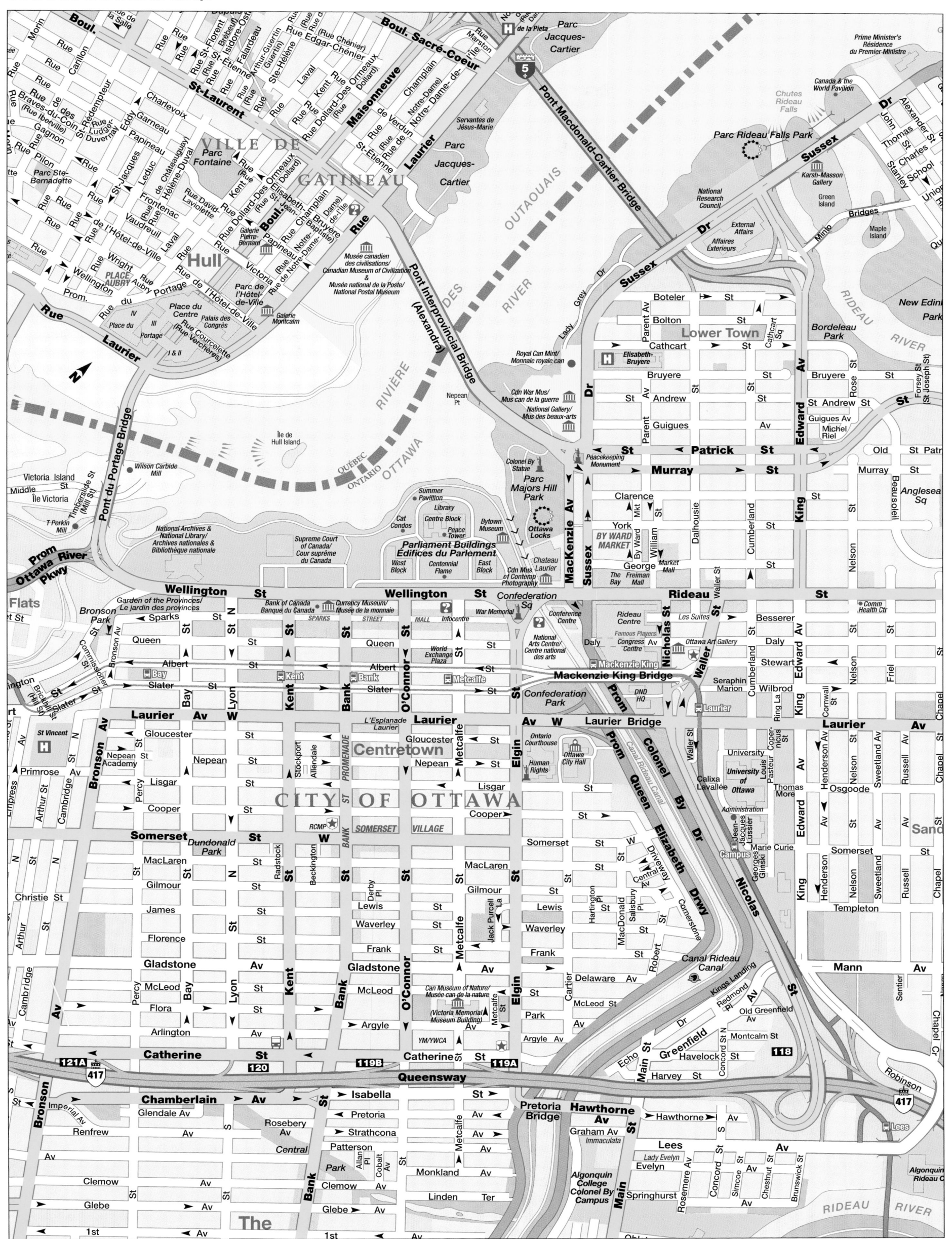

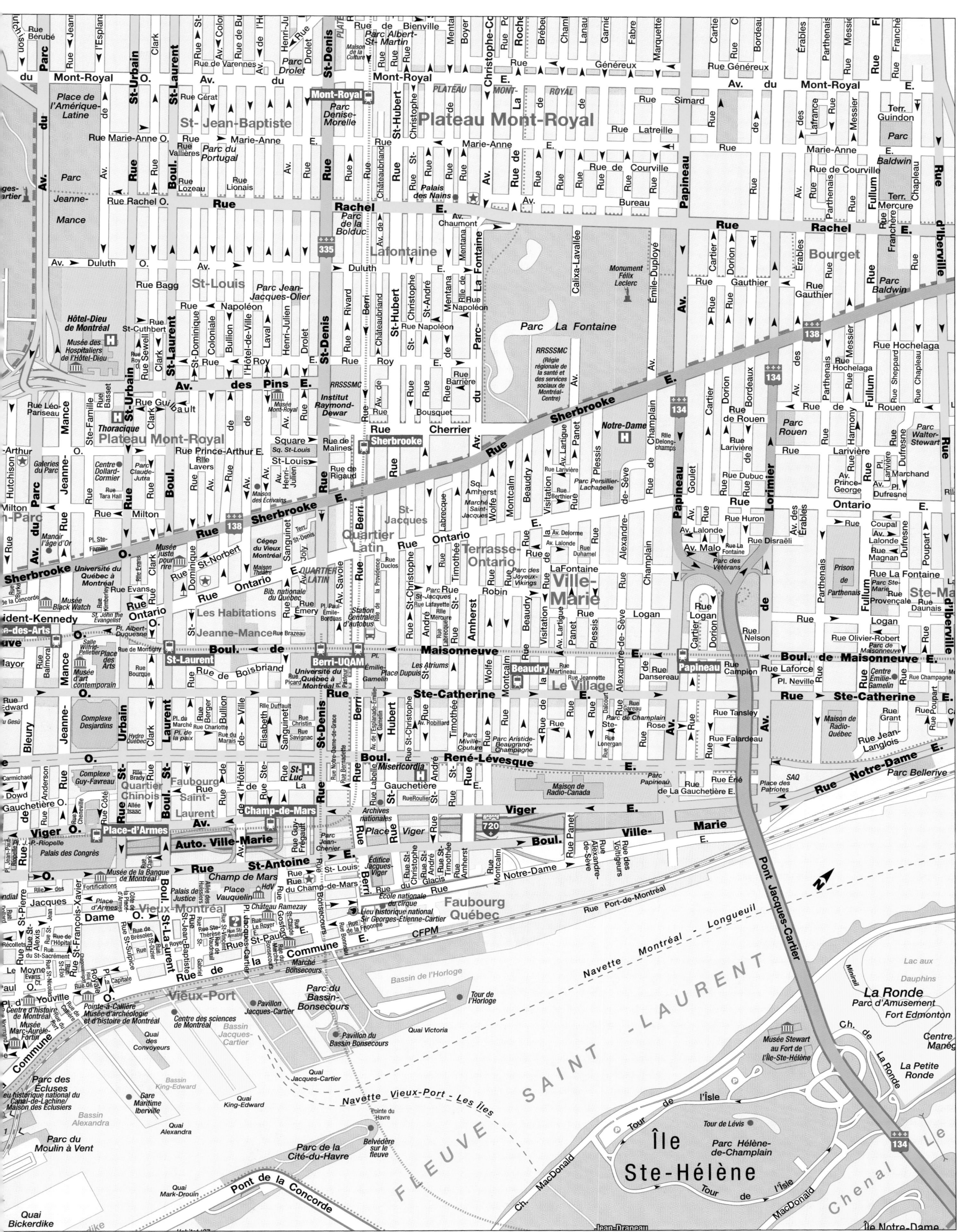

145

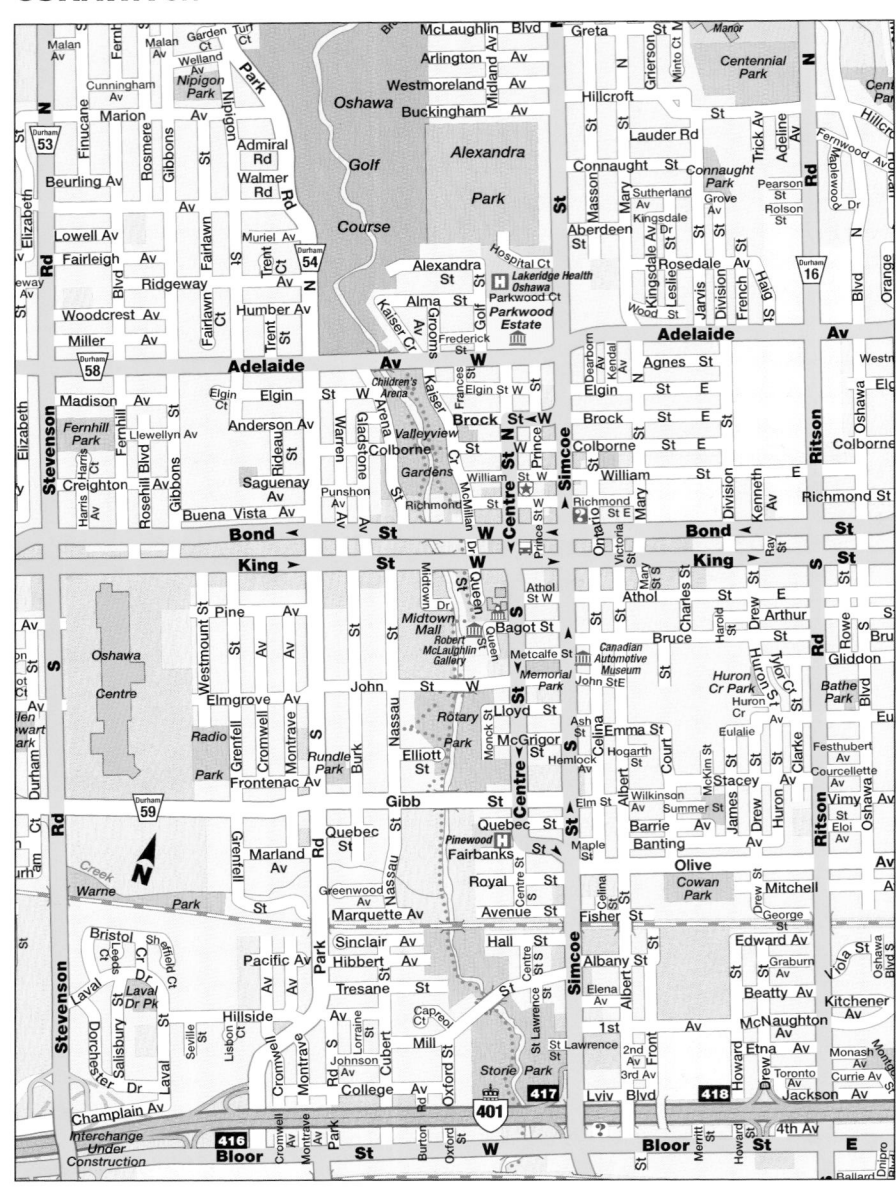

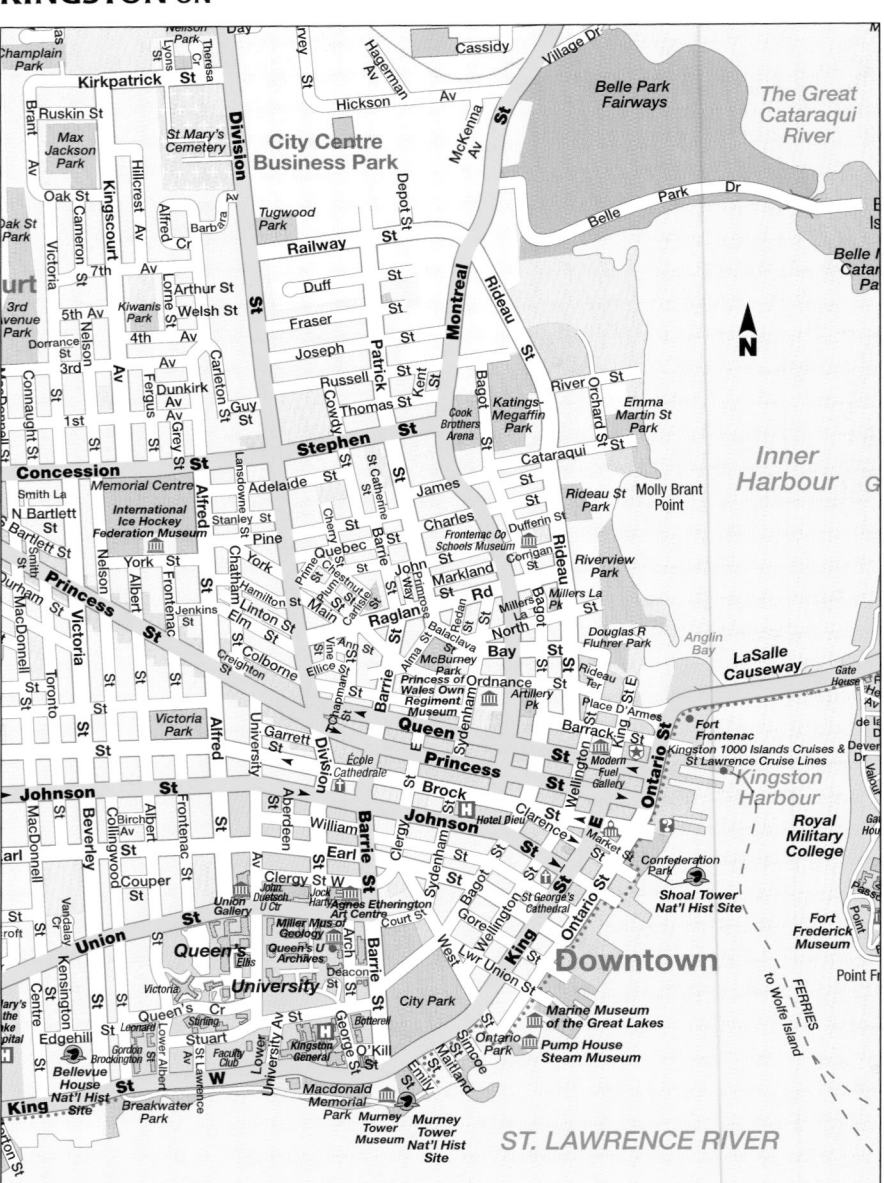

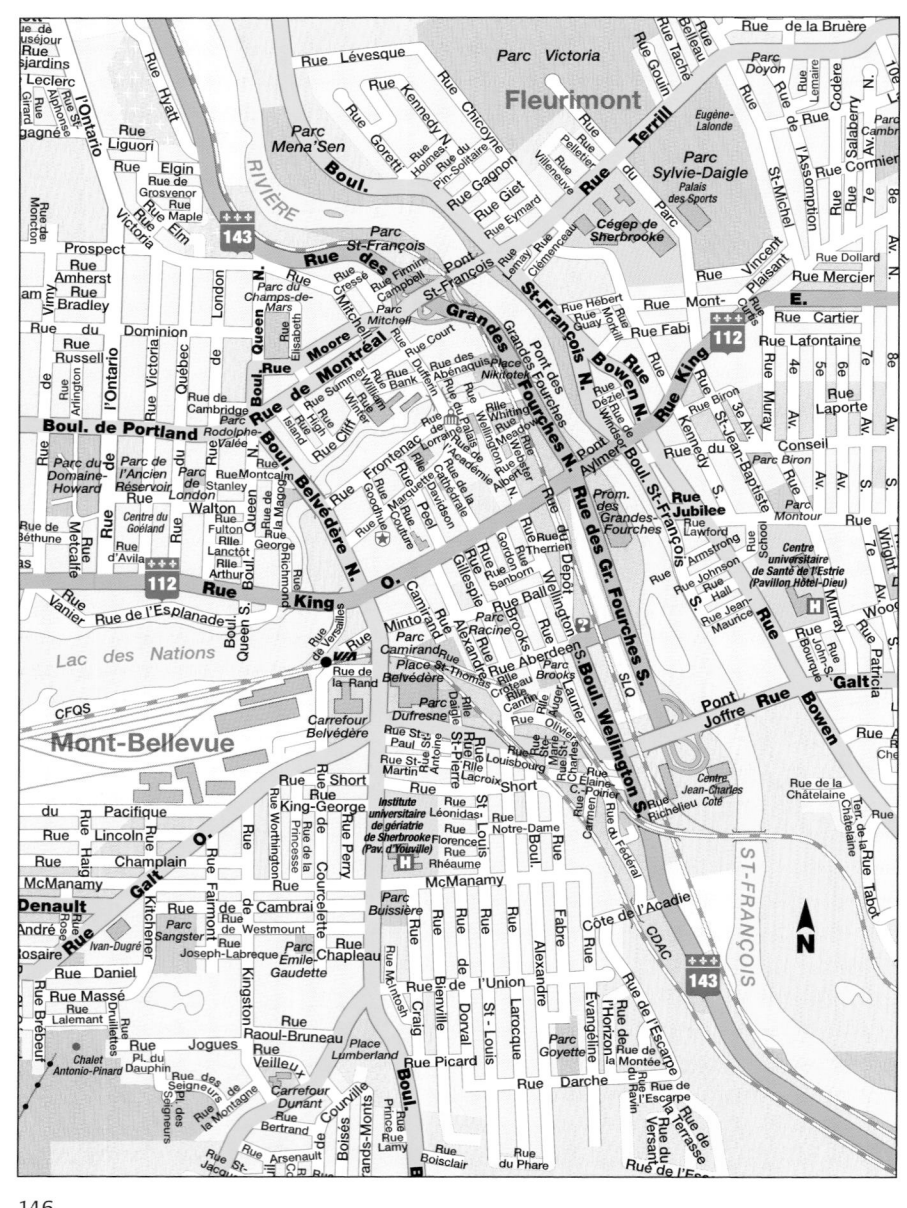

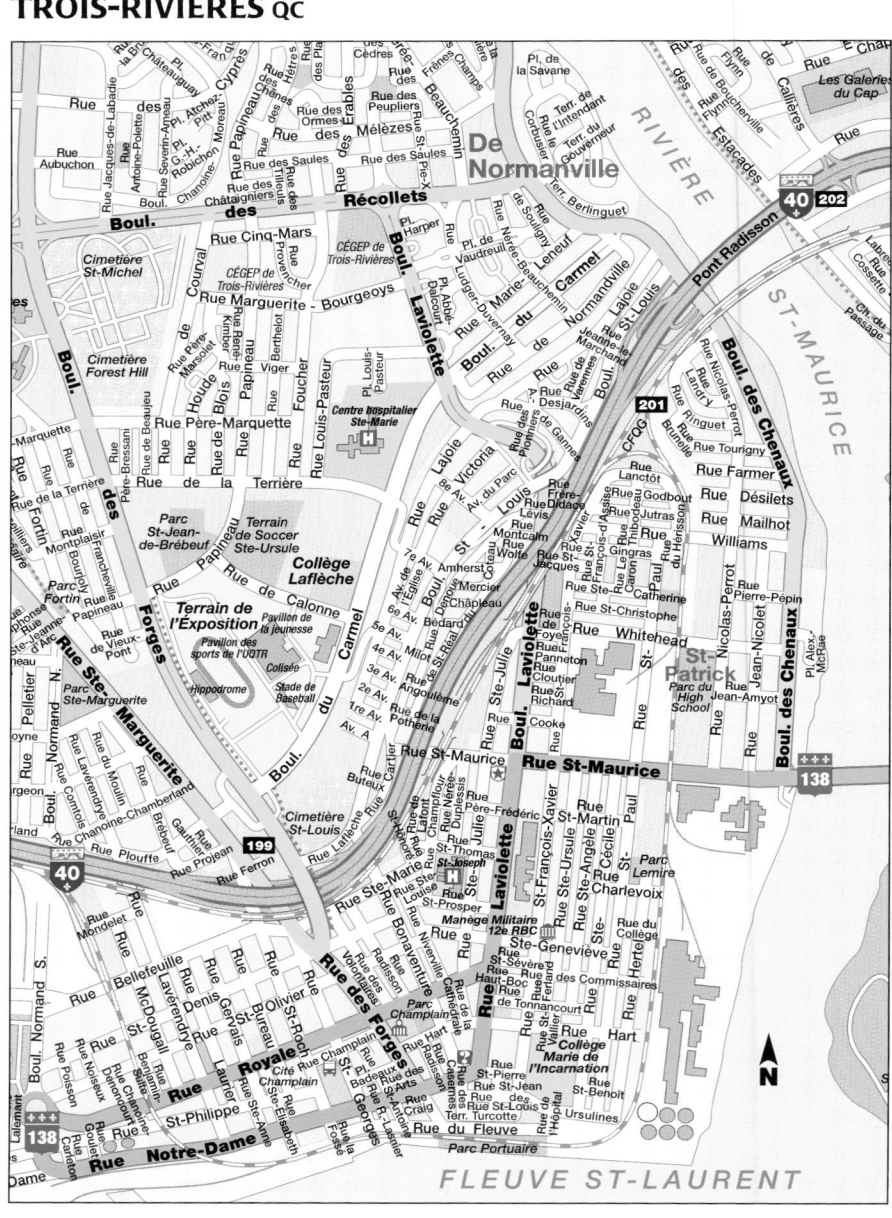

QUÉBEC QC

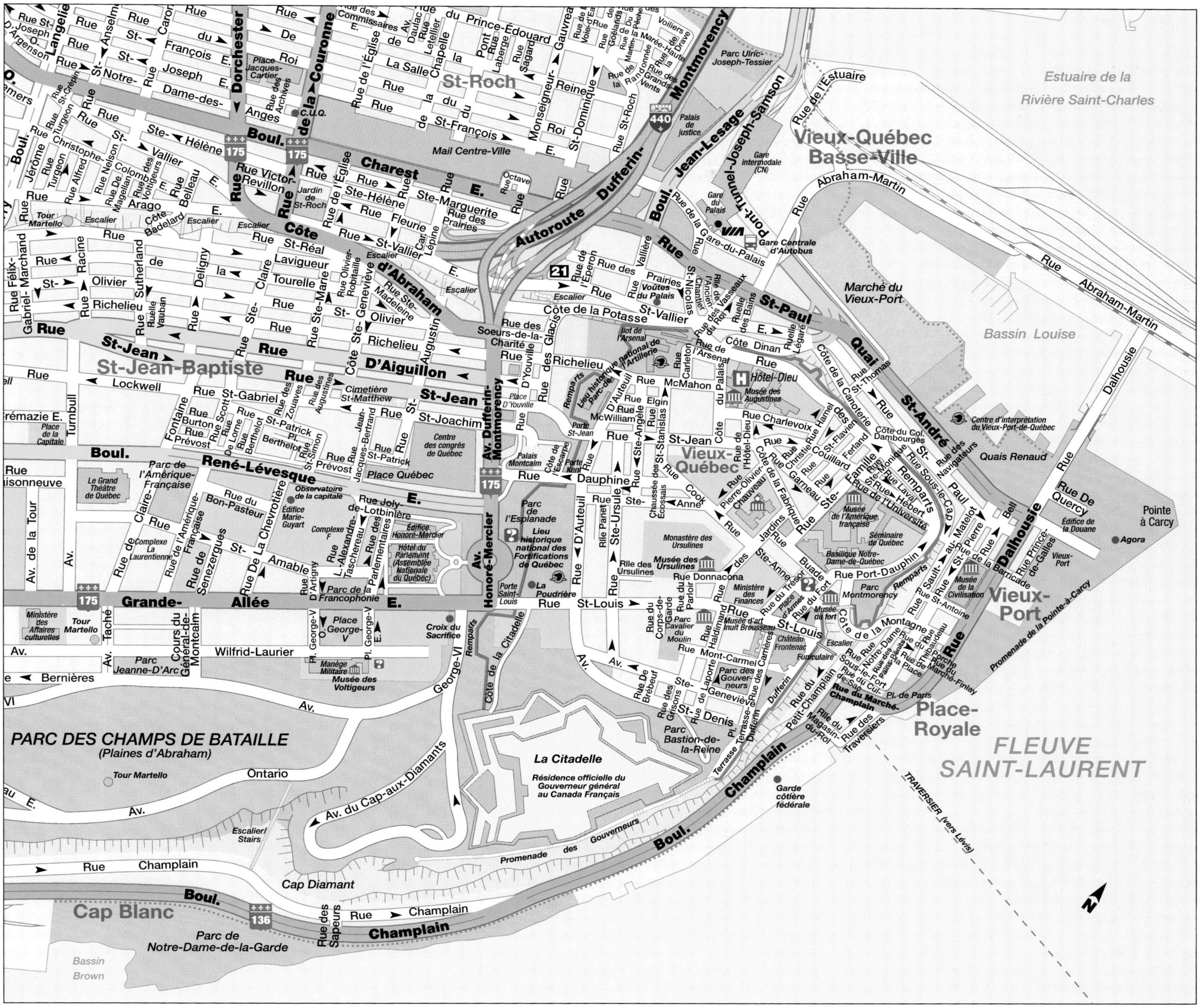

SAGUENAY QC

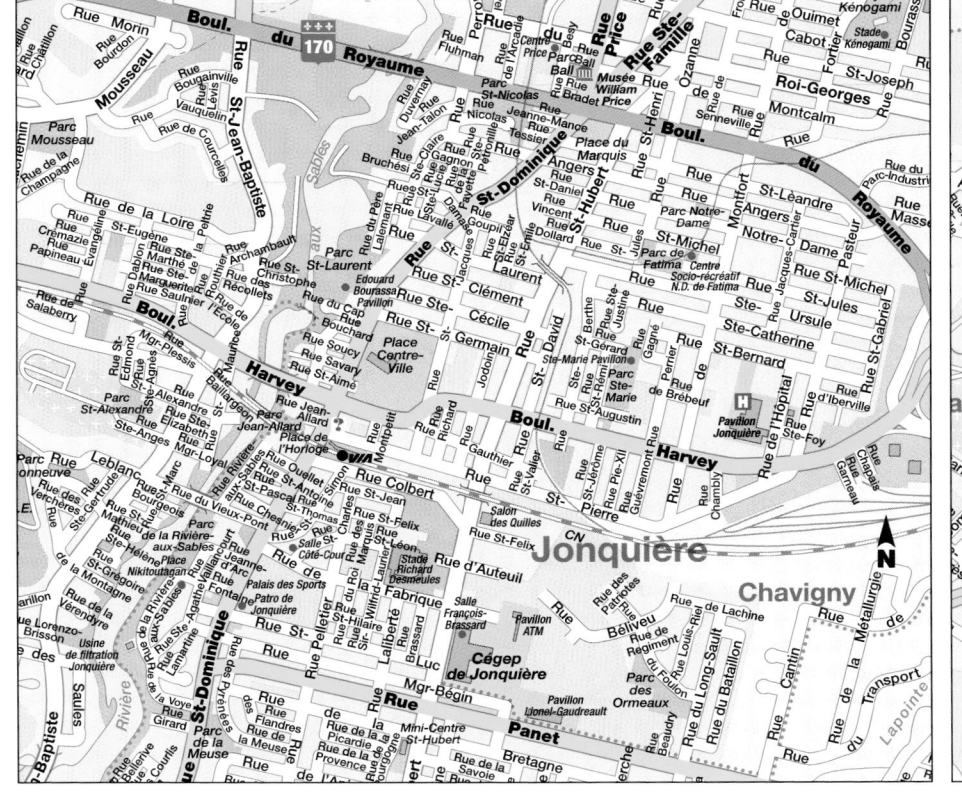

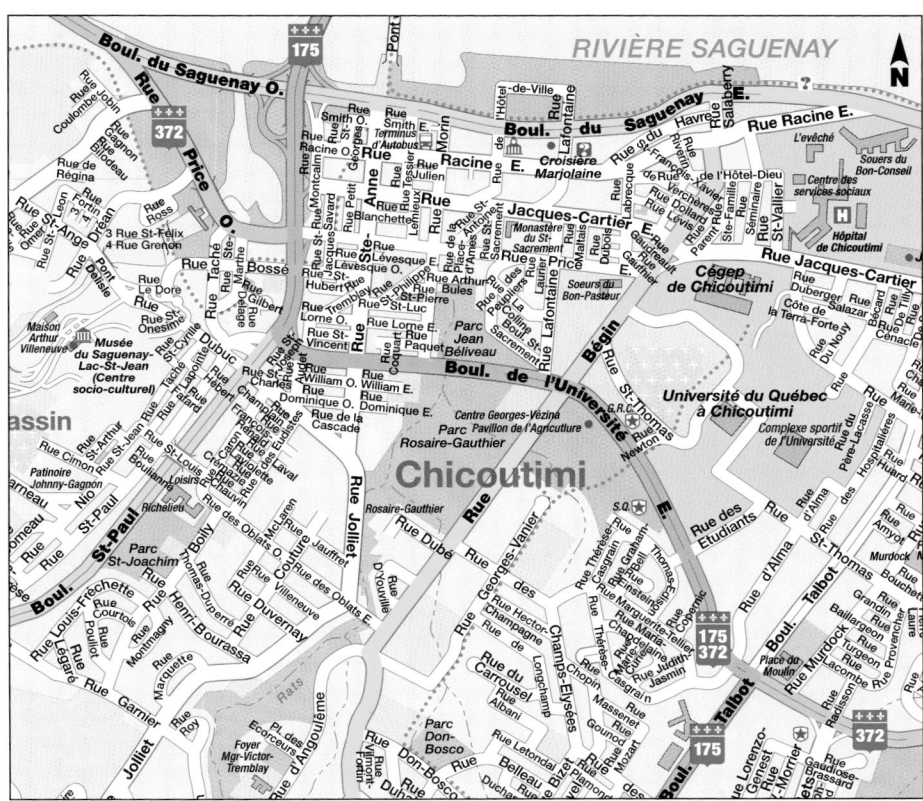

FREDERICTON NB

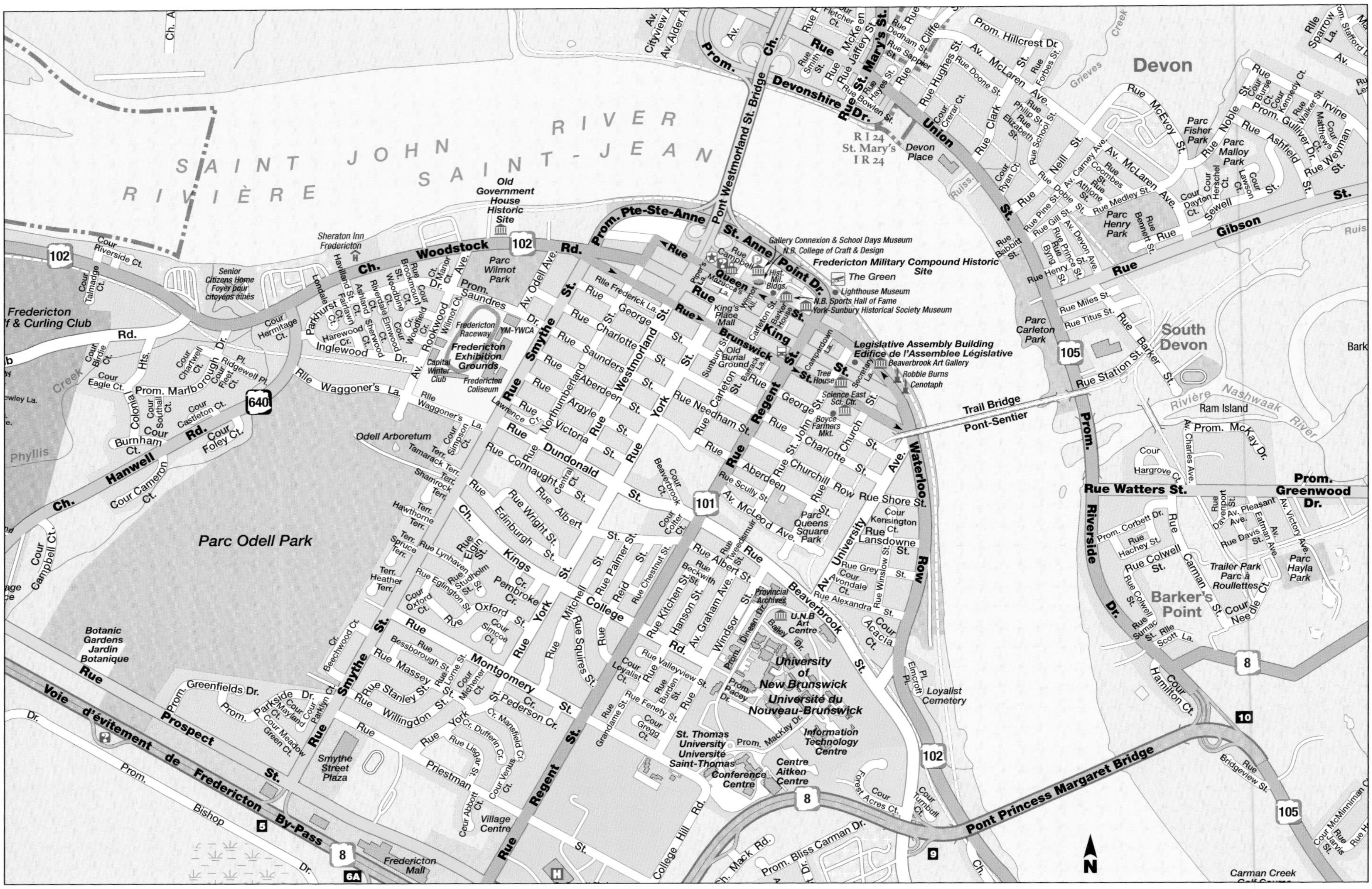

CHARLOTTETOWN PE

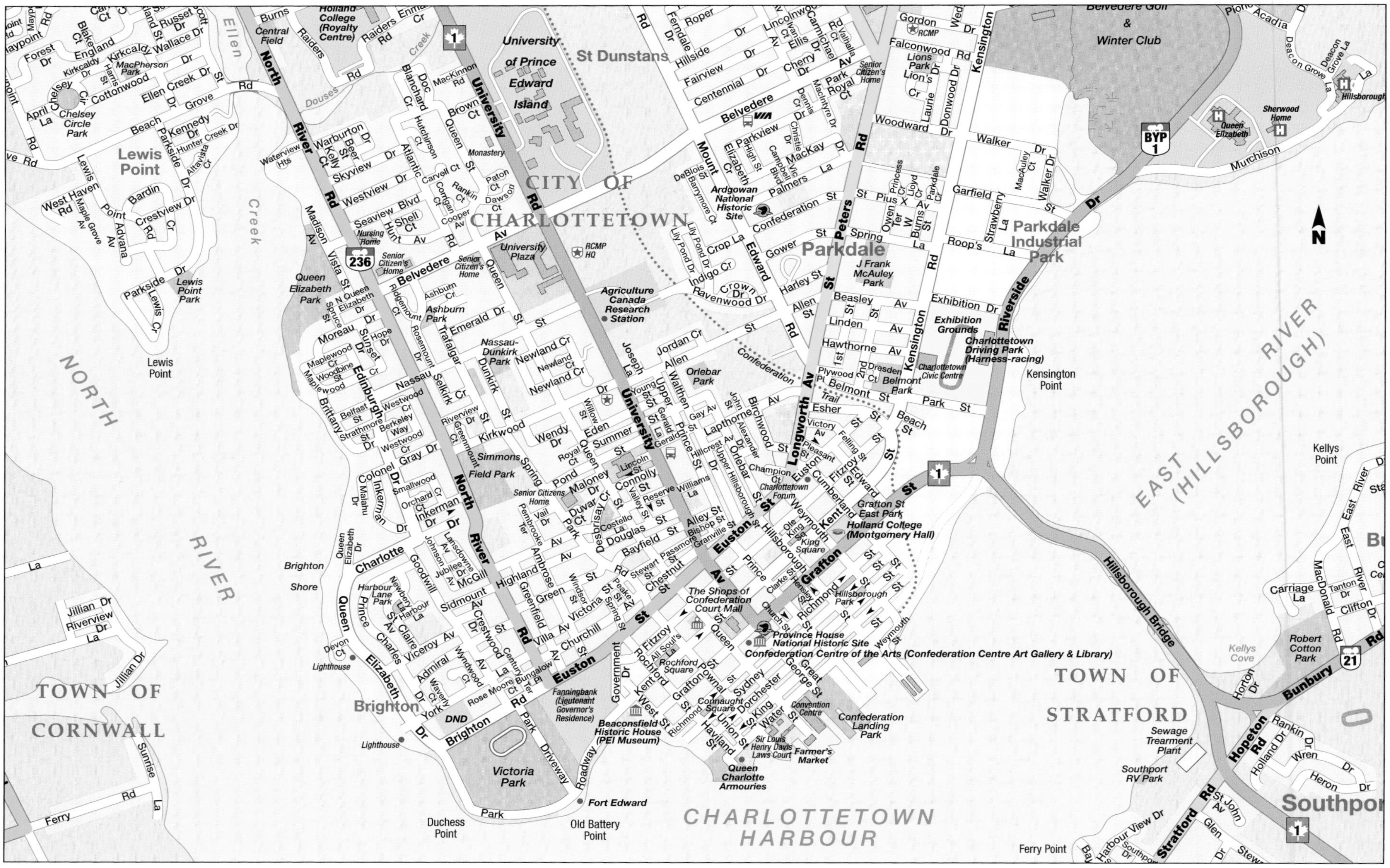

IQALUIT NU

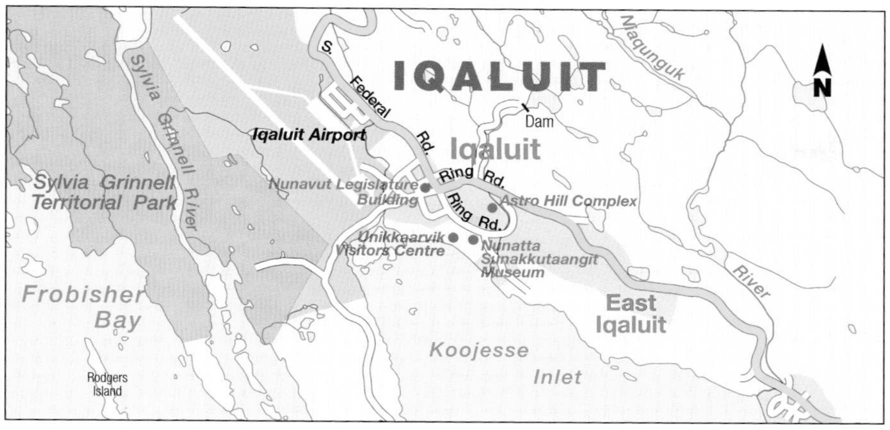

HALIFAX NS

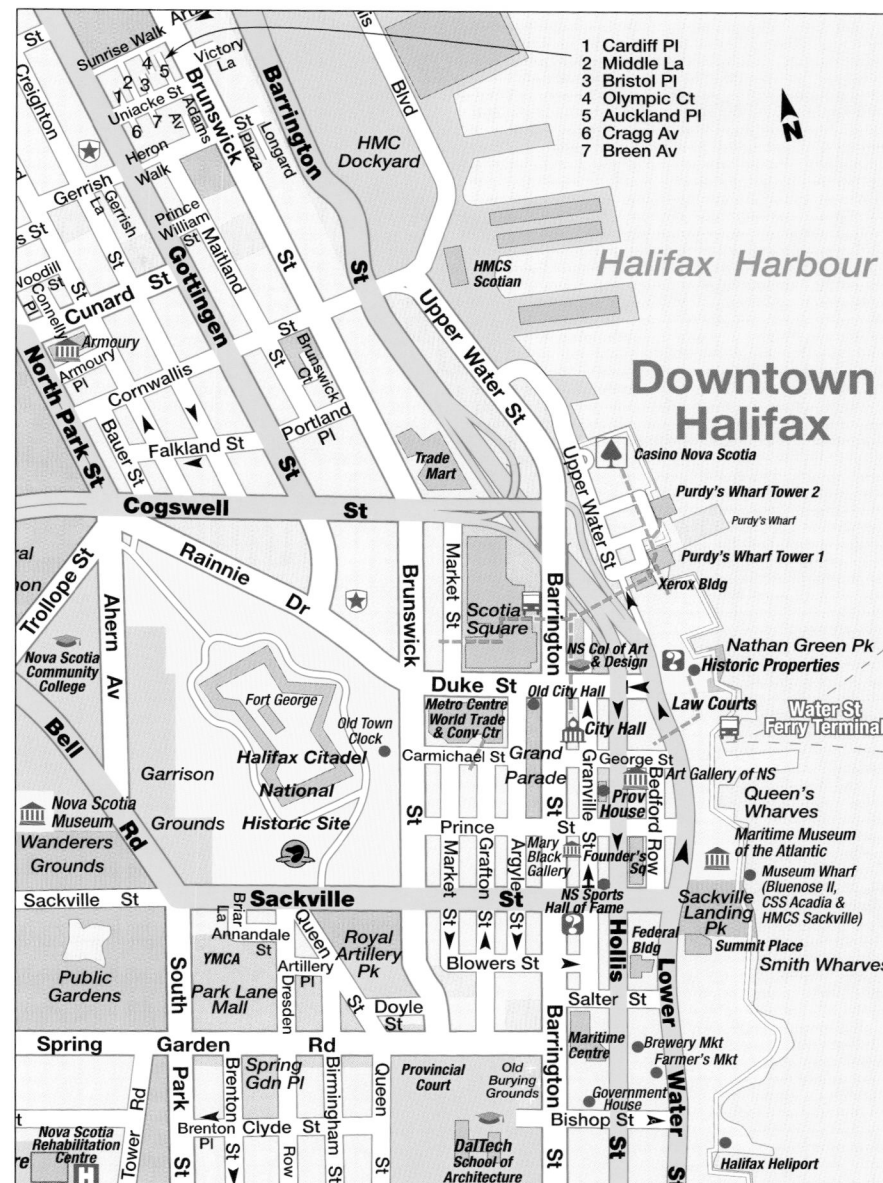

MONCTON NB

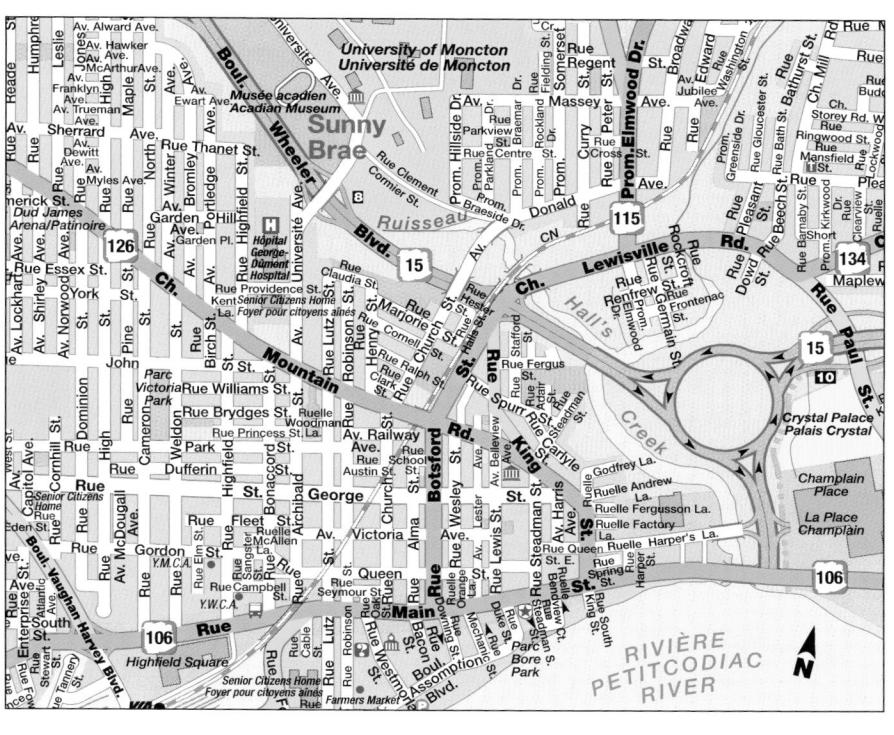

ST. JOHN'S NL

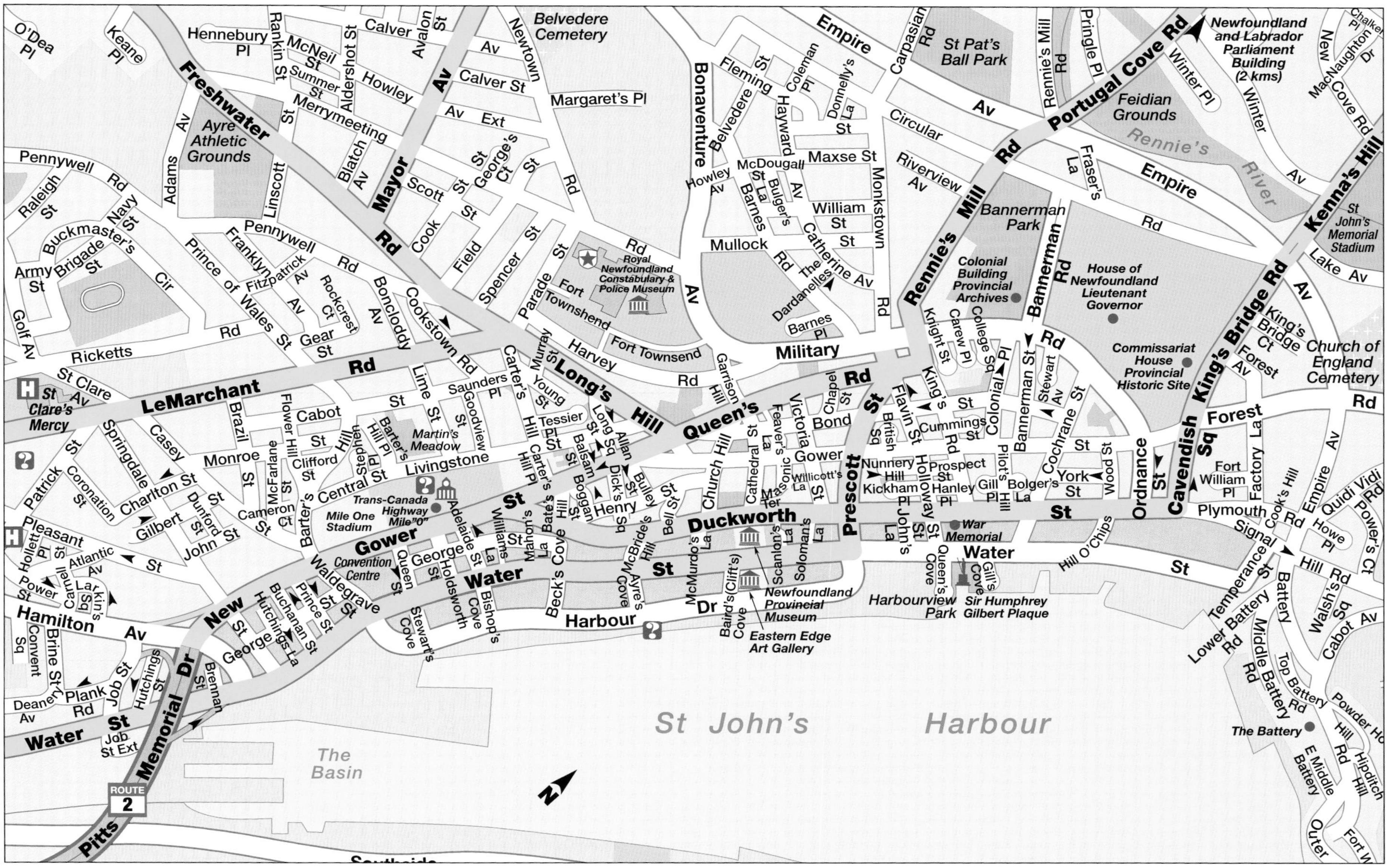

MAP INDEX

Below: *Winter scene, Jasper National Park*

HOW TO USE THE MAP INDEX

This index lists the names of more than 33,000 populated places, physical features, and other points of interest that appear on the preceding map plates. Each name is followed by a bold figure (the map plate number) and the map coordinates—the numbers at the top and bottom of the map plate, and the letters at each side of the map plate. When searching for a specific place name, use the coordinates to find the appropriate area on the map plate, then scan the area for the name. Aberdeen *NS*, for example, has the following references: **34** C6. After noting the references, go to Map Plate 34 and use the coordinates C6 to pinpoint the map area where Aberdeen appears.

ABBREVIATIONS

Provinces and Territories

AB	Alberta
BC	British Columbia
MB	Manitoba
NB	New Brunswick
NL	Newfoundland and Labrador
NS	Nova Scotia
NT	Northwest Territories
NU	Nunavut
ON	Ontario
PE	Prince Edward Island
QC	Quebec
SK	Saskatchewan
YT	Yukon

Other abbreviations used in the map index

APP	aire protégée provinciale		PP	Provincial Park/ parc provincial
CA	Conservation Area		PPA	Provincial Protected Area
HS	Historic Site		PRA	Provincial Recreation Area
LHN	lieu historique national		PRS	Provincial Recreation Site
MPP	Marine Provincial Park		PWP	Provincial Wilderness Park
NHS	National Historic Site		RF	réserve faunique
NMBS	National Migratory Bird Sanctuary		RNF	réserve nationale de faune
NMP	National Marine Park		RNOM	refuge national d'oiseaux migrateurs
NP	National Park		RNP	réserve naturelle provinciale
NPR	National Park Reserve		RPN	réserve de parc national
NWA	National Wildlife Area		SH	site historique
PCA	Provincial Conservation Area		TP	Territorial Park
PGA	Provincial Game Area		WMA	Wildlife Management Area
PHS	Provincial Historic Site		ZCP	zone de conservation provinciale
PMN	parc marin national		ZEC	zone d'exploitation contrôlée
PMP	Provincial Marine Park		ZPF	zone provinciale de faune
PN	parc national			
PNR	Provincial Nature Reserve			

Altario AB **11** C2
Alticane SK **11** B4
Alton NS **33** B6
Alton NS **34** F2
Alton ON **19** A7
Altona BC **3** D8
Altona BC **9** D1-2
Altona MB **13** F3
Altona MB **14** F4
Alvanley ON **18** F5
Alvanley ON **20** A4
Alvena SK **11** C4-5
Alvinston ON **19** C3
Alvinston ON **20** F2
Amable du Fond PP ON **22** D-E2
Amable du Fond River PP ON **22** D-E2
Amable du Fond River PP ON **17** F8
Amadjuak Lake NU **40** A5
Amadjuak Lake NU **41** C8
Amaranth MB **13** E3
Amaranth MB **14** C3
Amateewakea River ON **18** A-B6
Amazon SK **11** D5
Amberley ON **20** B3
Amber River AB **9** A-B2
Amedroz Island ON **18** B3
Amelia AB **8** B7
Ameliasburg ON **21** E5
Amery MB **15** D5
Amery, Mount AB **6** C3
Amery, Mount AB **8** F2
Amesdale ON **13** E6
Amesdale ON **16** D3
Ames Lake NT **10** B7
Amherst NS **32** C7
Amherstburg ON **19** E1
Amherst Cove NL **37** B7
Amherst Cove PE **35** E3
Amherst Head NS **35** F2
Amherst Point NS **32** C7
Amherst Point, RNOM NS **9** E6
Amherst Point NMBS NS **32** C7
Amherst Pointe ON **19** E1
Amherst Shore NS **35** F2
Amherst Shore PP NS **32** B-C8
Amherst Shore PP NS **34** D1
Amherst Shore PP NS **35** F2-3
Amherstview ON **21** E6
Amisk AB **6** C7
Amisk AB **11** B1-2
Amisk Lake SK **11** A7
Amisk Lake SK **12** F7
Amisk Lake SK **12** F7
Amisk Lake SK **15** F1
Amisk Lake SK **15** F1
Amisk PP Reserve MB **15** D3-4
Amisk River AB **6** A6
Amisk River AB **8** A8
Amisk River AB **9** F6
Amisk River AB **12** F1
Ammerman Mountain YT **5** B-C2
Amos QC **23** D4
Amour Point NL **38** A3
Amphitrite Point BC **1** F5
Amphitrite Point BC **2** E1
Amqui QC **29** C4-5
Amsterdam SK **11** C7
Amulet SK **11** F5
Amulree ON **19** B5
Amulree ON **20** D4
Amund Ringnes Island NU **42** C4
Amundsen, golfe d' NT **41** A1-2
Amundsen, golfe d' NT **42** E1
Amundsen Gulf NT **41** A1-2
Amundsen Gulf NT **42** E1
Amyot ON **17** C-D3
Anabusko River MB **15** D7
Anacla BC **2** E2
Anagance NB **32** C5
Anaharea Lake ON **17** C3-4
Analta AB **8** A6
Ancaster ON **19** B7
Ancaster ON **20** F2
Anchor Harbour NL **37** A6-7
Anchor Harbour NL **38** F8
Anchor Point NL **38** B4
Anchor Point NL **39** D7-8
Anctil, lac QC **26** D5
Anderson Bay PMP BC **2** D3
Anderson Creek YT **4** D6-7
Anderson Lake BC **1** E7
Anderson Lake BC **2** B6
Anderson Lake NT **10** B6
Anderson River BC **2** C-D7
Anderson River NT **5** A6 B2
Anderson River NT **5** C7-8
Anderson River Delta NMBS NT **5** A-B6
Anderson Road NB **31** D3
Andersons Cove NL **36** D6-7
Andersons Cove NL **37** D4-5
Anderson Settlement NB **35** E1
Andesite Peak BC **1** A4

Andou, lac QC **22** B6
Andreasen, Cape NT **42** C2
André Lake NL **39** A3
Andrew AB **6** A6
Andrew AB **8** B8
Andrew AB **11** A1
Andrew Island NS **34** E6
Andrew Lake AB **9** A7
Andrew Lake AB **12** A2-3
Andrew River NT **5** C6-7
Andrews Head NS **33** E4
Andys Corners ON **19** D6
Andys Corners ON **20** F5
Aneroid SK **11** F4
Anfield NB **31** D3
Angekum Lake ON **13** B6
Angels Cove NL **36** F8
Angels Cove NL **37** F6
Angers QC **24** D2-3
Angijak Island NU **11** A5
Angikuni Lake NU **41** E4
Anglais, rivière des QC **24** E5-6
Anglais, rivière des QC **27** F1
Anglemont BC **6** D2
Angliers QC **17** E8
Angliers QC **22** A2
Angliers QC **23** F2
Angling Lake MB **13** A6
Angling Lake ON **13** A8
Angling River MB **15** D5-6
Anglin Lake SK **11** A4-5
Anglin Lake PRS SK **11** A5
Anglo Tignish NB **34** A1
Anglo Tignish PE **35** A2
Anguille Mountains NL **36** D1
Angus ON **20** B6
Angus Ridge AB **8** D6-7
Angusville MB **11** D8
Angusville MB **13** D1
Anicet, lac QC **24** A5
Anicet, lac QC **28** E5
Anima Nipissing Lake ON **17** E7
Anishinabi Lake ON **16** C2
Anjigami Lake ON **17** D4
Annaheim SK **11** C5
Annan BC **2** D5
Annan ON **18** F5-6
Annan ON **20** A4
Annandale PE **34** C4
Annandale PE **35** E7
Annapolis Basin NS **32** E-F5
Annapolis Basin NS **32** F5
Annapolis Basin NS **33** C2
Annapolis River NS **32** F5
Annapolis Royal NS **32** F5
Annapolis Royal NS **33** C2-3
Annapolis Valley NS **32** E5-6
Annapolis Valley NS **33** B-C3
Annie Lake ON **21** A3
Annieopsquotch Mountains NL **37** C1-2
Annieoqsquotch Mountains NL **36** C3-4
Anola MB **13** E4
Anola MB **14** E6
Anse-Bleue NB **30** E1
Anse-Bleue NB **31** B7
Anselmo AB **8** B4
Anse Sloop, LHN de l' MB **15** B5
Anson Creek ON **21** C2
Anstruther Lake ON **21** C3
Anten Mills ON **18** F7
Anten Mills ON **20** A6
Anthony ON **17** C7
Anthony PP NS **32** D8
Anthony PP NS **33** A5-6
Anthony PP NS **34** E1-2
Anticline, Mount YT **4** D3-4
Anticosti, île d' QC **30** A5-6
Anticosti, île d' QC **39** F4
Anticosti, parc d' QC **40** F7
Anticosti, parc d' QC **30** A5
Anticosti, parc d' QC **39** F4
Anticosti, RF QC **30** A4-5
Anticosti, RF QC **30** A7
Anticosti, RF QC **39** F4
Anticosti Island QC **30** A5-6
Anticosti Island QC **39** F4-5
Anticosti Island QC **40** F7
Antigonish NS **34** D4-5
Antigonish Harbour NS **34** D5
Antique, lac QC **24** A5
Antique, lac QC **25** E5
Antler SK **11** F7-8
Antler SK **13** F1
Antler River SK **11** F7
Antoine, lac QC **22** D6
Antoine Creek ON **22** D2
Antoine Lake NT **10** D-E3
Antonid AB **6** A7
Antostagnan, lac QC **22** C6
Antrim NS **33** B6
Antrim NS **34** F2

Antrim ON **21** B7
Antrim ON **24** D1
Anville, lac QC **26** B3
Anvil Range YT **4** C4
Anzac AB **9** D7
Anzac AB **12** D2
Apeganau Lake MB **12** E8
Apeganau Lake MB **15** E2
Apex, Mount YT **4** C2
Apex Mountain PRA BC **5** D5-6
Apodaca PP BC **2** D4
Apohaqui NB **32** C5
Appalaches, monts QC **29** C5-6 D5
Appalachian Mountains QC **29** C5-6 D5
Appin ON **19** C-D4
Appin ON **20** F3
Appin Road PE **35** E4
Applegrove BC **6** E-F3
Apple Hill ON **24** E4
Apple River BC **2** A2
Apple River NS **32** B6
Apple River NS **33** A4
Appleton NL **36** B7
Appleton NL **37** B5
Appleton ON **21** B7
Appleton ON **24** E1
Apps Lake ON **16** A2
Apsley ON **21** C4
Apto ON **18** F7
Apto ON **20** A6
Aquadeo SK **11** B3
Aquaforte NL **37** E8
Arabella SK **11** C7
Arabella SK **13** C1
Arabian, lac QC **39** E6
Araignées, lac aux QC **27** E6-7
Ara Lake ON **16** C7
Ara Lake ON **17** A1
Arbakka MB **14** F6
Arborfield SK **11** B6
Arborg MB **13** D4
Arborg MB **14** B5
Arbor Vitae ON **13** F6
Arbuthnot SK **11** E4
Arbutus Grove PP BC **2** D-E3
Arcadia NS **33** E1
Archambault, lac QC **24** D3
Archambault, lac QC **25** F5
Archer Fiord NU **42** A5
Archerwill SK **11** C6
Archibald NB **31** B6
Archibald, Mount YT **4** E2
Archibald Lake SK **9** A-B8
Archipel-de-Mingan, LHN de l' QC **40** F7
Archipel-de-Mingan, RPN de l' QC **39** F3-4
Arcola NL **11** F7
Arctic Bay NU **41** A5
Arctic Bay NU **42** D-E5
Arctic Circle NT **5** D8
Arctic Circle YT **5** D2
Arctic Ocean NT **41** A1
Arctic Ocean NT **42** A2-3 B1-2
Arctic Ocean NT **42** A2-3 B1-3
Arctic Ocean YT **5** A3-4
Arctic Pacific Lakes PP BC **1** A7-8
Arctic Pacific Lakes PP BC **1** A7-8
Arctic Red River NT **5** E5 F6
Arctic Red River NT **5** D4-5
Arctique, cercle YT **5** D8
Arctique, cercle YT **5** D2
Arctique, océan NT **41** A1
Arctique, océan NT **42** A3 B1-3
Arctique, océan NU **42** A3 B1-3
Arctique, océan YT **5** A3-4
Ardath SK **11** D4
Ardbeg ON **18** C7
Arden MB **13** E2
Arden MB **14** D2
Arden ON **21** C6
Ardenode AB **6** D6
Ardenode AB **7** B5
Ardenville AB **7** E4-5
Ardill SK **11** E4-5
Ardley AB **8** F7
Ardmore AB **9** F7
Ardmore AB **11** A2
Ardmore AB **12** F2
Ardoch ON **21** C6
Ardoise NS **32** E8
Ardoise NS **33** B5
Ardoise NS **34** F1
Ardrossan AB **6** B6
Ardrossan AB **8** C7
Ardtrea ON **18** F8
Ardtrea ON **20** A7
Ardtrea ON **21** D1
Arelee SK **11** C3-4
Argent, rivière d' QC **22** C8
Argent, rivière d' QC **25** D2-3
Argenta BC **6** C3
Argentia NL **36** E8
Argentia NL **37** E6
Argile, lac de l' QC **22** F8

Argile, lac de l' QC **24** C2
Argo Bay NT **5** A8
Argolis ON **17** C4
Argonaut Mountain BC **6** C2-3
Argos, Cape NS **34** E6
Argyle MB **13** E3-4
Argyle MB **14** D4-5
Argyle NB **31** E3
Argyle NB **32** A2
Argyle NS **33** E2
Argyle ON **20** A8
Argyle ON **21** D2
Argyle Shore PE **34** C2
Argyle Shore PE **35** E4
Argyle Shore PP PE **32** B8
Argyle Shore PP PE **34** C1-2
Argyle Shore PP PE **35** E4
Arichat NS **34** E6
Arisaig NS **34** D4
Arisaig PP NS **34** D4
Ariss ON **19** A6
Ariss ON **20** D5
Aristazabal Island BC **1** B-C3
Arizona MB **13** E2
Arizona MB **14** E2-3
Arjay, lac QC **39** A3
Arkell ON **19** B7
Arkell ON **20** D6
Arkona ON **19** C3
Arkona ON **20** E2
Arkwright ON **18** F5
Arkwright ON **20** A3-4
Arlington ON **19** A6
Arlington Beach SK **11** D5
Arm River SK **11** D5
Armada AB **7** C5-6
Armagh QC **27** B7
Armagh, rivière QC **27** B7
Armena AB **6** B6
Armena AB **8** D7
Armi Lake NT **10** B6
Armit SK **11** B7
Armit SK **13** B1
Armit River MB **11** B7
Armit River MB **13** B1
Armley NL **11** B6
Armstrong BC **6** E2
Armstrong ON **16** D6
Armstrong ON **40** F1
Armstrong, lac QC **24** B3
Armstrong, lac QC **25** F3
Armstrong, Mount YT **4** B5
Armstrong Lake MB **13** C4
Arnaud MB **13** F4
Arnaud MB **14** F5
Arnaud, rivière QC **40** C5
Arnaud, rivière QC **41** E8
Arnes MB **13** D4
Arnes MB **14** C5
Arnold's Cove NL **36** D8
Arnold's Cove NL **37** D6
Arnot MB **15** A4
Arnoux, lac QC **23** D2
Arnprior ON **21** A7
Arnprior ON **24** D1
Arnstein ON **18** B7
Arntfield QC **17** D8
Arntfield QC **23** D2
Aroland ON **16** D7
Aroland ON **17** A2
Aroma Lake SK **11** C3
Aroostook NB **31** D3
Arran SK **11** C7
Arran SK **13** C1
Arran-Elderslie ON **18** F5
Arran-Elderslie ON **20** A4
Arran Lake ON **18** F5
Arran Lake ON **20** A3-4
Arras BC **3** E8
Arras BC **9** E1-2
Arrow Creek BC **6** F3
Arrowhead Lake AB **9** A5
Arrowhead Lake AB **10** F7
Arrowhead Lake AB **12** A1
Arrowhead Peninsula, RNP ON **16** F4
Arrowhead Peninsula PNR ON **16** F4
Arrowhead PP ON **18** D8
Arrowhead PP ON **21** B1
Arrowhead River NT **10** E3 F2
Arrow Lake ON **16** E3
Arrow Lake PP ON **16** F4-5
Arrow Lakes PP BC **6** E3
Arrow Park BC **6** E3
Arrow River MB **11** E8
Arrow River MB **13** E1
Arrow River ON **16** F5
Arrowstone PPA BC **2** A7
Arrowwood AB **6** E6
Arrowwood AB **7** C5
Artesian Lake SK **9** E-F8
Arthur ON **19** A6
Arthur ON **20** D5
Arthur, lac QC **39** E4
Arthurette NB **31** D3
Arthur Fiord NU **42** C4
Arthurville AB **8** B5
Artillery Lake NT **41** D2-3
Artland SK **9** B8
Artland SK **11** B2
Artlish Caves PP BC **1** E4
Arundel QC **24** C4
Arva ON **19** C4

Arva ON **20** E3
Arvia'juaq HS NU **41** E5
Arviat NU **40** B1
Arviat NU **41** E5
Arvida QC **28** C6
Arvila AB **8** B6
Arviliquiat NU **41** C5
Arviliquiat NU **42** F4-5
Asbestos QC **27** D4
Ascension PE **35** A2
Ascot QC **27** E4
Ascot, rivière QC **27** E4
Ascot Corner QC **27** E5
Asessippi PP MB **11** D7-8
Asessippi PP MB **13** D1
Ashburn ON **20** C8
Ashburn ON **21** E2
Ashby Lake ON **21** B5
Ash Creek BC **3** A-B3
Ash Creek BC **4** F6
Ashcroft BC **1** E8
Ashcroft BC **2** A7
Ashcroft BC **6** D1
Ashdale NS **34** E4-5
Ashern MB **13** D3
Ashern MB **12** F7
Asheweig River ON **40** E1-2
Ashgrove ON **19** A7
Ashgrove ON **20** D6
Ashmont AB **6** A7
Ashmont AB **11** A1
Ashmont AB **12** F1
Ashnola River BC **2** E8
Ashnola River BC **6** E1
Ash River BC **2** D2
Ashton ON **21** B7
Ashton ON **24** E1-2
Ashton PE **35** D7
Ashuanipi NL **39** C2
Ashuanipi Lake NL **39** C2
Ashuanipi Lake NL **40** E6-7
Ashuanipi River NL **39** B-C2
Ashuapmushuan, lac QC **26** C5-6
Ashuapmushuan, lac QC **26** C6 D7
Ashuapmushuan, RF QC **28** A1
Ashuapmushuan, RF QC **26** D6-7
Ashuapmushuan, RF QC **28** A1-B2
Ashuapmushuan, rivière QC **26** C6 D7
Ashville MB **11** D8
Ashville MB **13** D2
Ashville MB **14** B1
Asipoquobah Lake ON **13** A7
Aspen NS **34** E4
Aspen Beach PP AB **6** E5-6
Aspen Beach PP AB **8** E5-6
Aspen Brook NL **36** B5-6
Aspen Brook NL **37** B3-4
Aspen Cove NL **38** E-F7
Aspen Grove BC **1** E8
Aspen Grove BC **2** C8
Aspen Grove BC **6** E1
Aspotogan NS **32** F8
Aspotogan NS **33** C5
Aspy Bay NS **34** A7
Asquith SK **11** C4
Assaikwatamo River MB **15** D7
Assean Lake MB **15** E4
Assemetquagan, rivière QC **29** C6 D5-6
Assignу Lake SK **11** E6
Assiniboia SK **11** F4-5
Assiniboine, Mount AB **6** D6-8
Assiniboine, Mount AB **7** C2
Assiniboine River MB **13** F2-3
Assiniboine River MB **14** E3
Assinica, lac QC **40** F4
Assinica, parc QC **40** F4-5
Assomption, rivière l' QC **24** B5-6
Assomption, rivière l' QC **27** C1-2
Astle NB **31** E5
Astle NB **32** A3-B3
Aston-Jonction QC **24** B8
Aston-Jonction QC **27** C4
Astorville ON **17** F8
Astorville ON **18** B8
Astorville ON **22** D1-2
Astray NL **39** A2
Aswapiswanan Lake MB **13** A5
Aswapiswanan Lake MB **15** F5
Athabasca AB **6** A6-7
Athabasca AB **9** F6
Athabasca, lac AB **41** F2
Athabasca, lac SK **9** A8
Athabasca, lac SK **12** A3
Athabasca, lac SK **41** F2
Athabasca, Lake **12** A3 B2
Athabasca, Lake AB **41** F2
Athabasca, Lake SK **9** A8
Athabasca, Lake SK **12** A3
Athabasca, Lake SK **41** F2
Athabasca, rivière AB **6** A4
Athabasca, rivière AB **8** A5
Athabasca, rivière AB **8** B3
Athabasca, rivière AB **9** F5
Athabasca, rivière AB **12** C1
Athabasca, rivière AB **41** F1
Athabasca Landing AB **6** A6-7

Athabasca Landing AB **9** F6
Athabasca Pass NHS BC **6** C3
Athabasca Pass NHS BC **8** E1
Athabasca River AB **6** A3
Athabasca River AB **8** A5
Athabasca River AB **8** B-C6
Athabasca River AB **9** B-C6
Athabasca River AB **9** E6
Athabasca River AB **12** C1-2
Athabasca River AB **12** D-E1
Athabasca River AB **41** F1
Athabasca Sand Dunes PWP SK **9** A-B8
Athabasca Sand Dunes PWP SK **12** B3
Athabasca Sand Dunes PWP SK **41** F2
Athalmer BC **6** E4
Athalmer BC **7** C1
Athapap MB **11** A7
Athapap MB **12** F7
Athapap MB **15** F1
Athelstan QC **24** E5
Athenia Lake NT **10** B7-8
Athens ON **21** C7
Athens ON **24** F1-2
Atherley ON **18** F8
Atherley ON **20** A7
Atherley ON **21** D1
Athlone Island BC **1** C3
Athlow Bay NT **1** A1
Athol NS **32** C7
Athol Bay QC **21** F5
Atholville NB **29** D6
Atholville NB **31** A4
Aticonipi, lac QC **39** D6
Atikaki PWP MB **13** D5
Atikaki PWP MB **14** A7
Atikaki PWP MB **16** B1
Atikameg AB **9** F6
Atikameg Lake MB **11** A8
Atikameg Lake MB **13** A2
Atikamisis Lake AB **9** E4
Atik Lake MB **11** A7
Atik Lake MB **15** E-F5
Atikokan ON **13** F7
Atikokan ON **40** F1
Atikokan ON **16** E3-4
Atikonak Lake NL **39** C3
Atikonak Lake NL **40** E7
Atikonak River NL **39** C3
Atik River MB **15** E5
Atikwa Lake ON **13** F6
Atikwa Lake ON **16** D-E2
Atkinson (Fox) Lake MB **16** A7
Atkinson Point NT **5** A5
Atlantic Ocean NS **32** E2-3 F2-3
Atlantic Ocean NS **37** F1-2
Atlantic Ocean NS **33** D6-7 E6-7
Atlantic Ocean NS **34** E7-8 F7-8
Atlantic Ocean NS **33** E7 F6-8
Atlantique, océan NL **32** F3-4
Atlantique, océan NL **37** F4-6
Atlantique, océan NL **33** E7 F6-8
Atlantique, océan NL **34** F6-7
Atlantique, océan NL **33** B-C5
Atlantique, océan NL **34** F4-7
Atlee AB **7** B-C8
Atlin BC **3** A1
Atlin BC **4** F4
Atlin Lake BC **3** A1
Atlin Lake BC **4** F4
Atlin PP BC **3** A1
Atlin PP BC **4** F4
Atlin PRA BC **3** A1
Atlin PRA BC **4** F4
Atmore AB **9** F6
Atmore AB **12** F1
Attawapiskat AB **40** E3
Attawapiskat Lake ON **16** A7
Attawapiskat Lake ON **40** E2
Attawapiskat River ON **16** A7-8
Attawapiskat River ON **40** E2
Attercliffe ON **19** C8
Attercliffe ON **20** E7
Attica SK **11** D7
Attihamagan Lake NL **40** D6
Attikamagen Lake NL **39** A2
Attwood Lake ON **16** D-C6
Atwater SK **11** D7
Atwood NS **19** A5
Atwood ON **20** C4
Atwood Brook NL **33** F2
Aubergé-de-Ski-Skoki, LHN de l' AB **6** D4-5
Aubergé-de-Ski-Skoki, LHN de l' AB **7** A1-2
Aubichon Arm SK **9** C7
Aubigny MB **14** E5
Aubinadong River PP ON **17** E4-5
Aubrey Falls PP ON **17** E4-5
Aubry Lake NT **5** D7
Auburn NS **32** E6
Auburn NS **33** B3
Auburn ON **19** A4
Auburn ON **20** C3

Auburn PE **35** E5
Auclair QC **29** E2-3
Auclair QC **31** E1
Auden ON **16** D7
Auden ON **17** A1
Audet QC **27** E6-7
Augusta, lac QC **28** B8
Augustine Cove PE **34** C1
Augustine Cove PE **35** E3
Augustines, lac des QC **22** A8
Augustines, lac des QC **23** F8
Augustines, lac des QC **25** C2
Aujuittuq (Grise Fiord) NU **42** C5
Aulac NB **32** C7
Aulac NB **35** F1
Aulavik NP NT **42** D1
Aulds Cove NS **34** D5
Auleneau Peninsula ON **13** F5-6
Auleneau Peninsula ON **16** E1-2
Aumont, lacs QC **22** C5
Aupaluk QC **40** C5
Aurora ON **20** B-C7
Aurora ON **21** E1
Ausable River ON **19** B4 C3
Ausable River ON **20** D3 E2
Ausable River ON **20** D-E3
Austin MB **13** E3
Austin MB **14** D3
Austin QC **24** E8
Austin QC **27** F4
Australia Mountain YT **4** B2
Authier QC **17** C8
Authier QC **23** C3
Authier, lac QC **23** C3
Authier, rivière QC **23** C3
Authier-Nord QC **17** C8
Authier-Nord QC **23** C3
Auyuittuq NP NU **41** B8
Auyuittuq NP NU **42** F7-8
Avaugour, lac QC **26** D6
Avaugour, lac QC **28** A1
Avening ON **18** A6
Aventure, lac de l' QC **26** B5
Avola BC **6** C2
Avon BC **19** C5
Avon ON **20** E4
Avondale NL **36** E2-3 F2-3
Avondale NL **37** F1-2
Avondale NL **32** F4
Avondale NS **33** B5
Avondale NS **34** E4
Avondale NS **34** F1
Avondale PE **35** E6
Avonhurst SK **11** E6
Avonlea SK **11** E5
Avonmore ON **24** E3
Avonport NS **32** E7
Avonport NS **33** B5
Avon River NS **32** E7
Avon River NS **33** B-C5
Avon River ON **19** B5
Avon River ON **20** D4
Avonton ON **19** B5
Avonton ON **20** D4
A.W. Campbell CA ON **19** C3
A.W. Campbell CA ON **20** F2
Awenda PP ON **18** E7
Awry Lake NT **10** C6
Axe Lake ON **18** D8
Axe Lake ON **21** A-B1
Axe Lake ON **22** F1
Axel Heiberg Island NU **42** B4
Ayer's Cliff QC **27** F4
Aylen Lake ON **21** A4
Aylen Lake ON **22** F4
Aylesbury SK **11** D5
Aylesford NS **32** E6
Aylesford NS **33** B4
Aylesford Lake NS **32** E6-7
Aylesford Lake NS **33** B4
Aylmer ON **19** D5
Aylmer ON **20** F4
Aylmer ON **21** A8
Aylmer, lac QC **27** D5
Aylmer, Mount AB **6** D4-5
Aylmer, Mount AB **7** B2
Aylmer Lake NT **41** D2-3
Aylmer Sound QC **39** F4
Aylsham SK **11** B6
Aylwin QC **22** F5
Aylwin QC **24** C1
Ayr ON **19** B6
Ayr ON **20** D5
Ayton ON **20** B4
Azilda ON **17** F6
Azilda ON **18** A5
Azure AB **7** C4
Azure Lake BC **6** C2
Azure Mountain BC **6** C2

B

Babbage River YT **5** B3
Babine Lake BC **3** F5
Babine Lake (Pendleton Bay) PMP BC **1** A5
Babine Lake (Pendleton Bay) PMP BC **3** F5
Babine Lake (Smithers Landing) PMP BC **3** F5
Babine Mountains PP BC **3** F4-5
Babine Range BC **3** F4-5
Babine River BC **3** E4-5
Babine River Corridor PP BC **3** E4-5 F5
Baby, lac QC **22** A2
Baby, lac QC **23** F2
Baccalieu Island NL **37** C8
Baccalieu Island Ecological Reserve NL **37** C8
Baccaro NS **33** F2-3
Baccaro Point NS **33** F2-3
Bachelor, lac QC **23** A2-3
Bachelor, lac QC **26** B2
Bachois, lac QC **26** B5-6
Back MB **15** C5
Back Bay NB **32** E2
Backbone Ranges NT **4** A6 B7
Back Cove NL **38** E4
Back Harbour NL **38** E4
Back Lake NT **10** B8
Back River NU **41** C4
Back River NU **41** D3
Backus CA ON **19** D6
Backus CA ON **20** F5
Bacon Cove NL **37** E7-8
Bacon Ridge MB **13** D2
Bacon Ridge MB **14** B2
Badajoz, lac QC **25** D4
Baddeck NS **34** C7
Baddow ON **20** A8
Baddow ON **21** D2
Baden MB **11** B7
Baden MB **13** B1
Baden ON **19** B5-6
Baden ON **20** D4-5
Badesdawa Lake ON **16** D5
Badgeley Island ON **18** B4
Badgeley Point ON **18** B4
Badger MB **13** F4-5
Badger MB **14** F4
Badger NL **36** B5
Badger NL **37** B3
Badger NL **36** A6
Badger NL **37** A4
Badger Bay NL **36** B5-6
Badger Bay NL **37** A4
Badger Bay NL **38** F5
Badger's Corners ON **18** B4
Badger's Quay NL **37** A6-7
Badger's Quay NL **38** F8
Bad Hills SK **11** D3
Badjeros ON **20** B5
Bad Lake SK **11** D3
Badwater River SK **12** B5
Bad Weather Cape NU **42** B4
Baezaeko River BC **1** B8
Baezaeko River BC **6** A-B1
Baffin, baie de NU **41** A8
Baffin, baie de NU **42** D7-8
Baffin, île de NU **41** A7-8 B7-8
Baffin, île de NU **42** E6-7 F7-8
Baffin, île de **41** A7-8 B7-8
Baffin, île de NU **42** E6-7 F7
Baffin Bay NU **41** A7
Baffin Bay NU **42** D6-7
Baffin Island NU **40** A5-6
Baffin Island NU **41** A7-8 B7-8
Baffin Island NU **42** E6-7 F7
Bagg, Mount BC **1** B8
Bagg, Mount BC **6** A-B1
Bagot MB **13** E3
Bagot MB **14** D3
Bagotville QC **28** C6
Baie, lac de la QC **26** B4
Baie-Brador, RNOM de la QC **38** B3
Baie-Brador, RNOM de la QC **39** D-E4
Baie-Comeau QC **29** A3-4
Baie-Comeau QC **39** F1
Baie-Comeau QC **40** F6
Baie-de-l'Isle-Verte, RNF de la QC **29** D1
Baie de l'Isle-Verte NWA QC **29** D1
Baie-de-Percé, parc de la QC **30** C3
Baie-des-Bacon QC **29** C1-2
Baie-des-Ha! Ha! QC **38** C1
Baie-de-Shawinigan QC **24** A7
Baie-de-Shawinigan QC **25** F7
Baie-de-Shawinigan QC **27** B2-3
Baie-des-Homards QC **39** F2
Baie-des-Loups, RNOM de la QC **39** F2
Baie-des-Loups NMBS QC **39** F2
Baie-des-Rochers QC **28** D8
Baie-des-Sables QC **29** B4
Baie-du-Febvre QC **24** B8
Baie-du-Febvre QC **27** C3
Baie-du-Renard QC **30** A7
Baie-Egmont PE **35** D2
Baie-James QC **17** B8
Baie-Johan-Beetz QC **39** F4
Baie-Johan-Beetz QC **40** F7

153

Baies, lac des QC 22 B4
Baies, lac des QC 28 A7
Baies, lac des QC 29 D2
Baie-Sainte-Anne NB 31 D8
Baie-Sainte-Catherine QC 28 D8
Baie-Sainte-Catherine QC 29 D1
Baie-Sainte-Claire QC 39 F3
Baie-Saint-Paul 28 E-F7
Baie-Trinité QC 39 F1-2
Baie-Trinité QC 40 F6
Baie Verte NB 32 B7
Baie Verte NB 34 D1
Baie Verte NB 35 F2
Baie Verte NB 38 E4
Baie Verte NL 39 F8
Baie Verte Peninsula NL 38 E4-5
Baie Verte Road NB 35 F2
Baie-Wellers, RNF de QC 21 E4-5
Baikie Lake NL 39 B-C3
Bailey Brook NS 34 D4
Bailey Creek ON 20 B-C6
Bailey Lake SK 12 A6
Bailey Lake SK 15 A1
Bailey Point NT 42 D2
Bailieboro ON 21 E3
Baillairgé, lac QC 28 B1
Baillargé, lac QC 26 D5
Baillie NB 32 D1
Baillie-Hamilton Island NU 42 C-D4
Bailly, lac QC 23 C8
Bailly, lac QC 26 D3
Baine Harbour NL 36 E7
Baine Harbour NL 37 E5
Bain Lake MB 12 A-B8
Bain Lake MB 15 B3
Bainsville ON 24 E4
Baird Peninsula NU 41 B7
Baird Peninsula NU 42 F6
Baird Point BC 1 A2
Bajo Point BC 1 F4
Baker BC 7 E2
Baker, lac NB 29 F2
Baker, Mount YT 4 C1
Baker Brook NB 29 F3
Baker Brook NB 31 C1
Baker Creek BC 1 B7
Baker Lake NU 40 A1
Baker Lake NU 41 D4
Baker Lake (Qamanit'uaq) NU 41 D4-5
Bakers Brook NL 38 E2
Baker Settlement NS 33 D4
Bakers Narrows MB 12 F7
Bakers Narrows MB 15 F1
Bakers Narrows PP MB 12 F7
Bakers Narrows PP MB 15 F1
Bala ON 18 E8
Bala ON 21 C1
Balaclava ON 18 F5-6
Balaclava ON 20 A4
Balaclava ON 21 A-B6
Balaton Beach MB 14 B5
Balcarres SK 11 D6
Balderson ON 21 C7
Balderson ON 24 E1
Baldock Lake MB 15 D3
Baldonnel BC 3 B8
Baldonnel BC 9 D1-2
Bald Peak NB 29 F5
Bald Peak NB 31 C4
Baldur MB 13 F2
Baldur MB 14 F2
Baldwin ON 20 B7
Baldwin ON 21 E1
Baldwin Mills QC 27 F4
Baldwinton SK 6 B8
Baldwinton SK 11 B2-3
Baldy Hughes BC 1 B7
Baldy Mountain MB 11 D8
Baldy Mountain MB 13 D1
Baleine, rivière à la QC 40 C-D6
Balfour BC 6 F3
Balgonie SK 11 E6
Balijennie SK 11 C3
Balinhard, Mount AB 6 B3
Balinhard, Mount AB 8 E1-2
Ball, Cape BC 1 A1-2
Ballantine AB 8 B5
Ballantrae ON 20 B7
Ballantrae ON 21 E1
Ballantyne Bay SK 12 F6
Ballantyne River SK 11 A1
Ballantyne River SK 12 F6
Ballantynes Cove NS 34 D4
Ballantynes Strait NT 42 C2
Ballard, Cape NL 37 F8
Ballinafad ON 19 A7
Ballinafad ON 20 C6
Ballingall Islets Nature PP BC 2 E4
Ball's Falls CA ON 19 C8
Ball's Falls CA ON 20 E7
Ballymote ON 19 C4-5
Ballymote ON 20 E3-4
Balm Beach ON 18 F7
Balm Beach ON 20 A6
Balmertown ON 13 D6
Balmertown ON 16 C2
Balmoral MB 13 E3-4
Balmoral MB 14 D5
Balmoral ON 19 C7
Balmoral ON 20 E6
Balmoral Mills NS 34 D-E2

Balmoral Mills PP NS 34 E2-3
Balmy Beach MB 14 F2
Balmy Beach ON 18 F5
Balmy Beach ON 20 A4
Balsam Creek ON 17 F8
Balsam Creek ON 22 D2
Balsam Lake ON 20 A8
Balsam Lake ON 21 D2
Balsam Lake PP ON 20 A8
Balsam Lake PP ON 21 D2
Baltic PE 34 B4
Baltic PE 35 D8
Baltimore ON 21 E3-4
Balzac ON 6 D5
Balzac AB 7 B4
Bamaji Lake ON 13 D7
Bamaji Lake ON 16 B-C4
Bamberg ON 19 B5-6
Bamberton PP BC 2 F4
Bambridge Lake MB 12 A8
Bambridge Lake MB 15 A3
Bamfield BC 2 E2
Banc-de-Paspébiac, SH du QC 30 D1-2
Bancroft ON 21 B4
Banff AB 6 D4
Banff AB 7 B2
Banff, PN AB 6 D4
Banff, PN AB 7 A1-2
Banff, PN AB 7 A1-2 B2
Banff, PN AB 6 D4
Banff NP AB 7 A1
Banff Park Museum NHS AB
Bangall, lac QC 22 E7-8
Bangall, lac QC 24 C2
Bangle Lake MB 12 A8
Bangle Lake MB 15 A3
Bangor NS 33 D1
Bangor PE 34 C3
Bangor PE 35 D6
Bangor SK 11 D7
Bangs Falls NS 33 D4
Bankeir BC 2 C8
Bankeir BC 6 E1
Bankend SK 11 D6
Banks, île NT 41 A1-2
Banks, île NT 42 D1
Banks Island BC 1 B2-3
Banks Island NT 41 A1-2
Banks Island NT 42 D1
Banner ON 19 C5
Banner ON 20 E4
Bannock SK 11 B6
Bannockburn ON 21 D4
Bannock Lake SK 12 B6
Bannock Lake SK 15 B1
Banting Lake MB 36 A8
Banting Lake NL 37 A6
Banting Lake NL 38 F7
Bantry AB 7 C7
Bapaume SK 11 B3-4
Baptiste ON 21 B4
Baptiste, lac QC 28 D4
Baptiste Lake AB 8 A6
Baptiste Lake ON 21 B4
Barachois QC 30 C3
Barachois Brook NL 36 C2
Barachois Harbour NS 34 C4
Barachois Pond PP NL 36 C2
Baralzon Lake MB 15 A3
Barbara Lake ON 16 E7
Barbara Lake ON 17 B1
Barbel, lac QC 39 D1
Barber Lake SK 11 D3
Barbers Bay ON 17 C7
Barclay ON 20 B6-7
Barclay ON 21 D1
Bardanes, lac des QC 26 C5
Bardanes, lac des QC 28 A1
Bar-de-Cocagne AB 31 E8
Bar-de-Cocagne NB 32 A6-7
Bardoux, lac QC 39 D-E1
Bareneed NL 37 D7
Barge Bay NL 38 A4
Barge Bay NL 39 D8
Barich AB 8 B7
Baril Lake AB 9 B6
Baril Lake AB 12 B2
Barkerville BC 1 B8
Barkerville BC 6 B1
Barkerville PP BC 1 B8
Barkerville PP BC 6 B1
Barkfield MB 14 F5-6
Bark Lake ON 17 E5
Bark Lake ON 21 A4
Bark Lake ON 21 C3
Bark Lake ON 22 F4
Barkley Sound BC 2 E1-2
Barkway ON 20 B6-7
Bar Lake SK 12 E4
Barlow, lac QC 26 A4
Barma, lac QC 28 A7
Barnaby River NB 31 D-E7
Barnaby River NB 32 A6
Barnard, lac QC 24 A7
Barnard, lac QC 25 E7
Barnard, lac QC 27 B2
Barnard Point NT 41 A3
Barnard Point NT 42 D1
Barnes Crossing SK 11 A3
Barnes Lake MB 15 D3
Barnesville NB 32 D4-5

Barnesville NB 33 A2
Barneys Brook NL 36 A5
Barneys Brook NL 37 A3
Barneys Brook NL 38 F4
Barneys River NS 34 E4
Barnsley NB 14 E4
Barnston QC 27 F4-5
Barnwell AB 6 F6-7
Barnwell AB 7 E6
Barons AB 6 E6
Barons AB 7 D5
Barra Head NS 34 D7
Barraute QC 23 D5
Barr'd Harbour NL 38 C3
Barr'd Harbour NL 39 E8
Barr'd Islands NL 38 E7
Barré, cap QC 30 A1
Barrel Lake ON 13 F-F7
Barrel Lake ON 16 D4
Barren Island NS 34 F5
Barrhead AB 6 A5
Barrhead AB 8 B5
Barrie ON 21 D1
Barriefield ON 21 E7
Barrie Island ON 18 B2
Barrière BC 6 D1-2
Barrière, lac QC 23 E2
Barrière, lac QC 24 C3
Barrier Ford SK 11 C6
Barrier Islands ON 14 E8
Barrington NS 33 F2
Barrington QC 24 E6
Barrington QC 27 F1-2
Barrington Lake MB 12 D8
Barrington Lake MB 15 D2
Barrington Passage NS
Barrington River MB 12 D8
Barrington River MB 15 D2
Barron ON 21 A4
Barron River ON 22 E4-5
Barron River PP ON 22 E4-5
Barronsfield NS 32 C7
Barrow Bay ON 18 E5
Barrow Bay ON 18 E5
Barrows MB 11 B7
Barrows MB 13 B1
Barrow Strait NU 42 D4
Barry, lac QC 23 C8
Barry, lac QC 26 D7
Barrydowne ON 18 A5-6
Barry's Bay ON 21 A4
Barrys Bay ON 21 A4
Barry's Bay ON 22 F4
Barry's Bay ON 22 F4
Barryvale ON 21 B6
Barryville NB 29 F8
Barryville NB 31 C7
Barss Corner NS 32 F6-7
Barss Corner NS 33 C4
Barthel NB 6 A8
Barthel SK 11 A3
Barthélémy, lac QC 26 B5-6
Bartholomew River NB 31 E5-6
Bartholomew River NB 32 A4
Bartibog NB 29 F8
Bartibog NB 31 C7
Bartibog Bridge NB 31 C7
Bartibog River NB 29 F8
Bartibog River NB 31 C6-7
Bartlett Lake NT 10 C4-5
Bartletts Beach NS 33 E1
Bartletts Harbour NL 38 B3
Bartletts Harbour NL 39 E7-8
Barton NS 32 F4
Barton NS 33 C2
Barton, lac QC 22 C6
Barton Lake ON 13 C6
Barton Lake ON 16 A2
Barville QC 23 D5
Barwick ON 16 E-F2
Bas-Caraquet NB 30 E2
Bas-Caraquet NB 31 B8
Bashaw AB 6 C6
Bashaw AB 8 E7
Basin & Middle Lakes, RNOM ON 11 C5
Basin & Middle Lakes NMBS ON 11 C5
Basin Head PP PE 34 B4-5 C4-5
Basin Head PP PE 35 B8
Basin Lake SK 11 C5
Baskatong, réservoir QC 22 C8
Baskatong, réservoir QC 24 A2
Baskatong, réservoir QC 25 E2
Basket Lake ON 16 D3
Basler Lake NT 10 B6
Basques, île aux QC 29 D1
Bas-Saint-Laurent, ZEC QC 29 C3 D4
Bassano AB 6 D6
Bassano AB 7 C6
Basse-Aboujagane NB 35 E1
Basserode, lac QC 23 C1
Basses-Terres-du-Manitoba (projet), PN des MB 13 B2-3
Basses-Terres-du-Manitoba (projet), PN des MB 13 D4-5

Basses-Terres-du-Manitoba (projet), PN des MB 14 B6-7
Bassett Island ON 19 D2
Bassett Island ON 20 F1
Bassin QC 30 F7
Bass Lake ON 18 B3-4
Bass Lake ON 18 F5
Bass Lake ON 20 A4
Bass Lake ON 21 C7
Bass Lake ON 24 F1
Bass Lake PP ON 18 F8
Bass Lake PP ON 20 A7
Bass River NB 31 B-C6
Bass River NB 31 E7
Bass River NS 32 A5
Bass River NS 34 E1
Bass River NS 32 D8
Bass River NS 33 A5
Basswood MB 11 E8
Basswood MB 13 E2
Basswood MB 14 D1
Basswood Lake ON 18 A1
Basswood Ridge NB 32 D1-2
Bastille, lac QC 39 D5
Bat, lac QC 22 B2
Bataille-de-Fish Creek, LHN de la SK 11 C4
Bataille-de-la-Châteauguay, LHN de la QC 24 E5
Bataille-de-la-Ristigouche, LHN de la QC 29 D5-6
Bataille-de-la-Ristigouche, LHN de la QC 31 A3-4
Bataille-du-Moulin-à-Vent, LHN de la ON 24 F2-3
Batawa ON 21 E4
Batchawana Bay ON 17 F3-4
Batchawana Bay ON 17 E4
Batchawana Bay PP ON 17 F3-4
Batchawana River ON 17 E4
Batchawana River PP ON 17 E4
Bateau Cove NL 38 D2-3
Bateau Island ON 18 D6
Bateman NS 11 E4
Bates Beach BC 2 C2
Bates Lake YT 4 E2
Bath NB 31 E3
Bath NB 32 A1
Bath ON 21 E6
Bathing Lake NT 5 C5
Bathurst NB 29 E8
Bathurst NB 31 B6-7
Bathurst, Cape NT 41 A1
Bathurst Inlet NT 41 C3
Bathurst Inlet NU 42 F2
Bathurst Inlet (Kingaok) NU 41 C3
Bathurst Island NU 42 D3
Bathurst Mines NS 29 F7
Bathurst Mines NS 31 C6
Batiscan QC 24 B8
Batiscan QC 25 F8
Batiscan QC 27 B-C4
Batiscan, lac QC 28 D8
Batiscan, rivière QC 25 D-E8
Batiscan, rivière QC 27 A4
Batiscan-Neilson, ZEC QC 28 F4-5
Batiscan-Neilson, ZEC QC 27 A4
Batoche SK 11 B4
Batoche, LHN de SK 11 C4-5
Batoche NHS SK 11 C4-5
Batteau Harbour NL 39 B8
Battersea ON 21 D7
Battery PP NS 34 D6-7
Battle Creek SK 11 F2
Battleford SK 11 B3
Battle Harbour NL 39 C8
Battle River AB 6 B4 C6-7
Battle River AB 11 B-C1
Battle River AB 8 B-C8
Battle of Fish Creek NHS SK
Battle of the Chateauguay NHS QC 24 E5
Battle of the Restigouche NHS QC 29 D5-6
Battle of the Restigouche NHS QC 31 A3-4
Battle of the Windmill NHS ON 24 F2-3
Battrum SK 11 D3
Batty Lake MB 12 F7-8
Batty Lake MB 15 F2
Baude, lac QC 25 D6
Bauer, lac QC 27 A2
Bauline NL 37 D8
Bauline East NL 37 E8
Baumann Fiord NU 42 C4-5
Bawlf AB 6 B6
Bawlf AB 8 D7-8
Bawlf AB 11 B1
Baxter ON 20 B6
Baxter Harbour NS 32 D6-7
Baxter Harbour NS 33 B4
Baxters Corner NB 32 D4
Baxters Corner NB 33 A3
Bay, lac QC 22 A3-4
Bayard SK 11 E5
Bay Bulls NL 37 E8

Bay de l'Eau NL 36 D7
Bay de l'Eau NL 37 D5
Bay de Loup Brook NL 36 D3
Bay de Loup Brook NL 37 D1
Bay de Verde NL 37 D7
Bay du Nord River NL 36 D6
Bay du Nord River NL 37 D4
Bay du Nord Wilderness Reserve NL 36 C7
Bay du Nord Wilderness Reserve NL 37 C5
Bay du Vin NB 31 D7
Bay du Vin Island, ZCP NB 31 C7-8 D7-8
Bay du Vin Island PCA NB 31 C7-8 D7-8
Bay du Vin River NB 31 D7
Bayend MB 14 B2-3
Bayeuville, lac QC 28 A7
Bayfield NB 32 B8
Bayfield NB 34 C1
Bayfield NS 35 E3
Bayfield NS 34 D5
Bayfield ON 19 A4
Bayfield ON 20 D3
Bayfield PE 35 D7-8
Bayfield, île QC 38 B1-2
Bayfield Beach PP NS 34 D5
Bayfield Inlet ON 18 C6
Bayfield River ON 19 A-B4
Bayfield River ON 20 D3
Bayfield Sound ON 18 B2
Bayham ON 19 D5
Bayham ON 20 F4
Bayhead NS 34 D2
Bay L'Argent NL 36 D-E7
Bay L'Argent NL 37 D-E5
Bayly Lake MB 15 F5
Baynes Lake BC 7 F2
Bayonne, rivière QC 24 B-C6
Bayonne, rivière QC 27 C2
Bay Roberts NL 37 D7
Bays, Lake of ON 13 E7
Bays, Lake of ON 16 D4
Bays, Lake of ON 21 B2
Bay Shore Heights MB 14 C5
Bayshore Village ON 20 A7
Bayshore Village ON 21 D1
Bayside NB 32 E2
Bayside NB 35 F2
Bayside NB 38 F3-4
Bayside NS 32 F8
Bayside NS 33 C5-6
Bayside PE 35 C2
Baysville ON 21 B2
Bayswater NB 32 D4
Bayswater NB 33 A1
Bayswater NS 32 F7
Bayswater NS 33 C5
Bayswater Beach PP NS 32 F8
Bayswater Beach PP NS 33 C-D5
Baytona NL 36 A7
Baytona NL 37 A5
Baytona NL 38 F6
Bay Trail SK 11 C5
Bay Tree AB 3 E8
Bay Tree AB 9 E2
Bay View NL 36 F7
Bay View NL 37 F5
Bay View NL 38 E6
Bayview Escarpment, RNP ON 18 F6
Bayview Escarpment, RNP ON 20 A4-5
Bayview Escarpment PNR ON 18 F6
Bayview Escarpment PNR ON 20 A4-5
Bazil, lac QC 39 A2
Bazile, lac QC 28 D7
Bazin, rivière QC 24 B4
Bazin, rivière QC 25 F4
Bazinet, lac QC 25 C4
Beau Lac QC 29 F2
Beacham SK 11 C4
Beach Meadows NS 33 E4
Beachburg ON 21 A6
Beachburg ON 22 F5-6
Beach O'Pines ON 19 B3
Beach O'Pines ON 20 D2
Beach Point NS 34 D4
Beach Point PE 35 F7
Beach PP NB 32 B7
Beachside NL 38 E5
Beachville ON 19 C5
Beachville ON 20 E4
Beacon Hill SK 6 A8
Beacon Hill SK 9 F8
Beacon Hill SK 11 A2
Beacon Hill SK 12 F2
Beaconia MB 14 C5-6
Beadle SK 11 D3
Beale, Cape BC 2 E2
Bealton ON 19 C6-7
Bealton ON 20 E5-6
Beamsville ON 19 C8
Beamsville ON 20 E7
Beausejour MB 13 E4
Beausejour MB 14 C6
Beausejour, lac QC 28 A-B6
Beaubassin, lac QC 23 D4
Beauchastel, lac QC 23 D2
Beauchêne, lac QC 17 F8
Beauchêne, lac QC 22 C2
Beaucours, lac QC 26 C4
Beaudry QC 17 D8
Beaudry, lac QC 23 E2
Beaudry, lac QC 23 E7
Beaudry PP MB 13 E3
Beaufield SK 11 C3
Beaufort (océan Arctique), mer de YT 5 A3-4
Beaufort Sea (Arctic Ocean) YT 5 A3-4
Beauharnois QC 24 E5
Beauharnois QC 27 F1
Beaulac QC 27 D5
Beau Lac QC 29 F2
Beaulieu, rivière QC 28 C8
Beaulieu, rivière QC 29 C1
Beaulieu River NT 10 C7
Beaumesnil, lac QC 23 F2
Beaumier, lac QC 25 C6
Beaumier, lac QC 28 F1-2
Beaumont AB 6 B6
Beaumont AB 8 C6-7
Beaumont NL 38 E6
Beaumont PP BC 1 A6
Beaumont Unit (Gulf Islands NPR) BC 2 F4-5
Beauport QC 27 A6
Beaupré QC 27 A6
Beaupré Lake SK 11 A4
Beaupré Lake SK 12 F4
Beauregard, lac QC 25 D4 E3-4
Beaurivage, rivière QC 27 B5-6
Beausejour MB 13 E4
Beausejour MB 14 C6
Beausejour, lac QC 28 A-B6
Beaussier, lac QC 25 C5
Beauvais Lake PP AB 6 F5
Beauvais Lake PP AB 7 F4
Beauval SK 12 B4
Beauvallon AB 6 A7

Beauvallon AB 11 A1
Beaux-Rivages QC 24 A3
Beaux-Rivages QC 25 E3
Beaver AB 12 F2
Beaverbank NS 32 E-F8
Beaverbank NS 33 C6
Beaverbank Villa NS 32 E8
Beaverbank Villa NS 33 B5-6
Beaver Brook NL 38 B4
Beaver Brook NS 33 A6
Beaver Brook NS 34 E-F2
Beaver Brook Station NB 29 F8
Beaver Brook Station NB 31 C6-7
Beaver Cove BC 1 E4
Beaver Cove NL 36 A7-8
Beaver Cove NL 37 A5-6
Beaver Cove NL 38 F7
Beaver Creek ON 21 C5-6
Beaver Creek YT 4 C1
Beaver Creek PP MB 13 C4
Beaver Creek PP MB 14 A5
Beaver Crossing AB 6 A7-8
Beaver Crossing AB 9 F7-8
Beaver Crossing AB 11 A2
Beaver Crossing AB 12 F2
Beaverdam AB 6 A7-8
Beaverdam AB 11 A2
Beaver Dam NB 32 C2-3
Beaverdell BC 6 F2
Beaver Harbour NB 32 E3
Beaver Harbour NS 33 B8
Beaver Harbour NS 34 C2
Beaver Harbour NS 34 F4
Beaverhill AB 8 B-C7
Beaverhill Lake AB 8 C7
Beaverhill Lake AB 11 A-B1
Beaverhill Lake MB 13 A5
Beaver Island SK 12 C7
Beaver Island SK 15 C1
Beaver Lake AB 8 A1
Beaver Lake AB 9 F7
Beaver Lake AB 12 F1
Beaver Lake NT 10 E5
Beaver Lake ON 18 C7
Beaver Lake ON 21 D6
Beaver Lake SK 9 A8
Beaver Lake SK 12 A3
Beaton BC 6 D-E3
Beaton BC 18 D8
Beatrice ON 21 B1
Beattie PP BC 3 E8
Beatton River BC 9 D1
Beatton River BC 3 D8
Beatton River BC 9 C-D1
Beatton River PP BC 3 D-E8
Beatton River PP BC 9 D1-2
Beatty SK 11 B5
Beatty Creek BC 3 B2
Beatty Lake AB 9 A3
Beatty Lake AB 10 F5
Beatty Saugeen River ON 20 B4-5
Beaubier SK 11 F6
Beaucanton QC 17 C8
Beaucanton QC 23 B2
Beauceville QC 27 C6
Beauchamp, lac QC 23 D4
Beauchastel, lac QC 23 D2
Beau Bois NL 36 E7
Beau Bois NL 37 E5
Beaucage Point NL 22 D1
Beaver Mines AB 6 F5
Beaver Mines AB 7 E-F4
Beaver Mountain PP NS 33 A8
Beaver Mountain PP NS 34 E4
Beaver River AB 6 A6-7
Beaver River AB 9 F6-7
Beaver River AB 12 F1
Beaver River NL 39 B5
Beaver River ON 20 A-B5
Beaver River ON 21 D-E2
Beaver River SK 11 A3
Beaver River SK 12 F3
Beaver River YT 4 A4
Beaverstone Bay ON 18 B5
Beaver Stone River ON
Beaverton ON 20 A7
Beaverton ON 21 D1-2
Beaver Valley SK 11 C3
Beazer AB 6 F6
Beazer AB 7 F5
Bécancour QC 24 B8
Bécancour QC 27 C3-4
Bécancour, rivière QC 27 C5
Bechard SK 11 E6
Becher ON 19 D2
Becher ON 20 F1
Beckett Creek, RNOM ON 24 D3
Beckett Creek NMBS ON 24 D3
Beckwith Island ON 18 E7
Bedard Aspen PP BC 1 D7-8
Bedard Aspen PP BC 2 A6
Bedard Aspen PP BC 2 A6-7 B6-7
Bede Creek AB 9 B3-4
Bedeque PE 34 B8
Bedeque PE 35 D2
Bedeque Bay PE 34 C1
Bedeque Bay PE 35 D2
Bedford NS 32 F8
Bedford NS 33 C6
Bedford QC 24 E7
Bedford QC 27 F2-3
Bedford Island SK 12 C6
Bedford Island SK 15 C1
Bedwell River BC 2 D1
Bedwell Sound BC 2 D1
Beecham Settlement NS
Beechey Island NU 42 D4
Beechey Island HS NU

Beechey Island HS NU 42 D4
Beech Hill NS 32 F7
Beech Hill NS 33 C4
Beech Hill NS 33 E4
Beech Point PE 35 D3
Beechwood NB 31 E3
Beechwood NB 32 A1
Beechy SK 11 D3-4
Beersville NB 31 E7
Beersville NB 32 A5
Beeton ON 20 B6
Beetz, lac QC 39 E5
Bégin QC 28 B5
Beirnes, Mount NL 3 D4
Beirnes Lake NT 10 B8
Beiseker AB 6 D6
Beiseker AB 7 A4
Bekanon ON 18 C6
Bekere Lake NT 5 B7
Belair MB 13 E4
Belair MB 14 C6
Béland, lac QC 26 D6-7
Béland, lac QC 28 B2
Béland, rivière QC 29 B8
Béland, rivière QC 30 B1
Belanger Bay ON 18 B1-2
Bélanger River MB 13 B4
Bélanger River SK 12 E4-5
Belbutte SK 11 B3-4
Belcarra BC 2 C5
Belcher NB 15 C5-6
Belcher Channel NU 42 C4
Belcher Islands NU 40 D3-4
Belcourt QC 23 D5
Belfountain ON 19 A7
Belfountain ON 20 C6
Belgrave ON 19 A4
Belgrave ON 20 C3
Belinge, rivière QC 22 A-B7
Belinge, rivière QC 25 C1-2
Belisle, lac QC 26 D5
Belize Inlet BC 1 C4
Bell, rivière QC 23 A5 B6
Bell, rivière QC 30 A6
Bella Bella BC 1 C4
Bella Coola BC 1 C4-5
Bella Coola River BC 1 C5
Bella Lake ON 21 A2
Bella Lake ON 21 B2
Bell Bay PP ON 21 A4
Bell Bay PP ON 22 F4
Bellburns NL 38 D2-3
Bellburns NL 39 E7
Belldowns Point NL 38 E7
Belle-Anse QC 30 C3
Belledune NB 29 D7
Belledune NB 31 A6
Belledune, pointe NB 29 D7
Belledune, pointe NB 31 A6
Belledune Point NB 29 D7
Belledune Point NB 31 A6
Bellefeuille QC 24 C5
Bellefleur NB 29 F4
Bellefleur NB 31 D2
Bellefond NB 31 C6
Bellegarde SK 11 F7-8
Bellegarde SK 13 F1
Belle Isle NL 39 D8
Belle Isle NS 32 F5
Belle Isle, détroit de NL 38 A4
Belle Isle, détroit de NL 39 D8
Belle Isle, Strait of NL 38 A4
Belle Isle, Strait of NL 39 D8
Belleisle Bay NB 32 C-D4
Belleisle Bay NB 33 A2
Belleisle Creek NS 32 C4
Belle Marche NS 34 B6
Bellemare, lac QC 28 A4
Belleoram NL 36 E6
Belleoram NL 37 E4
Belle Plaine SK 11 E5
Belle River ON 19 E2
Belle River ON 20 F1
Belle River PE 34 D3
Belle River PE 35 F5-6
Bellerive-sur-le-Lac QC 24 B3-4
Bellerive-sur-le-Lac QC 25 F3-4
Belleterre QC 17 E8
Belleterre QC 22 A3
Belle Vallée ON 17 D8
Belle Vallée ON 23 F1
Belleville NB 31 F3
Belleville NS 32 B1
Belleville ON 21 E5
Bellevue AB 6 F5
Bellevue AB 7 E3
Bellevue NL 37 D6-7
Bellevue, lac QC 25 B5
Bellevue, lac QC 26 F5
Bellevue, lac QC 28 D1
Bellevue House NHS ON 21 E6-7
Bell Ewart ON 20 B7
Bell Ewart ON 21 D-E1
Bellhouse PP BC 2 E4-5
Bell-Irving River BC 3 D3
Bellis AB 6 A6
Bellis AB 8 B8
Bellis AB 11 A1
Bellis AB 12 F1
Bell Island NL 37 D8

Bell Island NL 38 C5
Bell Island NL 39 E8
Bell Island NT 10 A4
Belliveau NS 33 D1-2
Bell Lake ON 16 D4
Bell Lake PP MB 11 C7
Bell Lake PP MB 13 B-C1
Bellmans Cove NL 36 B-C2
Bellot Strait NU 41 A4-5
Bellot Strait NU 42 E4
Bell Peninsula NU 40 A3
Bell Peninsula NU 41 D6-7
Bell Point PE 35 F5-6
Bell River YT 5 D3
Bellrock ON 21 D6
Bellsite MB 11 C8
Bellsite MB 13 B1
Bells Point PE 35 E3
Bell II BC 3 D3
Belly River AB 6 F6
Belly River AB 7 E5
Belmont MB 13 F2
Belmont MB 14 E-F2
Belmont NS 33 A6
Belmont NS 33 B5
Belmont NS 34 E2
Belmont NS 34 F1
Belmont ON 19 C5
Belmont ON 20 E4
Belmont PE 32 A8
Belmont PE 34 B1
Belmont PE 35 D3
Belmont Lake ON 21 D4
Belmont PP PE 32 A8
Belmont PP PE 34 B1
Belmont PP PE 35 C-D2
Belmore ON 20 C4
Belnan NS 33 B6
Belnan NS 34 F2
Beloeil QC 24 D6-7
Beloeil QC 27 E2
Belot, lac NT 5 D7
Beloud Post YT 4 E3
Belsham Lake MB 12 B8
Belsham Lake MB 15 B3
Belwood ON 19 A6
Belwood ON 20 C5
Belwood, Lake ON 19 A6
Belwood, Lake ON 20 C5
Belwood Lake CA ON 19 A6
Belwood Lake CA ON 20 C5
Benalto AB 6 C5
Benalto AB 8 F5
Benbow AB 8 B2-3
Benbow Junction AB 8 B3
Bender AB 14 C4-5
Bender SK 11 E7
Benedict, Mount NL 39 A6
Ben Eoin NS 34 C7
Ben Eoin PP NS 34 C7
Bengough SK 11 F5
Beniah Lake NT 10 B-C7
Benito MB 11 C7
Benito MB 13 C1
Benjamin, lac QC 25 A3-4
Benjamin, lac QC 26 F3-4
Benjamin Lake NT 10 C7
Benjamin Point BC 1 C2
Benjamin River NB 29 D6-7
Benjamin River NB 31 A5-6
Ben Lomond MB 32 D4
Ben Lomond NB 33 A1-2
Benmiller ON 19 A4
Benmiller ON 20 C3
Bennett Bay & Georgeson Island Unit (Gulf Islands NPR) BC 2 E4-5
Bennett Lake MB 13 B4
Bennett Lake ON 21 C6-7
Bennett Lake YT 4 E4
Bennetts Bay NS 32 B7
Bennetts Bay NS 33 A-B4
Benny ON 17 F6
Benoit's Cove NL 36 B2-3
Benoit's Cove NL 40 F8
Benson SK 11 F6
Benson Bay MB 13 A5-6
Bentick Cove PE 35 D2-3
Bentley AB 6 C5
Bentley AB 8 E6
Bentley Lake SK 12 B6
Bentley Lake SK 15 B1
Benton NB 32 C1
Benton NL 36 B7-8
Benton NL 37 B5-6
Bent River ON 18 D8
Bent River ON 21 B1
Bérard, lac QC 25 B5
Bérard, lac QC 26 F5
Berens Island MB 13 C3-4
Berens Lake ON 16 B2
Berens River MB 13 C4
Berens River MB 13 C4
Berens River MB 16 A1
Berens River ON 16 A-B2
Berens River ON 16 B2-3
Beresford MB 11 E8
Beresford MB 13 E-F2
Beresford NB 14 E1
Beresford NB 29 E8
Beresford NB 31 B6
Beresford Bay BC 1 A1
Berford Lake ON 18 E5
Bergen AB 6 D5
Bergen AB 7 A3
Bergeron, lac QC 28 A6
Bergland ON 13 F5-6
Bergland ON 16 E1-2
Berkeley ON 20 B5
Berland River AB 6 A3

Berland River AB 8 B1
Berlinguet, lac QC 26 E5
Berlinguet Inlet NU 41 A-B6
Berlinguet Inlet NU 42 E5
Berlo MB 14 C5
Bermen, lac QC 39 B1
Bernabé, lac QC 28 B5
Bernard Lake ON 18 C8
Bernard Lake ON 22 E-F1
Bernic Lake MB 14 C7
Bernier, lac QC 23 D8
Bernier, lac QC 25 A3
Bernier, lac QC 26 E3
Bernier Bay NU 41 B5
Bernier Bay NU 42 E4-5
Bernierville QC 27 C5
Berry, monts QC 29 C7
Berry, rivière QC 23 C4
Berry Creek AB 6 D7
Berry Creek AB 7 A-B7
Berry Creek AB 9 A-B5
Berry Head NL 36 B-C2
Berry Head NS 34 F6
Berry Hill Pond NL 36 C4
Berry Hill Pond NL 37 C4
Berry Mills NB 31 F7-8
Berry Mills NB 32 B6
Berté, lac QC 39 E1
Berthelot, lac QC 23 D7
Berthelot, lac QC 26 E1-2
Berthier-sur-Mer QC 27 A6
Berthierville QC 24 C7
Berthierville QC 27 C2
Bertrand NB 30 E1-2
Bertrand NB 31 B7
Bertwell SK 11 C7
Bervie ON 20 B3
Berwick NS 32 C5
Berwick NS 32 E6
Berwick NS 33 B4
Berwick ON 24 E3
Berwyn AB 9 E3
Besace, lac de la QC 26 E5
Besa River BC 3 C-D6
Besnard Lake SK 12 E5
Bessonne, ZEC de la QC 25 C8
Bessonne, ZEC de la QC 28 E-F3
Best's Harbour NL 36 B8
Best's Harbour NL 37 B6
Bétaux, lac QC 39 D6
Betchouane, RNOM de QC 39 F4
Betchouane NMBS QC 39 F4
Beteau Lake ON 16 A7
Béthanie QC 24 D8
Béthanie QC 27 E3
Bethany MB 13 E2
Bethany MB 14 D2
Bethany ON 21 E2-3
Bethel PE 35 E5
Bethune SK 11 E5
Bethune Memorial House NHS ON 21 C1-2
Betsiamites QC 29 B3
Betsiamites, pointe de QC 29 B3
Betsiamites, rivière QC 29 A-B2
Betsiamites, rivière QC 40 F6
Beulah MB 11 E8
Beulah MB 13 E1
Beverley SK 11 A6
Bevin, lac QC 24 C4
Bevin Lake ON 18 A5
Bewdley ON 21 E3
Bewley Lake ON 16 A7
Beynon AB 7 B5
Bezanson AB 9 E-F3
Bible Hill NS 33 A6
Bible Hill NS 34 E2
Bic, île du QC 29 C2
Bic, parc du QC 29 C2
Bic, rivière du QC 29 C2
Bickerdike AB 8 C3
Bickleigh SK 11 D3
Bicknor, Cape NL 42 A4
Bide Arm NL 38 C4
Bide Arm NL 39 E8
Biderford PE 32 A8
Bideford PE 34 B1
Bideford PE 35 C2
Biefeld MB 11 D8
Biefeld MB 13 D1
Biencourt QC 29 D2-3
Biencourt QC 31 A1
Bienfait SK 11 F6-7
Bienville, lac QC 40 D5
Big Bald Mountain NB 29 F6
Big Bald Mountain NB 31 C5
Big Bar Creek BC 1 D7
Big Bay BC 1 E5-6
Big Bay BC 2 B2
Big Bay ON 18 E5
Big Bay ON 21 E5
Big Bay Point ON 18 F8
Big Bay Point ON 20 A-B7
Big Bay Point ON 21 D1
Big Beaver SK 11 F5
Big Bend CA ON 19 D4
Big Bend CA ON 20 F3
Big Bight NL 39 A6
Big Birch Island MB 13 D3
Big Birch Island MB 14 C3
Big Black River MB 13 B4
Big Blue Hill Pond NL 36 D7
Big Blue Hill Pond NL 37 D5

Big Bras d'Or NS 34 C7
Big Brook NL 38 A4
Big Brook NL 39 D8
Big Bunsby MPP BC 1 E7
Big Burnt Island ON 18 B4
Big C SK 9 D8
Big C SK 12 D3
Big Campbell Creek YT 4 D6
Big Canyon Lake ON 16 D2
Big Caribou Lake ON 18 B7
Big Cedar ON 21 D3
Big Chute ON 18 E8
Big Cove NB 31 E8
Big Cove NB 32 A6
Big Creek BC 1 D7
Big Creek BC 1 D7
Big Creek ON 19 C-D6
Big Creek ON 19 E2
Big Creek ON 20 E-F5
Big Creek YT 4 C3
Big Creek Marsh ON 19 D6
Big Creek Marsh ON 20 F5
Big Creek NWA ON 19 D6
Big Creek NWA ON 20 F5
Big Creek PP BC 1 D7
Big Crow Lake ON 22 E3
Big East River ON 21 A2
Big East River ON 22 F1-2
Big East River PP ON 21 A2
Big Falls NL 36 A4
Big Falls NL 37 A2
Big Falls NL 38 F3
Big Fish River YT 5 C3-4
Big Flat Lake MB 12 B-C8
Big Forks Brook NB 31 E6
Big Forks Brook NB 32 A4-5
Big Glace Bay Lake NMBS NS 34 C8
Big Grassy River ON 13 F6
Big Grassy River ON 16 E1-2
Big Gull Lake ON 21 C6
Big Harbour NS 34 C7
Big Harbour Island NS 34 D6
Bighead SK 6 A8
Bighead SK 9 F8
Bighead SK 11 A2-3
Bighead SK 12 F2-3
Bighead River ON 18 F6
Bighead River ON 20 A4-5
Big Hill NS 34 C7
Big Hill Springs PP AB 6 B8
Big Hill Springs PP AB 7 B3
Bighorn Wildland PRA AB 6 C4
Big Island MB 12 E-F7
Big Island MB 15 E1-2 F1-2
Big Island NT 10 E5
Big Island NU 40 B5
Big Island NU 41 D8
Big Island ON 13 F5
Big Island ON 14 F8
Big Island ON 16 E1
Big Island SK 12 F1
Big Jolicure, lac NB 31 F1
Big Jolicure Lake NB 31 F1
Big Kalzas Lake YT 4 B4
Big Knife PP AB 6 C6-7
Big Knife PP AB 8 E8
Big Knife PP AB 11 B-C1
Big LaHave Lake NS 32 F6
Big LaHave Lake NS 33 C3
Big Lake AB 6 B7
Big Lake AB 11 B1
Big Lake NT 10 A7
Big Lake NT 10 D5
Big Lake ON 13 D7
Big Lake Ranch BC 1 C7-8
Big Marsh NL 34 D4-5
Big Mink Lake ON 21 B3
Big Molly Upsim Lake NS 32 F7
Big Moon Lake ON 18 A2
Big Mossy Point MB 13 A3
Big Muddy Badlands SK 11 F5
Big Muddy Lake SK 11 F5
Big Narrows Island ON 14 E8
Bigot, lac QC 39 E2-3
Big Otter Creek ON 19 C-D5
Big Otter Creek ON 20 F4
Big Point PE 35 F5
Big Pond NL 37 E7
Big Pond NS 34 D7
Big Prairie AB 9 E4
Big Quill Lake SK 11 C5-6
Big Rideau Lake ON 21 C7
Big Rideau Lake ON 24 F1
Big River NB 39 A6
Big River SK 11 B4
Big River SK 11 A4
Big River (FN) SK 11 A-B4
Big Salmon YT 4 D4
Big Salmon Lake YT 4 D4-5
Big Salmon Range YT 4 D4 E4-5
Big Salmon River NB 32 D5
Big Salmon River NB 33 A2-3
Big Sand Island YT 4 D4
Big Sand Lake MB 12 C8
Big Sand Lake MB 15 C2-3
Big Sand Lake SK 12 A6
Big Sandy Lake SK 11 B-C2

Big Sandy Lake SK 12 F6
Bigsby Island ON 13 F5
Bigsby Island ON 14 F8
Bigsby Island ON 16 E1
Big Shell SK 11 B4
Big Shell Lake ON 16 B1-2
Big Smith Creek NT 10 A2
Big Spruce River MB 15 A-B2
Big Stick Lake SK 6 E8
Big Stick Lake SK 11 E2
Big Stone AB 7 B7
Big Stone AB 8 D6
Bigstone AB 11 D1
Bigstone Lake MB 13 A6
Big Stone Lake NT 5 D5
Bigstone Lake SK 12 F5
Bigstone River MB 15 E5
Big Tancook Island NS 32 F7
Big Tancook Island NS 33 D5
Big Timber Creek YT 4 C5-6
Big Timothy Mountain BC 1 C8
Big Timothy Mountain BC 6 C1
Big Tracadie River NB 30 F1
Big Tracadie River NB 31 B-C7
Big Trout Lake ON 13 A8
Big Trout Lake ON 13 A8
Big Trout Lake ON 21 A2
Big Trout Lake ON 22 E-F2
Big Trout Lake ON 40 E1
Big Trout Lake ON 40 D1
Big Valley AB 6 C6
Big Valley AB 8 F7
Big Whiteshell Lake MB 13 E5
Big Whiteshell Lake MB 14 D7
Bigwind Lake PP ON 21 B2
Bigwood ON 17 F7
Bigwood ON 18 B6
Bijoux Falls PP BC 3 F7
Billings, Mount YT 4 D7
Billtown NS 32 E6
Billtown NS 33 B4
Billy Lake NT 5 B8
Binamé Lake NT 5 B8
Binbrook ON 19 C7
Binbrook ON 20 E6
Binbrook CA ON 19 C7
Binbrook CA ON 20 E6
Bindloss AB 6 D8
Bindloss AB 11 D2
Bingley AB 8 E5
Binscarth MB 11 E7-8
Binscarth MB 13 E1
Bircham AB 7 A5
Birch Bay AB 8 E6
Birch Bay MB 14 A3
Birch Cove AB 6 A5
Birch Cove AB 8 B5
Birch Grove NS 34 C8
Birch Head NS 35 F3
Birch Hill PE 35 C2
Birch Hills SK 11 B6
Birch Island BC 6 C-D2
Birch Island MB 11 C8
Birch Island MB 13 C2
Birch Island MB 18 B4
Birch Island PP Reserve MB 11 C8
Birch Island PP Reserve MB 13 C2
Birch Lake AB 6 B7
Birch Lake AB 11 B1
Birch Lake NT 10 D5
Birch Lake ON 13 D7
Birch Lake ON 16 B3
Birch Lake SK 11 A3
Birch Lake SK 11 B3
Birch Mountains AB 9 C1
Birch Mountains AB 41 F1
Birch Mountains Wildland SK 11 F5
Birch Mountains Wildland PP AB 9 C6
Birch Mountains Wildland PP AB 12 C1
Birch Mountains Wildland PP AB 12 D3-4
Birch Point ON 21 D3
Birch Point PP MB 13 F5
Birch Point PP MB 14 F7
Birch Portage SK 12 F6-7
Birch Portage SK 15 F1
Birch River AB 9 B-C5
Birch River AB 41 F1
Birch River MB 13 C1
Birch River MB 19 C3 D3
Birch River ON 21 E5-6
Birchton QC 27 E5
Birchtown NS 33 F3
Birchy Bay NL 36 A7
Birchy Bay NL 37 A5
Birchy Bay NL 38 F6
Birchy Head NS 36 A3
Birchy Head NL 37 A1
Birchy Head NL 38 F2
Birchy Lake NL 36 A4
Birchy Lake NL 37 A4
Birchy Lake NL 38 F4
Bird MB 15 D5
Bird Cove NL 38 B3
Bird Cove NL 39 E7-8
Birdinia 14 B-C2

Bird River MB 14 C7-8
Bird River ON 13 E5
Bird Rocks NMBS QC 30 E8
Bird's Hill MB 14 D5
Bird's Hill PP MB 14 D5
Bird's Point SK 11 E7
Birds Creek ON 21 B4
Birdtail MB 11 D-E8
Birdtail MB 13 D1
Birdtail Creek MB 11 D8
Birdtail Creek MB 13 D1
Birken BC 1 E7
Birkenhead Lake PP BC 1 E7
Birkenhead Lake PP BC 2 B5
Birkenhead River BC 2 B5
Birmingham SK 11 D6
Birnie MB 13 E2
Birnie MB 14 C2
Birr ON 20 E3
Birsay SK 11 D4
Birtle MB 11 E8
Birtle MB 13 E1
Biscayan Cove NL 37 D8
Biscay Bay NL 37 F7
Biscay Bay NL 37 F7
Biscay Bay River NL 37 F7
Biscotasi Lake ON 17 E6
Biscotasi Lake PP ON 17 E5-6
Biscotasing ON 17 E6
Bishop River BC 1 D6
Bishop River BC 2 A3
Bishop River PP BC 2 A3
Bishop's Falls NL 36 A6
Bishop's Falls NL 37 A4
Bishop's Falls NL 38 F5
Bishopsgate ON 19 C6
Bishopsgate ON 20 E5
Bishops Mills ON 21 C8
Bishops Mills ON 24 F2
Bishopton QC 27 E5
Bishopville NS 32 E7
Bismarck ON 19 C7
Bismarck ON 20 E7
Bison Lake AB 9 D4
Bissett MB 13 D5
Bissett MB 14 B7
Bissett MB 16 B-C1
Bissett Creek ON 22 D-E3 D4
Bissett Creek ON 22 D3-4
Bissett Creek PP ON 22 E3
Bissett Lake ON 22 E3-4
Bistcho Lake AB 9 A3
Bistcho Lake AB 10 F5
Bitobi, lac QC 22 E7
Bitobi, lac QC 24 B2
Bitter Lake SK 6 E8
Bitter Lake SK 11 E2
Bittern Lake AB 6 B6
Bittern Lake AB 8 D7
Bittern Lake AB 8 D7
Bittern Lake SK 11 A5
Bittern Lake SK 11 A5
Bjorkdale SK 11 C7
Bjorne Peninsula NU 42 C4
Black, Mount YT 4 D4
Black Angus Creek SK 3 A4
Black Angus Creek BC 4 F7
Black Bank ON 20 B6
Black Bay NL 38 A4
Black Bay ON 16 F6
Black Bay QC 22 E5
Black Bay SK 9 A8
Black Bay SK 12 A3
Black Bay SK 12 E3
Black Bay Peninsula ON 16 F6
Black Beach NS 32 E4
Black Beach NB 33 B1
Black Bear Island Lake SK 12 E5
Black Bear River NL 39 B-C8
Blackbear River NL 13 A7-8
Black Birch Lake ON 16 A1
Black Birch Lake SK 12 D3-4
Blackburn ON 21 A8
Black Cape QC 29 D7
Black Cape QC 31 A6
Blackcomb Glacier PP BC 1 E7
Blackcomb Glacier PP BC 2 C5
Black Creek BC 1 E-F5
Black Creek BC 2 C2
Black Creek ON 19 C3 D3
Black Creek ON 20 D4
Black Creek ON 20 F1-2
Black Creek ON 21 E5-6
Black Diamond MB 14 D5
Black Diamond AB 6 C4
Black Diamond AB 7 C4
Black Donald Lake ON 21 B6
Black Duck NL 36 B2
Black Duck Brook NL 37 C6
Black Duck Cove NL 36 F7
Black Duck Cove NL 37 F5
Black Duck Cove NL 38 B3

Black Duck Cove NL 39 D7-8
Black Duck Cove PP NS 34 E6
Black Duck River ON 15 D8
Black Duck River ON 40 D2
Black Fox Creek YT 5 C2-3
Black Hawk ON 16 E2
Blackhead NL 37 D6-7-8
Blackhead NL 37 B7
Blackheath ON 19 C7
Blackheath ON 20 E6
Blackie AB 6 E6
Blackie AB 7 C4
Black Island MB 13 D4
Black Island MB 14 B6
Black Island NL 39 A-B7
Black Lake QC 27 D5
Black Lake SK 12 A5
Black Lake SK 12 B5
Black Lake YT 4 D6
Blackland NB 29 D6-7
Blackland NB 31 A5
Black Lichen Lake NT 10 B5-6
Black Point NB 29 D7
Black Point NB 31 A6
Black Point NS 32 F8
Black Point NS 33 C5
Black Point SK 9 D8
Black Point SK 12 A5
Black Pond, RNOM de PE 35 C4-5
Black Pond, RNOM de PE 35 D-E8
Black Pond NMBS PE 34 C4-5
Black Pond NMBS PE 35 D-E8
Black River MB 13 D4
Black River MB 14 C7
Black River NB 14 B-C6
Black River NB 31 D7
Black River NS 32 D7
Black River NS 33 B2
Black River NS 33 B4
Black River NS 34 C5-6
Black River ON 16 F-E8
Black River ON 20 B7
Black River ON 21 B-C2
Black River YT 3 D-E6
Black River Lake NS 32 E6
Black River Lake NS 33 B4
Black River Pond NL 36 C-D8
Black River Pond NL 37 C-D6
Black River PPA NB 31 D7
Black River Pond NL 14 A6-7
Black Rock NS 33 A6
Black Rock NS 34 E2
Blacks Corners ON 21 B7
Blacks Corners ON 24 E1
Blacks Harbour NB 32 E3
Blackstock ON 20 B8
Blackstock ON 21 E2
Blackstone Lake ON 16 C4
Blackstone Lake ON 18 D7
Blackstone River NT 10 E3
Blackstone River YT 5 F3
Blackstone TP NT 10 E2-3
Blackstrap PP SK 11 C-D4
Black Sturgeon MB 12 D7-8
Black Sturgeon MB 15 D2
Black Sturgeon Lake ON 16 E6
Black Sturgeon River ON 16 E6
Black Sturgeon River PP ON 16 E6
Black Tickle NL 39 B8
Blackville NB 31 E6
Blackville NB 32 A4
Blackwater ON 20 B7
Blackwater ON 21 A6
Blackwater Lake NT 10 B2
Blackwater Mountain BC 6 D3
Blackwater River NT 10 B2-3
Bladworth SK 11 D4-5
Blaine Lake SK 11 B4
Blaine Lakes SK 11 B4
Blainville QC 27 E1
Blair ON 19 B6
Blair ON 20 D5
Blairmore AB 6 F5
Blairmore AB 7 E3
Blairton ON 21 D4
Blakeney AB 8 B7
Blakeney ON 24 E1
Blaketown NL 37 E7
Blanc, cap QC 30 C3
Blanc, lac QC 29 A4-5
Blanc, mont QC 29 B5-6
Blanc, réservoir QC 25 B-C7
Blanc, réservoir QC 28 E2

Blanchard Road NS 33 A8
Blanchard Road NS 34 E4
Blanche NS 33 F3
Blanche QC 22 F8
Blanche QC 24 C3
Blanche, lac la QC 24 C3
Blanche, rivière QC 24 A8
Blanche, rivière QC 24 B8
Blanche, rivière QC 25 E8
Blanche, rivière QC 27 B4
Blanche, rivière QC 27 C4
Blanche River QC 22 C3
Blanchet Island NT 10 D7
Blanc-Sablon QC 39 D7-8
Blanc-Sablon QC 40 E8
Blandford NS 32 F7-8
Blandford NS 33 C-D5
Blanket Creek PP BC 6 D-E3
Blavet, lac QC 23 F7
Blavet, lac QC 25 B1
Bleu, mont NB 29 E7
Bleue, rivière QC 29 E2
Bleuets, rivière aux QC 25 B-C3
Blevins Lake MB 12 A8
Blevins Lake MB 15 A3
Blezard Valley ON 17 F6
Bligh Island PP BC 1 F4-5
Blind Bay BC 6 D2
Blind Bay MB 14 A-B5
Blind Bay PP NS 32 F8
Blind Bay PP NS 33 C5
Blind Channel BC 1 E5
Blind Channel BC 2 B1-2
Blind Creek YT 4 C5
Blind River ON 17 F5
Blind River ON 18 A2
Blind River ON 18 B2
Blissfield NB 31 E6
Blissfield NB 32 A4
Bliss Landing BC 1 E6
Bliss Landing BC 2 C2-3
Blissville NB 32 D3
Blockhouse NS 33 D4
Bloedel BC 1 E5
Bloedel BC 2 C2
Blomidon NB 31 D7
Blomidon NS 32 D7
Blomidon NS 33 B4
Blomidon PP NS 32 D7
Blomidon PP NS 33 A4-5
Blood Indian Creek AB 6 D7
Blood Indian Creek AB 7 B7
Blood Indian Creek AB 11 D1
Bloodsucker Lake ON 15 C3-4
Bloodsucker Lake ON 11 A6-7
Bloodvein MB 13 C4
Bloodvein MB 14 A5-6
Bloodvein River MB 13 D4-5
Bloodvein River MB 16 B1
Bloodvein River MB 14 A6-7
Bloody Falls (Kuklok) TP NU 41 B-C2
Bloomfield NB 31 F2-3
Bloomfield NB 32 B1
Bloomfield NB 33 D4-5
Bloomfield NL 33 A2
Bloomfield NL 36 E8
Bloomfield NL 37 C6
Bloomfield NL 37 F4
Bloomfield PE 35 B1
Bloomfield Corner PE 35 B1-2
Bloomfield PP PE 35 B1
Bloomfield Ridge NB 31 E5
Bloomfield Ridge NB 32 A3
Bloomingdale ON 19 B6
Bloomingdale ON 20 D5
Blooming Point PE 35 D5
Bloomington NS 32 E-F6
Bloomington NS 33 C4
Bloomsbury AB 6 A5
Bloomsbury AB 8 B5
Blossom Park ON 21 B8
Blossom Park ON 24 D2
Blouin, lac QC 23 E5
Blow Me Down PP NL 36 A2
Blow Me Down PP NL 38 F1
Blow River YT 5 B3
Blubber Bay BC 1 E-F6
Blubber Bay BC 2 C2
Blue, Cape NS 34 D5
Blueberry Mountain AB 9 E2
Blueberry River BC 3 D7-8
Blueberry River BC 9 E1
Blue Cove NL 38 B3
Blue Earth Lake PP BC 1 E8
Blue Earth Lake PP BC 6 C-D1
Bluefish River NT 5 D7
Bluefish River PP NT 5 C-D2
Blue Hills of Couteau NL 36 D3
Blue Hills of Couteau NL 37 D1

Blue Island SK 12 B6
Blue Jay Creek ON 18 C3
Blue Jay Creek PP ON 18 C3
Blue Lake PP ON 13 E6
Blue Lake PP ON 16 D2
Blue Mountain NB 29 E7
Blue Mountain NL 38 B3
Blue Mountain NS 33 A8
Blue Mountain NS 34 E4
Bluenose Lake NU 41 B1
Bluenose Lake NU 42 F1
Blue Point ON 19 C3
Blue Point ON 20 E2
Blue Ridge AB 6 A4-5
Blue Ridge AB 8 B4
Blue River AB 8 C3
Blue River NS 34 F6
Blue River BC 6 C2
Blue River Black Spruce PP BC 6 C2
Blue River Pine PP BC 6 C2
Blue Rocks NS 33 D5
Bluerock Wildland PP AB 6 E5
Bluerock Wildland PP AB 7 C3
Blue Sea QC 22 E7
Blue Sea QC 24 B1
Blue Sea, lac QC 22 E7
Blue Sea, lac QC 24 B1
Bluesky AB 9 E3
Blues Mills NS 34 D6
Bluets, rivière aux QC 27 D6
Bluevale ON 19 A4-5
Bluevale ON 20 C3-4
Bluewater ON 19 B4
Bluewater ON 20 D3
Bluewater Beach ON 18 F7
Bluewater Beach ON 20 A6
Bluff, Cape NL 39 E8
Bluff Creek MB 14 C3
Bluff Head NB 36 B2
Bluff Head Cove NL 39 A7
Bluff Point ON 19 D7
Bluff Point ON 20 F6
Bluffton AB 6 B5
Bluffton AB 8 E5
Bluffy Lake ON 13 D6-7
Bluffy Lake ON 16 C3
Blumenfeld MB 11 F3
Blumenfeld MB 14 F4
Blumenheim SK 11 E4
Blumenhof SK 11 E3-4
Blumenort MB 13 F4
Blumenort MB 14 E5-6
Blumenthal SK 11 C4
Blunt Mountain BC 3 F4
Blyth ON 19 A4
Blyth ON 20 C3
Blyth Brook ON 19 A4
Blyth Brook ON 20 C3
Blytheswood ON 19 E2
Boat Harbour NL 36 E7
Boat Harbour NL 37 E5
Boat Harbour NL 38 A4-5
Boat Harbour NL 39 D8
Boat Lake ON 18 F5
Boat Lake ON 20 A4
Bobby Cove NL 38 E5
Bobcaygeon ON 21 D2-3
Bob Creek Wildland PP AB 6 E5
Bob Creek Wildland PP AB 7 D3-4 E3-4
Bob-Grant, lac QC 25 B-C6
Bob-Grant, lac QC 28 E1
Bobinet, lac QC 22 B3-4
Bob Lake ON 21 C2
Bob Quinn Lake BC 3 D3
Bobs Lake ON 21 C6
Bobtail Mountain PP BC 1 B6
Bocabec NB 32 E2
Bocock Peak PP BC 3 E7
Bodmin SK 11 A4
Bodo AB 6 C8
Bodo AB 11 C2
Boggy Creek MB 11 D7
Boggy Creek MB 13 C-D1
Bognor ON 18 F6
Bognor ON 20 A5
Boharm SK 11 E5
Bohier, lac QC 39 E5
Bohn Lake AB 12 E2
Boiestown NB 31 E5
Boiestown NB 32 A3
Boileau QC 24 C4
Boileau QC 28 D6
Bois, lac des QC 10 F5
Bois, lac des QC 13 F5
Bois, lac des QC 16 E1
Bois, lac des QC 22 A2
Bois-Blanc NB 30 F-F1
Bois-Blanc NB 31 B7
Bois Blanc Island Lighthouse NHS ON 19 E1
Boisbriand QC 24 D5
Boisbriand QC 27 E1
Bois Brûlé, cap du QC 30 C3
Boischatel QC 27 A5-6
Boisdale NS 34 C7
Bois-des-Filion QC 24 D5
Bois-des-Filion QC 27 E1
Bois-Franc QC 22 D7-8
Bois-Franc QC 24 A4
Bois-Franc QC 25 F2
Bois-Gagnon NB 31 B7

Boishébert, LHN *NB*
  31 D5-6
Boishébert NHS *NB* **31** D5-6
Bois Island *NL* **36** D6
Bois Island *NL* **37** D4
Boissevain *MB* **11** F8
Boissevain *MB* **13** F2
Boissevain *MB* **14** F1
Boisvert, lac *QC* **26** B5
Boisvert, pointe de *QC* **29** B2
Boiteuse, lac de la *QC*
  28 B5-6
Boland Brook *NB* **29** E5
Bolingbroke *ON* **21** C6-7
Bolney *SK* **11** B3
Bolsover *ON* **20** A7-8
Bolsover *ON* **21** D2
Bolton *ON* **19** A7-8
Bolton *ON* **20** C6
Bolton Centre *QC* **24** E8
Bolton Centre *QC* **27** F4
Bolton Creek *ON* **21** C6
Bolton Lake *MB* **13** A5
Bolton Lake *NB* **32** C1
Bolton-Ouest *QC* **24** E8
Bolton-Ouest *QC* **27** F3-4
Bolton River *MB* **13** A4-5
Bompas Lake *SK* **12** A5
Bon Accord *AB* **6** A5-6
Bon Accord *AB* **8** B6
Bon Accord *NB* **31** E3
Bon Accord *NB* **32** A1
Bonanza *AB* **3** E8
Bonanza *AB* **9** E2
Bonaparte Lake *BC* **1** D8
Bonaparte Lake *BC* **6** D1
Bonaparte PP *BC* **1** D8
Bonaparte PP *BC* **6** D1
Bonaparte River *BC* **1** D8
Bonaparte River *BC* **2** A7
Bonaparte River *BC* **6** D1
Bonar *AB* **7** A7
Bonarlaw *ON* **21** D4-5
Bonaventure *QC* **29** D8
Bonaventure *QC* **30** D1
Bonaventure *QC* **31** A7
Bonaventure, île *QC* **30** C3
Bonaventure, pointe *QC*
  29 D8
Bonaventure, pointe *QC*
  30 D1
Bonaventure, pointe *QC*
  31 A6-7
Bonaventure, rivière *QC*
  29 B-C8
Bonaventure, rivière *QC*
  30 C1
Bonaventure, rivière *QC*
  31 A6-7
Bonaventure-et-du-Rocher-
  Percé, parc *QC* **30** C3
Bonaventure Ouest, rivière
  QC **29** B-C8
Bonavista *NL* **37** B7
Bonavista, baie de *NL*
  37 B7
Bonavista, Cape *NL* **37** B7
Bonavista Bay *NL* **37** B7
Bonavista Peninsula *NL*
  37 C6-7
Bon-Désir, cap de *QC* **29** C1
Bond Head *ON* **20** B6
Bond Head *ON* **21** F3
Bondiss *AB* **6** A6
Bondiss *AB* **8** A7
Bondiss *AB* **9** F6
Bondiss *AB* **12** F1
Bond Sound *BC* **1** D5
Bond Sound *BC* **2** A1
Bondy, lac *QC* **22** B8
Bondy, lac *QC* **25** D2
Bon Echo PP *ON* **21** C5
Bonfield *ON* **17** F8
Bonfield *ON* **22** D2
Bongard, lac *QC* **23** D8
Bongard, lac *QC* **25** A2
Bongard, lac *QC* **26** E2
Bonheur River Kame, RNP
  ON **16** E4
Bonheur River Kame PNR
  ON **13** F8
Bonheur River Kame PNR
  ON **16** E4
Bonilla Island *BC* **1** B2
Bonin, lac *QC* **25** A6
Bonin, lac *QC* **26** F6
Bonin, lac *QC* **28** C1
Bonjour, rivière *QC* **29** B5-6
Bonne Bay *NL* **38** E2
Bonne Bay Big Pond *NL*
  36 A3
Bonne Bay Big Pond *NL*
  37 A1
Bonne Bay Big Pond *NL*
  38 F2
Bonnechere *ON* **21** A4
Bonnechere *ON* **22** F4
Bonnechere *ON* **24** E1
Bonnechere PP *ON* **22** F4-5
Bonnechere River *ON*
  21 A6
Bonnechere River *ON*
  22 F5-6
Bonnechere River PP *ON*
  21 A4
Bonnechere River PP *ON*
  22 F4
Bonnechere Valley *ON*
  21 A5
Bonne-Espérance *QC* **39** D7
Bonnet Plume Range *YT*
  4 A5
Bonnet Plume River *YT*
  5 F4-5

Bonnie Doone *ON* **19** C3
Bonnie Doone *ON* **20** E2
Bonny River *NB* **32** E2
Bonnyville *AB* **9** F7
Bonnyville *AB* **11** A2
Bonnyville *AB* **12** F2
Bonokoski Lake *SK* **12** A6
Bonokoski Lake *SK* **15** A1
Bonsecours *QC* **24** D8
Bonsecours *QC* **27** E4
Bon Secours Beach *ON*
  20 B7
Bon Secours Beach *ON*
  21 D1
Bonshaw *PE* **34** C2
Bonshaw *PE* **35** E4
Bonshaw PP *PE* **34** C2
Bonshaw PP *PE* **35** E4
Bonville *ON* **24** E4
Boom Creek *ON* **22** D3
Booth, lac *QC* **22** C3
Boothia, golfe de *NU*
  41 A-B5
Boothia, golfe de *NU*
  42 E-F4
Boothia, Gulf of *NU*
  41 A-B5
Boothia, Gulf of *NU* **42** E-F4
Boothia, péninsule de *NU*
  41 B4-5
Boothia, péninsule de *NU*
  42 F4
Boothia Peninsula *NU*
  41 B4-5
Boothia Peninsula *NU*
  42 F4
Booth Island *NT* **5** A7-8
Booth Islands *NT* **5** A7-8
Booth Lake *MB* **12** A8
Booth Lake *MB* **15** A3
Booth Lake *ON* **21** A3
Booth Lake *ON* **22** F3
Boot Island National
  Wilderness Area *NS*
  32 E7
Boot Island National
  Wilderness Area *NS*
  33 B5
Borden *PE* **35** E3
Borden *SK* **11** C4
Borden-Carleton *PE* **32** B8
Borden-Carleton *PE* **34** C1
Borden Island *NT* **42** B3
Borden Lake *ON* **17** D5
Borden Peninsula *NU* **41** A6
Borden Peninsula *NU*
  42 E5
Border Lake PP *BC* **3** D2-3
Border Ranges *AB* **6** F5
Border Ranges *AB* **7** F3
Bornholm *ON* **19** B5
Bornholm *ON* **20** D4
Borups Corners *ON* **13** F7
Borups Corners *ON*
  16 D-E3
Boston *ON* **19** C6
Boston *ON* **20** E5
Boston Bar *BC* **1** E8
Boston Bar *BC* **2** C7
Boston Creek *ON* **17** D7
Boston Creek *ON* **20** E5-6
Boston Creek *ON* **23** E1
Bostonnais, rivière *QC*
  25 B-C8
Bostonnais, rivière *QC*
  28 D4 E3
Boswarlos *NL* **36** B2
Boswell *BC* **6** F4
Boswell River *YT* **4** D4
Bosworth Lake *ON* **16** A6
Botha *AB* **6** C6
Botha *AB* **8** F8
Bothwell *ON* **19** D3
Bothwell *ON* **20** F2
Bothwell *PE* **35** D8
Bothwell's Corners *ON*
  18 F5-6
Bothwell's Corners *ON*
  20 A4-5
Botsford Portage *NB*
  35 E1-2
Bottle Cove *NL* **36** A2
Bottle Cove *NL* **38** F1
Bottle Point *ON* **16** F7
Bottle Point *ON* **17** C2
Bottrell *AB* **6** D5
Bottrell *AB* **7** A-B3
Botwood *NL* **36** A6
Botwood *NL* **37** A4
Botwood *NL* **38** F5
Boucher *QC* **25** D-E7
Boucher *QC* **27** A3
Boucher, lac *QC* **22** E7-8
Boucher, lac *QC* **24** B2
Boucher, lac *QC* **25** D7
Boucher, lac *QC* **27** A3
Boucher, lac *QC* **28** F2
Boucher, rivière *QC* **29** A2
Boucherville *QC* **24** D6
Boucherville *QC* **27** E2
Bouchette *QC* **22** E7-8
Bouchette *QC* **24** B2
Bouchette *QC* **25** F2
Bouchette, lac *QC* **22** B8
Bouchette, lac *QC* **23** F6
Bouchette, lac *QC* **25** A8
Bouchette, lac *QC* **25** C1
Bouchette, lac *QC* **26** F8
Bouchie Lake *BC* **1** B7
Bouctouche *NB* **31** E8
Bouctouche *NB* **32** A6
Boyle *AB* **9** F6

Bouctouche, dune de *NB*
  31 E8
Bouctouche, dune de *NB*
  32 A6
Bouctouche-Baie *NB* **31** E8
Bouctouche-Baie *NB* **32** A6
Boudreau *NB* **35** E1
Bouffard, lac *QC* **39** E1
Boughey Bay *BC* **2** B1
Boughton Bay *PE* **34** C4
Boughton Bay *PE* **35** E7
Boughton Island *PE* **34** C4
Boughton Island *PE* **35** E7
Boughton Point *PE* **35** E7
Boughton River *PE* **35** E7
Boulain, lac *QC* **39** E5
Boularderie *NS* **34** C7
Boularderie *NS* **34** C7
Boularderie Island *NS*
  34 B-C7
Boulder Bay *MB* **14** C5-6
Boulder Creek *BC* **1** F7
Boulder Lake *NT* **10** B7
Boulé, ZEC *QC* **25** D4
Bouleau, rivière au *QC*
  39 E3
Boullé, rivière *QC* **42** E-F4
Boulogne Lake *NT* **10** E6
Boulter *ON* **21** B4
Boundary Bay *BC* **1** F7
Boundary Bay *BC* **2** E5
Boundary Bay *BC* **2** E5
Boundary Creek PP *BC* **6** F2
Boundary Dam *SK* **11** F6
Boundary Island *SK* **12** C7
Boundary Island *SK* **15** C1
Bounty *NL* **11** D4
Bourbeau, lac *QC* **26** A5
Bourbon, rivière *QC*
  27 C4-5
Bourdon, lac *QC* **39** A3
Bourgat, lac *QC* **26** B6
Bourgeois *NB* **35** D1
Bourgeois Mills *NB* **32** B7
Bourgeois Mills *NB* **35** E1
Bourget *ON* **24** D3
Bourgmont, lac *QC* **23** E8
Bourgmont, lac *QC* **25** A2
Bourgmont, lac *QC* **26** E2
Bourkes *ON* **17** D7
Bournemouth *SK* **11** B3
Bousso River *NT* **10** B-C6
Bouteille, lac de la *QC*
  24 A4
Bouteille, lac de la *QC*
  26 E6
Bouteille, lac de la *QC*
  27 B1
Bouteroute, lac *QC* **26** C5
Boutiliers Point *NS* **32** F7-8
Boutiliers Point *NS* **33** C5
Bow City *BC* **7** C6
Bowden *BC* **6** C5
Bowden *AB* **8** F5-6
Bowell *AB* **7** D8
Bowen Island *BC* **1** F7
Bowen Island *BC* **2** D4
Bowen Island *BC* **2** D4
Bowers Beach *ON* **18** F7
Bowers Beach *ON* **20** A6
Bowes Point *NU* **41** C4
Bowes Point *NU* **42** F3
Bow Island *AB* **6** E7
Bow Island *AB* **7** E7
Bow Island *AB* **11** E1
Bowland's Bay *ON* **17** F7
Bowman *QC* **22** E-F8
Bowman *QC* **24** B2
Bowman Bay *NU* **40** A4-5
Bowman Bay *NU* **41** C7-8
Bowmanton *AB* **7** D8
Bowmanville *ON* **20** C8
Bowmanville *ON* **21** F2
Bowmanville Creek *ON*
  20 C8
Bowmanville Creek *ON*
  21 E2
Bowron Lake *BC* **1** B8
Bowron Lake *BC* **6** B1
Bowron Lake PP *BC* **1** B8
Bowron Lake PP *BC* **6** B1
Bowron River *BC* **1** A7 B8
Bowron River *BC* **6** A-B1
Bowser *BC* **1** F6
Bowser *BC* **2** D3
Bowser Lake *BC* **3** D-E3
Bowsman *MB* **11** C7
Bowsman *MB* **14** D1
Bow Valley PP *AB* **6** D5
Bow Valley Wildland PP *AB*
  6 D4-5
Bow Valley Wildland PP *AB*
  7 B2
Boxey Harbour *NL* **36** E6
Boxey Harbour *NL* **37** E4
Box Lake *NT* **10** B8
Box Lake *SK* **12** A4
Boya Lake PP *BC* **3** A-B4
Boya Lake PP *BC* **4** F7
Boyd *MB* **15** E4
Boyd's Cove *NL* **36** A7
Boyd's Cove *NL* **37** A5
Boyd's Cove *NL* **38** F6
Boyer *AB* **9** B4
Boyer, rivière *QC* **27** B6
Boyer River *AB* **9** B3-4
Boyle *AB* **6** A6
Boyle *AB* **8** A7
Boyle *AB* **9** F6

Boyle *ON* **19** C8
Boyle *ON* **31** E8
Boyle Point PP *BC* **2** D3
Boylston *NS* **34** E5-6
Boylston PP *NS* **34** E5
Boyne Lake *AB* **6** A7
Boyne Lake *AB* **8** A-B8
Boyne Lake *AB* **11** A1
Boyne Lake *AB* **12** F1
Boyne River *ON* **20** B6
Boyne Valley PP *ON*
  20 B5-6
Brabant Lake *SK* **12** E6
Brabant Lake *SK* **12** E6
Bracebridge *ON* **21** B1-2
Bracebridge *ON* **21** C1-2
Bracken *SK* **11** F3
Brackendale *BC* **1** F7
Brackendale *BC* **2** C4-5
Bracken Lake *MB* **11** A5
Bracken Lake *MB* **13** A2
Brackett Lake *NT* **5** F8
Brackett Lake *NT* **10** A2
Brackley *PE* **34** C2
Brackley *PE* **35** E5
Brackley Beach *PE* **34** C2
Brackley Beach *PE* **35** D4
Braddock *SK* **11** E4
Bradford *ON* **20** B6-7
Bradford *ON* **21** E1
Bradford West Gwillimbury
  ON **20** B6-7
Bradford West Gwillimbury
  ON **21** E1
Bradian *BC* **2** A5
Bradleys Cove *NL* **37** D7-8
Bradner *AB* **2** E6
Brador *QC* **38** A3
Brador *QC* **39** D7
Brador Bay NMBS *QC* **38** B3
Brador Bay NMBS *QC*
  39 D-E7
Bradshaw Lake *MB* **15** C5
Bradwardine *MB* **11** E8
Bradwardine *MB* **13** E1-2
Bradwardine *MB* **14** D1
Bradwell *SK* **11** C4
Brae Harbour *PE* **35** C1-2
Braeside *ON* **24** D1
Bragg Creek *AB* **6** D5
Bragg Creek *AB* **7** B3
Bragg Creek PP *AB* **6** D5
Bralorne *BC* **1** E7
Bralorne *BC* **2** A5
Bramalea *ON* **19** A7
Bramalea *ON* **20** C6
Bramber *NS* **32** E7
Bramber *NS* **33** B5
Bramber *NS* **34** F1
Bramham Island *BC* **1** D4
Brampton *ON* **19** A7
Brampton *ON* **20** C6
Brancepeth *SK* **11** B5
Branch *NL* **36** F8
Branch *NL* **37** F6
Branch LaHave *NS* **33** D4
Branchton *ON* **19** B6-7
Branchton *ON* **20** D5
Brandon *MB* **11** E8
Brandon *MB* **13** E2
Brandon *MB* **14** E1
Brandon Hills *MB* **14** E1
Brandywine Falls PP *BC*
  1 E6-7
Brandywine Falls PP *BC*
  2 C5
Brant *AB* **6** E6
Brant *AB* **7** C4
Brant CA *ON* **19** C6
Brant CA *ON* **20** E5
Brant County *ON* **19** C6
Brant County *ON* **20** E5
Brantford *ON* **19** C6
Brantford *ON* **20** E5
Branton-Cundick, parc *ON*
  19 D2-3
Branton-Cundick, parc *ON*
  20 F1-2
Branton-Cundick Park
  (St. Clair Parkway) *ON*
  19 D2-3
Branton-Cundick Park
  (St. Clair Parkway) *ON*
  20 F1-2
Brantville *NB* **30** F2
Brantville *NB* **31** C8
Bras, rivière le *QC* **27** B6
Bras Coupé, lac du *QC*
  22 D7
Bras Coupé, lac du *QC*
  24 A1
Bras Coupé, lac du *QC*
  25 E1
Bras Coupé, lac du *QC*
  26 B4
Bras-Coupe-Désert, ZEC *QC*
  22 D7
Bras-Coupe-Désert, ZEC *QC*
  24 A1
Bras-Coupe-Désert, ZEC *QC*
  11 A-B3
Bras-d'Apic *QC* **27** A7-8
Bras d'Henri, rivière *QC*
  27 B5-6
Bras d'Or, lac *NS* **34** D6-7
Bras d'Or Lake *NS* **34** D6-7
Bras d'Or Lake *NT* **10** D6
Bras d'Or PP *NS* **34** B7
Bratnober, Mount *YT* **4** E3
Bray Island *NU* **41** B7

Bray Island *NU* **42** F6
Bray Lake *ON* **18** B8
Bray Lake *ON* **22** E1
Brazeau, Mount *AB* **6** C3
Brazeau, Mount *AB* **8** E2
Brazeau Canyon Wildland
  PP *AB* **6** B3-4
Brazeau Canyon Wildland
  PP *AB* **8** D-E2
Brazeau Dam *AB* **8** D4
Brazeau Reservoir *AB* **6** B4
Brazeau Reservoir *AB* **8** D-E3
Brazeau River *AB* **6** B4
Brazeau River *AB* **8** D-E3
Brazza, lac *QC* **28** A7
Breadalbane *PE* **32** A8
Breadalbane *PE* **34** C2
Breadalbane *PE* **35** D3-4
Breakeyville *QC* **27** B6
Bristol's Hope *NL* **37** D7-8
Breau-Village *NB* **31** E8
Breau-Village *NB* **32** A-B6
Brébeuf *QC* **24** C4
Brébeuf, lac *QC* **28** D7
Brécourt, lac *QC* **23** D8
Brécourt, lac *QC* **25** A2-3
Brécourt, lac *QC* **26** E2-3
Brechin *ON* **20** A7-8
Brechin *ON* **21** D2
Breckenridge *QC* **24** D2
Breckenridge *QC* **27** A7-8
Brecon *SK* **11** B2
Bredenbury *MB* **11** D7
Bredenbury *SK* **13** D1
Breeches, lac *QC* **27** D5
Bréhault, lac *QC* **25** D6
Bréhault, lac *QC* **27** A1
Bréhault, lac *QC* **28** F1
Breme, Point *NL* **36** F8
Breme, Point *NL* **37** F6
Brennan Lake *ON* **16** C5
Brent *ON* **22** E3
Brent Cove *NL* **36** D5
Brent Cove *NL* **37** D3
Brenton *NS* **33** E1
Brent's Cove *NL* **38** E5
Brentwood *ON* **20** B6
Bresaylor *SK* **11** B3
Breslau *ON* **19** B6
Breslau *ON* **20** D5
Breslayor *SK* **6** B8
Bressani, lac *QC* **25** D3
Brest *NB* **31** E8
Brest *NB* **32** A6
Breton *AB* **6** B5
Breton *AB* **8** D5
Breton, Cape *NS* **34** C8
Bretona *AB* **8** C6
Breton Cove *NS* **34** B7
Brett, Mount *AB* **6** D4
Brett, Mount *AB* **7** B1
Brevoort Island *NU* **40** A6
Brew, Mount *BC* **1** E7-8
Brew, Mount *BC* **2** B6
Brewer *SK* **11** D7
Breynat *AB* **9** F6
Breynat *AB* **12** E1
Brick, rivière du *QC* **30** A4
Brickley *ON* **21** E4
Briçonnet, lac *QC* **39** D6
Bridal Veil Falls PP *BC* **1** F8
Bridal Veil Falls PP *BC* **2** C7
Bridesville *BC* **6** F2
Bridge Creek *BC* **1** D8
Bridge Creek *BC* **6** C1
Bridgeford *SK* **11** D4
Bridge Lake *BC* **1** D8
Bridge Lake *BC* **6** D1
Bridge Lake PP *BC* **1** D8
Bridge Lake PP *BC* **6** C-D1
Bridgenorth *ON* **21** D3
Bridgeport *NL* **38** E6
Bridgeport *ON* **20** D5
Bridge River *BC* **1** D-E7
Bridge River *BC* **2** A4
Bridge River *BC* **2** A6
Bridgetown *NS* **33** C3
Bridgetown *PE* **34** C3
Bridgeville *NS* **33** A7
Bridgeville *NS* **34** E3
Bridgeville *QC* **30** C3
Bridgewater *NS* **33** D4
Briercrest *SK* **11** E5
Briereville *AB* **8** A8
Brier Island *NS* **33** E1
Brierly Brook *NS* **34** D-E4
Brig Bay *NL* **38** B3
Brigden *ON* **19** C2-3
Brigden *ON* **20** F1-2
Brigham *QC* **24** E7
Brigham *QC* **27** F3
Bright *ON* **19** B6
Bright *ON* **20** D5
Bright Lake *ON* **18** A1
Brighton *NL* **38** E5
Brighton *NS* **32** F4
Brighton *NS* **33** C2
Brighton *ON* **21** E4
Brightsand Lake *SK* **6** A8
Brightsand River PP *ON*
  13 E-F8
Brightsand River PP *ON*
  16 D4-5 E4
Brightside *ON* **21** B6-7
Brightstone *MB* **14** D6
Brightwater Lake *SK*
  11 C-D4
Brigus *NL* **37** D7

Brigus Junction *NL* **37** E7
Brigus South *NL* **37** E8
Bringadin, lac *QC* **39** A2
Brinston *ON* **24** F2-3
Brion, île *QC* **30** E8
Briquet, lac *QC* **22** C6
Brisay *QC* **40** E5
Brisbane *ON* **19** A7
Brisbane *ON* **20** C6
Brisco *BC* **6** D4
Brisco *BC* **7** C1
Bristol *NB* **14** E5
Bristol *NB* **31** E3
Bristol *NB* **32** A1
Bristol *PE* **35** D6
Bristol *QC* **21** A6-7
Bristol *QC* **24** D1
Britannia *NL* **37** C7
Britannia Beach *BC* **1** F7
Britannia Beach *BC* **2** D5
British Mountains *YT* **5** B2
Britt *ON* **18** C6
Britton *ON* **19** A5
Britton *ON* **20** C4
Britton Lake *MB* **12** E7
Britton Lake *MB* **15** E1
Broadacres *SK* **6** C8
Broadacres *SK* **11** C2-3
Broadback, rivière *QC*
  40 F4
Broad Cove *NL* **36** B2
Broad Cove *NL* **37** D7
Broad Cove *NS* **33** D4
Broad River *NS* **15** C6
Broad River *NS* **33** E3
Broad Valley *MB* **13** D3
Broad Valley *MB* **14** B4
Broadview *SK* **11** E7
Brochet *MB* **12** B7
Brochet, lac *MB* **12** B7
Brochet, lac *MB* **15** B2
Brochet, lac *QC* **24** C4
Brochet, lac du *QC* **26** C8
Brochet (Barren Lands) *MB*
  12 C7
Brochet (Barren Lands) *MB*
  15 C1-2
Brochet Bay *MB* **12** C7
Brochet Bay *MB* **15** C1-2
Brock *SK* **11** D3
Brocket *AB* **6** F6
Brocket *AB* **7** E4
Brockington *SK* **11** B5
Brock Island *NT* **42** C2
Brockton *ON* **20** B4
Brockton *PE* **35** B1
Brockville *ON* **21** D8
Brockville *ON* **24** F2
Brockway *SK* **32** D2
Broderick *SK* **11** D4
Brodeur Island *ON* **16** F6
Brodeur Peninsula *NU*
  41 A5
Brodeur Peninsula *NU*
  42 E4-5
Brodhagen *ON* **19** A5
Brodhagen *ON* **20** D4
Brodtkorb, lac *QC* **22** D6
Brodtkorb, lac *QC* **24** A1
Brodtkorb, lac *QC* **25** E-F1
Broken Group Islands *BC*
  2 E2
Brokenhead *MB* **13** E4
Brokenhead *MB* **14** D6
Broken Skull River *NT*
  4 B-C7
Brome *QC* **24** E8
Brome *QC* **27** F3
Brome, lac *QC* **24** E8
Brome, lac *QC* **27** F3
Bromhead *SK* **11** F6
Bromley Rock PP *BC* **6** F2
Bromont *QC* **24** E7
Bromont *QC* **27** F3
Brompton, lac *QC* **27** E4
Bromptonville *QC* **27** E4-5
Bronson Forest PRS *SK*
  6 A8
Bronson Forest PRS *SK*
  11 A2
Bronson Lake *SK* **6** A8
Bronson Lake *SK* **11** A2
Bronte *ON* **19** B7-8
Bronte *ON* **20** D6
Bronte Creek PP *ON*
  19 B7-8
Bronte Creek PP *ON*
  20 D6-7
Brooch, lac *QC* **39** E1
Brookdale *MB* **13** E2
Brookdale *MB* **14** D2
Brooke *ON* **18** F5
Brooke *ON* **20** A4
Brookfield *NL* **37** A7
Brookfield *NL* **38** F4
Brookfield *NS* **33** B5
Brookfield *NS* **34** F2
Brookfield *PE* **34** C2
Brookfield *PE* **35** D4
Brooklin *ON* **20** C8
Brooklin *ON* **21** E2
Brooklyn *NB* **35** F1
Brooklyn *NL* **37** C6
Brooklyn *NS* **32** E7-8
Brooklyn *NS* **33** B5
Brooklyn *NS* **33** E4
Brooklyn *NS* **34** C5
Brooklyn *NS* **34** F1

Brooklyn *PE* **35** F6
Brooklyn Road *NB* **35** F1
Brookmere *BC* **2** C8
Brooks *AB* **6** E7
Brooks *AB* **7** C7
Brooks *AB* **11** D1
Brooks Aqueduct PHS *AB*
  6 E7
Brooks Aqueduct PHS *AB*
  7 C7
Brooks Aqueduct PHS *AB*
  11 E1
Brooks Bay *BC* **1** E4
Brooks Brook *YT* **4** E5
Brooksby *SK* **11** B6
Brookside *NL* **36** E7
Brookside *NL* **37** E5
Brookside *NS* **32** F8
Brookside *NS* **33** C6
Brooks Mill *ON* **21** A2
Brooks Peninsula *BC* **1** E4
Brooks Peninsula PP *BC*
  1 E4
Brookvale Provincial Ski
  Park *PE* **32** B8
Brookvale Provincial Ski
  Park *PE* **34** C2
Brookvale Provincial Ski
  Park *PE* **35** E4
Brook Village *NS* **34** C6
Brookville *ON* **19** B7
Brookville *ON* **20** D6
Broomfield *MB* **11** F8
Broomhill *MB* **13** F1
Brossard *QC* **24** D6
Brossard *QC* **27** E2
Brosseau *AB* **6** A7
Brosseau *AB* **11** A1
Brougham *ON* **20** C7-8
Brougham *ON* **21** E2
Broughton Archipelago
  PMP *BC* **1** D-E4
Broughton Island *BC* **1** D4
Broughton Island
  (Qikiqtarjuaq) *NU*
  42 E-F8
Broughton Station *QC*
  27 C6
Brouillard, lac du *QC* **25** B3
Brouillard, lac du *QC* **26** F3
Brouseville *ON* **21** C8
Brouseville *ON* **24** F2-3
Brown *MB* **14** F3-4
Brown, lac *QC* **25** E7
Brown, lac *QC* **27** A2
Brown Creek *ON* **19** C3
Brown Creek *ON* **20** E2
Brownell Lake *SK* **12** F6
Brownfield *AB* **6** C7
Brownfield *AB* **11** C1
Brown Hill *ON* **20** B7
Brown Hill *ON* **21** E1
Browning *SK* **11** F7
Browning Entrance *BC* **1** A2
Brown Lake *NT* **10** B-C7
Brownlee *SK* **11** E4
Brown Lowery PP *AB* **6** D5
Brown Lowery PP *AB* **7** C3
Brown Passage *BC* **1** A2
Brown's Arm *NL* **36** A6-7
Brown's Arm *NL* **37** A4-5
Brown's Arm *NL* **38** F6
Brown's Cove *NL* **38** E3
Brownsburg *QC* **24** D4-5
Brown's Flat *NB* **32** D3-4
Browns Flat *NB* **33** A1
Browns Lake *ON* **19** C3
Brownstone Lake *MB* **12** A8
Brownstone Lake *MB* **15** A3
Brownsville *ON* **19** C5
Brownsville *ON* **20** F4
Brownvale *AB* **9** B3
Broxburn *AB* **7** E5-6
Broye, Cape *NL* **37** E8
Bruce *AB* **8** D8
Bruce, lac *QC* **22** C5
Bruce, péninsule de *ON*
  18 D4-5 E5
Bruce Mines *ON* **17** F4
Bruce Peninsula *ON*
  18 D-E5
Bruce Peninsula NP *ON*
  18 D4-5
Brucedale *ON* **19** A6
Brucedale *ON* **20** C5
Brucedale CA *ON* **18** F4
Brucedale CA *ON* **20** A3
Brucefield *ON* **19** B4
Brucefield *ON* **20** D3
Bruce Lake *ON* **13** D6
Bruce Lake *ON* **16** C2
Bruce Lake *ON* **16** C2
Bruce Mines *ON* **17** F4
Brucefield *ON* **19** B4
Bruderell *ON* **21** A5
Brudenell *PE* **35** D4
Brudenell *PE* **35** E6
Brudenell River PP *PE*
  34 C3-4
Brudenell River PP *PE* **35** E6
Bruderheim *AB* **6** B6
Bruderheim *AB* **8** B7
Brule *AB* **6** B3
Brule *AB* **8** C1
Brûlé, lac *QC* **22** C5
Brûlé, lac *QC* **24** C5
Brûlé, lac *QC* **28** E4

Brûlé, lac *QC* **39** D3-4
Brule Lake *ON* **21** B5
Brumlie *MB* **14** D1
Bruneau Lake *MB* **15** F3
Brunette Island *NL* **36** E5-6
Brunette Island *NL* **37** E3-4
Brunkild *MB* **13** F3
Brunkild *MB* **14** E4
Brunner *ON* **19** B5
Brunner *ON* **20** D4
Bruno *SK* **11** C5
Brunswick Lake *ON* **17** E5
Brunswick Mines *NB* **29** E7
Brunswick Mines *NB* **31** C6
Brussels *ON* **19** A4-5
Brussels *ON* **20** C3-4
Bruxelles *MB* **13** F2-3
Bruxelles *MB* **14** E2-3
Bryanston *ON* **19** C4-5
Bryanston *ON* **20** E3-4
Bryant *SK* **11** F6
Bryant's Cove *NL* **37** D7-8
Bryde, Mount *YT* **3** A1-2
Bryde, Mount *YT* **4** E-F5
Bryson *QC* **21** A6
Bryson *QC* **22** F6
Bryson, lac *QC* **22** D6
B-Say-Tah *SK* **11** D-E6
Buade, lac *QC* **26** C-D5
Buccaneer Bay PP *BC* **2** D3
Buchanan *SK* **11** D6-7
Buchanan, Mount *NB* **1** A8
Buchanan, Mount *BC*
  6 A1-2
Buchanan Bay *NU* **42** B5
Buchanan Point *PE* **35** F5
Buchan Gulf *NU* **41** A7
Buchan Gulf *NU* **42** E6
Buchan Lake *AB* **9** A4-5
Buchan Lake *AB* **10** F6
Buchans *NL* **36** B4
Buchans *NL* **37** B2
Buchans Junction *NL* **36** B5
Buchans Junction *NL* **37** B3
Buchans Lake *NL* **36** B4
Buchans Lake *NL* **37** B2
Buck Creek *AB* **8** B5
Buck Creek *AB* **8** D4-5
Buck Creek *BC* **1** A5
Buckfield *NS* **33** D3
Buckham Lake *NT* **10** C-D7
Buckhorn *ON* **21** D3
Buckhorn Lake *ON* **21** D3
Buckingham *QC* **21** A8
Buckingham *QC* **24** D2-3
Buckingham Island *NU*
  42 C4
Buckinghorse River *BC*
  3 D6-7
Buckinghorse River
  Wayside PP *BC* **3** C-D7
Buck Lake *AB* **6** B5
Buck Lake *AB* **8** D5
Buck Lake *AB* **8** D5
Buck Lake *ON* **18** D8
Buck Lake *ON* **21** A1
Buck Lake *ON* **21** D7
Buck Lake *ON* **22** F1
Buckland *QC* **27** B7
Buckland Lake *MB* **15** E4
Bucklaw *NS* **34** C6
Buckley Lake *BC* **3** C2
Buckshot Lake *ON* **21** C5
Budd *MB* **11** A8
Budd *MB* **13** A2
Buena Vista *SK* **11** E5
Buffalo *AB* **6** D7-8
Buffalo *AB* **7** C8
Buffalo *AB* **11** D1-2
Buffalo Bay *MB* **13** F5
Buffalo Bay *MB* **14** F7-8
Buffalo Bay *MB* **16** E1
Buffalo Creek *AB* **1** C-D8
Buffalo Creek *BC* **6** C1
Buffalo Gap *SK* **11** F5
Buffalo Head Hills *AB*
  9 C-D4
Buffalo Head Prairie *AB*
  9 C4
Buffalo Head Prairie *AB*
  41 F1
Buffalo Lake *AB* **6** C6
Buffalo Lake *AB* **8** E7
Buffalo Lake *AB* **9** E2
Buffalo Lake *NT* **10** F6
Buffalo Lake *NT* **41** E1
Buffalo Lake Estates *AB*
  8 E7
Buffaloland PP *PE* **34** C3
Buffaloland PP *PE* **35** E5
Buffalo Narrows *SK* **9** E8
Buffalo Narrows *SK* **12** E3
Buffalo Point *MB* **13** F5
Buffalo Point *MB* **14** F8
Buffalo Point *MB* **16** E1
Buffalo Pound Lake *SK*
  11 E5
Buffalo Pound PP *SK* **11** E5
Buffalo River *NT* **10** E-F6
Buffalo River *NT* **10** F6-7
Buffalo River *SK* **9** E8
Buffalo River *SK* **12** E3
Buffer Lake *SK* **11** C5
Bugaboo PP *BC*
  6 D3-4 E3-4
Buick *BC* **3** D7-8
Buick *AB* **9** D1
Buies, lac *QC* **25** A3
Buies, lac *QC* **26** F3
Buit, lac *QC* **39** E4
Bulkley Junction PP *BC* **3** F4

Bulkley River *BC* 3 F4
Bull Arm *NL* 36 D8
Bull Arm *NL* 37 D6
Bull Canyon PP *BC* 1 C6-7
Bull Cove *NL* 36 F7
Bull Cove *NL* 37 F5
Buller *BC* 21 C2
Bull Island *NL* 37 D6-7
Bullmoose Mountain *BC* 3 F7-8
Bullmoose Mountain *BC* 9 E-F1
Bullpond Creek *AB* 6 D7
Bullpond Creek *AB* 7 B6-7
Bullpound Creek *AB* 11 D1
Bull River *BC* 6 E-F5
Bull River *BC* 7 E2
Bull River *BC* 7 E2
Bull River *BC* 7 D8
Bullshead *AB* 7 D8
Bulmer Lake *NT* 10 C3-4
Bulstrode, rivière *QC* 27 C4-5 D5
Bulwer *QC* 27 E5
Bulyea *AB* 11 D5
Bulyea River *SK* 9 A8
Bulyea River *SK* 12 A4
Bummers Roost *ON* 18 C8
Bunclody *MB* 14 E1
Bunyans Cove *NL* 36 C8
Bunyans Cove *NL* 37 C6
Burbank *ON* 8 E-F6
Burchell Lake *ON* 16 F4
Burdett *AB* 6 E7
Burdett *AB* 7 E7
Burdett *AB* 11 E1
Burditt Lake *ON* 13 F6
Burditt Lake *ON* 13 F6
Burditt Lake *ON* 16 E2
Burford *ON* 19 C6
Burford *ON* 20 E5
Burge Lake PP *MB* 12 D7-8
Burge Lake PP *MB* 15 D2
Burgeo *NL* 36 D3
Burgeo *NL* 37 D1
Burgeo Islands *NL* 36 E3
Burgeo Islands *NL* 37 E1
Burges & James Gadsden PP *BC* 6 D3
Burgess, Mount *YT* 5 E2
Burgessville *ON* 19 C6
Burgessville *ON* 20 E5
Burgis *SK* 11 D7
Burgoyne *ON* 18 F5
Burgoyne *ON* 20 A3-4
Burgoynes Cove *NL* 37 C7
Burin *NL* 36 F7
Burin *NL* 37 F5
Burin Island *NL* 36 F7
Burin Island *NL* 37 F5
Burin Peninsula *NL* 36 F6
Burin Peninsula *NL* 37 F4
Burke, Mount *AB* 6 E5
Burke, Mount *AB* 7 D3
Burke Channel *BC* 1 C4
Burketon *ON* 20 B8
Burketon *ON* 21 E2
Burk's Falls *ON* 18 C8
Burk's Falls *ON* 21 A1
Burk's Falls *ON* 22 F1
Burleigh Falls *ON* 21 D3
Burlington *NL* 38 E4-5
Burlington *NL* 39 F8
Burlington *NS* 32 E6
Burlington *NS* 33 B4
Burlington *ON* 19 B7-8
Burlington *ON* 20 D6-7
Burman River *BC* 2 D1
Burmis *AB* 6 F5
Burmis *AB* 7 E3-4
Burnaby *BC* 1 F7
Burnaby *BC* 2 E4-5
Burnaby *ON* 19 C8
Burnaby *ON* 20 F7
Burnaby Island *BC* 1 C2
Burnett Lake *SK* 12 B6
Burnett Lake *SK* 15 A1
Burnie Lake *MB* 12 B8
Burnie Lake *MB* 15 B2
Burnley Creek *ON* 21 E4
Burnside *MB* 14 D3
Burnside *NL* 37 B6-7
Burnside *NS* 33 A7
Burnside *NS* 34 E-F3
Burnside River *NU* 41 C2-3
Burns Lake *BC* 1 A5
Burns Lake PP *BC* 1 A5
Burnstown *ON* 21 A-B6
Burnsville *NB* 29 E8
Burnsville *NB* 30 E1
Burnsville *NB* 31 B7
Burnt Berry Brook *NL* 36 A5
Burnt Berry Brook *NL* 38 F4
Burntbush River *ON* 17 B7
Burntbush River *ON* 23 A1-2
Burnt Cape Ecological Reserve *NL* 38 A5
Burnt Church *NB* 31 C7
Burntcoat Head *NS* 32 D8
Burntcoat Head *NS* 33 A5
Burntcoat Head *NS* 34 E-F1
Burnt Cove *NL* 37 E8
Burnt Creek *NL* 39 A2
Burnt Flat *BC* 6 F3
Burnt Island *ON* 18 B7
Burnt Island *ON* 18 B7
Burnt Island Lake *ON* 21 A2
Burnt Island PP *NS* 34 D2-6
Burnt Island River *NL* 36 D2
Burnt Islands *NL* 36 D-E1
Burnt Lake *NL* 36 A7
Burnt Lake *NL* 36 B6

Burnt Lake *NL* 37 A5
Burnt Lake *NL* 38 F6-7
Burnt Lake *NT* 5 D7
Burnt Lands, RNP *ON* 21 B7
Burnt Lands Alvar PP *ON* 24 E1
Burnt Lands PNR *ON* 21 B7
Burnt Point *NL* 37 C-D8
Burnt Point *PE* 35 D3
Burnt Pond *NL* 36 C3
Burnt Pond *NL* 37 C1-2
Burnt River *ON* 20 A8
Burnt River *ON* 21 C2-3
Burnt River *ON* 21 C2-3
Burntroot Lake *ON* 22 E2-3
Burntwood Island *AB* 9 B7
Burntwood Island *AB* 12 B2
Burntwood Lake *MB* 12 B8
Burntwood Lake *MB* 15 E2
Burpee, ZPF de *NB* 31 F5
Burpee, ZPF de *NB* 32 B5-6
Burpee PGA *NB* 31 F5
Burpee PGA *NB* 32 B3
Burr *SK* 11 C5
Burrage Creek *BC* 3 D3
Burrard Inlet *BC* 1 F7
Burrard Inlet *BC* 2 D4-5
Burritts Rapids *ON* 21 B-C8
Burritts Rapids *ON* 24 E2
Burrows *BC* 7 F1
Burrows Lake *ON* 16 D7-8
Burrows Lake *ON* 17 B2
Burstall *SK* 6 D-E8
Burstall *SK* 11 E2
Burstall Lake *AB* 9 A7
Burstall Lake *AB* 12 A2-3
Burtch *ON* 19 C6
Burtch *ON* 20 E5
Burton *BC* 6 E3
Burton *NB* 32 C3
Burton *PE* 35 B1
Burtonsville *AB* 8 C5
Burtts Corner *NB* 31 F4
Burtts Corner *NB* 32 B2
Burwash *ON* 17 F6-7
Burwash *ON* 18 A5-6
Burwash Lake *ON* 17 E8
Burwash Landing *YT* 4 D2
Bury *QC* 27 E5
Busby *AB* 6 A5
Busby *AB* 8 B6
Bushell Park *SK* 11 E5
Bushe River *AB* 9 B4
Bush Island PP *NS* 33 D4-5
Bustard Island *AB* 9 B7
Bustard Island *AB* 12 B2
Bustard Islands *ON* 18 B5
Buteux, lac *QC* 28 D8
Buteux-Bas-Saguenay, ZEC *QC* 28 D8
Butler Lake, RNP *ON* 16 D3
Butler Lake PNR *ON* 16 D3
Butler Lake PP *ON* 13 F6-7
Butler Ridge PP *BC* 3 E7
Butler Ridge PP *BC* 9 D1
Butler's Barracks NHS *ON* 20 D7-8
Butlerville *NL* 37 D7
Butnau River *MB* 15 E5
Butter Cove *NL* 37 E8
Butterfly Lake *MB* 13 A4
Butterfly Lake *MB* 15 F4
Butter Pot PP *NL* 37 E8
Butt Lake *ON* 21 A2
Butt Lake *ON* 22 F2
Buttle Lake *BC* 1 F5
Buttle Lake *BC* 2 C1
Button *MB* 15 F3
Button Bay *MB* 15 B5
Button Islands *NU* 40 B6
Buttonville *ON* 19 A8
Buttonville *ON* 20 C7
Buttonville *ON* 21 F1
Buttress *SK* 11 E5
Buzwah *ON* 18 C4
Byam Martin Island *NU* 42 D3
Bydand Bay *NT* 5 E8
Byemoor *AB* 6 C6
Byemoor *AB* 8 F8
Bylot *MB* 15 B5
Bylot Island *NU* 41 A6
Bylot Island *NU* 42 D5-6
Bynd, Mount *YT* 4 E4
Byng *ON* 19 F3
Byng *ON* 20 E-F7
Byng CA *ON* 19 C7-8
Byng Inlet *ON* 18 C6
Byng Inlet *ON* 18 C6
Byrd, lac *QC* 22 B6
Byrnes Road *PE* 35 E4
Byron Bay *NL* 39 A7

C

Caamaño Passage *BC* 3 F2
Caamaño Sound *BC* 1 B3
Cabana *SK* 11 A3
Cabano *QC* 29 E2
Cabano *QC* 31 B1
Cabano, rivière *QC* 29 E2
Cabbage Island Unit (Gulf Islands NPR) *BC* 2 E5
Cabin Creek *YT* 4 E6
Cable Head *PE* 35 D6
Cabonga, réservoir *QC* 22 B6-7
Cabonga, réservoir *QC* 25 C-D1
Cabot, détroit de *NL* 36 E1
Cabot, détroit de *NS* 34 A8
Cabot, lac *QC* 39 A3
Cabot Beach PP *PE* 32 A8
Cabot Beach PP *NB* 31 E8
Cabot Beach PP *PE* 35 C3
Cabot Head *ON* 18 D5
Cabot Head, RNP *ON* 18 D4-5
Cabot Head PNR *ON* 18 D4-5
Cabot's Landing PP *NS* 34 A7
Cabot Strait *NL* 36 E1
Cabot Strait *NS* 34 A8
Cabri *SK* 11 E3
Cabri Lake *SK* 6 D8
Cabri Lake *SK* 11 D2
Cacaoui, lac *QC* 39 E2
Cache Bay *ON* 17 F7
Cache Bay *ON* 18 A7
Cache Creek *BC* 1 D8
Cache Creek *BC* 2 A7
Cache Creek *NT* 4 A6-7
Cache Lake *ON* 10 C8
Cache Lake *ON* 18 A7
Cache Lake *ON* 21 A3
Cacouna *QC* 28 D8
Cacouna *QC* 29 D1
Cactus Lake *SK* 6 C8
Cactus Lake *SK* 11 C2
Cacuscanus, lac *QC* 28 A7
Caddy Lake *MB* 14 E7
Caddy Lake *MB* 15 D7
Cadillac *QC* 23 E3
Cadillac *QC* 23 E6
Cadillac, lac *QC* 23 D4
Cadman, pointe *NB* 35 E2
Cadman Corner *NB* 35 E2
Cadman Point *NB* 35 E2
Cadogan *AB* 6 C7-8
Cadogan *AB* 8 F8
Cadomin *AB* 6 B3
Cadomin *AB* 8 D1-2
Cadotte Lake *AB* 9 D4
Cadotte Lake *AB* 9 D4
Cadotte River *AB* 9 D3-4
Cadurcis *NT* 14 D1
Caen Lake *NT* 10 D5
Caesarea *ON* 20 B8
Caesarea *ON* 21 E2
Cain, Mount *BC* 1 E5
Cains River *NB* 31 E6
Cains River *NB* 32 A3
Cainsville *ON* 19 C6-7
Cainsville *ON* 20 E5-6
Cairn-de-Glengarry, LHN du *ON* 24 E4-5
Cairngorm *ON* 19 C4
Cairngorm *ON* 20 E3
Cairns Lake *ON* 13 C5-6
Cairns Lake *ON* 14 A8
Cairns Lake *ON* 16 B2
Cairns Lake *SK* 12 C6
Cairns Lake *SK* 15 C1
Cairo *ON* 19 D3
Cairo *ON* 20 F2
Caissie Cape *NB* 32 B6-7
Caistor Centre *ON* 19 C8
Caistor Centre *ON* 20 E6-7
Caistorville *ON* 19 C7-8
Caistorville *ON* 20 E6-7
Caithness *BC* 7 F2
Cajeux, lac du *QC* 26 E5
Calabogie *ON* 21 B6
Calabogie Lake *ON* 21 B6
Calahoo *AB* 6 A5
Calahoo *AB* 8 C5
Calais Lake *NT* 10 E5
Calder *SK* 11 D7
Calder *SK* 13 D1
Calderbank *SK* 11 E4
Calders Dock *MB* 13 D4
Calders Dock *MB* 14 A5
Caledon *ON* 19 A7
Caledon *ON* 20 C6
Caledon East *ON* 19 A7
Caledon East *ON* 20 C6
Caledonia *NS* 33 A8
Caledonia *NS* 33 D3
Caledonia *NS* 34 E4
Caledonia *ON* 19 C7
Caledonia *ON* 20 E6
Caledonia *PE* 34 C3
Caledonia *PE* 35 E6
Caledonia Gorge, APP *NB* 32 C6
Caledonia Gorge PPA *NB* 35 D8
Caledonia Mills *NS* 34 E5
Caledon Lake PP *BC* 6 C2
Caliento *MB* 14 D7
Caliento *MB* 14 F6
Caligata Lake PP *BC* 6 C2
Caliper Lake *ON* 13 D7
Caliper Lake *ON* 16 B2
Caliper Lake PP *ON* 13 D7
Caliper Lake PP *ON* 16 E2
Calixa-Lavallée *QC* 24 C6-7
Calixa-Lavallée *QC* 27 D2

Callaghan Lake PP *BC* 1 E6-7
Callaghan Lake PP *BC* 2 B-C5
Callaghan's Rapids CA *ON* 21 D4
Callander *ON* 17 F8
Callander *ON* 18 B8
Callander *ON* 22 D1
Calling Lake *AB* 9 F5-6
Calling Lake *AB* 9 F6
Calling Lake PP *AB* 9 F5-6
Call Inlet *BC* 3 F4-5
Call Lake PP *BC* 3 F4-5
Callum *ON* 17 F7
Callum *ON* 18 A6
Calm Lake *ON* 13 F7
Calm Lake *ON* 16 E3
Calm Lake *ON* 16 E3
Calmar *AB* 6 B5
Calmar *AB* 8 D6
Calstock *ON* 17 B4
Calton *ON* 19 D5
Calton *ON* 20 F4
Calumet *QC* 24 D4
Calvert *NL* 37 E8
Calvert, Cape *BC* 1 D4
Calvert Island *BC* 1 D4
Camachigama, lac *QC* 23 F7
Camachigama, lac *QC* 25 B1
Camatose, lac *QC* 22 B6
Camborne *ON* 21 E3
Cambrai, lac *QC* 23 E6
Cambrai, lac *QC* 25 B1
Cambrai, lac *QC* 26 F1
Cambray *ON* 20 A8
Cambray *ON* 21 D2
Cambria *NS* 7 A6
Cambria Icefield *BC* 3 E3
Cambridge *NS* 32 E6-7
Cambridge *NS* 33 B4
Cambridge *ON* 34 F1
Cambridge *ON* 19 B6-7
Cambridge *ON* 20 D5-6
Cambridge *PE* 35 F7
Cambridge Bay (Iqaluktuuttiaq) *NU* 41 B3-4
Cambridge Bay (Iqaluktuuttiaq) *NU* 42 F3
Cambridge-Narrows *NB* 32 C4
Camden *NS* 33 A4-7
Camden *NS* 34 E2
Camden *NS* 34 E2
Camden East *ON* 21 D6
Camden Lake *ON* 21 D6
Camelsfoot Range *BC* 1 D7
Camelsfoot Range *BC* 2 A5-6
Cameron *ON* 20 A8
Cameron *ON* 21 D2
Cameron, lac *QC* 24 B4
Cameron Creek *ON* 18 A3
Cameron Falls *ON* 16 E6
Cameron Hills *AB* 9 A3
Cameron Hills *AB* 10 F5
Cameron Island *NU* 42 C2
Cameron Lake *ON* 17 C4
Cameron River *BC* 2 E3
Cameron River *NT* 10 B-C7
Cameron River *NT* 10 C7
Cameron River *NB* 31 F6-7
Cameron River TP *NT* 10 C7
Camerons Brook PP *NS* 33 D3-4
Cameron Settlement *NS* 33 A8
Cameron Settlement *NS* 34 E3-4
Camilla *ON* 20 C6
Camille Pouliot, mont *QC* 28 E5
Camlachie *ON* 19 C3
Camlachie *ON* 20 E2
Campania Island *NL* 1 B3
Campbellcroft *ON* 21 E3
Campbellford *ON* 21 D4
Campbell Island *BC* 1 C3-4
Campbell Island *NT* 5 A6
Campbell Lake *BC* 1 E5
Campbell Lake *BC* 2 C1-2
Campbell Lake *MB* 15 D5
Campbell Lake *NT* 5 C5
Campbell Lake *SK* 12 E5
Campbell Point *ON* 16 F8
Campbell Point *ON* 17 C2
Campbell River *BC* 1 E5
Campbell River *BC* 2 C1-2
Campbell's Bay *QC* 21 A6
Campbell's Bay *QC* 22 F6
Campbell's Cove PP *PE* 34 B4
Campbells Cove PP *PE* 35 D8
Campbells Creek *NL* 36 C1-2
Campbell Settlement *NB* 31 F3
Campbell Settlement *NB* 32 B-C1
Campbellton *NB* 29 D6
Campbellton *NB* 31 A5
Campbellton *NL* 36 A7
Campbellton *NL* 37 A5
Campbellton *PE* 35 B1
Campbellvale *ON* 19 C4
Campbellvale *ON* 20 E3
Campbellville *ON* 19 B7
Campbellville *ON* 20 D6
Camp Boggy *NL* 36 D6
Camp Boggy *NL* 37 D4
Camp Creek *AB* 6 A5

Camp Creek *AB* 8 B5
Campden *ON* 19 C8
Campden *ON* 20 E7
Camper *MB* 13 D3
Camper *MB* 14 B3-4
Camperville *MB* 11 C8
Camperville *MB* 13 D3
Camp Farewell *NT* 5 B4
Camping Islands *MB* 13 C3
Camping Islands *MB* 13 C2
Camp Morton *MB* 13 E4
Camp Morton *MB* 14 C5
Camp Morton PP *MB* 13 D-E4
Camp Morton PP *MB* 14 C5-6
Camp Robinson *ON* 13 E6
Camp Robinson *ON* 16 D2-3
Camp Robinson *ON* 13 E6
Campsie *AB* 8 B5
Camrose *AB* 6 B6
Camrose *AB* 8 D7
Camsell Bend *NT* 10 D2
Camsell Lake *NT* 10 B8
Camsell Portage *SK* 9 A7
Camsell Portage *SK* 12 A1
Camsell Range *NT* 10 C-D2
Camsell River *NT* 10 A5
Canaan *NB* 31 F7
Canaan *NB* 32 B5-6
Canaan *NS* 32 E7
Canaan *NS* 32 F7
Canaan *NS* 33 B4
Canaan *NS* 33 C5
Canaan Bog PPA *NB* 31 F7
Canaan Bog PPA *NB* 32 B5
Canaan Forks *NB* 31 F6-7
Canaan Forks *NB* 32 B4-5
Canaan River *NB* 31 F6-7
Canaan River *NB* 32 B5
Canaan Road *NB* 31 F7
Canaan Road *NB* 32 B5
Canada Bay *NL* 38 C4
Canada Bay *NL* 39 E8
Canada Creek *NS* 32 E6
Canada Creek *NS* 33 B4
Canada Harbour *NL* 39 E8
Canadian Rocky Mountain Parks *AB* 6 C3-4
Canadian Rocky Mountain Parks *AB* 7 A1-2
Canadian Rocky Mountain Parks *AB* 8 E1-2
Canadian Rocky Mountain Parks *BC* 7 B1-2
Canairiktok River *NL* 40 D7
Canal-de-St. Peters, LHN du *NS* 34 D6-7
Canal Flats *BC* 6 E4
Canal Flats *BC* 7 D1
Canal Flats PP *BC* 6 E4
Canal Flats PP *BC* 7 D1-2
Canal Lake *ON* 20 A8
Canal Lake *ON* 21 D2
Canal Lake Concrete Arch Bridge NHS *ON* 20 A8
Canal Lake Concrete Arch Bridge NHS *ON* 21 D2
Canard *NS* 32 E7
Canard *NS* 33 B4
Canard, lac au *QC* 27 D5
Canard, lac du *QC* 26 D7
Canard, lac du *QC* 27 B2
Canard de Mer, lac *QC* 26 C6-7
Canard de Mer, lac *QC* 28 A2
Canards, lac aux *QC* 28 D7
Cana *NL* 36 C1-2
Candiac *QC* 27 E1-2
Candiac *QC* 11 E6
Candle Lake *SK* 11 A5
Candle Lake *SK* 11 A5
Candle Lake PP *SK* 11 A5
Cando *SK* 11 C3
Canfield *ON* 19 C7
Canfield *ON* 20 E6
Caniapiscau, réservoir *QC* 40 E6
Caniapiscau, rivière *QC* 40 C-D6
Canim Beach PP *BC* 1 C-D8
Canim Beach PP *BC* 6 C1
Caniminia, lac *QC* 22 B5-6
Canim Lake *BC* 1 C8
Canim Lake *BC* 6 C1
Canim Lake *BC* 6 C1
Canim Lake *BC* 6 C1
Canmore *AB* 6 D4
Canmore *AB* 7 B2
Cann *QC* 25 B5
Cann *QC* 28 D1
Cannes-de-Roches *QC* 30 C2-3
Cannifton *ON* 21 E5
Canning *NS* 32 E7
Canning *NS* 33 B4
Canning, Mount *BC* 3 A1
Canning, Mount *BC* 4 F1
Canning Lake *ON* 21 C2-3
Cannings Cove *NL* 36 C8
Cannings Cove *NL* 37 C6
Cannington *ON* 20 B8
Cannington *ON* 21 D2
Cannington Manor Provincial Historic Park *SK* 11 F7
Canoe *BC* 6 D2
Canoe Cove *PE* 35 E4

Canoe Cove *PE* 35 E4
Canoe Lake *ON* 21 A2
Canoe Lake *ON* 21 D6
Canoe Lake *SK* 12 F3
Canoe Lake *SK* 12 F3
Canoe Narrows *SK* 12 F3
Canoe Reach *BC* 6 B2
Canoe River *BC* 6 B2
Canoose Flowage, APP *NB* 35 D1-2
Canoose Flowage PPA *NB* 32 D2
Canoose Lake *NB* 32 D2
Canopus *SK* 11 F4
Canora *SK* 11 D7
Canot, rivière du *QC* 25 C3
Canot Lake *SK* 5 D6-7
Canots, lac aux *QC* 28 A6
Canots, lac des *QC* 26 A6
Canso *NS* 34 E6
Canso, Cape *NS* 34 E6
Canso, Strait of *NS* 34 D5-6 E6
Cantal *QC* 11 F7
Cantara Bay *SK* 9 A8
Cantara Bay *SK* 12 A3
Canterbury *NB* 32 C1
Cantin Lake *MB* 13 B5
Cantley *QC* 21 A8
Cantley *QC* 22 F8
Cantley *QC* 24 D2
Cantuar *SK* 11 E3
Canuck *SK* 11 F3
Canwood *SK* 11 B4
Canyon *BC* 6 F4
Canyon *BC* 7 F1
Canyon Creek *AB* 9 E4-5
Canyon Creek *AB* 4 E3
Canyon Creek *YT* 5 E3-4
Canyon Hot Springs *BC* 6 D3
Canyon Ranges *NT* 4 A7 A-B8
Canyon Ranges *NT* 10 A-B1
Cap, Mount *YT* 4 E4
Cap-à-l'Aigle *QC* 28 E7-8
Cap-au-Renard *QC* 29 A7
Cap-aux-Meules *QC* 30 A7
Cap aux Meules, île du *QC* 30 F7
Cap-Breton, île du *NS* 34 D7-8
Cap-Brulé *NB* 35 E1
Cap-Chat, pointe du *QC* 29 A6
Cap-Chat, rivière du *QC* 29 A-B6
Cap-de-la-Madeleine *QC* 24 B8
Cap-de-la-Madeleine *QC* 25 F8
Cap-de-la-Madeleine *QC* 27 C3-4
Cap-des-Caissie *NB* 32 B6-7
Cap-d'Espoir *QC* 30 C3
Cap-des-Rosiers *QC* 30 B3
Cape Anguille *NL* 36 D1
Cape Breton *NS* 34 C7
Cape Breton Highlands *NS* 34 B6-7
Cape Breton Highlands NP *NS* 34 A6-7
Cape Breton Highlands *NS* 34 B6 C5-6
Cape Broyle *NL* 37 E8
Cape Charles *NL* 39 C7
Cape Chignecto PP *NS* 32 D6
Cape Chignecto PP *NS* 33 A3
Cape Cove *NL* 38 E7
Cape Croker *ON* 18 E5
Cape Dorset (Kinngait) *NU* 40 A4
Cape Dorset (Kinngait) *NU* 41 D7-8
Cape Enrage *NB* 32 C-D6
Cape Enrage *NB* 35 D6
Cape Freels *NL* 37 A7
Cape Freels *NL* 38 E8
Cape Freels North *NL* 37 A7
Cape Freels North *NL* 38 F8
Cape Freels South *NL* 37 A7
Cape Freels South *NL* 38 F8
Cape George *NS* 34 D5
Cape George Harbour *NS* 34 D6
Cape Hurd *ON* 18 D4
Cape Jourimain NWA *NB* 34 C1-2
Cape Jourimain NWA *NB* 35 E3
Cape LaHave Island *NS* 33 D4-5
Cape la Hune *NL* 36 D-E4
Cape la Hune *NL* 37 D-E2
Cape Merry NHS *MB* 15 B5-6
Cape Negro Island *NS* 33 E7
Cape North *NS* 34 A7
Cape Onion *NL* 38 A5
Cape Parry *NT* 5 A8
Cape Parry NMBS *NT* 5 A8

Cape Race *NL* 37 F8
Cape Ray *NL* 36 D1
Cape St. George *NS* 36 C1
Cape St. George *NL* 40 F8
Cape St. Mary *NS* 33 E1
Cape St. Mary's Ecological Reserve *NL* 36 F8
Cape St. Mary's Ecological Reserve *NL* 37 F6
Cape Scott PP *BC* 1 D3-4
Cape Smokey PP *NS* 34 B7
Cape Spear *NB* 32 B8
Cape Spear *NB* 34 C1
Cape Spear NHS *NL* 37 D8
Cape Tatnam WMA *MB* 15 C7
Cape Tormentine *NB* 32 B8
Cape Tormentine *NB* 34 C1
Cape Tormentine *PE* 35 E3
Cape Traverse *PE* 35 E3
Cape Walsingham *NU* 42 F8
Cape Wolfe *PE* 32 A7
Cape Wolfe *PE* 35 B1
Capimit, lac *QC* 25 B5
Capitachouane, lac *QC* 23 E7-8
Capitachouane, lac *QC* 25 C2
Capitachouane, lac *QC* 26 F2
Capitachouane, rivière *QC* 23 E-F7 F6
Capitachouane, rivière *QC* 25 B1
Capitachouane, ZEC *QC* 23 F6-7
Capitachouane, ZEC *QC* 25 B1
Cap-Jourimain, RNF du *NB* 34 C1-2
Cap-Jourimain, RNF du *NB* 35 E3
Cap Le Moine *NS* 34 B6
Caplan *QC* 29 D7-8
Caplan *QC* 31 A6
Caplin Cove *NL* 37 C8
Caplin Cove *NL* 37 C-D7
Cap-Lumière *NB* 31 E8
Cap-Lumière *NB* 32 A6
Cap-Merry, LHN du *MB* 15 B5-6
Capot Blanc, lac *NT* 10 B8
Cappahayden *NL* 37 F8
Cap-Parry, RNOM du *NT* 5 A8
Cap-Pelé *NB* 32 B7
Cap-Pelé *NB* 35 E1
Cappon *QC* 7 B8
Capreol *ON* 17 F6-7
Caprona *AB* 8 F7
Cap-Rouge *QC* 27 B5
Cap-Saint-Ignace *QC* 27 A7
Cap-Santé *QC* 27 B4-5
Cap-Seize *QC* 29 B6
Cap-Spear, LHN du *NL* 37 D8
Capstan, Cape *NS* 32 D6
Capstan, Cape *NS* 33 A3
Capstick *NS* 34 A7
Cap-Tourmente *QC* 27 A6
Cap-Tourmente, RNF du *QC* 27 A6-7
Cap-Tourmente, RNF du *QC* 28 F6-7
Cap Tourmente National Wilderness Area *QC* 27 A6-7
Cap Tourmente National Wilderness Area *QC* 28 F6-7
Capucins *QC* 29 A5-6
Caramat *ON* 16 D8
Caramat *ON* 17 B2
Caraquet *NB* 30 E1-2
Caraquet *NB* 31 B7
Caraquet, rivière *NB* 30 E1
Caraquet, rivière *NB* 31 B7
Caraquet Island *NB* 30 E1
Carberry *MB* 13 E2
Carberry *MB* 14 E2
Carbon *AB* 6 D6
Carbon *AB* 7 A5
Carbondale *AB* 8 C6
Carbonear *NL* 37 D7
Carcajou *AB* 9 C4
Carcajou Lake *NT* 5 D6-7
Carcajou River *NT* 5 D6
Carcross *YT* 4 E4
Cardale *MB* 11 E8
Cardale *MB* 13 E1
Cardale *MB* 14 D1
Cardiff *ON* 21 C4
Cardigan *PE* 34 C3
Cardigan *PE* 35 E6
Cardigan Bay *PE* 35 E7
Cardigan Head *PE* 35 E6
Cardigan Point *PE* 35 E7
Cardigan River *PE* 35 E6-7
Cardinal *MB* 13 F3
Cardinal *MB* 14 E3
Cardinal *ON* 24 E3
Cardinal Lake *AB* 9 E3
Card Lake PP *NS* 33 C5
Cardross *PE* 35 E6

Cardross *SK* 11 E5
Card's Harbour *NL* 38 F5
Cardston *AB* 6 F6
Cardston *AB* 7 F5
Careen Lake *SK* 9 D8
Careen Lake *SK* 12 D3
Carey *MB* 13 F4
Carey *MB* 14 E5
Carghill Island *MB* 15 F5
Carheil, lac *QC* 39 C2
Cariboo Mountain *BC* 1 B8
Cariboo Mountain *BC* 6 B1
Cariboo Mountains *BC* 6 B1-2
Cariboo Mountains PP *BC* 6 B1
Cariboo Mountains PP *BC* 6 B1
Cariboo Nature PP *BC* 1 C8
Cariboo River *BC* 1 B8
Cariboo River *BC* 6 B1
Cariboo River PP *BC* 1 B8
Cariboo River PP *BC* 6 B1
Caribou *NS* 34 D3
Caribou, lac du *MB* 41 F3
Caribou, lac du *QC* 24 A7
Caribou, lac du *QC* 25 E7
Caribou, lac du *QC* 26 C7
Caribou, lac du *QC* 27 B2
Caribou, lac du *SK* 12 C7
Caribou, Mount *YT* 4 D4-5
Caribou Falls *ON* 13 E5
Caribou Falls *ON* 14 D8
Caribou Falls *ON* 16 C-D1
Caribou Gold Mines *NS* 33 B7
Caribou Gold Mines *NS* 34 F3
Caribou Islands *NS* 34 D3
Caribou Islands *NT* 10 D7
Caribou Lake *MB* 15 A4
Caribou Lake *SK* 5 C5
Caribou Lake *ON* 16 C5-6 E5
Caribou Lakes *YT* 5 E4
Caribou Mountains *AB* 9 B4-5
Caribou Mountains *AB* 41 E1
Caribou Mountains Wildland PP *AB* 9 A-B5
Caribou Mountains *AB* 9 B4-5
Caribou Mountains *AB* 41 E1
Caribou-Munroe's Island PP *NS* 34 D3-4
Caribou Point *NT* 10 E6
Caribou Pond *NL* 36 B7
Caribou Pond *NL* 37 B5
Caribou River *MB* 15 A4-5
Caribou River *MB* 40 B1
Caribou River *MB* 41 F4
Caribou River *NS* 33 D2
Caribou River *NS* 34 D3
Caribou River *NT* 4 D8
Caribou River *NT* 10 E1
Caribou River *YT* 5 E4
Caribou River, PP *MB* 40 B1
Caribou River PP *MB* 41 F4
Caribou River Wilderness PP *MB* 15 A4
Carievale *SK* 11 F7-8
Carievale *SK* 13 F1
Carignan *QC* 24 D6
Carignan *QC* 25 D7
Carignan *QC* 27 E2
Carignan *QC* 28 F3
Carignan, lac *QC* 25 D7
Carignan, lac *QC* 28 F3
Carillon *QC* 24 D4-5
Carillon, canal de *QC* 24 D4
Carillon Barracks NHS *QC* 24 D4-5
Carillon Canal *QC* 24 D4
Carlea *SK* 11 B6
Carleton *NS* 33 E2
Carleton *PE* 32 A7
Carleton *PE* 35 E3
Carleton *QC* 29 D7
Carleton *QC* 31 B2
Carleton, mont *NB* 29 F5-6
Carleton, mont *NB* 31 C4
Carleton, Mount *NB* 29 F5-6
Carleton, Mount *NB* 31 C4
Carleton Cove *PE* 35 E3
Carleton Martello Tower NHS *NB* 32 D4-5
Carleton Martello Tower NHS *NB* 33 B1-2
Carleton Place *ON* 21 B7
Carleton Place *ON* 24 E1
Carling *ON* 18 D7
Carlingford *NB* 31 E3
Carlingford *NB* 32 A1
Carlingford *ON* 19 B5
Carlingford *ON* 20 D4
Carlisle *ON* 19 B7
Carlisle *ON* 19 C4
Carlisle *ON* 20 E3
Carlos *AB* 8 E5
Carlow *ON* 19 A4
Carlow *ON* 20 C3
Carlowrie *MB* 13 F4
Carlowrie *MB* 14 F5
Carlsbad Springs *ON* 21 A8
Carlsbad Springs *ON* 24 D2-3
Carlsruhe *ON* 20 B4
Carluke *ON* 19 C7
Carluke *ON* 20 E6
Carl Wilson Lake *ON* 22 E2-3

Carlyle *SK* **11** F7
Carmacks *YT* **4** C3
Carman *MB* **13** F3
Carman *MB* **14** E4
Carmanah Walbran PP *BC* **2** F2-3
Carmangay *AB* **6** E6
Carmangay *AB* **7** D5
Carmanville *NL* **36** A8
Carmanville *NL* **37** A6
Carmanville *NL* **38** F7
Carmel *SK* **11** C5
Carmen, lac *QC* **25** A3-4
Carmen, lac *QC* **26** F3-4
Carmichael *SK* **11** E3
Carnarvon *AB* **21** B2-3
Carnduff *SK* **11** F7
Carnegie *MB* **14** D1
Carnwath River *NT* **5** C6-7
Carnwood *AB* **6** B5
Carnwood *AB* **8** D5
Caroline *AB* **6** C5
Caroline *AB* **8** F5
Caron *SK* **11** E4-5
Caron, lac *QC* **23** E2-3
Caron, lac *QC* **23** F2
Caron Brook *NB* **29** F2
Caron Brook *NB* **31** C1
Caronport *SK* **11** E5
Carp *ON* **21** B7
Carp *ON* **24** D1
Carpe, lac à la *QC* **28** C4
Carpenter *AB* **2** A7
Carpenter *SK* **11** C4-5
Carpenter Lake *BC* **1** D7
Carpenter Lake *BC* **2** A5
Carp Lake *BC* **3** F7
Carp Lake PP *BC* **3** F6
Carp River *ON* **24** D1-2
Carquile *BC* **2** A7
Carragana *SK* **11** C6
Carr-Harris Lake *MB* **14** B5
Carrick *MB* **13** F4
Carrick *MB* **14** F6
Carrière, lac *QC* **22** B5
Carroll *MB* **11** F8
Carroll *MB* **13** F2
Carroll *MB* **14** E1
Carroll Lake *ON* **13** C6
Carroll Lake *ON* **14** B7-8
Carroll Lake *ON* **16** B1
Carrolls Crossing *NB* **31** E5-6
Carrolls Crossing *NB* **32** A3-4
Carrot Creek *AB* **6** A-B4
Carrot Creek *AB* **8** C3
Carrot River *SK* **11** B6-7
Carruthers CA *ON* **20** B6
Carrying Place *ON* **21** E4-5
Carry The Kettle *AB* **11** E6
Carseland *AB* **6** D6
Carseland *AB* **7** C4-5
Carson *BC* **6** F2
Carson Lake *ON* **21** A4
Carson Lake *ON* **22** F4
Carson Lake *ON* **24** F3
Carson Lake PP *ON* **21** A4
Carson Lake PP *ON* **24** F3
Carson-Pegasus PP *AB* **6** A4
Carson-Pegasus PP *AB* **8** A-B4
Carsonville *NB* **32** C5
Carstairs *AB* **6** D5
Carstairs *AB* **7** A4
Carswell Lake *MB* **12** D7
Carswell Lake *MB* **15** D1
Carswell Lake *SK* **9** B7-8
Carswell Lake *SK* **12** B-C4
Cartage Mountains *NT* **10** B-C4
Carter Basin *NL* **39** B6
Carter's Cove *NL* **38** E6
Carthage *ON* **19** A5
Carthage *ON* **20** C4
Cartier *ON* **17** F6
Cartier, lac *QC* **22** A5
Cartier Lake *ON* **22** E4-5
Cartmel, Mount *BC* **3** C3
Cartwright *MB* **13** F2
Cartwright *MB* **14** F2
Cartwright *NL* **39** B7-8
Cartwright *NL* **40** D8
Cartwright Junction *NL* **39** D8
Cartwright Sound *BC* **1** B1
Cartyville *NL* **36** C2
Carway *AB* **6** F6
Carway *AB* **7** F5
Casault, lac *QC* **29** C5
Casault, ZEC *QC* **29** C5
Cascade *BC* **6** F3
Cascade Inlet *BC* **1** C4
Cascade PRA *BC* **1** F8
Cascade PRA *BC* **2** D7-8
Cascade PRA *BC* **6** F1
Cascade Range *BC* **2** D-F7
Cascapédia, branche du lac *QC* **29** B6-7 C6
Cascapédia, rivière *QC* **29** B2-3
Cascumpec *PE* **34** B1
Cascumpec *PE* **35** B2
Cascumpec Bay *PE* **34** B1
Cascumpec Bay *PE* **35** B2
Cascumpec Sand Hills *PE* **35** B2
Caserne-de-Carillon, LHN de la *ON* **24** D4-5
Casernes-de-Butler, LHN des *ON* **20** D7-8
Casey *ON* **25** B5
Cashel Lake *ON* **21** C5
Casimir *ON* **17** F7

Casimir *ON* **18** A6
Caslan *AB* **8** A7
Caslan *AB* **9** F6
Caslan *AB* **12** F1
Casselman *ON* **24** D-E3
Cassels Lake *ON* **22** B1
Cassiar Mountains *BC* **3** B-C4
Cassiar Mountains *YT* **4** E5-6
Cassidy *NL* **1** F6
Cassidy *BC* **2** E3-4
Cassils *AB* **7** C6
Castagnier, lac *QC* **23** C5
Castagnier, rivière *QC* **23** C5
Castalia *NB* **32** F3
Castalia PP *NB* **32** F3
Castelveyre, lac *QC* **25** D5
Castle Creek, RNP *ON* **16** F4-5
Castle Creek PNR *BC* **16** F4-5
Castledale *BC* **7** B1
Castleford *ON* **21** A6
Castleford *ON* **22** F6
Castlegar *BC* **6** F3
Castle Hill, LHN de *NL* **36** E8
Castle Hill, LHN de *NL* **36** E8
Castle Mountain *AB* **6** D4
Castle Mountain *AB* **7** B1-2
Castle Mountain *YT* **4** A4
Castle Hill NHS *NL* **36** E8
Castle Hill NHS *NL* **37** E6
Castle Rock Hoodoos PP *BC* **1** D8
Castle Rock Hoodoos PP *BC* **2** A7
Castle Rock Hoodoos PP *BC* **6** D1
Castleton *ON* **21** E4
Castonguay, lac *QC* **23** C7
Castonguay, lac *QC* **26** D1
Castor *AB* **6** C6-7
Castor *AB* **8** F8
Castor Blanc, lac *QC* **22** D8
Castor Blanc, lac *QC* **24** A8
Castor Blanc, lac *QC* **25** F2
Castor Lake *NT* **10** A5-6
Castor River *ON* **24** E3
Castors River *NL* **38** C3-4
Castors River *NL* **38** C3
Castors River *NL* **39** E8
Catala Island PMP *BC* **1** E4
Catalina *NL* **37** B7-8
Catalone *NS* **34** C8
Cataract Lake *ON* **18** A1-2
Cataracts PP *NL* **37** E7
Cataraqui River *ON* **21** D7
Cat Arm *NL* **38** D3
Cat Arm River *NL* **39** E7-8 F8
Catchacoma *ON* **21** C3
Catchacoma Lake *ON* **21** C3
Cater *SK* **11** B3
Catfish Creek *ON* **19** D5
Catfish Creek *ON* **20** F4
Catfish Lake *ON* **22** B1
Cathcart *ON* **19** D5
Cathcart, parc *ON* **19** D2-3
Cathcart, parc *ON* **20** F1-2
Cathcart Park (St. Clair Parkway) *ON* **19** D2-3
Cathcart Park (St. Clair Parkway) *ON* **20** F1-2
Cathedral PP *BC* **6** F1
Catherwood *SK* **11** C4
Cat Lake *ON* **13** C7
Cat Lake *ON* **13** C7
Cat Lake *ON* **16** B4
Cat Lake *ON* **16** B4
Cat Lake *ON* **16** B4
Cauchon Lake *MB* **15** E4
Cauchy, lac *QC* **39** E5
Caumont, lac *QC* **39** E5
Caupichigau, lac *QC* **26** A3
Causapscal *QC* **29** C5
Causapscal, rivière *QC* **29** C5
Caution, Cape *BC* **1** D4
Cavan *ON* **21** E3
Cavan, lac *QC* **26** B4
Cave and Basin, LHN *AB* **7** B1-2
Cave and Basin NHS *AB* **7** B1-2
Cavell *ON* **16** D7
Cavell *ON* **17** A1-2
Cavell *SK* **11** C3
Cavendish *AB* **7** C8
Cavendish *NL* **37** D7
Cavendish *PE* **34** B2
Cavendish *PE* **35** D4
Caverlys Landing *ON* **21** B5
Cavern Lake, RNP *ON* **16** E6
Cavern Lake PNR *ON* **16** E6
Cavers *ON* **16** E7
Cavers *ON* **17** C1
Cawaja Beach *ON* **18** F7
Cawaja Beach *ON* **20** A6
Cawasachouane, lac *QC* **22** A4
Cawatose, lac *QC* **22** B6
Cawston *BC* **6** F1-2
Caycuse *BC* **2** E3
Cayer *MB* **13** D2
Cayer *MB* **14** B2
Cayley *AB* **6** E5

Cayley *AB* **7** C4
Cayoose Creek *BC* **2** B6
Cayoosh Creek *BC* **1** E7-8
Cayoosh Creek *BC* **2** B6
Cayuga *ON* **19** C7
Cayuga *ON* **20** E6
Cazaville *QC* **24** E4-5
Cazeneuve, lac *QC* **26** B-C6
Ceba *SK* **11** B7
Ceba *SK* **13** B1
Cecebe *ON* **18** C8
Cecebe *ON* **21** A1
Cecebe *ON* **22** F1
Cecebe Lake *ON* **18** C8
Cecebe Lake *ON* **21** A1
Cecebe Lake *ON* **22** F1
Cecebe Lake *ON* **23** C5
Cecil *AB* **7** D7
Cecil Lake *BC* **3** E8
Cecil Lake *BC* **9** D2
Cécile, lac *QC* **26** E5-6
Cécile, lac *QC* **28** C1
Cedar *BC* **1** F6
Cedar *BC* **2** E4
Cedar Beach *ON* **19** F1
Cedar Beach CA *ON* **19** E1
Cedar Dunes PP *NL* **34** A6-7
Cedar Dunes PP *PE* **35** C1
Cedarhurst Park *ON* **19** F1-2
Cedar Lake *MB* **11** B8
Cedar Lake *MB* **13** B2
Cedar Lake *MB* **13** E6
Cedar Lake *MB* **22** E3
Cedar Lake *ON* **31** B6
Cedar Mills *ON* **19** A7
Cedar Mills *ON* **20** C6
Cedar Point PP *BC* **1** C7-8
Cedar Springs *ON* **19** E3
Cedarvale *BC* **3** F3-4
Cedoux *SK* **11** E6
Cèdres, lac aux *QC* **23** D8
Cèdres, lac aux *QC* **25** A2
Cèdres, lac aux *QC* **26** E2
Cèdres, lac des *MB* **11** B8
Cèdres, lac des *MB* **13** B2
Cèdres, lac des *QC* **22** D7
Cèdres, lac des *QC* **24** B1
Cèdres, lac des *QC* **25** F1
Cèdres, rivière des *QC* **29** B1
Ceepeecee *BC* **1** E4
Celista *BC* **6** D2
Centennial Lake *ON* **21** B5
Centennial Lake PP *ON* **21** B5-6
Centennial Range *YT* **4** D1
Central *AB* **8** D7
Central Argyle *NS* **33** E-F2
Central Bedeque *PE* **34** B8
Central Bedeque *PE* **35** D3
Central Blissville *NB* **32** C-D3
Central Butte *SK* **11** D-E4
Central Economy *NS* **32** D8
Central Economy *NS* **33** A5
Central Economy *NS* **34** E1
Central Elgin *ON* **19** D5
Central Elgin *ON* **20** F4
Central Grove *NS* **33** D1
Central Grove PP *NS* **33** D1
Centralia *ON* **19** B4
Centralia *ON* **20** C4
Central Lot 16 *PE* **35** D2
Central Onslow *NS* **33** A6
Central Onslow *NS* **34** E2
Central Patricia *ON* **13** D8
Central Patricia *ON* **16** B5
Central Saanich *BC* **2** F4
Central West River *NS* **33** A7
Central West River *NS* **34** E3
Centre Burlington *NS* **32** E7
Centre Burlington *NS* **33** B5
Centre Burlington *NS* **34** F1
Centre-d'Accueil-du-Parc-Jasper, LHN du *AB* **8** D1
Centre-d'Inscription-de-l'Entrée-Est, LHN du *MB* **14** C2-3
Centre Dummer *ON* **21** D4
Centre Island *ON* **18** B4
Centrelea *NS* **32** F5
Centrelea *NS* **33** C3
Centre Musquodoboit *NS* **33** B7
Centre Musquodoboit *NS* **34** F2-3
Centre Napan *NB* **31** D7
Centreton *ON* **21** E4
Centre Village *NB* **35** F1
Centreville *NB* **31** E3
Centreville *NB* **32** A1
Centreville *NB* **32** C5
Centreville *NL* **36** A8
Centreville *NL* **37** A6
Centreville *NL* **37** A6
Centreville *NS* **32** E6-7
Centreville *NS* **33** B4
Centreville *NS* **33** F2
Centreville *NS* **33** C1
Centreville *ON* **19** C5
Centreville *ON* **20** E4
Centreville *ON* **21** D6

Cerf, lac du *QC* **24** B2-3
Cerf, lac du *QC* **25** F2-3
Cessford *AB* **6** D7
Cessford *AB* **7** B7
Cessford *AB* **11** D1
Ceylon *ON* **20** B5
Ceylon *SK* **11** F5
Chaatl Island *BC* **1** B1
Chachukew Lake *SK* **12** B6-7
Chaffeys Locks *ON* **21** D7
Chaffeys Locks *ON* **24** F1
Chagoness *NS* **11** C6
Chaigneau, lac *QC* **39** A1-2
Chain Lakes *ON* **18** C8
Chain Lakes PP *AB* **7** D4
Chaleur, lac de la *QC* **26** A3
Chaleur Bay *NB* **29** E8
Chaleur Bay *QC* **30** E1
Chaleur Bay *QC* **31** A6-7
Chaleurs, baie de *QC* **29** D8
Chaleurs, baie de *QC* **30** E1
Chaleurs, baie de *QC* **31** A6
Chalk River *ON* **22** E5
Chaloupe, rivière de la *QC* **30** A6
Chambeaux, lac *QC* **39** B1
Chambeaux, lac *QC* **40** E6
Chamberlain *SK* **11** D5
Chamberlain Settlement *NB* **31** B6
Chambers Corners *ON* **19** C8
Chambers Corners *ON* **20** E7
Chambly *QC* **24** D6
Chambly *QC* **27** E2
Chamcook *NB* **32** E3
Champagne *YT* **4** E3
Champdoré, lac *QC* **40** D6
Champion *AB* **6** E6
Champion *AB* **7** D5
Champion Lakes PP *BC* **6** F3
Champlain *QC* **24** B8
Champlain *QC* **25** F8
Champlain *QC* **27** C4
Champlain, lac *QC* **27** F2
Champlain, rivière *QC* **24** B8
Champlain, rivière *QC* **25** F8
Champlain, rivière *QC* **27** B3-4 C3
Champlain Park *ON* **18** B8
Champneuf *QC* **23** D5
Champney's *NL* **37** C7
Chance Cove *NL* **37** E8
Chance Cove *NL* **37** D6-7
Chance Cove Head *NL* **37** D6-7
Chance Cove PP *NL* **37** E8
Chance Creek *YT* **5** D-E3
Chance Harbour *NS* **32** E3-4
Chancellor *AB* **7** B5
Chanceport *NL* **38** E6
Chandindu River *YT* **4** A2
Chandler *QC* **30** D2
Chandos Lake *ON* **21** C4
Change Island *NL* **38** E7
Change Islands *NL* **38** E7
Chan Lake TP **10** D5-6
Chan Lake TP *NT* **41** D1
Charlton Island *NU* **40** E3-4
Channel Island *MB* **11** B8
Channel Island *MB* **13** B2
Channel-Port aux Basques *NL* **36** E1
Chantrey Inlet *NU* **41** C4
Chantry *ON* **21** C7
Chantry *ON* **24** F1
Chantry Island NMBS *ON* **18** F4
Chantry Island NMBS *ON* **20** A3
Chapais *QC* **26** B4
Chapais *QC* **40** F4-5
Chapais, ZEC *QC* **28** F8
Chapais, ZEC *QC* **29** F1
Chapeau *QC* **22** E5
Chapeau, lac *QC* **25** D6
Chapeau, lac *QC* **22** A1
Chapeau, lac *QC* **28** F1
Chapeau-de-Paille, ZEC *QC* **25** E6
Chapeau-de-Paille, ZEC *QC* **27** A1-2
Chapel Arm *NL* **37** D-E7
Chapel Island *NL* **36** A7
Chapel Island *NL* **36** D-E6
Chapel Island *NL* **37** D-E4
Chapel Island *NL* **38** F6
Chapel Island *NL* **37** A6
Chapels Cove *NL* **37** E7-8
Chapin Bay *MB* **13** A5
Chapiteau, lac *QC* **39** A4
Chapleau *ON* **17** D4
Chapleau, réserve de chasse de la Couronne *ON* **17** C4
Chapleau Crown Game Preserve *ON* **17** C4
Chapleau-Nemegosenda River PP *ON* **17** D5
Chapleau River *ON* **17** D5
Chaplin *SK* **11** E4
Chaplin Lake *SK* **11** E4
Cerf, lac du *QC* **22** E8

Chapman, Cape *NU* **41** B5
Chapman, Cape *NU* **42** F5
Chapman Lake *MB* **15** D3
Chapmans *BC* **2** C7
Chapmans Corner *NB* **35** E2
Chapman Settlement *NS* **32** C7-8
Chapman Settlement *NS* **34** D1
Chapman Settlement *NS* **35** F2
Chaput *ON* **23** E1
Chaput Hughes *ON* **17** D7
Charbonneau River *SK* **9** B8
Charbonneau River *SK* **12** B3
Charcoal Lake *SK* **12** B6-7
Charcoal Lake *SK* **15** B1
Chard *AB* **9** E7
Chard *AB* **12** E2
Charest, rivière *QC* **24** A8
Charest, rivière *QC* **25** E8
Charest, rivière *QC* **27** B4
Charette *QC* **24** B7
Charette *QC* **25** F7
Charette *QC* **27** C2-3
Charette, lac *QC* **23** D7
Charette, lac *QC* **26** D-E1
Charing Cross *ON* **19** E3
Charland, lac *QC* **25** E5
Charlebois *MB* **15** D5
Charlemagne *QC* **24** D6
Charlemagne *QC* **27** D-E2
Charles *MB* **12** E7
Charles *MB* **15** E1-2
Charlesbourg *QC* **27** A5
Charles Island *NU* **40** B5
Charles Island *NU* **41** B5
Charles Lake *AB* **9** A7
Charles Lake *AB* **10** F8
Charles Lake *AB* **12** A7
Charleston *NS* **33** D4
Charleston *ON* **21** D7-8
Charleston *ON* **24** F1
Charleston Lake *ON* **21** D7-8
Charleston Lake PP *ON* **21** D7-8
Charlesville *NS* **33** F2
Charlie Lake *BC* **3** E8
Charlie Lake *BC* **9** D1
Charlie Lake *NU* **15** A2
Charlie Lake PP *BC* **3** E8
Charlie Lake PP *BC* **9** D1
Charlo *NB* **29** D6-7
Charlo *NB* **31** A5
Charlo River *NB* **29** D-E6
Charlo River *NB* **31** A-B5
Charlos Cove *NS* **34** E6
Charlotte, Lake *NS* **33** B7
Charlotte Lake *BC* **1** C5
Charlottetown *NL* **36** C8
Charlottetown *NL* **37** C6
Charlottetown *NL* **39** C8
Charlottetown *PE* **34** C2
Charlottetown *PE* **35** E5
Charlottetown Harbour *PE* **35** E5
Charlton *ON* **17** D7
Charlton *ON* **20** B5
Charlton Island *NU* **40** E3-4
Charny *QC* **27** B5-6
Charron, lac *QC* **26** B5
Charron Lake *MB* **13** B-C5
Charteris *QC* **21** A6-7
Charteris *QC* **22** F6-7
Charteris *QC* **24** C-D1
Chartersville *NB* **31** F8
Chartersville *NB* **32** B6
Chartier Lake *ON* **18** C7
Chartierville *QC* **27** F6
Chase *BC* **6** D2
Chasm *BC* **1** D8
Chasm *BC* **6** D1
Chasm PP *BC* **1** D8
Chasm PP *BC* **6** D1
Chassaigne, lac *QC* **25** A2-3
Chassaigne, lac *QC* **26** F2-3
Chasse, rivière la *QC* **39** E-F4
Chassignolle, lac *QC* **23** D3
Chastrier, lac *QC* **39** B1
Chaswood *NS* **33** B6
Chaswood *NS* **34** F2
Chateau *NL* **39** D8
Château Bay *NL* **39** D8
Châteauguay *QC* **24** D5-6
Châteauguay *QC* **27** E1
Châteauguay, rivière *QC* **24** E5
Châteauguay, rivière *QC* **27** F1
Château-Richer *QC* **27** A6
Châteauvert, lac *QC* **25** C5
Châteauvert, lac *QC* **28** E1
Chateh (Assumption) *AB* **9** B3
Chater *MB* **13** E2
Chater *MB* **14** E1
Chatfield *MB* **13** D3
Chatfield *MB* **14** C4
Chatham *ON* **19** D2-3
Chatham (Miramichi East) *NB* **31** D7
Chatham Head *NB* **31** D7
Chatham-Kent *ON* **19** D2-3
Chats, lac des *QC* **21** A6-7
Chatsworth *ON* **18** F5-6
Chatsworth *ON* **20** A4

Chatsworth *ON* **20** A4
Chatwin Lake *MB* **12** B7
Chatwin Lake *MB* **15** B2
Chaud, lac *QC* **24** B4
Chaud, lac *QC* **25** F4
Chaudière, lac *QC* **26** B5
Chaudière, rivière *QC* **27** B-C6
Chaudière, rivière *QC* **27** D7
Chaume, lac *QC* **23** F8
Chaume, lac *QC* **25** B2
Chaumont, lac *QC* **25** B7
Chaumont, lac *QC* **28** C7
Chauvin *AB* **6** B8
Chauvin *AB* **11** B2
Chauvin, ZEC *QC* **28** C7
Chavannes, lac *QC* **22** C5
Chazel *QC* **17** C8
Chazel *QC* **23** C2
Cheadle *AB* **7** B4-5
Cheapside *ON* **19** C7
Cheapside *ON* **20** E6
Chebogue Point *NS* **33** F1
Chebogue Point *NS* **33** E-F1
Checleset Bay *BC* **1** E4
Chédabouctou, baie de *NS* **34** E6
Chedabucto Bay *NS* **34** E6
Cheddarville *AB* **8** F4-5
Cheecham Hills *AB* **9** E7
Cheecham Hills *AB* **12** D-E2
Cheekamus *BC* **1** E-F7
Cheekamus *BC* **2** C5
Cheekamus River *BC* **2** C5
Cheeseman Lake *ON* **16** E5-6
Chef, rivière du *QC* **26** B6
Chegoggin *NS* **33** E1
Chehalis Lake *BC* **2** D6
Chelan *SK* **11** C6
Chelaslie River *BC* **1** B5
Chelmsford *NB* **31** D6-7
Chelmsford *ON* **17** F6
Chelmsford *ON* **18** A5
Chelsea *NS* **33** D4
Chelsea *QC* **21** A8
Chelsea *QC* **22** F8
Chelsea *QC* **24** D2
Cheltenham *ON* **19** A7
Cheltenham *ON* **20** C6
Chelton *PE* **35** D3
Chelton Beach PP *PE* **32** B8
Chelton Beach PP *PE* **34** C1
Chelton Beach PP *PE* **35** E2-3
Chemainus *BC* **1** F6
Chemainus *BC* **2** E4
Chemainus River PP *BC* **2** E3-4
Chemawawin *MB* **13** B4
Chemong *SK* **11** B7
Chemong *SK* **13** B1
Chemong Lake *ON* **21** D3
Chenaux *ON* **21** A6
Chenaux *ON* **22** F6
Chêne, rivière du *QC* **27** B4 C5
Chênes, île des *MB* **13** F4
Chênes, île des *MB* **14** E5
Chênevert, lac *QC* **25** A2
Chênevert, lac *QC* **26** F2
Chénéville *QC* **24** C3
Chénier *QC* **22** E7
Chénier *QC* **24** D1
Chenil, lac *QC* **39** D6
Chenon, lac *QC* **22** A-B3
Chepstow *PE* **35** D3
Chepstow *PE* **35** D-E8
Chepstow Point *PE* **35** E8
Cherhill *AB* **6** A5
Cherhill *AB* **8** B-C5
Cherrier, lac *QC* **23** E8
Cherrier, lac *QC* **26** D2
Cherrington Lake *ON* **16** A1
Cherryfield *NS* **32** E6
Cherryfield *NS* **33** C3-4
Cherry Grove *AB* **6** A7-8
Cherry Grove *AB* **9** F7-8
Cherry Grove *AB* **11** A2
Cherry Hill *NS* **33** D4
Cherry Point *AB* **3** E8
Cherry Point *AB* **9** B2
Cherry River *QC* **27** E-F4
Cherry Valley *ON* **21** E5
Cherry Valley *PE* **34** C3
Cherry Valley South *PE* **35** E5
Cherryville *BC* **6** E2
Chertsey *QC* **24** C5
Chertsey *QC* **27** D1
Cheslatta Lake *BC* **1** A-B5
Chesley *ON* **20** B4
Chesley Lake *ON* **18** F5
Chesley Lake *ON* **20** A3-4
Chesnay *NS* **15** C5
Chester *NS* **32** F7
Chester *NS* **33** D5
Chester Basin *NS* **32** F7
Chester Basin *NS* **33** D5
Chesterfield Bridge *SK* **6** D8
Chesterfield Bridge *SK* **11** D2
Chesterfield Inlet *NU* **40** A2
Chesterfield Inlet *NU* **41** D5
Chesterfield Inlet (Igluligaarjuk) *NU* **9** C-D2
Chesterfield Inlet (Igluligaarjuk) *NU* **40** A2

Chesterfield Inlet (Igluligaarjuk) *NU* **41** D-E5
Chester Grant *NS* **32** F7
Chester Grant *NS* **33** C4-5
Chestermere *AB* **6** D5-6
Chestermere *AB* **7** B4
Chesterville *ON* **24** E3
Chesterville *ON* **27** D5
Chéticamp *NS* **34** B6
Chéticamp Island *NS* **34** B6
Chéticamp Lake *NS* **34** B7
Chéticamp River *NS* **34** B6-7
Chetwynd *BC* **9** E1
Cheverie *NS* **32** E7
Cheverie *NS* **33** B5
Cheverie *NS* **34** F1
Chevery *QC* **39** E6
Cheves Creek *BC* **3** B6-7
Chevreuil, lac du *QC* **24** B-C3
Chevreuils, rivière aux *QC* **27** C5
Chevrillon, lac *QC* **26** A4-5
Chezacut *BC* **1** C6
Chiasson *NB* **31** B8
Chiblow Lake *ON* **17** F5
Chiblow Lake *ON* **18** A2
Chibouet, rivière *QC* **24** C7
Chibouet, rivière *QC* **32** C-D1
Chibougamau *QC* **26** A-B5
Chibougamau *QC* **40** F5
Chibougamau, lac *QC* **26** A-B5
Chibougamau, rivière *QC* **26** A3
Chic-Chocs, monts *QC* **29** B6-7
Chic-Chocs, RF des *QC* **29** B7-8
Chic-Chocs, RF des *QC* **29** A-B7
Chichester *QC* **22** E5
Chicken Lake *MB* **15** D2
Chick Lake *NT* **5** E7
Chicobi, lac *QC* **23** C3
Chicomo, lac *QC* **39** D2
Chicotte, rivière *QC* **30** A5
Chicoutai, pointe *QC* **39** F5
Chicoutimi *QC* **28** C6
Chicoutimi, rivière *QC* **28** C6
Chief Mountain *AB* **6** F5-6
Chief Mountain *AB* **7** F4
Chiefs Point *ON* **18** F4-5
Chiefs Point *ON* **20** A3
Chien, lac du *QC* **25** A6
Chien, lac du *QC* **26** F6
Chien, lac du *QC* **28** C2
Chienne, lac à la *QC* **25** D6
Chienne, lac à la *QC* **27** A1
Chiens, lac aux *QC* **26** A8
Chiganois River *NS* **33** A6
Chiganois River *NS* **34** E2
Chignecto *NS* **32** C7
Chignecto, Cape *NS* **32** D6
Chignecto, Cape *NS* **33** A3-4
Chignecto Bay *NB* **32** C-D6
Chignecto Bay *NB* **33** A3-4
Chignecto Bay *NS* **33** A3-4
Chignecto Game Sanctuary *NS* **32** C7
Chignecto Game Sanctuary *NS* **33** A4
Chignecto Isthmus *NB* **32** B7
Chignecto National Wilderness Area *NS* **32** C7
Chignectou, baie de *NB* **32** C-D6
Chignectou, baie de *NB* **33** A3-4
Chignectou, baie de *NS* **33** A3-4
Chignecto, RNF *NS* **32** C7
Chigoubiche, lac *QC* **26** C6
Chigoubiche, lac *QC* **28** A1-2
Chilako River *BC* **1** B6-7
Chilanko Forks *BC* **1** C6
Chilanko River *BC* **1** C6
Chilco Lake *BC* **1** D6
Chilcotin Ranges *BC* **1** D7
Chilcotin River *BC* **1** C6
Chilkat Pass *BC* **4** F3
Chilko Lake *BC* **1** C6
Chilko River *BC* **1** C6
Chilliwack *BC* **1** F8
Chilliwack *BC* **2** E6
Chilliwack Lake *BC* **2** E6
Chilliwack Lake PP *BC* **1** F8
Chilliwack Lake PP *BC* **2** E6-7
Chilliwack River *BC* **2** E6-7
Chilliwack River PP *BC* **2** E6-7
Chimney Corner *NS* **34** B5-6
Chin *AB* **7** E6
Chin, Cape *ON* **18** D5
Chinchaga River *AB* **9** C3
Chinchaga Wildland PP *AB* **9** C-D2
Chiniguchi Lake *ON* **17** E7

Chiniguchi Waterway PP *ON* **17** E7
Chin Lakes *AB* **6** F6-7
Chin Lakes *AB* **7** E6
Chinook *AB* **6** D7
Chinook *AB* **7** A8
Chinook *AB* **11** A1
Chinook Valley *AB* **9** D3
Chipewyan Lake *AB* **9** D5
Chipewyan Lake *AB* **9** D5
Chipewyan Lake *MB* **15** C3
Chip Lake *AB* **6** A4
Chip Lake *AB* **8** C4
Chipman *AB* **6** A6
Chipman *AB* **8** C7
Chipman *AB* **11** A1
Chipman *NB* **31** F6
Chipman *NB* **32** B4
Chipman Brook *NS* **32** E6
Chipman Brook *NS* **33** B4
Chipman Lake *ON* **16** D8
Chipman Lake *ON* **17** B2
Chipman River *SK* **12** A6
Chippawa *ON* **20** E7-8
Chippawa Creek CA *ON* **19** C8
Chippawa Creek CA *ON* **20** E7
Chippawa Hill *ON* **18** F4-5
Chippawa Hill *ON* **20** A3
Chiputneticook Lakes *NB* **32** C-D1
Chisasibi *QC* **40** E3-4
Chisel Lake *MB* **12** F8
Chisel Lake *MB* **15** F2
Chisholm *AB* **9** F5
Chiswick *ON* **17** F8
Chitek Lake *MB* **13** C2
Chitek Lake *SK* **11** A3
Chitek Lake *SK* **11** A3
Chitek Lake PP Reserve *MB* **13** C2
Chitek Lake PRS *SK* **11** A3
Chitek River *SK* **11** A3
Chochocouane, rivière *QC* **23** F5-6
Choelquoit Lake *BC* **1** D6
Choiceland *SK* **11** B5-6
Choiseul, lac *QC* **23** E8
Choiseul, lac *QC* **25** A2
Choiseul, lac *QC* **26** F2
Chokio *AB* **7** E4
Chopaka *BC* **6** F1-2
Chopin, lac *QC* **25** D3
Choquette, lac *QC* **25** B3
Chorkbak Inlet *NU* **40** A4-5
Chorkbak Inlet *NU* **41** C7-8
Chorlitz *MB* **14** F4
Chorney Beach *SK* **11** C6
Chouart, rivière *QC* **23** F8
Chouart, rivière *QC* **25** B2-3 C3
Chowade River *BC* **3** D7
Chown, Mount *AB* **8** B2
Christensen Beach *MB* **14** B2
Christian Channel *ON* **18** E7
Christian Island *ON* **18** E7
Christian Island *ON* **20** A5
Christie Bay *NT* **10** C8
Christie Bay *NT* **41** D1
Christie Beach CA *ON* **18** E7
Christie Beach CA *ON* **20** A5
Christie CA *ON* **19** B7
Christie CA *ON* **20** D6
Christie Islet NMBS *BC* **2** C4
Christie Lake *MB* **15** D4
Christie Lake *ON* **21** C6-7
Christie Memorial PP *BC* **6** F2
Christie Pass *NT* **4** B6
Christina Lake *AB* **12** E2
Christina Lake *BC* **6** E2
Christina Lake *BC* **6** F2-3
Christina Lake PP *BC* **6** F2-3
Christina Range *AB* **12** E2
Christina River *AB* **9** E7
Christina River *AB* **12** E2
Christmas Island *NS* **34** C7
Christopher Cross *PE* **35** A2
Christopher Lake *SK* **11** B5
Chub, lac *QC* **22** C8
Chub, lac *QC* **25** D3
Chubb Crater (cratère du Nouveau-Québec) **40** B4-5
Chubb Crater (cratère du Nouveau-Québec) **41** A1
Chub Lake *ON* **18** A1
Chuchi Lake *BC* **3** F6
Chu Chua Cottonwood PP *BC* **6** D1-2
Chukachida River *BC* **3** C4
Churchbridge *SK* **11** D7
Churchill *MB* **15** B5-6
Churchill *MB* **40** C1
Churchill *MB* **41** C1
Churchill *ON* **20** B7
Churchill *ON* **21** E1
Churchill *PE* **35** E4
Churchill, Cape *MB* **15** B6
Churchill, Cape *MB* **40** C1
Churchill, Cape *MB* **41** C5
Churchill, fleuve *NL* **39** C6
Churchill, fleuve *NL* **40** E7-8
Churchill, rivière *MB* **15** C5
Churchill, rivière *MB* **12** D8

Churchill, rivière *MB* **15** D2
Churchill, rivière *MB* **40** C1
Churchill, rivière *MB*
41 F4-5
Churchill, rivière *SK*
12 E6-7
Churchill, rivière *SK* **15** E1
Churchill Falls *NL* **39** B3-4
Churchill Falls *NL* **40** E7
Churchill Lake *ON* **13** D7-8
Churchill Lake *SK* **9** D-E8
Churchill Lake *SK* **12** E3
Churchill Lake *SK* **12** D-E3
Churchill Peak *BC* **3** C5
Churchill River *MB* **12** E8
Churchill River *MB* **15** C4-5
Churchill River *MB* **15** E2
Churchill River *MB* **40** C1
Churchill River *MB* **41** F4-5
Churchill River *NL* **39** B-C4
Churchill River *NL* **40** E7
Churchill River *SK* **12** E6
Churchover *NS* **33** F2-3
Church Point *NS* **33** D1
Churchview *NS* **34** C6
Churchville *NS* **33** A7-8
Churchville *NS* **34** E3
Churn Creek PPA *BC* **1** D7
Chute-aux-Outardes *QC*
29 A3
Chute-aux-Outardes *QC*
39 F2
Chute-Panet *QC* **27** A4
Chutes PP *ON* **17** F5-6
Chutes PP *ON* **18** A3
Chute-Saint-Philippe *QC*
24 E2
Chute-Saint-Philippe *QC*
25 E3
Chutes-de-Saint-Ursule,
parc des *QC* **25** F6-7
Chutes-de-Shawinigan,
parc des *QC* **25** F7-8
Chutine River *BC* **3** C2
Cigar Lake Mine *SK* **12** C5
Cinconsine *SK* **12** C4
Cinconsine, lac *QC* **25** C7
Cinconsine, lac *QC* **28** F2
Cinnemousun Narrows PP
BC **2** B7
Cinq Cerf River *NL* **36** D2-3
Cinq Cerf River *NL* **37** D1
Cinq Milles, lac des *QC*
25 A4
Cinq Milles, lac des *QC*
26 E4
Ciquart *NB* **31** B-C2
Circle River *ON* **23** B1
Cirque Mountain *NL* **40** C7
Clair *NB* **29** F2
Clair *NB* **31** C1
Clair *SK* **11** C6
Clair, lac *QC* **26** B8
Clairambault, lac *QC* **39** A1
Claire, lac *AB* **9** B6
Claire, lac *AB* **41** F1
Claire, Lake *AB* **9** B6
Claire, Lake *AB* **12** B1-2
Claire, Lake *AB* **41** F1
Claire Lake *YT* **4** D4
Clairmont *AB* **9** E2-3
Clairmont PP *NS* **32** E6
Clairmont PP *NS* **33** B3-4
Clairvaux, lac *QC* **26** E7
Clairvaux, lac *QC* **28** C3
Clairville *NB* **31** E7
Clairville *NB* **32** A5
Clam Bay *NS* **33** C7
Clam Harbour *NS* **33** C7
Clam Harbour Beach PP *NS*
33 C7
Clam Lake *SK* **12** E5
Clandeboye *MB* **13** E4
Clandeboye *MB* **14** D5
Clandeboye *ON* **19** B4
Clandeboye *ON* **20** E3
Clandonald *AB* **6** A-B7
Clandonald *AB* **11** A2
Clanwilliam *MB* **11** E8
Clanwilliam *MB* **13** E2
Clanwilliam *MB* **14** D1-2
Clapperton *BC* **2** B7
Clapperton Channel *ON*
18 B3
Clapperton Island *NT* **5** A8
Clapperton Island *ON* **18** B3
Clappison's Corners *ON*
19 B7
Clappison's Corners *ON*
20 D6
Clare *NS* **33** D1-2
Claremont *ON* **20** C7-8
Claremont *ON* **21** E2
Clarence *NS* **32** E5
Clarence *NS* **33** B3
Clarence *ON* **24** D3
Clarence, Cape *NU* **41** A5
Clarence, Cape *NU* **42** D4
Clarence Creek *ON* **24** D3
Clarence-Steepbank Lakes
PWP *SK* **11** A5
Clarendon *NS* **32** B3
Clarendon Station *ON*
21 C6
Clarenville *NL* **36** C8
Clarenville *NL* **37** C6
Clare River *ON* **21** D5
Claresholm *AB* **6** E6
Claresholm *AB* **7** D4
Clarington *ON* **20** B8
Clarington *ON* **21** E2-3
Clark, Point *ON* **20** B2-3
Clark Bay *SK* **12** B6

Clark Bay *SK* **15** B1
Clarke Bridge *SK* **11** F5
Clarke City *QC* **39** E2
Clarke Lake *SK* **11** A4
Clarke Lake *SK* **12** F4
Clarke's Beach *NL* **37** D7
Clarkes Head *NL* **36** A7
Clarkes Head *NL* **37** A5
Clarkes Head *NL* **38** F6-7
Clarkin *PE* **34** C3
Clarkin *PE* **35** E5-6
Clarkleigh *MB* **13** E3
Clarkleigh *MB* **14** C4
Clarksburg *ON* **18** F6
Clarksburg *ON* **20** A5
Clarkson *ON* **19** B8
Clarksville *NS* **32** E8
Clarksville *NS* **33** B5
Clarksville *NS* **34** F1
Clarkville *NB* **31** F3-4
Clarkville *NB* **32** B1-2
Clashmoor *SK* **11** B6
Clavering *ON* **18** F5
Clavering *ON* **20** A4
Claude, rivière à *QC* **29** A7
Claud Elliott Lake PP *BC*
1 E4
Clavet *SK* **11** C4
Claybank *SK* **11** E5
Claydon *SK* **11** F2-3
Clayhurst *BC* **3** E8
Clayhurst *BC* **9** D2
Clay Lake *ON* **13** E6
Clay Lake *ON* **16** D2
Clayoquot Arm *BC* **2** E1
Clayoquot Arm PP *BC* **1** F5
Clayoquot Arm PP *BC* **2** E1
Clayoquot Plateau PP *BC*
1 F5
Clayoquot Plateau PP *BC*
2 D1-2
Clayoquot Sound *BC* **1** E5
Clayoquot Sound *BC* **2** E1
Clayton *ON* **21** B7
Clayton *ON* **24** E1
Clayton Lake *ON* **21** B7
Clayton Lake *ON* **24** E1
Clear, Lake *ON* **21** A5
Clear Bay *MB* **11** B8
Clear Bay *MB* **13** B2
Clear Creek *ON* **19** D6
Clear Creek *ON* **20** F5
Clear Creek *YT* **4** B3
Cleardale *AB* **9** D1
Clearfield *SK* **11** F6
Clear Hills *AB* **9** D2
Clear Lake *MB* **11** D8
Clear Lake *MB* **13** D2
Clear Lake *MB* **14** C1
Clear Lake *ON* **21** B-C2
Clear Lake *ON* **21** A1-2
Clear Lake *ON* **21** C2
Clear River *AB* **9** D2
Clear Springs *PE* **35** D7
Clearwater *BC* **6** C1-2
Clearwater *MB* **13** F2
Clearwater *MB* **14** F2
Clearwater Bay *ON* **13** F5
Clearwater Bay *ON* **14** E8
Clearwater Bay *ON* **14** E8
Clearwater Bay *ON* **16** D1
Clearwater Brook *NB*
31 D-E4
Clearwater Creek *NT*
4 C-D8
Clearwater Creek *NT* **10** D1
Clearwater Lake *BC* **1** C8
Clearwater Lake *BC* **6** C1
Clearwater Lake *MB* **11** A7
Clearwater Lake *MB* **11** A8
Clearwater Lake *MB* **13** A1
Clearwater Lake *MB*
13 A-B2
Clearwater Lake PP *MB*
11 A7-8
Clearwater Lake PP *MB*
13 A1
Clearwater River *AB* **6** C5
Clearwater River *AB* **8** F4-5
Clearwater River *AB* **9** D7
Clearwater River *AB* **12** D2
Clearwater River *BC* **6** C1
Clearwater River *SK* **9** C8
Clearwater River
12 C-D3
Clearwater River PP *SK*
9 D8
Clearwater River PP *SK*
12 D3
Clearwater West Lake *ON*
13 F7
Clearwater West Lake *ON*
16 E3-4
Clemenceau *SK* **11** B6-7
Clemenceau Icefield *BC*
6 C3
Clemenceau Icefield *BC*
8 F1
Clemens *SK* **11** B5-6
Clementsport *NS* **32** F5
Clementsport *NS* **33** C2
Clementsvale *NS* **32** F5
Clementsvale *NS* **33** C2
Clemville *QC* **30** D1-2
Clemville *QC* **31** A7
Clendinning PP *BC* **1** E6
Clendinning PP *BC* **2** B4

Cléricy *QC* **23** D3
Clerke Point *BC* **1** E4
Clermont *PE* **35** D3
Clermont *QC* **17** C8
Clermont *QC* **23** C2
Clermont *QC* **28** E7
Clerval *QC* **17** C8
Clerval *QC* **23** C2
Cleveland *NS* **34** D6
Cleveland Beach PP *NS*
32 F8
Cleveland Beach PP *NS*
33 C5
Clifford *ON* **20** C4
Cliff Point *BC* **1** E4
Clifton *NB* **29** E8
Clifton *NB* **30** E1
Clifton *NB* **31** B7
Clifton *NL* **37** C7
Clifton Lake *MB* **12** B8
Clifton Lake *MB* **15** B3
Climax *SK* **11** F3
Clinton *BC* **1** D8
Clinton *BC* **2** A7
Clinton *ON* **19** A4
Clinton *ON* **20** C-D3
Clinton *PE* **32** A8
Clinton *PE* **34** B-C2
Clinton *PE* **35** D3
Clinton Creek *YT* **4** A1
Clinton Creek *YT* **5** F1
Clinton-Colden Lake *NT*
41 D3
Clisbako River *BC* **1** C6
Clive *BC* **6** C6
Clive *AB* **8** E6
Clive Lake *NT* **10** C4
Cloan *SK* **11** C3
Clode Sound *NL* **36** B-C8
Clode Sound *NL* **37** B-C6
Clo-oose *BC* **2** F2
Clore River *BC* **1** A4
Cloridorme *QC* **30** C4
Close Lake *SK* **12** C5
Close to the Edge PP *BC*
1 A8
Close to the Edge PP *BC*
6 A1
Cloud Bay *NT* **10** A3
Cloud Bay *ON* **16** F5
Cloud River *NL* **38** C4
Clouston *SK* **11** B4-5
Cloutier *QC* **17** D8
Cloutier *QC* **23** E2
Clova *QC* **25** A3
Clova *QC* **26** F3
Cloverdale *BC* **2** E5
Cloverdale *NB* **31** F3-4
Cloverdale *NB* **32** B2
Cloverdale *NS* **33** B6-7
Cloverdale *NS* **34** F2-3
Clover Lake *ON* **22** F4
Cloverlawn *AB* **8** D7
Cloverleaf *MB* **14** D6
Clowhom Lake *BC* **2** C4
Cloyne *ON* **21** C5
Club Island *ON* **18** C4
Cluculz Creek *BC* **1** A-B6
Cluculz Lake *BC* **1** A6
Cluny *AB* **6** D6
Cluny *AB* **7** C5
Clute *ON* **17** C6
Clyde *AB* **6** A5-6
Clyde *AB* **8** B6
Clyde *ON* **19** B6
Clyde *ON* **20** D5
Clydegale Lake *ON* **21** A-B3
Clydegale Lake *ON* **22** F3
Clyde Inlet *NU* **41** A7
Clyde Inlet *NU* **41** A7
Clyde River *AB* **9** E6-7
Clyde River *AB* **12** E1
Clyde River *NS* **33** F2
Clyde River *NS* **33** F2-3
Clyde River *ON* **21** B6
Clyde River *PE* **35** E4
Clyde River
(Kangiqtugaapik) *NU*
41 A8
Clyde River
(Kangiqtugaapik) *NU*
42 E7
Clydesdale *NS* **34** D4
Clymont *AB* **8** C6
C.M. Wilson CA *ON* **19** E3
Coachman's Cove *NL*
38 D4-5
Coal Branch *NB* **31** E7
Coal Branch *NB* **32** B5
Coal Brook *NL* **36** D1
Coal Creek *NB* **31** F6
Coal Creek *NB* **32** B4
Coaldale *AB* **6** F6
Coalhurst *AB* **7** E5
Coalmont *BC* **2** D8
Coalmont *BC* **6** F1
Coal River *BC* **3** A5
Coal River *BC* **4** C6
Coal River *YT* **4** D-E7
Coal River Springs TP *YT*
3 A4-5
Coal River Springs TP *YT*
4 E7-8
Coalspur *AB* **6** B3-4
Coalspur *AB* **8** D2
Coast Mountains *BC*
1 C-D5 E6

Coast Mountains *BC*
2 A2-3 B3-4
Coast Mountains *BC*
3 B-C2 D3
Coast Mountains *BC* **4** F4
Coast Mountains *YT* **4** E3
Coates Creek *ON* **20** A-B6
Coates Mills *NB* **31** E8
Coates Mills *NB* **32** A-B6
Coaticook *QC* **27** F5
Coaticook, rivière *QC*
27 F4-5
Coats Island *NU* **40** B3
Coats Island *NU* **41** E6
Coatsworth *ON* **19** E2
Cobalt *ON* **17** B8
Cobalt *ON* **22** A1
Cobbs Arm *NL* **38** E6-7
Cobden *ON* **21** A6
Cobden *ON* **22** F6
Cobden *ON* **24** A6
Cobequid Bay *NS* **32** D8
Cobequid Bay *NS* **33** A6
Cobequid Bay *NS* **34** E2
Cobequid Mountains *NS*
32 D8
Cobequid Mountains *NS*
34 E1-2
Cobham River *MB* **13** B5-6
Coboconk *ON* **20** A8
Coboconk *ON* **21** C-D2
Cobourg *ON* **21** E3-4
Coburg *NB* **35** F2
Coburg Island *NU* **42** C5
Coburn *SK* **32** C2
Cocagne *NB* **31** E8
Cocagne *NB* **32** B6
Cocagne, île *NB* **31** E8
Cocagne, île *NB* **32** A6-7
Cocagne Island *NB* **31** E8
Cocagne Island *NB* **32** A6-7
Cocagne River *NB* **31** F8
Cocagne River *NB* **32** B6
Cochenour *ON* **13** D6
Cochenour *ON* **16** C2
Cochin *SK* **11** B3
Cochrane *ON* **6** D5
Cochrane *AB* **7** B3
Cochrane *ON* **17** C6
Cochrane Bay *MB* **13** A5
Cochrane Ranche PHS *AB*
6 D5
Cochrane Ranche PHS *AB*
7 B3-4
Cochrane River *MB* **12** B-C7
Cochrane River *MB* **15** B-C2
Cochrane River *MB* **41** E3
Cockburn Island *ON* **18** B1
Cockfield, Mount *YT* **4** C2
Cockram Strait *NU* **41** B7
Cockram Strait *NU* **42** F6
Coddles Harbour *NS* **34** F5
Coderre *SK* **11** E4
Codes Corner *ON* **21** D7
Codette *SK* **11** B6
Codette Lake *SK* **11** B5
Codner *AB* **8** E5
Codrington *ON* **21** E4
Codroy Pond *NL* **36** C-D2
Codroy Valley Beach PP *NL*
36 D1
Codville Lagoon PMP *BC*
1 C4
Cody Caves PP *BC* **6** F3-4
Cody Creek *ON* **21** B7
Cody Creek *YT* **5** D2
Codys *NB* **32** C4
Coe Hill *ON* **21** C4
Cœur, lac en *QC* **24** C3
Coffee Cove *NL* **38** E5
Coffee Creek *YT* **4** B2
Coghill *AB* **8** E6-7
Coghlan Lake *YT* **4** F3
Cogmagun River *NS*
32 E7-8
Cogmagun River *NS* **33** B5
Cogmagun River *NS* **34** F1
Coin-du-Banc *QC* **30** C3
Col-Athabasca, LHN du *AB*
6 C3
Col-Athabasca, LHN du *BC*
8 E1
Colborne *ON* **21** E4
Colchester *ON* **19** F1
Cold Brook *NS* **36** B2
Coldbrook *NS* **32** E6
Coldbrook *NS* **34** F1
Coldbrook PP *NS* **32** E6-7
Coldbrook PP *NS* **33** B4
Cold Creek *ON* **21** E4
Cold Lake *AB* **9** F7
Cold Lake *AB* **12** F2
Cold Lake *SK* **9** F8
Cold Lake *SK* **12** F2
Cold Lake PP *AB* **9** F7-8
Cold Lake PP *AB* **12** F2
Coldspring Head *NS*
35 F2-3
Cold Spring Pond *NL* **36** C5
Cold Spring Pond *NL* **37** C3
Cold Springs *ON* **21** E3
Coldstream *BC* **6** E2
Coldstream *BC* **31** F3
Coldstream *NB* **32** B1
Coldstream *NS* **33** B6
Coldstream *NS* **34** F2
Coldstream *ON* **20** E3
Coldwater *ON* **18** F8
Coldwater *ON* **20** A7
Coldwater *ON* **21** C1
Coldwater River *BC* **1** E8
Coldwater River *BC* **2** C7-8

Coldwater River *BC* **6** E1
Coldwater River PP *BC*
1 E-F8
Coldwater River PP *BC*
2 C7-8
Coldwater River PP *BC* **6** E1
Coldwell *ON* **16** E8
Coldwell *ON* **17** C2
Cole Bay *SK* **9** E8
Cole Bay *SK* **12** F3
Colebrook *ON* **21** D6
Cole Harbour *NS* **33** C6
Cole Harbour *NS* **33** C6
Cole Harbour *NS* **34** E6
Coleman *AB* **6** F5
Coleman *AB* **7** E3
Coleman *PE* **32** A7
Coleman *PE* **35** E2
Coleman Lake *AB* **7** A6
Colen Lakes *MB* **13** A5
Colen Lakes *MB* **25** A8
Coleraine *QC* **27** D5-6
Coles Island *NB* **32** C4
Coles Pond *NL* **36** B4-5
Coles Pond *NL* **39** B4-5
Coleville *SK* **6** C8
Coleville *SK* **11** C2
Colfax *SK* **11** E6
Colgate *SK* **11** F6
Colin Archer Peninsula *NU*
42 C4
Colin-Cornwall Lakes
Wildland PP *AB* **9** A7
Colin-Cornwall Lakes
Wildland PP *AB* **12** A2
Colinet *NL* **37** E7
Colinet River *NL* **37** E7
Colinton *AB* **6** A7
Colinton *AB* **8** A7
Colinton *AB* **9** F6
Colin Lake *AB* **12** A2-3
Colin Lake *SK* **12** E6
Coliseum *SK* **11** A3
Colla *NB* **32** C4
Colleen *AB* **8** A7
College Bridge *NB* **32** B-C7
College Heights *AB* **6** C5
College Heights *AB* **8** E6
Collette *NB* **31** D7
Collicutt *AB* **7** B4
Colliers *NL* **37** E7
Collin, ZEC *QC* **24** A5
Collin, ZEC *QC* **25** E5
Collina *NB* **32** C4
Colline Saddle *QC* **27** E6
Collingwood *ON* **18** F7
Collingwood *ON* **20** A5-6
Collingwood Corner *NS*
32 C8
Collingwood Corner *NS*
34 E1
Collingwood Cove *AB* **8** C7
Collins *ON* **16** D5
Collins, lac *NB* **35** E2
Collins Bay *ON* **21** E6
Collins Bay *SK* **12** B6
Collins Cove *NL* **36** F7
Collins Cove *NL* **37** F5
Collins Inlet *ON* **18** B5
Collins Lake *NB* **35** E2
Collins Lake *ON* **21** E6
Colnet *AB* **8** E5
Colombier *QC* **29** B2
Colombier, cap *QC* **29** B2
Colombourg *QC* **17** C8
Colombourg *QC* **23** C2
Colonel Roscoe
Vanderwater CA *ON*
21 D5
Colonsay *SK* **11** C5
Colpitts Settlement *NB*
31 F8
Colpitts Settlement *NB*
32 C6
Colpoys Bay *ON* **18** E5
Colpton *NS* **32** F7
Colpton *NS* **33** D3-4
Colton *ON* **21** E4
Columbia, Cape *NU* **42** A5
Columbia, fleuve *BC* **6** D-E3
Columbia, fleuve *BC* **7** C1
Columbia, fleuve *BC* **6** D4
Columbia, Mount *AB* **8** F1
Columbia, Mount *BC* **8** F2
Columbia Icefield *AB* **8** F2
Columbia Lake *BC* **6** E4
Columbia Lake *BC* **7** D1-2
Columbia Lake PP *BC* **6** E4
Columbia Lake PP *BC*
7 D1-2
Columbia Mountains *BC*
6 B-C2
Columbia Reach *BC* **6** C3
Columbia River *BC* **6** D3
Columbia River *BC* **6** D4
Columbia River *BC* **7** B1
Columbus *ON* **20** C8
Columbus *ON* **21** E2
Colville Bay *PE* **35** E7-8
Colville Lake *NL* **39** B2-3
Colville Lake *NT* **5** D7-8
Colville Lake *NT* **10** C8
Colwell *ON* **20** B6
Colwood *BC* **2** F4
Col-Yellowhead, LHN du *AB*
6 B2-3
Comber *ON* **19** E2
Combermere *ON* **21** B4-5
Combermere, Cape *NU*
42 C5

Come By Chance *NL* **36** D8
Come By Chance *NL* **37** D6
Comeau Hill *NS* **33** F1
Comeauville *NS* **33** D1
Comet *ON* **19** E1
Comet Mountain *BC* **1** B4
Comfort, Cape *NU* **40** A3
Comfort, Cape *NU* **41** D6
Comfort Cove *NL* **36** A7
Comfort Cove *NL* **37** A5
Comfort Cove *NL* **38** F6
Comma Island *NL* **39** A5-6
Commanda *ON* **18** B8
Commanda *ON* **22** E1
Commanda Creek *ON*
18 B-C8
Commanda Lake *ON* **18** B7-8
Commanda Lake *ON* **22** E1
Commissaires, lac des *QC*
25 A8
Commissaires, lac des *QC*
26 F8
Commissaires, lac des *QC*
28 D4
Commissioner Island *MB*
13 C4
Committee Bay *NU*
41 B5-6 C6
Committee Bay *NU* **42** F5
Commodore, Cape *ON*
18 E5-6
Commonwealth Lake *MB*
15 A4
Como *QC* **24** D5
Comox *BC* **1** F6
Comox *BC* **2** D2
Comox Lake *BC* **1** F5
Comox Lake *BC* **2** D2
Compass *SK* **11** A3
Compeer *AB* **6** C8
Compeer *AB* **11** C2
Compton *QC* **27** F5
Compton, lac *QC* **26** C4
Compton Névé *BC* **1** E6
Compton Névé *BC* **1** E6
Complexe-Historique-de-
Dawson, LHN du *YT* **4** A2
Compulsion Bay *SK* **12** C6
Compulsion Bay *SK* **15** C1
Conception Bay *NL* **37** D8
Conception Bay South *NL*
37 D7-8
Conception Harbour *NL*
37 E7
Concession Discovery, LHN
de la *YT* **4** A1-2
Concession *NS* **33** D1-2
Conche *NL* **38** C5
Conche *NL* **39** E8
Condor *AB* **6** C5
Condor *AB* **8** E5
Cone Head *BC* **1** A-B1
Conche *NL* **38** C5
Conestogo *ON* **19** B6
Conestogo *ON* **20** D5
Conestogo Lake *ON* **19** A6
Conestogo Lake *ON* **20** C4
Conestogo Lake CA *ON*
19 A5
Conestogo Lake CA *ON*
20 C4
Conestogo River *ON*
19 A5-6 B6
Conestogo River *ON*
20 C4-5 D5
Confédération (péage), pont
de la *NS* **35** E3
Confederation CA *ON*
20 B7-8
Confederation CA *ON* **21** E1
Confederation Island *MB*
13 A5
Confederation Lake *ON*
13 D7
Confederation Lake *ON*
16 B-C3
Confusion Bay *NL* **38** D-E5
Congresbury *AB* **8** F4-5
Congress *SK* **11** F4
Conibear Lake *AB* **9** A5-6
Conibear Lake *AB* **12** A1
Coniston *ON* **17** F7
Coniston *ON* **18** A5-6
Conjuring Creek *AB* **8** D6
Conklin *AB* **9** E7
Conklin *AB* **12** E2
Conn *ON* **20** C5
Connaigre Bay *NL* **36** E5-6
Connaigre Bay *NL* **37** E3-4
Connaigre Head *NL* **36** E5-6
Connaigre Head *NL* **37** E3-4
Connaught *ON* **17** C6
Connell Creek *SK* **11** B6
Connelly, lac *QC* **28** A4
Connemara *BC* **7** C6
Conne River *NL* **36** D6
Conne River *NL* **37** D4
Conn Mills *NS* **32** C8
Conn Mills *NS* **34** D1
Connor *ON* **20** C6
Connor Creek *AB* **8** B4
Connors *NB* **29** F2
Connors *NB* **31** C1
Conquerall Mills *NS* **33** D4
Conquest *SK* **11** D4
Conrad *AB* **7** E7
Conrad Beach/Island PP *NS*
33 C6-7
Conrich *AB* **6** D5-6
Conrich *AB* **7** B4
Conroy Creek *BC* **3** C7

Conroy Creek *BC* **9** C1
Consecon *ON* **21** E4-5
Consecon Lake *ON* **21** E5
Consort *AB* **6** C7
Consort *AB* **11** C2
Constance Bay *ON* **21** A7
Constance Bay *ON* **22** F7
Constance Bay *ON* **24** D1
Constance Lake *ON* **17** B4
Constant Lake *ON* **21** A5-6
Consul *SK* **6** F8
Consul *SK* **11** F2
Contact Creek *BC* **3** A4
Contact Creek *BC* **4** F7
Contrecoeur *QC* **24** D5-7
Contrecoeur *QC* **27** D2
Contwoyto Lake *NU* **41** C2
Conway *ON* **21** E6
Conway *PE* **32** A7
Conway *PE* **34** B1
Conway *PE* **35** E2
Conway Narrows *PE* **35** E2
Cook, Cape *BC* **1** E4
Cook, Mount *YT* **4** E1
Cookshire *QC* **27** E5
Cook's Harbour *NL* **38** A5
Cook's Harbour *NL* **39** D8
Cooks Brook *NS* **33** B6
Cooks Brook *NS* **34** E5
Cooks Cove *NS* **34** E5-6
Cooks Mills *ON* **17** F8
Cooks Mills *ON* **18** A8
Cookstown *ON* **20** B6
Cookville *NB* **35** F1
Cookville PP *NS* **33** D4
Coombes Cove *NL* **36** E6
Coombes Cove *NL* **37** E4
Coombs *BC* **1** F6
Coombs *BC* **2** D3
Cooper Creek *MB* **15** C5-6
Copeland *SK* **11** D5-6
Copeland Islands PMP *BC*
1 E5
Copeland Islands PMP *BC*
2 C2
Copeland Lake *MB* **12** A8
Copeland Lake *MB* **15** B3
Copenhagen *ON* **19** D5
Copenhagen *ON* **20** F4
Copetown *ON* **19** B7
Copetown *ON* **20** E6
Copp Lake *NT* **10** F6-7
Copper, mont *QC* **29** B8
Copper, mont *QC* **30** B7
Copper Bay *BC* **1** B1-2
Copper Cliff *ON* **17** F6
Copper Cliff *ON* **18** A5
Copper Creek *BC* **1** D8
Copper Creek *BC* **2** A8
Copper Creek *BC* **6** D1
Coppermine (Kugluktuk) *NU*
41 B2
Coppermine (Kugluktuk) *NU*
42 F1
Coppermine River *NT*
10 A7-8
Coppermine River *NU*
41 C2
Coppin's Corners *ON*
20 B7-8
Coppin's Corners *ON* **21** E1
Copp Lake *NT* **10** F6-7
Coqualeetza *SK* **11** B4
Coques, anse aux *QC* **29** C3
Coquihalla Canyon PP *BC*
2 D5
Coquihalla Canyon PRA *BC*
1 F8
Coquihalla Canyon PRA *BC*
6 F1
Coquihalla River *BC* **2** D5
Coquihalla River PP *BC* **1** F8
Coquihalla River PP *BC* **6** F1
Coquihalla River PRA *BC*
1 F8
Coquihalla Summit PRA *BC*
2 D7
Coquihalla Summit PRA *BC*
6 F1
Coquille, lac de la *QC* **26** B5
Coquitlam *BC* **1** F7
Coquitlam *BC* **2** D5
Coquitlam Lake *BC* **2** C5
Coral Harbour (Salliq) *NU*
40 A3
Coral Harbour (Salliq) *NU*
41 D6-7
Coral Rapids, RNP *ON* **17** A6
Coral Rapids PNR *ON* **17** A6
Corbeau, lac du *QC* **24** E8
Corbeil *ON* **17** F8
Corbeil *ON* **18** A-B8
Corberrie *NS* **33** D2
Corbett *ON* **19** B4
Corbett *ON* **20** D-E3
Corbett Lake *AB* **12** A8
Corbett Lake *MB* **15** A3
Corbin *BC* **6** F5
Corbin *BC* **7** E3
Corbin *NL* **36** F7
Corbin *NL* **37** F5
Corbin, Cape *NL* **38** F7

Corbin Bay *NL* **36** D6
Corbin Bay *NL* **37** D4
Corbyville *ON* **21** E5
Cordova *MB* **14** D2
Cordova Bay *BC* **2** F4
Cordova Lake *ON* **21** D4
Cordova Mines *ON* **21** D4
Coriander *SK* **11** F3-4
Corinne *SK* **11** E5
Corinth *ON* **19** C5
Corinth *ON* **20** F4
Cork *NB* **32** C2
Corkscrew Island *ON*
14 E8
Cormac *ON* **21** A5
Cormac *ON* **22** F5
Cormack *NL* **36** A3-4
Cormack *NL* **37** A1-2
Cormack *NL* **38** F3
Cormack *NL* **39** F7
Cormack Lake *NT* **10** E3
Cormier, lac *QC* **39** E5
Cormier-Village *NB* **35** E1
Cormierville *NB* **31** E8
Cormierville *NB* **32** A6
Cormoran, pointe au *QC*
30 B7
Cormorant *MB* **11** A8
Cormorant *MB* **12** F8
Cormorant *MB* **13** A2
Cormorant Channel PMP *BC*
1 E4-5
Cormorant Lake *MB* **11** A8
Cormorant Lake *MB* **13** A2
Corneille Lake *SK* **12** F6-7
Cornelius Grinnell Bay *NU*
40 A6
Corner Brook *NL* **36** B2-3
Corner Brook Pond *NL*
36 B3
Corner Brook Pond *NL*
37 B1
Cornhill *NB* **32** C5
Corning *SK* **11** E6-7
Cornwall *ON* **24** E4
Cornwall *PE* **34** C2
Cornwall *PE* **35** E4-5
Cornwall Hills PP *BC* **1** E8
Cornwall Hills PP *BC* **2** A7
Cornwall Island *NU* **42** D4
Cornwall Lake *AB* **9** A7
Cornwall Lake *AB* **12** A2
Cornwallis *NS* **32** F5
Cornwallis Island *NU* **42** D4
Coronach *SK* **11** F5
Coronado *AB* **8** B6-7
Coronation *AB* **6** C7
Coronation *AB* **11** C1
Coronation Gulf *NU*
41 B2
Coronation Gulf *NU* **42** F1-2
Corralville *PE* **35** E6
Cortes Island *NL* **1** E6
Cortes Island *BC* **2** C2
Corunna *ON* **19** C2
Corunna *ON* **20** E1
Cosgrave Lake *ON* **16** F6-7
Cosgrave Lake *ON* **17** B1
Cosine *SK* **6** C8
Cosine *SK* **11** C2
Cosmo *AB* **8** B5
Costebelle, lac *QC* **39** E4-5
Costello Lake *MB* **12** D8
Costello Lake *MB* **15** D2
Costes Lake *MB* **13** A4
Costigan Lake *SK* **12** D5
Cote *SK* **11** D7
Cote *SK* **13** C1
Coteau Beach *SK* **11** D4
Côteau-du-Lac, LHN du *QC*
24 D5
Coteau Hills *SK* **11** D3
Côteau-du-Lac PRA *QC*
24 E4-5
Côteau-du-Lac NHS *QC*
24 E4-5
Côte-Nord-du-Golfe-du-
Saint-Laurent *QC* **38** B2
Côte-Nord-du-Golfe-du-
Saint-Laurent *QC*
39 E5-6
Côte-Saint-Luc *QC* **24** D6
Côte-Saint-Luc *QC* **27** E1
Côtière, chaîne *BC*
1 D4-5 E5
Côtière, chaîne *BC* **2** A2 B3
Côtière, chaîne *BC* **3** B-D2
Côtière, chaîne *BC* **4** F4
Cotswold *ON* **19** A5
Cotswold *ON* **20** C4
Cottage Cove PP *NS* **32** E5
Cottage Cove PP *NS*
33 B2-3
Cottage-Hawthorne, LHN
du *NL* **37** D7
Cottam *ON* **19** E2
Cottel Island *NL* **37** B7
Cottle's Island *NL* **38** E-F6
Cotton Lake *MB* **15** F4
Cottonwood *BC* **1** B7
Cottonwood House
Provincial Historic Park
BC **1** B7
Cottonwood River *BC* **1** B7
Cottonwood River *BC* **4** F6
Cottonwood River PP *BC*
1 B7
Cottonwoods *MB* **13** E2
Cottonwoods *MB* **14** E2
Cottrell's Cove *NL* **38** F6
Couchiching *ON* **16** F2

Couchiching, Lake *ON* **18** F8
Couchiching, Lake *ON* **20** A7
Couchiching, Lake *ON* **21** C-D1
Coucou, lac du *QC* **25** C5-6
Coucou, lac du *QC* **28** E1
Coudres, île aux *QC* **28** F7
Coughlan *NB* **31** D7-E6
Coughlan *NB* **32** A4
Coulonge, rivière *QC* **22** D-E6
Coulonge Est, rivière *QC* **22** D6
Coulson *ON* **18** F8
Coulson *ON* **20** A7
Coulson *ON* **21** D1
Coulter *MB* **11** F8
Coulter *MB* **13** F1
Countess *AB* **7** C6
Country Harbour *NS* **34** E5 F5
Country Harbour Cross Roads *NS* **34** E5
Country Harbour Mines *NS* **34** E5
Country Road *NL* **37** C7-8
Courageous Lake *NT* **10** B8
Courcelles *QC* **27** D6
Courtenay *NB* **1** F5
Courtenay *BC* **2** D2
Courtice *ON* **20** C8
Courtice *ON* **21** D2
Courtin Island *PE* **35** C-D3
Courtland *ON* **19** C5-6
Courtland *ON* **20** F4-5
Courtright *ON* **19** C2
Courtright *ON* **20** F1
Courval *SK* **11** E4
Cousacouta, lac *QC* **25** D6
Cousacouta, lac *QC* **28** F1
Cousins Lake *MB* **15** D3
Coutts *AB* **6** F7
Coutts *AB* **7** F7
Couture, lac *QC* **40** C4-5
Couture, lac *QC* **41** E8
Covedell *NB* **31** C7-8
Covehead Road *PE* **35** D5
Coverdale *NB* **31** D4
Coverdale *NB* **32** B6
Cove Road *NS* **32** D8
Cove Road *NS* **33** A5
Cove Road *NS* **34** E1
Covey Hill *QC* **24** E5
Covey Hill *QC* **27** F1
Cowan *MB* **11** C8
Cowan *MB* **13** C2
Cowan Lake *SK* **11** A4
Cowan River *SK* **11** A3-4
Cowansville *QC* **24** E7
Cowansville *QC* **27** F3
Cow Bay *NS* **33** C6
Cowessess *SK* **11** E6-7
Cow Head *NL* **38** E2
Cow Head *NL* **39** F7
Cowichan Bay *BC* **2** F4
Cowichan Lake *BC* **1** F6
Cowichan Lake *BC* **2** E3
Cowichan River *BC* **2** E3-4
Cowichan River PP *BC* **2** E3-4 F3-4
Cowley *AB* **6** F5
Cowley *AB* **7** E4
Coxipi, lac *QC* **39** D7
Coxipi, rivière *QC* **39** D7
Cox Island *BC* **1** D3
Cox Point *BC* **1** F5
Cox Point *BC* **2** E1
Cox's Cove *NL* **36** A3
Cox's Cove *NL* **37** A1
Cox's Cove *NL* **38** F2
Cox's Cove *NL* **40** F8
Coxvale *ON* **21** C6
Crabb Brook *NL* **36** A2-3
Crabb Brook *NL* **38** F2
Crabbs River *NL* **36** C-D2
Crabtree *QC* **24** C6
Crabtree *QC* **27** D2
Cracroft Bay *NT* **5** A8
Cracroft Island *BC* **1** E5
Craddock *AB* **7** B4-5
Craigdu *AB* **7** B4-5
Craigellachie *BC* **6** D2-3
Craighurst *ON* **18** F7-8
Craighurst *ON* **20** A6-7
Craigleith *ON* **18** F6-7
Craigleith *ON* **20** A5
Craigleith PP *ON* **18** F6-7
Craigleith PP *ON* **20** A5
Craigmore *NS* **34** D5
Craigmyle *AB* **6** C-D6
Craigmyle *AB* **7** A6
Craig River *BC* **3** D2
Craig's Pit, RNP *ON* **16** F8
Craig's Pit, RNP *ON* **17** C2
Craig's Pit PNR *ON* **16** F8
Craig's Pit PNR *ON* **17** C2
Craik *SK* **11** D5
Crammond *AB* **8** F5
Crampe, lac de la *QC* **26** E6
Crampe, lac de la *QC* **28** C2
Cranberry Junction *BC* **3** E4
Cranberry Lake *ON* **21** C2
Cranberry Lake *ON* **21** C8
Cranberry Lake, RNP *ON*
Cranberry Lake PNR *ON*
16 E1
Cranberry Portage *MB*
12 F7
Cranberry Portage *MB*
15 F1

Cranberry River *BC* **3** E-F4
Cranbrook *BC* **6** F4
Cranbrook *BC* **7** E2
Cranbrook *ON* **19** A4-5
Cranbrook *ON* **20** C3-4
Crandall *MB* **11** E8
Crandall *MB* **13** E1
Crane Lake *SK* **11** E2
Crane Narrows *MB* **13** C-D2
Crane Narrows *MB* **14** A2
Crane River *MB* **13** D2-3
Crane River *MB* **14** A2
Crane River (O-Chi-Chak-Ko-Sipi) *MB* **13** D2
Crane River (O-Chi-Chak-Ko-Sipi) *MB* **14** A2
Crane Valley *SK* **11** F5
Cranford *AB* **6** E6-7
Cranford *AB* **7** E1
Cranswick River *NT* **5** E-F5
Crapaud *PE* **32** B8
Crapaud *PE* **34** C2
Crapaud *PE* **35** E3
Crauford, Cape *NU* **41** A5-6
Crauford, Cape *NU* **42** A6-7
Craven *SK* **11** E5
Crawford Bay *BC* **6** F3-4
Crawford Lake CA *ON*
19 B7
Crawford Lake CA *ON*
20 D6
Crawford Park *MB* **14** C1
Crean Lake *SK* **11** A4
Crediton *ON* **19** B4
Crediton *ON* **20** D3
Credit River *ON* **19** A7
Credit River *ON* **20** C6
Crediteville *ON* **19** C6
Crediteville *ON* **20** E5
Creekland *AB* **8** D5
Cree Lake *SK* **12** C5
Cree Lake *SK* **12** C4
Cree Lake *SK* **41** F2
Creelman *SK* **11** E6
Creemore *ON* **20** B5-6
Cree River *SK* **33** B5 C4-5
Cree River *SK* **41** F2-3
Creighton *SK* **12** F7
Creighton *SK* **15** F1
Creighton Heights *ON*
21 E3-4
Creignish *NS* **34** D5
Cremona *AB* **6** D5
Cremona *AB* **7** A3
Crescent Beach *NS* **33** D4
Crescent Island *NU* **42** C3-4
Crescent Lake *SK* **11** D7
Crescent Spur *MB* **1** B8
Crescent Spur *BC* **6** A1
Crestomere *AB* **6** B-C5
Crestomere *AB* **8** E6
Creston *BC* **6** F4
Creston *NL* **36** E7
Creston *NL* **37** E5
Crestview *YT* **4** E4
Crestwynd *SK* **11** E5
Creswell Bay *NU* **41** A5
Creswell Bay *NU* **42** A4
Crevier, lac *QC* **22** B8
Crevier, lac *QC* **25** D2
Crewsons Corners *ON*
19 A6-7
Crewsons Corners *ON*
20 D5-6
Crichton *SK* **11** E-F3
Criddle/Vane Homestead PP Reserve *MB* **13** F2
Criddle/Vane Homestead PP Reserve *MB* **14** E2
Crieff *ON* **19** B7
Crieff *ON* **20** D6
Crilly *ON* **16** E3
Crimson Lake *AB* **8** E4
Crimson Lake PP *AB* **6** C4-5
Crimson Lake PP *AB* **8** E4
Crippsdale *AB* **8** B6-7
Crisafy, lac *QC* **26** C4
Cristal, lac *QC* **25** C-D5
Cristal, lac *QC* **28** F1
Croche, rivière *QC* **25** A-B7
Croche, rivière *QC* **28** D3
Croche, ZEC de la *QC*
25 B-C7
Croche, ZEC de la *QC* **28** F1
Crockets Corner *NB* **32** C5
Crofton *BC* **2** E4
Croil Lake *MB* **12** A8
Croil Lake *MB* **15** A3
Croix, lac à la *QC* **29** C4
Croix, lac en *QC* **24** A5
Croix, lac en *QC* **25** F5
Croix, pointe à la *QC* **29** A4
Croix, lac la *ON* **16** F3
Croker, Cape *ON* **18** E5
Croker Island *ON* **18** B3
Croll *MB* **14** F1
Cromarty *MB* **15** C5-6
Cromarty *ON* **19** B4
Cromarty *ON* **20** D3
Cromer *MB* **11** F7-8
Cromer *MB* **13** E1
Cromwell *MB* **13** E4
Crooked Bay *ON* **18** E7-8
Crooked Creek *AB* **9** F3
Crooked Lake *ON* **16** F3-4
Crooked Lake *ON* **18** B6
Crooked Lake PP *SK*
11 E6-7
Crooked River *NL* **39** B5
Crooked River *SK* **11** B6
Crooked River PP *BC* **1** A7

Croque *NL* **39** E8
Crosby *ON* **21** C7
Crosby *ON* **24** F1
Cross *SK* **11** E2
Cross Bay *MB* **11** B8
Cross Bay *MB* **13** B2
Cross Creek *NB* **31** F4-5
Cross Creek *NB* **32** B3
Crossfield *AB* **6** D5
Crossfield *AB* **7** A4
Crosshill *ON* **19** B5
Crosshill *ON* **20** D4
Cross Island *NS* **33** D5
Cross Lake *MB* **15** F3-4
Cross Lake *MB* **15** F3
Cross Lake *ON* **17** E7
Cross Lake *ON* **21** A-B4
Cross Lake PP *AB* **8** A6
Cross Lake PP *AB* **9** F5
Crossley Lakes *NT* **5** B6
Cross Point *ON* **22** D1
Crossroads *AB* **8** F6-7
Crossroads Lake *NL* **39** A3
Crotch Lake *ON* **21** A3
Crotch Lake *ON* **21** C6
Croton *ON* **19** D3
Croton *ON* **20** F2
Crouse *NL* **38** B5
Crouse *NL* **39** E8
Crousetown *NS* **33** D4
Crowduck Lake *MB* **12** E5
Crowduck Lake *MB* **14** D7
Crowe Bridge CA *ON*
21 D4
Crowe Lake *NL* **36** B6
Crowe Lake *NL* **37** B4
Crowe Lake *ON* **21** A4
Crowe Lake *ON* **21** D4
Crowell *NS* **33** F2
Crowe River *ON* **21** C-D4
Crowe River *ON* **21** D4
Crowes Landing *ON* **21** D4
Crowfoot *AB* **7** C6
Crow Head *NL* **38** E6
Crow Lake *ON* **13** F6
Crow Lake *ON* **16** E2
Crow Lake *ON* **21** C2
Crow Lake *ON* **21** C6
Crow Lake PP *AB* **9** E6
Crow Lake PP *AB* **12** E1
Crown Hill *ON* **18** F8
Crown Hill *ON* **20** A7
Crown Hill *ON* **21** D1
Crown Lake *ON* **22** F2
Crown Point *PE* **35** E5
Crown Prince Frederik Island *NU* **41** B5
Crown Prince Frederik Island *NU* **42** F5
Crow River *YT* **3** A5-6
Crow River *YT* **10** F1
Crows Nest *NS* **33** A8
Crows Nest *NS* **34** F4
Crowsnest Pass *AB* **6** F5
Crowsnest Pass *AB* **6** F5
Crowsnest Pass *AB* **7** E3
Crowsnest PP *BC* **6** F5
Crowsnest PP *BC* **7** E3
Croydon Station *BC* **6** A1
Crozier Channel *NT* **42** C2
Cruikshank *ON* **18** F5
Cruikshank *ON* **20** A4
Cruikshank *SK* **11** E3-4
Crumlin *ON* **19** C5
Crumlin *ON* **20** E4
Crump *AB* **7** A4
Crutwell *SK* **11** B4
Cry Lake *BC* **3** B3-4
Crysler *ON* **24** E3
Crystal Beach *ON* **20** F7-8
Crystal City *MB* **13** F2-3
Crystal City *MB* **14** F2-3
Crystal Cliffs *NS* **34** D5
Crystal Crescent Beach PP *NS* **33** D6
Crystal Falls *ON* **17** F7-8
Crystal Falls *ON* **18** A7
Crystal Lake *ON* **21** C3
Crystal Lake *ON* **21** C3
Crystal Lake *SK* **11** C7
Crystal Springs *ON* **21** B3
Crystal Springs *SK* **11** B5
Cuddle Lake *MB* **15** E4
Cudworth *SK* **11** C5
Cullen *SK* **11** F6-7
Culloden *NS* **32** F4
Culloden *NS* **33** C2
Culloden *ON* **19** C5
Culloden *ON* **20** E4
Culloden *PE* **35** F6
Culls Harbour *NL* **36** B8
Culls Harbour *NL* **37** B6
Cull's Island *NL* **38** E5
Culotte, lac à la *QC* **25** C4
Culross *MB* **13** F3
Culross *MB* **14** E4
Cultus *ON* **19** D6
Cultus *ON* **20** F5
Cultus Lake *BC* **2** E6
Cultus Lake *BC* **2** E6
Cultus Lake PP *BC* **1** F8
Cultus Lake PP *BC* **2** E6
Cumberland *BC* **1** F5
Cumberland *BC* **2** D2
Cumberland *ON* **24** D3
Cumberland *PE* **35** E4-5
Cumberland *SK* **11** B5
Cumberland Basin *NB*
32 C7
Cumberland Basin *NS*
32 C7
Cumberland Bay *NB* **31** F6
Cumberland Bay *NB* **32** B4
Cumberland Beach *ON*
18 F8

Cumberland Beach *ON*
20 A7
Cumberland Beach *ON*
21 C1
Cumberland Cove *PE* **35** E3
Cumberland House *SK*
11 A7
Cumberland House *SK*
13 A1
Cumberland House Provincial Historic Park *SK* **11** A6-7
Cumberland Lake *SK*
11 A6-7
Cumberland Peninsula *NU*
42 F8
Cumberland Sound *NU*
42 F8
Cumines Island *SK* **12** C6-7
Cumines Island *SK* **15** C1
Cumins Lake *SK* **9** B8
Cumins Lake *SK* **12** E3
Cummings Lake *ON* **18** A1
Cummins Lake PP *BC*
6 C2-3
Cummins Lake PP *SK* **8** F1
Cumshewa Head *BC* **1** B2
Cumshewa Inlet *BC* **1** B1-2
Cunning Bay *SK* **12** B6
Cunningham Lake *BC* **1** A5
Cupar *SK* **11** D6
Cupids *NL* **37** D7
Curières, lac *QC* **24** A4
Curières, lac *QC* **25** E4
Curlew *AB* **7** A5
Curran *AB* **24** D3
Currie *NB* **31** D3
Currie *NB* **32** A1
Currie Lake *MB* **15** C4
Curve Lake *ON* **21** D3
Curzon *BC* **7** F1
Curzon Village *NL* **38** E2
Cushing *QC* **24** D4
Cuslett *NL* **36** F8
Cuslett *NL* **37** F6
Cutarm Creek *SK* **11** D-E7
Cutarm Creek *SK* **13** D-E1
Cutbank *SK* **11** D4
Cut Beaver Lake *SK* **11** A6
Cut Knife *SK* **6** B8
Cut Knife *SK* **11** B3
Cutler *ON* **17** F5
Cutler *ON* **18** A-B3
Cuvillier, lac *QC* **23** C6-7
Cuvillier, lac *QC* **26** D1
Cuvillier, rivière *QC* **23** C6
Cygnes, lac des *QC* **26** A7
Cygnet *AB* **8** F6
Cygnet Lake *MB* **15** D3
Cymric *SK* **11** D5
Cynthia *AB* **8** B4
Cynthia *AB* **8** D4
Cypress Hills *AB* **6** E7-8
Cypress Hills *AB* **7** E8
Cypress Hills *AB* **11** E1
Cypress Hills *SK* **6** E8
Cypress Hills Interprovincial Park *AB* **6** E-F8
Cypress Hills Interprovincial Park *AB* **11** E-F2
Cypress Hills Interprovincial Park *SK* **6** E-F8
Cypress Hills Interprovincial Park *SK* **11** E-F2
Cypress Lake *SK* **6** F8
Cypress Lake *SK* **11** F2
Cypress Point *AB* **9** A7
Cypress Point *AB* **12** A2-3
Cypress PP *BC* **1** F7
Cypress PP *BC* **2** D4-5
Cypress River *MB* **13** F2
Cypress River *MB* **14** E2
Cyril River *MB* **15** E5
Czar *AB* **6** C7
Czar *AB* **11** C2

D

Daaquam *QC* **27** B7-8
Dacre *ON* **21** B6
Dafoe *SK* **11** C5
Dafoe Lake *MB* **15** E4
Dafoe River *MB* **15** E4-5
Dagenais, rivière *QC* **23** C2
Daggitt Lake *MB* **15** B3
Dahdinni River *NT* **4** A8
Dahdinni River *NT* **10** B-C2
Dahinda *NL* **11** F5
Dahl Lake PP *NU* **1** A6-7
Dain City *ON* **19** C8
Dain City *ON* **20** E7
Dakota Tipi *MB* **13** E3
Dakotah *MB* **14** E4
Dalcourt, lac *QC* **26** A7
Dale *ON* **21** E3
D'Alembert *QC* **17** D8
D'Alembert *QC* **23** D2-3
Dalemead *AB* **6** D5-6
Dalemead *AB* **7** C4
Dalem Lake PP *NU* **34** C7
Dalesville *QC* **24** D4-5
Dalewood CA *ON* **19** C5
Dalewood CA *ON* **20** F4
Dalhousie *NB* **29** D6-7
Dalhousie *NB* **31** A5

Dalhousie, Cape *NT* **5** A6
Dalhousie Junction *NB*
29 D6
Dalhousie Junction *NB*
31 A5
Dalhousie Lake *ON* **21** C6
Dalhousie Mills *ON* **24** E4
Dalhousie Road *NS* **32** F6
Dalhousie Road *NS* **33** A1
Dalhousie Station *ON* **24** E4
Dalkeith *ON* **24** D4
Dalmeny *SK* **11** C4
Dalny *MB* **11** F8
Dalny *MB* **13** F1
Dalroy *AB* **6** D6
Dalroy *AB* **7** B4
Dalrymple *ON* **20** A8
Dalrymple *ON* **21** D2
Dalston *ON* **18** F8
Dalston *ON* **20** A6-7
Dalston *ON* **21** D1
Dalton *ON* **17** D4
Dalton Post *YT* **4** E2-3
Dalum *AB* **6** D6
Dalum *AB* **7** B5-6
Dalvay-by-the-Sea Hotel NHS *PE* **34** B2-3
Dalvay-by-the-Sea Hotel NHS *PE* **35** D5
Daly Lake *SK* **12** D5
Damascus *ON* **19** A6
Damascus *ON* **20** C5
Damdochax PPA *BC* **3** D-E4
Damville, lac *QC* **26** C7
Damville, lac *QC* **28** A2
Dana *SK* **11** C5
Dana-Jowsey Lakes PP *ON*
1 F4-5
Dancing Point *MB* **13** C3
Dand *MB* **13** F1-2
Dandurand, lac *QC* **23** D2
Dane *ON* **17** D7
Daners *NS* **33** D2
Danford Lake *QC* **22** E7
Danford Lake *QC* **24** C1
Daniel's Cove *NL* **37** C7-8
Daniel's Harbour *NL* **38** D2
Daniel's Harbour *NL* **39** F7
Danielson PP *SK* **11** D4
Danish Strait *NU* **42** C3
Danube *AB* **8** A7
Danville *QC* **27** D4
Daphne *SK* **11** C5-6
Dapp *AB* **6** A5
Dapp *AB* **8** A6
Dapp *AB* **9** F5
Darby Harbour *NL* **36** E7
Darby Harbour *NL* **37** E5
Darch Island *ON* **18** B3
Dhaw Gro Habitat Protection Area *YT*
4 B3-4
D'Arcy *BC* **1** E7
D'Arcy *BC* **2** B5
D'Arcy *SK* **11** D3
D'Arcy, lac *QC* **22** B5
D'Arcy Island Unit (Gulf Islands NPR) *BC* **2** F4
Darfield *BC* **6** D1-2
Dark Cove *NL* **36** B8
Dark Cove *NL* **37** B6
Darke Lake PP *BC* **6** F1
Darlans, rivière *QC* **23** E3
Darlingford *MB* **13** F3
Darlingford *MB* **14** F3
Darling Lake *NS* **33** E1
Darlington PP *ON* **21** F2-3
Darmody *SK* **11** E4
Darnley *PE* **32** A8
Darnley *PE* **34** B1
Darnley *PE* **35** C3
Darnley Basin *PE* **35** C3
Darnley Bay *NT* **5** A8
Darnley Bay *NT* **41** A1
Darrell Bay *BC* **1** F7
Darrell Bay *BC* **2** D5
Dartford *ON* **21** E4
Dartmouth *NS* **33** C6
Dartmouth, rivière *QC*
30 B2
Dartmouth Point *NS* **33** D1
Darwell *AB* **6** A-B5
Darwell *AB* **8** C5
Darwin Sound *BC* **1** B1
Dashwood *ON* **19** B4
Dashwood *ON* **20** D3
Dashwood Pond *NL*
36 C-D2
Dasserat, lac *QC* **23** D2
D.A. Tiffin CA *ON* **20** B6
Daulnay *NB* **29** F8
Daulnay *NB* **30** F1
Daulnay *NB* **31** C7
Dauphin *MB* **11** D8
Dauphin *MB* **13** D2
Dauphin *MB* **14** B1
Dauphin, Cape *NS* **34** B7
Dauphin Beach *ON* **19** C7
Dauphin Lake *MB* **11** D8
Dauphin Lake *MB* **13** D2
Dauphin Lake *MB* **14** B1-2
Dauphin River *MB* **13** C3
Dauphin River *MB* **13** C3
D'Auteuil, lac *QC* **39** E5
Davéluyville *QC* **24** B8
Daveluyville *QC* **27** C4

David, lac *QC* **22** D7
David, lac *QC* **24** A1
David, lac *QC* **24** A3
David, lac *QC* **25** E3
David, lac *QC* **25** F1
David, rivière *QC* **24** B6
David, rivière *QC* **24** C7
David, rivière *QC* **25** F6
David, rivière *QC* **27** C1
David, rivière *QC* **27** D3
Davidson *QC* **22** E6
Davidson *SK* **11** D5
Davidson Lake *MB* **11** A8
Davidsville *NL* **36** A7-8
Davidsville *NL* **37** A5-6
Davidsville *NL* **38** F7
Davin *SK* **11** E6
Davin Lake *SK* **12** D6
Davis *SK* **11** B5
Davis, détroit de *NU* **40** A7
Davis, détroit de *NU* **42** D-E8
Davis Cove *NL* **36** D8
Davis Cove *NL* **37** D6
Davis Inlet *NL* **40** D7
Davis Lake PP *BC* **2** D6
Davis River *BC* **3** B4
Davis Strait *NU* **40** A7-8
Davis Strait *NU* **42** D-E8
Davy, rivière *QC* **23** C4
Davy Lake *SK* **9** B8
Davy Lake *SK* **12** B3-4
Dawes Pond *NL* **36** A5
Dawes Pond *NL* **37** A3
Dawes Pond *NL* **38** F4
Dawley Passage PMP *BC*
1 F4-5
Dawley Passage PMP *BC*
2 D-E1
Dawson, Mount *BC* **6** D3
Dawson Bay *MB* **11** B8
Dawson Bay *MB* **13** B1-2
Dawson Bay *MB* **13** B1-2
Dawson Bay *MB* **13** B1-2
Dawson City *YT* **4** A2
Dawson Creek *BC* **3** E8
Dawson Creek *BC* **9** D3
Dawson Inlet *NU* **40** B1-2
Dawson Inlet *NU* **41** E5
Dawson Landing *NT*
10 E6-7
Dawson Range *YT* **4** C2-3
Dawson River *ON* **13** B6
Dawsons Landing *BC* **1** D4
Dawsonville *NB* **29** D5-6
Dawsonville *NB* **31** A4
Daylesford *SK* **11** C5
Day Mills *ON* **17** F5
Day Mills *ON* **18** A1
Dayohessarah Lake *ON*
17 C3
Day Point *BC* **1** C3
Daysland *AB* **6** B6
Daysland *AB* **8** D8
Daysland *AB* **11** B1
Dayspring *NS* **33** D4
Day Star *SK* **11** D6
Dayton *NS* **33** E1

Deacon *ON* **21** A5
Deacon *ON* **7** B4
Deacons Corner *MB* **14** E5
Deadhorse Lake *AB* **6** D6
Deadhorse Lake *AB* **7** B6
Deadman River *BC* **1** D8
Deadman River *BC* **2** A7
Deadman River *BC* **6** E1
Deadman's Bay *NL* **37** A6-7
Deadman's Bay *NL* **38** F7
Deadman's Bay PP *NL*
36 A8
Deadman's Bay PP *NL*
37 A6
Deadman's Bay PP *NL*
38 B3-4
Deadman's Cove *NL*
38 B3-4
Dead Tree Point *BC* **1** A-B1
Deadwood *AB* **9** D3
Deadwood River *BC* **3** B4
Deadwood River *BC* **4** F7
Dealtown *ON* **19** E3
Dean *NS* **33** B7
Dean *NS* **34** F3
Dean Channel *BC* **1** C4
Dean Lake *ON* **18** A1
Dean Lake *ON* **18** A1
Deanlea Beach *ON* **18** F7
Deanlea Beach *ON* **20** A6
Dean River *BC* **1** B5
Dease Arm *NT* **41** C1
Dease Lake *BC* **3** B3
Dease Lake *BC* **3** B3
Dease River *BC* **3** A4 B3
Dease River *BC* **4** F7
Dease Strait *NU* **41** B3
Dease Strait *NU* **42** F2
Deauville *QC* **27** E4
Debden *SK* **11** B4
Debec *NB* **31** F3
Debec *NB* **32** B1
Debert *NS* **32** D8
Debert *NS* **33** A6
Debert *NS* **34** E2
Debert River *NS* **34** E2
DeBolt *AB* **9** E-F3
Decker *MB* **11** E8
Decker *MB* **13** E1
Decoigne, lac *QC* **39** D1
Deep Bay *NL* **38** E7
Deep Bay *NT* **10** E5
Deep Bay *ON* **14** F8
Deep Bay *SK* **12** D6
Deep Bay *SK* **15** D1
Deep Bight *NL* **36** C8
Deep Bight *NL* **37** C6
Deep Brook *NS* **32** F5
Deep Brook *NS* **33** C2
Deep Cove *NS* **32** F7
Deep Cove *NS* **33** C5
Deep River *ON* **22** E4-5
Deepwater Point *ON* **22** D1
Deer Bay *NU* **42** B3
Deer Bay *ON* **18** B6
Deerbrook *ON* **19** E2
Deer Cove *NL* **38** D2-3
Deerfield *NS* **33** E1-2
Deerhorn *MB* **13** D3
Deerhorn *MB* **14** C4
Deer Island *MB* **13** D4
Deer Island *MB* **14** B6
Deer Island *NB* **32** E2
Deer Island *MB* **13** B8
Deer Island *NB* **36** D3-4
Deer Island *NB* **37** B6
Deer Island *NB* **37** D1-2
Deer Lake *NL* **36** A3
Deer Lake *NL* **36** A3
Deer Lake *NL* **37** A1
Deer Lake *NL* **38** F2-3
Deer Lake *NL* **38** F2-3
Deer Lake *NL* **39** F7
Deer Lake *NL* **39** F7
Deer Lake *ON* **13** B5-6
Deer Lake *ON* **16** A2
Deer Lake *ON* **16** A2
Deer Lake *ON* **18** E8
Deer Lake *ON* **21** C1
Deer Lake *ON* **22** E1
Deer Pond *NL* **37** C-B5
Deer Pond *NL* **37** B-C5
Deer River *ON* **17** C3
Deerwood *MB* **13** F3
Deerwood *MB* **14** F3
Defeat Lake *NT* **10** D7
Defoy *QC* **27** C4
Dégelis *QC* **29** E2-3
Dégelis *QC* **31** B1
Deguire, lac *QC* **36** C1
De La Bidière, lac *QC*
25 C3
Delacour *AB* **6** D5
Delacour *AB* **7** B4
Delâge, lac *QC* **25** A4
Delâge, lac *QC* **26** F4
Delahey, lac *QC* **22** C6
Delaps Cove *NS* **33** C2
Delaronde Lake *SK* **11** A4
Delaware *ON* **19** C4
Delaware *ON* **20** E3
Delayee Lake *YT* **4** E4
Del Bonita *AB* **6** F6
Del Bonita *AB* **7** F5
Delburne *AB* **6** C6
Delburne *AB* **8** E7
Déléage *QC* **22** D7-8
Déléage *QC* **24** B2
Déléage *QC* **25** F2
Deleau *MB* **11** F8
Deleau *MB* **13** F1
DeLesseps Lake *ON* **13** D8
De Lesseps Lake *ON* **16** C4
Delhi *ON* **19** C6
Delhi *ON* **20** F5
Delia *AB* **6** D6
Delia *AB* **7** B5
Déline *NT* **10** A2
Déline Fishery and Fort Franklin HS *NT* **10** A2-3
Delisle *QC* **28** B5
Delisle *SK* **11** C4
Delisle, rivière *QC* **24** E4
Delisle, rivière *QC*
21 A7-8 B8
Delisle, lac *QC* **28** D7-8
Delmas *SK* **11** B3
Delmer *ON* **19** C5
Delmer *ON* **20** F4
Deloraine *MB* **11** F8
Deloraine *MB* **13** F1
Deloro *ON* **21** D4
Deloro *ON* **23** B6
Delph *AB* **8** B7
Delson *QC* **24** D6
Delta *BC* **1** F7
Delta *ON* **21** D7
Delta *ON* **24** F1
Delta (Ladner) *BC* **2** E5
Delta Beach *MB* **13** E3
Delta Beach *MB* **14** D3
Delta-de-la-Rivière-Anderson, RNOM du *NT*
5 A-B6
Delta-de-la-Rivière-Nisutlin, RNF du *YT* **4** E5

Deltaform Mountain *AB*
6 D4
Deltaform Mountain *AB*
7 B1
Delta Marsh *MB* **13** E3
Delta Marsh *MB* **14** D3
Delta Peak *BC* **3** D3
Demaine *SK* **11** D4
Demarcation Point *YT* **5** A2
Demay *AB* **9** E2
Demers-Centre *QC* **22** E5
Demmitt *AB* **9** E1
De Montigny, lac *QC* **23** D4
De Morbihan, lac *QC* **29** D4
Demorestville *ON* **21** E5
Dempsey Corners *NS* **32** F6
Dempsey Corners *NS* **33** B3
Demuth *BC* **6** E-F1
Denain *QC* **23** E-F6
Denare Beach *SK* **12** F7
Denare Beach *SK* **15** F1
Denault *QC* **26** C6
Denault, lac *QC* **28** A1
Denbeigh Point *MB* **11** B8
Denbeigh Point *MB* **13** B2
Denbigh *ON* **21** B5
Dencross *MB* **14** D6
Denetiah Lake *BC* **3** B4
Denetiah PP *BC* **3** B4
Denfield *ON* **19** C4
Denholm *QC* **22** F8
Denholm *QC* **24** C2
Denman Island *BC* **1** F5
Denman Island *BC* **2** D2-3
Denman Island *BC* **2** D2
Denmark *NS* **34** D2-3
Dennis Lake *MB* **14** C5
Denny Island *BC* **1** C4
Dension Lake *MB* **12** C8
Dension Lake *MB* **15** C2-3
Densmore Mills *NS* **32** D8
Densmore Mills *NS*
33 A5-6
Densmore Mills *NS* **34** E1
Denver *NS* **33** A8
Denver *NS* **34** E4
Denwood *AB* **6** B7
Denwood *AB* **11** B1-2
Denys Basin *NS* **34** D6
Denzil *SK* **6** C8
Denzil *SK* **11** C3
Departure Lake *ON* **17** B6
De Pas, rivière *QC* **39** A3
Depot Lake CA *ON* **21** D6
Derby *NB* **31** D6
Derby *PE* **32** A7
Derby *PE* **35** C1-2
Derby Junction *NB* **31** D6-7
Dering *MB* **11** A8
Dering *MB* **12** F8
Dering *MB* **13** A2
Dering *MB* **15** F2
Deroche, Point *PE* **35** D5
Derwent *AB* **6** A7
Derwent *AB* **11** A1-2
DeSable *PE* **32** B8
DeSable *PE* **34** C2
DeSable *PE* **35** E4
Desaulniers *ON* **17** F7
Désaulniers *ON* **18** A7
Desautels, lac *QC* **26** C6-7
Desbarats *ON* **17** F4
Desbiens *QC* **28** C4
Desboro *ON* **18** F5
Desboro *ON* **20** A4
Desboues, rivière *QC* **23** C4
Deschaillons-sur-Saint-Laurent *QC* **27** B4
Deschambault *QC* **27** B4-5
Deschambault Lake *SK*
12 F6
Deschambault Lake *SK*
12 F6
Deschambault River *SK*
12 F6
Deschamps, lac *QC* **26** E3
Descharme Lake *SK* **9** C-D3
Descharme Lake *SK*
12 C-D3
Descharme River *SK* **9** C7-8
Descharme River *SK* **12** C3
Deschênes, lac *QC*
21 A7-8 B8
Deschênes, lac *QC* **28** D7-8
D'Escousse *NS* **34** D6
Deseronto *ON* **21** E5-6
Désert, lac *QC* **22** D7
Désert, lac *QC* **24** A1
Désert, lac *QC* **25** E1
Deserters Peak *BC* **3** D6
Déserteur, lac du *QC* **26** E5
Desforges, lac *QC* **23** C3
Desforges, lac *QC* **26** D1
Desgly, lac *QC* **26** E5
Desgly, lac *QC* **28** A1-2
Desjardins, lac *QC* **23** B6
Desjardinsville *QC* **22** F8
Deskenatlata Lake *NT*
27 E1-2
Desliens Lake *NL* **39** A-B2
Desmarais, pointe *NT* **10** E6
Desmaraisville *QC* **23** A7-8
Desmaraisville *QC* **26** B2
Desmaraisville *QC* **40** F4-5
Des Méloizes, rivière *QC*
23 B2
Desolation Sound PMP *BC*
1 E6
Desolation Sound PMP *BC*
2 C3

des-Passes, ZEC *QC* 28 A4-5
Desperation Lake *NT* 10 C7
Despinassy *QC* 23 C5
Despinassy, lac *QC* 23 D7
Desrivières, lac *QC* 22 C-D7
Desrivières, lac *QC* 24 A1
Desrivières, lac *QC* 25 E1
Des Ruisseaux *QC* 22 D8
Des Ruisseaux *QC* 24 A2
Des Ruisseaux *QC* 25 F2
Desteffany Lake *NT* 10 A7
Destor *QC* 17 C8
Destor *QC* 23 D2-3
Destruction Bay *YT* 4 D2
Detlor *ON* 21 B4
Detroit River *ON* 19 E1
Dettah *NT* 10 D6-7
Dettah *NT* 41 D1
Deux Loutres, lac aux *QC* 39 D4
Deux-Montagnes *QC* 24 D5
Deux-Montagnes, lac des *QC* 24 D5
Deux Orignaux, lac des *QC* 26 A3
Deux Rivieres *ON* 22 D3
Devastation Channel *BC* 1 A3
De Veber, Mount *AB* 6 A2
Devenyns, lac *QC* 25 D5
Devenyns, lac *QC* 27 A1
Devil Lake *ON* 21 D7
Devil Lake *ON* 24 F1
Devil's Glen PP *ON* 20 B5
Devine *BC* 2 B5-6
Devlin *ON* 16 F2
Devlin, lac *QC* 22 A3
Devlin, lac *QC* 23 F3
Devon *AB* 6 B5
Devon *AB* 7 C4
Devon, île *NU* 42 D4-5
Devon Island *NU* 42 D4-5
Devon Road Mesa, RNP *ON* 16 F5-6
Devon Road Mesa PNR *ON* 16 F5-6
Dewar Lake *SK* 6 C-D8
Dewar Lake *SK* 11 D2
Dewar Peak *BC* 3 D5
Dewberry *AB* 6 A7
Dewberry *AB* 11 A2
Dewdney *BC* 1 F7-8
Dewdney *BC* 2 E6
De Winton *AB* 6 D5
De Winton *AB* 7 C4
Dewittville *QC* 24 E5
Dezadeash Lake *YT* 4 E3
Dezadeash River *YT* 4 E3
Diable, cap *NU* 38 A4
Diable, cap au *QC* 28 E8
Diable, rivière du *QC* 24 A-B4
Diable, rivière du *QC* 24 C4
Diable, rivière du *QC* 25 F4
Diamant, lac *QC* 25 D5
Diamant, lac *QC* 28 F1
Diamond City *AB* 6 E-F6
Diamond City *AB* 7 E5-6
Diamond Cove *NL* 36 D-E2
Diamond Jenness Peninsula *NT* 41 A2
Diamond Jenness Peninsula *NT* 42 E1
Diana Lake PP *BC* 1 A3
Dickey Lake *ON* 21 C4
Dickson *AB* 6 C5
Dickson *AB* 8 F5
Dickson, lac *QC* 24 C1
Dickson Lake *ON* 22 E-F3
Didsbury *AB* 6 D5
Didsbury *AB* 7 A3-4
Diefenbaker, Lake *SK* 11 D4
Dieppe *NB* 31 F8
Dieppe *NB* 32 B6
Dieppe, lac *QC* 25 D3
Dieppe Lake *NT* 10 D5-6
Digby *NS* 32 F4
Digby *NS* 33 C2
Digby *NS* 33 D2
Digby Neck *NS* 33 C-D1
Digdeguash *NB* 32 E2
Digdeguash River *NB* 32 D2
Digges *MB* 15 B6
Digges Islands *NU* 40 B4
Digges Islands *NU* 41 D7
Dildo *NL* 37 D7
Dildo Run PP *NL* 38 E6-7
Diligent River *NS* 32 D7
Diligent River *NS* 33 A4
Dilke *NB* 11 D4
Dillberry Lake PP *AB* 6 C7-8
Dillberry Lake PP *AB* 11 C2
Dillon *ON* 18 D6
Dillon *SK* 9 E8
Dillon *SK* 12 E3
Dillon Lake *SK* 9 E8
Dillon Lake *SK* 12 E3
Dillon River *SK* 9 E7-8
Dillon River *SK* 12 E2-3
Dinant *AB* 7 D6
Dingwall *NS* 34 A7
Dingwells Mills *PE* 34 C3-4
Dingwells Mills *PE* 35 D7
Dinorwic *ON* 13 F7
Dinorwic *ON* 16 D3
Dinorwic Lake *ON* 13 F7
Dinorwic Lake *ON* 16 D3
Dinosaur *AB* 6 D7
Dinosaur *AB* 7 C7
Dinosaur *AB* 11 D1
Dinosaur PP *AB* 6 D7
Dinosaur PP *AB* 7 C7
Dinosaur PP *AB* 11 D1

Dinsmore *SK* 11 D4
Dionisio Point PP *BC* 1 F6-7
Dionisio Point PP *BC* 2 E4
Dipper Harbour *NB* 32 E3
Dipper Harbour *NB* 33 B1
Dipper Lake *SK* 12 E4
Dipper Lake *SK* 12 E4
Disappointment Lake *NL* 39 B4
Discovery Island PMP *BC* 2 F4-5
Discovery, Cape *BC* 2 F4-5
Discovery Claim NHS *YT* 4 A1-2
Discovery Passage *BC* 2 B-C2
Disley *SK* 11 E5
Disraeli *QC* 27 D5
Ditidaht *BC* 2 E3
Divide *SK* 11 F2
Divide Ridge, RNP *ON* 16 F5
Divide Ridge PNR *ON* 16 F5
Dividing Lake, RNP *ON* 21 B2-3
Dividing Lake PNR *ON* 21 B2-3
Dixie Cove PP *BC* 1 E4
Dix Milles, lac *QC* 22 C4
Dix Milles, lac des *QC* 25 F4
Dixon *SK* 11 C5
Dixon, détroit de *BC* 1 A1-2
Dixon, détroit de *BC* 3 F1-2
Dixon Entrance *BC* 1 A1-2
Dixon Entrance *BC* 3 F1-2
Dixonville *AB* 9 D3
Dixville *QC* 27 F5
Dizzy Creek *AB* 9 A4
Doak, lieu historique provincial *NB* 31 E6
Doak, lieu historique provincial *NB* 32 A4-5
Doak PHS *NB* 31 E6
Doak PHS *NB* 32 A4-5
Doaktown *NB* 31 E5
Doaktown *NB* 32 A3-4
Dobbin Bay *NU* 42 B5
Dobbinton *ON* 18 F5
Dobbinton *ON* 20 A4
Dobie *ON* 17 D7-8
Dobie Lake *ON* 13 D8
Dobie Lake *ON* 16 B4
Dobie River *ON* 13 C-D8
Dobie River *ON* 16 B4-5
Dobleau, lac *QC* 26 B6
Dock Cove *NL* 37 B7
Doctor Island *NS* 34 D3
Doctors Brook *NS* 34 D4
Doctors Harbour *NL* 36 D6
Doctors Harbour *NL* 37 D4
Doda, lac *QC* 26 C3
Dodd Lake *BC* 2 C3
Dodds *AB* 8 D7-8
Dodge Lake *SK* 12 A5
Dodsland *SK* 11 C3
Doe Lake *ON* 18 C8
Doe Lake *ON* 18 C8
Doe Lake *ON* 21 A1
Doe Lake *ON* 21 A1
Doe Lake *ON* 22 F1
Doe River *BC* 3 E8
Doe River *BC* 9 E2
Dog Creek *BC* 1 D7-8
Dog Creek *BC* 1 D7
Dog Creek *MB* 13 D3
Dog Creek *MB* 14 B3
Dogface Lake *NT* 10 F4
Dog Islands *NL* 39 A6
Dog Lake *MB* 13 D3
Dog Lake *MB* 14 B3
Dog Lake *ON* 16 E5
Dog Lake *ON* 17 D4
Dog Lake *ON* 21 D7
Dogpound *AB* 7 A3
Dog River *ON* 16 E5
Dogskin Lake *MB* 14 A7
Dogskin Lake *ON* 13 C5
Dogskin River *MB* 13 C5
Dogtooth Lake *ON* 13 F6
Dogwood Creek *BC* 2 D7
Doig River *BC* 3 D8
Doig River *BC* 9 D2
Dokis *ON* 18 B7
Dokis Point *ON* 18 A7
Dolbeau *QC* 26 D8
Dolbeau *QC* 28 B4
Dollard *SK* 11 E3
Dollard Brook *NL* 36 D5
Dollard Brook *NL* 37 D3
Dollar Lake PP *NS* 33 B6
Dollars Lake *SK* 18 B7
Dollier, lac *QC* 26 B5
Dolly Bay *NL* 37 B7
Dolphin and Union Strait *NT* 41 B2
Dolphin and Union Strait *NT* 42 E-F1
Dolphin and Union Strait *NU* 41 B2
Dolphin and Union Strait *NU* 42 E-F1
Dolu, lac *QC* 39 A1
Domain *MB* 13 F3-4
Domain *MB* 14 E5
Dome *ON* 17 D6
Dome Creek *BC* 1 A8
Dome Creek *BC* 6 A1
Dominion *NS* 34 C8
Dominion, Cape *NU* 41 C7

Dominion Beach PP *NS* 34 C8
Dominion City *MB* 13 F4
Dominion City *MB* 14 F5
Dominion Lake *NL* 39 C5
Domino *NL* 39 B8
Domogaya Lake *NL* 39 D3
Domremy *SK* 11 B5
Domville *ON* 21 C8
Domville *ON* 24 F2
Donagh *PE* 34 C3
Donagh *PE* 35 E5
Donald *BC* 6 D3
Donald *ON* 21 C3
Donalda *AB* 6 C6
Donalda *AB* 8 E7-8
Donald Lake *ON* 13 D5
Donald Lake *ON* 14 B8
Donaldston *PE* 35 D5
Donavon *SK* 11 C4
Donegal *ON* 19 A5
Donegal *ON* 20 A4
Donegal *SK* 11 C2
Don Getty Wildland PP *AB* 6 D5
Don Getty Wildland PP *AB* 6 E5
Don Getty Wildland PP *AB* 7 A-B2
Don Getty Wildland PP *AB* 7 B-C3
Don Getty Wildland PP *AB* 7 C3
Don Getty Wildland PP *AB* 7 D3
Donjek *YT* 4 C1
Donjek River *YT* 4 C1-2
Donkin *NS* 34 C8
Donnacona *QC* 27 B5
Donnegana Lake *ON* 17 E6
Donnelly *AB* 9 E3-4
Donnelly Lake *ON* 16 A4
Donnelly River *NT* 5 E7
Don Peninsula *BC* 1 C4
Donwell *SK* 11 C7
Donwood *ON* 21 D3
Doolittle, lac *QC* 22 C6
Doran Lake *ON* 13 D8
Dorcas Bay *ON* 18 D4
Dorchester *NB* 32 C6-7
Dorchester *ON* 19 C5
Dorchester *ON* 20 C4
Dorchester, Cape *NU* 41 B6
Dorchester, Cape *NU* 41 C7
Dorchester Crossing *NB* 31 F8
Dorchester Crossing *NB* 32 B6-7
Doré, lac *QC* 22 A7
Doré, lac *QC* 25 C1
Doré, Lake *ON* 21 A5
Doré, Lake *ON* 22 F5
Dore Lake *SK* 12 F4
Doré Lake *SK* 12 F4
Doré River *SK* 12 F3-4
Dorés, lac aux *QC* 26 A3-5
Dorintosh *SK* 11 A3
Dorintosh *SK* 12 F3
Dorion *ON* 16 E6
Doris *AB* 8 A5
Dorking *ON* 19 A5
Dorking *ON* 20 C4
Dorland *ON* 21 E5-6
Dormans Cove *NL* 36 A7
Dormans Cove *NL* 37 A5
Dormans Cove *NL* 38 F6-7
Dornoch *ON* 20 B4
Dorothy *AB* 6 D6-7
Dorothy *AB* 7 C6
Dorset *ON* 21 B2
Dory Point TP *NT* 38 F7-8
Dosquet *QC* 27 B-C5
Doting Cove *NL* 38 F7-8
Douaire, lac *QC* 22 B8
Douaire, lac *QC* 25 D2
Double Island *NL* 39 A6-7
Double Mer *NL* 39 B6
Doucet, lac *QC* 28 A4
Doucet Lake *NT* 10 B8
Doucetville *NS* 33 D2
Douglas *MB* 13 E2
Douglas *MB* 14 E2
Douglas *NB* 31 F4-5
Douglas *NB* 32 C3
Douglas *ON* 21 A6
Douglas *ON* 22 F6
Douglas Channel *BC* 1 A-B3
Douglas Harbour *NB* 32 C4
Douglas Lake *BC* 2 C8
Douglas Lake *BC* 2 B8
Douglas Lake *BC* 6 E1
Douglas Point *SK* 11 D4
Douglas River *SK* 9 B7
Douglas River *SK* 12 A3
Douglastown *NB* 31 D6-7
Doumic, lac *QC* 28 B6
Douro *ON* 21 D3
Dover *NL* 36 B8
Dover *NL* 37 B6
Dover *PE* 35 F6
Dover Bay *NL* 34 E6
Dover Centre *ON* 19 D2
Dover Centre *ON* 20 F1
Dowager Island *BC* 1 C3
Dowling *ON* 17 F6
Dowling *ON* 18 A5
Dowling Lake *AB* 6 C7
Dowling Lake *AB* 7 A6
Dowling Lake *AB* 11 C1

Downe *SK* 11 C3
Downeyville *ON* 21 D3
Downing PP *BC* 1 D8
Downing PP *BC* 2 A6
Downton, Mount *BC* 1 C6
Downton *BC* 2 A5
Doyles *NL* 36 D1
Dozois, réservoir *QC* 22 A5
Drag Lake *ON* 21 B3
Drague-Numéro-Quatre, LHN de la *YT* 4 A1-2
Drake *SK* 11 C-D5
Draney Inlet *BC* 1 D4
Drapeau *QC* 29 D7
Drapeau *QC* 31 A5
Drayton *ON* 19 A6
Drayton *ON* 20 A5
Drayton Valley *AB* 6 B4-5
Drayton Valley *AB* 8 D4-5
Dredge No. 4 NHS *YT* 4 A1-2
Dresden *ON* 19 D3
Dresden *ON* 20 F2
Drew Station *ON* 20 C4
Drewry Point PP *BC* 6 F3-4
Driftpile *AB* 9 E4
Driftpile River *AB* 9 F4
Drift River *MB* 15 B4
Driftwood *ON* 17 D6
Driftwood Canyon PP *BC* 3 F4
Driftwood Lake *MB* 11 A8
Driftwood Lake *MB* 13 A1-2
Driftwood PP *ON* 22 D4
Driftwood River *YT* 5 C3
Drinkwater *SK* 11 E5
Driscol Lake *SK* 11 F3
Drizzle Lake *BC* 21 B3
Droit, lac du *QC* 25 B6
Droit, lac du *QC* 26 F6
Droit, lac du *QC* 28 D1
Drolet, lac *QC* 27 D6
Dromore *ON* 20 B5
Dromore *PE* 35 E6
Dropmore *MB* 11 D7
Dropmore *MB* 13 D1
Drowning River *ON* 17 A2-3
Droxford *SK* 11 E4
Druid *SK* 11 C3
Drumbeg PP *BC* 2 E4
Drumbo *ON* 19 B6
Drumbo *ON* 20 C5
Drumhead *NS* 34 F5
Drumheller *AB* 6 D6
Drumheller *AB* 7 A5-6
Drummond *NB* 31 D3
Drummondville *QC* 24 C8
Drummondville *QC* 27 D3-4
Drury Inlet *BC* 1 D4
Drury Lake *YT* 4 C4
Dryberry Lake *ON* 13 F6
Dryberry Lake *ON* 16 D2
Drybones Lake *NT* 10 B7
Drybrough *MB* 12 D7
Drybrough *MB* 15 D2
Dry Creek *YT* 4 C1
Dry Gulch PP *BC* 6 E1
Dry Gulch PP *BC* 7 C1-2
Dry Island Buffalo Jump PP *AB* 6 C6
Dry Island Buffalo Jump PP *AB* 8 F7
Drylake *NL* 39 C2
Dry River *MB* 14 F2-3
Dryson, lac *QC* 22 C6
Drywood *AB* 7 F4
Duagh *AB* 8 C6-7
Duart *ON* 19 D3-4
Duart *ON* 20 F2-3
Dubawnt *NT* 41 E3
Dubawnt Lake *NU* 41 D-E4
Dublin *ON* 19 B4
Dublin *ON* 20 D3
Dublin Shore *NS* 33 D5
Dubois, lac *QC* 26 D4-5
Duborne, Lake *ON* 18 A2
Dubose, Mount *AB* 1 A-B4
Dubreuilville *ON* 17 D3-4
Dubuc *SK* 11 E7
Dubuisson *QC* 23 E4
Ducharme, lac *QC* 26 C5
Ducharme, lac *QC* 28 A1
Duchastel, lac *QC* 25 C3
Duchenier, réserve *QC* 31 A1
Duchesnay *QC* 27 A5
Duchesnay Creek *ON* 18 A6
Duchess *AB* 6 D-E7
Duchess *AB* 7 C7
Duck Bay *MB* 11 C8
Duck Bay *MB* 13 C1
Duck Lake *BC* 6 E2
Duck Lake *MB* 15 F3
Duck Lake *ON* 17 D5
Duck Lake *ON* 18 C8
Duck Lake *ON* 21 A1
Duck Lake PP *BC* 2 C8
Duck Mountain *MB* 11 C-D7
Duck Mountain *MB* 13 C1
Duck Mountain PP *MB* 11 C-D8
Duck Mountain PP *MB* 13 C-D1
Duck Mountain PP *SK* 11 C-D7
Duck Mountain PP *SK* 13 C-D1

Duck River *MB* 11 C8
Duck River *MB* 13 C2
Duclos *QC* 21 A7
Duclos *QC* 22 F7
Duclos *QC* 24 C1
Duclos Point *ON* 20 B7
Duclos Point *ON* 21 D1
Duclos Point, RNP *ON* 20 B7
Duclos Point, RNP *ON* 21 D1
Duclos Point PNR *ON* 20 B7
Dudidontu River *BC* 3 B2
Dudley *QC* 22 D8
Dudley, lac *QC* 24 B2-3
Dudley, lac *QC* 25 F2-3
Dudswell *QC* 27 E5
Dufault, lac *QC* 23 D2-3
Dufay, lac *QC* 23 E2
Duff *SK* 11 D6
Duffey Lake PP *BC* 1 E7
Duffey Lake PP *BC* 2 B5-6
Duffield *AB* 6 B5
Duffield *AB* 8 C5
Duffin Lake *MB* 15 B3
Duff Roblin PP *MB* 14 D5
Duff's Corners *ON* 19 C7
Duff's Corners *ON* 21 F1
Duffys Corner *NB* 31 F5-6
Duffys Corner *NB* 32 B4
Dufresne *AB* 8 B5
Dufresne *MB* 14 E5-6
Dufresne, lac *QC* 26 C7
Dufresnoy, lac *QC* 23 D2-3
Dufrost *MB* 13 F4
Dufrost *MB* 14 F5
Du Gas, lac *QC* 39 E6
Dugald *MB* 13 E4
Dugald *MB* 14 E5
Dugué, lac *QC* 25 A4
Dugué, lac *QC* 26 F4
Dugué, rivière *QC* 27 A8
Duguayville *NB* 31 B7
Duhamel *AB* 8 D7
Duhamel *QC* 24 C3
Duke River *YT* 4 D2
Dulwich *SK* 11 B3
Dumais, lac *QC* 25 B4
Dumas *SK* 11 E7
Dumas, lac *QC* 26 A3-4
Dumbarton *NB* 32 D2
Dumbell Lake *NL* 39 C3
Dumfries *NB* 32 C2
Dummer *SK* 11 E5
Dumoine, lac *QC* 22 C-D4
Dumoine, rivière *QC* 22 C-D4
Dumoine, ZEC *QC* 22 D4
Dumoine River *ON* 22 C-D4
Dumond River *ON* 13 C8
Dumont, lac *QC* 24 B-C1
Dumoulin, lac *QC* 25 C3
Dumoulin, lac *QC* 28 E2-3
Dunblane *SK* 11 D4
Duncairn *SK* 11 E3
Duncairn, RNOM de la réserve *SK* 11 E3
Duncairn Reserve NMBS *SK* 11 E3
Duncan *BC* 2 E4
Duncan, lac *QC* 22 C4
Duncan Crevice Caves, RNP *ON* 18 F6
Duncan Crevice Caves, RNP *ON* 20 B7
Duncan Crevice Caves PNR *ON* 18 F6
Duncan Crevice Caves PNR *ON* 20 B7
Duncan Lake *BC* 6 E3-4
Duncan Lake *NT* 10 C7
Duncan Lake *ON* 17 D6
Duncan River *BC* 6 D3
Dunchurch *ON* 18 C7
Dundalk *ON* 20 B5
Dundas *ON* 19 B7
Dundas *ON* 20 D6
Dundas *PE* 34 C4
Dundas *PE* 35 E7
Dundas, Cape *ON* 18 E5
Dundas Island *NL* 39 A6
Dundas Peninsula *NT* 42 D2
Dundas Valley CA *ON* 19 B7
Dundas Valley CA *ON* 20 D6
Dundee *NB* 31 A5
Dundee *NS* 34 D6
Dundee *QC* 24 E4
Dundee Centre *QC* 24 E4
Dundee PP *NS* 34 D6
Dundela *ON* 24 F3
Dundonald *ON* 21 E4
Dundurn *SK* 11 C4
Dune Za Keyih PP *BC* 3 B-C5
Dunfield *NL* 37 C7
Dungannon *ON* 19 A4
Dungannon *ON* 20 B4
Dungarvon River *NB* 31 E5-6
Dungarvon River *NB* 32 A3-4
Dungeon PP *NL* 37 B7-8
Dunham *QC* 24 E7
Dunham *QC* 27 F3
Dunière, RF de *QC* 29 C6
Dunière, RF de *QC* 29 C6
Dunkerron *ON* 20 B6
Dunkirk *ON* 11 E5

Dunkirk River *AB* 9 D6
Dunkirk River *AB* 12 C-D1
Dunk River *PE* 35 D3
Dunleath *SK* 11 D7
Dunlevy PRA *BC* 3 E7
Dunlop *ON* 19 A4
Dunlop *ON* 20 C3
Dunlop Lake *ON* 18 A2
Dunmaglass *NS* 34 D4
Dunmore *AB* 6 E8
Dunmore *AB* 7 D8
Dunmore *AB* 11 E1-2
Dunning Lake *SK* 9 C8
Dunnottar *MB* 14 C5
Dunns Brook *NL* 36 D7
Dunns Brook *NL* 37 D5
Dunns Valley *ON* 17 F4-5
Dunns Valley *ON* 18 A1
Dunnville *ON* 19 C8
Dunnville *ON* 20 E7
Dunphy Lakes *MB* 12 D7
Dunphy Lakes *MB* 15 D1
Dunraven *ON* 21 A6
Dunraven *ON* 22 F6
Dunrea *MB* 13 F2
Dunrea *MB* 14 E-F1
Dunrobin *ON* 21 A7
Dunrobin *ON* 24 D1
Dunsheath Lake *MB* 15 D2
Dunstable *AB* 8 B5
Dunstaffnage *PE* 35 D-E5
Dunster *BC* 6 A2
Dunsterville, Cape *NU* 42 B-C5
Duntroon *ON* 18 F7
Duntroon *ON* 20 A-B6
Duntura *NL* 37 B7
Duparquet *QC* 17 C8
Duparquet *QC* 23 D2
Duparquet, lac *QC* 23 D2
Duperow *SK* 11 C3
Duplessis, lac *QC* 25 D4
Duplessis, lac *QC* 28 E1
DuPont, RNP *ON* 24 F3-4
DuPont PNR *ON* 24 F3-4
Dupuis, lac *QC* 25 C6
Dupuis, lac *QC* 28 F1
Dupuis Corner *NB* 32 B7
Dupuy *QC* 17 C7
Dupuy *QC* 23 C2
Durban *MB* 11 C7
Durban *MB* 13 C1
Durban, lac *QC* 25 D5
Durban, lac *QC* 28 E2-3
Durell Point *PE* 35 E7
Durham *ON* 20 B4
Durham CA *ON* 20 D6
Durham Bridge *NB* 31 F5
Durham Bridge *NB* 32 B3
Durham Centre *NB* 29 D7
Durham Centre *NB* 31 A6
Durham-Sud *QC* 24 D8
Durham-Sud *QC* 27 E4
Durocher, lac *QC* 39 E5
Durrell *NL* 38 E6
Durward *AB* 7 D4
Dusey River *ON* 16 C8
Dusterlo, lac *QC* 39 C1
Dusty River *YT* 4 E2
Dutch Valley *NB* 32 C5
Dutility, lac *QC* 26 A6
Duti River *BC* 3 D4
Dutton *ON* 19 D4
Dutton *ON* 20 F3
Duttona Beach *ON* 19 D4
Duttona Beach *ON* 20 F3
Dutton/Dunwich *ON* 19 D4
Dutton/Dunwich *ON* 20 F3
Duval *SK* 11 D5
Duval, lac *QC* 26 D2
Duval, rivière *QC* 29 C-D8
Duval, rivière *QC* 30 D1
Duvernay *AB* 6 A7
Duvernay *AB* 11 A1
Duvernay *QC* 24 D5
Dwight *ON* 21 B2
Dyce *MB* 11 A8
Dyce *MB* 13 A2
Dyce *MB* 15 F2
Dyer, Cape *NU* 42 E6
Dyer's Bay *ON* 18 D5
Dyke Lake *NL* 39 A2
Dyment *ON* 13 F7
Dymond *ON* 17 E7-8
Dymond *ON* 22 A1
Dymond *ON* 23 F1
Dysart *SK* 11 D6

E

Eabamet Lake *ON* 16 B7
Eabametoong (Fort Hope) *ON* 16 B6
Eades *ON* 17 C7
Eager Lake *MB* 12 D7
Eager Lake *MB* 15 D1
Eagle *ON* 19 D4
Eagle *ON* 20 F3

Eagle Bay *BC* 6 D2
Eagle Creek *BC* 1 C8
Eagle Creek *BC* 1 C8
Eagle Creek *BC* 6 C1
Eagle Creek *BC* 6 C1
Eagle Creek *SK* 11 C-D3
Eagle-Dogtooth PP *ON* 13 F6
Eagle-Dogtooth PP *ON* 16 D2
Eagle Head *NS* 33 E4
Eaglehead Lake *ON* 16 E5
Eagle Island *MB* 13 A3
Eagle Island *ON* 18 B3
Eagle Lake *BC* 7 B5
Eagle Lake *BC* 1 C8
Eagle Lake *ON* 13 F6
Eagle Lake *ON* 13 F6
Eagle Lake *ON* 16 C1
Eagle Lake *ON* 16 D2-3
Eagle Lake *ON* 18 C8
Eagle Lake *ON* 21 B3
Eagle Lake *ON* 22 E1
Eagle Peak *BC* 1 A4
Eagle Plain *YT* 5 D-E3
Eagle Plains *YT* 5 E3-4
Eagle Point *ON* 18 D4
Eagle River *BC* 6 D2
Eagle River *NL* 39 C6-7
Eagle River *NL* 40 D-E8
Eagle River *ON* 13 F6
Eagle River *ON* 16 D2
Eagle River *YT* 5 E3
Eaglesham *AB* 9 E3
Eagles Lake *SK* 11 A3-4
Eakin Creek Canyon PP *BC* 6 D1-2
Eakin Creek Floodplain PP *BC* 6 D1-2
Earchman *MB* 15 F3
Eardley *QC* 21 A7
Eardley *QC* 22 F7
Eardley *QC* 24 C1
Eardley Lake *MB* 13 C4-5
Ear Falls *ON* 13 E6
Ear Falls *ON* 16 C2
Ear Lake *SK* 6 C8
Earl Grey *SK* 11 D5
Earl Island *NL* 39 B7
Earl Rowe PP *ON* 20 B6
Earls Cove *BC* 1 F6
Earls Cove *BC* 2 C3-4
Earlton *ON* 17 D7
Earlton *ON* 23 F1
Earltown *NS* 33 A6
Earltown *NS* 34 E2
Earnscliffe *PE* 34 C2-3
Earnscliffe *PE* 35 E5
East Angus *QC* 27 E5
East Apple River *NS* 32 D6
East Apple River *NS* 33 A3-4
East Archibald Lake *SK* 9 A8
East Archibald Lake *SK* 12 A-B3
East Arm *BC* 1 C8
East Arm *BC* 6 C1
East Arm *YT* 4 D6
East Bay *NL* 36 D2-3
East Bay *NS* 34 C7
East Bay *NS* 34 C7
East Bideford *PE* 35 C2-3
East Braintree *MB* 13 F5
East Braintree *MB* 14 F7
East Braintree *MB* 16 E1
East Broughton *QC* 27 C6
East Castor River *ON* 24 E2-3
East Channel *NT* 5 B4
East Chester *NS* 32 F7
East Chester *NS* 33 C5
East Chezzetcook *NS* 33 C6-7
East Clifton *QC* 27 F5
East Coulee *AB* 6 D6
East Coulee *AB* 7 B6
East Dalhousie *NS* 32 F6
East Dalhousie *NS* 33 C3-4
East Dover *NS* 32 F7
East Dover *NS* 33 C-D5
East Earltown *NS* 34 E2-3
Eastend *SK* 11 F3
East English River PP *ON* 16 C2
Easter Head *SK* 9 A7-8
Easter Head *SK* 12 A3
Eastern Blue Pond *NL* 38 D5
Eastern Island *NL* 38 D5
Eastern Meelpaeg *NL* 36 C7
Eastern Meelpaeg *NL* 38 C1
Eastern Passage *NS* 33 C6
Easterville *MB* 11 B8
Easterville *MB* 13 B2
East Farnham *QC* 24 E7-8
East Farnham *QC* 27 F3
East Ferry *NS* 33 D1
East Gate Registration Complex NHS *MB* 14 C2-3
East Gore *NS* 32 E8
East Gore *NS* 33 B6
East Gore *NS* 34 F2
East Grant Island *ON* 18 B1
East Green Harbour *NS* 33 F3
East Gwillimbury *ON* 20 B7
East Gwillimbury *ON* 21 E1

East Hants *NS* 32 E8
East Hants *NS* 33 B6
East Hants *NS* 34 F1-2
East Hereford *QC* 27 F5
East Jeddore *NS* 33 C7
East Jordan Station *NS* 33 E3
East Kemptville *NS* 33 E2
East LaHave *NS* 33 D4-5
East Lake *ON* 21 F5
East Lake Ainslie *NS* 34 C6
East Linden *NS* 35 F3
East Linton *ON* 18 F5
East Linton *ON* 20 A4
Eastmain, rivière *QC* 40 E4-5 F5
Eastman *QC* 24 E8
Eastman *QC* 27 F4
East Mapleton *NS* 32 D7
East Mapleton *NS* 33 A5
East Margaree *NS* 34 B6
East Margaretsville *NS* 32 E5-6
East Margaretsville *NS* 33 B3
East Medway *NS* 33 D4
East Mines Station *NS* 32 D8
East Mines Station *NS* 33 A6
East Mines Station *NS* 34 E2
East New Annan *NS* 34 E2
East Niska Lake *MB* 15 E6
East Oro *ON* 18 F8
East Oro *ON* 20 A7
East Paint Lake PP Reserve *MB* 15 E3-4
East Pine PP *BC* 3 E8
East Pine PP *BC* 9 E1
East Point *NS* 33 F3
East Point *PE* 34 B4
East Point *PE* 34 B4
East Point *PE* 35 D8
East Point *PE* 35 D8
Eastport *NL* 37 B6-7
East Port l'Hébert *NS* 33 E3-4
East Prairie River *AB* 9 E-F4
East Pubnico *NS* 33 F2
East Quoddy *NS* 33 B8
East Quoddy *NS* 34 F5
East Redonda Island *BC* 2 B2-3
East Redonda PP *BC* 1 E6
East Redonda PP *BC* 2 B2-3
East River *NL* 38 E3
East River *NS* 33 A7-8
East River *NS* 34 C5
East River *NS* 34 E3-4
East River PP *NS* 33 A8
East River PP *NS* 34 C5
East River St. Marys *NS* 33 A8
East River St. Marys *NS* 34 E4
East River Sheet Harbour *NS* 33 B8
East River Sheet Harbour *NS* 34 F4
East Riverside-Kingshurst *NB* 32 C5
East Riverside-Kingshurst *NB* 33 A1-2
East Rous Island *ON* 18 B3
East Selkirk *MB* 13 E4
East Selkirk *MB* 14 D5
East Sister Island *ON* 19 F1
East Sister Island PP *ON* 19 F1
East Thurlow Island *BC* 2 B2
East Trout Lake *SK* 11 A5
East Trout Lake *SK* 12 F5
East Uniacke *NS* 33 B5-6
East Uniacke *NS* 34 F1
Eastville *NS* 33 A7
Eastville *NS* 34 F1
East Walton *NS* 33 A5
East Walton *NS* 34 F1
Eastwood *ON* 19 C5-6
Eastwood *ON* 20 E4-5
East York *ON* 19 C7
East York *ON* 20 C7
East York *ON* 21 F1
Eaton, rivière *QC* 27 E5
Eatonia *SK* 6 D8
Eaton Nord, rivière *QC* 27 E5
Eatonville *NS* 32 D6
Eau, lac d' *QC* 24 E7-8
Eau Claire, lac à l' *QC* 24 A7
Eau Claire, lac à l' *QC* 25 C5-6
Eau Claire, lac à l' *QC* 25 F7
Eau Claire, lac à l' *QC* 28 D2
Eau Claire, lac à l' *QC* 40 D5
Eau-Claire Lake *NL* 39 C2-3
Eau Jaune, lac à l' *QC* 26 B4

Eau Noire, lac à l' *QC* **26** A3-4
Ebb & Flow *MB* **13** D3
Ebb & Flow *MB* **14** B2-3
Ebb and Flow Lake *MB* **13** D2-3
Ebb and Flow Lake *MB* **14** B2
Ebbutt Hills *NT* **10** D3
Ebenezer *PE* **34** C2
Ebenezer *SK* **11** D7
Eberts *ON* **19** D3
Eberts *ON* **20** F2
Ebor *MB* **11** E7-8
Ebor *MB* **13** E1
Écarté, lac *QC* **25** B8
Écarté, lac *QC* **26** F8
Écarté, lac *QC* **28** D4
Écho, lac *QC* **22** E8
Écho, lac *QC* **23** E2
Écho, lac *QC* **24** C2
Echo Bay *BC* **1** E5
Echo Bay *NT* **41** C1
Echo Bay *ON* **17** F4
Echo Bay *SK* **11** B4
Echo Bay PMP *BC* **1** D-E5
Echoing Lake *ON* **13** A7
Echoing Lake *ON* **15** F7
Echoing River *MB* **15** E-F7
Echo Lake *NS* **33** C6
Echo Lake *ON* **17** F4
Echo Lake *ON* **21** B2
Echo Lake *SK* **11** C6
Echo Lake PP *BC* **6** E2
Échouani, lac *QC* **22** A8
Échouani, lac *QC* **23** F8
Échouani, lac *QC* **25** B2
Echo Valley PP *SK* **11** D5-6
Eckville *AB* **6** C5
Eckville *AB* **8** E-F5
Eclipse Sound *NU* **41** A6
Eclipse Sound *NU* **42** D-E5
Écorce, lac de l' *QC* **22** B7
Écorce, lac de l' *QC* **25** D1
Écorces, lac aux *QC* **22** B4
Écorces, lac aux *QC* **28** D5
Écorces, lac des *QC* **22** A8
Écorces, rivière aux *QC* **28** C-D5
Ecstall River *BC* **1** A3
Ecum Secum *NS* **33** B8
Ecum Secum *NS* **34** F4
Eda Lake *NT* **10** A7
Edam *SK* **6** B8
Edam *SK* **11** D4
Edberg *AB* **6** B6
Edberg *AB* **8** E7
Ed Bird-Estella Lakes PP *BC* **3** D5-6
Eddies Cove *NL* **38** A4
Eddies Cove *NL* **39** D8
Eddies Cove West *NL* **38** C3
Eddies Cove West *NL* **39** E7-8
Eddontenajon Lake *BC* **3** C3
Eddystone *MB* **13** D2-3
Eddystone *MB* **14** B2
Edehon Lake *NU* **40** B1
Edehon Lake *NU* **41** F4
Eden *MB* **13** E2
Eden *MB* **14** C-D2
Eden *ON* **19** D5
Eden *ON* **20** F4
Eden Lake *MB* **12** D8
Eden Lake *MB* **15** D2
Eden Lake *NS* **33** A8
Eden Lake *NS* **34** F4
Eden Mills *ON* **19** A-B7
Eden Mills *ON* **20** D5-6
Edenvale *ON* **18** F7
Edenvale *ON* **20** A6
Edenwold *SK* **11** E5-6
Edgar *ON* **18** F8
Edgar *ON* **20** A6-7
Edgar *ON* **21** D1
Edge Hills PP *BC* **1** D7-8
Edge Hills PP *BC* **2** A6
Edgeley *SK* **11** E6
Edgell Island *NU* **40** F6
Edgerton *AB* **6** B7
Edgerton *AB* **11** B2
Edgewater *BC* **6** E4
Edgewater *BC* **7** C1
Edgewood *BC* **6** E2-3 F2-3
Edillen *MB* **14** A2
Edith Cavell, Mount *BC*
Edith Cavell, Mount *AB* **8** E1
Edmonton *AB* **6** B5-6
Edmonton *AB* **8** C6
Edmonton Beach *AB* **6** B5
Edmonton Beach *AB* **8** C5-6
Edmund Lake *MB* **15** F6
Edmundston *NB* **29** F3
Edmundston *NB* **31** C1-2
Edmund Walker Island *NU* **42** C5
Édouard, lac *QC* **25** C8
Édouard, lac *QC* **28** E4
Edrans *MB* **13** E2
Edrans *MB* **14** D2
Edson *AB* **6** B4
Edson *AB* **8** C3
Edward *AB* **8** B8
Edward Island *ON* **16** F6
Edward Island, RNP *ON* **16** F6
Edward Island PNR *ON* **16** F6
Edwards *ON* **21** B8
Edwards *ON* **24** C2-3
Edwards, Mount *YT* **4** B4
Edwin *MB* **14** D-E3

Edys Mills *ON* **19** D3
Edys Mills *ON* **20** F2
Edziza, Mount *BC* **3** C2-3
Edzo *NT* **10** C5-6
Edzo *NT* **41** D1
Eel Ground *NB* **31** D6
Eel River *NB* **29** D6
Eel River Bar *NB* **29** D6-7
Eel River Bar *NB* **31** A5
Eel River Bridge *NB* **31** D7
Eel River Crossing *NB* **29** D6-7
Eel River Crossing *NB* **31** A5
Eels Creek *ON* **21** C3-4
Eels Lake *ON* **21** C3
Effingham *ON* **19** C8
Effingham *ON* **20** E7
Effingham Inlet *BC* **2** E2
Effingham Lake *ON* **21** C5
Egan Chutes PP *ON* **21** B4
Egan Creek *ON* **21** C4
Eganville *ON* **21** A5
Eganville *ON* **22** F5
Egenolf Lake *MB* **12** A8
Egenolf Lake *MB* **15** B2
Egerton *NS* **34** E3-4
Egg Lake *MB* **12** F7
Egg Lake *MB* **13** A1
Egg Lake *MB* **15** F1
Egg Lake *SK* **11** A7
Egg Lake *SK* **12** F5
Egg Lake *SK* **13** A1
Eglington *PE* **34** C4
Eglington *PE* **35** E7
Eglington Bay *PE* **35** E7
Eglinton Island *NT* **42** C2
Egmondville *ON* **19** A-B4
Egmondville *ON* **20** D3
Egmont *BC* **1** F6
Egmont *BC* **2** C4
Egmont, cap *PE* **32** A7
Egmont Bay *PE* **32** A7
Egmont Bay *PE* **35** C1
Egremont *AB* **6** A6
Egremont *AB* **8** B7
Egypt Road *NS* **34** B6
Ehdacho HS *NT* **41** C1
Eholt *BC* **6** F2
Eighteen Mile Island *ON* **18** B6
Eight Island Lake *NS* **34** E5
Ekati Mine *NT* **10** A8
Ekati Mine *NT* **41** D2
Ekka Island *NT* **41** C1
Ekwan River *ON* **40** E2-3
Ekwi River *NT* **4** A-B7
Elaho River *BC* **1** E6-7
Elaho River *BC* **2** B-C4
Elak Dase *SK* **12** E4
Elbow *SK* **11** D4
Elbow Creek *ON* **18** A6
Elbow Lake *MB* **12** F7
Elbow Lake *MB* **15** F2
Elbow Lake *SK* **11** C7
Elbow River *AB* **6** D-E5
Elbow River *AB* **7** C3
Elbow-Sheep Wildland PP *AB* **6** D-E5
Elbow-Sheep Wildland PP *AB* **7** C3
Elcan *AB* **7** E6
Eldee *ON* **17** F8
Eldee *ON* **22** C2
Elderbank *NS* **33** B6-7
Elderbank *NS* **34** F2-3
Elderbank PP *NS* **33** B6-7
Elderbank PP *NS* **34** F2-3
Eldersley *SK* **11** B6
Eldon *PE* **34** C3
Eldon *PE* **35** E7
Eldorado *ON* **21** D4-5
Eldorado *SK* **9** A8
Eldorado *SK* **12** A3-4
Eldorena *AB* **8** B7
Eldred *SK* **11** B4
Eleanor Island National Wilderness Area *ON* **18** E8
Eleanor Island National Wilderness Area *ON* **21** C1
Elephant Hill PP *BC* **2** A7
Elephant Lake *ON* **21** B3
Elfrida *ON* **19** C7
Elfrida *ON* **20** E6
Elfros *SK* **11** C6
Elgin *MB* **11** F8
Elgin *MB* **13** F2
Elgin *MB* **14** E1
Elgin *NB* **32** C5
Elgin *NS* **33** A7
Elgin *NS* **34** E3
Elgin *ON* **21** D7
Elgin *ON* **24** F1
Elgin, lac *QC* **27** D5-6
Elginburg *ON* **21** D6
Elginfield *ON* **19** C4
Elginfield *ON* **20** E3
Elgin Mills *ON* **19** A8
Elgin Mills *ON* **20** C7
Elie *MB* **14** D-E3
Elimville *ON* **19** B4
Elimville *ON* **20** D3
Elizabeth, Mount *NB* **29** F6
Elizabeth Bay *ON* **18** B2
Elizabeth Bay *ON* **18** C2
Elizabeth Bay *ON* **21** E3
Elk Falls PP *BC* **1** E5
Elk Falls PP *BC* **2** C1-2

Elkford *BC* **6** E5
Elkford *BC* **7** D3
Elkhorn *MB* **11** E7-8
Elkhorn *MB* **13** E1
Elk Island *MB* **14** C6
Elk Island *MB* **15** F6
Elk Island, PN *AB* **6** A-B6
Elk Island NP *AB* **8** C7
Elk Island NP *AB* **6** A-B6
Elk Island NP *AB* **8** C7
Elk Island Provincial Heritage Park *MB* **13** D-E4
Elk Island Provincial Heritage Park *MB* **14** C6
Elk Lake *ON* **17** D7
Elk Lakes PP *BC* **6** E5
Elk Lakes PP *BC* **7** C2
Elko *BC* **6** F5
Elko *BC* **7** F2
Elk Point *AB* **6** A7
Elk Point *AB* **11** A2
Elk Ranch *MB* **14** C2
Elk River *BC* **6** E5
Elk River *BC* **7** D-E3
Elkton *AB* **6** D5
Elkton *AB* **7** A3
Elk Valley *AB* **7** B3
Elk Valley PP *BC* **6** E5
Elk Valley PP *BC* **7** E2-3
Elkwater *AB* **6** E-F8
Elkwater *AB* **11** F2
Ella Lake *ON* **18** A4
Ellard Lake *ON* **15** F7
Ellef Ringnes Island *NU* **42** B-C3
Elleh Creek *BC* **3** B7
Elleh Creek *BC* **9** B1
Ellen, lac *QC* **22** E6
Ellen, lac *QC* **24** C1
Ellenwood PP *NS* **33** E1-2
Ellershouse *NS* **33** B5
Ellerslie *AB* **8** C6
Ellerslie *PE* **32** A7
Ellerslie *PE* **34** B1
Ellerslie *PE* **35** C2
Ellesmere, île d' *NU* **42** A-C5
Ellesmere Island *NU* **42** A-C5
Ellesmere Island (Quttinirpaaq) NP *NU* **42** A-C5
Ellice Island *NT* **5** B4
Ellice River *NU* **41** C3
Ellington Lake *NT* **10** A5
Elliot Lake *MB* **13** B5
Elliot Lake *ON* **17** F5
Elliot Lake *ON* **18** A2
Elliot Lake *ON* **18** A2-3
Elliotvale *PE* **35** E6
Ellison PP *ON* **6** E1-2
Elliston *NL* **37** B7-8
Ellscott *AB* **6** A6
Ellscott *AB* **8** A7
Ellscott *AB* **9** F6
Ellscott *AB* **12** F1
Ells River *AB* **9** C-D6
Ells River *AB* **12** C1
Ellwood, lac *QC* **25** B5
Ellwood, lac *QC* **26** F5
Elma *MB* **13** E5
Elma *MB* **14** D7
Elm Creek *MB* **13** F3
Elm Creek *MB* **14** E4
Elmira *ON* **19** A6
Elmira *ON* **20** D5
Elmira *PE* **34** B4
Elmira *PE* **35** D8
Elmira Railway Station Provincial Heritage Site *PE* **35** D8
Elmsdale *NS* **33** B6
Elmsdale *NS* **34** F2
Elmsdale *PE* **35** B2
Elmstead *ON* **19** E1
Elmsvale *NS* **33** B7
Elmsvale *NS* **34** F3
Elmsville *NB* **32** D2
Elmvale *ON* **18** F7
Elmvale *ON* **20** A6
Elmwood *NB* **31** F4
Elmwood *NB* **32** B1
Elmwood *ON* **20** B4
Elmwood *ON* **21** E6
Elmwood *PE* **35** E4
Elmworth *AB* **9** F2
Elnora *AB* **6** C6
Elnora *AB* **8** F7
Elora *ON* **19** A6
Elora *ON* **20** C5
Elora Gorge CA *ON* **19** A6
Elora Gorge CA *ON* **20** C5
Elphin *ON* **21** C6
Elphinstone *MB* **11** E8
Elphinstone *MB* **13** E2
Elphinstone *MB* **14** C1
Elrose *SK* **11** D3
Elsa *YT* **4** A-B4
Elsas *ON* **17** C5
Elsie Lake *BC* **2** D2
Elson, lac *QC* **39** A3
Elspeth *AB* **8** A7
Elstow *SK* **11** C4-5
Eltrut Lake *ON* **16** E3
Elva *MB* **11** F8
Elva *MB* **13** F1
Emaril *NL* **39** C2
Emar Lakes PP *BC* **1** D8
Emar Lakes PP *BC* **6** D1
Embar *NL* **39** C3
Embarras, lac *QC* **22** C6-7

Embarras, lac *QC* **25** E1
Embleton Lake *MB* **15** C-D5
Embree *NL* **36** A7
Embree *NL* **37** A5
Embree *NL* **38** F6
Embro *ON* **19** C5
Embro *ON* **20** E4
Embrun *ON* **24** E3
Embury *MB* **14** C3
Embury Lake *MB* **12** F7
Embury Lake *MB* **15** F1
Emerald *PE* **35** D3
Emerald Estates *QC* **24** C5
Emerald Isle *NT* **42** C2
Emerald Lake *ON* **18** A2
Emerson *MB* **13** F4
Emerson *MB* **14** E4
Emeryville *ON* **19** E1-2
Emile River *NT* **10** B6
Emily PP *ON* **21** D3
Emma Lake *SK* **11** B5
Emma Lake PRS *SK* **11** A4
Emmeline Lake *SK* **12** F4
Emmerson Lake *ON* **15** F7
Emmuraillé, lac *QC* **28** A-B7
Emo *ON* **16** F2
Emory Creek PP *BC* **1** F8
Emory Creek PP *BC* **2** D7
Empire Corners *ON* **19** C7
Empire Corners *ON* **20** E6
Empress *AB* **6** D8
Empress *AB* **11** D2
Emsdale *ON* **18** C8
Emsdale *ON* **21** A1
Emsdale *ON* **22** F1
Ena Lake *ON* **13** E5
Ena Lake *ON* **14** D8
Ena Lake *ON* **14** D8
Ena Lake *SK* **9** A8
Ena Lake *SK* **12** A4
Enatik Lake *MB* **12** C-D8
Enatik Lake *MB* **15** D2
Enchant *AB* **6** E6
Enchant *AB* **7** D6
Endako *BC* **1** A5-6
Endeavour *SK* **11** C6-7
Enderby *BC* **6** E1
Endiang *AB* **6** C6-7
Endiang *AB* **8** F8
Endiang *AB* **11** C1
Eneas Lakes PP *BC* **6** E1
Enemy, Lake of the *NT* **10** B8
Enfield *NS* **33** B6
Enfield *NS* **34** F2
Engemann Lake *SK* **12** C4
Engen *BC* **1** A6
Englee *NL* **38** C4-5
Englee *NL* **39** E8
Englefeld *SK* **11** B5
Englefield, Cape *NU* **41** B6
Englefield, Cape *NU* **42** F5
Englefield Bay *BC* **1** B1
Englehart *ON* **17** D7-8
Englehart *ON* **23** E1
Englehart R. Fine Sand Plain & Waterway PP *ON* **17** D7
Engler Lake *SK* **12** B4
English Chief River *NT* **10** C2
English Harbour *NL* **37** C7
English Harbour East *NL* **36** D7
English Harbour East *NL* **37** D5
English Harbour West *NL* **36** E6-7
English Harbour West *NL* **37** E4-5
Englishman River Falls PP *BC* **1** F6
Englishman River Falls PP *BC* **2** D-E3
Englishmans Range *YT* **4** E5
English River *ON* **13** E5
English River *ON* **13** E6
English River *ON* **13** E-F7
English River *ON* **14** C-D8
English River *ON* **16** C-D1 D2
English River *ON* **16** D-E4
English River *ON* **16** E4
Englishtown *NS* **34** C7
Enilda *AB* **9** E4
Enmore *PE* **34** A7
Enmore *PE* **34** B1
Enmore *PE* **35** C2
Ennadai Lake *NU* **41** E3-4
Ennett Turnerville *NL* **19** D3
Ennett Turnerville *NL* **20** F1-2
Ennishone *NB* **31** D3
Enniskillen *NB* **32** D3
Enniskillen *ON* **19** B7
Enniskillen *ON* **21** E3
Enniskillen CA *ON* **20** C8
Enniskillen CA *ON* **21** E2
Ennismore *ON* **21** C3
Ennotville *ON* **19** A6
Ennotville *ON* **20** C5
Enon *NS* **34** D7
Enrage, Cape *NB* **32** C-D6
Enrage, Cape *NB* **33** A3-4
Ensign *AB* **6** E6

Ensign *AB* **7** C5
Enterprise *NT* **10** E-F6
Enterprise *NT* **41** E1
Enterprise *ON* **21** D6
Entiako PP *BC* **1** B5
Entiako River *BC* **1** B5
Entrance *AB* **6** B3
Entrance *AB* **8** C1
Entrelacs *QC* **24** B5
Entwistle *AB* **6** B5
Entwistle *AB* **8** C4
Environ *SK* **11** C4
Épaule, lac à l' *QC* **28** F5
Epper Passage PP *BC* **1** F5
Epper Passage PP *BC* **2** D1
Epsom *ON* **20** B8
Epsom *ON* **21** E2
Epsom PP *ON* **1** E8
Epsom PP *ON* **2** B7
Epworth *NL* **36** F6-7
Epworth *NL* **37** F4-5
Equity *AB* **7** A5
Erables Lake *ON* **22** E2
Eramosa *ON* **19** A6
Eramosa *ON* **20** D5
Eramosa River *ON* **19** A7
Eramosa River *ON* **20** C-D6
Erg Mountain PP *NS* **38** A6
Erg Mountain PP *BC* **6** A-B1
Eric, lac *QC* **39** D2
Éric, lac *QC* **39** D2-3
Erickson *BC* **6** F4
Erickson *BC* **7** F1
Erickson *MB* **13** E2
Erickson *MB* **14** C1
Erie *BC* **6** F3
Erie *ON* **19** C7
Erie *ON* **20** E6
Érié, lac *QC* **19** E4-6
Érié, lac *ON* **20** F6-7
Erie, Lake *ON* **19** E5-6
Erie, Lake *ON* **20** F3-4
Erieau *ON* **19** E3
Erie Beach *ON* **19** E3
Erie Creek PP *BC* **6** F3
Eriksdale *MB* **13** D3
Eriksdale *MB* **14** B4
Erin *ON* **19** A7
Erin *ON* **20** C6
Erinferry *SK* **11** A4
Erinsville *ON* **21** D5
Erinview *MB* **13** E3
Erinview *MB* **14** D4-5
Erinville *NS* **34** E5
Erith *AB* **8** C2
Ermine *SK* **11** C3
Ernest *AB* **8** C4
Ernestown *ON* **21** E6
Ernfold *SK* **11** D3
Errington *BC* **2** D3
Errol *ON* **19** C3
Errol *ON* **20** E2
Erskine *AB* **6** C6
Erskine *AB* **8** E-F7
Ervick *AB* **8** D7
Erwood *SK* **11** B7
Erwin, lac *QC* **22** F8
Escalier, réservoir l' *QC* **22** F8
Escalier, réservoir l' *QC* **24** C2
Esclaves, rivière des *AB* **9** A-B6
Esclaves, rivière des *AB* **12** A-B2
Escott *ON* **21** D7-8
Escoumins, rivière des *QC* **28** C8
Escoumins, rivière des *QC* **29** C8
Escuminac *NB* **31** D8
Escuminac *QC* **29** D6
Escuminac *QC* **31** A7
Escuminac, Point *NB* **31** D8
Escuminac, rivière *QC* **29** C-D6
Esden, lac *QC* **22** A7
Esden, lac *QC* **23** F7
Esden, lac *QC* **25** C1-2
Esk *SK* **11** C5
Eskasoni *NS* **34** C7
Eskbank *SK* **11** E4
Esker Lakes PP *ON* **17** D7
Esker Lakes PP *ON* **23** D1
Esker Siding *NL* **39** B2
Eskers PP *SK* **11** A4
Eskimo Lakes *NT* **5** A-B5
Eskimo Point *NS* **15** B5
Esli Dodge CA *ON* **19** C3
Esli Dodge CA *ON* **20** E2
Esnagami Lake *ON* **16** C-D7
Esnagami Lake *ON* **17** A2
Esnagami River *ON* **16** C7
Esnagami River *ON* **17** A2
Esnagi Lake *ON* **17** C4
Espanola *ON* **17** F6
Esperanza *BC* **1** E4
Esperanza Inlet *BC* **1** E-F4
Espinosa Inlet *BC* **1** E4
Espoir, Bay d' *NL* **36** D6
Espoir, Bay d' *NL* **37** D3-4
Espoir, cap d' *QC* **30** C3
Esprit-Saint *QC* **29** D3
Esprit-Saint *QC* **31** A1
Essex *ON* **19** E1
Essipit *QC* **29** C1
Essonville *ON* **21** B3
Esox Lake *ON* **13** F6
Espanola *ON* **17** F6
Est, île de l' *QC* **30** E8
Est, lac de l' *QC* **24** E7
Est, lac de l' *QC* **28** F1

Est, pointe de l' *QC* **30** A7
Estabrook Lake *NT* **5** C8
Estaire *ON* **17** F7
Estaire *ON* **18** A6
Estcourt *QC* **29** E1-2
Esten Lake *ON* **18** A2
Estérel *QC* **24** C5
Estérel *QC* **27** D1
Esterhazy *SK* **11** E7
Esterhazy *SK* **13** D1
Esther *AB* **6** C8
Esther *AB* **11** C2
Esther, lac *QC* **23** B6
Esther, lac *QC* **26** B-C1
Estlin *SK* **11** E5-6
Estmere *NS* **34** D6
Estevan *SK* **11** F7
Estevan Group *BC* **1** B3
Estevan Sound *BC* **1** B3
Estuary *SK* **11** D8
Estuary *SK* **11** D8
Établissement Melanson, LHN de l' *NS* **32** F3-5
Établissement Melanson, LHN de l' *NS* **33** C1-2
Établissement-Ryan, LHN de l' *NL* **37** B7
Étagaulet Bay *NL* **39** B6
Étagaulet Point *NL* **39** B6
Etamamiou *QC* **39** E6
Étamamiou, rivière *QC* **39** E6
Étang-des-Caps *QC* **30** F7
Étang Morice *NB* **35** E1-2
Étang-Sand, RNF de l' *NS* **33** E2
Etawney Lake *MB* **15** C4
Etchemin, lac *QC* **27** C7
Etchemin, rivière *QC* **27** B6-7
Éternité, lac *QC* **28** C7
Ethel *ON* **19** A5
Ethel *ON* **20** C4
Ethelbert *MB* **11** D8
Ethelbert *MB* **13** D1-2
Ethelbert *MB* **14** A1
Ethel F. Wilson Memorial PP *BC* **1** A5
Ethel Lake *YT* **4** B3
Ethelton *SK* **11** B5
Etobicoke *ON* **19** A8
Etobicoke *ON* **20** C7
Etobicoke *ON* **21** F1
Etomami River *SK* **11** C7
Eton-Rugby *ON* **13** E6
Eton-Rugby *ON* **16** E3
Etter's Beach *SK* **11** D5
Etthen Island *NT* **10** D7-8
Ettington *SK* **11** E4-5
Etzikom *AB* **6** F7
Etzikom *AB* **7** E8
Etzikom *AB* **11** F1
Etzikom Coulee *AB* **6** F7
Etzikom Coulee *AB* **7** E-F7
Euchiniko River *BC* **1** B6
Euchu Reach *BC* **1** A5
Eudistes, lac des *QC* **39** E3
Eugenia *ON* **20** B5
Eugenia Falls CA *ON* **20** B5
Eugenia Lake *ON* **20** B5
Eulas Lake *SK* **12** E5
Eureka *NS* **33** A7
Eureka *NS* **34** E3
Eureka *NU* **42** B4
Eureka River *AB* **9** D2-3
Eureka River *AB* **9** D2-3
Eureka Sound *NU* **42** B4
Eutsuk Lake *BC* **1** B5
Éva, lac *QC* **26** A5-6
Évain *QC* **17** D8
Évain *QC* **23** D2
Eva Lake *AB* **9** B5
Evandale *NB* **32** D4
Evandale *NB* **33** A1
Evangeline *NB* **31** B8
Evangeline *NB* **31** F8
Evangeline *NB* **32** B6
Evans, Cape *NU* **42** A4
Evansburg *AB* **6** B4-5
Evansburg *AB* **8** C4
Evans Strait *NU* **40** A3
Evans Strait *NU* **41** D6-7
Evanston *NS* **34** D6
Evansville *ON* **18** A1
Evansville *ON* **18** C2
Evarts *AB* **8** F5
Evesham *SK* **6** C8
Evesham *SK* **11** C2
Excel *AB* **7** A8
Exchamsiks River PP *BC* **1** A3
Exeter *ON* **19** B4
Exeter *ON* **20** D3
Exeter Lake *NT* **10** A7-8
Exmoor Lake *NT* **10** A5-6
Expanse *SK* **11** E4
Exploits, Bay of *NL* **38** E6
Exploits Islands *NL* **38** E6
Exploits River *NL* **37** B3-4
Exshaw *AB* **6** D4-5
Exshaw *AB* **7** B2
Eyapamikama Lake *ON*
Eyebrow *SK* **11** D4-5

Eyehill Creek *AB* **6** B-C8
Eyehill Creek *AB* **11** B-C2
Eyrie Lake *MB* **12** C7
Eyrie Lake *MB* **15** C2

F

Faber Lake *NT* **10** B5
Faber Lake *NT* **41** D1
Fabre *QC* **17** E8
Fabre *QC* **22** B1-2
Fabyan *AB* **6** B7
Fabyan *AB* **11** B1
Factorydale *NS* **32** E6
Factorydale *NS* **33** B4
Faden *NL* **39** B2
Fagan, pointe *NB* **35** E1-2
Fagan Point *NB* **35** E1-2
Faguy, lac *QC* **26** E5
Faillon, lac *QC* **23** D6-7
Faillon, lac *QC* **25** A1
Faillon, lac *QC* **26** E1
Fairbairn Lake *NT* **10** D8
Fairbank Lake *ON* **18** A4-5
Fairbank PP *ON* **17** F6
Fairbank PP *ON* **18** A4
Fairchild Creek *ON* **19** B6 C6-7
Fairchild Creek *ON* **20** E5-6
Fairfax *MB* **11** F8
Fairfax *MB* **13** F2
Fairfax *MB* **14** E1
Fairfield *NB* **32** D4-5
Fairfield *NB* **33** A1
Fairfield *NB* **35** F1
Fairfield *PE* **35** D8
Fairford *MB* **13** D3
Fairford *MB* **14** A3
Fairground *ON* **19** D5-6
Fairground *ON* **20** F4-5
Fair Harbour *BC* **1** E4
Fairhaven *NB* **32** E2
Fair Haven *NL* **36** D8
Fair Haven *NL* **37** D6
Fairholme *ON* **18** C7
Fairholme *SK* **11** B3
Fairisle *NB* **29** F8
Fairisle *NB* **30** F1
Fairisle *NB* **31** C7
Fairlight *SK* **11** E7
Fairlight *SK* **13** E1
Fairmont Hot Springs *BC* **6** E4
Fairmont Hot Springs *BC* **7** D1
Fair Ness *NU* **40** A5
Fair Ness *NU* **41** B6
Fairvale *NB* **32** D4
Fairvale *NB* **33** A1
Fairview *AB* **9** E3
Fairweather Lake *YT* **4** B5
Fairweather Mountain *YT* **4** F2
Fairy Glen *SK* **11** B5
Fairy Lake *ON* **21** B1-2
Falaise, lac *QC* **39** A1
Falaise Lake *NT* **10** E6
Falconbridge *ON* **17** F7
Falconbridge *ON* **18** A6
Falcon Island *ON* **16** F2
Falcon Island *ON* **14** E-F8
Falcon Lake *MB* **13** F5
Falcon Lake *MB* **14** E7
Falcon Lake *MB* **14** E7
Falcon Lake *ON* **16** D1
Falher *AB* **9** E3
Falkland *BC* **6** E2
Falkland *ON* **19** C2
Falkland *ON* **20** E5
Falkland Ridge *NS* **32** C7
Falkland Ridge *NS* **33** C4
Falkland Ridge *NS* **33** C4
Fallaize Lake *NT* **5** B8
Fallbrook *ON* **21** C6-7
Fallbrook *ON* **24** E1
Fall Caribou Crossing HS *NU* **41** D4
Fallis *AB* **6** B5
Fallis *AB* **8** C5
Fallison *MB* **14** F3
Falloon Lake *SK* **15** A4
Fall River *NS* **32** E8
Fall River *NS* **33** C6
Falls Brook *NB* **29** E4
Falls Creek *BC* **2** C7
Falls Lake PP *NS* **32** E8
Falls Reserve CA *ON* **19** A4
Falls Reserve CA *ON* **20** C3
Falmouth *NS* **32** E7
Falmouth *NS* **33** B5
False Bay *BC* **1** F6
False Bay *BC* **2** D3
False Cape *NL* **39** A7
False Detour Channel *ON* **18** D3-4
Falun *AB* **6** B5
Falun *AB* **8** D6
Famham, Mount *BC* **6** E4
Family Lake *MB* **13** C5
Family Lake *MB* **16** A1
Famine, rivière *QC* **27** C7
Fancy Lake PP *NS* **33** D4
Fanning Brook *PE* **35** D-E6
Fanny Bay *BC* **1** F5-6
Fanny Bay *BC* **2** D2
Fannystelle *MB* **13** F3
Fannystelle *MB* **14** E4

Fanshawe CA *ON* **19** C5
Fanshawe CA *ON* **20** E4
Fanshawe Lake *ON* **19** C4-5
Fanshawe Lake *ON* **20** E3-4
Fanson Lake *SK* **12** A6
Fanson Lake *SK* **15** A1
Faraud Lake *SK* **12** A5
Farbus, lac *QC* **22** A7
Farbus, lac *QC* **23** F7
Farbus, lac *QC* **25** B2
Farewell *NL* **38** E7
Farewell *ON* **19** A5-6
Farewell *ON* **20** C4-5
Faribault, lac *QC* **40** C5
Faride, lac *QC* **39** E6
Farley *QC* **22** D-E7
Farley *QC* **24** B2
Farley *QC* **25** F2
Farmer Island *NU* **40** C3
Farmer Island *NU* **41** F7
Farmington *BC* **3** E8
Farmington *BC* **9** E2
Farmington *NS* **32** F6
Farmington *NS* **32** D8
Farmington *NS* **33** C4
Farmington *NS* **33** A5
Farmington *NS* **34** E1
Farmington *PE* **35** D7
Far Mountain *BC* **1** B5-6
Farnham *QC* **24** E7
Farnham *QC* **27** F2-3
Farnham-Centre *QC* **24** E7
Farnham-Centre *QC* **27** F2-3
Faro *YT* **4** C4-5
Farrant Island *BC* **1** B3
Farrell Lake *AB* **6** C6
Farrellton *QC* **22** F7-8
Farrellton *QC* **24** C2
Farrow *AB* **7** C5
Farwell Lake *MB* **12** F8
Farwell Lake *MB* **15** F2
Fassett *QC* **24** D3
Fassifern *NS* **34** D4
Father, lac *QC* **26** C3
Fathom Five, PMN *ON* **18** D3-4
Fathom Five NMP *ON* **18** D3-4
Fatima *QC* **30** F7
Faucon, lac du *QC* **22** F8
Faucon, lac du *QC* **24** C3
Faulkner *MB* **13** D3
Faulkner *MB* **14** A3
Fauquier *BC* **6** E3
Fauquier *ON* **17** C7
Faust *AB* **9** E-F4
Favourable Lake *ON* **13** B6
Fawcett *AB* **6** A5
Fawcett *AB* **8** A5-6
Fawcett Lake *AB* **9** E3
Fawcett Lake *AB* **9** E-F5
Fawcett Lake *ON* **13** D7
Fawcett Lake *ON* **16** B4
Fawcett Lake *ON* **21** E5
Fawcettville *ON* **21** E5
Fawnie Range *BC* **1** B5-6
Fawn Lake *NT* **10** D5
Fawn River *NT* **40** D1-2
Fawn River PP *ON* **13** A8
Fay Lake *MB* **12** F7-8
Fay Lake *MB* **15** F2
Featherstone Point *ON* **19** F6
Featherstone Point *ON* **20** F6
Fecteau, île *QC* **39** E6-7
Felix, Cape *NU* **41** B4
Felix, Cape *NU* **42** F3
Felix Cove *NL* **36** C2
Felton, rivière *QC* **27** D-E6
Feltz South *NS* **33** D5
Fenella *ON* **21** E3-4
Fenelon Falls *ON* **20** A8
Fenelon Falls *ON* **21** D2
Fenn *AB* **6** C6
Fenn *AB* **8** F7
Fennell *ON* **20** B7
Fennell *ON* **21** E1
Fenton *SK* **11** B5
Fenton Lake *NT* **10** C7
Fenwick *NS* **32** C7
Fenwick *ON* **19** C8
Fenwick *ON* **20** E7
Fenwood *SK* **11** D6-7
Fergus *ON* **19** A6
Fergus *ON* **20** C5
Ferguson Creek *MB* **15** F3
Fergusonvale *ON* **18** F7
Fergusonvale *ON* **20** A6
Ferintosh *AB* **6** B6
Ferintosh *AB* **8** E6
Ferland *ON* **16** D6
Ferland *ON* **17** A1
Ferland *QC* **28** E7
Ferland *SK* **11** F4
Ferme MacDonald, lieu historique provincial *NB* **31** D7
Ferme-Neuve *QC* **22** C8
Ferme-Neuve *QC* **24** A2-3
Ferme-Neuve *QC* **25** E2-3
Fermeuse *NL* **37** F8
Fermont *QC* **39** C2
Fermont *QC* **40** E6
Fern Creek *AB* **8** D5
Ferndale *MB* **14** E4
Ferndale *ON* **18** E5
Fernie *BC* **6** F5
Fernie *BC* **7** E2-3

Fernleigh ON 21 C5-6
Fernwood PE 32 B8
Fernwood PE 34 C1
Fernwood PE 35 D2-3
Ferré, lac QC 29 C-D3
Ferrer Point BC 1 F4
Ferrier AB 8 E4
Ferris PP ON 21 D-E4
Ferryland NL 37 E8
Ferry Road NB 31 D7
Fertile ON 11 F7-8
Fertile SK 13 F1
Fesserton ON 18 F8
Fesserton ON 20 A6-7
Fesserton ON 21 C1
Festubert, lac QC 23 F7-8
Festubert, lac QC 25 B2
Festubert, lac QC 26 F2
Festubert, rivière QC 23 F7-8
Festubert, rivière QC 25 B2
Festubert, ZEC QC 23 F8
Festubert, ZEC QC 25 B2
Feu, lac du QC 39 E6
Feudal SK 11 C3
Feuilles, lac aux QC 40 C5
Feuilles, rivière aux QC 40 C5
Feuquières, lac QC 26 C5
Feversham ON 20 B5
Feversham Gorge CA ON 20 B5
F.H. Barber PP BC 2 D7
Fichot Islands NL 38 B5
Fidler AB 8 D2-3
Fidler Bay SK 12 B-C6
Fidler Bay SK 15 B1
Fidler-Greywillow Wildland PP AB 9 A-B7
Fidler-Greywillow Wildland PP AB 12 A-B2
Fidler Lake MB 15 D4
Fidler Point AB 9 A-B7
Fidler Point AB 12 A-B2
Fiedmont, lac QC 23 D5
Field BC 6 D4
Field BC 7 A-B1
Field ON 17 F7
Field ON 18 A7
Fielding NB 31 E3
Fielding NB 32 A1
Fielding SK 11 C3-4
Fife Island SK 12 B6
Fife Island SK 15 B1
Fife Lake SK 11 F4
Fife Lake SK 14 F8
Fifteenmile River YT 4 A2
Fifteenmile River YT 5 F2
Fifth Lake NB 32 C1
Fifty Point CA ON 19 B8
Fifty Point CA ON 20 E7
Fighting Ile ON 19 E1
Fiji Island NT 5 A8
Fildegrand, rivière QC 22 D4
File Lake MB 12 F8
File Lake MB 15 F2
Filkars, rivière QC 27 C5-6
Fillmore SK 11 E6
Fillongley PP BC 1 F5-6
Fillongley PP BC 2 D2-3
Filndt River ON 16 C5
Fil River AB 12 E-F8
Fil River MB 15 F2
Fils, lac du QC 22 C3-4
Finch ON 24 E3
Findlater SK 11 E5
Findlayson River YT 4 D6
Fingal ON 19 D4
Fingal ON 20 F3
Finger MB 11 A8
Finger MB 13 A2
Finger Lake BC 1 B6
Finger Lake ON 13 B6
Finger-Tatuk PP BC 1 B6
Finlay Ranges BC 3 D5
Finlay Ranges BC 3 D-E6
Finlay Reach BC 3 C5
Finlay River BC 3 C-D5
Finlay-Russel PP BC 3 C5
Finlayson NS 34 C6
Finlayson Channel BC 1 C3
Finlayson Lake YT 4 D6
Finlayson Lake Territorial Historic Site YT 4 D6
Finlayson Point PP ON 17 E7
Finlayson Point PP ON 22 B1
Finmark ON 16 F5
Finn Creek PP BC 6 C2
Finnegan AB 6 D6-7
Finnegan AB 7 C7
Finnie SK 11 D6
Finns MB 14 B-C5
Fintry PP BC 6 E1-2
Fiord, Bay NU 42 B4-5
Fiordland PRA BC 1 B4
Firdale MB 13 E2-3
Firdale MB 14 D2
Firebag River AB 9 C7
Firebag River AB 12 C2
Firefight Lake SK 6 E8
Firefight Lake SK 11 E2-3
Fire River ON 17 C4
Fireside BC 3 A4
Fireside BC 4 F7-8
Firesteel River BC 3 D4-5
Fir Island SK 12 A-B5
Fir Mountain SK 11 F4
Fir River SK 11 B7
1st Burnt Pond NL 36 B8
1st Burnt Pond NL 37 B6

First Eel Lake NB 32 C1
First Lac NB 29 E3
First Oil Well in Western Canada NHS AB 7 F3-4
Firth River YT 5 B2
Firvale BC 1 C5
Fischells NL 36 C2
Fischells Brook NL 36 C2
Fishburn AB 7 F4
Fish Cove Point NL 39 B7
Fish Creek PP AB 6 D5
Fish Creek PP AB 7 B4
Fish Egg Inlet BC 1 C-D4
Fisher, Mount BC 6 F4
Fisher, Mount BC 7 E2
Fisher Bay MB 13 D4
Fisher Bay MB 14 A5
Fisher Bay PP Reserve MB 13 D4
Fisher Bay PP Reserve MB 14 A5
Fisher Branch MB 13 D3
Fisher Branch MB 14 B4
Fisher Channel BC 1 C4
Fisher Home AB 6 B5
Fisher Home AB 8 D5
Fisher Lake NS 32 F5
Fisher Lake NT 3 A3
Fisherman Lake NT 3 A6
Fisherman Lake NT 10 F2
Fisherman's Haven PP PE 34 A1
Fisherman's Haven PP PE 35 B2-3
Fisher Peak AB 6 D5
Fisher Peak AB 7 C2-3
Fisher River MB 13 D3-4
Fisher River MB 14 A-B4
Fisher River MB 14 A4-5
Fishers Glen ON 19 D6
Fishers Glen ON 20 F5
Fisher Strait NU 40 A3
Fisher Strait NU 41 D-E6
Fisherton MB 14 B4
Fisherville ON 19 C7
Fisherville ON 20 E-F6
Fishing Branch YT 5 E2
Fishing Branch Ecological Territorial Reserve YT 5 D2
Fishing Branch Wilderness TP Reserve YT 5 D-E2
Fishing Eagle Lake MB 15 F5
Fishing Lake MB 13 C5
Fishing Lake MB 16 A1
Fishing Lake SK 11 C5
Fishing Lake SK 11 C6
Fishing River MB 11 D8
Fishing River MB 13 D2
Fishing River MB 14 A5
Fishing Ships Harbour NL 39 C8
Fish Island PE 35 C3
Fish Lake NT 10 C3
Fish Lake YT 4 E4
Fish Lake YT 4 E5
Fishog Lake ON 21 C2
Fish Point, RNP ON 19 F2
Fish Point PNR ON 19 F2
Fishtail Lake ON 21 B1
Fishtrap Lake ON 16 A7-8
Fiske SK 11 D3
Fitch Bay QC 24 E8
Fitch Bay QC 27 F4
Fitton, Mount YT 5 B-C3
Fitton Point NT 5 A7
Fitzallen AB 8 C8
Fitzgerald AB 9 A6
Fitzgerald AB 10 F8
Fitzgerald AB 12 A2
Fitz Hugh Sound BC 1 D4
Fitzmaurice SK 11 D6
Fitzroy Harbour ON 24 D1
Fitzroy Harbour ON 21 A4
Fitzroy PP ON 21 A4
Fitzroy PP ON 24 D1-2
Fitzwilliam Channel ON 18 D4
Fitzwilliam Island ON 18 C4
Fitzwilliam Strait NT 42 C2-3
Five Finger Rapids Territorial Recreation Site YT 4 C3
Five Fingers NB 29 E7
Five Fingers NB 31 B3-4
Five Islands NS 32 D7
Five Islands NS 33 A5
Five Islands NS 33 A5
Five Islands NS 34 E1
Five Islands PP NS 32 D7
Five Islands PP NS 33 A5
Five Islands PP NS 34 E1
Five Mile BC 3 A1
Five Mile BC 4 F4
Five Mile Bay ON 18 D7
Five Mile Lake NS 32 E-F8
Five Mile Lake NS 33 C5
Five Mile Lake PP ON 17 E4-5
Flamand, lac QC 25 C4
Flamand, lac QC 28 E2
Flamand, rivière QC 25 C4
Flamand, rivière QC

Flanders ON 16 E3
Flat Bay NL 36 C2
Flat Bay Brook NL 36 C2
Flatbush AB 8 A5
Flatbush AB 9 F5
Flat Creek ON 19 B4-5
Flat Creek ON 20 D3-4
Flathead BC 6 F5
Flathead BC 7 F3
Flat Island NL 38 C3
Flat Lake AB 6 A7
Flat Lake AB 8 A7
Flat Lake AB 9 F6
Flat Lake MB 11 A2
Flat Lake NT 10 B7
Flat Lake PP BC 1 D8
Flat River NT 4 D7
Flat River PE 34 D3
Flat River PE 35 F5
Flat Rock NL 37 D8
Flatlands NB 29 D5-6
Flatlands NB 31 A4
Flat Sound NU 42 B4
Flatstone Lake SK 12 D3-4 E4
Flat Valley SK 11 A3
Flat Valley SK 12 F3
Flavrian, lac QC 23 D2
Flaxcombe SK 6 D8
Flaxcombe SK 11 D2
Fleet AB 6 C7
Fleet AB 11 C1
Fleming SK 11 E7
Fleming SK 13 E1
Fleming Lake ON 16 D7
Fleming Lake ON 17 A2
Flesherton ON 20 B5
Flesherton ON 21 D1
Fletcher ON 19 E2
Fletcher Hall AB 8 D6
Fletcher Lake ON 17 E7
Fleur de Lys NL 38 D4-5
Fleur de Lys NL 39 E8
Fleur-de-May Lake NL 39 D3
Fleurimont QC 27 E4-5
Flin Flon MB 12 F7
Flin Flon MB 15 F1
Flintoft SK 11 F4
Flinton ON 21 C5
Floradale ON 19 A6
Floradale ON 20 C-D5
Floral SK 11 C4
Floral Park ON 18 F5
Floral Park ON 20 A7
Floral Park ON 21 C1
Florence NS 34 C7-8
Florence ON 19 D3
Florence ON 20 F2
Florence Lake ON 17 E7
Florenceville NB 31 E3
Florenceville NB 32 A1
Florencia Bay BC 2 E1
Flores Island BC 1 F5
Flores Island BC 2 D1
Flores Island PP BC 1 F4-5
Florimond, lac QC 26 C5
Florio, lac QC 22 C4
Florze MB 14 F7
Flotten Lake SK 9 F8
Flotten Lake SK 12 F3
Flowerpot Island ON 18 C4
Flower's Cove NL 38 B3-4
Flower's Cove NL 39 D7-8
Flower Station ON 21 B6
Flowing Well SK 11 E4
Fluery Bight NL 38 E5
Fluor Island BC 16 F6-7
Fluor Island ON 17 C1
Flying Dust SK 11 A3
Flying Point NS 34 E-F6
Flyingshot Lake AB 9 F2
Fly Lake MB 15 F5
Foam Lake SK 11 C-D6
Foam Lake SK 11 D6
Foch Lagoon BC 1 A3
Fogo NL 38 E7
Fogo, Cape NL 38 E7
Fogo Head NL 38 E7
Fogo Island NL 38 E7
Fogo Island Region NL 38 E7
Foins, lac aux QC 22 B4
Foisy AB 6 A7
Foisy AB 11 A1
Foldens ON 19 C5
Foldens ON 20 E4
Foley Island NU 41 B8
Foley Island NU 42 F6
Foley Mountain CA ON 21 C7
Foley Mountain CA ON 24 F1
Folger ON 21 B6
Folly Lake NS 32 D7
Folly Lake NS 33 A6
Folly Lake NS 34 E1-2
Fond du Lac SK 12 A4
Fond-du-Lac SK 12 A4
Fond du Lac River SK 41 F2-3
Fond du Lac River SK 12 A4-5
Fond du Lac River SK 41 F1
Fond du Lac River SK 12 B5-6
Fontainebleau QC 27 D-E5
Fontas River BC 3 C7-8
Fontas River BC 9 B1-2
Fontbonne, lac QC 23 D3
Fonteneau, lac QC 39 D5

Fonthill ON 19 C8
Fonthill ON 20 E7
Foothills AB 6 B4
Foothills AB 8 A5
Foothills Creek NT 5 A-B8
Footprint Lake MB 15 E3
Foot's Bay ON 18 E7-8
Forant, lac QC 22 D5
Forbes, lac QC 24 A5
Forbes, lac QC 25 F5
Forbes Point NS 33 F2
Ford Bay NT 5 E8
Fords Mills NB 31 E7
Fords Mills NB 32 A5-6
Fordwich ON 19 A5
Fordwich ON 20 C4
Foremost AB 6 F7
Foremost AB 7 E7
Foremost AB 11 F1
Forest ON 19 C3
Forest ON 20 E2
Forestburg AB 6 C6-7
Forestburg AB 8 E8
Forest City NB 32 C1
Forester Lake ON 16 A5
Foresters Falls ON 21 A6
Foresters Falls ON 23 F6
Forest Estates ON 19 B-C6
Forest Estates ON 20 E4-5
Forest Field NL 37 E7
Forest Gate SK 11 B5
Forest Glen NS 33 B6
Forest Glen NS 34 F2
Forest Grove BC 1 C8
Forest Grove BC 6 C1
Forest Hill NS 34 E5
Forest Home ON 18 F8
Forest Home ON 20 A7
Forest Home ON 21 D1
Forest Mills ON 21 D5
Forestville ON 19 D6
Forestville ON 20 F5
Forestville QC 29 B2
Forestville, ZEC de QC 28 A8
Forestville, ZEC de QC 29 A1-2
Forfar ON 21 C7
Forfar ON 24 F1
Forgan SK 11 D3
Forges-du-Saint-Maurice, LHN des QC 24 B7
Forges-du-Saint-Maurice, LHN des QC 25 F7
Forges-du-Saint-Maurice, LHN des QC 27 C2-3
Forges du Saint-Maurice NHS QC 24 B7
Forges du Saint-Maurice NHS QC 25 F7
Forges du Saint-Maurice NHS QC 27 C2-3
Forget SK 11 F7
Forillon, PN QC 40 F7
Fork Lake AB 6 A7
Fork Lake AB 8 A8
Fork Lake AB 9 F7
Fork Lake AB 12 F1
Fork River MB 11 D8
Fork River MB 13 D2
Fork River MB 14 A5
Forks Baddeck NS 34 C6-7
Forks Branch NB 29 E3
Forks of the Credit PP ON 19 A6-7
Forks of the Credit PP ON 20 C5-6
Formosa ON 20 B3-4
Forrest MB 11 E8
Forrest MB 13 E2
Forrest MB 14 D1
Forresters Point NL 38 B3
Forresters Point NL 39 D-E8
Forrest Lake SK 9 D5
Forrest Lake SK 12 C3
Forshee AB 8 E5-6
Forshee AB 9 F7
Forsythe QC 23 E7
Forsythe QC 26 F1
Forsyth Lake SK 9 A8
Forsyth Lake SK 12 A4
Forsyth Lake SK 12 B5
Fort Albany ON 40 E3
Fort Alexander MB 13 E4
Fort Alexander MB 14 C6
Fort Anne & Scots Fort NHS NS 32 F3-5
Fort Anne & Scots Fort NHS NS 33 C1-2
Fort-Anne et du Fort-Scots, LHN du NS 32 F3-5
Fort-Anne et du Fort-Scots, LHN du NS 33 C1-2
Fort Assiniboine AB 6 A5
Fort Assiniboine AB 8 A5
Fort Assiniboine Sandhills Wildland PP AB 6 A5
Fort Assiniboine Sandhills Wildland PP AB 8 A5
Fort Assiniboine Sandhills Wildland PP AB 9 F5
Fort Augustus PE 34 C3
Fort Augustus PE 35 E5
Fort Babine BC 3 F4-5
Fort-Battleford, LHN du SK 11 B-C3
Fort Battleford NHS SK 11 B-C3
Fort-Beauséjour, LHN du NB 32 C7
Fort-Beauséjour, LHN du NS 35 F1-2

Fort Beauséjour NHS NB 32 C7
Fort Beauséjour NHS NS 35 F1-2
Fort Black SK 12 E3-4
Fort Carlton Provincial Historic Park SK 11 B4
Fort-Chambly, LHN du 24 D6-7
Fort-Chambly, LHN du 27 E2-3
Fort Chambly NHS QC 24 D6-7
Fort Chambly NHS QC 27 E2-3
Fort Chipewyan AB 9 B7
Fort Chipewyan AB 12 A2
Fort Chipewyan AB 41 F1-2
Fort Churchill SK 15 B5-6
Fort-Coulonge QC 22 F6
Fort-Edward, LHN du NS 32 E7-8
Fort-Edward, LHN du NS 33 B5
Fort-Edward, LHN du NS 34 F1
Fort Edward NHS NS 32 E7-8
Fort Edward NHS NS 33 B5
Fort Edward NHS NS 34 F1
Forteresse-de-Louisbourg, LHN de la NS 34 C-D8
Fort Erie ON 20 E8
Fort-Espérance, LHN du SK 11 E7
Fort Espérance NHS SK 11 E7
Fort Frances ON 16 F2
Fort Fraser BC 1 A6
Fort-Gaspareaux, LHN du NB 32 B7-8
Fort-Gaspareaux, LHN du NB 34 D1-2
Fort-Gaspareaux, LHN du NB 35 F2-3
Fort Gaspareaux NHS NB 32 B7-8
Fort Gaspareaux NHS NB 34 D1
Fort Gaspareaux NHS NB 35 F2-3
Fort-George, LHN QC 20 D7-8
Fort George/Buckingham House PHS QC 6 A7
Fort George/Buckingham House PHS AB 11 A2
Fort George Canyon PP BC 1 B7
Fort George NHS ON 20 D7-8
Fort Good Hope (Radilih Koe) NT 5 E7
Forthton ON 21 C8
Forthton ON 24 F2
Fort Hope (Eabametoong) ON 16 B6
Fortier MB 13 E3
Fortier MB 14 D4
Fortier, rivière QC 25 B3
Fortierville QC 27 B8
Forties Settlement NS 32 F7
Forties Settlement NS 33 C4
Fortin, lac QC 28 C7
Fortin, lac QC 39 E1
Fortin Lake YT 4 C-D6
Fort Irwin ON 21 B3
Fort Kent AB 6 A7
Fort Kent AB 9 F7
Fort Kent AB 11 A2
Fort Kent AB 12 F1
Fort-Kitwanga, LHN du 3 F3-4
Fort-Langley, LHN du BC 2 E5-6
Fort Langley NHS BC 2 E5-6
Fort Lawrence NS 32 C7
Fort-Lennox, LHN du QC 24 E6-7
Fort-Lennox, LHN du QC 27 F2
Fort Lennox NHS QC 24 E6-7
Fort Lennox NHS QC 27 F2
Fort Liard NT 3 A6
Fort Liard NT 10 F2
Fort-Livingstone, LHN du 11 C7
Fort-Livingstone, LHN du 13 C1
Fort Livingstone NHS SK 11 C7
Fort Livingstone NHS SK 13 C1
Fort MacKay AB 9 C-D6
Fort MacKay AB 12 C-D2
Fort MacKay AB 41 F1
Fort Macleod AB 6 F6
Fort Macleod AB 7 E4
Fort McLeod Provincial Historic Park BC 3 F7
Fort McMurray AB 9 D6-7
Fort McMurray AB 12 D2
Fort McPherson (Tetlit Zheh) NT 5 D4

Fort-Malden, LHN du ON 19 E1
Fort Malden NHS ON 19 E1
Fort-Mississauga, LHN du ON 20 D7-8
Fort Mississauga NHS ON 20 D7-8
Fort Nelson BC 3 B7
Fort Nelson BC 9 B1
Fort Nelson River BC 3 B6-7
Fort-Pelly, LHN du SK 11 C7-8
Fort-Pelly, LHN du SK 13 C1
Fort Pelly NHS SK 11 C7-8
Fort Pelly NHS SK 13 C1
Fort Pitt Provincial Historic Park SK 6 A-B8
Fort Pitt Provincial Historic Park SK 11 A3
Fort-Prince-de-Galles, LHN du MB 15 B5
Fort-Prince-de-Galles, LHN du MB 40 C1
Fort Providence NT 10 E5
Fort Providence NT 41 D-E1
Fort Providence TP NT 10 E5
Fort Providence TP NT 41 D-E1
Fort Qu'Appelle SK 11 D6
Fort Resolution NT 41 E1-2
Fort Resolution (Deninoo) NT 10 E7
Fort Rodd Hill et du Phare-de-Fisgard, LHN de BC 2 F4-5
Fort Rodd Hill & Fisgard Lighthouse NHS BC 2 F4-5
Fort St. James BC 1 A6
Fort-St. James, LHN du BC 1 A6-7
Fort St. James NHS BC 1 A6-7
Fort St. John BC 3 E8
Fort St. John BC 9 D1
Fort-St. Joseph, LHN du ON 17 F4-5
Fort St. Joseph NHS ON 17 F4-5
Fort San SK 11 D6
Fort Saskatchewan AB 6 A6
Fort Saskatchewan AB 8 C7
Fort Selkirk YT 4 C2-3
Fort Selkirk Territorial Historic Site YT 4 C3
Fort Severn ON 15 F8
Fort Severn ON 40 D2
Fort Simpson NT 10 D3
Fort Simpson TP NT 10 D3
Fort Smith NT 9 A6
Fort Smith NT 10 F8
Fort Smith NT 41 E2
Fort Steele BC 6 F4
Fort Steele BC 7 E2
Fort Stewart ON 21 B4
Fort-Témiscamingue, LHN du QC 17 E8
Fort-Témiscamingue, LHN du QC 22 A1-2
Fort-Témiscamingue NHS QC 17 E8
Fort-Témiscamingue NHS QC 22 A1-2
Fort Vermilion AB 9 B4
Fort Vermilion AB 41 F1
Fort-Walsh, LHN du SK 6 F8
Fort-Walsh, LHN du SK 11 F1-2
Fort Walsh NHS SK 6 F8
Fort Walsh NHS SK 11 F1-2
Fort Ware BC 3 C-D5
Fort-Wellington, LHN du 21 C8
Fort-Wellington, LHN du 24 F2
Fort Wellington NHS ON 21 C8
Fort Wellington NHS ON 24 F2
Fort William ON 16 F5-6
Fort William QC 22 E5
Fortymile YT 4 A2
Forty Mile, Fort Cudahy and Fort Constantine HS YT 4 A2
Forty Mile Flats BC 3 C3
Fortymile River YT 4 A1
Fosheim Peninsula NU 42 B4-5
Fossambault-sur-le-Lac QC 27 A5
Fossli PP BC 2 D-E2
Fosston SK 11 C6

Foster QC 24 E8
Foster QC 27 F3
Foster, mont QC 24 E8
Foster, mont QC 27 F3
Foster, Mount BC 4 F3-4
Foster River SK 12 D-E5
Fosterton SK 11 E3
Fosterville NB 32 C1
Fountain Valley BC 1 E8
Fountain Valley BC 2 A6
Fourche, lac de la QC 23 F8
Fourche, lac de la QC 25 B2
Fourche, lac de la QC 26 F2
Fourché Harbour NL 39 E8
Fourches, lac des QC 22 A3
Fourches, lac des QC 23 F3
Fourchu NS 34 D7-8
Fourchu Bay NS 34 D8
Four Falls NB 31 D3
Four Mile Lake ON 20 A8
Four Mile Lake ON 21 C2
Four Mile River BC 3 B3
Fourmel, lac QC 38 A2
Fourmont Lake NL 39 D5-6
Fournel, lac QC 39 D7
Fournier ON 24 D3-4
Fournier SK 11 E8
Fournière, lac QC 39 D2-3
Fournière, lac QC 23 E4
Four Roads NB 31 B8
Fowlers Corners ON 21 D3
Foxboro ON 21 E5
Fox Cove NL 36 E7
Fox Cove NL 37 E5
Fox Cove-Mortier NL 36 E-F7
Fox Cove-Mortier NL 37 E-F5
Fox Creek AB 6 A3-4
Fox Creek AB 8 A2-3
Fox Creek AB 9 F3-4
Fox Creek NB 32 B6
Fox Harbour NL 36 E8
Fox Harbour NL 37 E6
Fox Harbour NS 34 D2
Fox Harbour PP NS 34 D2
Fox Island SK 31 C7-8
Fox Island ON 18 B3
Fox Island Harbour NL 36 D4
Fox Island Harbour NL 37 D2
Fox Island Main NL 36 E6
Fox Island River NL 36 B2
Fox Lake AB 9 B5
Fox Lake MB 15 D5-6
Fox Lake ON 21 B1
Fox Lake ON 18 D8
Foxley Bay PE 35 B-C2
Foxley River PE 35 C2
Fox Mine MB 12 D7
Fox Mine MB 15 D1
Fox Mount YT 4 D4-5
Fox Point NS 33 C5
Fox Point NT 10 A3
Fox Point NT 21 B2
Fox Point SK 12 F5
Fox River NL 36 F6
Fox River NL 15 E5-6
Fox River NL 36 F2
Fox River NL 33 A4
Fox Roost NL 36 E1
Fox Valley SK 6 E8
Fox Valley SK 11 E2
Foxwarren MB 11 E7
Foxwarren MB 13 E1
Foymount ON 21 A5
Foy PP ON 21 A5
Foy PP ON 22 F5
Fraleigh Lake, RNP ON 16 F5
Fraleigh Lake PNR ON 16 F5
Framboise NS 34 D7-8
Framboise Cove NS 34 D7-8
Framnes MB 14 B5
Français, rivière des ON 17 F7
Français, rivière des ON 18 B7
Français, rivière des ON 18 B6
France, lac QC 26 A5
Frances Lake YT 4 D6
Frances Lake Territorial Campground & Historic Site YT 4 D6-7
Frances River YT 4 A7
Franchere AB 6 A7
Franchere AB 8 A8
Franchere AB 11 A1-2
Franchere AB 12 F1
Franchère, lac QC 24 A3-4
Franchère, lac QC 25 E3-4
Francis, Lake MB 14 D4
Francis Lake YT 4 D6
Francis Lake ON 18 F5
Francis Point PP BC 2 D3
Francoeur, lac QC 25 A4
Francoeur, lac QC 26 F4
François NL 36 D-E4
François NL 37 D-E2
François, lac QC 24 B3

François, lac QC 25 F3
François, lacs à QC 39 D2
François Lake BC 1 A5
François Lake BC 1 A5
François Lake NT 10 C7
François Lake PP BC 1 A5-6
Franey Corner NS 32 F6
Franey Corner NS 33 C4
Frank BC 6 F5
Frank AB 7 E3
Frank Lake AB 7 C4
Frank Lake YT 4 D4
Frankford ON 21 E4-5
Franklin MB 13 E2
Franklin MB 14 D2
Franklin QC 24 E5
Franklin QC 27 F1
Franklin Bay NT 5 A7
Franklin Bay NT 41 A1
Franklin Beach ON 20 B7
Franklin Beach ON 21 D1
Franklin Island ON 18 D6
Franklin Lake SK 12 A6
Franklin Lake SK 15 A1
Franklin Strait NU 41 A-B4
Franklin Strait NU 42 E4
Franklyn Arm BC 1 D6
Frank Mackie Glacier BC 3 E3
Frank Slide PHS AB 7 E3-4
Franks Pond NL 37 E7
Franktown ON 21 B7
Franktown ON 24 E1-2
Frankville NS 34 D5
Frankville ON 21 C7
Frankville ON 24 F1
Franquelin QC 29 A4
Franquelin QC 39 F1
Franz ON 17 C-D4
Fraser BC 3 A1
Fraser BC 4 F4
Fraser, fleuve BC 1 D7
Fraser, fleuve BC 1 B7
Fraser, fleuve BC 1 B8
Fraser, fleuve BC 1 F8
Fraser, fleuve BC 2 B6
Fraser, fleuve BC 2 C-D7
Fraser, fleuve BC 2 E6
Fraser, fleuve BC 6 B1-2
Fraserburg ON 21 B1-2
Fraserdale ON 17 B6
Fraserdale ON 40 F3
Fraser Lake BC 1 A5
Fraser Lake NL 39 A-B4
Fraser Lake ON 21 B4
Fraser Mills NS 34 E5
Fraser River BC 1 A-B8
Fraser River BC 1 D7
Fraser River BC 1 E8
Fraser River BC 2 A6
Fraser River BC 2 C7
Fraser River BC 2 E6
Fraser River BC 6 A-B1
Fraser River NL 40 C-D7
Fraser River PP BC 1 B7
Fraserville NS 32 D6
Fraserville NS 33 A4
Fraserville ON 21 E3
Fraserwood MB 13 E4
Fraserwood MB 14 C5
Frater ON 17 E4
Fraxville NS 32 F6-7
Fraxville NS 33 C4
Frazer Bay ON 18 B4
Frazer Lake ON 18 E6
Fréchette, rivière QC 39 D3
Frederick ON 17 C6
Frederick Arm BC 2 B2
Frederick House Lake ON 17 C6
Frederick House Lake, RNP ON 17 C6
Frederick House Lake PNR ON 17 C6
Frederick Lake YT 4 D3
Fredericton NL 36 A8
Fredericton NL 37 A6
Fredericton NB 38 F7
Fredericton NB 32 C2-3
Fredericton PE 32 A8
Fredericton PE 34 C2
Fredericton PE 35 D4
Fredericton Junction NB 32 C3
Fredericton Road NB 31 F7
Fredericton Road NB 32 B5
Fred Henne TP NT 10 C6
Fred Henne TP NT 41 D1
Freedale MB 14 A2
Freedom AB 8 B5
Freeland PE 32 A7-8
Freeland PE 34 B1
Freeland PE 35 C2
Freels, Cape NL 37 F7
Freelton ON 19 B7
Freelton ON 20 D6
Freeman Lake ON 16 C4
Freeman River AB 8 A4-5
Freeman River AB 8 A5
Freeman River AB 9 F4-5
Freeport NS 33 D1
Freetown PE 32 A8
Freetown PE 34 C1-2
Freetown PE 35 D3
Frelighsburg QC 24 E7
Frelighsburg QC 27 F3
French Beach PP BC 2 F3
French Cove NS 38 C5
French Cove NS 39 E8
French Cove NS 34 C6
French Creek MB 15 D6

Frenchfort PE **35** E5
French Lake NB **32** C3
Frenchman Butte SK **6** A8
Frenchman Butte SK **11** A-B2
Frenchman Butte, LHN de SK **6** A8
Frenchman Butte, LHN de SK **11** A2
Frenchman Butte NHS SK **6** A8
Frenchman Butte NHS SK **11** A2
Frenchman River SK **11** F3-4
Frenchman's Cove NL **35** E5
Frenchman's Cove NL **36** E6
Frenchman's Cove NL **37** E4
Frenchman's Cove NL **38** F2
Frenchman's Cove PP NL **36** E6
Frenchman's Cove PP NL **37** E4
Frenchmans Island NL **39** B-C8
French Portage ON **13** F5
French Portage ON **14** E8
French River NS **34** B7
French River NS **34** C4
French River ON **17** F7
French River ON **17** F7
French River ON **18** B6
French River ON **18** B6
French River ON **18** B6
French River PE **32** A8
French River PE **34** B2
French River PE **35** D3
French River North Channel ON **18** B6
French River PP ON **18** B5-6
French River PP ON **18** B6-7
French River PP ON **18** B7
French Village NS **32** F8
French Village NS **33** C5
French Village PE **35** D5
Frenchville SK **11** F3
Freshfield Icefield BC **6** B-C8
Freshford NB **11** A7-8
Freshford MB **13** A1
Freshwater NL **36** E8
Freshwater NL **37** D7
Freshwater NL **37** B6
Freshwater Bay NL **36** B8
Freshwater Bay NL **37** B6
Freshwater Lake SK **6** B-C8
Freshwater Lake SK **11** C2
Freshwater Point NL **37** F7
Friars Cove NL **36** D1
Friedensfeld MB **14** E6
Friedensruh MB **14** F4
Frigon, lac QC **23** E8
Frigon, lac QC **25** A2-3
Frigon, lac QC **26** E2-3
Frisco SK **8** E5
Froatburn ON **24** E3
Frobisher SK **11** F7
Frobisher Bay NU **40** A-B6
Frobisher Lake SK **12** D3-4
Frodsham Lake NT **10** B7
Frog Lake AB **6** A7
Frog Lake AB **11** A2
Frog Lake AB **11** A2
Frog Lake Massacre NHS AB **6** A7-8
Frog Lake Massacre NHS AB **11** A2
Frogmore ON **19** D6
Frogmore ON **20** F5
Frog River BC **3** C4
Froid, lac QC **24** A4
Froid, lac QC **25** E4
Froid, lac QC **26** D4
Frome ON **19** D4
Frome ON **20** F3
Frontenac QC **27** E6-7
Frontenac, lac QC **26** D5
Frontenac, lac QC **28** B1
Frontenac, parc de QC **27** D6
Frontenac PP ON **21** D6-7
Frontier SK **11** F3
Froomfield ON **19** C2
Froomfield ON **20** E1
Frosty Hollow NB **32** C7
Froude SK **11** F6
Frozen Ocean Lake NL **36** B6
Frozen Ocean Lake NL **37** B4
Frozen Strait NU **41** D5
Fruitland ON **19** B-C7
Fruitland ON **20** E6
Fruitvale BC **6** F3
Fryatt, Mount AB **6** C3
Fryatt, Mount AB **11** A5
Fry Lake NL **13** D7-8
Fry Lake ON **16** B4
Frys SK **11** F7
Frys SK **13** F1
Fugèreville QC **17** E8
Fugèreville QC **22** A2
Fulda SK **11** C5
Fulford QC **24** E8
Fulford QC **27** F3
Fulford Harbour BC **2** E-F4
Fullarton ON **19** B5
Fullarton ON **20** D4
Fullerton, Cape NU **40** A2
Fullerton, Cape NU **41** D5
Fulmore Lake BC **2** B1
Fulton ON **19** C7-8

Fulton ON **19** C7-8
Fulton ON **20** E6-7
Fulton River BC **3** F5
Fulton River Dam BC **3** F5
Fundy, baie de NB **32** E5-6
Fundy, baie de NB **33** B2-3
Fundy, Bay of NB **32** E-F4
Fundy, Bay of NB **33** B1-2 C1
Fundy, îles de NB **32** F3
Fundy, PN NB **32** D6
Fundy, PN NB **33** A3
Fundy (péage), route d'accès du sentier NB **32** D5
Fundy (péage), route d'accès du sentier NB **33** A2-3
Fundy Escarpment PP (proposed/projeté) NB **32** D5-6
Fundy Escarpment PP (proposed/projeté) NB **33** A3
Fundy NP NB **32** C-D6
Fundy Trail Parkway (toll) NB **32** D5
Fundy Trail Parkway (toll) NB **33** A2-3
Funk Island NL **38** E8
Funk Island Ecological Reserve NL **38** E8
Furby's Cove NL **36** D6
Furby's Cove NL **37** D4
Furman AB **7** D4
Furnace Falls ON **21** C3
Furness SK **6** B8
Furness SK **11** B2
Fury and Hecla Strait NU **41** B6
Fury and Hecla Strait NU **42** F5
Fushimi Lake PP ON **17** B6
Fusilier SK **6** C8
Fusilier SK **11** C2

G

Gabarus NS **34** D8
Gabarus, Cape NS **34** D8
Gabarus Bay NS **34** D8
Gabarus Lake NS **34** D8
Gable Mountain BC **6** F2
Gabriel, lac QC **26** C4
Gabriel Strait NU **40** B6
Gabriola BC **1** F6
Gabriola BC **2** E4
Gabriola Island BC **2** E4
Gabriola Sands PP BC **2** E4
Gadsby AB **6** C6
Gadsby AB **8** F8
Gads Hill ON **19** B5
Gads Hill ON **20** D4
Gagamo, lac QC **22** C6
Gagamo, lac QC **24** A1
Gage, Cape PE **35** B1
Gagetown NB **32** C3-4
Gagnon QC **39** D1
Gagnon QC **40** E6
Gagnon, lac QC **24** B3
Gagnon Lake NT **10** D8
Gagnon-Siding QC **23** E7
Gagnon-Siding QC **25** A1-2
Gagnon-Siding QC **26** F1-2
Gaillarbois, lac QC **39** D2
Gainford AB **6** B5
Gainsborough SK **11** F7-8
Gainsborough SK **13** F1
Gairloch NS **33** A7
Gairloch NS **34** E3
Galahad AB **6** C6-7
Galahad AB **8** E8
Galahad AB **11** B1
Galarneau, lac QC **22** E6
Galarneauville AB **7** B6-7
Gale, lac QC **22** C6
Galeairy Lake ON **21** A3
Galeairy Lake ON **22** F3
Galena Bay BC **6** E3
Galetta ON **21** A7
Galetta ON **24** D1
Galeville NL **38** E4
Galiano Island BC **2** E4
Galibois, îles QC **39** E-F6
Galifet, lac QC **25** D5
Galifet, lac QC **28** F1
Galilee NS **11** E5
Galinée, lac QC **26** D5
Galinée, lac QC **28** B1
Gallants NL **36** B2
Gallant Settlement NB
Gallas Point PE **34** C2-3
Gallas Point PE **35** E5
Gallet, lac QC **38** A1
Gallichan QC **17** C8
Gallichan QC **23** C2
Gallivan SK **11** B3
Gallix QC **39** F2
Galloway AB **8** C2
Galloway BC **6** F4-5
Galloway BC **7** F2
Gallows Cove NL **37** E7
Gamart, lac QC **39** C1
Gambier Island BC **2** D4
Gambo NL **36** B8
Gambo NL **37** B6
Gambo Pond NL **36** B7-8

Gambo Pond NL **37** B5-6
Gamebridge ON **20** A7
Gamebridge ON **21** D1-2
Gameland ON **16** E1
Gamsby River BC **1** B4
Gananoque ON **21** D7
Gananoque Lake ON **21** D7
Ganaraska River ON **21** E3
Gander NL **36** A-B7
Gander NL **37** A-B5
Gander Bay NL **36** A7-8
Gander Bay NL **38** E-F7
Gander Bay NL **38** F7
Gander Lake NL **36** B7
Gander Lake NL **37** B5
Gander River NL **36** A7
Gander River NL **37** A5
Gander River NL **38** F6-7
Ganges BC **2** E4
Gang Ranch BC **1** D7
Gannet Islands Ecological Reserve NL **39** B8
Gaotanaga, lac QC **22** A5
Gaotanaga, lac QC **23** F5
Gap NL **37** B7
Garancières, lac QC **25** A4
Garancières, lac QC **26** E4
Garden Bay NL **37** C7
Garden Bay BC **2** D3-4
Garden Cove NL **36** D8
Garden Cove NL **37** D6
Garden Hill MB **13** A5
Garden Hill ON **21** E3
Garden Lake ON **16** E5
Garden of Eden NS **33** A8
Garden of Eden NS **34** E4
Garden Plain AB **8** F8
Garden River ON **17** F4
Garden River ON **17** F4
Garden River SK **11** A-B5
Gardenton MB **13** F4
Gardenton MB **14** F6
Garden Village ON **17** F7-8
Garden Village ON **18** A7
Gardenville ON **21** E4-5
Gardiner ON **17** B6
Gardiner Dam SK **11** D4
Gardiner Lakes AB **9** C6
Gardiner Lakes AB **12** C1
Gardiner Mines NS **34** C8
Gardner, lac QC **22** B5
Gardner AB **7** B5
Gardner Canal BC **1** B4
Gardner Creek NB **32** D4-5
Gardner Creek NB **33** B2
Garfield AB **7** A3
Garfield PE **34** C3
Garfield PE **35** F5-6
Gargamelle NL **38** C3
Gargantua, Cape ON **17** E3
Garia Bay NL **36** D2
Garia Brook NL **36** D2
Garibaldi BC **1** E7
Garibaldi BC **2** C5
Garibaldi, Mount BC **1** E7
Garibaldi, Mount BC **2** C5
Garibaldi Heights BC **1** F7
Garibaldi Heights BC **2** C5
Garibaldi Lake BC **1** E7
Garibaldi Lake BC **2** C5
Garibaldi PP BC **1** E-F7
Garibaldi PP BC **2** C5
Garin, rivière QC **29** C8
Garin, rivière QC **30** D1
Garland MB **11** C-D8
Garland MB **13** C-D2
Garland MB **14** A1
Garner Lake PP AB **6** A6-7
Garner Lake PP AB **8** B8
Garner Lake PP AB **11** A1
Garner Lake PP AB **12** F1
Garnet ON **20** E6
Garnett Settlement NB **32** D4-5
Garnett Settlement NB **33** A2
Garnish NL **36** E6
Garnish NL **37** E4
Garrett Lake ON **13** D7
Garrick SK **11** B5-6
Garrington AB **8** F5
Garry Island NT **5** A-B4
Garry Lake NU **41** C4
Garson MB **14** D5
Garson ON **17** F7
Garson ON **18** A5
Garson Lake AB **9** D7
Garson Lake AB **12** D2
Garson Lake SK **9** D7-8
Garson Lake SK **12** D2
Garth AB **8** E-F4
Gartly AB **7** A5-6
Gascoigne SK **11** D2
Gascoigne Cove PE **35** F5
Gaskiers NL **37** F7
Gaskiers-Port La Haye NL **37** F7
Gasline ON **19** C8
Gasline ON **20** E7
Gaspard Creek BC **1** D7
Gaspé QC **30** B2
Gaspé QC **30** B2-3
Gaspé, baie de QC **30** B-C3
Gaspé, baie de QC **40** F7
Gaspé, cap de QC **30** B3
Gaspereau NS **33** B4-5
Gaspereau, rivière QC **35** E1-2 F2
Gaspereau Forks NB **31** F6
Gaspereau Forks NB **32** B4
Gaspereau Lake NS **32** E6-7

Gaspereau Lake NS **33** B4
Gaspereau River NB **31** E-F6
Gaspereau River NB **32** B4
Gaspereau River NB **35** E-F2
Gaspereaux PE **34** C4
Gaspereaux PE **35** F7
Gaspésie, parc de la QC **29** B6-7
Gassend Lake NT **5** C7
Gaston, lac QC **25** A4
Gaston, lac QC **26** E6
Gaston, lac QC **26** E4
Gaston, lac QC **28** C2
Gataga Mountain BC **3** B5
Gataga River BC **3** B5
Gateshead Island NU **41** B4
Gateshead Island NU **42** E3
Gathto Creek BC **3** C6
Gatineau QC **21** A8
Gatineau QC **22** F8
Gatineau QC **24** D2
Gatineau, parc de la QC **21** A7
Gatineau, parc de la QC **22** F7
Gatineau, parc de la QC **24** D1-2
Gatineau, rivière QC **17** A7-8
Gatineau, rivière QC **22** C-D7
Gatineau, rivière QC **22** F7-8
Gatineau, rivière QC **24** A-B2
Gatineau, rivière QC **24** C-D2
Gatineau, rivière QC **25** E-F2
Gatineau Park QC **21** A7-8
Gatineau Park QC **22** F7
Gatineau Park QC **24** D2
Gaudois, lac QC **22** A7
Gaudois, lac QC **25** C2
Gauer Lake MB **15** D3-4
Gaultois NL **36** D6
Gaultois NL **37** D4
Gauvin, lac QC **24** A3
Gauvin, lac QC **25** F3
Gayford AB **7** B5
Gayna River NT **5** F6
Gays River NS **33** B6
Gays River NS **34** F2
Gaythorne NB **31** C7
Geary NB **32** C3
Gegogan Harbour NS **34** F5
Gegogan Harbour NS **34** F5
Geikie Island ON **16** D6
Geikie River SK **12** C5-6
Gelert ON **21** C3
Gem AB **6** D6
Gem AB **7** B6
Genelle BC **6** F3
Genesee AB **6** B5
Genesee AB **8** C5
Genévriers, île des QC **39** D-E7
Génier, lac QC **22** D8
Génier, lac QC **24** A2
Génier, lac QC **25** E2
Gensart, lac QC **39** C1
Gens de Terre, rivière QC **22** B-C7
Gens de Terre, rivière QC **25** D1
Gens de Terre, rivière QC **22** B-C7
Gentilly QC **24** B8
Gentilly QC **25** E8
Gentilly QC **27** C4
Geoffrion, lac QC **25** D6
Geoffrion, lac QC **28** F2
George, Cape BC **1** A2
George, Cape NS **34** D5
George, Lake NB **32** C2
George, rivière QC **40** C-D6
George Evanoff PP BC **1** A8
George Evanoff PP BC **6** A1
Georgefield NS **33** B6
Georgefield NS **34** F2
George Island MB **13** B3-4
George Island NL **39** B7
George Island ON **18** B4
George Island SK **12** B6
George Island SK **15** B1
George Lake SK **12** D-E4
George Richards, Cape NU **42** E7
Georges Brook NL **36** C8
Georges Brook NL **37** C6
Georges Lake NL **36** B2-3
Georges Lake NL **36** B2
Georges River NS **34** C7
Georgetown ON **19** A7
Georgetown ON **20** C6
Georgetown PE **34** C3-4
Georgetown PE **35** E7
Georgetown Harbour PE **34** C3-4
Georgetown Harbour PE **35** E7
Georgeville NS **34** D4
Georgeville QC **24** E8
Georgeville QC **27** F4
Georgia, Strait of BC **1** F6
Georgia, Strait of BC **2** C-D3
Georgian Bay ON **18** C5
Georgian Bay ON **20** A5
Georgian Bay Islands NP ON **18** E6-7
Georgian Bay Islands NP ON **18** E7
Georgian Beach ON **18** F6

Georgian Beach ON **20** A5
Georgie, détroit de BC **1** F6
Georgie, détroit de BC **2** D3-4
Georgienne, baie ON **18** D5-6
Georgienne, baie ON **20** A5-6
Georgina ON **20** B7
Georgina ON **21** E1
Georgina Island ON **18** F8
Georgina Island ON **20** A7
Georgina Island ON **20** B7
Georgina Island ON **21** D1
Georgina Island ON **21** D1
Gerald SK **11** E7
Gerald SK **13** D1
Geraldton ON **16** D7
Geraldton ON **17** B2
Geraldton ON **40** F2
Geraldton East ON **16** D7-8
Geraldton East ON **17** B2
Germaine, lac QC **39** C1
Germaine River NT **10** C6
Germansen Lake BC **3** E6
Germansen Landing BC **3** E6
Gesgapegiag QC **29** D7
Gesgapegiag QC **31** A6
Gesto ON **19** E1
Geyser MB **14** B5
Ghost Lake AB **6** D5
Ghost Lake AB **7** B3
Ghost Lake NT **10** B6
Ghost River ON **13** E7-8
Ghost River ON **16** D6
Ghost River ON **23** C-D1
Ghost River Provincial Wilderness Area AB **7** A-B2
Ghost River Wilderness Area AB **6** D4-5
Giant Lake NS **34** E5
Giants Tomb Island ON **18** E7
Gibbons AB **6** A6
Gibbons AB **8** B6
Gibbs Fiord NU **41** A7
Gibbs Fiord NU **42** E6-7
Gibraltar BC **3** C4
Gibraltar Rock PP Reserve NS **33** B6-7
Gibson Lake ON **18** E7-8
Gibson PMP BC **1** F4-5
Gibson River, RNP QC **18** E8
Gibson River, RNP QC **21** C1
Gibson River PNR ON **18** E8
Gibson River PNR ON **21** C1
Gibsons BC **1** F6
Gibsons BC **2** D4
Gift Lake AB **9** E4
Gilbert Bay NL **39** C8
Gilbert Cove NS **32** F4
Gilbert Plains MB **11** D8
Gilbert Plains MB **13** D1-2
Gilbert River NL **39** C7-8
Gilbertville ON **19** C6
Gilbertville ON **20** F5
Gilby AB **8** E5
Gilchrist Bay ON **21** D4
Giles Lake SK **12** B5
Gilford ON **20** B7
Gilford ON **21** E1
Gil Island BC **1** B3
Gillam MB **15** D5
Gillam MB **40** F1
Gillams NL **36** A-B3
Gillams NL **37** A-B1
Gillespie Settlement NB **31** D2-3
Gillies ON **17** E7
Gillies Bay BC **1** F6
Gillies Bay BC **2** D3
Gillis Point PE **35** C3
Gilltoyees Inlet BC **1** A3
Gilmour ON **21** C4
Gilmour ON **22** B3
Gilnockie PP BC **6** F4-5
Gilnockie PP BC **7** F2
Gimli MB **13** E4
Gimli MB **14** C5
Gin Cove NL **36** C8
Gin Cove NL **37** C6
Gingolx (Kincolith) BC **3** F3
Ginoogaming ON **16** D8
Ginoogaming ON **17** B2
Gipsy Lake AB **9** D7
Gipsy Lake AB **12** D2
Gipsy Lake Wildland PP AB **9** D7
Gipsy Lake Wildland PP AB **12** D2
Girardville QC **26** D7-8
Girardville QC **28** A3
Girène, lac QC **25** C3
Girouard, lac QC **23** D7
Girouard, lac QC **26** E1
Giroux MB **13** F4
Giroux, lac QC **22** B5
Giroux Lake ON **13** E4
Girouxville AB **9** E3
Girvin SK **11** D4-5
Gisborne Lake NL **36** D7
Gisborne Lake NL **37** D5
Gitanmaax BC **3** F4
Gitchie River ON **13** C8 D7
Gitnadoix River PRA BC **1** A3
Gitwinksihlkw BC **3** F3

Gjoa Haven (Uqsuqtuuq) NU **41** B-C4
Gjoa Haven (Uqsuqtuuq) NU **42** F3-4
Glace Bay NS **34** C8
Glacial Mountain BC **3** C3
Glacier BC **6** D3
Glacier Lake NB **31** C1
Glacier NP BC **6** D3
Glaciers, PN des BC **6** D3
Glade BC **6** F3
Gladmar SK **11** F5-6
Gladstone AB **8** D8
Gladstone MB **13** E3
Gladstone MB **14** D3
Gladstone ON **19** C5
Gladstone ON **20** E4
Gladstone Creek YT **4** D2
Gladstone PP BC **6** F2-3
Gladstone Valley AB **7** F3-4
Gladue Lake SK **11** A3
Gladue Lake SK **12** F3
Gladwin BC **2** B7
Gladys AB **7** C4
Gladys Lake BC **3** A2
Gladys Lake BC **4** F5
Gladys River BC **3** A2
Gladys River BC **4** F5
Glamis SK **11** D3
Glamis ON **20** B3
Glamor Lake ON **21** C3
Glanford ON **19** B-C7
Glanworth ON **19** C5
Glanworth ON **20** F4
Glasgow Station ON **21** A6
Glaslyn SK **11** B3
Glasnevin SK **11** F5
Glass MB **14** D5-6
Glassville NB **31** B4
Glassville NB **32** A1-2
Gledhow SK **11** C4
Gleichen AB **6** D6
Gleichen AB **7** C5
Glen Alda ON **21** C4
Glen Allan ON **19** A6
Glen Allan ON **20** C5
Glen Alpine NS **34** E4-5
Glenarm ON **20** A8
Glenarm ON **21** D2
Glenavon SK **11** E6
Glen Becker ON **24** F3
Glenbain SK **11** E4
Glenboro MB **13** F2
Glenboro MB **14** E2
Glenbourne MB **14** B3
Glenbow AB **7** B3
Glenburnie NL **36** A3
Glenburnie NL **37** A1
Glenburnie NL **38** F2
Glenburnie NL **39** F7
Glenburnie ON **21** D6-7
Glenburnie-Birchy Head-Shoal Brook NL **36** A3
Glenburnie-Birchy Head-Shoal Brook NL **37** A1
Glenburnie-Birchy Head-Shoal Brook NL **38** F2
Glenbush SK **11** B3
Glencairn MB **14** C2
Glencairn ON **20** B6
Glencoe NB **29** D6
Glencoe NS **31** A4
Glencoe NS **33** A8
Glencoe ON **19** D3-4
Glencoe NS **34** E5
Glencoe ON **20** F2-3
Glencoe Mills NS **34** C-D6
Glencoe Station NS **34** D5-6
Glendale NS **34** D5-6
Glendale Cove BC **1** E5
Glendale Cove BC **2** A-B1
Glendon AB **6** A7
Glendon AB **11** A2
Glendon AB **12** F2
Glendyer NS **34** C6
Glenelg NS **33** A8
Glenelg NS **34** E4
Glenella MB **13** E2
Glenella MB **14** C2
Glenevis AB **6** A5
Glenevis AB **8** C5
Glen Ewen SK **11** F7
Glenfanning PE **35** E6-7
Glenfinnan PE **35** E5
Glenford AB **8** B5
Glenforsa MB **14** C1
Glengarry NS **33** A7
Glengarry NS **34** E3
Glengarry Cairn NHS ON **24** E4-5
Glen Haffy CA ON **19** A7
Glen Haffy CA ON **20** C6
Glen Harbour SK **11** D5
Glen Haven NS **32** F8
Glen Haven NS **33** C5
Glenholme NS **33** B6
Glenholme NS **34** F2
Glenholme NS **34** E2
Gleniffer Lake AB **8** F5-6
Glenister AB **9** D7
Glen Kerr SK **11** E4
Glenlea NB **31** A4
Glen Levit NB **31** A4
Glenlily BC **7** F1
Glen Livet NB **29** D5-6
Glenlyon Peak YT **4** C4
Glenlyon Range YT **4** C4
Glen McPherson SK **11** D3
Glen Margaret NS **32** F8
Glen Margaret NS **33** C5
Glen Meyer ON **19** D5-6

Glen Meyer ON **20** F4-5
Glen Miller ON **21** E4-5
Glen Mont NS **32** E6-7
Glenmoor MB **14** D6
Glen Morris ON **19** B6
Glen Morris ON **20** D-E5
Glenmount ON **21** B2
Glen Nevis ON **24** E4
Glenora BC **2** E-F4
Glenora BC **3** C2
Glenora MB **13** F2
Glenora MB **14** F2
Glenora ON **21** E5
Glen Orchard ON **18** E8
Glen Orchard ON **21** B1
Glenorchy ON **16** E3
Glen Park AB **8** D6
Glen Robertson ON **24** D4
Glen Ross ON **21** E4
Glenside SK **11** D4
Glen Sutton QC **24** E8
Glen Sutton QC **27** F3
Glen Tay QC **21** C7
Glen Tay ON **24** F1
Glentworth SK **11** F4
Glen Valley PE **35** D4
Glenview NL **36** D8
Glenview NL **37** D6
Glen Walter ON **24** E4
Glenwilliam ON **20** B3
Glen Williams ON **19** A7
Glen Williams ON **20** C6
Glenwood AB **6** F5-6
Glenwood AB **7** F4-5
Glenwood NB **31** B4
Glenwood NB **32** D4
Glenwood NL **36** A7
Glenwood NL **37** A5
Glenwood NS **32** F3
Glenwood PE **32** A7
Glenwood PE **35** C1
Glenwood PP NS **33** E2
Glidden SK **6** D6
Glidden SK **11** D2-3
Gloucester ON **21** A-B8
Gloucester ON **24** D2
Gloucester Pond NL **36** B3
Gloucester Pond NL **37** B1
Glover Island NL **36** B3
Glover Island NL **37** B1
Glovers Harbour NL **36** A6
Glovers Harbour NL **37** A4
Glovers Harbour NL **38** F5
Glovertown NL **36** B8
Glovertown NL **37** A1
Glowworm Lake NT **10** A8
Glundebery Creek BC **3** B2
Glyvelde Lake NL **39** C4
Gnadenthal MB **14** F4
Goatfell BC **7** F1
Goathard Mountain YT **4** E2
Goat Island BC **1** E6
Goat Island BC **2** C3
Goat Lake BC **2** C3
Goat Range PP BC **6** E4
Goat River BC **6** F4
Goat River BC **7** F1
Gobles ON **19** C6
Gobles ON **20** E5
Goblin NL **36** D5
Goblin NL **37** D3
Godbout QC **39** D1
Godbout QC **40** F6
Godbout, rivière QC **39** F1
Goderich ON **19** A3-4
Goderich ON **20** C2-3
Godfrey ON **21** D6
Gods Lake MB **15** F5
Gods Lake MB **15** F5-6
Gods Lake Narrows MB **13** A5
Gods Lake Narrows MB **15** F5
Gods Mercy, Bay of NU **40** A3
Gods Mercy, Bay of NU **41** D6
God's Pocket PMP BC **1** D4
Gods River MB **15** F5
Gods River MB **15** F5-6
Gods River MB **40** D1
Goeland, lac au QC **40** F4
Goéland, lac au QC **23** A6
Goéland, lac au QC **26** A-B1
Goéland, lac du QC **22** C5
Goélands, lac aux QC **25** A6
Goélands, lac aux QC **26** A6
Goélands, lac aux QC **26** E6
Goélands, lac aux QC **28** C1
Goélands, lac aux QC **28** D3
Goffs NS **33** B6
Gogama ON **17** D-E6
Go Home Lake ON **18** E7
Go Home River ON **18** E7
Goldboro NS **34** F5
Gold Bridge BC **1** D7
Gold Bridge BC **2** A5
Goldburn AB **11** B6
Gold Centre ON **17** D6
Gold Creek BC **6** F4
Gold Creek BC **7** F2
Golden BC **6** D3-4
Golden Ears PP BC **1** F7
Golden Ears PP BC **2** D5-6
Golden Hinde BC **1** E5
Golden Hinde BC **2** C-D1
Golden Lake ON **21** A5
Golden Lake ON **21** A5
Golden Lake ON **22** F5

Golden Lake ON **22** F5
Golden Prairie SK **6** E8
Golden Prairie SK **11** E2
Goldenville NS **34** F4
Gold Lake ON **21** C3
Gold Muchalat PP BC **1** E5
Goldpan PP BC **1** E8
Goldpan PP BC **2** B7
Goldpan PP BC **6** E1
Goldpines ON **13** E6
Goldpines ON **16** C2-3
Gold River BC **1** E5
Gold River BC **1** F5
Gold River BC **2** C1
Gold River NS **32** F7
Gold River NS **32** F7
Gold River NS **33** C4
Gold River NS **33** C4
Gold Rock ON **13** F7
Gold Rock ON **16** E3
Goldsand Lake MB **12** D7
Goldsand Lake MB **15** D2
Goldsmith Channel NU **41** A3
Goldsmith Channel NU **42** D-E3
Gold Spur AB **7** B8
Goldstone ON **19** A6
Goldstone ON **20** C5
Goldstream PP BC **2** F4
Golet, lac QC **39** D5
Goletas Channel BC **1** D4
Golspie ON **19** C5
Golspie ON **20** E4
Gondola Point NB **32** D4
Gondola Point NB **33** A1-2
Goobies NL **36** D8
Goobies NL **37** D6
Gooderham ON **21** C3
Goodeve SK **11** D6
Goodfare AB **9** E2
Goodfish Lake AB **6** A6-7
Goodfish Lake AB **8** A8
Goodfish Lake AB **9** F6-7
Goodfish Lake AB **11** A1
Goodfish Lake AB **12** F1
Good Hope AB **8** C7
Good Hope BC **7** F2
Good Hope Bay NT **5** E8
Good Hope Lake BC **3** A-B3
Good Hope Lake BC **4** F6-7
Good Hope Mountain BC **1** D6
Good Hope Mountain BC **2** A3
Goodlands MB **11** F8
Goodlands MB **13** F1-2
Goodlow BC **3** E8
Goodlow BC **9** D2
Goodrich Loomis CA ON **21** E4
Goodridge AB **6** A7
Goodridge AB **9** F7
Goodridge AB **11** A1-2
Goodridge AB **12** F1-2
Goodsir, Mount BC **6** D4
Goodsir, Mount BC **7** B1
Goodsoil SK **6** A8
Goodsoil SK **9** F8
Goodsoil SK **11** A3
Goodsoil SK **12** F3
Good Spirit Lake SK **11** D6-7
Good Spirit Lake PP SK **11** D6-7
Goodwater SK **11** F6
Goodwin Corner AB **9** F3
Goodwin Mill NB **31** B6-7
Goodwood ON **20** B7
Goodwood ON **21** E1
Goodwood, rivière QC **39** A2
Goose Bay NL **39** B-C6
Gooseberry Cove NL **36** E8
Gooseberry Cove NL **37** E6
Gooseberry Cove PP NL **36** E-F8
Gooseberry Cove PP NL **37** E-F6
Gooseberry Lake PP AB **6** C7-8
Gooseberry Lake PP AB **11** C2
Goose Cove NL **36** B8
Goose Cove NL **37** C7
Goose Cove NL **37** D6
Goose Cove NL **38** B5
Goose Cove NL **38** D8
Goose Harbour River NS **34** E5-6
Goose Island BC **1** C3
Goose Island NS **34** F5
Goose Islands PP Reserve MB **13** B2
Goose Lake MB **11** A7
Goose Lake MB **12** F7
Goose Lake MB **13** A1
Goose Lake MB **13** A5
Goose Lake MB **15** F1
Goose Lake NU **12** A7-8
Goose Lake NU **15** A2
Goose Lake ON **13** C6
Goose Lake SK **11** A-B7
Goose Lake SK **13** A1
Goose Lake SK **11** A1
Goose Pond NL **36** A4
Goose Pond NL **37** A2
Goose River NL **38** F3
Goose River NL **39** B5
Goose River PE **35** D6-7
Gordon SK **11** D5-6
Gordon Bay ON **18** D7
Gordon Bay PP BC **2** E3

Gordondale AB 9 E2
Gordon Horne Peak BC 6 C-D2
Gordon Lake AB 9 D7
Gordon Lake AB 12 D2
Gordon Lake NT 10 C7
Gordon Lake SK 12 E4
Gordon Pittock Reservoir ON 19 B-C5
Gordon Pittock Reservoir ON 20 E4
Gordon Point PE 35 E3
Gordon River BC 2 E-F3
Gordonsville NB 31 E3
Gordonsville NB 32 A1
Gore NS 32 E8
Gore NS 33 B5-6
Gore NS 34 F1
Gore Bay ON 18 B3
Gore Bay ON 18 B3
Gores Landing ON 21 E3
Gorge de Jacquet River, APP NB 29 E7
Gorge de Jacquet River, APP NB 31 B5-6
Gorgotton, lac QC 28 B7-8
Gorlitz SK 11 D7
Gorman, lac QC 25 D3
Gormanville NS 32 D8
Gormanville NS 33 A5-6
Gormanville NS 34 F1-2
Gormley ON 19 A8
Gormley ON 20 C7
Gormley ON 21 E1
Gorrie ON 19 A5
Gorrie ON 20 C4
Goschen Island BC 1 A2
Gosford, mont QC 27 F6
Goshawk Lake ON 14 D8
Goshen NB 32 C5
Goshen NS 34 E5
Gosnell BC 6 C2
Gosport ON 21 E6
Gossage River NT 5 D6
Gosselin, lac QC 25 B3
Gosselin, lac QC 28 B7
Goudreau ON 17 D4
Goudreau, lac QC 26 B4
Gouffre, rivière du QC 28 E7
Gough Lake AB 6 C6
Gough Lake AB 8 F8
Gouin, réservoir QC 26 E3-4
Goulais Bay ON 17 F4
Goulais Point ON 18 A7
Goulais River ON 17 F4
Goulais River PP ON 17 E-F4
Gould QC 27 E5
Gould Lake CA ON 21 D6
Goulds NL 37 E8
Gouldtown SK 11 E4
Goulet Lake MB 15 E4
Gounamitz River NB 29 E4
Gounamitz River NB 31 B2-3
Goupil, lac QC 39 C1
Gourlay Lake ON 17 C3
Gournet, lac QC 25 A7
Gournet, lac QC 26 F7
Gournet, lac QC 28 C2
Gousvaris, baie QC 23 B7
Gousvaris, baie QC 26 C1
Gouverneur SK 11 E-F3
Govan SK 11 D5
Govenlock SK 6 F8
Govenlock SK 11 F2
Government Landing ON 16 E2
Governor Lake NS 33 A-B7
Governor Lake NS 34 F3
Governors Island PE 35 E5
Gowan River MB 15 E5
Gowanstown ON 19 A5
Gowanstown ON 20 C4
Gowgaia Bay BC 1 C1
Gowganda ON 17 D6-7
Gowganda Lake ON 17 D-E7
Gow Lake SK 12 D5
Gowlland Tod PP BC 2 F4
Goyelle, lac QC 39 E5
Gracefield QC 22 E7
Gracefield QC 24 B1-2
Grady Harbour NL 39 B8
Grafton NB 31 F3
Grafton NB 32 B1
Grafton NS 32 E6
Grafton NS 33 B4
Grafton ON 21 E4
Graham ON 13 F8
Graham ON 16 E4-5
Graham, île BC 1 A1
Graham Corner NB 32 C1
Grahamdale MB 13 D3
Grahamdale MB 14 A3
Graham Head PE 35 D2
Graham Island BC 1 A1
Graham Island NU 42 C4
Graham Lake AB 9 D5
Graham Lake NT 10 C7
Graham Lake ON 21 D8
Graham Lake ON 24 F2
Graham Laurier PP BC 3 D6
Graham Moore, Cape NU 41 A6-7
Graham Moore, Cape NU 42 D6
Graham Moore Bay NU 42 D3
Graham River BC 3 D6-7 E7
Grahams Road PE 35 D3-4
Grainfield NB 31 D6
Grainger AB 7 A5
Grainger River NT 10 E2
Graminia AB 8 C6
Granada QC 23 E2
Granary Lake ON 18 A2
Granby QC 24 D7
Granby QC 27 E3
Granby PP BC 6 E2-3
Granby River BC 6 F2
Grand, lac QC 21 A8
Grand, lac QC 22 F8
Grand, lac QC 24 D2
Grand Anse NS 34 D6
Grand Bank NL 36 E6
Grand Bank NL 37 E4
Grand-Barachois NB 32 B7
Grand Bay NB 32 D4
Grand Bay NB 32 D3-4
Grand Bay NB 33 A1
Grand Bay NB 33 A1
Grand Bay NL 36 D1
Grand Bay East NL 36 D-E1
Grand Beach MB 13 E4
Grand Beach MB 14 C5-6
Grand Beach NB 36 E6
Grand Beach NB 37 E4
Grand Beach PP MB 13 E4
Grand Beach PP MB 14 C5-6
Grand Bend ON 19 B3-4
Grand Bend ON 20 D2-3
Grand Bruit NL 36 D2
Grand Calumet, île du ON 21 A6
Grand Calumet, île du QC 22 F6
Grand Canyon of the Stikine BC 3 C2-3
Grand Castor, lac du QC 25 B5
Grand Castor, lac du QC 26 F5
Grand Castor, lac du QC 28 D1
Grand Centre AB 6 A7-8
Grand Centre AB 9 F7-8
Grand Centre AB 11 A2
Grand Centre AB 12 F2
Grand Coulee SK 11 E5
Grande-Aldouane NB 31 D-E8
Grande-Aldouane NB 32 A6
Grande-Anse NB 29 E8
Grande-Anse NB 30 E1
Grande-Anse NB 31 B7
Grande-Anse QC 25 D7
Grande-Anse QC 27 A3
Grande-Anse QC 28 F3
Grande Cache AB 6 A2
Grande-Cascapédia QC 29 D7-8
Grande-Cascapédia QC 31 A6
Grande-Clairière MB 11 F8
Grande-Clairière MB 13 F1
Grande Décharge, lac de la QC 28 B5
Grande-Digue NB 31 E-F8
Grande-Digue NB 32 B6-7
Grande Digue Point PE 35 C1-2
Grande-Entrée QC 30 F8
Grande Entrée, île de la QC 30 E-F8
Grande Greve NS 34 D6
Grande-Île QC 24 E5
Grande Île, la QC 39 E-F3
Grande Passe, île de la QC 38 B1
Grande Passe, île de la QC 39 E7
Grande Pointe MB 13 F4
Grande Pointe MB 14 E5
Grande Pointe ON 19 D2
Grande Prairie AB 9 F2-3
Grande Rivière NB 29 F4
Grande Rivière NB 31 C3
Grande Rivière QC 30 C2
Grande-Rivière QC 30 C2
Grande Rivière, la QC 40 E4
Grande Rivière, la QC 40 E5
Grande-Rivière, ZEC de la QC 30 C2
Grande rivière de la Baleine QC 40 D4-5
Grandes-Bergeronnes QC 28 C8
Grandes-Bergeronnes QC 29 C1
Grandes-Piles QC 24 A7-8
Grandes-Piles QC 25 E7-8
Grandes-Piles QC 27 B3
Grandes Pointes, lac aux QC 28 A5
Grand Etang NS 34 B6
Grande-Vallée QC 30 A1
Grande-Vallée QC 39 F3
Grande-Vallée QC 40 F7
Grand Falls NB 31 D3
Grand Falls-Windsor NL 36 B6
Grand Falls-Windsor NL 37 B4
Grand-Fonds QC 28 E7
Grand Forks BC 6 F2
Grand Harbour NB 32 F2-3
Grandin AB 8 A8
Grandin, lac NT 10 B4
Grandin, lac NT 41 D1
Grandin, rivière NT 10 B4-5
Grand Island MB 11 B8
Grand Island MB 13 B2
Grand Island ON 20 A8
Grand Island ON 21 D2
Grand Island PP Reserve MB 11 B8
Grand Island PP Reserve MB 13 B2
Grand Jardin NL 36 C1
Grand lac Bostonnais QC 25 B8
Grand lac Bostonnais QC 28 D4
Grand lac Caotibi QC 39 E1
Grand Lac de l'Ours NT 41 C1
Grand Lac des Esclaves NT 10 E6-7
Grand Lac des Esclaves NT 41 E1
Grand lac Germain QC 39 D2
Grand lac Jourdain QC 26 A-B8
Grand lac Squatec QC 29 E2-3
Grand lac Squatec QC 31 B1
Grand lac Victoria QC 22 A5
Grand-Lac-Victoria QC 22 A5
Grand-Lac-Victoria QC 23 F5
Grand Lake NB 31 F6
Grand Lake NB 32 B-C4
Grand Lake NB 32 C1
Grand Lake NL 36 A-B4
Grand Lake NL 37 A-B2
Grand Lake NL 39 B5
Grand Lake NL 40 D7-8
Grand Lake NS 32 E8
Grand Lake NS 33 B6
Grand Lake ON 22 E4
Grand Lake Meadows PPA NB 31 F5-6
Grand Lake Meadows PPA NB 32 B-C4
Grand Lake Road NB 31 E6
Grand Lake Road NB 32 A4
Grand Lake Road NS 34 C8
Grand Le Pierre NL 36 D7
Grand Le Pierre NL 37 D5
Grand Manan NB 32 F2-3
Grand Manan, chenal du NB 32 F2
Grand Manan, île du NB 32 F3
Grand Manan, RNOM NB 32 F3
Grand Manan Channel NB 32 F2
Grand Manan Island NB 32 F2-3
Grand Manan NMBS NB 32 F3
Grand Marais MB 13 E4
Grand Marais MB 14 C5-6
Grand-Mère QC 24 A7
Grand-Mère QC 25 E7
Grand-Mère QC 27 B3
Grandmesnil, lac QC 39 D1
Grand-Métis QC 29 B3-4
Grandmother's Bay SK 12 E5-6
Grand Mountain BC 6 D3
Grand Narrows NS 34 C-D7
Grandois NL 38 B5
Grandois NL 39 D8
Grandora SK 11 C4
Grand Pabos, rivière du QC 30 C1-2
Grand Pacific Glacier BC 4 F2
Grand Portage, rivière du QC 26 D6-7
Grand Portage, rivière du QC 28 A2 B1-2
Grand Pré NS 32 E7
Grand Pré NS 33 B4
Grand-Pré, LHN de NS 32 E7-8
Grand-Pré, LHN de NS 33 B4-5
Grand Pré NHS NS 32 E7-8
Grand Pré NHS NS 33 B4-5
Grand Rapids MB 13 B2-3
Grand Rapids (IR) MB 13 B2-3
Grand Rapids PP MB 13 B2
Grand Rapids Wildland PP AB 9 D-E6
Grand Rapids Wildland PP AB 12 D1
Grand-Remous QC 22 D7-8
Grand-Remous QC 24 A2
Grand-Remous QC 25 E2
Grand River NS 34 D7
Grand River NS 34 D7
Grand River ON 19 A-B6
Grand River ON 19 C7
Grand River ON 20 B5
Grand River ON 20 C-D5
Grand River ON 20 E6
Grand River PE 35 D2
Grand-Saint-Esprit QC 24 B8
Grand-Saint-Esprit QC 27 C3
Grand-Sault NB 31 D3
Grands-Jardins, parc des QC 28 E6
Grand Sudbury ON 17 F7
Grand Sudbury ON 18 A5
Grand Tracadie PE 35 D5
Grand Valley ON 19 A6
Grand Valley ON 20 C5
Grandview AB 8 D6
Grandview MB 11 D8
Grandview MB 13 D1
Grandview PE 35 E-F6
Grandys Brook NL 36 D3
Grandys Brook NL 37 D1
Granet, lac QC 23 F5
Granet Lake NT 5 B7-8
Granger, Mount YT 4 E3-4
Granger Lake SK 12 B5
Granisle BC 3 F5
Granite Bay BC 1 E5
Granite Bay BC 2 B2
Granitehill Lake ON 17 C3
Granite Lake NL 36 C4
Granite Lake NL 37 C2
Granite Point NL 38 C4
Granite Point NL 39 E8
Graniteville QC 24 E8
Graniteville QC 27 F4
Grantham AB 7 D6-7
Grantham BC 2 C2
Grant Lake NT 10 A5
Granton NS 34 E3
Granton ON 19 B4
Granton ON 20 E3
Grant Point NT 10 D7
Grant's Creek PP ON 22 E4
Grantville NS 34 D6
Granum AB 6 E6
Granum AB 7 E5
Granville Centre NS 32 F5
Granville Centre NS 33 C2
Granville Ferry NS 32 F5
Granville Ferry NS 33 C2
Granville Lake MB 12 D8
Granville Lake MB 12 D7-8
Granville Lake MB 15 D2
Granville Lake MB 15 D-E2
Gras, lac QC 39 C2
Gras, lac de NT 10 A8
Gras, lac de NT 41 D2
Grasmere BC 6 F5
Grasmere BC 7 F2-3
Grassberry River SK 11 A6
Grassdale SK 11 F6
Grassie ON 19 C7-8
Grassie ON 20 E6-7
Grassland AB 9 F6
Grassland AB 12 F1
Grasslands NP SK 11 F3-4
Grassmere ON 21 B1-2
Grass River MB 14 C2-3
Grass River MB 15 E3-4
Grass River PP MB 12 F7-8
Grass River PP MB 15 F2
Grassy Island Fort NHS NS 34 E6-7
Grassy Island Lake AB 6 C7-8
Grassy Island Lake AB 11 C2
Grassy Lake AB 6 E7
Grassy Lake AB 7 E7
Grassy Lake AB 11 E1
Grassy Narrows ON 13 E6
Grassy Narrows ON 16 D2
Grassy Plains BC 1 A5
Grassy River-Mond Lake Lowlands & Ferris Lake Uplands, RNP ON 17 D6
Grassy River-Mond Lake Lowlands & Ferris Lake Uplands PNR ON 17 D6
Grates Cove NL 37 C8
Gravelbourg SK 11 E4
Gravel Hill NB 31 A5
Gravel Hill ON 24 E3
Gravelle BC 1 B7
Gravel River, RNP ON 16 E7
Gravel River, RNP ON 17 C1
Gravel River PNR ON 16 E7
Gravel River PNR ON 17 C1
Gravenhurst ON 18 E8
Gravenhurst ON 21 C1
Graves Island PP NS 32 F7
Graves Island PP NS 33 C5
Gray SK 11 E5-6
Gray Creek BC 6 F4
Grayling Fork YT 5 E1-2
Grayling River BC 3 A5
Grayling River BC 10 F1
Grayson SK 11 D-E7
Grayson Lake ON 16 C5
Graysville MB 13 F3
Graysville MB 14 E3-4
Grease River SK 12 A4
Greasy Lake NT 10 C3
Great Barasway NL 36 E8
Great Barasway NL 37 E6
Great Barren Lake NS 33 E2
Great Bay de l'Eau NL 36 E6
Great Bay de l'Eau NL 37 E4
Great Bear Lake NT 41 C1
Great Bear River NT 5 F8
Great Bear River NT 10 A2
Great Bona NL 36 E8
Great Bona NL 37 E6
Great Bras d'Or NS 34 C7
Great Brehat NL 38 A5
Great Brehat NL 39 D8
Great Central Lake BC 1 F5
Great Central Lake BC 2 D2
Great Codroy NL 36 D1
Great Colinet Island NL 37 F7
Great Duck Island ON 18 C2
Greater Napanee ON 21 E6
Greater Sudbury ON 17 F6-7
Greater Sudbury ON 18 A5
Great Falls MB 13 E4-5
Great Falls MB 14 C6-7
Great Glacier BC 3 D2
Great Glacier PP BC 3 D2
Great Gull Lake NL 36 C6
Great Gull Lake NL 37 C4
Great Gull Pond NL 36 A5
Great Gull Pond NL 37 A3
Great Gull Pond NL 38 F4
Great Harbour Deep NL 39 E8
Great La Cloche Island ON 18 B4
Great Paradise NL 36 E7-8
Great Paradise NL 37 E5-6
Great Plain of the Koukdjuak NU 41 C7-8
Great Pubnico Lake NS 33 F2
Great Rattling Brook NL 36 B6
Great Rattling Brook NL 37 B4
Great Slave Lake NT 10 D6-7
Great Slave Lake NT 41 E1
Great Snow Mountain BC 3 D6
Great Village NS 32 D8
Great Village NS 33 A6
Great Village NS 34 E1-2
Greaves Island BC 1 D4
Greece's Point QC 24 D4
Greely ON 21 B8
Greely ON 24 E2
Greely Fiord NU 42 B5
Green, lac QC 22 B6
Green, Mount YT 4 D5
Green Acres SK 11 B5
Greenan SK 11 D3
Greenbank ON 20 B8
Greenbank ON 21 E2
Green Bay MB 14 D6
Green Bay NL 38 E5
Green Bay NS 33 D4
Greenbough Lake ON 22 D3
Greenbush ON 21 C8
Greenbush ON 24 F2
Greenbush SK 11 B6
Greenbush Lake ON 13 D8
Greenbush Lake ON 16 C5
Green Court AB 6 A5
Green Court AB 8 B4
Green Cove BC 1 F6
Green Cove BC 2 E2
Greene Island ON 18 B1
Greene Valley PP AB 9 E3-4
Greenfield NS 32 E7
Greenfield NS 33 A7
Greenfield NS 33 B4-5
Greenfield NS 33 D4
Greenfield NS 34 E2-3
Greenfield ON 24 E4
Greenfield PE 35 E6
Green Gables House PE 35 D4
Green Hill NS 33 A7
Green Hill NS 34 E3
Green Hill PP NS 34 E3
Greenhurst-Thurstonia ON 21 D2-3
Green Inlet PMP BC 1 B3-4
Green Island Brook NL 38 B4
Green Island Cove NL 38 B4
Green Island Cove NL 39 D8
Green Lake BC 1 D8
Green Lake BC 6 D1
Green Lake SK 11 A3
Green Lake SK 11 A3-4
Green Lake SK 12 F3-4
Green Lake PP BC 1 D8
Green Lake PP BC 6 D1
Greenland NS 32 F5
Greenland NS 33 C2
Green Lane ON 24 D4
Greenly, île QC 38 B3
Greenmount PE 34 A1
Greenmount PE 35 B2
Green Oaks NS 33 A6
Green Oaks NS 34 F2
Greenoch Swamp CA ON 20 B3
Greenock ON 20 B3-4
Greenough Point ON 18 F4
Green Park PP PE 32 A8
Green Park PP PE 34 B1
Green Park PP PE 35 C2-3
Green Point NS 33 F3
Green River BC 2 B-C5
Green River NB 29 E3
Green River NB 31 B2
Green River ON 19 A8
Green River ON 20 C7
Green River ON 21 E-F1
Green River YT 4 E7
Green Road NB 31 F3
Green Road NB 32 B1
Greens Brook NS 33 A8
Greens Brook NS 34 E4
Greens Corner ON 19 D6
Greens Corner ON 20 F5
Green's Harbour NL 37 D7
Greenspond NL 37 A7
Greenspond NL 38 F8
Greenstone ON 16 D7
Greenstone ON 16 D7-8
Greenstone ON 17 B1
Greenstone ON 17 A2
Greenstone Mountain PP BC 1 E8
Greenstone Mountain PP BC 2 B8
Greenstone Mountain PP BC 6 E1
Greenstreet SK 6 B8
Greenstreet SK 11 B2
Greensville ON 19 B7
Greensville ON 20 D6
Greenvale PE 35 D4
Green Valley ON 24 E4
Greenville Station NS 32 C8
Greenville Station NS 34 E1
Greenwald MB 14 D6
Greenwater Lake ON 16 F5
Greenwater Lake PP SK 11 C6
Greenwater PP ON 17 C6-7
Greenway MB 14 F2
Greenway ON 19 B3-4
Greenway ON 20 D2-3
Greenwich PE 34 B3
Greenwich PE 35 D6
Greenwich Hill NB 32 D3-4
Greenwich Hill NB 33 A1
Greenwich Lake ON 16 E-F6
Greenwood BC 6 F2
Greenwood NS 32 E6
Greenwood NS 33 B3-4
Greenwood ON 20 C7-8
Greenwood ON 21 E2
Gregg MB 14 D2
Gregoire Lake AB 9 D7
Gregoire Lake AB 12 D2
Gregoire Lake PP AB 9 D7
Gregoire Lake PP AB 12 D2
Grégoires Mill ON 17 B-C6
Grenfell SK 11 E6-7
Grenville QC 24 D4
Grenville Channel BC 1 A-B3
Grenville Lake NT 10 A6
Gresham Lake MB 15 C5
Gretna MB 13 F3
Gretna MB 14 F4
Grew Lake AB 9 D6
Grew Lake AB 12 C-D1
Grey Island SK 12 C4
Grey Islands NL 38 C5
Grey River NL 36 C-D4
Grey River NL 36 D4
Grey River NL 37 C-D2
Grey River NL 37 D2
Greywood NS 32 F5
Greywood NS 33 C2-3
Gribbell Island BC 1 B3
Grief Point BC 1 B2
Griesbach AB 8 C6-7
Griffin SK 11 F6
Griffith ON 21 B5
Griffith Island NU 42 D4
Griffith Island ON 18 E5
Grimes Lake MB 12 B8
Grimes Lake MB 15 B3
Grimsby ON 19 C7-8
Grimsby ON 20 E6-7
Grimsby Beach ON 19 B8
Grimsby Beach ON 20 E7
Grimshaw AB 9 E3
Grimsthorpe ON 18 C3
Grindrod BC 1 E8
Grindstone, lac QC 22 C2-3
Grindstone Point MB 13 D4
Grindstone Point MB 14 A5-6
Grinnell Peninsula NU 42 C4
Griquet NL 38 A5
Griquet NL 39 D8
Grise Fiord (Aujuittuq) NU 42 C5
Griswold MB 11 E8
Griswold MB 13 E1-2
Griswold MB 14 E1
Grizzle Bear Lake NT 10 B7
Grizzly Bear Hills SK 9 E8
Grizzly Bear Hills SK 12 E2-3
Grizzly Ridge Wildland PP AB 9 F5
Groais Island NL 38 B-C5
Groais Island NL 39 E8
Grog Brook NB 29 E5
Grohman Narrows PP BC 6 F3
Grole NL 36 E5
Grole NL 37 E3
Grollier Lake SK 12 A5
Grondines QC 27 B4
Gronick, lac QC 26 C7
Gronick, lac QC 28 A2
Gronlid SK 11 B5-6
Gros Cap ON 17 F3-4
Gros-Cap QC 30 F7
Gros-Mécatina QC 39 E6
Gros Mécatina, île du QC 39 E6-7
Gros Mécatina, rivière du QC 39 E6
Gros-Mécatina, RNOM de QC 38 C1
Gros-Mécatina, RNOM de QC 39 E6-7
Gros-Mécatina NMBS QC 38 C1
Gros-Mécatina NMBS QC 39 E6-7
Gros Morne NL 38 E2
Gros Morne NL 38 E2-3
Gros Morne NL 39 F6-7
Gros-Morne NL 39 F6-7
Gros-Morne QC 29 A8
Gros-Morne QC 30 A1
Gros-Morne QC 39 F2-3
Gros-Morne, PN du NL 38 E2-3
Gros-Morne, PN du NL 39 F7
Gros Morne NP NL 38 E2-3
Gros Morne NP NL 39 F7
Gros Ours, lac du QC 25 A4-5
Gros Ours, lac du QC 26 F4-5
Gros Pate NL 38 D3
Grosse Île QC 27 A6
Grosse-Île QC 30 E7-8
Grosse Île, la QC 30 E7
Grosse Île, lac de la QC 25 B6
Grosse Île, lac de la QC 26 F6
Grosse Île, lac de la QC 28 D1
Grosse Île and the Irish Memorial NHS QC 27 A6-7
Grosse-Île-et-le-Memorial-des-Irlandais, LHN de la QC 27 A6-7
Grosse-Île-Nord QC 30 E7-8
Grosse Isle MB 13 E3-4
Grosse Isle MB 14 D5
Grosses Coques NS 33 D1
Grosses-Roches QC 29 B5
Groswater Bay NL 39 B7
Groswater Bay NL 40 D8
Grouard AB 9 E4
Grouard Lake NT 10 A5
Grouard Mission AB 9 E4
Groulx, monts QC 40 E-F6
Groundbirch BC 3 E8
Groundbirch BC 9 E1
Groundhog River ON 17 B6
Groundhog River ON 17 D5-6
Groundhog River PP ON 17 C5
Grouse Island SK 11 A8
Grouse Island SK 12 A3
Grovedale AB 9 F2
Groves Point PP NS 34 C7
Gruenthal SK 11 C4
Grund MB 14 E2
Grundy Lake PP ON 18 B6
Grunthal MB 13 F4
Grunthal MB 14 F5
Guay, lac QC 22 B2
Gué, rivière du QC 40 D5
Guéguen, lac QC 23 E5
Guelph ON 19 B6-7
Guelph ON 20 D5-6
Guelph Lake CA ON 19 A6
Guelph Lake CA ON 20 D5
Guénette QC 24 A3
Guénette QC 25 F3
Guénette, lac QC 25 B5
Guérin QC 17 D8
Guérin QC 23 F2
Guernesé, lac QC 38 A2
Guernsey SK 11 C5
Guernsey Cove PE 35 F7
Guernsey Island NL 36 A2
Guernsey Island NL 38 F1
Guichon Creek BC 1 E8
Guichon Creek BC 2 B8
Guichon Creek BC 6 E1
Guilds ON 19 E3
Guillaume-Delisle, lac QC 40 D4
Guillemot, lac QC 39 B1
Guines Lake NL 39 D5
Guiquac River NB 31 D4
Gulch NL 37 F7
Gulch Cove NL 36 E4
Gulch Cove NL 37 E2
Gulf of Georgia Cannery, LHN BC 2 E4-5
Gulf of Georgia Cannery NHS BC 2 E4-5
Gulf Shore NS 32 C8
Gulf Shore NS 34 D2
Gulf Shore PP NS 32 C8
Gulf Shore PP NS 34 D1-2
Gulf Shore PP NS 35 F3
Gulf River ON 16 D5
Gull Bay ON 16 D6
Gull Harbour MB 13 D4
Gull Harbour MB 14 B6
Gull Island NL 37 C-D7
Gull Island NL 40 E7
Gull Island Point NL 37 F7
Gulliver Lake ON 16 E4
Gullivers Cove NS 32 F4
Gullivers Cove NS 33 C1-2
Gull Lake AB 6 C5
Gull Lake AB 6 C5
Gull Lake AB 8 E6
Gull Lake AB 8 E6
Gull Lake MB 15 D-E5
Gull Lake NL 39 C5
Gull Lake ON 13 D7
Gull Lake ON 21 C2
Gull Lake SK 11 E3
Gull Pond NL 36 B8
Gull Pond NL 37 B6
Gull Pond NL 38 E4
Gull River ON 21 C2
Gull River PP ON 16 D5-6 E5
Gullrock Lake ON 13 D6
Gullrock Lake ON 16 C2
Gullwing Lake ON 13 E7
Gundahoo River BC 3 B5
Gundahoo River BC 4 F8
Gunisao Lake MB 13 B4
Gunisao River MB 13 A3-4 B4
Gunn AB 6 A5
Gunn AB 8 C5
Gunners Cove NL 38 A5
Gunning Cove NS 33 F3
Gunnworth SK 11 D3
Gunter ON 21 C5
Gunton MB 13 E3-4
Gunton MB 14 D5
Gurneyville AB 6 A7
Gurneyville AB 11 A2
Guthrie ON 18 F8
Guthrie ON 20 A7
Guthrie ON 21 D1
Guthrie Lake MB 12 E7-8
Guthrie Lake MB 15 E2
Guy AB 9 E3
Guyenne QC 23 C3
Guynemer MB 14 A2-3
Guysborough NS 34 E5
Guysborough NS 34 E5-6
Gwaii Haanas, RPN et site du patrimoine haïda BC 1 C1-2
Gwaii Haanas (projetée), réserve d'aire marine nationale BC 1 C2
Gwaii Haanas National Marine Reserve (proposed) BC 1 B-C2
Gwaii Haanas NPR and Haida Heritage Site BC 1 B1-2 C1-2
Gwich'in TP NT 5 C5
Gwillim, lac QC 26 A4-5
Gwillim Lake PP BC 3 F8
Gwillim Lake PP BC 9 E1
Gwynne AB 6 B6
Gwynne AB 8 D7
Gypsum Point NT 41 D-E1
Gypsumville MB 13 C3
Gypsumville MB 14 A3
Gyrfalcon Islands NU 40 C6

H

Habay AB 9 B3
Hache, lac de la QC 25 E3
Hache, lac la BC 1 C8
Hache, lac la BC 1 C8
Hache, lac la BC 6 C1
Hache, lac la BC 6 C1
Hackett AB 8 F8
Hackett, lac QC 25 D8
Hackett, lac QC 27 A3
Hacketts Cove NS 32 F8
Hacketts Cove NS 33 C5
Hadashville MB 13 F4
Hadashville MB 14 E6-7
Haddock AB 8 B3
Hadley Bay NU 41 A3
Hadley Bay NU 42 D-E2
Hafford SK 11 B4
Hagar ON 17 F7
Hagar ON 18 A6
Hagen SK 11 B5
Hagensborg BC 1 C5
Hagersville ON 19 C7
Hagersville ON 20 E6
Hague SK 11 C4
Hagwilget BC 3 F4
Ha! Ha!, baie des QC 28 C6
Ha! Ha!, lac QC 28 D6
Haida BC 1 A1
Haig BC 1 F8
Haig BC 2 D7
Haig, Mount BC 6 F5
Haig, Mount BC 7 F3
Haig Lake AB 9 D4
Haight AB 8 C8
Haig-Thomas Island NU 42 C4
Haileybury ON 17 E7-8
Haileybury ON 22 A1
Haileybury ON 23 F1
Haines Junction YT 4 E2-3
Haines Lake ON 18 D7
Hainsfield NS 32 F4
Hainsfield NS 33 C-D2
Hainsville ON 21 C8
Hainsville ON 24 F2-3
Hairy Hill AB 6 A6
Hairy Hill AB 8 C8
Hairy Hill AB 11 A1
Hakai Passage BC 1 C3-4
Hakai PRA BC 1 C3-4 D3-4
Halbrite SK 11 F6
Halbstadt MB 14 F5
Halcomb NB 31 D6
Halcrow MB 11 A8
Halcrow MB 13 A2
Haldimand QC 30 B3
Haldimand CA ON 19 C-D7
Haldimand CA ON 20 F6
Haldimand County ON 19 C7
Haldimand County ON 20 E6
Hale Lake MB 15 D4
Haley Lake NMBS NS 33 E3
Haley Station ON 21 A6
Haley Station ON 22 F6
Halfmoon Bay BC 1 F6
Halfmoon Bay BC 2 D4
Halfway Cove NS 34 E6
Halfway Lake AB 8 B6-7
Halfway Lake MB 15 F3
Halfway Lake PP ON 17 E6
Halfway Point NL 36 B2-3
Halfway Pond NL 37 B-C7
Halfway River BC 3 D-E7
Halfway River MB 15 F3
Halfway River NS 32 D7
Halfway River NS 33 A4-5

Haliburton ON **21** B3
Haliburton PE **35** B1
Haliburton Lake ON **21** B3
Halifax NS **32** F8
Halifax NS **33** B7
Halifax NS **33** C6
Halifax NS **34** F2-3
Halifax Harbour NS **33** C6
Halkett Bay PP BC **2** D4
Halkirk AB **6** C6
Halkirk AB **8** F8
Halkirk AB **11** C1
Hall, rivière QC **29** D8
Hall, rivière QC **30** D1
Hall, rivière QC **31** A7
Hallam Peak BC **6** C2
Hall Beach (Sanirajak) NU **41** B6
Hall Beach (Sanirajak) NU **42** F6
Hallboro MB **13** E2
Hallboro MB **14** D2
Hallebourg ON **17** B4-5
Hall Glen ON **21** D3
Hall Lake BC **3** A2
Hall Lake NU **4** F5
Hall Lake NU **41** B6
Hall Lake NU **42** F5
Hall Peninsula NU **40** A6
Halls Bay NL **38** E-F5
Halls Harbour NL **32** D6 E6-7
Halls Harbour NS **33** B4
Halls Hill NB **35** F2
Halls Lake ON **21** B2
Hallville ON **21** B8
Hallville ON **24** E2
Halvorgate SK **11** E4
Hamber PP BC **6** C3
Hamber PP BC **8** E1
Hamer Bay ON **18** D7
Hamill Creek BC **6** E4
Hamilton ON **19** B-C7
Hamilton ON **20** D-E6
Hamilton ON **20** E6
Hamilton PE **32** A8
Hamilton PE **34** B1
Hamilton PE **35** D3
Hamilton, lac QC **22** C3
Hamilton Inlet NL **39** B6-7
Hamilton Inlet NL **40** D8
Hamilton Sound NL **38** E7
Hamiota MB **11** E8
Hamiota MB **13** E1-2
Hamlin AB **6** A6
Hamlin AB **8** B8
Hamlin AB **11** A1
Hammond ON **24** D3
Hammonds Plains NS **32** F8
Hammonds Plains NS **33** C5
Hammondvale NB **32** D5
Hammondvale NB **33** A2
Hammone, lac QC **38** A3
Hammone, lac QC **39** D7
Ham-Nord QC **27** D5
Hampden NL **38** E3
Hampden NL **39** F8
Hampstead NB **32** D4
Hampstead NB **33** A1
Hampton NB **32** D4
Hampton NB **33** A1-2
Hampton NS **32** E5
Hampton NS **33** B2-3
Hampton ON **20** C8
Hampton PE **21** E2
Hampton PE **32** B8
Hampton PE **34** C2
Hampton PE **35** E4
Hampton Lake ON **16** A2
Ham-Sud QC **27** D5
Hamton SK **11** D7
Hamtown Corner NB **31** F4-5
Hamtown Corner NB **32** B3
Hanbury River NU **41** D3
Hanceville BC **1** C7
Hanctin, lac QC **39** C1
Hand Branch River NB **29** F5
Hand Branch River NB **31** C-D4
Handel SK **11** C3
Handsworth SK **11** E6-7
Haneytown NB **32** C3
Hanford Brook NB **32** D5
Hanford Brook NB **33** A2
Hanley NS **11** D4
Hanmer ON **17** F7
Hanna AB **6** C6-7 D6-7
Hanna AB **7** A6
Hanna AB **11** C1
Hannah Bay ON **40** F3
Hannah Lake SK **12** B6
Hanover ON **18** F5
Hansen Harbour NT **5** A4-5
Hansford NS **32** E7
Hansford NS **34** D1
Hanson Lake SK **12** F6
Hant's Harbour NL **37** C7
Hantsport NS **32** E7
Hantsport NS **33** B7
Hantzsch River NU **41** B8
Hantzsch River NU **42** F7
Hanwell NB **32** C2-3
Happy Adventure NL **37** D-E7
Happy Isle Lake ON **22** F3
Happy Valley TP NT **5** C5
Happy Valley-Goose Bay NL **39** C5

Happy Valley-Goose Bay NL **40** E7-8
Hara Lake SK **12** A-B7
Hara Lake SK **15** A1
Harbour Breton NL **36** E6
Harbour Breton NL **37** E4
Harbour Buffett NL **36** D-E8
Harbour Buffett NL **37** D-E6
Harbour Deep NL **38** D4
Harbour Grace NL **37** D7
Harbour Le Cou NL **36** D2
Harbour Main NL **37** E7-8
Harbour Mille NL **36** D7
Harbour Mille NL **37** D5
Harbour Round NL **38** E5
Harbourville NS **34** C-D5
Harbourville NS **32** E6
Harbourville NS **33** B3-4
Harcourt NB **31** F7
Harcourt NB **32** A5
Harcourt NL **36** C8
Harcourt NL **37** C6
Harcourt ON **21** B3
Harcus MB **14** C3
Harding MB **11** E8
Harding MB **13** E1-2
Harding Lake MB **15** D-E3
Harding Lake NT **10** D7
Hardings Point MB **32** D3-4
Hardings Point NB **33** A1
Hardisty AB **6** B7
Hardisty AB **11** B1
Hardisty Lake NT **10** A5
Hardwicke NB **31** D7-8
Hardwicke Island NL **1** E5
Hardwicke Island BC **2** B1
Hardwood Lake ON **21** B5
Hardwood Lands NS **32** E8
Hardwood Lands NS **33** B3
Hardwood Lands NS **34** F2
Hardwood Ridge NB **31** E8
Hardwood Ridge NB **32** B4
Hardy NB **35** F2
Hardy SK **11** F5
Hardy Channel PE **35** C2-3
Hardy Inlet BC **1** C-D4
Hardy Lake NT **10** A8
Hardy Lake PP ON **18** E7-8
Hare Bay NL **36** B8
Hare Bay NL **36** D5
Hare Bay NL **37** B6
Hare Bay NL **37** D3
Hare Bay NL **38** B5
Hare Bay NL **38** E7
Hare Bay NL **39** D8
Hare Bay Islands Ecological Reserve NL **38** A6
Hare Fiord NL **42** B4
Hare Hill NL **36** B2-3
Hare Indian River NT **5** E7-8
Hares Islands NL **39** A6
Hargrave MB **11** E8
Hargrave MB **13** D1
Hargrave Lake MB **11** A8
Hargrave Lake MB **12** F8
Hargrave Lake MB **13** A2
Hargrave Lake MB **15** F2
Hargwen AB **8** C2
Harley ON **19** C6
Harley ON **20** E5
Harlowe ON **21** C5
Harmattan AB **6** C5
Harmattan AB **7** A3
Harmon Lake ON **13** E8
Harmon Lake ON **16** D5
Harmony NS **33** A6-7
Harmony MB **13** F3
Harmony MB **14** E2
Harmony PE **35** C2
Harmony Islands PMP BC **1** E6
Harmony Islands PMP BC **2** C3-4
Harmony Mills NS **33** D3
Haro Strait BC **2** F4
Harpellville NS **34** F5
Harper ON **21** C7
Harper ON **24** F1
Harper, Mount YT **5** F2
Harper Creek AB **9** C5
Harper Lake SK **9** A7
Harper Lake SK **12** A3
Harperville MB **14** C4
Harp Lake NL **39** A5
Harptree SK **11** F5
Harpurhey ON **19** A4
Harpurhey ON **20** D3
Harricana, rivière QC **23** A3 B4
Harricana, rivière QC **23** D-E4
Harricana, rivière QC **40** F4
Harricott NL **37** E7
Harrietsfield NS **32** F8
Harrietsfield NS **33** C6
Harrietsville ON **19** C5
Harrietsville ON **20** E4
Harrigan Cove NS **33** B8
Harrigan Cove NS **34** F4
Harrington ON **19** B5
Harrington ON **20** E4
Harrington PE **35** D4-5
Harrington, îles QC **39** E6
Harrington Harbour QC **39** E6
Harriott Lake SK **12** D-C6
Harriott Lake SK **15** D1
Harriott River SK **12** D6
Harriott River SK **15** D1
Harris SK **11** C3-4
Harris, lac QC **22** A6
Harris, lac QC **23** F6

Harris, lac QC **25** B1
Harris Bay ON **21** A-B4
Harrisburg ON **19** B6
Harrisburg ON **20** E5
Harrisburg ON **18** F8
Harris Lake ON **18** C6
Harrison, Cape NL **39** A7
Harrison, Cape NL **40** D8
Harrison Hot Springs BC **1** F7-8
Harrison Hot Springs BC **2** D6
Harrison Islands NU **41** B5
Harrison Islands NU **42** F4
Harrison Lake BC **1** F7-8
Harrison Lake BC **2** C-D6
Harrison Mills BC **1** F7-8
Harrison Mills BC **2** D6
Harrison River BC **2** D6
Harrisons Corner ON **24** E3
Harrison Settlement NS **33** A4
Harrison Settlement NS **32** D7
Harriston ON **19** A5
Harriston ON **20** C4
Harrogate BC **6** D4
Harrogate BC **7** B1
Harrop BC **6** F3
Harrop Lake MB **13** C4-5
Harrow ON **19** E1
Harrowby MB **11** D7-8
Harrowby MB **13** D1
Harrowby Bay NT **5** A6-7
Harrowsmith ON **21** D6
Harry Lake ON **21** A3
Harry Lake ON **22** F3
Harry Lake Aspen PP NL **1** D-E8
Harry Lake Aspen PP NL **2** A7
Harry's Harbour NL **38** E5
Harry's Harbour NL **39** F8
Hart SK **11** F5
Hart, Mount YT **4** A1
Harte MB **14** D2
Hartell AB **7** C4
Hartford ON **21** D6
Hart Jaune, rivière QC **39** D1
Hartland NB **31** F3
Hartland NB **32** B1
Hartlen Point NS **33** C6
Hartley ON **20** A8
Hartley ON **21** D2
Hartley Bay BC **1** B3
Hartleyville AB **7** F4
Hartney MB **11** F8
Hartney MB **13** F1-2
Hart Ranges BC **3** E-F7
Hart River YT **5** F3
Hartshorn AB **8** F8
Hartsmere ON **21** B4-5
Hartsville PE **35** D-E4
Harty ON **17** B5
Harvey NB **32** C2
Harvey NB **32** C6
Harvey, anse QC **30** A6
Harvey Lake NB **32** C2
Harvie Heights AB **6** D4-5
Harvie Heights AB **7** B2
Harwill MB **13** D3
Harwill MB **14** B4
Harwood ON **21** E3
Harwood Island BC **2** C2-3
Harwood Islands ON **18** A7
Hasbala Lake NT **8** A8
Hasbala Lake MB **15** A1-2
Haskett MB **13** F3
Haskett MB **14** F4
Haslam Lake BC **2** C3
Hassel Sound NU **42** B-C3 C4
Hassett NS **33** D1-2.1
Hasté, lac QC **39** C1
Hastings NS **32** C7
Hastings NS **34** D1
Hastings ON **21** D4
Hastings Arm BC **3** E3
Hatchet Cove NL **36** C8
Hatchet Cove NL **37** C6
Hatchet Lake SK **12** B6
Hatchet Lake SK **12** B-C6
Hatchet Lake SK **15** B1
Hat Creek BC **2** A7
Hatfield NS **11** D5
Hatfield Point NB **32** C4
Hatfield Point NB **33** A1-2
Ha'thayim PMP BC **2** B-C2
Hatley QC **27** F4
Hatton SK **6** E8
Hatton SK **11** E2
Hattonford AB **8** C3-4
Haultain AB **8** D7
Haultain Lake SK **12** D4
Haultain River SK **12** D-E4
Haut-Bouctouche NB **31** E8
Haut-Bouctouche NB **32** A6
Haute-Aboujagane NB **35** E1
Hautes-Gorges-de-la-Rivière-Malbaie, parc des QC **28** D7
Hautes-terres-du-Cap-Breton, PN des NS **34** A6-7 B6-7
Hauteurs-de-Queenston, LHN des ON **20** E8
Haut-Pokemouche NB **30** E2
Haut-Pokemouche NB **31** B7-8
Haut-Shippagan NB **31** B8

Havelock NB **31** F7
Havelock NB **33** B-C5
Havelock NS **33** D1-2
Havelock ON **21** D4
Havelock QC **24** E5
Havelock QC **27** F1
Havelock Lake ON **21** B3
Havre, île du QC **39** F4
Havre-Aubert QC **30** F7
Havre Aubert, île du QC **30** F7
Havre-aux-Maisons QC **30** F7
Havre aux Maisons, île du QC **30** F7
Havre Boucher NS **34** D5
Havre-Saint-Pierre QC **39** E3-4
Havre-Saint-Pierre QC **40** F7
Hawarden SK **11** D4
Hawke Harbour NL **39** C4
Hawke Hill Ecological Reserve NL **37** E7-8
Hawke Island NL **39** C8
Hawke River NL **39** C7-8
Hawke's Bay NL **38** C3
Hawke's Bay NL **39** E7-8
Hawkesbury ON **24** D4
Hawkesbury Island BC **1** A-B3
Hawkestone ON **18** F8
Hawkestone ON **20** A7
Hawkestone ON **21** D1
Hawkesville ON **19** A6
Hawkesville ON **20** D5
Hawkins Corner ON **18** F8
Hawkins Corner ON **20** A7
Hawkins Corner ON **21** C1
Hawkins Lake SK **12** D4
Hawk Junction ON **17** D4
Hawkrock River SK **12** B5
Hawks, Cape NU **42** B5
Hawks Creek BC **1** C7
Hawkshaw NB **32** C2
Hawthorne Cottage NHS NL **37** D-E7
Hawtrey ON **19** C6
Hawtrey ON **20** E5
Hay Bay ON **21** E5-6
Hay Camp AB **9** A6
Hay Camp AB **10** F8
Hay Camp AB **12** A2
Hay Cove NS **34** D7
Hay Island ON **18** E5
Hay Lake AB **9** B3
Hay Lake ON **21** B3
Hay Lakes AB **6** B6
Hay Lakes AB **8** D7
Hay River AB **9** A4
Hay River AB **41** E1
Hay River AB **3** B8
Hay River BC **9** B2
Hay River NT **10** E6
Hay River NT **41** E1
Hay River TP NT **10** E6
Hay River TP NT **41** E1
Haydon ON **20** B8
Haydon ON **21** E2
Hayes Bay ON **21** B6-7
Hayes Bay ON **24** D1
Hayes Creek BC **2** D8
Hayes Peak YT **4** E4-5
Hayes River MB **13** A4
Hayes River MB **15** E6
Hayes River MB **40** C-D1
Hayes River NU **41** B-C5
Hayesville NB **31** E5
Hayesville NB **32** A3
Hayfield MB **14** E1
Haynes AB **8** F6
Haynes Point PP BC **6** F2
Hays AB **6** E7
Hays AB **7** D7
Hays AB **11** E1
Haysville ON **19** B5-6
Haysville ON **20** D4-5
Hayter AB **6** C8
Haywood MB **13** F3
Haywood MB **14** E3
Haywood Lake NT **10** B8
Hay-Zama Lakes Wildland PP AB **9** B2-3
Hazel MB **13** E4
Hazel MB **14** E6
Hazel Dell SK **11** C6
Hazel Hill NS **34** E6
Hazelbrook PE **35** E5
Hazeldean NB **31** D3
Hazeldean ON **24** E2
Hazelglen MB **13** E4
Hazelglen MB **14** D6
Hazell AB **7** E3
Hazelridge MB **13** E4
Hazelridge MB **14** D5-6
Hazelton BC **3** F4
Hazelton Mountains BC **3** E-F4
Hazelton Mountains BC **7** C-D2
Hazen, Lake NU **42** A5
Hazenmore SK **11** F4
Hazel Strait NT **42** C2-3
Hazen Strait NU **42** C2-3
Hazlet SK **11** E3
Headingley MB **14** E5
Head Lake ON **20** A8
Head Lake ON **21** C1
Head Lake QC **21** C2
Head Lake QC **21** C2
Head of Amherst NS **34** D1
Head of Bay d'Espoir NL **36** D6

Havelock NB **31** F7
Head of Bay d'Espoir NL **37** D4
Head of Chezzetcook NS **33** C6-7
Head of Hillsborough PE **35** D6
Head of Jeddore NS **33** C7
Head of Loch Lomond NS **34** D7
Head of Millstream NB **32** C5
Head of St. Margarets Bay NS **32** F7-8
Head of St. Margarets Bay NS **33** C5
Headquarters CA ON **20** B4
Head River ON **20** A7-8
Head River ON **21** C1-2
Head-Smashed-In **6** F5-6
Head-Smashed-In AB **7** E4
Hendrickson Island NT **5** B6
Head-Smashed-In Buffalo Jump PHS AB **6** E5-6 F5-6
Head-Smashed-In Buffalo Jump PHS AB **7** E4
Healey Lake ON **18** D7
Heaman MB **12** E7
Heaman MB **15** E1-2
Hearne SK **11** E5
Hearne Lake NT **10** D7
Hearst ON **17** B4
Hearst ON **40** F2-3
Heart Lake AB **9** F7
Heart Lake AB **12** F1-2
Heart River AB **9** E4
Heart's Content NL **37** D7
Hearts Content Cable Station PHS NL **37** D7
Heart's Delight NL **37** D7
Heart's Delight-Islington NL **37** D7
Heart's Desire NL **37** D7
Hearts Hill SK **6** C8
Hearts Hill SK **11** C2
Heaslip MB **14** E1
Heaslip ON **17** D7-8
Heatburg AB **8** F7
Heath AB **6** B7
Heath AB **11** B2
Heathcote ON **18** F6
Heathcote ON **20** A5
Heatherbrae AB **8** D7
Heatherdale PE **35** F6
Heather-Dina Lakes PP BC **3** E-F6
Heatherdown AB **8** C5-6
Heatherton NL **36** C2
Heatherton NS **34** E5
Heath Steele NB **29** F7
Heath Steele NB **31** C5-6
Hebbs Cross NS **33** D4
Hebbville NS **33** D4
Héber Down CA ON **20** C8
Heber Down CA ON **21** E2
Hébert NB **31** F8
Hébert NB **32** B6
Hébert, lac QC **23** E1-2
Hébert, lac QC **26** C3
Hébert, River NS **32** C-D7
Hébertville QC **28** C4-5
Hébertville-Station QC **28** C4-5
Hecate BC **1** D4
Hecate, détroit d' BC **1** B-C2 C3
Hecate Island BC **1** D4
Hecate Strait BC **1** A-B2
Heckman Pass BC **1** C6
Heckston ON **21** C8
Heckston ON **24** E2
Hecla MB **13** D4
Hecla MB **14** B5-6
Hecla, Cape NU **42** A5
Hecla and Griper Bay NT **42** C2
Hecla/Grindstone PP MB **13** D4
Hecla/Grindstone PP MB **14** B5-6
Hecla Island MB **13** D4
Hecla Island MB **14** B5-6
Hedley BC **6** F1
Hedley BC **2** E8
Heddery Lake SK **12** D4
Hedley BC **6** F1
Heffley Creek BC **2** A8
Heffley Creek BC **6** D1
Heidelberg ON **19** B6
Heidelberg ON **20** D5
Height of the Rockies PP BC **6** E4-5
Height of the Rockies PP BC **7** C-D2
Heinsburg AB **6** A7
Heinsburg AB **11** A2
Heisler AB **6** B-C6
Heisler AB **8** E8
Heisler AB **11** B1
Hekkla ON **18** D8
Hekkla ON **21** B1
Helena Island NU **42** C3
Helen Bay ON **18** B4
Helen Lake ON **21** C2
Helen Lake ON **17** C1
Helliwell PP BC **2** D3

Helmer Lake SK **9** A8
Helmer Lake SK **12** A4
Helston MB **14** D2
Hemaruka AB **7** A7
Hemer PP BC **2** E4
Hemford NS **32** F6
Hemford NS **33** C3-4
Heming Lake MB **12** F7-8
Heming Lake MB **15** F1-2
Hemlo ON **16** E-F8
Hemlo ON **17** C2
Hemmingford QC **24** E6
Hemmingford QC **27** F1
Hemphill, Cape NT **42** C2
Henday AB **8** F6
Henderson ON **21** C6
Henderson Corner YT **4** A2
Henderson Lake BC **1** F5-6
Henderson Lake BC **2** C2
Hendon SK **11** C6
Heney, lac QC **22** E7
Heney, lac QC **24** C2
Hen Island ON **19** F1
Henley Harbour NL **39** D8
Hennepin Island ON **18** B1
Henri, rivière QC **27** C5
Henribourg SK **11** B5
Henrietta Island NL **39** B6-7
Henrietta Maria, Cape ON **40** D3
Henry House, LHN NB **8** D1
Henry House NHS AB **8** D1
Henry Island SK **34** D5
Henry Kater, Cape NL **42** E7
Henry Kater, Cape NU **42** E7
Henryville QC **24** E7
Henryville QC **27** F2
Hensall ON **19** B4
Hensall ON **20** D3
Henvey Inlet ON **18** B-C6
Henvey Inlet ON **18** B-C6
Hepburn SK **11** C4
Hepburn Lake SK **12** E4
Hepworth ON **18** F5
Hepworth ON **20** A4
Herald PP BC **6** D2
Herbert SK **11** E4
Herbert Inlet BC **1** F5
Herbert Inlet BC **2** D1
Herbert Lake SK **12** A5
Herbert Point BC **1** D3-4
Herb Lake MB **12** F8
Herb Lake MB **15** F2
Herb Lake Landing MB **12** F8
Herb Lake Landing MB **15** F2
Herblet Lake MB **12** F8
Herblet Lake MB **15** F2
Herchmer MB **15** C5-6
Herdman QC **24** E5
Hereford, mont QC **27** F5
Heriot Bay BC **1** E5
Heriot Bay BC **2** C2
Hermanville PE **34** B4
Hermitage PE **35** F6
Hermitage Bay NL **36** D5-6
Hermitage Bay NL **37** D3-4
Hermitage-Sandyville N **36** E5-6
Hermitage-Sandyville NL **37** E3-4
Hermon ON **21** B4
Hernando Island BC **2** C2
Heron, île NB **31** A5
Heron Bay ON **16** F8
Heron Bay ON **17** C2
Heron Island NB **29** D7
Heron Island NB **31** A5
Herronton AB **7** C5
Herschel SK **11** D3
Herschel Island YT **5** A3
Herschel Island TP YT **5** A3
Hertel, lac QC **26** A7
Hertzberg Island NL **18** D6
Hervey-Jonction QC **25** E8
Hervey-Jonction QC **27** A-B3
Hesketh AB **7** A5
Hesquiat BC **1** F5
Hesquiat Lake PP BC **1** F5
Hesquiat Peninsula PP BC **1** F5
Hess Mountains YT **4** B5
Hesson ON **19** A6
Hesson ON **20** C-D4
Hess River YT **4** B4-5
Heward SK **11** F6
Heyden ON **17** F4
Heydon Lake BC **2** B2
Heywood Island ON **18** B4
Hiawatha ON **21** D3
Hibbard QC **25** B5
Hibbard QC **28** D1

Hibben Island BC **1** B1
Hibbs Cove NL **37** D7-8
Hibernia NS **33** D3
Hickey, lac QC **22** E6
Hickman's Harbour NL **37** C7
Hickory Corner ON **19** C4
Hickory Corner ON **20** E3
Hicks-Oke Bog, RNP ON **17** C6
Hicks-Oke Bog PNR ON **17** C6
Hickson ON **19** B5
Hickson ON **20** D4
Hickson Lake SK **12** C6
Hidden Bay SK **12** C6
Hidden Lake TP NT **10** C7
Higginsville NS **33** B7
Higginsville NS **34** F3
High Bank PE **34** D3
High Bank PE **35** F6
High Beach NL **36** F6
High Beach NL **37** F4
High Bluff MB **13** E3
High Bluff MB **14** D4
High Falls ON **17** F6
High Falls ON **18** A4
Highgate ON **19** D3
Highgate ON **20** F2
Highgate ON **20** C5-6
High Hill Lake MB **15** E4-5
High Hill River MB **15** E5
High Lakes Basin PP BC **6** D1
Highland Creek ON **21** E2
Highland Grove ON **21** B3-4
Highland Lake NT **10** C3
Highland Valley BC **2** B7
Highlands BC **2** F4
Highlands NL **36** C1
High Level AB **9** B3
High Prairie AB **9** E4
Highridge AB **8** B5-6
High River AB **6** E5
High River AB **7** C4
Highrock MB **12** E8
Highrock MB **15** E2
Highrock Lake MB **12** E8
Highrock Lake SK **12** D5
Highvale AB **8** C5
Highwater QC **27** F3-4
Highway AB **8** E4
Highwood House AB **7** D3
Highwood River AB **6** E5
Highwood River AB **7** C-D3
Hilbre MB **13** D3
Hilbre MB **14** A3
Hilda AB **6** E8
Hilda AB **11** E2
Hilden NS **33** A6
Hilden NS **34** E2
Hillandale NB **31** E3
Hillandale NB **32** A1
Hillcrest Mines AB **6** F5
Hillcrest Mines AB **7** E3
Hillgrade NL **38** E6
Hillgrove NS **32** F4
Hillgrove NS **33** C2
Hilliard AB **8** C7
Hilliard AB **8** C7-8
Hilliard AB **11** A1
Hilliard's Bay PP AB **9** D4
Hilliardton AB **17** D8
Hilliardton AB **23** F1
Hillier ON **21** E5
Hilliers BC **2** D3
Hillman Marsh CA ON **19** E2
Hillmond SK **6** B8
Hillmond SK **11** B2
Hillsborough NB **32** C6
Hillsborough Bay PE **34** C2
Hillsborough Bay PE **35** E5
Hillsborough River PE **34** C2-3
Hillsborough River PE **35** E5
Hillsburgh ON **19** A7
Hillsburgh ON **20** D6
Hillsburn NS **32** F5
Hillsdale MB **12** D7
Hillsdale NS **33** A2
Hillsdale ON **18** F7
Hillsdale ON **20** A6
Hillside AB **8** D6
Hillside NS **32** C6
Hillside NS **34** C8
Hillside NS **34** F1
Hillsport ON **16** E8
Hillsport ON **17** B3
Hillspring AB **6** F6
Hillsvale NS **32** E8
Hillsvale NS **33** B5
Hillsvale NS **34** F1
Hilltop MB **14** C1
Hilltop Lake NT **10** B8
Hillview NL **36** C8
Hillview NL **37** C6
Hilton ON **21** B4
Hilton Beach ON **17** F4
Hilton Falls CA ON **19** B7
Hilton Falls CA ON **20** D6
Hinchcliffe SK **11** C6-7
Hinds Brook NL **36** A4
Hinds Brook NL **37** A2
Hinds Lake NL **36** B4
Hinds Lake NL **37** B2

Hines Creek AB **9** D3
Hinton AB **6** B3
Hinton AB **8** C1-2
Hippa Island BC **1** A1
Hirsch SK **11** F7
Hislop Lake NT **10** C5
Hitchcock SK **11** F6
Hitchcock Bay SK **11** D4
Hitchie Creek PP BC **2** B7
Hixon BC **1** B7
Hjalmarson Lake MB **12** C7
Hjalmarson Lake MB **15** C2
Hnausa MB **13** D4
Hnausa MB **14** B5
Hnausa PP MB **13** D4
Hnausa PP MB **14** B5
Hoadley AB **6** B5
Hoadley AB **8** D-E5
Hoards ON **21** D4
Hoare Bay NU **42** F8
Hobbema AB **6** B6
Hobbema AB **8** E6
Hobson Lake BC **6** C1-2
Hochfeld MB **13** F3
Hochfeld MB **14** F4
Hochstadt MB **14** F4
Hockin AB **15** E-F3
Hocking Lake SK **12** B4-5
Hockley ON **20** C6
Hockley Valley, RNP ON **20** C5-6
Hockley Valley PNR ON **20** C5-6
Hodge's Cove NL **37** C-D7
Hodges Hill NL **36** A6
Hodges Hill NL **37** A4
Hodges Hill NL **38** F5
Hodgeville SK **11** E4
Hodgson AB **13** D3-4
Hodgson MB **14** B4
Hodson NS **34** D3
Hoey SK **11** B5
Hoffer SK **11** F6
Hogan Lake ON **22** E3
Hogem Ranges BC **3** E5
Hogg, Mount YT **4** D5
Hogg Lake MB **15** C4
Hog Island NS **32** A8
Hog Island PE **34** B1
Hog Island PE **35** D2
Hokanson Point MB **14** B5
Holbein SK **11** B4-5
Holberg BC **1** E3-4
Holberg Inlet BC **1** E3-4
Holborn AB **8** C5
Holbrook ON **19** C5-6
Holbrook ON **20** E4-5
Holden AB **6** B6
Holden AB **8** D8
Holden AB **11** B1
Holdfast SK **11** D5
Hole-in-the-Wall PP BC **3** F7-8
Hole-in-the-Wall PP BC **9** E-F1
Holgar Lake SK **12** C5
Holiday Beach CA ON **19** E1
Holinshead Lake ON **16** D5
Holland MB **13** F2-3
Holland MB **14** E3
Holland Centre ON **18** F6
Holland Centre ON **20** A4-5
Holland Landing ON **20** B7
Holland Landing ON **21** E1
Holland Landing Prairie, RNP ON **20** B7
Holland Landing Prairie, RNP ON **21** E1
Holland Landing Prairie PNR ON **20** B7
Holland Landing Prairie PNR ON **21** E1
Holland River ON **20** B6-7
Holland River ON **21** E1
Hollow Lake AB **8** B7
Hollow Water MB **13** D4
Hollow Water MB **14** B6
Holly ON **20** B6
Holman NT **41** A2
Holman NT **42** E1
Holmes Lake MB **15** D4
Holmes River BC **6** B2
Holmesville NB **31** E3
Holmesville NB **32** A1
Holmesville ON **19** A4
Holmesville ON **20** C3
Holmfield MB **13** F2
Holmfield MB **14** F2
Holstein ON **18** B4-5
Holt ON **20** B7
Holt ON **21** E1
Holton NL **39** A7
Holton Island NL **39** A7
Holtville NB **31** E5
Holtville NB **32** A3
Holtyre ON **17** C-D7
Holyrood NL **37** E7-8
Holyrood ON **20** D3
Holyrood Pond NL **37** F7
Homards, baie des QC **39** F2
Homathko Estuary PP BC **1** D6
Homathko Estuary PP BC **2** A2-3
Homathko Icefield BC **1** D6
Homathko River BC **1** D6
Homathko River BC **2** A2
Homathko River-Tatlayoko PPA BC **1** D6
Home Bay NU **41** B8
Home Bay NU **42** E7
Homebrook MB **13** C2-3
Homebrook MB **14** A2-3

Homefield *SK* **11** D6
Homeglen *AB* **8** E6
Home Pond *NL* **36** A8
Home Pond *NL* **37** E6
Homestead-Motherwell,
LHN du *SK* **11** E6
Homeville *NS* **34** C8
Homewood *MB* **13** F3
Homewood *MB* **14** E4
Homfray Channel *BC* **1** E6
Homfray Channel *BC*
2 B-C3
Hondo *AB* **9** F5
Hone *MB* **12** D7
Hone *MB* **15** D1-2
Honey Harbour *ON* **18** E7
Honeydale *NB* **32** D2
Honeymoon Bay *BC* **2** E3
Honeywell Corners *ON*
21 E5
Honeywood *ON* **20** B5
Honfleur *QC* **27** B6
Honguedo, détroit d'
30 A3-4
Honguedo, détroit d'
39 F3
Honguedo, détroit d' *QC*
40 F7
Honguedo, Strait of *QC*
39 F3
Honguedo, Strait of *QC*
40 F7
Honora *ON* **1** F8
Hood River *NU* **41** C2-3
Hooker Lake *ON* **13** E8
Hoole Canyon *YT* **4** D5
Hoole River *YT* **4** D5
Hooper Island *NT* **5** A4
Hooping Harbour *NL* **38** C4
Hoosier *SK* **6** C7
Hoosier *SK* **11** C-D2
Hootalinqua *YT* **4** D4
Hope *BC* **1** F8
Hope *BC* **2** D7
Hopeall *NL* **37** D7
Hope Bay *ON* **18** E5
Hope Bay Forest, RNP *ON*
18 E5-6
Hope Bay Forest PNR *ON*
18 E5-6
Hopedale *NL* **39** A6
Hopedale *NL* **40** D7
Hopedale Mission NHS *NL*
39 A5-6
Hopedale Mission NHS *NL*
40 D7
Hopefield *PE* **35** F6
Hope Island *ON* **18** E7
Hope Mill CA *ON* **21** E3
Hope Town *QC* **30** D1
Hope Town *QC* **31** A7
Hopes Advance, cap *QC*
40 B5-6
Hopetown *ON* **21** B7
Hopetown *ON* **24** E1
Hopeville *ON* **20** B5
Hopewell *NS* **33** C5
Hopewell *NS* **34** E3
Hopewell Cape *NB* **32** C6
Hopewell Islands *NU* **40** C4
Hopewell Islands *NU*
41 F7-8
Hopewell Rocks PP *NB*
32 C6
Horburg *AB* **8** E4
Horen *AB* **8** C5
Horizon *SK* **11** F5
Hornaday River *AB* **9** A6
Hornaday River *AB* **10** F8
Hornaday River *AB* **12** A2
Hornaday River *NT* **41** B1
Hornby *ON* **19** A-B7
Hornby *ON* **20** D6
Hornby Island *BC* **1** F6
Hornby Island *BC* **2** D2-3
Hornby Island *BC* **2** D3
Hornby Lake *ON* **16** A2
Horndean *AB* **14** F4
Horne Lake *BC* **2** D2-3
Horne Lake PP *BC* **2** D2
Horneline Creek PP *BC* **3** B4
Hornell Lake *NT* **10** D4
Hornepayne *ON* **17** C3
Hornepayne *ON* **40** F7
Horner Creek *ON* **19** B-C6
Horner Creek *ON* **20** D-E5
Horne Settlement *NS* **32** E8
Horne Settlement *NS* **33** B6
Horne Settlement *NS* **34** E2
Hornes Road *NS* **34** C8
Horning's Mills *ON* **20** B5
Horn Lake *ON* **18** C8
Horn Lake *ON* **18** D8
Horn Lake *ON* **21** A1
Horn Lake *ON* **21** B1
Horn Lake *ON* **22** F1
Horn Mountains *NT* **10** D4
Horn Peak *YT* **4** B5-6
Horn River *NT* **10** D4-5
Horn River *NT* **41** D1
Horsefly *BC* **1** C8
Horsefly *BC* **6** C1
Horsefly Lake PP *BC* **1** C8
Horsefly Lake PP *BC* **6** C1
Horsefly Lake Reservoir *AB*
6 F7
Horsefly Lake Reservoir *AB*
7 E6-7
Horsefly Lake Reservoir *AB*
11 E1
Horsefly River *BC* **1** C8
Horsefly River *BC* **6** C1
Horse Island *MB* **13** B3
Horse Islands *NL* **38** D5

Horse Islands *NL* **39** E8
Horse Lakes *SK* **11** D6
Horse River *AB* **9** D6-7
Horse River *AB* **12** D1
Horseshoe Bay *AB* **6** A7
Horseshoe Bay *AB* **11** A1
Horseshoe Bay *BC* **1** F6-7
Horseshoe Bay *BC* **2** D4
Horseshoe Lake *AB* **8** F6
Horseshoe Lake *BC* **2** C3
Horseshoe Lake *ON* **13** C8
Horseshoe Lake *ON*
16 A4-5
Horseshoe Lake *ON* **18** A3
Horseshoe Lake *ON*
18 D7-8
Horseshoe Lake *ON* **21** C2
Horseshoe Slough
Territorial Habitat
Protection Area *YT* **4** B4
Horseshoe Valley *ON* **18** F8
Horseshoe Valley *ON* **20** A7
Horseshoe Valley *ON* **21** D1
Horsethief Creek *BC* **6** E4
Horsethief Creek *BC* **7** C1
Horsham *SK* **6** E8
Horsham *SK* **11** E2
Horton *NS* **32** E7
Horton *NS* **33** B4-5
Horton, rivière *QC* **29** D2-3
Horton Lake *NT* **41** B1
Horton River *NT* **5** B7-8
Horton River *NT* **41** A-B1
Horwood *NL* **38** F7
Horwood Lake *ON* **17** D5
Hoselaw *AB* **6** A7
Hoselaw *AB* **11** A2
Hoselaw *AB* **12** F2
Hosmer *BC* **6** F5
Hosmer *BC* **7** E3
Hotchkiss *AB* **9** D3
Hôtel-Dalvay-by-the-Sea,
LHN de l' *PE* **34** C2-3
Hôtel-Dalvay-by-the-Sea,
LHN de l' *PE* **35** D5
Hotham Sound *BC* **1** E6
Hotham Sound *BC* **2** C3
Hottah Lake *NT* **10** A4
Hottah Lake *NT* **41** C1
Hough Lake *ON* **18** A2-3
Houghton Centre *ON*
19 D5-6
Houghton Centre *ON*
20 F4-5
Houghton Lake *SK* **11** C5
House Mountain *AB* **9** F4
House River *AB* **9** E6
House River *AB* **12** D-E1
Houseys Rapids *ON* **21** C1-2
Houston *BC* **1** A5
Houston Stewart Channel
BC **1** C2
Howard *NB* **31** E6
Howard *NB* **32** A4
Howard, lac *QC* **22** D8
Howard, lac *QC* **24** A2
Howard, lac *QC* **25** E2
Howden *MB* **14** D2
Howdenvale *ON* **18** E4-5
Howe Bay *PE* **35** E7
Howe Bay *PE* **35** F7
Howe Island *ON* **21** D7
Howell Point *MB* **13** B3
Howells River *NL* **39** A2
Howe Point *PE* **34** C4
Howe Point *PE* **35** E7
Howe Sound *BC* **1** F6-7
Howe Sound *BC* **2** D4
Howick *QC* **24** E5
Howick *QC* **27** F1
Howie *AB* **7** B7
Howie Centre *NS* **34** C7-8
Howlan *PE* **35** B1
Howley *NL* **36** A4
Howley *NL* **37** A2
Howley *NL* **38** F3
Howley *NL* **39** F8
Howse Pass *BC*
Howse Pass NHS *AB* **6** C3-4
Howser *BC* **6** E3
Howser Creek *BC* **6** E3-4
Hoyle *ON* **17** C6
Hoyt *NB* **32** D3
H.R. Frink CA *ON* **21** D-E5
Huallen *AB* **9** F2
Huard, lac *QC* **28** C6
Huards, lac aux *QC* **26** B8
Huards, lac aux *QC* **28** A6
Hubbard *SK* **11** D6
Hubbard, Mount *YT* **4** C7
Hubbards *NS* **32** F7
Hubbards *NS* **33** C5
Hubbards PP *NS* **32** F7
Hubbards PP *NS* **33** C5
Hubbart Point *MB* **15** A5
Huberdeau *QC* **24** C4
Hubert Lake Wildland PP *AB*
8 A5
Hubert Lake Wildland PP *AB*
9 F5
Hubley *NS* **32** F8
Hubley *NS* **33** C5
Huddersfield, lac *QC* **22** E6
Hudson *AB* **8** F7
Hudson *ON* **16** D3
Hudson *QC* **24** D5
Hudson, baie d' *MB* **15** C7-8
Hudson, baie d' *NU* **41** F5-7
Hudson, baie d' *QC* **40** C2-3
Hudson, détroit d' *NU*
40 A-B5
Hudson, détroit d' *NU*
41 D8
Hudson Bay *MB* **15** B7

Hudson Bay *NU*
41 E5-7 F5-7
Hudson Bay *QC* **40** B2-3
Hudson Bay *SK* **11** B7
Hudson's Hope *BC* **9** E1
Hudson's Hope *BC* **2** E1
Hudson Strait *NU* **40** A-B5
Hudson Strait *NU* **41** D8
Hudwin Lake PP Reserve
MB **13** B5
Hughenden *AB* **6** C7
Hughenden *AB* **11** C1-2
Hughes *ON* **23** E1
Hughes Brook *NL* **36** A-B3
Hughes Brook *NL* **37** A-B1
Hughes River *MB* **12** D7-8
Hughes River *MB* **15** D2
Hughton *SK* **11** D3
Huguenin, lac *QC* **25** A4
Huguenin, lac *QC* **26** E-F4
Hu Haven *AB* **8** B-C7
Huit, lac du *QC* **27** C-D6
Huit Chutes, lac des *QC*
28 B6
Hull *QC* **21** A8
Hull *QC* **24** D2
Humber Arm South *NL*
36 A-B2
Humber River *NL* **36** A-B3
Humber River *NL* **37** A-B1
Humber River *ON* **20** C6
Humber Village *NL* **36** B3
Humber Village *NL* **37** B1
Humboldt *SK* **11** C5
Hume *SK* **11** F6
Hume River *NT* **5** E-F6
Humphrey *ON* **18** D7-8
Humphrey Corner *NB* **31** F6
Humphrey Corner *NB* **32** B4
Humpy Lake *NT* **10** A6-7
Humqui, rivière *QC* **29** C4-5
Hungry Creek *YT* **5** E3-4 F3
Hungry Lake *ON* **21** C6
Hunta *ON* **17** C6
Hunter Bay *SK* **12** F5-6
Hunter Creek *SK* **12** B-C4
Hunter Island *BC* **1** C4
Hunter Island *ON* **16** F3-4
Hunter Lake *ON* **18** A4
Hunter Point *BC* **1** B1
Hunter River *PE* **34** C2
Hunter River *PE* **35** D4
Hunter River *PE* **35** D4
Hunters Mountain *NS* **34** C6
Hunters Point *NL* **17** E8
Hunter's Point *QC* **22** B2-3
Hunterstown *QC* **24** B7
Hunterstown *QC* **25** F7
Hunterstown *QC* **27** C2
Huntingdon *BC* **1** F7-8
Huntingdon *BC* **2** E6
Huntingdon *QC* **24** E5
Huntingdon Island *NL*
39 B7
Huntingdon Island *NL*
40 D8
Huntingford *ON* **19** B5
Huntingford *ON* **20** E4
Hunting River *MB* **15** D4
Huntingville *QC* **27** E4-5
Huntley *PE* **35** B2
Huntoon *SK* **11** F6
Hunt River *NL* **39** A5
Hunts Point *NS* **33** E4
Hunts Pond *NL* **36** B7
Hunts Pond *NL* **37** B5
Huntsville *ON* **18** D8
Huntsville *ON* **21** B1
Huntsville *ON* **21** B1-2
Hurd, Cape *ON* **18** D4
Hurds Lake *ON* **21** A6
Hurkett *ON* **16** E6
Hurley River *BC* **2** A-B5
Huron, lac *ON* **18** E2-3
Huron, lac *ON* **19** B3
Huron, lac *ON* **20** C1-2
Huron, Lake *ON* **18** D2-3
Huron, Lake *ON* **19** A2-3
Huron, Lake *ON* **20** B1-2
Huron, rivière *QC* **27** B5
Huron East *ON* **19** A4-5
Huron East *ON* **20** C3-4
Huron Heights *ON* **19** C3
Huron Heights *ON* **20** E2
Huron Park *ON* **19** B4
Huron Park *ON* **20** D3
Hurons, rivière des *QC*
27 A5
Husavik *MB* **14** C5
Husky Channel *NT* **5** C4
Hussar *AB* **6** D6
Hussar *AB* **7** B5
Hutchison Bay *NT* **5** A5
Hutch Lake *AB* **9** B3-4
Hutshi Lake *YT* **4** D3
Hutton Heights *ON* **19** A4
Hutton Heights *ON* **20** C3
Huttonville *ON* **19** A7
Huttonville *ON* **20** C6
Huxley *AB* **6** C6
Huxley *AB* **7** B4
Huxley Island *BC* **1** C2
Hyas *SK* **11** C7
Hyas *SK* **13** C1
Hyde Park *ON* **19** C4
Hyde Park *ON* **20** E3
Hyland River *YT* **4** D-E7
Hyland River *NT* **4** E7
Hyland River PP *NT* **4** F7
Hylo *AB* **8** A8
Hylo *AB* **9** F6
Hylo *AB* **12** F1

Hymers *ON* **16** F5
Hyndford *ON* **21** A5-6
Hyndford *ON* **22** F5-6
Hyndman Lake *NT* **5** C5-6
Hyperite Point *NU* **42** C4
Hythe *AB* **9** E2

I

Ian Lake *BC* **1** A1
Iberville *QC* **24** E6-7
Iberville *QC* **27** F2
Iberville, lac d' *QC* **40** D5
Iberville, ZEC d' *QC* **28** B8
Iberville, ZEC d' *QC* **29** B1
Ibex Valley *YT* **4** E3
Iceberg Point *NU* **42** B4
Ice Mountain *BC* **1** A8
Ice Mountain *BC* **6** A1
Ice Mountain *BC* **9** F1
Iconoclast Mountain *BC*
6 D3
Ida *ON* **21** E3
Ida, lac *QC* **26** A5
Iddesleigh *AB* **6** D7
Iddesleigh *AB* **7** C7
Iddesleigh *AB* **11** D1
Ideal *MB* **14** C4
Iffley *SK* **11** F6
Igloolik (Iglulik) *NU* **41** B6
Igloolik (Iglulik) *NU* **42** F5-6
Igluligaarjuk (Chesterfield
Inlet) *NU* **40** A2
Igluligaarjuk (Chesterfield
Inlet) *NU* **41** D-E5
Iglulik (Igloolik) *NU* **41** B6
Iglulik (Igloolik) *NU* **42** F5-6
Ignace *ON* **13** F7
Ignace *ON* **16** E3-4
Ijiraliq, parc territorial *NU*
40 A2
Ijiraliq, parc territorial *NU*
41 E5
Ijiraliq TP *NU* **40** A2
Ijiraliq TP *NU* **41** E5
Ikeda Point *BC* **1** C2
Ikpiarjuk (Arctic Bay) *NU*
41 A5
Ikpiarjuk (Arctic Bay) *NU*
42 D-E5
Ilford *MB* **15** E5
Ilderton *ON* **19** C4
Ilderton *ON* **20** E3
Île-à-la-Brume, RNOM de l'
QC **39** E5-6
Ile-à-la-Brume NMBS *QC*
39 E5-6
Île-à-la-Crosse *SK* **12** E3
Île-à-la-Crosse (IR) *SK*
12 E4
Île-à-la-Crosse, lac *SK*
12 E3-4
Île au Castor, lac de l' *QC*
38 A1
Île au Castor, lac de l' *QC*
40 D8
Île-aux-Canes, RNOM de l'
NL **38** C5-6
Île aux Canes NMBS *NL*
38 C5-6
Île-Bizard *QC* **24** D5
Île-Bizard *QC* **27** E1
Île-Boot, RNF de l' *NS* **32** E7
Île-Boot, RNF de l' *NS* **33** B5
Île-Brion, refuge écologique
QC **30** E8
Île-Chantry, RNOM de l'
NU **18** F4
Île-Chantry, RNOM de l'
NU **20** A3
Ile-Corossol NMBS *QC*
39 F2
Île-d'Entrée *QC* **30** D7
Île-de-Carillon, RNF de l' *QC*
24 D4-5
Île-de-Carillon National
Wilderness Area *QC*
24 D4-5
Île-de-Salaberry *QC* **24** E5
Île-de-Salaberry *QC* **27** F1
Île-du-Corossol, RNOM de l'
QC **39** F2
Île-Dupas *QC* **24** C7
Île-Dupas *QC* **27** D2
Île-du-Prince-Édouard, PN
de l' *PE* **34** B3
Île-du-Prince-Édouard, PN
de l' *PE* **34** B2-3
Île-du-Prince-Édouard, PN
de l' *PE* **35** D4-5
Île-Eleanor, RNF de l' *ON*
18 E8
Île-Eleanor, RNF de l' *ON*
21 C1
Île-Grassy, LHN de l' *ON*
34 E6-7
Île-Kendall, RNOM de l' *NT*
5 A3-4
Île-Machias-Seal, RNOM de
l' *NB* **32** F2
Île-Mohawk, RNF de l' *ON*
19 D8
Île-Mohawk, RNF de l' *ON*
20 F7
Île-Navy, LHN de l' *ON*
20 E8
Île-Portage, RNF de l' *NB*
31 C7-8
Iles, lac des *SK* **9** F8

Îles, lac des *ON* **16** E5
Îles, lac des *QC* **22** D8
Îles, lac des *QC* **24** B2
Îles, lac des *QC* **24** C5
Îles, lac des *QC* **25** F2
Îles, lac des *QC* **26** A7
Îles, lac des *QC* **28** D5-6
Îles, lac des *SK* **11** A3
Îles, lac des *SK* **12** F3
Île-Sainte-Croix, lieu
historique international
de l' *NB* **32** E1-2
Île-Scotch-Bonnet, RNF de l'
21 F4-5
Îles-de-Boucherville, parc
des *QC* **24** D6
Îles-de-Boucherville, parc
des *QC* **27** E1-2
Îles-de-Contrecœur, RNF
des *QC* **24** C6
Îles-de-Contrecœur, RNF
des *QC* **27** D2
Îles-de-Contrecœur
National Wilderness
Area *QC* **24** C6
Îles-de-Contrecœur
National Wilderness
Area *QC* **27** D2
Îles-de-la-Baie-Georgienne,
PN des *ON* **18** E6-7
Îles-de-la-Baie-Georgienne,
PN des *ON* **18** E7
Îles-de-la-Madeleine, les *QC*
30 F7
Îles-de-l'Estuaire, RNF des
QC **28** D8
Îles-de-l'Estuaire, RNF des
QC **29** D1
Îles-de-l'Estuaire NWA *QC*
28 D8
Îles-de-l'Estuaire NWA *QC*
29 D1
Îles-du-Saint-Laurent, PN
des *ON* **21** D-E7 D8
Île-Sea-Wolf, RNF de l' *NS*
34 B5-6
Île-Shepherd, RNOM de l'
NL **38** C5-6
Îles-Monger *QC* **38** B1
Îles-Sainte-Marie, RNOM
des *QC* **39** E-F6
Iles-Sainte-Marie NMBS *QC*
39 E-F6
Île Verte, lac de l' *QC* **28** A7
Ilford *MB* **15** E5
Ilfracombe *ON* **18** D8
Ilfracombe *ON* **21** A-B1
Îlots, lac aux *QC* **28** D7
Imperial *SK* **11** D5
Imperial Eagle Channel *BC*
2 D5
Imperial Mills *AB* **9** F6-7
Imperial Mills *AB* **12** F1
Imperial River *NT* **5** E7
Inchkeith *SK* **11** E7
Incomappleux River *BC*
6 D3
Inconnu, lac *QC* **26** A2
Independence *AB* **8** B6
Indian, lac *QC* **23** F7
Indian, lac *QC* **25** B1-2
Indian Arm *BC* **2** D5
Indian Arm PMP *BC* **2** D5
Indian Arm PP *BC* **1** F7
Indian Arm PP *BC* **2** D5
Indian Bay *NL* **14** E7-8
Indian Bay *NL* **36** A8
Indian Bay *NL* **37** A6
Indian Bay *NL* **38** F7-8
Indian Bay Pond *NL*
36 A7-8
Indian Bay Pond *NL* **37** A6
Indian Bay Pond *NL* **38** F7
Indian Brook *NS* **32** E8
Indian Brook *NS* **33** B6
Indian Brook *NS* **34** B7
Indian Brook *NS* **34** B7
Indian Brook *NS* **34** B7
Indian Brook *NS* **34** B7
Indian Cabins *AB* **9** A3-4
Indian Cabins *AB* **10** F5
Indian Creek *NB* **21** B6-7
Indian Creek *NB* **24** E1
Indian Gardens *NS* **33** D3
Indian Harbour *NL* **36** B2
Indian Harbour *NL* **39** A7
Indian Harbour *NS* **32** F8
Indian Harbour *NS* **33** C5
Indian Harbour *NS* **34** F5
Indian Harbour Lake *NS*
34 F5
Indian Head *SK* **11** E6
Indian Head, RNOM d' *SK*
11 E6
Indian Head NMBS *SK*
11 E6
Indian Hill Lake *NT* **10** B-C8
Indian Island *NB* **31** E8
Indian Island *NB* **32** A6
Indian Lake *NS* **32** E8
Indian Lake *NS* **33** C5
Indian Lake *ON* **13** E6
Indian Lake *ON* **13** F7
Indian Lake *ON* **16** E3-4
Indian Line CA *ON* **20** C6
Indian Mountain Lake *NT*
10 C8
Indian Point *NS* **32** F7
Indian Point *NS* **33** D4-5
Indian Point PP *ON* **20** B3
Indian Point PP *ON* **21** E1
Indian Point Sand Hills *PE*
35 C1
Indian Pond *NL* **36** B3
Indian Pond *NL* **37** B1

Indian River *BC* **2** D5
Indian River *NL* **36** A5
Indian River *NL* **37** A3
Indian River *NL* **38** F4
Indian River *NL* **39** F8
Indian River *ON* **21** A5
Indian River *ON* **21** B6-7
Indian River *ON* **21** D3-4
Indian River *ON* **22** F5
Indian River *ON* **24** E1
Indian River *PE* **35** D3
Indian River *YT* **4** B2
Indian Tickle *NL* **39** B8
Indienne, lac de l' *QC* **22** E6
Indin Lake *NT* **10** B6
Indin River *NT* **10** A-B6
Indus *AB* **7** B4
Inférieur, lac *QC* **29** C4
Ingebrigt Lake *SK* **6** F8
Ingebrigt Lake *SK* **11** E2
Ingelow *MB* **14** D2
Ingenika River *BC* **3** D5-6
Ingersoll *ON* **19** C5
Ingersoll *ON* **20** E4
Ingleside *ON* **24** E3
Inglestrom Lake *ON* **13** A7
Inglewood *ON* **19** A7
Inglewood *ON* **20** C6
Inglis *MB* **11** D8
Inglis *MB* **13** D1
Inglis Falls CA *ON* **18** F5
Inglis Falls CA *ON* **20** A4
Inglisville *NS* **32** E5-6
Inglisville *NS* **33** C3
Ingoldsby *ON* **21** C3
Ingol *ON* **13** F5
Ingol *ON* **14** E8
Ingol *ON* **16** D1
Ingomar *NS* **33** F3
Ingonish *NS* **34** B7
Ingonish Beach *NS* **34** B7
Ingonish Centre *NS* **34** B7
Ingonish Ferry *NS* **34** B7
Ingonish River *NS* **34** B7
Ingornachoix Bay *NL* **38** C3
Ingornachoix Bay *NL* **39** E7
Ingraham Bay *BC* **1** A1
Ingramport *NS* **32** F7-8
Ingramport *NS* **33** C5
Ingray Lake *NT* **10** B5
Ings River *YT* **4** D6
Inkaneep PP *BC* **6** F2
Inkerman *NB* **30** E2
Inkerman *NB* **31** B8
Inkerman *ON* **24** E3
Inkerman Ferry *NB* **31** B8
Inkerman Ferry *NB* **32** A6
Inkin River *BC* **3** B1-2
Inland *AB* **13** C8
Inland Lake *MB* **13** A5
Inner Bay *ON* **19** D6
Inner Bay *ON* **20** F5
Innerkip *ON* **19** B5-6
Innerkip *ON* **20** E4-5
Innes *SK* **11** F6
Innes Island *ON* **18** B3
Innisfail *AB* **6** C5
Innisfail *AB* **8** F6
Innisfil *ON* **20** B6-7
Innisfil *ON* **21** E1
Innisfil Heights *ON* **20** B6-7
Innisfil Heights *ON* **21** E1
Innisfree *AB* **6** B7
Innisfree *AB* **11** B1
Innismara *NL* **36** A2-3
Innismara *NL* **38** F2
Innisville *ON* **21** B7
Innisville *ON* **24** E1
Insinger *SK* **11** C7
Instow *SK* **11** E-F3
Intata Reach *BC* **1** B5
Intérieur, plaines de l' *BC*
3 C-D8
Intérieur, plaines de l' *BC*
9 C-D2
Intérieur, plateau de l' *BC*
1 B-C6
Interior Plains *BC* **3** B-C8
Interior Plains *BC* **9** B-C2
Interior Plateau *BC*
1 B6 C6-7
International Ridge PRA *BC*
2 E6
Intersection Mountain *AB*
6 A2
Intrepid Inlet *NT* **42** C2
Inukjuak *QC* **40** C4
Inukjuak *QC* **41** F8
Inuuvarjik TP *NU* **41** D4
Inuvik *NT* **5** C5
Inverarden House NHS *ON*
24 E4
Inverary Lake *ON* **15** A3
Inverary *ON* **21** D7
Inverhaugh *ON* **19** A6
Inverhaugh *ON* **20** D5
Inverhuron *ON* **20** B3
Inverhuron PP *ON* **20** B3
Invermay *ON* **20** A4
Invermay *SK* **11** C6
Invermere *BC* **6** E4
Invermere *BC* **7** C1
Inverness *NS* **34** C5-6
Inverness *NS* **34** B1
Inverness *PE* **32** A7
Inverness *QC* **24** A8
Inverness *QC* **25** F8
Inverness *QC* **27** C5
Inverness *QC* **37** C5
Inwood *MB* **14** C4-5
Inwood *ON* **19** C3
Inwood *ON* **20** F2
Inzana Lake *BC* **3** F6

Iona *NS* **34** C6
Iona *ON* **19** D4
Iona *ON* **20** F3
Iona *PE* **34** C3
Iona *PE* **35** F6
Iona Station *ON* **19** D4
Iona Station *ON* **20** F3
Iosegun Lake *AB* **6** A3-4
Iosegun Lake *AB* **8** A2
Iosegun Lake *AB* **9** F3-4
Iosegun River *AB* **8** A2-3
Iosegun River *AB* **9** F3-4
Ipperwash Beach *ON* **19** B3
Ipperwash Beach *ON*
20 D-E2
Iqaluit *NU* **41** A5-6
Iqaluktuutiaq (Cambridge
Bay) *NU* **41** B3
Iqaluktuutiaq (Cambridge
Bay) *NU* **42** F3
Ira Lake PP *ON* **18** E4
Ireland *NS* **33** A8
Ireland *NS* **34** E4
Ireland's Eye *NL* **37** C7
Irène, lac *QC* **26** B4
Ireton *AB* **8** D6
Iris *PE* **34** D3
Iris *PE* **35** F6
Irish Cove *NS* **34** D7
Irish Cove PP *NS* **34** D7
Irish Lake *ON* **20** B5
Irishtown *NB* **31** F8
Irishtown *NB* **32** B6
Irishtown *NL* **36** B3
Irishtown *NL* **37** B1
Irma *AB* **6** B7
Irma *AB* **11** B1
Ironbound Islands *NL* **39** A6
Iron Bridge *ON* **17** D5
Iron Bridge *ON* **18** A1
Irondale *ON* **21** C3
Irondale River *ON* **21** C3
Iron Hill *QC* **24** E8
Iron Hill *QC* **27** F3
Iron Island *ON* **18** A7
Iron Mines *NS* **34** D6
Iron River *AB* **6** A7
Iron River *AB* **9** F7
Iron River *AB* **12** F2
Irons Creek *YT* **3** A4
Irons Creek *YT* **4** E4
Iron Springs *AB* **6** D6
Iron Springs *AB* **7** D5-6
Iroquois *ON* **24** F3
Iroquois, lac *QC* **22** E8
Iroquois, lac *QC* **24** C2
Iroquois, pointe *ON* **28** E7-8
Iroquois, rivière aux *NB*
29 E3
Iroquois, rivière aux *QC*
26 E8
Iroquois, rivière aux *QC*
28 C3
Iroquois Falls *ON* **17** C7
Iroquois River *NT* **5** C-D6
Irricana *AB* **6** D6
Irricana *AB* **7** B4
Irvine *AB* **6** E8
Irvine *AB* **11** E2
Irvine Creek *YT* **4** E5-6
Irvines Landing *BC* **1** F6
Irvines Landing *BC* **2** D3
Isaac Creek *YT* **4** B-C2
Isabel, lac *QC* **22** F7
Isabel, lac *QC* **24** C1
Isabella *MB* **11** E8
Isabella *MB* **13** E1
Isabella Lake *ON* **18** D7
Isachsen, Cape *NU* **42** B3
Isachsen Peninsula *NU*
42 B3
Isbister River *MB* **13** B5
Isham *SK* **11** D3
Ishkheenickh River *BC* **3** F3
Isidore, lac *QC* **28** A8
Isidore, lac de *QC* **29** A1
Iskut *BC* **3** C3
Iskut Hot Springs PP *BC*
3 D3
Iskut River *BC* **3** D2-3
Iskwatam Lake *SK* **12** E6
Iskwatikan Lake *SK* **12** E5-6
Island Brook *QC* **27** E5
Island Cove *NL* **37** D6-7
Island Cove *NL* **37** E8
Island Falls *SK* **12** E1
Island Falls *SK* **15** E1
Island Grove *ON* **20** B7
Island Grove *ON* **21** D-E1
Island Harbour *NL* **38** E7
Island Lake *AB* **8** B7
Island Lake *MB* **13** A5
Island Lake *ON* **17** F4
Island Lake *ON* **18** C7
Island Lake CA *ON* **20** C6
Island Lake PP *BC* **2** C3
Island Lake South *AB* **9** F5-6
Island of Ponds *NL* **39** B8
Island Pond *NL* **36** A8
Island Pond *NL* **36** B-C3
Island Pond *NL* **36** C5
Island Pond *NL* **37** A6
Island River *NT* **3** A8
Island River *NT* **3** A8
Islands, Bay of *NL* **36** A2
Islands, Bay of *NL* **38** F2

Islands, Bay of *NL* **39** A6
Islands, Bay of *NL* **40** F8
Islands, Bay of *ON* **18** B4
Islay *AB* **6** B7
Islay *AB* **11** B2
Isle aux Morts *NL* **36** D-E1
Isle Pierre *BC* **1** A6-7
Isle Valen *NL* **36** E8
Isle Valen *NL* **37** E6
Islington *NL* **37** D7
Ispas *NL* **8** B8
Issac Lake *BC* **1** B8
Issac Lake *BC* **6** B1
Issac's Harbour *NS* **34** F5
Issac's Harbour North *NS*
34 E-F5
Issoudun *QC* **27** B5
Italy Cross *NS* **33** D4
Itcha Ilgachuz PP *BC* **1** C5-6
Ithingo Lake *AB* **6** B5
Itomamo, lac *QC* **28** A7
Ittatsoo *BC* **1** F5
Ittatsoo *BC* **2** E1-2
Ituna *SK* **11** D6
Ivanhoe *ON* **21** D5
Ivanhoe Lake PP *ON* **17** D5
Ivanhoe River *ON* **17** C-D5
Ivry, lac *QC* **38** B1
Ivry, lac *QC* **39** E6-7
Ivry-sur-le-Lac *QC* **24** C4
Ivujivik *QC* **40** B4
Ivujivik *QC* **42** D4
Ivvavik, PN *YT* **5** B2
Ivvavik NP *YT* **5** A2-3 B2-3
Ivy *ON* **20** B6
Ivy Lea *ON* **21** D7
Ivy Lea, parc *ON* **21** D7
Ivy Lea Park *ON* **21** D7
Ixworth, sanctuaire d' *QC*
28 F8

J

Jacam *MB* **15** D5
Jackfish Lake *MB* **13** E2-3
Jackfish Lake *MB* **14** C3
Jackfish Lake *SK* **11** B3
Jackfish Lake *SK* **11** B3
Jackfish Lake *YT* **4** D5
Jackfish River *AB* **9** A-B5
Jackfish River *AB* **12** A1
Jackhead *MB* **13** C4
Jack Lake *ON* **21** C4
Jack Lake *ON* **21** C4
Jackman Flats PP *BC* **6** B2
Jackpine River *AB* **6** A2
Jackson *NS* **32** C8
Jackson *NS* **33** A5
Jackson *NS* **34** E1
Jackson *ON* **18** F5
Jackson *ON* **20** A4
Jackson Creek *SK* **11** F7-8
Jackson Creek *SK* **13** E-F1
Jackson Narrows PMP *BC*
1 C3-4
Jackson's Arm *NL* **38** E3-4
Jackson's Arm *NL* **39** F8
Jackson's Cove *NL* **38** E4-5
Jacksons Point *ON* **20** B7
Jacksons Point *ON* **21** E1
Jacksonville *NB* **31** F3
Jacksonville *NB* **32** B1
Jacques, lac à *NT* **5** E7
Jacques-Cartier, détroit *QC*
30 D1
Jacques-Cartier, détroit de
QC **39** F3-4
Jacques-Cartier, lac *QC*
28 E6
Jacques-Cartier, mont *QC*
29 B7
Jacques-Cartier, mont *QC*
40 F7
Jacques-Cartier, parc de la
QC **28** F8
Jacques Cartier, Strait of *QC*
39 F4-5
Jacques Cartier, Strait of *QC*
40 F7
Jacques Cartier PP *PE*
34 A-B1
Jacques Cartier PP *PE* **35** B2
Jacques Fontaine *NL* **36** E7
Jacques Fontaine *NL* **37** E6
Jacquet River *NB* **29** E6-7
Jacquet River *NB* **29** D7
Jacquet River *NB* **31** B5
Jacquet River *NB* **31** A6
Jacquet River Gorge PPA *NB*
29 E7
Jacquet River Gorge PPA *NB*
31 B6
Jade City *BC* **3** B3
Jade City *BC* **3** B3
Jaffray *BC* **6** F5
Jaffray *BC* **7** F3
Jaffray Melick *ON* **16** D1-2
Jaffray Melick *ON* **13** E5-6 F5
Jake's Corner *YT* **4** E4
Jak TP *NT* **5** C5
J. Albert Bauer PP *ON*
21 A-B2
J. Albert Bauer PP *ON* **22** F2
Jalobert, lac *QC* **23** E8
Jalobert, lac *QC* **25** A2
Jalobert, lac *QC* **26** F2
Jalobert, lac *QC* **28** A3
James, baie *QC* **40** E3
James Bay *QC* **40** E3

James Chabot PP *BC* 6 E4
James Chabot PP *BC* 7 C1-2
James Creek *AB* 9 A4
James Lake *MB* 12 C8
James Lake *MB* 15 C2
James Lake *NT* 10 C5
Jameson, Cape *NU* 41 A7
Jameson, Cape *NU* 42 D-E6
James River 34 E4
James River Bridge *BC* 6 E5
James River Bridge *AB* 8 F5
James Ross Strait *NU* 41 B4
James Ross Strait *NU* 42 F4
James Smith *NL* 37 C6-7
Jamestown *NL* 37 C6-7
Jamet, rivière *QC* 24 A3-4
Jamet, rivière *QC* 25 F4
Jamie Lake *MB* 12 B8
Jamie Lake *MB* 15 B2
Jamière, lac *QC* 39 A3-4
Jamot *ON* 18 B6
Jamyn, lac *QC* 38 A2
Janet Lake *YT* 4 B4
Janetville *ON* 20 B8
Janetville *ON* 21 E2
Janeville *NB* 29 E8
Janeville *NB* 30 E1
Janeville *NB* 31 B7
Jan Lake *SK* 12 F6
Jans Bay *SK* 12 F3
Jansen *SK* 11 C5
Janvrin Island *NS* 34 E6
Jardins, lac des *QC* 22 C3
Jaro, ZEC *QC* 27 D7
Jaroslaw *MB* 14 B5
Jarratt *ON* 18 F8
Jarratt *ON* 20 A7
Jarratt *ON* 21 D1
Jarvie *AB* 6 A5
Jarvie *AB* 8 A6
Jarvie *AB* 9 F5
Jarvis *ON* 19 C7
Jarvis *ON* 20 F6
Jarvis Bay *AB* 8 E-F6
Jarvis Bay PP *AB* 8 E5-6
Jarvis River *YT* 4 D-E2
Jasmin *SK* 11 D6
Jasper *AB* 6 B3
Jasper *AB* 8 D1
Jasper *ON* 21 C7
Jasper *ON* 24 F2
Jasper, PN *AB* 6 B3
Jasper, PN *AB* 8 E1-2
Jasper House, LHN *AB*
  6 B2-3
Jasper House, LHN *AB* 8 D1
Jasper House NHS *AB*
  6 B2-3
Jasper House NHS *AB* 8 D1
Jasper NP *AB* 6 B2-3
Jasper NP *AB* 8 D1-2 E1-2
Jasper Park Information
  Centre NHS *AB* 8 D1
Jaux, lac *QC* 25 A3
Jaux, lac *QC* 26 F3
Jean, lac *ON* 17 B1
Jean, Lake *ON* 16 E6-7
Jean Cote *AB* 9 E3
Jean de Baie *NL* 36 E7
Jean de Baie *NL* 37 E3
Jeanette Bay *NL* 39 A7
Jean Maria River *NT*
  10 E3-4
Jean Marie River *NT* 10 E4
Jeanne Mance *NB* 31 C7
Jeanne-Mance *NB* 29 E8
Jeanne-Mance *NB* 30 F1
Jeannettes Creek *ON* 19 E2
Jeannotte, ZEC *QC* 25 C3
Jeannotte, ZEC *QC* 28 F4
Jean-Péré, lac *QC* 22 B-C6
Jean-Péré, lac *QC* 25 D1
Jedburgh *SK* 11 D6
Jeddore Cape *NS* 33 C7
Jeddore Harbour *NS* 33 C7
Jeddore Lake *NL* 36 C6
Jeddore Lake *NL* 37 C4
Jeddore Oyster Pond *NS*
  33 C7
Jedediah Island PMP *BC*
  2 D3
Jeffrey *AB* 8 B6
Jeffrey's *NL* 36 C1-2
Jeffries Corner *NB* 32 C5
Jeffries Corner *NB* 33 A2
Jellicoe *BC* 2 C-D8
Jellicoe *BC* 6 E-F1
Jellicoe *ON* 16 D7
Jellicoe *ON* 17 B1
Jemseg *NB* 32 C4
Jennejohn Lake *NT* 10 C-D7
Jenner *AB* 6 D7
Jenner *AB* 7 C7
Jenner *AB* 11 D1
Jennings River *BC* 3 A2-3
Jennings River *BC* 4 F5-6
Jenny Lind Island *NU*
  41 B-C4
Jenny Lind Island *NU* 42 F13
Jenpeg *MB* 13 A3
Jenpeg *MB* 15 F3
Jens Munk Island *NU* 41 B6
Jens Munk Island *NU*
  42 F5-6
Jenssen, lac *QC* 26 E7
Jenssen, lac *QC* 28 C2
Jérôme, lac *QC* 25 D6
Jérôme, lac *QC* 27 A1
Jerrys Nose *NL* 36 C1
Jersey Cove *NL* 30 B3
Jersey Harbour *NL* 36 E6
Jersey Harbour *NL* 37 E4
Jersey Mills *QC* 27 C6-7

Jerseyside *NL* 36 E8
Jerseyside *NL* 37 E6
Jerseyville *ON* 19 B6-7
Jerseyville *ON* 20 E5-6
Jervis Inlet *BC* 1 E6
Jervis Inlet *BC* 2 C4
Jesmond *BC* 1 D7-8
Jessups Falls CA *ON* 24 D3
Jésuite, lac du *QC* 25 E8
Jésuite, lac du *QC* 27 A3
Jetait *MB* 12 E7
Jetait *MB* 15 E1
Jeune, lac *QC* 25 B3
Jewel Lake PP *BC* 6 F2
Jewellville *ON* 21 B5
Jewett Lake *SK* 12 E5
Jim, lac *QC* 22 E6
Jim's Cove *NL* 38 E-F5
Jimsmith Lake PP *BC* 6 F4
Jimsmith Lake PP *BC* 7 E1
Joannès, lac *QC* 23 E3
Job Lake *ON* 16 B1
Job's Cove *NL* 37 C7
Jocko Point *ON* 22 D1
Jocko River *ON* 22 C2
Jocko River PP *ON* 17 F8
Jock River *ON* 21 B7
Jock River *ON* 24 E1-2
Joe Batt's Arm *NL* 38 E7
Joe Batt's Arm-Barr'd
  Islands-Shoal Bay *NL*
  38 E7-8
Joe Creek *YT* 5 B2
Joe Glodes Pond *NL* 36 B5
Joe Glodes Pond *NL* 37 B3
Joe Lake *AB* 21 A2
Joe Lake *ON* 23 B1-2
Joffre *AB* 6 C6
Joffre *AB* 8 F6
Joffre Lakes PP *BC* 1 E7
Joffre Lakes PP *BC* 2 B5-6
Joggins *NS* 32 C7
Jogues *ON* 17 B4
John, Cape *NS* 34 D2
John Dean PP *BC* 2 F4
John D'or Prairie *AB* 9 B5
John D'or Prairie *AB*
  41 E-F1
John Dyer, Cape *NU* 41 A4
John Dyer, Cape *NU* 42 D3
John E. Pearce PP *ON* 19 D4
John E. Pearce PP *ON*
  20 F3
John Island *ON* 18 B2
John Osborn Lake *MB*
  12 A8
John Osborn Lake *MB*
  15 A3
John Richardson Bay *NU*
  42 B5
Johnny Hoe River *NT* 10 B3
John's Beach *NL* 36 A2
Johnson, lac *QC* 22 F7
Johnson, lac *QC* 24 C1
Johnson Creek *YT* 5 D3
Johnson's Mills *NB* 32 C6-7
Johnsons Crossing *YT* 4 E5
Johnsons Landing *BC*
  6 E3-4
Johnstone Creek PP *BC* 6 F2
Johnstone Strait *BC* 1 E5
Johnstone Strait *BC* 2 B1
Johnston Harbour/Pine Tree
  Point PP *ON* 18 D-E4
Johnston Island *BC* 9 A8
Johnston Island *SK* 12 A3
Johnstown *NS* 34 D6-7
Johnstown *ON* 24 F2-3
Johnville *NB* 31 E3
Johnville *NB* 32 A1
Johnville *QC* 27 E-F5
Joint Lake *MB* 13 A5
Joint Lake *MB* 15 F5
Joinville, lac *QC* 24 B3
Joinville, lac *QC* 25 F3
Jo-Jo Lake *YT* 4 E3
Jolicure *NB* 32 C7
Jolicure *NB* 35 F1
Joliette *QC* 24 C6
Joliette *QC* 27 D1-2
Joly *QC* 27 B-C5
Jonathan's Pond *NL* 36 A7
Jonathan's Pond *NL*
  37 A5-6
Jonathan's Pond *NL* 38 F7
Jonchée, lac *QC* 39 D5
Jones *ON* 13 E6
Jones *ON* 16 D2
Jones Falls *ON* 21 D7
Jones Sound *NU* 42 C4-5
Jonquière *QC* 28 C5
Jordan *ON* 19 C8
Jordan *ON* 20 E7
Jordan Bay *NS* 33 F3
Jordan Bay *NS* 33 F3
Jordan Falls *NS* 33 E3
Jordan Lake *MB* 12 C8
Jordan Lake *MB* 15 C2
Jordan Lake *ON* 33 E3
Jordan River *BC* 2 F3
Jordan River *NS* 33 E3
Jordan River *ON* 21 C-D8
Jordan River *SK* 11 B6
Jordan Station *ON* 19 C8
Jordan Station *ON* 20 E7
Jordanville *NS* 34 F5
Joseph, lac *NL* 40 E7
Joseph, Lake *ON* 18 D7-8
Joseph, pointe *QC* 30 A6-7
Josephburg *AB* 6 A6
Josephburg *AB* 8 C7

Joseph Henry, Cape *NU*
  42 A5
Joseph Lake *NL* 39 C3
Jourimain, île *NB* 35 E2-3
Jourimain Island *NB*
  35 E2-3
Journois *NL* 36 C2
Joussard *AB* 9 E4
Joutel *AB* 17 B8
Joutel *QC* 23 A4
Joy, Mount *YT* 4 B5
Joy Bay *QC* 40 B5
Joy Bay *QC* 41 D8
Joyceville *ON* 21 D7
J.T. Cheeseman PP *NL*
  36 D1
Juan de Fuca, détroit de *BC*
  2 F2-3
Juan de Fuca, Strait of *BC*
  2 F3
Juan de Fuca PP *BC* 2 F3
Juan Perez Sound *BC*
  1 B-C2
Jubilee Island *MB* 13 A5
Jubilee Lake *NL* 36 C7
Jubilee Lake *NL* 37 C5
Jude Island *NL* 36 E7
Jude Island *NL* 37 E5
Judes Point *PE* 35 A2
Judge Daly Promontory *NU*
  42 A5
Judique *NS* 34 D5
Judith Island *NT* 5 F7
Judson *AB* 7 E6
Juillet, lac *QC* 39 A3-4
Julia Bay *ON* 18 B2
Jumonville, lac *QC* 26 A6
Jumpers Brook *NL* 36 A-B6
Jumpers Brook *NL* 37 A-B4
Junction Brook *NL* 36 A3-4
Junction Brook *NL* 37 A2
Junction Brook *NL* 38 F3
Junction Sheep Range PP
  BC 1 C7
Juniper *NB* 31 E3-4
Juniper *NB* 32 A2
Juniper Beach PP *BC* 1 D8
Juniper Beach PP *BC* 2 A7
Juniper Beach PP *BC* 6 D1
Juniper Station *NB* 31 E4
Juniper Station *NB* 32 A2
Juno *MB* 14 E7
Jupiter Bay *NT* 10 A3
Jura *BC* 2 D8
Jura *BC* 6 F1
Juskatla *BC* 1 A1
Justice *MB* 13 E2
Justice *MB* 14 D1-2

K

Kabamichigama Lake *ON*
  16 E7
Kabamichigama Lake *ON*
  17 C1
Kabania Lake *ON* 16 A6
Kabenung Lake *ON* 17 D3
Kabika River *ON* 17 B-C7
Kabinakagami Lake *ON*
  17 C4
Kabinakagami River *ON*
  17 B-C4
Kabinakagamisis Lake *ON*
  17 C4
Kabitotikwia River, RNP *ON*
  16 D6
Kabitotikwia River PNR *ON*
  16 D6
Kaboni *ON* 18 C4
Kaegudeck Lake *NL*
  36 C6-7
Kaegudeck Lake *NL*
  37 C4-5
Kagawong *ON* 18 B3
Kagawong, Lake *ON*
  18 B-C3
Kagianagami Lake *ON*
  16 C7
Kagiano Lake *ON* 16 E8
Kagiano Lake *ON* 17 B2
Kaglik Lake *NT* 5 B6
Kahkewistahaw *SK* 11 E7
Kahnawake *QC* 24 D5-6
Kahnawake *QC* 27 E1
Kahntah River *BC* 3 C8
Kahntah River *BC* 9 C1-2
Kahshe Lake *ON* 18 E8
Kahshe Lake *ON* 21 C1
Kahshe Lake *ON* 21 C1
Kahwin *BC* 8 B8
Kaiashk PP *ON* 16 B7
Kainairiktok Bay *NL* 39 A6
Kaipokok River *NL* 39 A5-6
Kakabeka Falls *ON* 16 F5
Kakabeka Falls *ON* 40 F1
Kakabeka Falls PP *ON*
  16 F5
Kakagi Lake *ON* 13 F6
Kakagi Lake *ON* 16 C-D4
Kakiddie Lake *BC* 3 C3
Kakisa *NT* 10 D5
Kakisa *NT* 41 E1
Kakisa Lake *ON* 10 E5
Kakisa Lake *ON* 41 E1
Kakisa River *NT* 10 E-F4
Kakisa River Bridge TP *NT*
  10 E5-6
Kakwa PP *BC* 1 A8
Kakwa PP *BC* 6 A1-2

Kakwa River *AB* 6 A2
Kakwa Wildland PP *AB* 6 A2
Kakweiken River *BC* 2 A1
Kaladar *ON* 21 C-D5
Kalamalka Lake *BC* 6 E2
Kalamalka Lake PP *BC* 6 E2
Kaleden *BC* 6 F2
Kaleida *MB* 13 F3
Kaleida *MB* 14 F3
Kaleland *AB* 8 C8
Kalliecahoolie Lake *MB*
  13 A5
Kalone Peak *BC* 1 C4-5
Kalzas River *YT* 4 B4
Kama Hills, RNP *ON* 16 E7
Kama Hills, RNP *ON* 17 C1
Kama Hills PNR *ON* 16 E7
Kama Hills PNR *ON* 17 C1
Kamaniskeg Lake *ON* 21 B4
Kamatsi Lake *SK* 12 F7
Kamatsi Lake *SK* 15 F-E7
Kaminak Lake *NU* 40 A1
Kaminak Lake *NU* 41 E4
Kaminuriak Lake *NU* 40 A1
Kaminuriak Lake *NU* 41 E4
Kamiskotia River *ON*
  17 C-D6
Kamloops *BC* 2 A8
Kamloops *BC* 6 D1-2
Kamloops Lake *BC* 1 E8
Kamloops Lake *BC* 2 A8
Kamloops Lake *BC* 6 D1
Kamouraska *QC* 28 E8
Kamouraska, îles de *QC*
  28 E8
Kamsack *SK* 11 D7
Kamsack *SK* 13 C1
Kamuchawie Lake *SK*
  12 D7
Kamuchawie Lake *SK*
  15 D1
Kanaaupscow, rivière *QC*
  40 D5 E4-5
Kanairiktok River *NL* 39 A5
Kanairiktok River *NL*
  39 A-B4
Kananaskis Village *AB*
  7 B-C2
Kananuta, rivière *QC* 23 D2
Kanata *ON* 21 B7
Kanata *ON* 24 D2
Kandahar *SK* 11 C6 D5-6
Kane *MB* 13 F3
Kane *MB* 14 F4
Kangiqiniq (Rankin Inlet)
  NU 40 A1-2
Kangiqiniq (Rankin Inlet)
  NU 41 E5
Kangiqsualujjuaq *QC* 40 C6
Kangiqsujuaq *QC* 40 B5
Kangiqsujuaq *QC* 41 D-E8
Kangiqtugaapik (Clyde
  River) *NU* 41 A8
Kangiqtugaapik (Clyde
  River) *NU* 42 E7
Kangirsuk *QC* 40 B-C5
Kano Inlet *BC* 1 B1
Kaotisinimigo Lake *ON*
  22 C1
Kapikik Lake *ON* 13 D7
Kapikik Lake *ON* 16 B3
Kapikog Lake *ON* 18 E-E7
Kapikotongwa River *ON*
  16 C7
Kapikotongwa River *ON*
  17 A1-2
Kapiskau River *ON* 40 C2-3
Kapiskong Lake *ON* 17 D6
Kapkichi Lake *ON* 16 B4-5
Kap-Kig-Iwan PP *ON* 17 D7
Kap-Kig-Iwan PP *ON* 23 E1
Kaposvar Creek *SK* 11 D-E7
Kapuskasing *ON* 17 B5
Kapuskasing *ON* 40 F3
Kapuskasing Lake *ON*
  17 C-D5
Kapuskasing River *ON*
  17 B5
Kapuskasing River *ON*
  17 C5
Kapusta Lake *MB* 12 A8
Kapusta Lake *MB* 15 A3
Karagamac, lac *QC* 24 A3-4
Karagamac, lac *QC* 25 E5-6
Karagamac, lac *QC* 27 B1
Karalash Corners *ON* 17 F4
Karloose River *MB* 15 E5-6
Karsakuwigamak Lake *MB*
  12 D8
Karsakuwigamak Lake *MB*
  15 D2
Karsdale *NS* 32 F4-5
Karsdale *NS* 33 C2
Kars Lake *NU* 41 F3
Kars *ON* 21 B8
Kars *ON* 24 E2
Kasakuwigamak Lake *MB*
  12 D8
Kaskattama River *MB*
  15 D7
Kaskattama River *MB*
  40 C-D1
Kaskawulsh River *YT* 4 E2
Kaslo *BC* 6 E3

Kasmere Lake *MB* 12 A7
Kasmere Lake *MB* 15 A2
Katannilik, parc territorial
  NU 40 A5
Katannilik TP *NU* 40 A5
Katannilik TP *NU* 41 D8
Katepwa Beach *SK* 11 E6
Katepwa Point PP *SK* 11 E6
Kathleen *AB* 9 E3-4
Kathleen Lake *YT* 4 E2-3
Kathryn *AB* 6 D5-6
Kathryn *AB* 7 B4
Katimik Lake *MB* 13 B2-3
Katrime *MB* 13 E2-3
Katrime *MB* 14 D3
Katrine *ON* 18 C8
Katrine *ON* 21 A1
Katrine *ON* 22 F1
Kattawagami Lake *ON*
  17 B7
Kattawagami River *ON*
  17 A7
Kavaksuatilik Island *NL*
  39 A6
Kavanagh *AB* 6 B6
Kavanagh *AB* 8 D6
Kawagama Lake *ON* 21 B2
Kawartha Highlands,
  réserve récréative
  provinciale de *ON*
  21 C-D3
Kawartha Highlands
  Provincial Recreation
  Reserve *ON* 21 C3
Kawartha Lakes *ON*
  20 A-B8
Kawartha Lakes *ON*
  21 C-D3
Kawartha Lakes *ON*
  21 D2-3
Kawartha Park *ON*
  21 D2-3
Kawawachikamach *QC*
  39 A2
Kawawachikamach *QC*
  40 D6
Kawawaymog Lake *ON*
  22 B2
Kawaweogama Lake *ON*
  13 E8
Kawaweogama Lake *ON*
  16 D5
Kaweenakumik Lake *MB*
  13 B2-3
Kaweenakumik Lake *MB*
  13 E1
Kawene *ON* 13 F8
Kawene *ON* 16 E4
Kawigamog Lake *ON*
  18 B6-7 C7
Kawinogans River *ON*
  13 D8
Kawinogans River *ON*
  16 B4
Kawnipi Lake *ON* 16 F4
Kaybob *AB* 8 A2
Kay Point *YT* 5 B3
Kayville *SK* 11 F5
Kazabazua *QC* 22 E7
Kazabazua *QC* 24 C1-2
Kazabazua, rivière *QC*
  22 E-F7
Kazabazua, rivière *QC*
  24 C1
Kazan Lake *SK* 9 E8
Kazan Lake *SK* 12 E3
Kazan River *NU* 41 D-E4
Kazchek Lake *BC* 3 F6
Keady *ON* 18 F5
Keady *ON* 20 A4
Kearney *ON* 21 A1-2
Kearney *ON* 22 F1-2
Kearns *ON* 17 D8
Kearns *ON* 23 E1
Keating, lac *QC* 39 A2
Keatley *SK* 11 B4
Keats Island *BC* 1 F6-7
Keats Island *BC* 2 D4
Kebaowek *QC* 17 E8
Kebaowek *QC* 22 C2
Kechika Ranges *BC* 3 B4 C5
Kechika River *BC* 3 B4 C5
Kecil Lake *ON* 18 A3
Kedgwick *NB* 29 E5
Kedgwick *NB* 31 B3-4
Kedgwick, RF de *NB*
  29 D-E4
Kedgwick Game Refuge *NB*
  29 E4
Kedgwick Provincial WMA
  NB 31 A-B3
Kedgwick River *NB* 29 E4
Kedgwick River *NB* 29 E4
Kedgwick River *NB* 31 B3
Kedgwick River *NB* 31 B3
Keefe Lake *SK* 12 C5
Keeler *SK* 11 E5
Keele Range *YT* 5 D1-2
Keele River *NT* 4 A-B7
Keeley Lake *SK* 12 E4
Keelor Lake *ON* 18 A2
Keels *NL* 37 B7
Keene *ON* 21 E3
Keeper River *ON* 16 A-B1
Keeper River *ON* 16 A-B1
Keeseekoose *SK* 11 D7
Keeseekoose *SK* 13 C1
Keeseekoowenin *MB* 11 E8

Keeseekoowenin *MB*
  13 E1-2
Keeseekoowenin *MB* 14 C1
Keewatin *ON* 13 F5
Keewatin *ON* 14 E8
Keewatin *ON* 16 D1
Keewatin River *MB* 12 D7-8
Keewatin River *MB* 15 D2
Kee-way-win *ON* 13 B6-7
Keezhik Lake *ON* 16 B6
Kegaska *QC* 39 E5
Kegaska, lac *QC* 39 E-F5
Keg Lake *SK* 12 E6
Keg River *AB* 9 C3
Keg River *AB* 9 C3
Kegworth *SK* 11 E6-7
Keho Lake *AB* 6 E6
Keho Lake *AB* 7 D5
Keith Arm *NT* 41 C1
Keith Island *NT* 10 D8
Keiver's Lake *SK* 12 E5
Kejick *ON* 13 F5
Kejick (Iskutewizaageegan)
  ON 13 F5
Kejick (Iskutewizaageegan)
  ON 16 D1
Kejimkujik, LHN *NS*
  33 D2-3
Kejimkujik, PN *NS* 33 D3
Kejimkujik (annexe côtière),
  PN *NS* 33 E4-5
Kejimkujik NHS *NS* 33 D2-3
Kejimkujik NP *NS* 33 D3
Kejimkujik NP (seaside
  adjunct) *NS* 33 E4-5
Kekek, rivière *QC*
  23 D8 E7-8
Kekek, rivière *QC* 25 A2
Kekek, rivière *QC* 26 F2
Kekuli Bay PP *BC* 6 E2
Keld *ON* 14 B1
Keldon *ON* 20 B5
Kelfield *SK* 6 C8
Kelfield *SK* 11 C3
Kelk Bridge *ON* 21 D4-5
Keller Lake *MB* 10 B3
Keller Lake *SK* 12 E4
Kellett Strait *NT* 42 C2
Kelley Cove *NS* 33 E1
Kelliher *SK* 11 D6
Kelloe *MB* 11 E8
Kelloe *MB* 13 E1
Kelly *QC* 29 D8
Kelly *QC* 30 D1
Kelly *QC* 31 A7
Kelly Lake *BC* 1 D8
Kelly Lake *BC* 2 A6
Kelly Lake *BC* 3 F8
Kelly Lake *BC* 9 E-F2
Kelly Lake *NT* 5 F8
Kelly Lake *NT* 41 A7
Kellys Corner *ON* 21 A5-6
Kellys Cross *PE* 32 B8
Kellys Cross *PE* 34 C2
Kellys Cross *PE* 35 A4
Kelowna *BC* 6 E2
Kelp Point *BC* 1 B2
Kelsey *AB* 6 B6
Kelsey *AB* 8 D7
Kelsey Bay *BC* 1 E5
Kelsey Bay *BC* 2 B1
Kelsey Creek *MB* 15 C6
Kelsey Lake *MB* 11 A-B7
Kelso *SK* 11 F8
Kelso *SK* 13 D1
Kelso CA *ON* 19 B7
Kelso CA *ON* 20 D6
Kelstern *SK* 11 E4
Kelvin *ON* 19 C6
Kelvin *ON* 20 E5
Kelvin Grove *PE* 35 D3
Kelvington *SK* 11 C6
Kelvin Island *ON* 16 D6
Kelwood *MB* 13 E2
Kelwood *MB* 14 C2
Kemano *BC* 1 B4
Kemble *ON* 18 E-F5
Kemble *ON* 20 A4
Kemkby *SK* 18 A3
Kemmay *MB* 11 E8
Kemnay *MB* 13 E2
Kemnay *MB* 14 D2
Kemp River *AB* 9 C3
Kemp River *AB* 9 C3
Kempt *NS* 33 D3
Kempt, lac *QC* 25 C5
Kempt, rivière *QC* 29 D6
Kemptown *NS* 33 A5-6
Kemptown *NS* 34 E2-3
Kempt Shore *NS* 32 E7
Kempt Shore *NS* 33 B5
Kempt Shore *NS* 34 F1
Kemptville *ON* 21 B8
Kemptville *ON* 24 E3
Kemptville Creek *ON* 21 B8
Kemptville Creek *ON*
  24 E-F2
Keefers *BC* 1 E8
Keefers *BC* 2 C7
Kenabeek *ON* 17 D-E7
Kenaston *SK* 11 D4-5
Kendal *ON* 20 B7-8
Kendal *ON* 21 E1
Kendall, Cape *NU* 41 D6
Kendall Island *NT* 5 A4
Kendall Island NMBS *NT*
  5 A3-4
Keno *NL* 37 B7

Kenilworth *ON* 20 C5
Kenmore *ON* 24 E3
Kennaway *ON* 21 B3
Kennebecasis River *NB*
  32 C-D4
Kennebec Lake *ON* 21 C5-6
Kennedy *SK* 11 E7
Kennedy, Mount *BC* 1 D5
Kennedy, Mount *BC* 2 A1
Kennedy, passage *NU*
  42 A5
Kennedy Bog PP *BC* 1 F5
Kennedy Bog PP *BC* 2 E1
Kennedy Channel *NU* 42 A5
Kennedy Lake *BC* 2 E1-2
Kennedy Lake PP *BC* 1 F5
Kennedy Lake PP *BC* 2 E1-2
Kennedy Lakes, APP *NB*
  31 D5
Kennedy Lakes PRA *BC*
  1 F5
Kennedy Lakes PP *ON*
  17 C6
Kennedy Lakes Wilderness,
  ZCP *NB* 31 D4-5
Kennedy Lakes Wilderness
  PCA *NB* 31 D4-5
Kennedy River *BC* 2 E2
Kennedy River *BC* 31 D5
Kennetcook *NS* 32 E8
Kennetcook *NS* 33 B5-6
Kennetcook *NS* 34 F1
Kennetcook River *NS*
  32 D-E8
Kennetcook River *NS*
  33 B5-6
Kennetcook River *NS*
  34 F1-2
Kenney, Mount *BC* 3 F3
Kenney Dam *BC* 1 B6
Kennisis Lake *ON* 21 B2
Kennisis Lake *ON* 21 B3
Kenny Forest, RNP *ON*
  17 F8
Kenny Forest, RNP *ON*
  22 C1
Kenny Forest PNR *ON* 17 F8
Kenny Forest PNR *ON*
  22 C1
Kénogami, lac *QC* 28 C5
Kenogami Lake *ON* 17 D7
Kenogami River *ON* 17 A4
Kenogami River *ON*
  17 A3-4
Kenogami River *ON* 40 F2
Kenogamissi Lake *ON*
  17 C6
Kenora *ON* 13 F5
Kenora *ON* 14 E8
Kenora *ON* 16 D1
Kenosee Lake *SK* 11 F7
Kenton *MB* 11 E8
Kenton *MB* 13 E1
Kent *BC* 1 F8
Kent *BC* 2 D6-7
Kent Bridge *ON* 19 D3
Kent Bridge *ON* 20 F2
Kent Junction *NB* 31 E7
Kent Junction *NB* 32 A5
Kent Lake *NB* 31 E7
Kent Lake *NB* 32 A5-6
Kent Peninsula *NU* 41 B3
Kent Peninsula *NU* 42 F2
Kent, Cape *PE* 32 B8
Kentucky-Alleyne PP *BC*
  2 C8
Kentucky-Alleyne PP *BC*
  6 E1
Kentvale *NS* 1 F6
Kentville *NS* 32 E6-7
Kentville *NS* 33 B4
Kenville *MB* 11 C7-8
Kenville *MB* 13 B1
Kenyon Lake *MB* 15 F6
Kenzieville *NS* 34 E4-5
Keoma *AB* 6 D6
Keoma *AB* 7 B4-5
Kepenkeck Lake *NL* 36 C7
Kepenkeck Lake *NL* 37 C5
Kepimits Lake *NL* 39 C5
Kerbodot, lac *QC* 39 B1
Kerbodot, lac *QC* 6 F1
Keremeos Columns PP *BC*
  6 F1-2
Kerensky *AB* 8 B7
Kergwenan *AB* 12 B4
Kerr Lake *MB* 14 C2
Kerrobert *SK* 6 C8
Kerrobert *SK* 11 C2-3
Kerry *MB* 14 E6
Kerry *QC* 7 A4
Kersey *AB* 6 F1
Kersley *BC* 1 B7
Kerwood *ON* 19 C3-4
Kerwood *ON* 20 E2-3
Keseechewun Lake *SK*
  12 A6
Keseechewun Lake *SK*
  15 A1
Kesagami Lake *ON* 17 A7
Kesagami Lake *ON* 40 F3
Kesagami PP *ON* 17 A7
Keswick *NB* 31 F4
Keswick *NB* 32 B2
Keswick *ON* 20 B7
Keswick *ON* 21 C1
Keswick Ridge *NB* 31 F4
Keswick Ridge *NB* 32 C2

Keswick River *NB* 31 F4
Keswick River *NB* 32 B2
Ketchen *SK* 11 C6-7
Ketch Harbour *NS* 33 C6
Ketsall River *BC* 4 F3
Kettleby *ON* 20 B-C7
Kettleby *ON* 21 E1
Kettle Cove *NL* 38 E6
Kettle Creek *ON* 19 C5
Kettle Creek *ON* 20 E-F4
Kettle Lake *MB* 15 E5
Kettle Lakes PP *ON* 17 C6
Kettle Point *ON* 19 B3
Kettle Point *ON* 19 B3
Kettle Point *ON* 20 E2
Kettle Point *ON* 20 E2
Kettle Rapids *MB* 15 D5
Kettle River *BC* 6 F2
Kettle River *MB* 15 E5
Kettle River *MB* 15 E5
Kettle River PRA *BC* 6 F2
Kettle Stones PP *MB* 11 C8
Kettle Stones PP *MB* 13 C2
Kettle Valley *BC* 6 F2
Kevisville *AB* 8 F5
Key *SK* 11 C7
Key *SK* 13 C1
Keyano *QC* 40 E5
Keyes *MB* 14 D2
Key Junction *ON* 18 B6
Key River *ON* 18 B6
Key River *ON* 18 B6
Keyser *ON* 19 C3-4
Keyser *ON* 20 E2-3
Keystone *AB* 8 D5
Kezik Lake *ON* 16 B4
Khartum *ON* 21 B5
Khedive *SK* 11 F5-6
Khutzeymateen Inlet *BC*
  3 F3
Khutzeymateen PP *BC* 3 F3
Khutzeymateen Valley
  Grizzly Bear Sanctuary
  BC 3 F3
Kiamika *QC* 24 B3
Kiamika *QC* 25 F3
Kiamika, réservoir *QC*
  25 E3
Kiamika, réservoir *QC*
  24 A3
Kianuko PP *BC* 6 F4
Kianuko PP *BC* 7 F1
Kicking Horse Pass *BC* 6 D4
Kicking Horse Pass *BC* 7 A1
Kicking Horse Pass NHS *BC*
  7 A1
Kickininee PP *BC* 6 E2
Kikiak (Rigolet) *NL* 39 B6-7
Kikino *AB* 6 A6
Kikino *AB* 8 A8
Kikino *AB* 9 F6
Kikino *AB* 12 F1
Kikiwissi, lac *QC* 22 B3
Kikkertavak Island *NL*
  39 A6-7
Kikomun Creek *BC* 6 F4-5
Kikomun Creek PP *BC* 7 F2
Kikupegh Pond *NL* 36 C5
Kikupegh Pond *NL* 37 C3
Kilbride *ON* 19 B7
Kilbride *ON* 20 D6
Kilburn *NB* 31 E3
Kilburn *NB* 32 A1
Kilby PP *BC* 1 F7-8
Kilby PP *BC* 2 D-E6
Kildala Arm *BC* 1 A3-4
Kildala River *BC* 1 A3-4
Kildare *PE* 34 A1
Kildare *PE* 35 B2
Kildare, Cape *PE* 34 A1
Kildare, Cape *PE* 35 B2
Kildare Capes *PE* 34 A1
Kildare Capes *PE* 35 B2
Kildare River *PE* 35 B2
Kildonan *BC* 1 F6
Kildonan *BC* 2 E2
Kilkale Lake *NT* 5 D8
Kilian Island *NU* 41 A3
Kilian Island *NU* 42 D2
Killala Lake *ON* 16 E8
Killala Lake *ON* 17 C2
Killaloe *ON* 21 A5
Killaloe *ON* 22 F5
Killally *SK* 11 D6-7
Killam *AB* 6 B7-8
Killam *AB* 8 D-E8
Killam *AB* 11 B1
Killarney *MB* 13 F2
Killarney *MB* 14 F1-2
Killarney *ON* 17 F7
Killarney *ON* 18 B4
Killarney *ON* 18 B4
Killarney *ON* 18 B4
Killarney Beach *BC* 6 E2
Killarney Coast and Islands
  PP *ON* 18 B4
Killarney Coast and Islands
  PP *ON* 18 B5
Killarney Lakelands and
  Headwaters PP *ON*
  17 F6-7
Killarney Lakelands and
  Headwaters PP *ON*
  18 B5
Killarney PP *ON* 17 F6
Killarney PP *ON* 18 B4
Killbear PP *ON* 18 D6-7
Killdeer *SK* 11 F4
Killiniq *NU* 40 B6
Killiniq Island *NL* 40 B6
Killmar *QC* 24 C4

Kilmuir *PE* **34** C3
Kilmuir *PE* **35** F6
Kilsyth *ON* **18** F5
Kilsyth *ON* **20** A4
Kilwinning *SK* **11** B4
Kilworth *ON* **19** C4
Kilworth *ON* **20** E3
Kilworthy *ON* **18** E8
Kilworthy *ON* **21** C1
Kimball *AB* **7** F5
Kimberley *BC* **6** F4
Kimberley *ON* **7** E1
Kimberley *ON* **18** F6
Kimberley *ON* **20** A-B5
Kimiwan Lake *AB* **9** E4
Kimmirut (Lake Harbour)
    NU **40** B5-6
Kimowin River *SK* **9** D8
Kimowin River *SK* **12** D3
Kimsquit *BC* **1** B4
Kimsquit River *BC* **1** B4
Kinaskan Lake *BC* **3** C3
Kinaskan Lake PP *BC* **3** C3
Kinbasket Lake *BC* **6** C2-3
Kinbasket Lake *BC* **8** F1
Kin Beach PP *BC* **2** C2
Kinbrook Island PP *AB*
    40 B5-6
Kinbrook Island PP *AB*
    6 E6-7
Kinbrook Island PP *AB*
    7 C6-7
Kinburn *ON* **19** A4
Kinburn *ON* **20** C3
Kinburn *ON* **21** A7
Kinburn *ON* **24** D1
Kincaid *SK* **11** F4
Kincardine *NB* **31** E3
Kincardine *NB* **32** A1
Kincardine *ON* **20** B3
Kindakun Point *BC* **1** B1
Kindersley *SK* **6** D8
Kindersley *SK* **11** B1
Kingaok (Bathurst Inlet) *NU*
    41 C3
King Christian Island *NU*
    42 C3
King City *ON* **20** C7
King City *ON* **21** E1
Kingcome Inlet *BC* **1** D4-5
Kingcome Inlet *BC* **1** D5
Kingcome Inlet *BC* **2** A1
Kingcome River *BC* **1** D5
Kingfisher *BC* **6** E2
Kingfisher Lake *ON* **13** B8
Kingfisher Lake *ON* **13** B8
King George, Mount *BC*
    6 E4
King George, Mount *BC*
    7 C2
King George IV Ecological
    Reserve *NL* **36** C3
King George IV Ecological
    Reserve *NL* **37** C1
King George IV Lake *NL*
    36 C3
King George IV Lake *NL*
    37 C1
King George Islands *NU*
    40 C-D4
King George Islands *NU*
    41 F8
King Island *BC* **1** C4
King Kirkland *ON* **17** D7
King Kirkland *ON* **23** D1
Kingkown Inlet *BC* **1** A-B2
King Lake *NT* **10** E8
Kingman *AB* **6** B6
Kingman *AB* **8** D7
Kingman's *NL* **37** E8
Kingnait Fiord *NU* **42** F8
King Peak *YT* **4** E1
King Point *YT* **5** B3
Kingross *SK* **34** B6
Kingsboro *PE* **34** B4
Kingsboro *PE* **35** D8
Kingsbridge *ON* **20** C2-3
Kingsburg *NS* **33** D5
Kingsbury *QC* **24** D8
Kingsbury *QC* **27** E4
Kings Castle PP *PE*
    34 C-D3 C-D4
Kings Castle PP *PE* **35** F6
Kingsclear *NS* **32** C2
Kingsclear (IR) *NB* **32** C2-3
Kingscote Lake *ON* **21** B3
King's Cove *NL* **37** B7
King's Cove *NL* **38** E5
Kingscroft *QC* **27** F4
Kingsey Falls *QC* **27** D4
Kingsford *SK* **11** C7
Kingsgate *BC* **6** F4
Kingsgate *BC* **7** F1
Kingsley *MB* **14** F3
Kingsley *NB* **31** F4
Kingsley *NB* **32** B2-3
Kingsley Lake *SK* **12** B7
Kingsley Lake *SK* **15** B1
Kingsmere Lake *SK* **11** A4
King's Point *NL* **38** E4
King's Point *NL* **39** F8
Kingsport *NS* **32** E7
Kingsport *NS* **33** B4-5
Kingston *NB* **32** C4
Kingston *NB* **33** A1
Kingston *NL* **37** D7-8
Kingston *NS* **32** E6
Kingston *NS* **33** B3
Kingston *ON* **21** D6-7
Kingston *ON* **24** F1
Kingston *PE* **34** C2
Kingston *PE* **35** E4
Kingston Lake *SK* **12** B6
Kingston Lake *SK* **15** B1
Kingston Mountain *YT* **4** D3
Kingsvale *BC* **2** C7-8

Kingsville *NS* **34** D6
Kingsville *ON* **19** E1-2
Kingwell *NL* **36** D8
Kingwell *NL* **37** D6
King William Island *NU*
    41 B4
King William Island *NU*
    42 F3-4
Kinhuron *ON* **20** B3
Kinikinik *BC* **8** A7
Kinistin *SK* **11** C6
Kinistino *SK* **11** B5
Kinkora *ON* **19** B5
Kinkora *ON* **20** D4
Kinkora *PE* **32** B8
Kinkora *PE* **34** C1-2
Kinkora *PE* **35** D3
Kinley *SK* **11** C4
Kinloch Lake *ON* **16** B4
Kinloss *ON* **20** B3
Kinlough *ON* **20** B3
Kinmount *ON* **21** C2-3
Kinnear, rivière *NB* **35** E1
Kinnear River *NB* **35** E1
Kinnear's Mills *QC* **27** C5
Kinngait (Cape Dorset) *NU*
    40 A4
Kinngait (Cape Dorset) *NU*
    41 D7
Kinojévis, lac *QC* **23** D4
Kinojévis, rivière *QC* **23** D3
Kinojévis, rivière *QC* **23** E3
Kinoosao *SK* **12** D7
Kinoosao *SK* **15** D1
Kinosota *MB* **13** D3
Kinosota *MB* **14** B3
Kinross *PE* **35** E6
Kinsale *ON* **20** C8
Kinsale *ON* **21** E2
Kinsella *AB* **6** B7
Kinsella *AB* **8** E6
Kinsman Corner *NS* **32** E6
Kinsman Corner *NS* **33** B4
Kinsman Lake *MB* **12** B8
Kinsman Lake *MB* **15** B3
Kintail *ON* **20** C3
Kintore *NB* **31** E3
Kintore *NB* **32** A1
Kintore *ON* **19** C5
Kintore *ON* **20** E3
Kinusisipi *MB* **13** A3
Kinuso *AB* **9** E4
Kinwow Bay *MB* **13** C3-4
Kinwow Bay PP Reserve *MB*
    13 C3-4
Kioshkokwi Lake *ON* **22** E2
Kiosk *ON* **17** F8
Kiosk *ON* **22** E1
Kiowana Beach *ON* **18** F6
Kiowana Beach *ON* **20** A5
Kipahigan Lake *MB* **12** E7
Kipahigan Lake *MB* **15** E1
Kipawa *ON* **17** E-F8
Kipawa, lac *QC* **22** C2
Kipawa, lac *QC* **17** E8
Kipawa, lac *QC* **22** B2
Kipawa, ZEC *QC* **22** B3
Kipling *SK* **11** E7
Kippase *BC* **1** E4
Kippen *ON* **19** B4
Kippen *ON* **20** D3
Kippens *NL* **36** C2
Kirby *ON* **21** E3
Kirbys Corner *ON* **17** F4
Kirkcaldy *AB* **6** E6
Kirkcaldy *AB* **7** D5
Kirk Cove *ON* **21** C6
Kirkella *MB* **11** E7
Kirkella *MB* **13** E1
Kirkfield *ON* **20** A8
Kirkfield *ON* **21** D2
Kirkhill *NS* **32** D7
Kirkhill *NS* **33** A4
Kirkland *QC* **24** D5
Kirkland *QC* **27** E1
Kirkland Centre *YT* **4** D3
Kirkland Lake *ON* **17** D7
Kirkland Lake *ON* **23** D1
Kirkness Lake *ON* **13** D6
Kirkness Lake *ON* **16** B2
Kirkpatrick Lake *AB* **6** C7
Kirkpatrick Lake *ON* **11** C1
Kirkpatrick Lake *ON* **17** F5
Kirkton *ON* **19** B4
Kirkton *ON* **20** D3
Kirriemuir *AB* **6** C8
Kirriemuir *AB* **11** C2
Kisbey *SK* **11** F7
Kisgegas Peak *BC* **3** E4
Kishikas Lake *ON* **16** A4
Kishkas River *ON* **13** B-C7
Kiskatinaw *BC* **9** E2
Kiskatinaw PP *BC* **9** E2
Kiskatinaw River *BC* **3** E-F8
Kiskatinaw River *BC* **9** E-F2
Kiski Lake *MB* **13** A3
Kiski Lake *MB* **15** F3
Kiskissink, lac *QC* **28** D4
Kiskitto Lake *MB* **13** A4
Kiskittogisu Lake *MB* **13** A3
Kispiox *BC* **3** F4
Kispiox River *BC* **3** F4
Kisseynew Lake *MB* **12** F7
Kisseynew Lake *MB* **15** F1
Kississing Lake *MB* **12** F7
Kississing Lake *MB* **15** E-F1
Kississing River *MB* **12** F7
Kississing River *MB* **15** E1-2
Kistigan Lake *MB* **16** F6
Kitamaat Village *BC* **1** A3-4
Kitchener *BC* **6** F4
Kitchener *BC* **7** F1
Kitchener *ON* **19** B6
Kitchener *ON* **20** D5
Kitchener Lake *BC* **3** D4

Kitchie Lake *ON* **16** A7-8
Kitigan *ON* **17** B5
Kitigan-Zibi *ON* **22** D7-8
Kitigan-Zibi *QC* **24** B2
Kitimat *BC* **1** A3
Kitimat Arm *BC* **1** A3
Kitimat Ranges *BC* **1** A-B3
Kitkatla *BC* **1** A2
Kitkiata Inlet *BC* **1** A3
Kitlope Heritage
    Conservancy Provincial
    Area *BC* **1** B4
Kitlope River *BC* **1** B4
Kitsakie *SK* **12** F5
Kitsault River *BC* **3** E3
Kitscoty *AB* **6** B7-8
Kitscoty *AB* **11** B2
Kitseguecla (Gitsegukla) *BC*
    3 F4
Kitson Island PMP *BC*
    1 A2-3
Kitsumkalum Lake *BC*
    3 F3-4
Kitsumkalum PP *BC* **3** F3
Kitsumkalum River *BC* **3** F3
Kitsumkaylum *BC* **1** A3
Kitsumkaylum *BC* **3** F3
Kittigazuit *NT* **5** B5
Kitty Coleman PP *BC* **2** C2
Kitwancool *BC* **3** F4
Kitwanga *BC* **3** F4
Kitwanga Fort NHS *BC*
    3 F3-4
Kitwanga Mountain PP *BC*
    3 F4
Kivimaa-Moonlight Bay *SK*
    11 A3
Kivimaa-Moonlight Bay *SK*
    11 D3
Kiyiu Lake *SK* **11** D3
Klappan River *BC* **3** D3
Klashwun Point *BC* **1** A1
Klastline River *BC* **3** C3
Klawli River *BC* **3** F6
Klaza, Mount *YT* **4** C2-3
Klaza River *BC* **4** C2-3
Kleanza Creek PP *BC* **3** F4
Kleczkowski, lac *QC* **39** E4
Kledo Creek PP *BC* **3** B6
Kleefeld *MB* **13** F4
Kleefeld *MB* **14** E5
Kleena Kleene *BC* **1** C6
Kleinburg *ON* **19** A7-8
Kleinburg *ON* **20** C6-7
Klemka, lac *QC* **28** A7
Klemtu *BC* **1** C3
Klewnuggit Inlet PMP *BC*
    1 A3
Klinaklini River *BC* **1** C-D5
Klinkit Creek *BC* **3** A2-3
Klinkit Creek *BC* **4** F5-6
Klin-se-za PP *BC* **3** E7
Klondike Historic Complex
    NHS *YT* **4** A2
Klondike River *YT* **4** A2-3
Kloo Lake *YT* **4** D2
Klotassin River *YT* **4** C2
Klotz, Mount *YT* **5** F2
Klotz Lake *ON* **16** D8
Klotz Lake *ON* **17** B3
Klua Lakes PPA *BC* **3** C7
Klua Lakes PPA *BC* **9** B1
Kluane *YT* **4** D1
Kluane, PN et réserve *YT*
    4 E1-2
Kluane Lake *YT* **4** D2
Kluane NP and Reserve *YT*
    4 D1-2
Kluane Ranges *YT*
    4 D1-2 D-E2
Kluane River *YT* **4** C-D2
Kluane Wildlife Sanctuary
    YT **4** D1
Kluane Wildlife Sanctuary
    YT **4** D2
Kluane Wildlife Sanctuary
    YT **4** E2-3
Kluane/Wrangell-St. Elias/
    Glacier Bay/Tatshen-
    shini-Alsek *BC* **4** E1-2
Kluane/Wrangell-St. Elias/
    Glacier Bay/Tatshen-
    shini-Alsek *YT* **4** E1-2
Kluayetz Creek *BC* **3** D4
Klueys Bay *ON* **18** E8
Klueys Bay *ON* **21** C1
Klukshu *YT* **4** E3
Kluksu Creek *YT* **4** D3
Kluskoil Lake PP *BC* **1** B6
Kluskus *BC* **1** B6
Klutlan Creek *YT* **4** C-D1
Kluziai Island *NT* **10** C8
Kneehill *AB* **7** A5
Kneehills Creek *AB*
    6 C5 D5-6
Kneehills Creek *AB* **7** A4-5
Knee Lake *MB* **15** F5
Knee Lake *SK* **12** E4
Knee Lake *SK* **12** E4
Knewstubb Lake *BC* **1** B6
Knife Lake *ON* **13** A7
Knight Inlet *BC* **1** E5
Knight Inlet *BC* **2** A1
Knights Cove *NL* **37** B7
Knightville *NB* **32** C5
Knorr Creek *YT* **5** F4-5
Knorr Range *YT* **5** F5
Knowlesville *NB* **31** E3-4
Knowlesville *NB* **32** A1-2
Knowlton *QC* **24** E8
Knowlton *QC* **27** F3
Knox, Cape *BC* **1** A1
Knox Lake *NL* **39** A3

Knox Lake *ON* **16** B2
Knoydart *NS* **34** D4
Knutsford *BC* **2** B8
Knutsford *BC* **6** E1
Knutsford *PE* **34** A7
Knutsford *PE* **35** C1
Koch Creek *BC* **6** F3
Koch Island *NU* **41** B7
Koch Island *NU* **42** E-F6
Kogaluc, rivière *QC* **40** C4
Kogaluc, rivière *QC* **41** E8
Kogaluk Bay *QC* **40** C4
Kogaluk Bay *QC* **41** F7-8
Kogaluk River *NL* **39** C7
Kohler *ON* **19** C7
Kohler *ON* **20** E6
Kohn Lake *SK* **12** A6
Kohn Lake *SK* **15** A1
Koidern *YT* **4** C1
Kokanee Creek PP *BC* **6** F3
Kokanee Glacier PP *BC*
    6 E-F3
Koko Lake *ON* **18** A4
Koksilah River PP *BC* **2** F4
Koksoak, rivière *QC* **40** C6
Kola *MB* **11** E8
Kola *MB* **13** E1
Kolbec *NS* **32** C8
Kolbec *NS* **34** D1
Komakuk Beach *YT* **5** A2
Komarno *MB* **13** E3-4
Komarno *MB* **14** C5
Komoka *ON* **19** C4
Komoka *ON* **20** E3
Komoka PP *ON* **19** C4
Komoka PP *ON* **20** E3
Kondiaronk, lac *QC* **22** C6
Koocanusa, Lake *BC* **7** F2
Kookatsoon Lake Territorial
    Recreation Site *YT* **4** E2
Koona Lake *MB* **12** A7
Koona Lake *MB* **15** A2
Kootenay, PN *BC* **6** D4
Kootenay, PN *BC* **7** B1
Kootenay Bay *BC* **6** F3-4
Kootenay Lake *BC*
    6 E3-4 F3-4
Kootenay Lake (Davis
    Creek) PP *BC* **6** F3-4
Kootenay Lake (Lost Ledge)
    PP *BC* **6** E3-4
Kootenay Lake (Midge
    Creek) PP *BC* **6** F4
Kootenay NP *BC* **6** D4
Kootenay NP *BC* **7** B1
Kootenay Pass *BC* **6** F3-4
Kootenay Ranges *BC*
    6 D-E4
Kootenay Ranges *BC* **7** B-C1
Kootenay River *BC* **6** E4
Kootenay River *BC* **7** E-F2
Kootenay River *BC* **7** F-F2
Kopka River *ON* **16** D5
Kopka River PP *ON* **16** D5
Kopp's Cove *SK* **11** B3
Kormak *ON* **17** E5
Kortright CA *ON* **19** A8
Kortright CA *ON* **20** C6-7
Kortright CA *ON* **21** F1
Koshlong Lake *ON* **21** C3
Koskaecodde Lake *BC*
    36 D6-7
Koskaecodde Lake *BC*
    37 D4-5
Kotaneelee River *NT*
    10 E-F2
Kotcho Lake *BC* **3** B7
Kotcho Lake *BC* **9** A-B1
Kotcho Lake Village Site PP
    BC **3** B8
Kotcho Lake Village Site PP
    BC **9** B1-2
Kouchibouguac *NB* **31** D7
Kouchibouguac, baie de *NB*
    31 D8
Kouchibouguac, PN *NB*
    31 D8
Kouchibouguac Bay *NB*
    31 D8
Kouchibouguac NP *NB*
    31 D8
Kouchibouguac River *NB*
    31 D7-8
Kouchibouguasis River *NB*
    31 E7
Kouchibouguasis River *NB*
    31 E7
Koukdjuak River *NU* **40** A5
Koukdjuak River *NU* **41** C8
Koukdjuak River *NU* **42** F7
Kovik Bay *QC* **40** B4
Kovik River *QC* **41** E7
Kowkash *ON* **16** C6-7
Krakow *AB* **8** C8
Kronau *SK* **11** E5-6
Kronsgart *MB* **14** F6
Krusenstern, Cape *NU*
    41 B2
Krusenstern, Cape *NU*
    42 E7 E7-8 F7-8
Krydor *SK* **11** B4
Ksi Hlginx *BC* **3** F3
Kugaluk River *NT* **5** B5-6
Kugluktuk (Coppermine) *NU*
    41 B2
Kugluktuk (Coppermine) *NU*
    42 F1
Kugmallit Bay *NT* **5** A5
Kukatush *ON* **17** D5-6
Kuklok (Bloody Falls) TP *NU*
    41 B-C2
Kuklok (Bloody Falls) TP *NU*
    42 F1
Kukukus Lake *ON* **13** E7

Kukukus Lake *ON* **16** D3-4
Kunga Island *BC* **1** B1
Kunghit Island *BC* **1** C2
Kunghit Island *BC* **1** C2
Kuper Island *BC* **1** F6-7
Kuper Island *BC* **2** E4
Kuroki *SK* **11** C6
Kurtzville *BC* **19** A5
Kurtzville *ON* **20** C4
Kusawa Lake *YT* **4** E3
Kusawa Lake Territorial
    Recreation Site and
    Campground *YT* **4** E3
Kushog Lake *ON* **21** B2
Kustra Lake *MB* **12** C8
Kustra Lake *MB* **15** C2
Kutawagan Lake *SK* **11** D5
Kutcho Creek *BC* **3** B-C4
Kuujjuaq *QC* **40** C6
Kuujjuarapik *QC* **40** D4
Kwadacha PRA *BC* **3** C5
Kwadacha PWP *BC* **3** C5-6
Kwadacha River *BC* **3** C5-6
Kwanta Inlet *BC* **1** C4
Kwejinne Lake *NT* **10** B6
Kwokullie Lake *BC* **3** B8
Kwokullie Lake *BC* **9** A2
Kyaska Lake *SK* **12** D6
Kyaska Lake *SK* **15** D1
Kyklo Creek *BC* **3** B7-8
Kyklo Creek *BC* **9** B1-2
Kyle *SK* **11** D3
Kylemore *SK* **11** C6
Kynoch *BC* **18** A1
Kynoch Inlet *BC* **1** B4
Kyuquot *BC* **1** E4
Kyuquot Sound *BC* **1** E4

L

La Baie *QC* **28** C6
La Baleine *QC* **28** F7
L'Abbé *QC* **24** E6
L'Abbé, lac *QC* **28** C1-2
Labelle *NS* **33** D4
Labelle *QC* **24** B4
Labelle *QC* **25** F4
Labelle, lac *QC* **24** B4
Labelle *QC* **25** F4
Laberge, Lake *YT* **4** D4
La Biche River *AB* **9** F6
La Biche River *AB* **12** E-F1
La Biche River *AB* **10** E1 F2
La Biche River Wildland PP
    AB **9** F6
La Bostonnais *QC* **25** C7-8
La Bostonnais *QC* **28** E3
Labouchere Channel *BC*
    1 C4
Labrador, mer du *NL*
    39 A7-8
Labrador (océan
    Atlantique), mer du *NL*
    37 A7-8
Labrador (océan
    Atlantique), mer du *NL*
    38 C6-8
Labrador (océan
    Atlantique), mer du *NL*
    39 B-C8
Labrador City *NL* **39** C2
Labrador City *NL* **40** E6
Labrador Sea *NL* **39** A-B8
Labrador Sea (Atlantic
    Ocean) *NL* **37** A7-8
Labrador Sea (Atlantic
    Ocean) *NL* **38** B6-8
Labrador Sea (Atlantic
    Ocean) *NL* **40** B7-8
La Branche *QC* **27** A5-6
Labrecque *QC* **28** B5
Labreque, lac *QC* **28** B5
Labrieville *QC* **29** A1-2
Labrieville-Sud *QC* **29** A1-2
La Broquerie *MB* **13** F4
La Broquerie *MB* **14** E6
La Bruère, lac *QC* **25** A6
La Bruère, lac *QC* **26** E6
La Bruère, lac *QC* **12** F1
La Butte Creek *AB* **9** A6-7
La Butte Creek *AB* **12** A2
La Butte Creek Wildland PP
    AB **9** A6-7
La Butte Creek Wildland PP
    AB **10** F8
La Butte Creek Wildland PP
    AB **12** A2
Lac, île du *QC* **39** F6
Lac-à-Beauce *QC* **25** D7-8
Lac-à-Beauce *QC* **28** F3
Lacadena *SK* **11** D3
Lac Le Jeune *BC* **2** B8
Lac Le Jeune *BC* **6** E1
Lac Le Jeune PP *QC* **27** E-F2
Lac Le Jeune PP *BC* **6** E1
L'Acadie *QC* **24** E6
L'Acadie, rivière *QC* **24** E6
La Cloche Creek *ON*
    18 B3-4
La Cloche Lake *ON* **18** B3
La Cloche Mountains *ON*
    18 B3-4
Laclu *ON* **13** E-F5
Laclu *ON* **14** E8
Laclu *ON* **16** D1
Lac-Manitou-Sud *QC* **24** C4
Lac-Marois *QC* **1** F6
Lac-Mégantic *QC* **27** E6
Lac-Mississippi, réserve
    naturelle de faune du *QC*
    21 B-C7
Lac-Mississippi, réserve
    naturelle de faune du *QC*
    24 E1

Lac-aux-Sables *QC*
    27 A3-4 B3-4
Lac-Baker *NB* **29** F2
Lac-Baker *NB* **31** C1
Lac-Beauport *QC* **27** A5-6
Lac-Big-Glace-Bay, RNOM
    NS **34** D8
Lac-Bouchette *QC* **28** C4
Lac-Brébeuf, ZEC du *QC*
    28 D7
Lac Brochet (Northlands)
    MB **12** B7
Lac Brochet (Northlands)
    MB **15** B2
Lac-Brome *QC* **24** E8
Lac-Brome *QC* **27** F3
Lac-Castagnier *QC* **23** C5
Lac-Cayamant *QC* **22** E7
Lac-Cayamant *QC* **24** B1
Lac-Connelly *QC* **24** C5
Lac-Connelly *QC* **27** D1
Lac-Daigle *QC* **39** E2
Lac-de-la-Boiteuse,
    ZEC du *QC* **28** B5-6
Lac-de-l'Achigan *QC* **24** C5
Lac-de-l'Achigan *QC* **27** D1
Lac-Delage *QC* **27** A5
Lac-des-Aigles *QC* **29** A3
Lac-des-Aigles *QC* **31** A1
Lac-des-Commissaires *QC*
    25 A8
Lac-des-Commissaires *QC*
    26 F8
Lac-des-Commissaires *QC*
    28 D4
Lac-des-Écorces *QC* **24** A3
Lac-des-Écorces *QC*
    25 E-F3
Lac-Désert *QC* **24** B3
Lac-Désert *QC* **25** F3
Lac-des-Îles *QC* **22** D8
Lac-des-Îles *QC* **24** B2
Lac-des-Îles *QC* **25** F3
Lac-des-Loups *QC* **21** A7
Lac-des-Loups *QC* **22** F7
Lac-des-Loups *QC* **24** C-D1
Lac-des-Plages *QC* **24** C3
Lac-des-Seize-Îles *QC*
    24 C4
Lac-Drolet *QC* **27** D6
Lac-du Bois PP *BC* **1** D8
Lac-du Bois PP *BC* **2** A8
Lac-du Bois PP *BC* **6** D1
Lac du Bonnet *MB* **13** E4-5
Lac du Bonnet *MB* **14** D6-7
Lac-du-Cerf *QC* **22** D8
Lac-du-Cerf *QC* **24** B2-3
Lac-du-Cerf *QC* **25** F2-3
Lac-Dufault *QC* **17** D8
Lac-Dufault *QC* **23** D2
Lac-Dufresne (Mai) *QC*
    39 D2-3
Lac-Écho *QC* **24** C5
Lac-Écho *QC* **27** D1
Lac-Édouard *QC* **25** C8
Lac-Édouard *QC* **28** E4
Lac-Etchemin *QC* **27** C7
Lac-Frontière *QC* **27** B7-8
Lac-Gatineau *QC* **22** D8
Lac-Gatineau *QC* **24** A2
Lac-Gatineau *QC* **25** E2
Lac-Grosleau *QC* **24** C3
Lac-Haley, RNOM du *NS*
    33 E3
Lachenaie *QC* **24** D6
Lachenaie *QC* **27** E1-2
Lachine *QC* **24** D5-6
Lachine *QC* **27** E1
Lac-Humqui *QC* **29** C4
Lachute *QC* **24** D5
Lac-John *QC* **39** A2
Lac-John *QC* **40** D6
Lacolle *QC* **24** E6
Lacolle *QC* **27** F2
Lacombe *AB* **6** C5-6
Lacombe *AB* **8** E6
La Conception *QC* **24** C4
La-Opuntia, RNOM du *SK*
    11 C4
La Corey *AB* **6** A7
La Corey *AB* **9** F7
La Corey *AB* **12** F2
La Corne *QC* **23** D4
Lacoste, lac *QC* **24** B4
Lacoste, lac *QC* **25** A2
La Coulee *AB* **6** A7
Lacoursière, lac *QC* **23** E8
Lacoursière, lac *QC* **25** A2
Lacoursière, lac *QC* **26** E-F2
Lac-Paré *QC* **24** B5
Lac-Paré *QC* **27** C1
Lac-Poulin *QC* **27** C6
Lac-Rapide *QC* **22** B2
Lac-Redberry, RNOM du *SK*
    11 C4
Lac-Richardson, RNOM du
    AB **9** B7
Lac-Richardson, RNOM du
    AB **12** B2
La Croche *QC* **25** C7-8
La Croche *QC* **28** E3
Lacroix, lac *QC* **26** C-D3
Lac-Saguay *QC* **24** A3
Lac-Saguay *QC* **25** F3
Lac-Saint-Charles *QC*
    27 A5
Lac-Sainte-Marie *QC* **22** E7
Lac-Sainte-Marie *QC* **24** B1
Lac-Saint-François, RNF du
    QC **24** E4-5
Lac-Saint-Joseph *QC* **27** A5
Lac-Saint-Paul *QC* **24** A3
Lac-Saint-Paul *QC* **25** E3
Lac-Sergent *QC* **24** A4-5
Lac-Simon *QC* **23** E5
Lac-Simon *QC* **24** C3
Lac-Spiers, RNF du *AB*
    6 C6-7
Lac-Spiers, RNF du *AB*
    8 F7-8
Lac-Spiers, RNF du *AB*
    11 C1
Lac-Supérieur *QC* **24** B4
Lac-Supérieur (projet), aire
    marine nationale de
    conservation du *ON*
    17 C-D1
Lac-Supérieur (projet), aire
    marine nationale de
    conservation du *ON*
    16 F6-7
Lacs-Waterton, PN des *AB*
    6 F5
Lacs-Waterton, PN des *AB*
    7 F4
Lac-Upper Roussay, RNOM
    du *SK* **11** D6-7
Lac-Labrie *QC* **39** E2
Lacusta Lake *NT* **12** A5
La Dauversière, lac *QC*
    26 B5
Ladder Lake Beach *SK*
    11 A4
Ladle Cove *NL* **38** E7
La Doré *QC* **26** D7-8
La Doré *QC* **28** B3
La Durantaye *QC* **27** B6
Lady Cove *NL* **37** C6-7
Lady Douglas Island *BC*
    1 C3
Lady Evelyn Falls *NT* **10** E5
Lady Evelyn Falls TP *NT*
    10 E5
Lady Evelyn Lake *ON* **17** E7
Lady Evelyn-Smoothwater
    PP *ON* **17** E7
Lady Franklin Bay *NU*
    42 A5
Lady Grey Lake *NT* **10** E8
Lady Lake *SK* **11** C7
Ladysmith *BC* **2** E3-4
Ladysmith *BC* **6** F1
Ladysmith *QC* **22** C1
Ladysmith *QC* **24** C1
Ladywood *MB* **13** E4
Ladywood *MB* **14** D6
Laferté *QC* **17** C8
Laferté *QC* **23** C-D3

Laferte River *NT* **10** D4-5
Laflamme, lac *QC* **28** B7
Laflamme, rivière *QC*
    23 B-C5
Lafleche *SK* **11** F4
Lafond *AB* **6** A7
Lafond *AB* **11** A1
Lafond Creek *AB* **9** D4
Lafontaine *ON* **18** E7
Lafontaine *ON* **20** A6
Lafontaine *QC* **24** C5
Laforce *QC* **17** E8
Laforce *QC* **22** A3
Laforce *QC* **23** F3
Laforge-Deux *QC* **40** D-E5
Lafrenay, lac *QC* **26** E6
Lafrenay *QC* **28** B-C1
La Galissonnière, lac *QC*
    39 D4-5
Laganière, lac *QC* **26** A6
La Garde, lac *QC* **24** D4
Laggan *ON* **24** D4
La Glace *AB* **9** E2
La Course, lac *SK* **11** C7
La Course, lac *SK* **13** C1
La Guadeloupe *QC* **27** D6
LaHave *NS* **33** D4-5
La Have River *NS* **32** F6
La Have River *NS* **33** C3-4
La Hêtrière *QC* **31** E4
La Hêtrière *QC* **32** B6
La Hune, Cape *NL* **36** E4
La Hune, Cape *NL* **37** E2
Laidlaw *BC* **1** F8
Laidlaw *BC* **2** D7
Laird *SK* **11** B4
Laird Island *SK* **9** A7-8
Laird Island *SK* **12** A3
La Jannaye, lac *QC* **39** B1-2
Lajoie, lac *QC* **25** A4
Lajoie, lac *QC* **26** F4
Lajoie, lac *QC* **11** E6
Lajoue, lac *QC* **25** C3
Lajord *SK* **11** E6
La Justone, lac *QC* **39** C1
Lake Abitibi Island PP *ON*
    17 C7-8
Lake Alma *SK* **11** F6
Lake Annis *NS* **33** E1
Lake Audy *NS* **14** C1
Lakeburn *NB* **31** F8
Lakeburn *NB* **32** B6
Lake Charlotte *NS* **33** C7
Lake Country *BC* **6** E2
Lake Cowichan *BC* **2** E3
Lake Creek *YT* **4** A3
Lake Creek *YT* **4** B3
Lake Dalrymple *ON* **20** A8
Lake Dalrymple *ON*
    21 C-D2
Lake Edward *NT* **31** D3
Lake Errock *BC* **2** D-E6
Lakefield *ON* **21** D3
Lakefield *ON* **24** C5
Lake Francis *MB* **13** E3
Lake Francis *MB* **14** D4
Lake George *NB* **32** C2
Lake George *NS* **32** E6
Lake George *NS* **33** E1
Lake George PP *NS* **32** E6-7
Lake George PP *NS* **33** B4
Lake Harbour (Kimmirut)
    NU **40** B5-6
Lake Helen *NS* **16** E6
Lake Helen *NS* **17** C1
Lake Huron Highland *ON*
    20 B3
Lakehurst *ON* **21** D3
Lakeland *MB* **13** E3
Lakeland *MB* **14** D4
Lakeland *NS* **32** E8
Lakeland *NS* **33** B5
Lakeland PP *AB* **9** F7
Lakeland PP *AB* **12** F1
Lakeland PRA *AB* **9** F7
Lakeland PRA *AB* **12** F1-2
Lakelands *NS* **32** D7
Lakelands *NS* **33** A4-5
Lake La Rose *NS* **32** E7
Lake La Rose *NS* **33** C2-3
Lake Laurentian CA *ON*
    18 A5-6
Lake Lenore *SK* **11** C5
Lakelet *ON* **20** C4
Lake Louise *AB* **6** C4
Lakelse Lake *BC* **1** A3
Lakelse Lake *BC* **1** A3-4
Lakelse Lake PP *BC* **1** A3-4
Lake Majeau *AB* **8** B5
Lake Midway PP *NS* **32** F4
Lake Midway PP *NS*
    33 C-D1
Lake Nipigon PP *ON* **16** E6
Lake Nipigon PP *ON*
    40 F3-4
Lake O'Brien *NL* **36** A7
Lake O'Brien *NL* **37** A5
Lake O'Brien *NL* **38** F7
Lake of the Woods PP *ON*
    13 F5-6
Lake of the Woods PP *ON*
    16 E1
Lake O'Law PP *NS* **34** C4-5
Lake on the Mountain PP
    ON **21** E5-6
Lake Paul *NS* **32** E6-7
Lake Paul *NS* **33** C4

Lakeport ON 21 E4
Lake Ramsay NS 32 F7
Lake Ramsay NS 33 C4
Lake St. George PP NL 13 D3-4
Lake St. George PP NL 14 A4
Lake St. Peter ON 21 B3-4
Lake St. Peter PP ON 21 B4
Lakeshore ON 19 E2
Lakeside NB 32 D4
Lakeside NB 33 A2
Lakeside NS 32 F8
Lakeside NS 33 C5-6
Lakeside ON 19 B5
Lakeside ON 20 E4
Lakeside Beach ON 20 B8
Lakeside Beach ON 21 E2
Lake Stream NB 31 F7
Lake Stream NB 32 B5
Lake Superior National Marine Conservation Area (proposed) ON 16 F6-7
Lake Superior National Marine Conservation Area (proposed) ON 17 C-D1
Lake Superior PP ON 17 E3-4
Laketon NB 31 D7
Lake Uist NS 34 D7
Lakevale NS 34 D5
Lake Valley SK 11 E4
Lakeview AB 8 F6-7
Lakeview NB 37 E7
Lakeview NS 32 F8
Lakeview NS 33 C6
Lakeview QC 24 C4
Lakeview Heights ON 24 E3-4
Lakeville NB 31 F3
Lakeville NB 31 F8
Lakeville NB 32 B1
Lakeville NB 32 B6
Lakeville NS 32 E6-7
Lakeville NS 33 B4
Lakeville PE 34 B4
Lakeville PE 35 D8
Lakeville Corner NB 32 C3-4
Lake Whittaker CA ON 19 C5
Lake Whittaker CA ON 20 E4
Laliberté, lac QC 25 C5
Laliberté, rivière QC 29 A2
La Loche SK 9 D8
La Loche SK 12 D3
La Loche, lac SK 9 D7-8
La Loche, lac SK 12 D2-3
La Loche River NT 10 D8
La Loche River SK 9 D8
La Loche River SK 12 D3
La Loche West SK 9 D7-8
La Loche West SK 12 D2-3
L'Alverne QC 29 D6
L'Alverne QC 31 A4-5
L'Amable ON 21 C4
La Macaza QC 24 B4
La Macaza QC 25 F4
La Malbaie QC 28 E7-8
La Malbaie QC 30 C3
Lamaline NL 36 F6
Lamaline NL 37 F4
La Manche NL 36 D8
La Manche NL 37 D6
La Manche NL 37 E8
La Manche PP NL 37 E8
Lamarche QC 28 B5
La Martre QC 29 A7
La Martre, lac NT 10 C5
La Martre, lac NT 41 D1
La Mauricie NP QC 24 A7
La Mauricie NP QC 25 E7
La Mauricie NP QC 27 B2-3
Lamb, lac QC 22 B5
Lambert, lac QC 22 A5
Lambert, lac QC 23 F5
Lambertville NB 32 B2
Lambertville ON 19 C4-5
Lambeth ON 20 E3-4
Lambton QC 27 D6
Lambton, Cape NT 41 A1
Lambton, Cape NT 42 E1
Lambton Shores ON 19 B-C3
Lambton Shores ON 20 E2
Lamèque NB 30 E2
Lamèque NB 31 B8
Lameque, île NB 30 E2
Lameque, île NB 31 B8
Lameque Island NB 30 E2-3
Lameque Island NB 31 B8
Lamerton AB 8 E7
La Minerve QC 24 B3-4
La Minerve QC 25 F3-4
La Minerve, lac QC 24 B3
La Minerve, lac QC 25 F3
Lamont AB 8 A6
Lamont AB 8 C7
Lamont AB 11 A1
La Morandière QC 23 D4-5
La Mothe, lac QC 28 B5-6
La Motte QC 23 D4
La Motte, lac QC 23 D4
La Motte Lake PP ON 17 D6
Lampman SK 11 F6-7
Lamprey ON 21 B7
La Muir, Lake ON 22 E2-3
Lamy QC 29 E2
Lanark ON 21 B7
Lanark ON 24 E1
Lancaster ON 24 E4

Lancaster Sound NU 42 D4-5
Lance, Point NL 36 F8
Lance, Point NL 37 B7-8
Lance Cove NL 37 D8
Lancer SK 11 D3
Landing SK 9 E8
Landing SK 12 E3
Landing Lake MB 15 E4
Landis SK 11 C3
Landmark MB 13 F4
Landmark MB 14 E5-6
Landrienne QC 23 D4
Landron, lac QC 22 A7
Landron, lac QC 23 F7
Landron, lac QC 25 C1
Landry, lac QC 25 B4
Landry, lac QC 39 E5
Landry Creek NT 4 B-C8
Landry Creek NT 10 C1-2
Landseer MB 14 E3
Land's End NT 42 C1
Lanezi Lake BC 1 B8
Lanezi Lake BC 6 B1
Lanfine AB 7 A8
Lang ON 21 E3
Lang SK 11 E5-6
Langara Island BC 1 A1
Langara Point BC 1 A1
Langbank SK 11 E7
Lang Bay BC 1 F6
Lang Bay BC 2 C3
Langdale BC 1 F6-7
Langdale BC 2 D4
Langdon AB 6 D6
Langdon AB 7 B4
L'Ange-Gardien QC 21 A8
L'Ange-Gardien QC 22 F8
L'Ange-Gardien QC 24 D2-3
L'Ange-Gardien QC 24 E7
L'Ange-Gardien QC 27 A5-6
Langenburg SK 11 D7
Langenburg SK 13 D1
Langford BC 2 F4
Langham SK 11 C4
Langlade QC 23 E8
Langlade QC 25 A2
Langlade QC 26 F2
Lang Lake ON 18 B4
Langley BC 1 F7
Langley BC 2 E5
Langruth MB 13 E3
Langstaff ON 19 A8
Langstaff ON 20 C7
Langstaff ON 21 F1
Langton ON 19 D6
Langton ON 20 F5
Langton Bay NT 5 A-B7
Languedoc QC 17 C8
Languedoc QC 23 C3
Laniel QC 17 E8
Laniel QC 22 B2
Lanigan SK 11 C5
Lanigan Creek SK 11 C5
L'Annonciation QC 24 B4
L'Annonciation QC 25 F4
Lanoraie QC 24 C6
Lanoraie QC 27 D2
Lansdowne NS 33 A7
Lansdowne ON 21 D7
Lansdowne House ON 40 E2
Lansdowne House (Neskantaga) ON 16 A6-7
Lansdowne Lake SK 12 C4
L'Anse-à-Beaufils QC 30 C3
L'Anse-à-Fugère QC 30 B3
L'Anse-à-Gilles QC 27 A7
L'Anse-à-Gilles QC 28 F7
L'Anse-à-la-Cabane QC 30 F6-7
L'Anse-Amour NL 38 A3-4
L'Anse-au-Clair NL 38 A3-4
L'Anse-au-Diable NL 38 A4
L'Anse-au-Griffon QC 30 B3
L'Anse-au-Loup NL 38 A3
L'Anse-au-Loup NL 39 D8
L'Anse-aux-Gascons QC 30 D2
L'Anse-aux-Gascons QC 31 A8
L'Anse aux Meadows, LHN de NL 38 A5-6
L'Anse aux Meadows, LHN de NL 39 D8
L'Anse aux Meadows NL 38 A5
L'Anse aux Meadows NL 39 D8
L'Anse aux Meadows NHS NL 38 A5-6
L'Anse aux Meadows NHS NL 39 D8
L'Anse-à-Valleau QC 30 B2-3
L'Anse-de-Roche QC 28 C8
L'Anse-Pleureuse QC 29 A8
L'Anse-Saint-Jean QC 28 C7
Lansing River YT 4 B5
Lantier QC 24 B5

Lantz NS 33 B6
Lantz NS 34 F2
Lantzville BC 1 F6
Lantzville BC 2 D3
Lanz Island BC 1 D3
La Passe ON 22 F6
La Patrie QC 27 E6
La Pêche QC 21 A7
La Pêche QC 22 F7
La Pêche QC 24 C-D1
La Pêche, lac QC 21 A7
La Pêche, lac QC 22 F7
La Pêche, lac QC 24 D1
La Pérouse MB 15 F3
La Pérouse Bay MB 15 B6
Lapeyrère, lac QC 25 D8
Lapeyrère, lac QC 27 A3
Lapeyrère, lac QC 28 F3
Lapie River YT 4 B5
La Plaine QC 24 C6
La Plaine QC 27 D1
Lapland NS 33 D4
LaPlante NB 31 B6
La Plonge SK 12 E-F4
La Plonge, lac SK 12 F4
La Pocatière QC 28 F8
La Poile NL 36 D2
La Poile Bay NL 36 D2
La Poile River NL 36 D2
Lapointe, lac QC 39 B1
Laporte SK 6 D8
Laporte SK 11 D2
La Potherie, lac QC 40 C5
La Prairie QC 24 D6
La Prairie QC 27 E2
La Présentation QC 24 D7
La Présentation QC 27 E2-3
L'Archevêque NS 34 D7
Larchwood ON 18 A5
Lardeau River BC 6 E3
Larder Lake ON 17 D7-8
Larder Lake ON 23 E1
Larder River PP ON 17 D7-8
Larder River PP ON 23 E1
La Rédemption QC 29 C4
Lareau, lac QC 25 A4
Lareau, lac QC 26 F6
Lareau, lac QC 28 D1
La Reine QC 17 C7-8
La Reine QC 23 C2
La Reine, rivière QC 23 B-C2
La Ribourde, lac QC 26 A3
Larive, lac QC 22 B5
La Riviere MB 13 F3
La Riviere MB 14 F3
Lark Harbour NL 36 A2
Lark Harbour NL 38 F1-2
Larkspur AB 8 A6
Larkspur AB 9 F5-6
La Rochelle MB 13 F4
La Rochelle MB 14 F5
La Romaine QC 39 F5
La Romaine QC 40 F8
La Ronde QC 26 B3
La Ronge SK 12 F5
La Ronge, lac SK 12 F5
Larouche QC 28 C5
Larrey, lac QC 28 B7-8
Larrys River NS 34 E5-6
Larsen Sound NU 41 B4
Larsen Sound NU 42 E-F3
Larus Lake ON 16 B2
La Salette ON 19 C6
La Salette ON 20 E5
La Salle MB 13 F3-4
La Salle MB 14 E5
La Salle ON 19 E1
LaSalle, lac QC 24 D6
LaSalle ON 27 E1
Lasalle, lac QC 28 E4
La Sarre QC 17 C8
La Sarre QC 23 C2
La Sarre, rivière QC 23 C2
La Saulx, lac QC 28 A7-8
L'Ascension QC 24 A4
L'Ascension QC 25 F4
L'Ascension QC 28 B5
L'Ascension-de-Patapédia QC 29 D5
L'Ascension-de-Patapédia QC 31 A3-4
La Scie NL 38 E5
La Scie NL 39 F8
Lashburn SK 6 B8
Lashburn SK 11 B2
Laskeek Bay BC 1 B2
La Société QC 29 D2
Lasqueti Island BC 1 F6
Lasqueti Island BC 2 D3
L'Assomption QC 24 C6
L'Assomption QC 27 D1-2
Lasswade ON 21 C4
Last Lake ON 18 C7
Last Mountain SK 11 D5
Last Mountain House Provincial Historic Park SK 11 E5
Last Mountain Lake SK 11 D5
Last Mountain Lake National Wilderness Area SK 11 D5

La Tabatière QC 40 E8
Latchford ON 17 E7
Latchford ON 22 A1
Latchford Bridge ON 21 B5
Laterrière QC 28 C6
Lathorn AB 7 C6
La Trinité-des-Monts QC 29 D3
La Trinité-des-Monts QC 31 A1
Latties Brook NS 32 D8
Latties Brook NS 33 A6
Latties Brook NS 34 F2
Latulipe QC 17 E8
Latulipe QC 22 A2
La Tuque QC 25 C7
La Tuque QC 28 F3
Lauder MB 11 F8
Lauder MB 13 F1
Launay QC 17 C8
Launay QC 23 C3
Launay, lac QC 25 D5
Launching PE 34 C4
Launching PE 35 E7
Launching Point PE 34 C4
Launière, rivière QC 28 E-F5
Laura SK 11 C4
Lauréat, lac QC 25 C6
Lauréat, lac QC 28 E2
Laurel ON 20 C5
Laurel QC 24 C4
Laurel Creek CA ON 19 B6
Laurel Creek CA ON 20 E6
Laurenceton NL 36 A6-7
Laurenceton NL 37 A4-5
Laurenceton NL 38 F6
Laurent QC 28 C6
Laurentia Beach MB 14 C4
Laurentian Mountains QC 25 E7 F6-7
Laurentian Mountains QC 27 A5-6
Laurentian Mountains QC 27 B2 C1-2
Laurentian Mountains QC 28 E5-6
Laurentian Mountains QC 29 A2 B1
Laurentian Mountains QC 40 F6
Laurentides QC 24 C5-6
Laurentides QC 27 D1
Laurentides, les QC 25 E6 F5-6
Laurentides, les QC 27 A4-5
Laurentides, les QC 27 B1-2
Laurentides, les QC 28 D5-6 E5
Laurentides, les QC 29 A1-2 B1
Laurentides, RF des QC 28 D-E5
Laurie Lake SK 12 D7
Laurie Lake SK 15 D1
Laurier PP NS 32 E8
Laurier PP NS 33 B6
Laurier MB 13 D2
Laurier MB 14 B2
Laurier, Mount SK 13 D6
Laurier River MB 12 D7
Laurier River MB 15 D2
Laurier-Station QC 27 B5
Laurierville QC 27 C5
Laurus Lake ON 13 C5
Laurus Lake ON 14 A-B8
Laussedat, Mount BC 6 D3
Lauzon QC 27 B5
Lava Forks PP BC 3 D2
Laval QC 24 D6
Laval QC 27 E1
Laval, lac QC 29 C4
Laval, rivière QC 29 A-B2
La Vallee ON 16 F2
Lavaltrie QC 24 C6
Lavaltrie QC 27 D2
Lavant ON 21 C4
Lavender Peak BC 3 E3
Lavenham MB 13 E-F3
Lavenham MB 14 E3
L'Avenir QC 24 C8
L'Avenir QC 27 D4
La Vérendrye, RF QC 22 A5-6 B5-6
La Verendrye PP ON 16 F4-5
Laverlochère QC 17 E8
Laverlochère QC 22 A2
Laverlochère QC 23 F2
La Vernière QC 30 F7
Lavesta AB 8 E5
Lavieille, Lake ON 22 E3
Lavigne ON 17 F7
Lavigne ON 18 A7
Lavigne, ZEC QC 24 B5-6
Lavigne, ZEC QC 25 F5-6
Lavigne, ZEC QC 27 B-C1
Lavillette NB 29 F8
Lavillette NB 30 F1
Lavillette NB 31 C7
Lavington BC 6 E2
Lavinia MB 11 E8
Lavinia MB 13 E3
La Visitation-de-Yamaska QC 24 C7-8
La Visitation-de-Yamaska QC 27 C3
Lavoie, lac QC 25 B6
Lavoie, lac QC 28 E1-2
Lavoie Bay SK 12 E4
Lavoy AB 6 B6-7
Lavoy AB 8 C8

Lavoy AB 11 A-B1
Lawford Lake MB 13 A4
Lawford Lake MB 15 A4
Lawford River MB 13 A4
Lawford River MB 15 A4
Lawledge MB 15 D5
Lawn NL 36 F6
Lawn NL 37 F4
Lawn Bay NL 36 F6
Lawn Bay NL 37 F4
Lawn Point NL 36 F6
Lawn Point NL 37 F4
Lawn Point PP BC 1 E3-4
Lawrence Bay SK 12 D6
Lawrence Bay SK 15 D1
Lawrence Lake AB 9 F5
Lawrence Lake ON 13 F6
Lawrence Lake ON 14 D6
Lawrence Station NB 32 D2
Lawrencetown NS 32 F5-6
Lawrencetown NS 33 C3
Lawrencetown NS 33 C6
Lawrencetown Beach PP NS 33 C6-7
Lawrenceville QC 24 D8
Lawrenceville QC 27 E4
Lawson NS 34 D4
Lawton AB 8 B5
Laxgalts'ap (Greenville) BC 3 F3
Lax Kw'alaams BC 3 F2
Layco SK 11 C-D6
Layland MB 13 E-F3
Layland MB 14 E3-4
Lazo BC 1 F6
Lazo BC 2 C-D2
Leach Island ON 17 E3
Leacross SK 11 B6
Lead Cove NL 37 C7-8
Leader SK 6 D8
Leader SK 11 D2
Leading Tickles NL 38 E-F5
Leaf Lake SK 11 B7
Leaf Lake SK 13 B1
Leaf Rapids MB 12 D8
Leaf Rapids MB 15 D2
Leahurst AB 8 E7-8
Leamington ON 19 E2
Leancoil BC 6 D3-4
Leander PNR ON 16 F6
Leander QC 24 C6
Leask SK 11 B4
Leaskdale ON 20 B7-8
Leaskdale ON 21 E2
Leavitt AB 6 F6
Leavitt AB 7 F5
Leavitt Bay SK 12 C-D4
Le Barrois, lac QC 26 E7
Le Barrois, lac QC 28 C3
Lebel-sur-Quévillon QC 23 B-C6
Lébel-sur-Quevillon QC 23 B-C6
Le Bic QC 29 C2-3
Leblanc NB 32 B7
Leblanc NB 35 E1
Leblanc, lac QC 25 A5
Leblanc, lac QC 26 F5
Le Blanc Lake SK 9 A8
Le Blanc Lake SK 11 D-E6
Lebret SK 11 D-E6
Le Breton, lac QC 28 B7
Lebrix Lake MB 13 A4
Leclercville QC 27 B4
Lecointre, lac QC 23 C6-7
Lecointre, lac QC 25 C2
Lecompte, lac QC 26 D1
Lecompte, rivière QC 23 C6
Le Croq, lac QC 28 A7
Le Doré, lac QC 39 D-E5
Leduc AB 6 B5-6
Leduc AB 8 D6
Ledwyn MB 14 F5
Lee AB 7 E3-4
Leeburn ON 17 F4
Leech NB 31 C7
Leechdale AB 8 E5
Leedale AB 8 E5
Lee River MB 14 D7
Lee River Falls MB 14 D6-7
Lefaivre ON 24 D3-4
Lefebvre QC 24 D8
Lefebvre QC 27 D-E3
Le Fer, lac QC 39 A2
Lefroy ON 20 B7
Lefroy ON 21 E1
Leftrook Lake MB 15 E3
Legal AB 6 A5-6
Legal AB 8 B6
Le Gardeur QC 24 C6
Le Gardeur QC 27 D1-2
Legend AB 7 E7
Legend Lake AB 9 C6
Legend Lake AB 12 C1
Le Gentilhomme, lac QC 39 C2
Légerville NB 31 F7-8
Légerville NB 32 B6
Leggat Lake ON 21 C6
Le Goulet NB 30 E2
Le Goulet NB 31 B8
Leg Pond NB 38 C3-4
Leifur MB 14 C3
Leinan SK 11 E3
Leipzig SK 11 C3
Leitch Collieries PHS AB 7 E3
Leith ON 18 F5
Leith ON 20 A4
Leitrim ON 21 B8
Leitrim ON 24 D2
Lejac BC 1 A6
Lejeune QC 29 E3

Lejeune QC 31 B1
Leland Lakes AB 9 A7
Leland Lakes AB 10 F8
Leluau, lac QC 25 C4
Lemain, lac QC 25 D3
Leman, lac QC 28 A8
Leman, lac QC 29 A1
Le Marié, lac QC 28 B7
Lemay, rivière QC 39 E1
Lemberg SK 11 D-E6
Lemieux ON 24 D3
Lemieux QC 24 B8
Lemieux QC 25 F8
Lemieux Islands NU 40 A6
Leminster NS 32 E-F7
Leminster NS 34 C4-5
Lemoine, lac QC 23 E4
Lemsford SK 11 D2-3
Lena MB 13 F2
Lena MB 14 F1-2
Leney SK 11 C3-4
Lennox ON 24 D3
Lennox Island PE 32 A8
Lennox Island PE 34 B1
Lennox Island PE 35 C2-3
Lennox Island PE 35 C2-3
Lennox Passage PP NS 34 D6-7
Lennoxville QC 24 E4-5
Lenore MB 11 E8
Lenore MB 13 E1
Lenore Lake SK 11 C5
Lenore Lake, RNOM QC 11 C5
Lenore Lake NMBS SK 11 C5
Lenôtre, lac QC 22 B7
Lenôtre, lac QC 25 C2
Lenswood MB 11 C8
Lenswood MB 13 C1-2
Lenvale SK 11 B5
Leo AB 8 F8
Leo Creek BC 3 F5
Leonardville NB 32 E2
Leonville SK 11 A4
Léoville PE 35 B2
Lepage, lac QC 25 A4
Lepage, lac QC 26 E4
Le Pate, RNP ON 16 F6
Le Pate PNR ON 16 F6
L'Épiphanie QC 24 C6
L'Épiphanie QC 27 D2
Lepreau NB 32 E3
Lepreau, Point NB 32 E3
Lepreau, ZPF de NB 32 D3
Lepreau PGA NB 32 D3
Lepreau River NB 32 D-E3
Le Précieux-Sang QC 24 B8
Le Précieux-Sang QC 27 C3
Le Rageois, lac QC 39 B1
Le Royer, lac QC 26 E5
Léry QC 24 E5-6
Léry QC 27 E1
Lesage, lac QC 24 B3
Lesage, lac QC 25 F3
Les Boules QC 29 A3
Les Buissons QC 29 A3
Les Caps QC 30 F7
Les Cèdres QC 24 E5
Lescarbot QC 28 E4
Lescot QC 22 B3
Les Coteaux QC 24 E4-5
Les Éboulements QC 28 E7
Les Escoumins QC 29 C1-2
Les-Hauteurs QC 29 C3-4
Les Îlets-Jérémie QC 29 B2-3
Les Islets-Caribou QC 29 F1-2
Les Jardins de la République PP NB 29 E2-3 F2-3
Les Jardins de la République PP NB 31 C2
Leslie SK 11 D6
Leslieville AB 6 C5
Leslieville AB 8 E5
Les Méchins QC 29 B5
Lesser Slave Lake AB 9 E4
Lesser Slave Lake PP AB 9 E5
Lesser Slave Lake Wildland PP AB 9 E4-5
Lesser Slave River AB 9 F5
Lester Beach MB 14 C5-6
Lester-Garland Premises PHS NL 37 C7-8
Le Sueur, ZEC QC 25 D3
Lestock SK 11 D6
L'Étang-du-Nord QC 30 F7
Letellier MB 13 F4
Letellier MB 14 F5
L'Etete NB 32 E2
Lethbridge AB 6 F6
Lethbridge AB 7 E5
Lethbridge NL 36 C8
Lethbridge NL 37 C6
Letondal, lac QC 25 D5
Le Tort, lac QC 39 D5
Lett SK 11 C3
Letty Harbour NT 5 A8
Leumat Lake NT 5 B7
Levack ON 17 F6
Levasseur, lac QC 25 A5
Levasseur, lac QC 26 E5

Levasseur, lac QC 28 C1
Leven MB 15 E4
Lever Lake NT 10 A5
Leville MB 14 D1
Leville NS 33 C4
Levine MB 14 D1
Lévis, lac NT 10 C5
Lévis QC 27 B6
Lewes PE 35 F6
Lewin's Cove NL 36 E6-7
Lewin's Cove NL 37 E4-5
Lewis MB 14 D6
Lewis, lac QC 23 A8
Lewis, lac QC 26 B2-3
Lewis Hill NL 36 B2
Lewis Hills NL 36 B2
Lewis Hills NL 40 F8
Lewis Lake PP NS 32 F8
Lewis Lake PP NS 33 C5-6
Lewisporte NL 37 A5
Lewisporte NL 38 F6
Lewvan SK 11 E6
Lexington NS 34 D6
Leyond River MB 13 C-D4
Leyond River MB 14 A3
Leyson Point NU 41 D7
L'Heureux, lac QC 22 B6-7
L'Heureux, lac QC 25 C2
L'hutsaw Wetland Territorial Habitat Protection Area YT 4 C3
Liard Canyon Territorial Recreation Site BC 3 A4
Liard Canyon Territorial Recreation Site BC 4 F7
Liard River AB 3 A6
Liard River BC 3 B5
Liard River BC 4 F8
Liard River BC 10 F2
Liard River YT 3 A3-4
Liard River YT 4 E6
Liard River Corridor PP BC 3 A5
Liard River Crossing River TP NT 10 D3-4
Liard River Hotsprings PP BC 3 A5
Liard River Hotsprings PP BC 4 F7
Libau MB 13 E4
Lichen, lac QC 23 B8
Lichen, lac QC 26 B2
Lichtenthal SK 6 D8
Liddon Gulf NT 42 D2
Liebenthal AB 6 E7
Liebenthal SK 11 E2
Liège, lac QC 25 B6
Liège, lac QC 28 E2
Liège River QC 9 C-D5
Lièvre, rivière du QC 22 F7
Lièvre, rivière du QC 24 C2 D2-3
Lièvre, ZEC de la QC 25 A7-8
Lièvre, ZEC de la QC 26 E7-8
Lièvres, île aux QC 28 D8
Lièvres, île aux QC 29 D1
Lighthouse Cove ON 19 E2
Lighthouse Point PNR ON 19 F1-2
Lighthouse Point, RNP ON 19 F1-2
Ligne, lac de la QC 23 C7
Ligne, lac de la QC 26 C1
L'Islet QC 28 F7
L'Islet QC 28 F7
L'Islet-sur-Mer QC 28 F7
L'Isle-Verte QC 29 D1-2
Lismore NS 34 D4
Listen Lake SK 11 A4
Listen Lake SK 12 F4
Lister BC 6 F4
Lister BC 7 F1
Listowel ON 19 A5
Listowel ON 20 C4
Listuguj QC 29 D6
Listuguj QC 31 A4
Litchfield NS 32 F5
Litchfield NS 33 C2
Little Abitibi Lake ON 17 B7
Little Abitibi PP ON 17 B6-7
Little Abitibi River ON 17 A-B8
Little Andrews Bay PMP BC 1 A4-5
Little Anse NS 34 E6-7
Little Ausable River ON 19 B-C4
Little Ausable River ON 20 D-E3
Little Barachois Brook NL 36 C2-3
Little Barachois Brook NL 36 D2
Little Baraway NS 34 E6
Little Barasway NL 36 E8
Little Barasway NL 36 D2
Little Bay NL 36 D5
Little Bay NL 36 E7
Little Bay NL 37 D3
Little Bay NL 38 E5
Little Bay East NL 36 F7
Little Bay East NL 37 E5
Little Bay Islands NL 38 E5
Little Bay West NL 36 E7
Little Bay West NL 37 E4

Limestone Mountain AB 8 F4
Limestone Point MB 13 A3
Limestone Point Lake MB 12 F8
Limestone Point Lake MB 15 F2
Limestone River MB 15 D5
Limoges ON 24 D3
Linacre SK 6 E8
Linacre SK 11 E2
Linaria AB 8 A5-6
Lincoln AB 8 A6
Lincoln NB 32 C3
Lincoln ON 19 C8
Lincoln ON 20 E7
Lincoln, mer de NU 42 A4
Lincoln Sea NU 42 A5
Lincolnville NS 34 E5
Lindale AB 6 B5
Lindale AB 8 D5
Lindbergh AB 6 A7
Lindbergh AB 11 A2
Lindbrook AB 6 B6
Lindbrook AB 8 C7
Linden AB 6 D6
Linden AB 7 A4-5
Linden NS 14 E5
Linden NS 32 C8
Linden NS 34 D1
Linden NS 35 F2-3
Linden Beach ON 19 E1-2
Lindsay NB 31 F3
Lindsay NB 32 B1
Lindsay ON 20 B8
Lindsay ON 21 D2-3
Lindsay, lac QC 26 D5
Linear Mounds NHS MB 11 F7-8
Linear Mounds NHS MB 13 F1
Lingan NS 34 C8
Lingham Lake ON 21 C5
Linière, rivière QC 27 D7
Link Lake ON 1 C4
Linkletter PE 35 D2
Linkletter PP PE 32 A7-8
Linkletter PP PE 35 D2
Lintlaw SK 11 C6
Linton QC 25 D8
Linton QC 28 F4
Linwood NS 34 D5
Linwood ON 19 A5
Linwood ON 20 D4
Lions Bay BC 1 F7
Lion's Head ON 18 E5
Lion's Head, RNP ON 18 E5
Lion's Head PNR ON 18 E5
Lipton SK 11 D6
Lisbon ON 19 B5
Lisbon ON 20 D4
Lisburn AB 8 B5
Liscomb NS 34 F5
Liscomb Game Sanctuary NS 33 A7-8 B7-8
Liscomb Game Sanctuary NS 34 F3-4
Liscomb Harbour NS 34 F5
Liscomb Island NS 34 F5
Liscomb Mills NS 33 B8
Liscomb Mills NS 34 F5
Liscomb River NS 33 B8
Liscomb River NS 34 F5
Lisieux SK 11 F4-5
Lisle ON 20 B6
L'Isle-aux-Grues QC 27 A7
L'Isle-aux-Grues QC 28 F7

Little Bear Lake SK 11 A5
Little Bear Lake SK 12 F5
Little Bear River NT 4 A7-8
Little Bear River NT 10 A1
Little Beaver River MB 15 C4-5
Little Black Bear 11 D6
Little Bluff CA ON 21 E6
Little Bolton Lake MB 13 A4
Little Bona NL 36 E7-8
Little Bona NL 37 E5-6
Little Bone SK 11 D7
Little Bow PP AB 6 E6
Little Bow PP AB 7 D5
Little Bow River AB 6 E6
Little Bow River AB 7 C4 D4-5
Little Bras d'Or NS 34 C7-8
Little Brehat NL 38 A5
Little Britain NL 20 B8
Little Britain ON 21 D-E2
Little Brook NS 33 D1
Little Buffalo AB 9 D4
Little Buffalo NT 10 E-F7
Little Buffalo River NT 41 E1
Little Buffalo River Crossing TP NT 10 E7
Little Buffalo River Falls TP NT 10 E7
Little Burnt Bay NL 36 A7
Little Burnt Bay NL 37 A5
Little Burnt Bay NL 38 F6
Little Cadotte River AB 9 D4
Little Catalina 37 B7-8
Little Cataraqui Creek CA ON 21 D6
Little Chiblow Lake ON 18 A1-2
Little Churchill River MB 15 C5 D4-5
Little Clarke Lake SK 11 A4
Little Clarke Lake SK 12 F4
Little Cornwallis Island NU 42 D4
Little Cove, RNP ON 18 D4
Little Cove PNR ON 18 D4
Little Current ON 18 B4
Little Current River ON 16 C8
Little Current River ON 40 F2
Little Current River PP ON 16 C7-8
Little Current River PP ON 17 A2-3
Little Cygnet Lake MB 15 D5
Little Doctor Lake NT 10 D2
Little Dover NS 34 E6
Little Duck Lake MB 15 A4
Little Duck Post MB 15 A3
Little Falls NL 36 A4
Little Falls NL 37 A2
Little Falls NL 38 F3
Little Fishing Lake SK 11 A2-3
Little Fish Lake PP AB 6 D6-7
Little Fish Lake PP AB 7 A6
Little Fish Lake PP AB 11 D1
Little Flatstone Lake SK 12 E3-4
Little Fogo Islands NL 38 E7
Little Forehead Lake NT 10 A6-7
Little Fort BC 6 D1
Little French River ON 18 B7
Little Friars Cove NL 36 D1
Little Gander Pond NL 36 C6-7
Little Gander Pond NL 37 C5
Little Gem AB 7 A8
Little Grand Lake NL 36 B3
Little Grand Lake NL 37 B1
Little Grand Rapids MB 13 C5
Little Grand Rapids MB 16 A1
Little Grand Rapids (IR) MB 13 C5
Little Grand Rapids (IR) MB 16 A1
Little Greenwater Lake, RNP ON 16 F4-5
Little Greenwater Lake PNR ON 16 F4-5
Little Harbour NL 36 A3-4
Little Harbour NL 36 D8
Little Harbour NL 36 E7
Little Harbour NL 37 A1-2
Little Harbour NL 37 C7
Little Harbour NL 37 D6
Little Harbour NL 37 E5
Little Harbour NL 38 F3
Little Harbour NS 33 C7
Little Harbour NS 33 F3
Little Harbour NS 34 E3-4
Little Harbour NS 35 D-E8
Little Harbour Deep NL 39 E8
Little Harbour East NL 36 D7
Little Harbour East NL 37 D5
Little Hawk Lake ON 21 B2-3
Little Heart's Ease NL 37 C-D7
Little Hills SK 12 F5

Little Hyland River YT 4 C-D7
Little Jackfish River ON 16 C-D6
Little Jackfish River ON 17 A1
Little Joseph Lake 39 C3
Little Judique NS 34 D5
Little Keele River NT 4 A7
Little Keele River NT 5 F7
Little Key River ON 18 B-C6
Little Klappan River BC 3 C-D3
Little Lake 19 B6-7
Little Lake CA ON 19 C6
Little Lake CA ON 20 E5
Little Limestone Lake MB 11 A8
Little Limestone Lake MB 13 A2
Little Limestone Lake MB 15 D5
Little Main Restigouche River NB 29 E4
Little Maitland River ON 19 A5
Little Maitland River ON 20 C4
Little Manitou Lake SK 11 D5
Little Marten Lake NT 10 A7
Little Mecatina River NL 39 C4-5
Little Missinaibi Lake ON 17 D4
Little Mississippi River ON 21 B4-5
Little Moose Island MB 14 A5
Little Nahanni River NT 4 C7
Little Narrows NS 34 C6
Little Paradise NL 36 E7-8
Little Paradise NL 37 E5-6
Little Partridge River MB 12 A7
Little Partridge River MB 15 A2
Little Pickerel River ON 18 B7-8 C7
Little Pic River ON 16 E8
Little Pine SK 6 B8
Little Pine SK 11 B3
Little Playgreen Lake MB 13 A3
Little Pond PE 34 C4
Little Pond PE 35 E7
Little Port l'Hébert NS 33 E3-4
Little Qualicum Falls PP BC 1 F6
Little Qualicum Falls PP BC 2 D-E3
Little Quill Lake SK 11 C6
Little Quirke Lake ON 18 A2-3
Little Rancheria River BC 3 A3
Little Rancheria River BC 4 C6
Little Rapids NL 36 A-B3
Little Rapids NL 37 A-B1
Little Rapids ON 17 F4
Little Rapids ON 18 A1
Little Red River SK 11 B4
Little Ridge NB 32 D-E1
Little River BC 1 F6
Little River NB 2 C2
Little River NB 29 F4
Little River NB 31 C3
Little River NB 32 B3
Little River NB 32 C6
Little River NB 32 C6
Little River NB 32 C8
Little River NB 33 D1
Little River NB 34 D1
Little River Harbour NS 33 F1
Little Sachigo Lake ON 13 A7
Little St. Lawrence NL 36 F6
Little St. Lawrence NL 37 F4
Little Salmon Lake YT 4 C4
Little Salmon River Gorge, ZCP NB 32 D5-6
Little Salmon River Gorge, ZCP NB 33 A2-3
Little Salmon River Gorge PCA NB 32 D5-6
Little Salmon River Gorge PCA NB 33 A2-3
Little Sand Lake MB 15 C3
Little Sand Lake ON 14 B8
Little Sands PE 34 D3
Little Sands PE 35 F6
Little Saskatchewan MB 13 D3
Little Saskatchewan MB 14 A3
Little Seal River MB 15 A-B5
Little Seldom NL 38 E7
Little Shemogue NB 34 C4
Little Shemogue NB 34 C1
Little Shemogue NB 35 E2
Little Shemogue, havre de NB 35 E2

Little Shemogue Harbour NB 35 E2
Little Smoky AB 9 F3
Little Smoky River AB 9 F3
Little Southwest Miramichi River NB 31 D5
Little Stull Lake MB 15 F1
Little Sturgeon River ON 18 A8
Little Sturgeon River ON 22 C-D1
Little Tobique River NB 31 C4
Little Tobique River (Nictor Branch) NB 29 F5
Little Tomiko River ON 18 A8
Little Traverse Bay ON 14 F8
Little Trout Lake ON 13 D6
Little Trout Lake ON 16 C2-3
Little Turtle Lake ON 16 E3
Little Turtle River ON 16 E3
Little Vermilion Lake ON 13 D6
Little Vermilion Lake ON 16 B2
Little White River ON 17 F5
Little White River PP ON 17 F5
Livelong SK 6 A-B8
Livelong SK 11 B3
Lively ON 17 F6
Lively ON 18 A5
Liverpool NS 33 E4
Liverpool Bay NT 5 A6
Liverpool Bay NT 41 A1
Livingston NL 39 B2
Livingstone Cove NS 34 D4
Livingstone Creek YT 5 D4
Livingstone Lake ON 21 B2
Livingstone Lake SK 12 B4
Livingstone Point PP ON 16 D6
Livingstone Point PP ON 17 B1
Lizotte QC 25 B8
Lizotte QC 26 F8
Llewellyn Glacier BC 3 A-B1
Lloyd Lake SK 9 C8
Lloyd Lake SK 12 C3
Lloydminster AB 6 B7-8
Lloydminster AB 11 B2
Lloyds Lake NL 36 C3
Lloyds Lake NL 37 C1
Lloyds River NL 36 C3-4
Lloyds River NL 37 C1-2
Lloydtown ON 20 C6
Lobo ON 19 C4
Lobo ON 20 B3
Lobster Cove NL 38 D-E4
Lobster Cove NL 38 E2
Lobster House NL 36 A4
Lobster House NL 37 A2
Lobster House NL 38 F3-4
Lobstick AB 6 A6
Lobstick AB 8 B7
Lobstick AB 11 A1
Lobstick Island SK 9 A7
Lobstick Island SK 12 A3
Lobstick Lake Provincial Reserve SK 11 A7
Lochaber NS 34 E4-5
Lochaber Lake NS 34 E4
Lochaber Mines NS 33 B7-8
Lochaber Mines NS 34 F3-4
Lochalsh ON 17 D4
Loch Alva NB 32 D4
Loch Alva, APP NB 32 D-E3
Loch Alva, APP NB 33 B1
Loch Alva PPA NB 32 D-E3
Loch Alva PPA NB 33 B1
Loch Broom NS 34 D-E3
Lochiel ON 24 D4
Lochiel Lake PP NB 34 E4-5
Lochinvar AB 8 E6
Lochlin ON 21 C3
Loch Lomond NB 32 D4
Loch Lomond NB 33 A2
Loch Lomond NS 34 D7
Loch Lomond NS 34 D7
Lockeport NS 33 F3
Lockhart AB 8 E5
Lockhart Beach PP BC 6 F4
Lockhart Creek PP BC 6 F4
Lockhart Lake NT 10 D7-8
Lockhart River NT 10 B8
Lockston NB 21 A2
Lockport MB 13 E4
Lockport MB 14 D5
Lock's Cove NL 36 D5
Lock's Cove NL 37 D3
Lockston NB 21 B1
Lockston Path PP NL 37 B7
Lockwood SK 11 D5
Lodge Bay NL 39 C8
Lodge Creek AB 6 F8
Lodge Creek AB 11 F2
Lodgepole AB 6 B4
Lodgepole AB 8 D4
Logan, mont QC 29 B6
Logan, Mount YT 4 E1
Logan Glacier YT 4 E1
Logan Island ON 16 D6
Logan Lake AB 9 F3
Logan Lake AB 12 E1-2
Logan Lake BC 1 E8
Logan Lake BC 2 B8
Logan Lake BC 6 E1
Logan Lake ON 21 B1
Logan Mountains YT 4 D7
Logans Point NS 34 D3

Loganville NS 34 E3
Loggieville NB 31 D7
Log Valley SK 11 D-E4
Logy Bay NL 39 D8
Logy Bay-Middle Cove-Outer Cove NL 37 D8
Lohbiee BC 1 D6
Lois, rivière QC 23 C2-3
Lois Lake BC 2 C3
Lokken, Mount YT 4 C-D4
Loks Land NL 40 B6
Lola Lake, RNP ON 16 D3
Lola Lake PNR ON 13 E7
Lola Lake PNR ON 16 D3
Lombardy ON 21 C7
Lombardy ON 24 F1
Lomond AB 6 E6
Lomond AB 7 D6
Londesborough ON 19 A4
Londesborough ON 20 C3
London ON 19 C4
London ON 20 B3
Londonderry NS 32 D8
Londonderry NS 33 A5-6
Londonderry NS 34 E1-2
Londonderry PP NS 33 A6
Londonderry PP NS 34 E2
Londonderry Station NS 32 D8
Londonderry Station NS 33 A6
Londonderry Station NS 34 E1-2
Lone Butte BC 1 D8
Lone Butte BC 6 D1
Lone Creek ON 22 E4
Lonely Bay NT 10 D6
Lonely Island NL 18 C4
Lonely Lake MB 13 D2
Lonely Lake MB 14 B2
Lonely Lake MB 14 B2
Lone Mountain YT 5 D2
Lone Pine AB 8 A-B4
Lone Prairie BC 3 E8
Lone Prairie BC 9 E1
Lone Ridge AB 8 B8
Lone Rock SK 6 B8
Lone Rock SK 11 B2
Lonesand AB 14 F6
Lone Spruce MB 14 C3
Long, lac QC 25 C4
Long, lac QC 28 C5
Long Bay ON 18 C3
Long Beach NL 37 B7-8
Long Beach NL 37 F8
Long Beach ON 19 C8
Long Beach ON 20 A8
Long Beach ON 20 F7
Long Beach ON 21 D2
Long Beach CA ON 19 C8
Long Beach CA ON 20 F7
Longbow Lake ON 13 F6
Longbow Lake ON 16 D2
Long Cove NL 37 D7
Long Creek NB 32 C4-5
Long Creek NB 35 E4
Long Creek SK 11 F6
Longford ON 18 F8
Longford ON 20 A7
Longford ON 21 C1
Long Harbour NL 36 D7
Long Harbour NL 37 D5
Long Harbour NL 37 E6-7
Long Harbour-Mount Arlington Heights NL 36 E8
Long Harbour-Mount Arlington Heights NL 37 E6
Long Island NL 36 D5-6
Long Island NL 36 D8
Long Island NL 37 D3-4
Long Island NL 37 D6
Long Island NS 38 E5
Long Island NS 33 D1
Long Island NU 40 D3-4
Long Island ON 16 D8
Long Island ON 17 B2
Longlac ON 16 D8
Longlac ON 17 B2
Longlac ON 40 F2
Longlegged Lake ON 13 D6
Longlegged Lake ON 16 C2
Longlegged River ON 16 C2
Long Plain NB 13 E3
Long Plain NB 14 E3
Long Point MB 13 B3
Long Point MB 15 D3
Long Point NB 32 E4
Long Point NB 33 A2
Long Point NB 32 B8
Long Point NL 36 D2
Long Point NL 34 D1
Long Point NL 35 F2
Long Point NL 39 B6

Long Point NL 39 B7
Long Point NL 34 A7
Long Point NS 34 D5
Long Point ON 19 D6-7
Long Point ON 19 D6
Long Point ON 20 F5-6
Long Point ON 21 E6
Long Point ON 21 E6
Long Point, RNF de ON 20 F5-6
Long Point Bay ON 19 D6-7
Long Point Bay ON 20 F5-6
Long Point NWA ON 19 D6-7
Long Point NWA ON 20 F5-6
Long Point PP NS 34 D5
Long Point PP ON 19 D6
Long Point PP ON 20 F5
Long Pond NL 36 D8
Long Pond NL 37 D6
Long Range Mountains NL 36 C2-3 D2
Long Range Mountains NL 38 C3-4 D-E3
Long Range Mountains NL 39 E8 F7-8
Long Reach NB 32 D4
Long Reach NB 32 D4
Long Reach NB 33 A1
Long Reach NB 33 A1
Long River PE 35 D3
Long Sault ON 24 E3
Long Sault ON 20 B8
Long Sault CA ON 21 E2
Long Schooner Lake ON 21 B6
Longs Creek NB 32 C2
Longtom Lake NT 10 A5
Longue-Pointe QC 40 F7
Longue-Pointe-de-Mingan QC 40 F7
Longueuil QC 24 D6
Longueuil QC 27 E1-2
Longview AB 6 E5
Longwoods Road CA ON 19 C4
Longwoods Road CA ON 20 E3
Longworth BC 1 A8
Longworth BC 6 A1
Lonira AB 8 B4
Looma AB 8 C7
Loomis SK 11 F3
Loon Bay NL 36 A7
Loon Bay NL 37 A5
Loon Bay NL 38 F6
Loonhaunt Lake ON 13 F6
Loonhaunt Lake ON 16 D2
Loonhead Lake MB 12 F7-8
Loonhead Lake MB 15 F2
Loon Lake NS 34 E5
Loon Lake NT 5 D-E6
Loon Lake ON 18 B1-2
Loon Lake ON 21 B-C3
Loon Lake SK 6 A8
Loon Lake SK 11 A3
Loon Lake SK 12 E7
Loon Lake SK 15 E1
Loon Lake PP BC 1 D8
Loon Lake PP BC 2 A7
Loon Lake PP BC 6 D1
Loon River AB 9 D4-5
Loon River MB 12 D-E7
Loon River MB 15 D-E1
Loos BC 1 B8
Loos BC 6 A1
Loranger Island SK 12 C6
Lord Creek YT 5 D2
Lords Cove NL 36 F5
Lord's Cove NL 36 F6
Lord's Cove NL 37 F4
Lord Selkirk PP PE 34 C2-3
Lord Selkirk PP PE 35 F5
Loreburn SK 11 D4
Lorens, lac QC 39 D5
Lorette MB 13 E4
Lorette MB 14 E5
Lorette, lac QC 25 B4
Lorette, lac QC 26 F4
Loretteville QC 27 B5
Loretto ON 20 B6
Lories NL 36 F5-6
Lories NL 37 F3-4
L'Orignal ON 24 D4
Lorimer Lake ON 18 C7
Loring ON 18 B7
Lorlie SK 11 D6
Lorne NB 29 E7
Lorne NB 31 B5
Lorne NS 33 A7
Lorne NS 34 D4
Lorne, Mount YT 4 E4
Lorne Beach ON 20 B3
Lorne Lake PP AB 6 A6
Lorne Lake PP AB 8 A7
Lorne Lake PP AB 9 F6
Lorne Lake PP AB 12 F1
Lorne C. Henderson CA ON 19 C2-3
Lorne C. Henderson CA ON 20 E2
Lorne Valley PE 35 E6
Lornevale NS 32 D8
Lornevale NS 33 A5-6
Lornevale NS 34 E1
Lorneville NB 32 E4
Lorneville NS 32 B8
Lorneville NS 33 B1
Lorneville NS 34 D1
Lorneville NS 35 F2
Lorneville ON 20 A8

Lorneville ON 21 D2
Lorraine QC 24 D6
Lorraine QC 27 E1
Lorrain Valley ON 17 E8
Lorrainville ON 17 E8
Lorrainville QC 22 A2
Lortie, lac QC 23 D7
Losier Settlement NB 30 F2
Losier Settlement NB 31 B6
Lost Channel ON 18 B6
Lost Lake ON 19 C5
Lost Reindeer Lakes NT 19 D6-7
Lost River QC 24 C4
Lotbinière QC 27 B5
Lots-Renversés QC 29 E2-3
Lots-Renversés QC 31 B1
Loudon Channel BC 2 E2
Loughborough Inlet BC 1 E5
Loughborough Inlet BC 2 B2
Loughborough Lake ON 21 D6-7
Lougheed AB 6 B7
Lougheed AB 8 B8
Lougheed Island NU 42 C3
Louis Creek BC 6 D1-2
Louisa QC 24 C4-5
Louisa, Lake ON 21 A3
Louisa, Lake ON 22 F3
Louisbourg NS 34 C8
Louisbourg Harbour NS 34 C-D8
Louisdale NS 34 D6
Louise, lac QC 24 C4
Louise, lac QC 27 D5
Louise Falls NT 10 F6
Louise Island BC 1 B1
Louise Lake YT 4 E2
Louiseville QC 24 B7
Louiseville QC 25 F7
Louiseville QC 27 C2
Louis Head NS 33 E3-4
Louison, lac QC 23 D7
Louis-XIV, pointe QC 19 C4
Louis-S.-St-Laurent, LHN QC 27 F5
Louis S. St. Laurent NHS QC 27 F5
Louisville ON 19 D3
Louisville ON 20 D6
Loup, rivière du QC 24 A6-7 B7
Loup, rivière du QC 25 E6 F7
Loup, rivière du QC 27 B-C2
Loup, rivière du QC 29 E1
Loup, rivière du QC 29 A3
Loup Marin, lac au QC 22 A1-2
Loups, lac des QC 22 B3
Loups Marins, rivière aux QC 30 A7
Lourdes NL 36 B1
Lourdes-de-Blanc-Sablon QC 38 A-B3
Lourdes-de-Blanc-Sablon QC 39 D7
Lousana AB 6 C6
Lousana AB 8 A7
Loutre, rivière de la QC 22 A1-2
Loutres, lac aux QC 25 A6
Loutres, lac aux QC 26 D2
Louvicourt QC 23 B5
Love SK 11 B6
Loveland Bay PP BC 1 E5
Loveland Bay PP BC 2 C1-2
Lovering, lac QC 27 F4
Loverna SK 6 C8
Loverna SK 11 C2
Low QC 22 F7
Low QC 24 C2
Low, Cape NU 40 A3
Low, Cape NU 41 D6
Lowbanks ON 19 C8
Lowbanks ON 20 F7
Low Bush River ON 17 C7
Lowe Farm MB 13 E3
Lowe Farm MB 14 F4
Lowe Inlet PMP BC 1 A-B3
Lower Argyle NS 33 F2
Lower Arrow Lake BC 6 F3
Lower Barneys River NS 34 D4
Lower Bedeque PE 35 D3
Lower Beverley Lake ON 24 F1
Lower Branch NS 33 D4
Lower Caledonia NS 33 A8
Lower Caledonia NS 34 E4
Lower Cove NL 36 C1
Lower East Pubnico NS 33 F2
Lower Economy NS 32 D8
Lower Economy NS 33 A5
Lower Economy NS 34 E1
Lower Eel Brook NS 33 E2
Lower Five Islands NS 32 D7
Lower Five Islands NS 33 A5
Lower Five Islands NS 34 E1
Lower Fort Garry, LHN de MB 14 D5-6
Lower Fort Garry NHS MB 14 D5-6

Lower Foster Lake SK 12 D5
Lower Freetown PE 35 D3
Lower Gulf Shore NS 35 E3
Lower Hainesville NB 31 F4
Lower Hainesville NB 32 B2
Lower Hay Lake ON 21 A-B3
Lower Hay Lake ON 22 F3
Lower Island Cove NL 37 C7
Lower Jemseg NB 32 C4
Lower Lance Cove NL 37 C7
Lower Madawaska River PP ON 21 B5
Lower Manitou Lake ON 13 F6
Lower Manitou Lake ON 16 E3
Lower Middle River NS 34 C6
Lower Montague PE 35 E6-7
Lower Newtown PE 35 F5-6
Lower Nicola BC 2 C7-8
Lower Nicola BC 6 E1
Lower Nimpkish PP BC 1 E4
Lower Ohio NS 33 E3
Lower Post BC 3 A4
Lower Post BC 4 C5
Lower Rockport NB 32 C6-7
Lower Sackville NS 32 E4
Lower Sackville NS 33 C5-6
Lower St. Esprit NS 34 D7
Lower Shinimicas NS 35 F2
Lower South Jaquet River NB 29 E7
Lower South River NS 34 D-E5
Lower Tsitika River PP BC 1 E4-5
Lower Twin Lake ON 16 D8
Lower Twin Lake ON 17 A2
Lower Wedgeport NS 33 F2
Lower West Pubnico NS 33 F2
Lower Windsor NB 31 E-F3
Lower Windsor NB 32 B1-2
Lower Woods Harbour NS 33 F2
Low Landing NS 33 D3
Low Point NL 37 C8
Lowther ON 17 B5
Lowther Island NU 42 D3-4
Lowville ON 19 B7
Lowville ON 20 D6
Loyalist PE 35 E4
Lozeau, lac QC 39 D3-4
Lubicon Lake AB 9 D4
Lubicon River AB 9 D4
Lucan ON 19 B4
Lucan ON 20 C3
Lucania, Mount YT 4 D1
Lucasville ON 19 C2-3
Lucasville ON 20 E1-2
Lucerne, lac QC 22 F7-8
Lucerne, lac QC 24 C2
Luceville QC 29 C3
Lucie, lac QC 17 A8
Lucien, lac QC 25 A6
Lucien, lac QC 26 E6
Lucien, lac QC 28 C2
Lucknow ON 20 C3
Lucky Lake SK 11 D4
Lucky Lake SK 11 D4
Lucky Man SK 11 B4
Ludger, lac QC 24 B5
Ludlow NB 31 E5
Ludlow NB 32 A3
Luke MB 15 D-E5
Luke's Arm NL 38 E6
Lumberton BC 7 F1
Lumby BC 6 E2
Lumsden NL 37 A7
Lumsden NL 38 F8
Lumsden SK 11 D5
Lumsden Pond PP NS 32 E7
Lumsden Pond PP NS 33 B7
Lund BC 1 E6
Lund BC 2 C2-3
Lundar MB 13 D4
Lundar MB 14 C4
Lundar Beach MB 13 D3
Lundar Beach MB 14 C4
Lundar Beach PP MB 13 E3
Lundar Beach PP MB 14 C3
Lundbreck AB 6 F5
Lundbreck AB 7 E5
Lundy NL 34 E5
Lunenburg NS 32 F6-7
Lunenburg NS 33 C4
Lunenburg NS 33 D4-5
Lunenburg Fisheries Museum of the Atlantic NS 33 D5
Lunenburg Harbour NS 33 D5
Lurgan SK 11 B6
Lurgan Beach ON 20 B3
Luscar AB 8 D1-2
Luseland SK 6 C8
Luseland SK 11 C2-3
Lushes Bight NL 38 E5
Lushes Bight-Beaumont-Beaumont North NL 38 E5
Lusignan, lac QC 24 A5
Lusignan, lac QC 25 E5
Luskville QC 22 F7
Luskville QC 24 D1
Lutes Mountain NB 31 F8
Lutes Mountain NB 32 B6
Luther Lake ON 19 A6

Luther Lake ON 20 C5
Luther Marsh CA ON 19 A6
Luther Marsh CA ON 20 C5
Luton ON 19 D5
Luton ON 20 F4
Lutose AB 9 A3-4
Lutsel K'e NT 10 C8
Lutsel K'e NT 41 E2
Luxana Bay BC 1 C2
Luzan AB 8 B8
Lyal Island ON 18 E4
Lyalta AB 6 D6
Lyalta AB 7 B4
Lyddal MB 15 F3
Lyddat MB 14 D6
Lyell Island BC 1 B1-2
Lyleton ON 11 F8
Lyleton MB 13 F1
Lymburn AB 9 E2
Lyn ON 21 D8
Lyn ON 24 F2
Lynch, lac QC 22 D5
Lynch, lac QC 24 A4
Lynch, lac QC 25 F4
Lynden ON 19 B6-7
Lynden ON 20 E5-6
Lynde Shores CA ON 20 C8
Lynde Shores CA ON 21 D7
Lyndhurst ON 21 D7
Lyndoch ON 19 D6
Lyndoch ON 20 F5
Lynn NS 32 D7
Lynn NS 33 A5
Lynn NS 34 E1
Lynn Lake MB 12 D7
Lynn Lake MB 15 D1-2
Lynnwood ON 19 D3
Lyn Valley CA ON 21 D8
Lynx Bay MB 13 C3-4
Lynx City YT 4 C2
Lynx Lake NT 41 E3
Lyon, Cape NT 5 A8
Lyon, Cape NT 41 A1
Lyons ON 19 C5
Lyons ON 20 F4
Lyons Brook NS 34 D3
Lyonshall MB 14 F1-2
Lyster QC 27 C5
Lyttleton NB 31 D6
Lytton ON 1 E8
Lytton BC 2 B7

M

Mabee's Corners ON 19 D6
Mabee's Corners ON 20 F5
Mabel Lake BC 6 E2
Mabel Lake PP BC 6 E2
Maberly NL 37 B7-8
Maberly ON 21 C6
Mabille, lac QC 39 D4
Mabou NS 34 C5
Mabou Harbour NS 34 C5
Mabou Harbour Mouth NS 34 C5
Mabou Mines NS 34 C5
Mabou PP NS 34 C5
McAdam NB 32 D2
Macallum Lake SK 9 F8
Macallum Lake SK 12 F3
MacAlpine Lake NU 41 C3
Macamic ON 17 C8
Macamic QC 23 C2-3
Macamic, lac QC 23 C2-3
Macamic, rivière QC 23 C3
McAras Brook NS 34 D4
McArthur Falls MB 13 E4-5
McArthur Falls MB 14 C6-7
McArthur Mills ON 21 B4
Macaulay Mountain CA ON 21 E5
McAuley MB 11 E7-8
McAuley MB 13 E1
Macaza, lac QC 24 B4
Macaza, lac QC 25 F4
McBean Harbour ON 18 B3
McBeth Fiord NU 41 B7-8
McBeth Fiord NU 42 E7
McBeth Point MB 13 C3-4
McBride BC 1 B8
McBride BC 6 B1
McBride Lake SK 11 C7
McBride River BC 3 C3
McCabe Creek YT 4 C3
McCallum NL 36 D3
McCallum NL 37 D3
McCallum Lake MB 12 E7
McCallum Lake MB 15 E1
McCallum Settlement NS 33 A6
McCallum Settlement NS 34 E2
MacCallums Point PE 35 D3
Maccan NS 32 C7
Maccan River NS 32 C7
McCarthy, lac QC 25 C8
McCarthy, lac QC 28 B2
McCauley Island BC 1 A2
McClarty Lake MB 12 A8
McClarty Lake MB 13 A2
McClarty Lake MB 15 F2
McClelland Lake AB 9 C6-7
McClelland Lake AB 12 A1
Maccles Lake NL 36 B8
Maccles Lake NL 37 B6
M'Clintock Channel NU 41 A-B4
M'Clintock Channel NU 42 E3

M'Clintock Inlet NU 42 A4-5
M'Clintock River YT 4 E4
M'Clure Strait NT 42 D2
McConnell MB 11 E8
McConnell MB 13 E1-2
McConnell Lake PP NU 2 B8
McConnell Lake PP BC 6 E1
McConnell Range NT 10 B2
McConnell River YT 4 D5
McCord SK 11 F4
MacCormack PP NS 34 C6-7
McCoy Lake ON 13 C7
McCoy Lake ON 16 A3
McCracken, lac QC 22 D2-3
McCrackens Landing ON 21 D3-4
McCraney Lake ON 21 A2
McCrea Lake ON 13 D8
McCrea Lake ON 16 C5
McCrea River NT 10 B-C7
McCreary MB 13 D2
McCreary MB 14 C2
McCulloch ON 6 E2
Macculloch, Cape NT 41 A7
Macculloch, Cape NU 42 D6
McCusker Lake ON 13 D5
McCusker Lake ON 14 A8
McCusker Lake ON 16 B1
McCusker Lake Provincial Ecological Reserve SK 9 E8
McCusker Lake Provincial Ecological Reserve SK 12 E-F3
McCusker River SK 9 E-F8
McCusker River SK 12 E3
Macdiarmid ON 16 E6
Macdiarmid ON 17 B1
Macdonald MB 13 E3
Macdonald MB 14 D3
McDonald, lac QC 24 C4
McDonald, lac QC 26 A2
McDonald Bay SK 12 E4
McDonald Creek PP BC 6 E3
MacDonald Farm PHS ON 31 D7
MacDonald Lake NT 10 D8
MacDonald Lake 12 A4
MacDonald Range BC 6 F5
MacDonald Range BC 7 F3
McDonalds Corners ON 21 C6
McDonalds Point NB 32 C4
McDonald Unit (Gulf Islands NPR) 2 F4
McDougal Creek SK 11 A5-6
MacDougall NB 31 F8
MacDougall MB 32 B6
McDougall Gulch NL 36 D1
McDougall Pass NT 5 C3-4
Macdowell MB 13 C6
McDowell ON 13 C6
McDowell ON 16 A3
MacDowell Lake ON 13 C7
MacDowell Lake ON 16 B1
MacDowell River ON 13 B-C7
MacDuff ON 17 C4
MacElmons Pond PP NS 32 D8
MacElmons Pond PP NS 33 A5-6
MacElmons Pond PP NS 34 E1-2
Maces Bay NB 32 E3
McEvoy, Mount BC 3 D4
McEvoy Lake YT 4 D6
McFarlane Lake ON 18 A5
McFarlane River SK 9 A-B8
McFarlane River SK 9 B8
MacFarlane River SK 12 B4
MacFarlane River SK 41 F2
McGavock Lake MB 12 D7
McGavock Lake MB 15 D1-2
McGee SK 11 D3
McGerrigle, monts QC 29 B7
McGillivray, lac QC 22 E5
McGillivray Range BC 6 F4
McGillivray Range BC 7 F1-2
McGivney NB 31 F5
McGivney NB 32 B3
McGowans Corner NB 32 C5
McGrath Cove NS 32 F8
McGrath Cove NS 33 C5
McGraw Brook NB 31 D5-6
MacGregor MB 13 D3
MacGregor MB 14 C2
McGregor ON 19 E1
McGregor, lac QC 21 A8
McGregor, lac QC 22 A8
McGregor, lac QC 22 E4
McGregor, lac QC 24 E2
McGregor, lac QC 24 D2
McGregor, lac QC 24 D2
McGregor Bay ON 18 B4
McGregor Bay ON 18 B4
McGregor Lake AB 6 E6
McGregor Lake AB 7 C5
MacGregor Point PP ON 18 F4
MacGregor Point PP ON 20 A3
McGregor River BC 1 A7-8
McGregor River BC 6 A1

McGregor River BC 9 F1
Machawaian Lake ON 16 B6-7
Machias Seal Island NMBS NB 32 F2
Machichi River MB 15 D7
Machmell River BC 1 D4-5
McInnes Lake ON 13 C6
McInnes Lake ON 13 C6
McInnis Lake ON 16 A2
McInnis Point PE 35 E-F5
McIntosh ON 13 E6
McIntosh ON 16 D2
McIntosh Lake ON 22 F2
McIntyre Bay BC 1 A1
McIntyre Lake SK 12 C5
McIver's NL 36 A3
McIver's NL 37 A1
McIver's NL 38 F2
McIvor River AB 9 C6
McIvor River AB 12 B-C1
MacIvors Point PE 35 E4
McKague SK 11 C6
McKaskill Lake ON 21 A4
McKaskill Lake ON 22 F4
MacKay AB 6 A-B4
MacKay AB 8 C4
McKay Creek AB 22 F5
McKay Lake NT 10 B8
Mackay Lake NT 41 D2
McKay Lake ON 17 B2
MacKay Lake NL 39 B2-3
McKay Lake ON 16 D-E8
McKay River AB 12 D1
McKay's NL 36 C2
McKays Corners ON 19 D3
McKean River NU 40 A5
McKeand River NU 41 C8
McKees Mills NB 31 E8
McKees Mills NB 32 A6
McKellar ON 18 C7
McKendrick NB 31 A5
Mackenzie AB 9 B4
Mackenzie BC 3 F6-7
Mackenzie, fleuve NT 4 A8
Mackenzie, fleuve NT 5 E6
Mackenzie, fleuve NT 10 A-B2
Mackenzie, fleuve NT 10 E4
Mackenzie, fleuve NT 41 D-E1
McKenzie, Mount YT 4 C4
Mackenzie Bay YT 5 B3-4
Mackenzie Bison Sanctuary NT 10 D6
McKenzie Creek ON 19 C6-7
McKenzie Creek ON 20 E5-6
Mackenzie Delta NT 5 C4
McKenzie Island ON 13 D6
Mckenzie Island 16 B-C2
Mackenzie King Island 42 C2-3
McKenzie Lake ON 21 B4
Mackenzie Lake SK 11 A7
Mackenzie Lake SK 12 C6
Mackenzie Lake SK 13 A1
Mackenzie Lake SK 15 C1
Mackenzie Lake SK 15 F1
Mackenzie Mountains NT 4 A6-7 B7-8 C8
Mackenzie Mountains NT 5 E-F4 F5-6
Mackenzie River NT 4 A8
Mackenzie River NT 5 D5-6
Mackenzie River NT 10 A1
Mackenzie River NT 10 D3-4
Mackenzie River NT 41 D-E1
MacKerracher Lake MB 12 C8
MacKerracher Lake MB 15 C3
McKerrow ON 17 F6
McKerrow ON 18 A4
Mackey ON 22 E4
McKiel Lake NB 31 E4
McKiel Lake NB 32 A2
McKillop, lac QC 22 C3
Mackin Creek BC 1 C7
McKinlay Lake NT 10 C8
McKinley Bay NT 5 A5-6
McKinleyville NB 31 D6-7
McKinnon TP NT 5 F7-8
McKinnon TP NT 10 A1
McKinnons Harbour NS 34 D6-7
Mack Lake MB 15 C5
Macklin SK 6 C8
Macklin SK 11 C2
McKnight Lake MB 12 E7
McKnight Lake MB 15 E2
McLachlin, lac QC 22 C3
McLaren, lac QC 25 D6
McLaren, lac QC 28 C4
McLaren, lac QC 28 F2
McLaughlin AB 6 B8
McLaughlin AB 11 B2
McLaughlin River MB 13 A4
McLean SK 11 E6
McLean Settlement NB 31 E7-8
McLean Settlement NB 32 A6
Maclean Strait NU 42 C3
McLeese Lake BC 1 C7
McLennan AB 9 E3-4

McLennan, lac QC 22 A8
McLennan, lac QC 25 C2
McLennan Lake SK 12 E5
McLeod Bay NT 41 D2
McLeod Lake BC 3 F7
McLeod Lake BC 3 F7
MacLeod Lake MB 15 B3-4
MacLeod PP ON 16 D7
MacLeod PP ON 17 B2
McLeod River AB 6 A4 B3-4
McLeod River AB 8 B-C3
McLeod River AB 8 C2
McLeod Valley AB 8 C3
McLure BC 2 A8
McLure BC 6 D1
McMahon SK 11 E3-4
McManus Hill, ZCP ON 32 C5-6
McManus Hill PCA ON 32 C6
McMaster Island, RNP ON 18 E4
McMaster Island PNR ON 18 E4
McMillan Lake AB 9 E6
McMillan Lake AB 12 E1
McMillan Lake MB 12 D7
McMillan Lake MB 15 D1
Macmillan Pass YT 4 B6
MacMillan PP BC 2 D-E2
Macmillan River YT 4 B4 C3-4
McMorran SK 11 D3
McMunn MB 13 F5
McMunn MB 14 E7
McMurdo BC 6 D3-4
McNab SK 7 F6
McNab Cove NS 34 D7
McNabs Island NS 33 C6
McNallie Creek TP NT 10 E5-6
McNamara, Mount BC 3 C4-5
McNamara Lake ON 16 E4
McNamee NB 31 E5
McNamee NB 32 A3
McNeil Lake YT 4 D5
McNeil River YT 4 D5
McNeills Mills PE 35 C2
MacNutt SK 11 D1
McNutt Lake NB 31 D7
McNutt Island NS 33 F3
Macoun SK 11 F6
Macoun Lake SK 12 D6
Macpès, lac QC 29 C3
McPhadyen River NL 39 B2
McPhee Corner NS 32 E8
McPhee Corner NS 33 B6
McPhee Corner NS 34 F2
McPherson Lake YT 4 D6
McPhersons Mills NS 33 A8
McPhersons Mills NS 34 A4
McQuaby Lake ON 22 E1
McQuesten YT 4 B3
McQuesten Lake YT 4 A4
McQuesten River YT 4 B3
McRae AB 6 A7
McRae AB 8 A4
McRae AB 9 F7
McRae AB 11 A1
McRae AB 12 F1
MacRae YT 4 E4
McRae Point PP ON 18 F8
McRae Point PP ON 20 A7
McRae Point PP ON 21 D1
Macrorie SK 11 D4
Mactaquac NB 32 C2
Mactaquac NB 32 B8
Mactaquac PP NB 32 C2
McTaggart SK 11 F6
McTaggart Lake SK 9 B-C8
McTaggart Lake SK 12 C3
McTavish MB 13 F4
McTavish MB 14 E4-5
McTavish Lake SK 12 E5
MacTier ON 18 D7-8
Macton ON 19 A5-6
Macton ON 20 D4-5
McVeigh MB 12 D7
McVeigh MB 15 D2
McVicar Arm NT 10 A3-4
McVicar Arm NT 41 C1
McVittie ON 17 F7
McVitties ON 18 A5-6
McWatters QC 17 D8
McWatters QC 23 E2-3
Madame, Isle QC 27 A6
Madame, Isle NS 34 E6
Madawaska ON 21 A4
Madawaska ON 22 F3-4
Madawaska, rivière QC 31 B-C1
Madawaska Lake ON 21 B3
Madawaska River ON 21 A3-4
Madawaska River ON 21 B5
Madden AB 6 D5
Madden AB 7 A3-4
Maddox Cove NL 37 D-E8
Madeira Park BC 2 D3-4
Madeleine, cap de la QC 29 A8
Madeleine, cap de la QC 30 A1
Madeleine, cap de la QC 34 E4-5
Madeleine, îles de la QC 30 F2-3
Madeleine, lac QC 23 B6
Madeleine, rivière QC 29 A-B8
Madeleine, rivière QC 30 B1
Madeleine-Centre QC 29 A8

Madeleine-Centre QC 30 A1
Madge Lake MB 11 C-D7
Madge Lake MB 13 C1
Madison SK 11 D3
Madoc ON 21 D5
Madran NB 31 B6
Mad River ON 20 B6
Madsen ON 13 D6
Madsen ON 16 C2
Maelpaeg Reservoir NL 36 C5
Maelpaeg Reservoir NL 37 C3
Mafeking MB 11 B8
Mafeking MB 13 B1
Magaguadavic QC 32 C2
Magaguadavic Lake NB 32 C1
Magaguadavic River NB 32 D2
Maganasipi, lac QC 22 D3
Maganasipi, ZEC QC 22 D3
Magill Lake ON 15 F5
Magiss Lake ON 13 B7
Magna Bay BC 6 D2
Magnet MB 14 B2
Magnetawan ON 18 C7-8
Magnetawan ON 22 F1
Magnetawan River ON 18 C6-7
Magnetawan River ON 18 C7
Magnetawan River ON 18 C8
Magnetawan River ON 21 A1
Magnetawan River PP ON 18 C6-7
Magnolia AB 8 C5
Magog QC 24 E8
Magog QC 27 E4
Magog, lac QC 27 E-F4
Magpie QC 39 E-F3
Magpie, baie de QC 39 E-F3
Magpie, lac QC 39 E3
Magpie, lac QC 40 F7
Magpie, rivière QC 39 E3
Magpie Ouest, rivière QC 39 D3
Magpie River ON 17 D3-4
Magrath AB 6 F6
Magrath AB 7 F5-6
Magundy NB 32 C2
Magundy River YT 4 C4-5
Maguse Lake NU 40 A-B1
Maguse Lake NU 41 E4
Mahaska AB 8 B3
Mahekun Lake MB 12 C7
Mahekun Lake MB 15 C2
Mahers NL 37 E7
Mahone Bay NS 32 F7
Mahone Bay NS 33 C-D5
Mahone Bay NS 33 D4-5
Mahony Lake NT 5 E8
Mahood Falls BC 1 C8
Mahood Falls BC 6 C1
Mahood Lake BC 1 C8
Mahood Lake BC 6 C1
Maicasagi, rivière QC 26 A2
Maiden Point NL 38 B5
Maidstone NL 36 C1
Maidstone ON 19 E1
Maidstone SK 6 B8
Maidstone SK 11 B3
Main-à-Dieu NS 34 C8
Main Brook NL 38 B4
Main Brook NL 39 D8
Main Centre SK 11 E4
Main Channel ON 18 D4
Main Dam SK 34 A4
Main Dam NL 37 A2
Main Dam NL 38 F3
Main Duck Island ON 21 E6
Maine, golfe du NS 32 F2-3
Maine, golfe du NS 33 E1
Maine, Gulf of NS 32 F1-2
Main Lake PP BC 2 B2
Mainland NL 36 B1
Mainland River MB 13 A5
Main Outlet NL 38 B5-6
Main Passage ON 18 D4
Main Point NL 36 A7-8
Main Point NL 37 A5-6
Main Point NL 38 F7
Main River NL 38 E3
Main River NL 39 F7-8
Main Topsail NL 36 A5
Main Topsail NL 37 A3
Main Topsail NL 38 F4
Mair SK 11 E-F7
Mair SK 13 E1
Maison-Commémorative-Bethune, LHN de la ON 21 C1-2
Maison de Pierre, lac de la NL 31 D7
Maison-de-Pierre, ZEC de la QC 23 E4
Maison Green Gables PE 35 C3
Maison-de-Sir-John-Johnson, LHN de la QC 24 E4-5
Maison-Inverarden, LHN de la ON 24 E4
Maison-Riel, LHN de la MB 14 E5
Maitland NS 32 D8

Maitland NS 33 A6
Maitland NS 33 D4
Maitland NS 34 E-F2
Maitland ON 21 C8
Maitland ON 24 F2
Maitland Bridge NS 33 D3
Maitland Island BC 1 A2
Maitland Point NT 5 A6
Maitland River ON 19 A4-5
Maitland River ON 20 C3
Maitland River ON 20 C4
Major SK 11 C2
Major, lac QC 22 C4
Major, lac QC 24 A3
Major, lac QC 25 E3
Majorville AB 7 C5-6
Makaroff AB 11 D7
Makaroff MB 13 D1
Makepeace AB 7 B6
Makinak MB 13 D2
Makinak MB 14 B2
Makinsons NL 37 D7
Makkovik NL 39 A6
Makkovik NL 40 D8
Makkovik Bay NL 39 A6
Makkovik River NL 39 A6
Makobe Lake ON 17 E7
Makobe-Grays River PP ON 17 E7
Makokibatan Lake ON 16 B7
Makoop River ON 13 A7
Makwa AB 6 A8
Makwa SK 11 A3
Makwa Lake PP SK 6 A8
Makwa Lake SK 11 A2-3
Makwa-Sahfahcan SK 6 A8
Makwa-Sahgaican SK 6 A8
Makwa-Sahgaican SK 11 A3
Makwa-Sahgaiehcan SK 11 A3
Malachi ON 13 E5
Malachi ON 14 D8
Malachi ON 16 D1
Malagash NS 34 D2
Malagash Point NS 34 D2
Malagash Station NS 34 D2
Malagawatch NS 34 D6
Malahat NS 2 F4
Malakwa BC 6 D2
Malartic QC 23 E4
Malartic, lac QC 23 D4
Malbaie, rivière QC 28 D-E7
Malbaie, rivière QC 30 C2-3
Malcolm Island BC 1 E4
Malcolm Island SK 12 D7
Malcolm Island SK 15 D1
Malcolm River YT 5 A2
Malden ON 35 E2
Malden Centre ON 19 E1
Mâle, du QC 26 E3
Maleb AB 6 F7
Maleb AB 7 E7-8
Maleb AB 11 F1-2
Malfait, lac NT 10 A4-5
Malfait, lac NT 28 A4
Maliceet NB 31 D3
Malignant Cove NS 34 D4
Maligne Lake AB 6 E1-2
Maligne River AB 8 B3
Maligne River AB 8 D1
Maligne River ON 16 F3-4
Maliotenam QC 39 E-F2
Mallaig AB 6 A7
Mallaig AB 11 A1
Mallard AB 13 C2
Mall Bay NL 37 F7
Mallery Lake NU 41 D4
Mallet River ON 19 A5-6
Mallet River ON 20 C4-5
Malloch, Cape NT 42 B3
Mallorytown ON 21 D7-8
Mallorytown Landing ON 21 D8
Mallow AB 7 B-C6
Mallow AB 8 E7
Malo, lac QC 26 B5
Malone ON 21 D4-5
Malonton MB 13 E4
Malonton MB 14 C5
Malouin, rivière QC 17 A8
Malpeque PE 32 A8
Malpeque PE 34 B3
Malpeque Bay PE 35 C3
Malpeque Bay NS 32 A8
Malpeque Bay PE 35 C2-3
Malpeque Harbour PE 35 C3
Maltais NB 29 D6
Maltais NB 31 A5
Maltampec NB 30 E1-2
Maltampec NB 31 B7
Malton ON 19 A7-8
Malton ON 20 C6-7
Malton ON 21 F1
Maniwaki QC 22 D1-2
Maniwaki QC 24 B1-2
Maniwaki QC 25 F1-2
Mankota SK 11 F4
Manly AB 8 C5-6

Mamakwash Lake ON 16 B3
Manners Sutton NB 32 C2
Manneville ON 17 C8
Manneville QC 23 D3
Mannheim ON 19 B6
Mannheim ON 20 D5
Manning AB 9 D3
Manning, Cape NT 42 C1
Manning Park PE 2 E8
Ma-Me-O Beach AB 8 D6
Mamiegowish Lake ON 16 B5
Mamozekel River NB 29 F5
Mamozekel River NB 31 C4
Man of War Point NL 38 F8
Manawan QC 25 D4
Manawan Lake SK 12 F7
Manche-d'Épée QC 29 A8
Manche-d'Épée QC 30 A1
Manchester ON 20 B8
Manchester ON 21 E2
Mandanna Lake YT 4 D3-4
Mandaumin ON 19 C3
Mandaumin ON 20 E2
Manful Bight NL 38 E5
Manhattan Beach MB 14 F2
Manic-Cinq QC 39 E1
Manic-Deux, barrage QC 29 A3
Manic-Deux, réservoir QC 39 F1
Manic-Un, barrage QC 29 A3
Manic-Un, barrage QC 40 F6
Manicouagan QC 40 F6
Manicouagan, péninsule de QC 29 A3
Manicouagan, pointe de QC 29 A3-4
Manicouagan, réservoir QC 39 D1
Manicouagan, réservoir QC 40 F6
Manicouagan, rivière QC 39 E-F1
Manicouagan, rivière QC 40 F6
Manic-Un, barrage QC 29 A3
Manilla ON 20 B8
Manilla ON 21 D2
Manigotagan MB 13 D4
Manigotagan MB 14 B6
Manigotagan River MB 13 D5
Manigotagan River MB 14 B7
Manigotagan River PP Reserve MB 13 D4-5
Manigotagan River PP Reserve MB 14 B6
Manilla ON 20 B8
Manipogo PP MB 13 D2
Manipogo PP MB 14 A2
Manitoba, lac MB 13 D3
Manitoba, lac MB 13 E3
Manitoba, lac MB 14 C3 D3-4
Manitoba, Lake MB 13 D2-3
Manitoba, Lake MB 13 D-E3
Manitoba, Lake MB 14 B3
Manitoba, Lake MB 14 C3
Manitoba Lowlands NP (proposed) MB 13 D4-5
Manitoba Lowlands NP (proposed) (Black & Deer Island Section) MB 14 B6
Manitoba Lowlands NP (proposed) (Limestone Bay Section) MB 13 A3
Manitoba Lowlands NP (proposed) (Long Point Section) MB 13 B2-3
Manitou MB 13 F3
Manitou MB 14 F3
Manitou QC 39 E-F3
Manitou, lac QC 39 E3
Manitou, Lake ON 18 C3-4
Manitou, rivière QC 39 E3
Manitou Beach SK 11 D5
Manitou Island NT 5 E6
Manitou Island PP ON 17 F7-8
Manitou Islands ON 18 A-B8
Manitou Islands ON 22 D1
Manitou Islands PP ON 18 B8
Manitou Islands PP ON 22 D1
Manitou Lake ON 17 F8
Manitou Lake ON 18 F2
Manitou Lake SK 6 B8
Manitou Lake SK 11 B2
Manitoulin, île ON 18 C2-3
Manitoulin Island ON 18 C2-3
Manitou River ON 18 C2-3
Manitouwabing Lake ON 18 C7
Manitouwadge ON 16 E8
Manitouwadge ON 17 C3
Manitouwadge ON 40 F2
Manitowaning ON 18 C4
Manitowaning Bay ON 18 B-C4
Manitowik Lake ON 17 D4
Maniwaki QC 22 D1-2
Maniwaki QC 24 B1-2
Maniwaki QC 25 F1-2
Mankota SK 11 F4
Manly AB 8 C5-6

Manning Park BC 6 F1
Manning PP BC 1 F8
Manning PP BC 2 F8
Manning PP BC 6 F1
Mann Settlement QC 29 D5
Mann Settlement QC 31 A4
Mannville AB 6 B5
Manola AB 6 A5
Manola AB 8 B5
Manor SK 11 F7
Manotick ON 21 B8
Manotick Station ON 21 B8
Manotick Station ON 24 E2
Manouane, lac QC 25 C5
Manseau QC 27 C4
Mansel Island NU 40 B4
Mansel Island NU 41 E7
Mansfield NS 32 C7-8
Mansfield NS 34 D1
Mansfield ON 20 B6
Mansfield ON 21 B7
Manson MB 11 E7-8
Manson MB 13 E1
Manson Creek BC 3 E6
Mansons Landing PP BC 2 C2
Mansonville QC 24 E8
Mansonville QC 27 F3-4
Mantago Lake MB 14 A4
Mantago River MB 13 D3
Mantago River MB 14 A4
Mantario SK 6 D8
Mantario SK 11 D2
Manuel Lake NT 5 D6
Manuels NB 31 B8
Manvers ON 21 E2-3
Manyberries AB 6 F8
Manyberries AB 11 F1-2
Many Islands Lake AB 6 E8
Maple ON 19 A8
Maple ON 20 C7
Maple ON 21 F1
Maple Bay BC 2 E4
Maple Beach MB 14 A3
Maple Creek SK 6 E8
Maple Creek SK 11 E2
Maple Glen NB 31 E4-5
Maple Grove NB 31 E4-5
Maple Grove NB 32 A3
Maple Grove QC 24 E5-6
Maple Grove QC 27 E-F1
Maple Island QC 18 C7
Maple Lake ON 18 D7
Maple Lake ON 21 B2-3
Maple Lake ON 22 F2
Maple Lake Park ON 18 D7
Maple Lake Park ON 20 E5
Maple Leaf ON 21 B4
Maple Plains PE 35 E3
Maple Ridge BC 1 F7
Maple Ridge BC 2 E4
Maple Ridge NB 31 F3-4
Maple Ridge NB 32 B2
Mapleton NS 32 C-D7
Mapleton NS 33 A5
Mapleton NS 34 E1
Mapleton ON 20 F4
Maple Valley ON 20 B5
Maplewood NS 32 F6-7
Maplewood NS 33 C4
Maplewood NS 33 D2
Mapova AB 2 A-B7
Maquereau, pointe au QC 30 D2
Maquinna PP BC 1 F4-5
Mar ON 18 E5
Mara ON 20 A7
Mara PP BC 6 D-E2
Mara PP ON 18 F8
Mara PP ON 20 A7
Mara PP ON 21 D1
Maraiche Lake SK 11 A7
Maraiche Lake SK 12 F7
Maraiche Lake SK 15 F1
Marathon ON 16 F8
Marathon ON 17 C2
Marathon Village ON 21 B7
Marble Canyon PP BC 1 D7-8
Marble Canyon PP BC 2 A6
Marble Island NU 40 A2
Marble Island NU 41 E5
Marble Lake NL 39 A-B2
Marble Mountain NS 34 D6
Marble Range PP BC 1 D8
Marble Range PP BC 6 D1
Marble River PP BC 1 E4
Marbleton QC 27 E5
Marceau, lac QC 39 D2
Marcelin SK 11 B4
Marchand MB 13 F4
Marchand MB 14 E6
Marches Point NL 36 C1

Marchington Lake ON 16 D4
March Water PE 35 C-D3
Marchwell SK 11 D7
Marchwell SK 13 D1
Marcil QC 30 D2
Marcil QC 31 A7
Marc Lake NL 39 C3-4
Marconi, LHN NS 34 C8
Marconi, Mount BC 6 E5
Marconi, Mount BC 7 D2
Marconi NHS NS 34 C8
Marden ON 19 A6
Marden ON 20 D5
Marécage, lac du QC 22 B3
Marengo SK 6 D8
Marengo SK 11 D2
Margaree NL 36 E1
Margaree Centre NS 34 C6
Margaree Forks NS 34 C6
Margaree Harbour NS 34 B6
Margaree River NS 34 B6
Margaree Valley NS 34 B6
Margaret MB 13 F2
Margaret MB 14 E-F1
Margaret, Cape NU 41 B5
Margaret, Cape NU 42 F4
Margaret Hamilton Lake NL 39 B2
Margaret Lake AB 9 B6
Margaret Lake NT 10 A-B5
Margaretsville NS 32 E5-6
Margaretsville NS 33 B3
Margate PE 35 D3
Margo SK 11 C6
Margry, lac QC 23 B8
Margry, lac QC 26 C2
Marguerite BC 1 C7
Marguerite, lac QC 22 B-C8
Marguerite, lac QC 25 D2
Marguerite River AB 9 C7
Marguerite River AB 12 C2
Marguerite River Wildland AB 41 F1-2
Marguerite River Wildland PP AB 9 C7
Marguerite River Wildland PP AB 12 C2
Maria QC 29 D7
Maria QC 31 A6
Maria Lake MB 13 F2
Maria Lake MB 15 B2
Maria Lake ON 13 B8
Maria Lake ON 16 A3
Mariana Lake AB 9 E6
Mariana Lake AB 12 E1
Marian Lake NT 10 C5-6
Marian River NT 10 B5
Marian River NT 10 C5
Mariapolis MB 13 F2-3
Mariapolis MB 14 F2-3
Maribelli Lake SK 12 D5-6 E5-6
Maricourt QC 24 D8
Maricourt QC 27 E4
Maricourt, lac QC 23 D7-8
Maricourt, lac QC 26 E2
Marie-Claire, lac QC 39 F5
Marie Joseph NS 34 F4-5
Marie Joseph PP NS 34 F4-5
Marie Lake AB 9 F7
Marie Lake AB 12 F2
Marie-Lefranc, lac QC 24 B3-4
Marie River PE 35 D6
Marie-Reine AB 9 E3-4
Marieval SK 11 E6-7
Marieville QC 24 D7
Marieville QC 27 E2
Mari Lake SK 15 E-F1
Mari Lake SK 22 C2
Marin, lac QC 22 C2
Marine (St. Clair Parkway), parc ON 19 D2
Marine (St. Clair Parkway), parc ON 20 F1
Marine Park (St. Clair Parkway) ON 19 D2
Marine Park (St. Clair Parkway) ON 20 F1
Marinette NS 33 B7
Marinette NS 34 F4
Maringouin, Cape NB 32 C6-7
Marion, lac QC 22 C4
Marion, lac QC 26 D6
Marion, lac QC 27 F2
Marion Bridge NS 34 C7-8
Marion Lake AB 6 A8
Marion Lake AB 8 F8
Marion Lake AB 11 C1
Marion Lake NT 5 D6
Marionville ON 21 B8
Marionville ON 24 E2-3
Mariposa Brook ON 20 B8
Mariposa Brook ON 21 D-E2
Markdale ON 20 B4-5
Markerville AB 6 C5
Markerville AB 8 F5
Markham ON 20 C7-8
Markham ON 21 F1-2
Markham Bay NU 40 A5
Markham Bay NU 41 E8
Markham River PP BC 1 E4
Markinch SK 11 D5-6
Markland MB 14 C4
Markland NL 37 E7
Mark S. Burnham PP ON 21 E2-3
Markstay ON 17 D-E7
Markstay ON 18 A6

Markstay-Warren ON 18 A6
Marktosis BC 1 F5
Marktosis BC 2 D1
Marlbank ON 21 D5
Marlboro AB 6 B4
Marlboro AB 8 C2
Marl Creek PP BC 6 D3
Marmion Lake ON 13 F8
Marmion Lake ON 16 E4
Marmora ON 21 D4
Marquette MB 13 E3
Marquette MB 14 D4
Marquette, lac QC 26 D5-6
Marquette, lac QC 28 B1
Marquis SK 11 E5
Marriott SK 11 C3
Marriotts Cove NS 32 F7
Marriotts Cove NS 33 C5
Mars, rivière à QC 28 C-D6
Marsboro QC 27 E6
Marsden SK 6 B8
Marshall SK 6 B8
Marshall SK 11 B2
Marshall Lake ON 16 C-D7
Marshall Lake ON 17 A1
Marshalltown NS 32 F4-5
Marshalltown NS 33 C2
Marshfield PE 35 E5
Marsh Lake ON 18 D7
Marsh Lake YT 4 E4
Marsh Lake YT 4 E4
Marshland Lake ON 18 A2
Marsh Point MB 15 D6-7
Marshville NS 34 D2-3
Marshy Hope NS 34 E4
Mars-Moulin, ZEC QC 28 D6
Marsoui QC 29 A7
Marsoui QC 39 F2
Marsoui, rivière QC 29 A7
Marsville ON 19 A6
Marsville ON 20 C5
Marteau, lac du QC 25 A5
Marteau, lac du QC 26 F5
Marteau, lac du QC 28 C1
Martel BC 2 B7
Marten Beach AB 9 E5
Marten Falls ON 16 B8
Marten Lake ON 22 C1
Marten River ON 17 F7
Marten River ON 22 C1
Marten River PP BC 17 F7-8
Marten River PP ON 22 C1
Martensville SK 11 C4
Martha Creek PP BC 6 D3
Martin ON 13 F8
Martin ON 16 E4
Martin, lac QC 23 D6
Martin House NT 5 D5
Martin River NT 10 D-E3
Martindale QC 22 F7-8
Martindale QC 24 C2
Martineau River SK 9 F8
Martineau River SK 12 F2
Martinenda Lake ON 17 F5
Martinique Beach PP NS 33 C7
Martins Point NS 32 F7
Martins Point NS 33 C5
Martins River NS 32 F7
Martins River NS 33 C-D4
Martintown ON 24 E4
Martinvale PE 35 E6
Martin-Valin, ZEC QC 28 B-C7
Martinville QC 27 F5
Martock NS 32 E7
Martock NS 33 B5
Martock NS 34 F1
Martres, lac des QC 28 D-E6
Marwayne AB 6 B7
Marwayne AB 11 B2
Mary Ann Lake NL 36 A6
Mary Ann Lake NL 37 A4
Mary Ann Lake NL 38 F5
Maryen, lac QC 39 D5-6
Maryfield SK 11 E7
Maryfield SK 13 E1
Maryhill ON 19 B6
Maryhill ON 20 D5
Mary Lake ON 18 D8
Mary Lake ON 21 B1
Marysburg SK 11 C5
Mary's Harbour NL 39 C8
Marys River ON 13 F8
Marystown NL 36 E6-7
Marystown NL 37 E4-5
Marysvale NL 37 D-E7
Marysville BC 7 E1
Marysville NB 31 F5
Marysville NB 32 C3
Marysville ON 21 D7
Marysville ON 21 E7
Maryville NS 34 E4
Mascouche QC 24 C-D6
Mascouche QC 27 D1
Mascouche, rivière QC 24 C-D5
Mascouche, rivière QC 27 D1
Masefield SK 11 F3
Maseres, lac QC 23 C8
Maseres, lac QC 26 D2
Mashkinonje PP ON 17 F7
Mashkinonje PP ON 18 B6
Mashteuiatsh QC 26 E8
Mashteuiatsh QC 28 C4
Masketsi, lac QC 25 D8

Masketsi, lac QC 27 A3
Maskinongé QC 24 B7
Maskinongé QC 25 F7
Maskinongé QC 27 C2
Maskinongé, lac QC 24 B6
Maskinongé, lac QC 25 F6
Maskinongé, lac QC 27 C2
Maskinonge Lake ON 17 E-F7
Mason Point NS 34 C6
Mason River NT 5 A7
Masquapit Lake NB 32 C3-4
Masquapit Lake NB 32 C3-4
Massawipi QC 27 F4
Massawippi, lac QC 27 F4
Masset BC 1 A1
Masset BC 1 A1
Masset Inlet BC 1 A1
Massey ON 17 F6
Massey ON 18 A3-5
Massey Drive NL 36 B3
Massey Drive NL 37 B1
Massey Island NU 42 C4
Massey Sound NU 42 B-C4
Massie ON 18 B6
Massie ON 20 A4-5
Masson QC 22 F8
Masson QC 24 D2-3
Masson, lac QC 24 C5
Masson, lac QC 27 D1
Masstown NS 32 D8
Masstown NS 33 A4
Masstown NS 34 E2
Massueville QC 24 C7
Massueville QC 27 D3
Mastigou, lac QC 24 A6
Mastigou, lac QC 25 E6
Mastigouche, lac QC 24 B6
Mastigouche, lac QC 25 F6
Mastigouche, RF QC 24 A6
Mastigouche, RF QC 25 E6
Mastigouche, RF QC 27 B1-2
Matachewan ON 17 D6-7
Matachewan (IR) ON 17 D6-7
Matagami ON 23 A5
Matagami QC 40 F4
Matago MB 15 E4
Matamec QC 39 E2
Matamec, lac QC 39 E2
Matane QC 29 B4
Matane, lac QC 29 B4
Matane, rivière QC 29 B-C5
Matane, RF de QC 29 B5-6
Matapédia QC 29 D5
Matapédia QC 31 A4
Matapédia, lac QC 29 C4-5
Matapédia, rivière QC 29 C-D5
Matapédia, rivière QC 31 A4
Matawatchan ON 21 B5
Matawatchan PP ON 21 B5
Matawin, lac QC 24 A5
Matawin, lac QC 25 E5
Matawin, rivière QC 25 E6-7
Matawin, rivière QC 27 A2
Matawin River PP ON 16 F4-5
Matchi-Manitou, lac QC 23 E6
Matchlee Bay BC 2 D1
Mates Corner NB 35 E2
Mather MB 13 F2
Mather MB 14 F2
Mathers Corners ON 21 E3
Matheson ON 17 D7
Matheson Island ON 13 C4
Matheson Island MB 14 A5
Mathieson Channel BC 1 B-C4
Mathieu, lac QC 26 D4
Matimek, ZEC QC 39 E2
Matimekosh QC 39 A2
Matimekosh QC 40 D6
Matinenda Lake ON 18 A2
Matinenda PP ON 17 F5
Matlock MB 13 E4
Matlock MB 14 C5
Matson Creek YT 4 B1-2
Matsqui BC 2 E6
Mattagami ON 17 D6
Mattagami Lake ON 17 D6
Mattagami River ON 17 B-C6
Mattawa ON 22 D2
Mattawa, rivière ON 22 D2
Mattawa River ON 22 D1-2
Mattawa River PP ON 17 F8
Mattawitchewan River ON 17 B-C6
Mattberry Lake NT 10 B6
Matthews NB 31 D6
Mattice ON 17 B5
Mattie Settlement NS 34 D-E5
Mattis Point NL 36 C2
Matty Island NU 41 B4
Matty Island NU 42 B4
Maude, lac QC 23 D7
Maude, lac QC 25 A1
Maude, lac QC 26 E1
Maugerville NB 32 C3
Maunoir, lac NT 5 C8
Maunoir, lac NT 41 B1
Maurice Point SK 9 A7
Maurice Point SK 12 A3

Mauricie, PN de la QC 24 A7
Mauricie, PN de la QC 25 E7
Mauricie, PN de la QC 27 A2-3 B2-3
Mauser, lac QC 25 B4
Mauves, lac des QC 24 B3-4
Mavillette NS 33 E1
Mavillette Beach PP NS 33 E1
Mawdesley Lake MB 11 A8
Mawdesley Lake MB 13 A2
Mawer SK 11 D-E4
Max Lake MB 13 A4-5
Max Lake MB 15 F4-5
Maxim SK 11 F6
Maxime, lac QC 22 B8
Maxime, lac QC 23 F8
Maximeville PE 35 D1-2
Maxstone SK 11 F4
Maxville ON 24 E3-4
Maxwell ON 20 B5
Maxwell Bay NU 42 D4-5
Maxwellton Station ON 33 D1-2
Medora MB 11 F8
May, Mount AB 6 A2
May Inlet NU 42 C3
May Lake SK 12 D6
Maybelle River Wildland PP AB 9 B7
Maybelle River Wildland PP AB 12 B2
Mayberry SK 11 E5
Maycroft AB 7 E3-4
Maymont SK 11 C3-4
Maynard ON 21 C8
Maynard ON 24 F2
Maynard Lake ON 13 E6
Maynard Lake ON 16 C-D2
Maynard Lake, RNP ON 16 C2
Maynard Lake PNR ON 13 E6
Maynard Lake PNR ON 16 C2
Mayne Bay BC 2 E2
Mayne Island BC 2 E4
Maynooth ON 21 B4
Maynooth Station ON 21 B4
Mayo QC 22 F8
Mayo QC 24 D2
Mayo YT 4 B3-4
Mayo River YT 4 B3-4
Mayson Lake SK 12 C4
Mayton AB 7 A4
Mazana, lac QC 25 D4
Mazana, ZEC QC 25 D4
Mazenod SK 11 E4
Mazenod Lake NT 10 B5
Mazeppa AB 7 C4
Mazinaw Lake ON 21 C5
Meacham SK 11 C5
Meachen BC 6 F4
Meachen BC 7 E1
Mead ON 17 B4
Meadow Bank PE 35 E4
Meadow Brook NB 32 B6-7
Meadow Creek BC 6 D3
Meadow Lake AB 6 A8
Meadow Lake SK 11 A3
Meadow Lake SK 11 A4
Meadow Lake SK 11 B4
Meadow Lake PP SK 9 F8
Meadow Lake PP SK 11 A2-3
Meadow Lea MB 14 D4
Meadow Portage MB 13 C2
Meadowlands MB 14 A2
Meadows MB 14 D4
Meadows NL 36 B3
Meadows NL 37 B1
Meadowside ON 18 A7
Meadowvale AB 14 E5-6
Meadowvale Village ON 19 A7
Meadowvale Village ON 20 D6
Meadowview AB 8 B5
Meadowville NS 34 D3
Meaford ON 18 D8
Meaford ON 20 A4-5
Meaford ON 20 A5
Meaghers Grant NS 33 B6-7
Meaghers Grant NS 34 F2-3
Mealy Mountains NL 39 B-C6
Mealy Mountains NL 40 D-E8
Meander River AB 9 B3
Meanook AB 8 A6-7
Meanook, RNF AB 8 A5-6
Meanook, RNF AB 8 A6
Meanook, RNF AB 9 F5-6
Meanook NWA AB 6 A5-6

Meanook NWA AB 8 A6
Meanook NWA AB 9 F5-6
Meares Island BC 1 F5
Meares Island BC 2 D1
Meat Cove NS 34 A7
Mechamego, lac QC 26 B3
Mechanic Settlement NB 32 C5-6
Meath Park SK 11 B5
Médard MB 15 F3
Médard QC 23 E8
Médéra, lac QC 25 A3
Médéra, lac QC 26 F3
Medford NS 32 E7
Medford NS 33 B4
Medicine Hat AB 6 E7-8
Medicine Hat AB 7 D8
Medicine River AB 11 E1-2
Medicine River AB 6 C5
Medicine River AB 8 E5
Medika MB 13 F4
Medika MB 14 E7
Medina ON 19 B5
Medina ON 20 E4
Medley AB 6 A7
Medley AB 9 F7
Medley AB 11 A2
Medley AB 12 F2
Medonnegonix Lake NL 36 D6
Medonnegonix Lake NL 37 D4
Medora MB 11 F8
Medora MB 13 F1
Medstead SK 11 B3
Meductic NB 32 C1
Meduxnekeag River NB 31 F3
Meduxnekeag River NB 32 B1
Medway Creek ON 19 C4
Medway Creek ON 20 E3
Medway Harbour NS 33 D4
Medway River NS 32 F5-6
Medway River NS 33 C-D3
Medway River NS 33 D4
Meek Lake ON 21 B3
Meeting Creek AB 6 B-C6
Meeting Creek AB 8 E6
Meeting Lake SK 11 B3-4
Meeting Lake SK 11 B4
Meetoos SK 11 A3
Meeyomot Lake SK 11 A5
Meeyomot Lake SK 12 F5
Mega River AB 9 B2
Mégantic, lac QC 27 E6
Mégantic, mont QC 27 E6
Megisan Lake ON 17 E4-5
Mégiscane, lac QC 23 D8
Mégiscane, lac QC 26 E2
Mégiscane, rivière QC 23 D-E6
Mégiscane, rivière QC 23 D7
Mégiscane, rivière QC 26 E1-2
Mehatl Creek PP BC 1 E7-8
Mehatl Creek PP BC 2 C6
Meighen Island NU 42 B3
Meikle River AB 9 C-6
Meister River YT 4 A5
Mékinac, lac QC 25 D7-8
Mékinac, lac QC 27 A3
Melanson NS 32 E7
Melanson NS 33 B4-5
Melanson Settlement NHS NS 32 F3-5
Melanson Settlement NHS NS 33 C1-2
Melaval SK 11 F4
Melbourne MB 14 E2
Melbourne ON 19 C4
Melbourne ON 20 F3
Melbourne ON 24 D8
Melbourne ON 27 E4
Melbourne Island NU 41 B-C3
Melbourne Island NU 42 F3
Melchett Lake ON 16 C7
Melchett Lake ON 17 A2
Meldrum Bay ON 18 B1-2
Meldrum Bay ON 18 B1
Meleb MB 13 D3
Meleb MB 16 E1
Mélèzes, rivière aux QC 40 C-D5
Melford NS 34 D6
Melford NS 34 E6
Melfort SK 11 B5-6
Melissa ON 18 D8
Melissa ON 21 B1
Melissa ON 22 F1
Melita MB 11 F8
Melita MB 13 F1
Mellowdale AB 8 B5
Melmerby Beach PP NS 34 D-E4
Melnice MB 14 C5
Melocheville QC 24 E1
Melrose NB 32 B7-8
Melrose NB 34 C1
Melrose NS 35 E2
Melrose NL 37 B7-8
Melrose NS 34 E4-5
Melrose ON 17 E6
Melrose ON 19 C4
Melrose ON 20 E4

Melville ON 21 E5
Melville PE 34 D3
Melville PE 35 F6
Melville SK 11 D7
Melville, île de NT 42 D2-3
Melville, île de NU 42 D2-3
Melville, Lake NL 39 B6
Melville, Lake NL 40 D8
Melville Beach SK 11 E7
Melville Hills NT 41 A-B1
Melville Island BC 1 A2
Melville Island NT 42 C-D2-3
Melville Island NU 42 C2 D2-3
Melville Peninsula NU 41 B-C6
Melville Peninsula NU 42 F5
Melville Sound NU 41 C3
Melville Sound NU 42 F2
Melville Sound ON 18 E5
Melvin Lake MB 12 C8
Melvin River AB 9 B3-4
Memesagamesing Lake ON 18 B7
Memewin, lac QC 22 D3
Memory Island PP BC 2 F4
Memphrémagog, lac QC 24 E8
Memphrémagog, lac QC 27 F4
Memramcook NB 31 F8
Memramcook NB 32 B6
Memramcook NB 32 B7
Memramcook, rivière NB 35 E-F1
Memramcook-Est NB 32 B7
Memramcook River NB 35 E-F1
Menaik AB 8 E6
Menako Lake ON 13 C8
Menako Lake ON 16 A5
Ménascouagama, lac QC 39 D-E5
Mendham SK 6 D8
Mendham SK 11 D2
Meneset ON 19 A3-4
Meneset ON 20 C2-3
Menihek NL 39 A2
Menihek Lakes NL 39 B2
Menihek Lakes NL 40 E6
Menisino MB 13 F4
Menisino MB 14 F6
Menneval NB 29 E5
Menneval NB 31 B3-4
Menzel Centennial, RNP ON 21 D5
Menzel Centennial PNR ON 21 D5
Menzie MB 11 E8
Menzie MB 13 E1-2
Meota SK 11 B3
Mephisto Lake ON 21 C4
Merasheen NL 36 E6
Merasheen NL 37 E6
Merasheen Island NL 36 D-E8
Merasheen Island NL 37 D-E6
Mercer's Cove NL 36 E6
Mercer's Cove NL 37 E4
Merchant Lake ON 22 E3
Mercier ON 24 E6
Mercier QC 27 E-F1
Mercoal AB 6 B3
Mercoal AB 8 D2
Mercutio Lake ON 16 E4
Mercy Bay NT 42 D1-2
Merid SK 6 D8
Merid SK 11 D2
Meridian Island NT 10 E5
Meridian River NB 15 D-E4
Merigomish NS 34 E4
Merlin ON 19 E3
Merlin ON 20 F1
Mermaid PE 34 C2-3
Mermaid PE 35 E5
Merrickville ON 21 C8
Merrickville ON 24 F2
Merrickville Blockhouse NHS ON 21 C7-8
Merrickville Blockhouse NHS ON 24 E1-2
Merrickville-Wolford ON 21 C8
Merrickville-Wolford ON 24 F2
Merrill, lac QC 26 B4
Merrill, Mount YT 3 A6
Merrill, Mount YT 10 F1
Merrill Lake ON 16 E4
Merritt BC 1 E8
Merritt BC 2 C8
Merritt BC 6 E1
Merritt's Harbour NL 38 E6
Mersey River NS 33 D3 E4
Meruimticook, lac QC 29 E2
Meruimticook, lac QC 31 C1
Merville BC 1 F5-6
Merville BC 2 E2
Mervin SK 11 B3
Mervin SK 11 B3
Mesa Lake NT 10 A6
Mesachie Lake BC 2 E4
Mesgouez, lac QC 40 F4
Mesilinka River BC 3 B6
Meskanaw SK 11 B5
Mesomikenda Lake ON 17 E6
Mespark Lake NS 33 E2

Mesplet, lac QC 23 C8
Mesplet, lac QC 26 D2
Mess Creek BC 3 C2
Messines QC 22 D-E7
Messines QC 24 B1-2
Messines QC 25 F1-2
Meta Incognita Peninsula NU 40 A5-6 B6
Meta Pond NL 36 C7
Meta Pond NL 37 C5
Métabetchouan QC 28 C4
Métabetchouane, lac QC 28 D4
Métabetchouane, rivière QC 28 C-E4
Métabetchouane Est, rivière QC 28 E4-5
Métascouac, lac QC 28 E4-5
Metagama ON 17 E6
Metcalf Bay MB 12 D7
Metcalf Bay MB 15 D2
Metcalfe ON 21 B8
Metcalfe ON 24 E2-3
Metchin River NL 39 B4
Metchosin BC 2 F4
Meteghan NS 33 D1
Meteghan River NS 33 D1
Meteghan Station NS 33 D1
Methley MB 14 B2
Methley Beach MB 14 B2
Metinota SK 11 B3
Metionga Lake ON 13 E8
Metiskow AB 6 C7-8
Metiskow AB 11 C2
Métis-sur-Mer QC 29 B3-4
Metlakatla BC 1 A2
Metlakatla BC 3 E3
Mewassin AB 8 C5
Mewatha Beach AB 12 F1
Mewatha Beach AB 9 F6
Meyersburg ON 21 E4
Meyronne SK 11 F4
Meziadin Junction BC 3 E3
Meziadin Lake BC 3 E3
Meziadin Lake PP BC 3 E3
Miami MB 13 F3
Miami MB 14 F3-4
Micksburg ON 21 A5-6
Micksburg ON 21 A5-6
Mica Creek BC 6 C2-3
Mica Dam BC 6 C3
Mica Peak BC 3 D-E6
Michael, Lake NL 39 A6
Michael, Lake NL 39 B7
Michael Lake MB 12 C6
Michael's Bay ON 18 C3
Michaels Harbour NL 36 A7
Michaels Harbour NL 38 F6
Michaud Point NS 34 D7
Michel BC 6 F5
Michel BC 7 E3
Michel SK 9 E8
Michel SK 12 E3
Michel, lac QC 22 D7-8
Michel, lac QC 24 B2
Michel, lac QC 25 F2
Michel, lac NL 38 D3
Michener, Mount AB 6 C4
Michener, Mount AB 8 F3
Michichi AB 6 D6
Michichi AB 7 A6
Michikamats Lake NL 39 A3
Michikinabish Lake MB 15 F5
Michipicoten Island ON 17 D2-3
Michipicoten Island PP ON 17 E2-3
Michipicoten Post PP ON 17 D3
Michipicoten River ON 17 D3
Michwacho, lac QC 26 A4
Micksburg ON 21 A5-6
Micksburg ON 21 A5-6
Midale SK 11 F6
Middle Arm NL 36 A3
Middle Arm NL 37 A1
Middle Arm NL 38 F2
Middle Arm NL 38 E4
Middle Bay QC 38 A2-3
Middle-Bay QC 39 D7
Middlebro MB 14 F7
Middlebro MB 16 E1
Middle Brook NL 36 B8
Middle Brook NL 37 B6
Middle Cape NL 34 D7
Middle Castor River ON
Middle Channel NT 5 C4
Middlechurch MB 14 D5
Middle Clyde River NS 33 E2-3
Middle East Pubnico NS 33 F2
Middle Falls PP ON 16 F5-6
Middlefield NS 33 D3-4
Middle Fiord NL 38 B4
Middle Foster Lake SK 12 D5
Middle Head NS 34 B7
Middle Island NL 19 F2
Middle Lake SK 36 C-D3
Middle Lake SK 37 C-D1
Middle Lake SK 11 C5
Middle Maitland River ON 19 A5
Middle Maitland River ON 20 C4

Middlemiss ON 19 D4
Middlemiss ON 20 F3
Middle Musquodoboit NS 33 B7
Middle Musquodoboit NS 34 F2-3
Middle New Cornwall NS 32 F7
Middle New Cornwall NS 33 C-D4
Middle Ohio NS 33 E3-4
Middle Pond NL 36 C8
Middle Pond NL 37 C6
Middleport ON 19 C7
Middleport ON 20 E6
Middle Ridge Pond NL 36 C6-7
Middle Ridge Pond NL 37 C4-5
Middle Ridge Wildlife Reserve NL 36 C6
Middle Ridge Wildlife Reserve NL 37 C4
Middle River BC 3 F5
Middle River NB 29 F7
Middle River NB 31 B6
Middle River NS 34 C6
Middle Sackville NB 32 C7
Middle Sackville NS 32 F1
Middle Sackville NS 33 C5-6
Middlesex NB 31 F7-8
Middlesex NB 32 C5-6
Middle Stewiacke NS 33 A6-7
Middle Stewiacke NS 34 F2
Middle Thames River ON 19 C5
Middle Thames River ON 20 D-E4
Middle West Pubnico NS 33 F1-2
Middlewood NS 33 D4
Midgell PE 35 D6
Midgic NB 32 B7
Midgic NB 35 F1
Midhurst ON 18 F7
Midhurst ON 20 A6
Midland ON 18 E-F7
Midland ON 20 A6
Midland PP AB 6 D6
Midland PP AB 7 A6
Midnight Lake SK 11 B3
Midway BC 6 F2
Mignault, lac QC 26 C5-6
Migration Lake NT 10 A8
Miguasha QC 29 D7
Miguasha QC 31 A5
Miguasha, parc de QC 29 D7
Miguasha, parc de QC 31 A5
Miguasha, pointe de QC 29 D7
Miguasha Ouest QC 31 A5
Miguasha-Ouest QC 29 D6-7
Miguels Lake NL 36 B6
Miguels Lake NL 37 B4
Mikado SK 11 D7
Mikisew PP ON 18 C8
Mikisew PP ON 22 E1
Mikkola ON 18 A5
Mikkwa River AB 9 C5
Mikkwa River AB 41 F1
Milan QC 27 E6
Milbanke Sound BC 1 C3
Milburn PE 35 C1
Milden SK 11 D3-4
Mildmay ON 20 B4
Mildred NL 11 B4
Mile Lake ON 21 B6
Miles Cove NL 38 E5
Milestone SK 11 E5-6
Militia Point NS 34 D6
Milk River AB 6 F6-7
Milk River AB 7 F6
Milk River AB 11 F1-2
Milk River Ridge Reservoir AB 7 F6
Mill Bay BC 2 F4
Mill Cove NS 32 F7-8
Mill Cove NS 33 C5
Mill Cove NS 34 C2-3
Mill Creek ON 20 B3
Mill Island NU 40 A4
Mill Island NU 41 D7
Mill Lake ON 18 D7
Mill of Kintail CA ON 21 B7
Mill of Kintail CA ON 24 E1
Mill Pond CA ON 21 C7
Mill Pond CA ON 24 F1

Mill Pond CA ON 24 F1
Mill River PE 35 B1
Mill River East PE 34 B1
Mill River East PE 35 B2
Mill River PP NE 35 B2
Mill Road NS 32 F7
Mill Road NS 33 C4
Mill Village NS 33 D4
Millar Hill ON 21 B2
Millar Hill ON 22 F2
Millarton ON 20 B3
Millarville AB 6 D-E5
Millarville AB 7 C3
Millbank ON 19 A5
Millbank ON 20 D4
Millbridge ON 21 C4
Millbrook NS 33 A6
Millbrook NS 34 E2
Millbrook ON 21 E3
Millburn PE 35 D6
Milldale ON 19 C6
Milldale ON 20 E5
Mille Îles, rivière des QC 24 D5-6
Mille Îles, rivière des QC 27 E1
Mille-Isles QC 24 C5
Mille Lacs, lac des ON 13 F8
Mille Lacs, lac des ON 16 E4-5
Mille Lacs, lac des ON 40 F1
Millerand ON 17 F7
Millerand ON 18 A7
Millerdale AB 6 C8
Millerdale SK 11 C2-3
Miller Lake ON 18 D4-5
Miller Lake ON 18 E4
Miller's Passage NL 36 E6
Miller's Passage NL 37 E4
Millerton NB 31 D6-7
Millertown NL 36 B5
Millertown NL 37 B3
Millertown Junction NL 36 A-B5
Millertown Junction NL 37 A3
Millet AB 6 B6
Millgrove ON 19 B7
Millgrove ON 20 D6
Millhaven ON 21 E6
Millicent AB 6 D7
Millicent AB 7 C7
Millicent AB 11 D1
Millieu, rivière du QC 24 A5
Millieu, rivière du QC 25 E5
Milligan Hills PP BC 3 C8
Milligan Hills PP BC 9 F1
Milliken ON 19 A8
Milliken ON 20 C7
Milliken ON 21 F1
Million MB 14 B2
Mills Lake NT 10 E5
Mills Lake NT 41 D1
Millstream River NB 32 C5
Milltown NL 36 D6
Milltown NL 37 D4
Milltown Cross PE 34 C3
Milltown Cross PE 35 F6
Milltown-Head of Bay d'Espoir NL 36 D6
Milltown-Head of Bay d'Espoir NL 37 D4
Millvale PE 35 D4
Millview PE 34 C3
Millview PE 35 E5-6
Millville NB 31 F4
Millville NB 32 B2
Millville NL 36 D1
Millville NS 33 B3-4
Millwood NL 11 E7
Millwood MB 13 D1
Milne Inlet NU 41 A6
Milne Inlet NU 42 E5
Milner Lake NL 39 B2
Milner Ridge MB 13 E4
Milner Ridge MB 14 D6
Milnerton AB 8 F6
Milnes Landing BC 2 F4
Milnet ON 17 E6
Milnikek, rivière QC 29 E7
Milo AB 6 E6
Milo AB 7 C5
Milo PE 32 A7
Milo PE 35 B1
Milton NL 36 C8
Milton NL 37 C6
Milton NS 33 E4
Milton ON 19 B7
Milton ON 20 D6
Milton PE 34 C2
Milton PE 35 E4
Milton Lake SK 12 C5
Milverton ON 19 A5
Milverton ON 20 D6
Miminegash PE 35 B1
Miminegash Pond PE 35 B1
Miminiska Lake ON 16 B6
Minago River MB 13 A3
Minahico ON 13 E8
Minahico ON 16 E1-2
Minaki ON 13 E8
Minaki ON 14 D8
Minaki ON 16 D1
Minaret AB 7 A4
Minas Basin NS 32 D7
Minas Basin NS 33 A5
Minas Basin NS 34 E-F1
Minas Channel NS 32 D6-7

Minas Channel NS 33 A4
Minasville NS 32 D8
Minasville NS 33 A5
Minasville NS 34 F1
Minburn AB 6 B7
Minburn AB 11 B1
Mindemoya ON 18 C3
Mindemoya Lake ON 18 C3
Minden ON 21 C2
Mine Centre ON 16 E-F3
Mine Mile Lake ON 18 C-D7
Miner River NT 5 B-C5
Miner River YT 5 E2
Miners Bay ON 21 C2
Miners Range YT 4 D3
Mines, bassin des NS 32 D7-8
Mines, bassin des NS 33 A5
Mines, bassin des NS 34 E-F1
Mines, chenal des NS 32 D6-7
Mines, chenal des NS 33 A-B4
Minesing ON 18 F7
Minesing ON 20 A6
Minett ON 18 D8
Mineville NS 33 C6
Mingan QC 39 E3
Mingan QC 40 F7
Mingan, rivière QC 39 E4
Mingan Archipelago NPR QC 39 F3-4
Mingo Lake NU 40 A5
Mingo Lake NU 41 C8
Mingré, lac QC 39 B-C2
Ming's Bight NL 38 D4-5
Ming's Bight NL 39 E-F8
Miniac, rivière QC 23 C4
Miniota MB 11 E8
Miniota MB 13 E1
Minipi Lake NL 39 C-D5
Minipi River NL 39 C5
Minisinakwa Lake ON 17 E6
Miniss Lake ON 13 D8
Miniss Lake ON 16 C4
Miniss River ON 16 C4
Ministic, lac QC 25 B7
Ministic, lac QC 26 F7
Ministic, lac QC 28 D2
Ministic Lake ON 18 A4
Ministik AB 8 C7
Ministik Lake AB 8 C7
Ministikwan SK 6 A8
Ministikwan SK 11 A2
Ministikwan Lake SK 6 A8
Ministikwan Lake SK 11 A2
Minitonas MB 11 C8
Minitonas MB 13 C1
Mink Cove NS 33 D1
Mink Lake AB 9 D5-6
Mink Lake NT 10 D5
Mink Lake ON 21 A5
Mink Lake ON 22 F5
Minnedosa MB 11 E8
Minnedosa MB 13 E2
Minnedosa MB 14 D1-2
Minnewakan MB 14 C3-4
Minnewanka, Lake AB 6 D4-5
Minnewanka, Lake AB 7 B2
Minnitaki ON 13 E6
Minnitaki ON 16 D2
Minnitaki Kames PP ON 13 E7
Minnitaki Kames PP ON 16 D3
Minnitaki Lake ON 13 E7
Minnitaki Lake ON 16 D3
Minoachak SK 11 D7
Minomaquam, lac QC 25 C7
Minomaquam, lac QC 28 E3
Minstrel Island BC 1 E5
Minstrel Island BC 2 A-B1
Minto MB 11 F8
Minto MB 13 F2
Minto MB 14 E-F1
Minto NB 31 F6
Minto NB 32 B4
Minto ON 19 A5
Minto ON 20 C4
Minto ON 21 D5
Minto YT 4 C3
Minto, lac QC 40 D4
Minto, lac QC 41 F8
Minto Head NU 41 A4
Minto Head NU 42 D-E3
Minto Inlet NT 41 A2
Minto Inlet NT 42 E1
Minton SK 11 F5
Minudie NS 32 C7
Miquelon QC 23 B7
Miquelon QC 26 B-C1
Miquelon Lake PP AB 6 B6
Miquelon Lake PP AB 8 D7
Mira NS 34 C8
Mira Bay NS 34 C8
Mirabel QC 24 D5
Miracle Beach PP BC 1 E6 F5-6
Miracle Beach PP BC 2 C2
Miracle Valley BC 2 D-E6
Mirage, lac au QC 25 B8
Mirage, lac au QC 26 F8
Mirage, lac au QC 28 D4
Miramichi NB 31 D6-7
Miramichi NB 31 D7
Miramichi, baie de NB 31 C7
Miramichi Bay NB 31 C7
Miramichi Lake NB 31 E4
Miramichi Lake NB 32 A2
Mirande, lac QC 23 E8
Mirande, lac QC 25 B2
Mirande, lac QC 26 F2
Mira River NS 34 C7-8 D7
Mira River PP NS 34 C8
Mira Road NS 34 C8
Mirepoix, lac QC 28 A6-7
Mirond Lake SK 12 F6-7
Mirond Lake SK 12 F6
Mirond Lake SK 15 F1
Mirond Lake SK 15 E1
Mirror AB 6 C6
Mirror AB 8 E7
Mirror Lake BC 6 E3
Mirror River SK 9 C8
Mirror River SK 12 C3
Misamikwash Lake ON 13 B8
Misaw Lake SK 12 A6
Misaw Lake SK 15 A1
Miscou, île NB 30 D2-3
Miscou, île NB 31 A8
Miscou Centre NB 30 E2
Miscou Centre NB 31 A8
Miscouche PE 32 A7-8
Miscouche PE 34 C1
Miscouche PE 35 D2
Miscou Harbour NB 30 E2-3
Miscou Island NB 30 D2-3
Miscou Island NB 31 A8
Misehkow River ON 16 B-C5
Misekumaw Lake SK 12 B6
Misery Bay, RNP ON 18 C2
Misery Bay PNR ON 18 C2
Mishibishu Lake ON 17 D3
Mishkeegogamang (New Osnaburgh) ON 16 B4-5
Mishwamakan River ON 13 A-B8
Misikeyask Lake ON 13 A8
Misinchinka Ranges BC 3 E6-7 F7
Miskwabi Lake ON 21 B3
Mispec NB 32 E4
Mispec NB 33 B1-2
Misquamaebin Lake ON 13 A7-8 B7
Missanabie ON 17 D4
Missawai Lake AB 9 F6
Missawai Lake AB 12 F1
Missi Lake SK 12 D6
Missinaibi Lake ON 17 D4
Missinaibi PP ON 17 A5
Missinaibi PP ON 17 C4-5
Missinaibi PP ON 17 D4
Missinaibi River ON 17 A5-6
Missinaibi River ON 17 B-C5
Missinipe SK 12 E5
Mission BC 1 F7
Mission BC 2 E6
Mission-de-Hopedale, LHN de la NL 39 A5-6
Mission-de-Hopedale, LHN de la NL 40 C6-7
Missionnaire, lac du QC 25 E8
Missionnaire, lac du QC 27 A3
Mission-Saint-Louis, LHN de la ON 18 F7-8
Mission-Saint-Louis, LHN de la ON 20 A6-7
Missipuskiow River SK 11 A6
Missisa Lake ON 16 A8
Missisquoi Nord, rivière QC 24 E8
Missisquoi Nord, rivière QC 27 E-F4 F3
Mississagagon Lake ON 21 C5
Mississagi Bay ON 18 B1-2
Mississagi Delta PNR ON 18 B2
Mississagi Delta PNR ON 17 F5
Mississagi Delta, RNP ON 17 F5
Mississagi Delta, RNP ON 18 B2
Mississagi Island ON 18 B2
Mississagi PP ON 17 F5
Mississagi River ON 17 E5
Mississagi River ON 17 F4-5
Mississagi River ON 17 F5
Mississagi River ON 18 A1
Mississagi River ON 18 A2
Mississagi River PP ON 17 E5
Mississagi Strait ON 18 B1
Mississagua Lake ON 21 C3
Mississagua River ON 21 C-D3
Mississauga ON 19 A-B8
Mississauga ON 20 D7
Mississauga ON 21 F1
Mississippi Lake ON 21 B7
Mississippi Lake ON 24 E1
Mississippi Lake National Wildlife Reserve ON 24 E1
Mississippi Lake NWA ON 21 B-C7
Mississippi Mills ON 21 B7
Mississippi River ON 21 B7
Mississippi River ON 21 C6
Mississippi River ON 24 D-E1
Mississippi Station ON 21 C6
Mista Lake MB 15 E6
Mistahayo Lake ON 15 D8
Mistaken Point Ecological Reserve NL 37 F7-8
Mistanipisipou, rivière QC 39 D4-5
Mistaouac, lac QC 17 B8
Mistaouac, lac QC 23 A3
Mistaouac, rivière QC 17 B8
Mistassibi, rivière QC 26 A8
Mistassibi, rivière QC 26 C8
Mistassibi, rivière QC 28 A4
Mistassini QC 28 B4
Mistassini, lac QC 40 F5
Mistassini, rivière QC 26 B7-8 C8
Mistassini, rivière QC 40 F5
Mistatim SK 11 B6
Mistawasis SK 11 B4
Mistigougèche, lac QC 29 D4
Mistigougèche, lac QC 31 A2-3
Mistigougèche, rivière QC 29 C3-4
Mistikokan River MB 15 D7
Mistinic, lac QC 39 C1
Mistinikon Lake ON 17 D6-7
Mistinippi Lake NL 39 A5
Mistissini QC 40 F5
Mistusinne SK 11 D4
Misty Icefield BC 1 E-F7
Misty Icefield BC 2 C5-6
Misty Lake MB 12 B7
Misty Lake MB 15 B2
Misty Lake ON 21 A2
Misty Lake ON 22 F2
Mitchell MB 13 F4
Mitchell MB 14 E5
Mitchell ON 19 B4-5
Mitchell ON 20 D3-4
Mitchell Lake BC 1 B8
Mitchell Lake BC 6 B1
Mitchell Lake ON 18 A8
Mitchell's Bay ON 19 D2
Mitchells Brook NL 37 E7
Mitchell's Corners ON 20 C8
Mitchell's Corners ON 21 E2
Mitchellton SK 11 E5
Mitchinamécus, lac QC 25 C3-4
Mitchinamécus, rivière QC 25 C4
Mitchinamécus, ZEC QC 25 C-D3
Mitford AB 7 B3
Mitis, rivière QC 29 C4
Mitishto River MB 12 F8
Mitishto River MB 13 A2
Mitishto River MB 15 F2
Mitlenatch Island PP BC 2 C2
Mittimatalik (Pond Inlet) NU 41 A6
Mittimatalik (Pond Inlet) NU 42 D-E6
Mizzen Topsail NL 36 A4
Mizzen Topsail NL 37 A2
Mizzen Topsail NL 38 F3-4
Moar Lake MB 13 C5
Moar Lake ON 16 A1
Moberly BC 6 D3-4
Moberly Lake BC 3 E7-8
Moberly Lake BC 9 E1
Moberly Lake PP BC 3 E7
Moberly Lake PP BC 9 E1
Moberly River BC 3 E7
Mobile NL 37 E8
Mobile Big Pond NL 37 E8
Mocassins, lac des QC 24 A4-5
Mocassins, lac des QC 25 E-F4
Mocauque de Canaan, APP NB 31 F7
Mocauque de Canaan, APP NB 32 B5
Mochelle NS 32 F5
Mochelle NS 33 C2-3
Mocodome, Cape NS 34 F5
Moe, rivière QC 27 F5
Moes River QC 27 F5
Moffat ON 19 B7
Moffat ON 20 D6
Moffat, lac QC 27 E5-6
Moffet QC 17 E8
Moffet QC 22 A2
Moffet QC 23 F2
Moha BC 2 A6
Mohawk Island NWA ON 19 D8
Mohawk Island NWA ON 20 F7
Mohawk Lake NT 10 B7
Mohawk Point ON 19 C8
Mohawk Point ON 20 F7
Moira ON 21 D5
Moira Lake ON 21 D5
Moira River ON 21 C4-5
Moira River ON 21 D5
Moira River ON 21 D-E5
Moisie QC 39 E-F2
Moisie QC 40 F7
Moisie, baie de QC 39 E2
Moisie, rivière QC 39 D-E2
Moisie, rivière QC 39 E2
Moisie, rivière QC 40 E6 F6-7
Mojikit Lake ON 16 C6
Mojikit Lake ON 17 A1
Molanosa SK 11 A5
Molanosa SK 12 F5
Molard, lac QC 26 E6
Molard, lac QC 28 B-C2
Molega NS 33 D3-4
Molega Lake NS 33 D4
Molesworth ON 19 A5
Molesworth ON 20 C4
Molliers NL 36 E-F6
Molliers NL 37 E-F4
Molly Ann Cove NL 36 B2
Mollyguajeck Lake NL 36 C7
Mollyguajeck Lake NL 37 C5
Molson MB 13 E4
Molson MB 14 D6
Molson Lake MB 13 A4
Molson Lake NL 39 B2
Molson River MB 13 A4
Moltke ON 20 B4
Molus River NB 31 E7
Molus River NB 32 A5
Momich Lakes PP BC 6 D2
Monaco, lac QC 26 C4
Monarch AB 6 E6
Monarch AB 7 E5
Monarch Icefield BC 1 C5
Monarch Mountain BC 1 C5
Monashee Mountains BC 6 C-D2
Monashee Mountains BC 6 E3 F2-3
Monashee PP BC 6 E3
Monastery NS 34 E5
Monchy NL 36 B7
Monchy NL 37 B5
Monchy SK 11 F3
Monck ON 20 C5
Monck PP BC 1 E8
Monck PP BC 2 B8
Monck PP BC 6 E1
Moncouche, lac QC 28 B6
Moncton NB 31 F8
Moncton NB 32 B6
Mondonac, lac QC 25 C5
Mondonac, lac QC 28 F1
Mondou SK 11 D3
Monet QC 23 E8
Monet QC 25 A2
Monet QC 26 F2
Monetville ON 17 F7
Monetville ON 18 B6-7
Money Creek YT 4 D6
Monger, îles QC 39 E7
Monitor AB 6 C7
Monitor AB 11 C2
Monkland ON 24 E3-4
Monkman PP BC 9 F1
Monkstown NL 36 D8
Monkstown NL 37 D6
Monkton ON 19 A5
Monkton ON 20 D4
Monmouth Mountain BC 1 D6-7
Monmouth Mountain BC 2 A4
Monnery River SK 6 A8
Monnery River SK 11 A2
Mono ON 20 B-C6
Mono Centre ON 20 B6
Mono Cliffs PP ON 20 B6
Mono Mills ON 20 C6
Monominto MB 14 E6
Mono Road ON 19 A7
Mono Road ON 20 C6
Monroe NL 37 C6-7
Montagnais, lac aux QC 28 D4-5
Montagnais Point NL 39 B6
Montague PE 34 C3
Montague PE 35 E6
Montague Harbour PMP BC 2 E4-5
Montague Lake SK 11 F5
Montague River PE 35 E-F6
Mont-Apica QC 28 D5
Montauban, lac QC 25 E8
Montauban, lac QC 27 A4
Montbeillard QC 17 D8
Montbeillard QC 23 E2
Montbeillard, lac QC 23 E2
Mont-Brun QC 23 D3
Mont-Carleton, APP NB 29 F6
Mont-Carleton, APP NB 31 C5
Mont-Carleton, PP NB 29 F6
Mont-Carleton, PP NB 31 C4
Mont-Carmel PE 32 A7
Mont-Carmel PE 34 C1
Mont-Carmel PE 35 D2
Mont-Carmel QC 28 E8
Montcerf QC 22 D7
Montcerf QC 24 A1-2
Montcerf QC 25 E-F1
Montcevelles, lac QC 39 E5-6
Montebello QC 24 D3
Monte Creek BC 6 E1-2
Monte Creek PP BC 6 E2
Monteith MB 14 E1
Monteith ON 17 C7
Monte Lake BC 6 E1-2
Monte Lake PP BC 6 E2
Montfort QC 24 C4
Mont-Gabriel QC 24 C5
Monticello ON 20 C5
Monticello PE 34 B3
Monticello PE 35 D7
Monticules-Linéaires, LHN des MB 11 F7-8
Monticules-Linéaires, LHN des MB 13 F1
Montjoie, lac QC 24 B3
Montjoie, lac QC 25 F3
Mont-Joli QC 29 C3
Mont-Laurier QC 22 D8
Mont-Laurier QC 24 A3
Mont-Laurier QC 25 F3
Mont-Lebel QC 29 C3
Mont-Louis QC 29 A8
Mont-Louis QC 39 F2
Mont-Louis QC 40 F7
Mont-Louis, rivière de QC 29 A7-8 B7
Montmagny QC 27 A6-7
Montmartre SK 11 E6
Mont-Mégantic, parc du QC 27 E6
Montmorency, rivière QC 28 F6
Mont Nebo SK 11 B4
Montney BC 3 D8
Montney BC 9 D1
Mont-Orford QC 24 E8
Mont-Orford, parc du QC 24 D-E8
Mont-Orford, parc du QC 27 E4
Montpellier QC 24 C3
Montréal QC 24 D5-6
Montréal QC 27 E1
Montréal, lac QC 26 C7
Montréal, lac QC 28 A3
Montreal Island ON 17 E3
Montreal Lake SK 11 A5
Montréal Lake SK 11 A5
Montréal-Nord QC 24 D6
Montréal-Nord QC 27 E1-2
Montreal Point MB 13 B3
Montreal River ON 17 D-E7
Montreal River ON 17 E4
Montreal River ON 17 E3-4
Montreal River ON 22 A-B1
Montréal River SK 12 F5
Montreal River PP ON 17 E3
Mont-Revelstoke, PN du BC 6 D2-3
Mont-Riding, PN du MB 11 D8
Mont-Riding, PN du MB 13 D1-2
Mont-Riding, PN du MB 14 C1-2
Mont-Rolland QC 24 C5
Montrose BC 6 F3
Montrose NS 32 D8
Montrose NS 33 A5-6
Montrose NS 34 E1
Montrose PE 34 A1
Montrose PE 35 B2
Mont-Saint-Bruno, parc du QC 24 D6
Mont-Saint-Bruno, parc du QC 27 E2
Mont-Sainte-Anne, parc du QC 27 A6
Mont-Sainte-Anne, parc du QC 28 F6
Mont-Saint-Grégoire QC 24 D6-7 E6-7
Mont-Saint-Grégoire QC 27 E2
Mont-Saint-Hilaire QC 24 D7
Mont-Saint-Hilaire QC 27 E2
Mont-Saint-Michel QC 24 A3
Mont-Saint-Michel QC 25 E3
Mont-Saint-Pierre QC 29 A7
Mont-Saint-Pierre, rivière de QC 29 A7
Monts-Torngat-et-de-la-Rivière-Koroc, parc des QC 40 C6
Monts-Valin, parc des QC 28 C6
Mont-Tremblant QC 24 B4
Mont-Tremblant QC 25 F4
Mont-Tremblant, parc du QC 24 A4-5 B4-5
Mont-Tremblant, parc du QC 25 F4-5
Mont-Tremblant-Village QC 24 B4
Mont-Wright QC 39 C1-2
Monument-Lefebvre, LHN du NB 32 B7
Monument Lefebvre NHS NB 32 B7
Moodie Island NU 40 A6
Moonbeam ON 17 B5-6
Moon Island ON 18 D7
Moon Lake AB 8 C5
Moon Lake NT 5 E-F7
Moonlight Bay AB 8 B5-6
Moon River ON 18 E7
Moonshine Lake PP AB 9 E2
Moonstone ON 18 F7-8
Moonstone ON 20 A6
Moore Dale MB 14 B2
Moore Falls ON 21 C2
Moore Lake ON 21 A4
Moore Park MB 13 E2
Moore Park MB 14 D1
Moore, lac QC 22 A2
Moorefield ON 19 A5-6
Moorefield ON 20 C4-5
Mooresburg ON 20 B4
Moores Mills NB 32 D2
Mooretown ON 19 C2
Mooretown ON 20 F1
Mooring Cove NL 36 E6-7
Mooring Cove NL 37 E4-5
Moose, baie QC 27 D5
Moose Bay MB 14 A1-2
Moose Brook NS 32 D8
Moose Brook NS 33 A5
Moose Brook NS 34 F1
Moose Channel YT 5 B3-4
Moose Creek ON 24 E3
Moose Factory ON 40 F3
Moosehead NS 33 B8
Moosehead NS 34 F4
Moose Heights BC 1 B7
Moosehorn MB 13 D3
Moosehorn MB 14 B3
Moose Horn River NT 4 B7
Moose Island MB 12 E-F7
Moose Island MB 13 C-D4
Moose Island MB 14 A5
Moose Island MB 15 E1
Moose Jaw SK 11 E5
Moose Jaw River SK 11 E5-6
Moose Lake MB 11 A8
Moose Lake MB 13 A2
Moose Lake MB 14 A5
Moose Lake MB 14 F7
Moose Lake MB 14 F7-8
Moose Lake ON 18 B4
Moose Lake (Mosakahiken) MB 11 A8
Moose Lake (Mosakahiken) MB 13 A2
Moose Lake PP AB 6 A7
Moose Lake PP AB 9 F7
Moose Lake PP AB 11 A1-2
Moose Lake PP AB 12 F1-2
Moose Lake PP MB 13 F5
Moose Lake PP MB 14 F7-8
Moose Lake PP MB 16 E1
Mooseland NS 33 B7
Mooseland NS 34 F3
Moose Mountain Creek SK 11 F6-7
Moose Mountain Lake SK 11 E6
Moose Mountain PP SK 11 E-F7
Moose Nose Lake MB 15 E4-5
Moose Point ON 18 E7
Moose Range SK 11 B6
Moose River NS 32 D7
Moose River NS 33 A8
Moose River NS 33 A5
Moose River NS 33 A8
Moose River NS 34 E4
Moose River NS 34 E4
Moose River NS 34 E1
Moose River ON 17 A6
Moose River ON 40 F3
Moose River Gold Mines NS 33 B7
Moose River Gold Mines NS 34 F3
Moose River Gold Mines PP NS 33 B7
Moose River Gold Mines PP NS 34 F3
Moose Valley SK 11 E5
Moose Valley PP BC 1 D7-8
Moose Woods SK 11 C4
Moosomin SK 11 B3
Moosomin SK 11 E7
Moosomin SK 13 E1
Moosonee ON 40 F3
Moraine Point NT 10 D6
Morass Point MB 13 C3
Moraviantown ON 19 D3
Moraviantown ON 20 F2
Morden MB 13 F3
Morden MB 14 F3-4
Morden NS 33 B3
Morden Colliery PP BC 2 E3-4
Morecambe AB 6 A7
Morecambe AB 11 A1
More Creek BC 3 D2-3
Moreland SK 11 F5-6
Morell PE 34 B-C3
Morell PE 35 D6
Morell (IR) PE 35 D6
Morell Bay SK 12 C6
Morell East PE 34 B-C3
Morell East PE 35 D6
Morell River PE 34 C3
Morell River PE 35 D-E6
Moresby, île BC 1 B1
Moresby Camp BC 1 B1
Moresby Island BC 1 B1
Moreton's Harbour NL 38 E6
Morewood ON 24 E3
Morgan Point BC 1 A1
Morganston ON 21 E4
Morganville NS 32 F5
Morganville NS 33 D2
Morice Lake BC 1 A4
Morice Pond NB 35 F1
Morice River BC 1 A4
Moricetown BC 3 F4
Morien, Cape NS 34 C8
Morin, lac QC 25 D8
Morin, lac QC 28 B6
Morin, lac QC 28 F4
Morin Heights QC 24 C4-5
Morin Lake MB 12 E7
Morin Lake MB 15 E1
Morin Lake SK 12 F5
Morinville AB 6 A5
Morinville AB 8 B-C6
Morisset-Station QC 27 C7
Morkill River BC 6 A1-2
Morley AB 6 D5
Morley AB 7 B3
Morley River YT 4 E5
Morley River Territorial Recreation Site YT 3 A2
Morley River Territorial Recreation Site YT 4 E-F5
Morleyville AB 7 B3
Morning Lake SK 12 E-F5
Morningside AB 6 C6
Morningside AB 8 E6
Morpeth ON 19 E3
Morrin AB 6 D6
Morrin AB 7 A5
Morris MB 13 F3-4
Morris MB 14 F5
Morris QC 30 B2
Morris Lake ON 16 B4
Morris Lake YT 4 E5
Morris River ON 13 C7-8
Morris Tract, RNP ON 19 A3-4
Morris Tract, RNP ON 20 C2-3
Morris Tract PNR ON 19 A3-4
Morris Tract PNR ON 20 C2-3
Morrisburg ON 24 F3
Morrisdale NB 32 D3-4
Morrisdale NB 33 A1
Morrison Lake MB 13 B2
Morrissey BC 7 F2-3
Morrissey PP BC 6 F5
Morrissey PP BC 7 F3
Morriston ON 19 B7
Morriston ON 20 D6
Morristown NS 32 E6
Morristown NS 33 B4
Morristown NS 34 D5
Morrisville NL 36 D6
Morrisville NL 37 D4
Morse SK 11 E4
Morson ON 13 F6
Morson ON 16 E1
Mort, lac du NT 10 C8
Mortier NL 36 E-F7
Mortier NL 37 E-F5
Mortimers Point ON 18 E8
Mortimers Point ON 21 B-C1
Mortlach SK 11 E4
Morton ON 21 D7
Morton Lake PP BC 1 E5
Morton Lake PP BC 2 B-C1
Morvan NS 33 A8
Morvan NS 34 E4
Morven ON 21 E6
Morweena MB 13 D3-4
Morweena MB 14 B4-5
Moscow ON 21 D6
Mose Ambrose NL 36 E6
Mose Ambrose NL 37 E4
Moser River NS 33 B8
Moser River NS 33 B8
Moser River NS 34 F4
Moser River NS 34 F4
Moses Inlet BC 1 C4
Mosher Lake NT 10 C6
Mosley Creek BC 1 D5-6
Mosque River BC 3 D4
Mosquic, lac QC 24 A4
Mosquic, lac QC 25 E4
Mosquito NL 36 D5
Mosquito NL 37 D3
Mosquito NL 37 F7
Mosquito Bay QC 40 B4
Mosquito Bay QC 41 E7
Mosquito Cove NL 37 D6-7
Mosquito Creek ON 18 A7-8
Mosquito-Grizzly Bear's Head-Lean Man SK 11 C3
Mossbank SK 11 E4
Moss Glen NB 32 D4
Moss Glen NB 33 A1
Mosside AB 8 B5
Moss Lake MB 12 C8
Moss Lake MB 15 C3
Mossleigh AB 6 E6
Mossleigh AB 7 C5
Mossley ON 19 C5
Mossley ON 20 E4
Moss Spur MB 14 D6
Mossy Point PE 35 C2
Mossy River SK 11 A6
Mostoos Hills SK 9 E8
Mostoos Hills SK 12 E3
Motase Peak BC 3 E4
Motherwell Homestead NHS SK 11 E6
Motion Head NL 37 E8
Mouat, Cape NU 42 C5
Mouche, rivière à la QC 38 A1
Moulin, lac QC 22 D4
Moulin, rivière du QC 29 D5
Moulin-Morneault NB 29 E3
Moulin-Morneault NB 31 C1-2
Mound AB 7 A3
Mountain ON 21 B8
Mountain ON 24 E2
Mountain Grove ON 21 C6
Mountain Lake ON 18 F5
Mountain Lake ON 20 A4
Mountain Lake ON 21 C2
Mountain Lake SK 12 E5
Mountain Park AB 6 B3
Mountain Park AB 8 D2
Mountain River NT 5 E6 F6-7
Mountain Road MB 13 E2
Mountain Road MB 14 C2
Mountainside MB 14 F1
Mountain View AB 6 F6
Mountain View AB 7 F4-5
Mountain View ON 21 E5
Mount Albert ON 20 B7
Mount Albert ON 21 E1
Mount Albion PE 35 E5
Mount Arlington Heights NL 36 E8
Mount Arlington Heights NL 37 E6
Mount Assiniboine PP BC 6 D4
Mount Assiniboine PP BC 7 B1-2
Mount Blanchet PP BC 3 F5
Mount Brydges ON 19 C4
Mount Brydges ON 20 E3
Mount Buchanan PE 34 C2-3
Mount Buchanan PE 35 F5
Mount Carleton PP NB 29 F6
Mount Carleton PP NB 31 C4
Mount Carleton PPA NB 29 F6
Mount Carleton PPA NB 31 C5
Mount Carmel NL 37 E7
Mount Carmel ON 19 B4
Mount Carmel ON 20 D3
Mount Carmel-Mitchell's Brook-St. Catherines NL 37 E7
Mount Carmel Pond NL 37 E8
Mount Chesney ON 21 D6-7
Mount Currie BC 1 E7
Mount Currie BC 2 B5
Mount Denson NS 32 E7
Mount Denson NS 33 B5
Mount Denson NS 34 F1
Mount Edziza PP BC 3 C2-3
Mount Edziza PRA BC 3 C3
Mount Elgin ON 19 C5-6
Mount Elgin ON 20 E4-5
Mount Elphinstone PP BC 2 D4
Mount Fernie PP BC 6 F4-5
Mount Fernie PP BC 7 E2
Mount Forest ON 20 B-C5 C4-5
Mount Hanley NS 33 B3
Mount Herbert PE 35 E5
Mount Hope ON 19 C7
Mount Hope ON 20 E6
Mount Judge Howay PP BC 2 D6
Mount Judge Howay PRA BC 1 F7-8
Mount Julian ON 21 D3
Mount Lorne YT 4 E4
Mount Maxwell PP BC 2 E4
Mount Mellick PE 35 E5
Mount Moriah NL 36 B2-3
Mount Pearl NL 37 D-E8
Mount Pelly (Uvajuq) TP NU 41 B3
Mount Pelly (Uvajuq) TP NU 42 F3
Mount Pleasant NB 31 E3
Mount Pleasant NB 32 A-B1
Mount Pleasant NS 32 C8
Mount Pleasant NS 34 D1
Mount Pleasant ON 19 C6
Mount Pleasant ON 20 E5
Mount Pleasant ON 21 E3
Mount Pleasant PE 32 A7
Mount Pleasant PE 34 B1
Mount Pleasant PE 35 C2
Mount Pope PP BC 1 A6
Mount Revelstoke NP BC 6 D2-3
Mount Richardson PP BC 2 D4
Mount Robson BC 6 B2
Mount Robson PP BC 6 B2
Mount Royal PE 35 C1
Mount St. Patrick ON 21 B6
Mount Salem ON 19 D5
Mount Salem ON 20 F4
Mount Savona PP BC 1 E8
Mount Savona PP BC 2 A-B8
Mount Savona PP BC 6 D-E1
Mountsberg CA ON 19 B7
Mountsberg CA ON 20 D6
Mount Seymour PP BC 1 F7
Mount Seymour PP BC 2 D5
Mount Stewart PE 34 C3
Mount Stewart PE 35 D5-6
Mount Terry Fox PP BC 6 B2
Mount Thom NS 33 A7
Mount Thom NS 34 E3
Mount Uniacke NS 32 E8
Mount Uniacke NS 33 B5
Mount Uniacke PP NS 32 E8
Mount Uniacke PP NS 33 B5-6
Mount Vernon ON 19 C6
Mount Vernon ON 20 E5
Mount View NB 35 F1

Mount Whatley NB 35 F1
Mourier, lac QC 23 E4
Mouton Island NS 33 E4
Mowachaht BC 1 F5
Mowachaht BC 2 C1
Mowbray MB 14 F3
Moyen, lac QC 39 A3-4
Moyie BC 6 F4
Moyie BC 7 F1-2
Moyie Lake PP BC 6 F4
Moyie Lake PP BC 7 F1
Moyie River BC 6 F4
Moyie River BC 7 F1
Moyre, lac QC 25 D5
Mozart SK 11 C6
Muchalat Inlet BC 1 F5
Muchalat Inlet BC 2 D1
Mud Island NS 33 F1
Mud Lake NL 39 C6
Mud Lake ON 18 C3
Mud Lake Delta PP BC 6 C2
Muddy Bay NL 39 B7
Muddy Hole Point NL 36 D3
Muddy Hole Point NL 37 D1
Muddy Lake SK 6 C8
Muddy Lake SK 11 C3
Mudge Bay ON 18 B3
Mudjatik River SK 12 D4
Mudzenchoot PP BC 3 F6
Muenster SK 11 C5
Muir MB 14 D3
Muirkirk ON 19 D3-4
Muirkirk ON 20 F2-3
Mukutuwa River MB 13 B4
Mulgrave NS 34 D5
Mulgrave, Lake NS 32 F5
Mulgrave, Lake NS 33 C2 D2-3
Mulhurst Bay AB 6 B5
Mulhurst Bay AB 8 D6
Mulligan Bay NL 39 B6
Mulligan River NL 39 B6
Mullingar SK 11 B4
Mullin Stream Lake NB 31 D5
Mulock Lake ON 18 A8
Mulvihill MB 13 D3
Mulvihill MB 14 B4
Mummery, Mount BC 6 C-D4
Muncey ON 19 C4
Muncey ON 20 F3
Muncho Lake BC 3 B5
Muncho Lake BC 3 B5
Muncho Lake BC 4 F8
Muncho Lake PP BC 3 B5
Mundare AB 6 B6
Mundare AB 8 C8
Mundare AB 11 A1
Munekun Lake ON 13 B7
Munk MB 15 E4-5
Munn Lake NT 10 B8
Munro Lake MB 15 F5
Munroe Lake MB 15 A-B3
Munsee-Delaware ON 19 C-D4
Munsee-Delaware ON 20 F3
Munson AB 6 D6
Munson AB 7 A5
Munster ON 21 B7-8
Munster ON 24 E2
Murchison Island ON 16 D6
Murchison Island ON 17 A1
Murchyville NS 33 B7
Murchyville NS 34 F3
Murdochville QC 29 B8
Murdochville QC 30 B1
Murdochville QC 40 F7
Murdock Lake MB 15 B3
Muriel Lake AB 6 A7
Muriel Lake AB 11 A2
Murillo ON 16 F5
Murky Lake NT 10 C-D8
Murphy, Mount YT 4 E4-5
Murphy Cove NS 33 C7
Murphys Point PP ON 21 C7
Murphys Point PP ON 24 F1
Murray, lac QC 25 D-E5
Murray, Mount YT 4 E7
Murray Beach PP NB 32 B7
Murray Beach PP NB 34 C1
Murray Beach PP NB 35 E2
Murray Corner NB 32 B7-8
Murray Corner NB 34 C1
Murray Corner NB 35 E2
Murray Harbour PE 34 C4
Murray Harbour PE 34 D3-4
Murray Harbour PE 35 F7
Murray Harbour PE 35 F7
Murray Harbour North PE 34 C4
Murray Harbour North PE 35 F7
Murray Head PE 34 C-D4
Murray Head PE 35 F7
Murray Islands PE 35 F7
Murray Lake MB 13 A5
Murray Lake MB 15 F5
Murray Lake SK 11 B3
Murray Lake, RNOM SK 11 B3
Murray Lake NMBS SK 11 B3
Murray River BC 3 E-F8
Murray River BC 9 E1
Murray River PE 34 D3
Murray River PE 35 F6
Murray Road NB 35 E2
Murray Road PE 35 C2
Murrin PP BC 1 F7
Murrin PP BC 2 D4
Murtle Lake BC 6 C2
Musclow Lake ON 13 D5
Musclow Lake ON 14 A7-8
Musclow Lake ON 16 B1
Muscote Bay ON 21 E5
Muscovite Lakes PP BC 3 E6
Musée-du-Parc-Banff, LHN du AB 7 B1-2
Musgrave Harbour NL 38 E7-8
Musgravetown NL 36 C8
Musgravetown NL 37 C6
Mushaboom NS 33 B8
Mush Lake YT 4 E2-3
Musidora AB 6 A7
Musidora AB 11 A1
Muskeg Lake ON 13 F8
Muskeg Lake ON 16 E5
Muskeg Lake SK 11 B4
Muskeg River NT 3 A7
Muskeg River NT 10 F2-3
Musket Island PMP BC 2 C-D3
Muskiki Lake SK 11 C5
Muskoday SK 11 B5
Muskoka, Lake ON 18 E8
Muskoka, Lake ON 21 B-C1
Muskoka Falls ON 18 E8
Muskoka Falls ON 21 C1
Muskosung Lake ON 18 A7
Muskowekwan SK 11 D6
Muskowpetung SK 11 D5-6
Muskrat Dam ON 13 B7
Muskrat Dam Lake ON 13 B7
Muskrat Lake ON 21 A6
Muskrat Lake ON 22 F6
Muskwa-Kechika Management Area BC 3 B4 D6-7
Muskwa-Kechika Management Area BC 4 F7-8
Muskwa Lake AB 9 E5
Muskwa Ranges BC 3 C5 D5-6
Muskwa River AB 9 E5
Muskwa River BC 3 B-C6
Muskwesi MB 15 C3
Muskwesi River MB 12 C8
Musquaro QC 39 F5
Musquaro, lac QC 39 E5
Musquaro, rivière QC 39 E5
Musquash NB 32 E3
Musquash NB 33 B1
Musquash Lake NB 32 C1
Musquodoboit Harbour NS 33 C7
Musquodoboit Harbour NS 33 C6-7
Musquodoboit River NS 33 B6-7
Musquodoboit River NS 34 F2-3
Musquodoboit Valley PP NS 33 B7
Musquodoboit Valley PP NS 34 F3
Mussel Inlet BC 1 B3-4
Musselman's Lake ON 20 B-C7
Musselman's Lake ON 21 E1-2
Mustard Lake NT 10 D4
Mutton Bay NL 37 F7
Mutton Bay QC 38 C1
Mutton Bay QC 39 E6
Muzon, Cape BC 3 F1
Mye, Mount YT 4 C5
Myers Cave ON 21 C5
Myles Bay ON 18 E4
Mynarski Lakes MB 12 D8
Mynarski Lakes MB 15 E3
Myre Lake MB 15 D5
Myrnam AB 6 A7
Myrnam AB 11 A1-2
Myrtle MB 13 F3
Myrtle MB 14 F4
Myrtle ON 20 C8
Myrtle ON 21 E2
Myrtle Station ON 20 B8
Myrtle Station ON 21 E2
Mystery Lake AB 8 B4-5
Mystery Lake NT 10 D7
Mystic QC 24 E7
Mystic QC 27 F2-3

N

Nabel Lake MB 15 A4
Nabisipi, pointe QC 39 F5
Nabisipi, rivière QC 39 E4
Nachvak Fiord NL 40 C7
Nackawic NB 31 F3-4
Nackawic NB 32 C1-2
Nackawic Stream NB 31 F3-4
Nackawic Stream NB 32 B2
Nacmine AB 6 D6
Nacmine AB 7 A5
Naco AB 7 A8
Nadaleen Range YT 4 A4-5
Nadaleen River YT 4 A5
Naden Harbour BC 1 A1
Nadina Lake BC 1 A4
Nadina River BC 1 A4-5
Naelin Lake MB 15 B4
Nagagami Lake ON 17 B3
Nagagami Lake PP ON 17 B3
Nagagami River ON 17 B3-4
Nagagamisis Lake ON 17 B3-4
Nagagamisis PP ON 17 B3-4
Nagas Point BC 1 C1-2
Nahanni NT 4 D8
Nahanni NT 10 E1
Nahanni, RPN NT 4 D8
Nahanni, RPN NT 10 E1-2
Nahanni Butte NT 10 E2
Nahanni NPR NT 4 D8
Nahanni NPR NT 10 D-E1
Nahanni Range NT 10 D-E2
Nahatlatch PP BC 1 E8
Nahatlatch PP BC 2 C6-7
Nahatlatch River BC 1 E7-8
Nahatlatch River BC 2 C6
Nahili Lake MB 12 A7
Nahili Lake MB 15 A2
Nahlin River BC 3 B2
Nahmint BC 1 F5-6
Nahmint BC 2 E2
Nahoni Range YT 5 E2
Naicam SK 11 C5-6
Naikoon PP BC 1 A1-2
Nail Pond PE 35 A2
Nail Pond PE 35 A2
Nain NL 40 C-D7
Nainlin Brook NT 4 A7
Nainlin Brook NT 10 B1
Nairn ON 20 E3
Nairn, lac QC 26 D5
Nairn Centre ON 17 F6
Nairn Centre ON 18 A4
Nairn Falls PP BC 1 E7
Nairn Falls PP BC 2 B5
Naiscoot River ON 18 C6
Najoua, lac QC 25 B5
Najoua, lac QC 26 F5
Najoua, lac QC 28 D1
Nakamun AB 8 B5-6
Nakaneet SK 11 E-F2
Nakina ON 16 D7-8
Nakina ON 17 A2
Nakina ON 40 F2
Nakina Moraine, RNP ON 16 D7-8
Nakina Moraine, RNP ON 17 B2
Nakina Moraine PNR ON 16 D7-8
Nakina Moraine PNR ON 17 B2
Nakina River BC 3 A2
Nakina River BC 4 F5
Nakusp BC 6 E3
Namaka AB 7 B5
Namakan River ON 16 F3
Namao AB 6 A6
Namao AB 8 C6
Nameigos Lake ON 17 C3
Nameless Cove NL 38 B3-4
Nameless Lake SK 12 D4
Namew Lake SK 11 A7
Namew Lake SK 13 A1
Nampa AB 9 E3-4
Namu BC 1 C4
Namur QC 24 C3
Namur Lake AB 9 C6
Namur Lake AB 12 C1
Nanaimo BC 1 F6
Nanaimo BC 2 E3
Nanaimo River BC 1 F6
Nanaimo River BC 2 E3
Nancy Greene PP BC 6 F3
Nancy's Cellar NS 33 A7-8
Nancy's Cellar NS 34 E3
Nango Lake ON 16 A4
Nango River ON 13 B-C7
Nanika Lake BC 1 A4
Nanika River BC 1 A4
Nanisivik NU 41 A6
Nanisivik NU 42 D5
Nanoose Bay BC 1 F6
Nanoose Bay BC 2 D3
Nan Sdins, LHN de BC 1 C1-2
Nan Sdins NHS BC 1 C1-2
Nansen, Mount YT 4 C3
Nansen Sound NU 42 A-B4
Nantais, lac QC 40 B4-5
Nantais, lac QC 41 E8
Nantes QC 27 E6
Nanticoke ON 19 D7
Nanticoke ON 20 F6
Nanticoke Creek ON 19 C6-7
Nanticoke Creek ON 20 E5-6
Nanton AB 6 E5
Nanton AB 7 D4
Nantyr Park ON 20 B7
Nantyr Park ON 21 D-E1
Naococane, lac QC 40 E5-6
Naosap Lake MB 12 F7
Naosap Lake MB 15 F1
Napadogan NB 31 E4
Napadogan NB 32 A2
Napanee ON 21 E6
Napanee River ON 21 D6
Napatak SK 12 F5
Napetipi, lac QC 38 A-B2
Napetipi, rivière QC 39 D7
Napier ON 19 C3
Napier ON 20 E2
Napierville QC 24 E6
Napierville QC 27 F2
Napinka MB 11 F8
Napinka MB 13 F1
Naples AB 8 B5
Napoiak Channel NT 5 B-C4
Napoléon, lac QC 25 D4-5
Nappan NS 32 C7
Naramata BC 6 F2
Narcisse MB 13 E3
Narcisse MB 14 C4-5
Narcosli Creek BC 1 C7
Nardin Lake NT 10 B7
Nares, détroit de NU 42 B5
Nares Strait NU 42 B5
Narraway River BC 6 A1-2
Narrow Hills PP SK 11 A5
Narrows Inlet BC 1 F6
Narrows Inlet BC 2 C4
Nash Creek NB 29 D7
Nash Creek NB 31 A6
Nashwaak Bridge NB 31 F5
Nashwaak Bridge NB 32 B3
Nashwaak River NB 31 F4-5
Nashwaak River NB 32 B2-3
Nashwaak Village NB 31 F5
Nashwaak Village NB 32 B3
Nasigon, lac QC 25 C3
Naskaupi River NL 39 B4-5
Naskaupi River NL 39 B5
Naskaupi River NL 40 D7
Nasoga Gulf BC 3 F3
Nasonworth NB 32 C2-3
Nasparti Inlet BC 1 E4
Nass Bay BC 3 F3
Nass Camp BC 3 F3
Nass River BC 3 E3
Nastapoca, rivière QC 40 D4-5
Nastapoca, rivière QC 41 F8
Nastapoka Islands NU 40 D4
Nastapoka Islands NU 41 F8
Natal BC 7 E3
Natalkuz Lake BC 1 B5
Natashquan QC 39 F5
Natashquan QC 40 F8
Natashquan, rivière QC 39 D5
Natashquan, rivière QC 39 E5
Natashquan, rivière QC 40 E7-8 F8
Natashquan Est, rivière QC 39 D-E5
Natashquan River NL 39 C-D4
Nathorst, Cape NU 42 C3
National Mills MB 11 B7
National Mills MB 13 B1
Nation River BC 3 F6
Nation River YT 5 E1-2
Native Bay NU 40 A3
Native Bay NU 41 D6
Natla River NT 4 B6-7
Nat River ON 17 C6
Natuashish NL 40 D7
Naufrage PE 35 D7
Naufrage, baie du QC 30 A7
Naufrage Harbour PE 35 D7
Naughton ON 17 F6
Naughton ON 18 A5
Naujaat (Repulse Bay) NU 41 C5-6
Nautley BC 1 A6
Nauwigewauk NB 32 D4
Nauwigewauk NB 33 A2
Navan ON 21 A8
Navan ON 24 D1
Navin MB 14 E5
Navy Board Inlet NU 41 A6
Navy Board Inlet NU 42 D5
Navy Island NHS ON 20 E8
Nazco BC 1 B6
Nazko BC 1 B6
Nazko Lake PP BC 1 C6
Nazko River BC 1 B6 C7
Neapolis AB 7 A4
Neault, lac QC 23 E8
Neault, lac QC 25 A-B3
Neault, lac QC 26 F3
Neawagank Lake ON 16 A5
Nechako Reservoir BC 1 A6-7
Nechako River BC 1 A-B6
Nechako River NMBS BC 1 A6
Necoslie BC 1 A6
Necum Teuch NS 33 B8
Necum Teuch NS 34 F4
Neddy Harbour NL 38 E2
Nédélec QC 17 D8
Nédélec QC 23 F2
Neeb SK 11 A3
Needles BC 6 E2-3
Neelin MB 13 F2
Neelin MB 14 F2
Neely Lake NMBS SK 11 B6-7
Neepawa MB 13 E2
Neepawa MB 14 D2
Neerlandia AB 6 A5
Neerlandia AB 8 A5
Negassa Lake MB 15 B3
Neguac NB 30 F1-2
Neguac NB 31 C7
Neguac, plage NB 30 F2
Neguac Beach NB 30 F2
Neguac Beach NB 31 C7-8
Negwazu Lake ON 17 D3
Neidpath SK 11 E4
Neiges, lac des QC 25 B3
Neiges, lac des QC 26 F3
Neiges, lac des QC 28 E6
Neiges, rivière des QC 28 F6
Neigette QC 29 C3
Neigette, rivière QC 29 C3
Neilburg SK 6 B8
Neilburg SK 11 B2
Neils Harbour NS 34 A7
Neiman Lake SK 9 A8
Neiman Lake SK 12 A4
Nejanilini Lake MB 15 A4
Nejanilini Lake MB 41 F4
Nekweaga Bay SK 12 C6
Nelles Corners ON 19 C7
Nelles Corners ON 20 E6
Nellie Lake ON 17 C6-7
Nelson BC 6 F3
Nelson, fleuve MB 15 D6
Nelson, fleuve MB 15 E4
Nelson, fleuve MB 40 C1
Nelson, lac QC 26 D4
Nelson Forks BC 3 A6
Nelson House MB 15 E3
Nelson Island BC 1 F6
Nelson Island BC 2 D3
Nelson Lake MB 12 E8
Nelson Lake MB 15 E2
Nelson-Miramichi NB 31 D7
Nelson Range BC 6 F3
Nelson River BC 9 A1
Nelson River MB 15 D6
Nelson River MB 15 E-F4
Nelson River MB 40 C1
Nelway BC 6 F3
Nemaiah Valley BC 1 D6
Nemebien River SK 12 E-F5
Nemegos ON 17 D5
Nemegosenda Lake ON 17 D5
Nemegosenda River ON 17 D5
Nemegosenda River Wetlands PP ON 17 D5
Némégousse, lac QC 26 C4
Nemeiben Lake SK 12 F5
Nemio, rivière QC 26 E4
Némiscachingue, lac QC 25 C4-5
Nemiscau QC 40 F4
Nemiskam AB 7 E7
Nepean ON 21 B8
Nepean ON 24 D-E2
Nepewassi Lake ON 17 F7
Nepewassi Lake ON 18 A6
Nephton ON 21 D4
Nepisiguit, baie de NB 29 E8
Nepisiguit, baie de NB 30 E1
Nepisiguit, baie de NB 31 B6-7
Nepisiguit Bay NB 29 E8
Nepisiguit Bay NB 30 E1
Nepisiguit Bay NB 31 B6-7
Nepisiguit River NB 29 F6-7
Nepisiguit River NB 31 C5-6
Neptune SK 11 F6
Nerepis NB 32 D3
Nerepis NB 33 A1
Nerepis River NB 32 C-D3
Nerepis River NB 33 A1
Neroutsos Inlet BC 1 E4
Nesbitt MB 13 F2
Nesbitt MB 14 E1
Neskantaga (Lansdowne House) ON 16 A6-7
Nesle, lac QC 39 E6
Nesselrode, Mount BC 3 B1
Nesslin Lake SK 11 A4
Nestaocano, rivière QC 26 A-B6
Nestleton ON 20 B8
Nestleton ON 21 E2
Nestleton Station ON 20 B8
Nestleton Station ON 21 E2
Nestor Falls ON 13 F6
Nestor Falls ON 16 E2
Nestorville ON 17 F4
Nestow AB 6 A5
Nestow AB 8 B6
Netalzul Meadows PP BC 3 F4-5
Netherhill SK 11 D3
Net Lake ON 22 B1
Netley MB 13 E4
Netley MB 14 D5
Netley Creek PP MB 13 E4
Netley Creek PP MB 14 D5
Netley Lake MB 14 D5
Netson Creek BC 3 B5
Nettell Lake SK 9 A7
Nettell Lake SK 12 A3
Nettilling Fiord NU 41 C8
Nettilling Fiord NU 42 F7-8
Nettilling Lake NU 41 C8
Nettilling Lake NU 42 F7
Nettogami River ON 17 A7
Neubergthal MB 13 F3-4
Neubergthal MB 14 F4-5
Neudorf SK 11 E6-7
Neuf Milles, lac des QC 22 B5
Neuhorst SK 11 C4
Neustadt ON 20 B4
Neuville QC 27 B5
Neveton MB 14 C4
Neville SK 11 E3-4
Nevins Lake SK 9 A8
Nevins Lake SK 12 A4
Nevis AB 6 C6
Nevis AB 8 E-F7
Nevoir, lac QC 26 B6
New Acadia PE 35 D7
New Aiyansh BC 3 F3
New Albany NS 32 F5-6
New Albany NS 33 C3
New Annan PE 35 D3
New Bandon NB 29 E8
New Bandon NB 30 E1
New Bandon NB 31 B7
New Bandon NB 31 E5
New Bandon NB 32 A3-4
New Bay Pond NL 36 A6
New Bay Pond NL 37 A4
New Bay Pond NL 38 F5
Newbliss ON 21 C7
Newbliss ON 24 F1-2
New Bonaventure NL 37 C7
Newboro ON 21 C7
Newboro ON 24 F1
Newboro Lake ON 21 D7
Newboro Lake ON 24 F1
New Bothwell MB 13 F4
New Bothwell MB 14 E5
New Bridge NL 37 E7
New Brigden AB 6 C7-8
New Brigden AB 7 A8
New Brigden AB 11 C2
New Brighton BC 1 F7
New Brighton BC 2 D4
Newbrook AB 6 A6
Newbrook AB 8 A7
Newbrook AB 9 F6
Newburg NB 31 F3
Newburg NB 32 B1
Newburgh ON 21 D6
Newburne NS 32 F7
Newburne NS 33 C4
Newbury ON 19 D3
Newbury ON 20 F2
New Campbellton NS 34 B-C7
New Canaan NB 31 F7
New Canaan NB 32 B5
New Canaan NS 32 D7
New Canaan NS 33 A5
New Carlisle QC 29 D8
New Carlisle QC 30 D1
New Carlisle QC 31 A7
Newcastle ON 21 E-F3
Newcastle (Miramichi West) NB 31 D6
Newcastle Bridge NB 31 F6
Newcastle Bridge NB 32 B4
Newcastle Creek NB 31 F6
Newcastle Creek NB 32 B4
Newcastle Island PMP BC 1 F6
Newcastle Island PMP BC 2 E4
New Chelsea NL 37 C7
New Chester NS 33 B8
New Chester NS 34 F4
Newcombville NS 33 D4
New Credit ON 19 C7
New Credit ON 20 E6
Newdale MB 11 E8
Newdale MB 13 E2
Newdale MB 14 D1
New Dayton AB 6 F6-7
New Dayton AB 7 E-F6
New Denmark NB 31 D3
New Denver BC 6 E3
New Dominion PE 34 C2
New Dominion PE 35 E4
New Dublin ON 21 C8
New Dublin ON 24 F2
New Dundee ON 19 B6
New Dundee ON 20 D5
New Durham ON 19 C6
New Durham ON 20 E5
New Edinburgh NS 33 D1
Newell, Lake AB 6 E6-7
Newell, Lake AB 7 C6-7
Newellton NS 33 F2
New Elm NS 33 D4
New Ferolle NL 38 B3
New Ferolle NL 39 E7-8
New Ferolle Peninsula NL 38 B3
New Fish Creek AB 9 E-F3
Newfoundland T'Railway PP NL 36 A7 B7-8
Newfoundland T'Railway PP NL 36 A4-5
Newfoundland T'Railway PP NL 36 C2 D1
Newfoundland T'Railway PP NL 37 A2-3
Newfoundland T'Railway PP NL 37 A-B5 B5-6
Newfoundland T'Railway PP NL 37 D6 E7
Newfoundland T'Railway PP NL 38 F3-4
New France NS 34 E5
Newgate BC 6 F4-5
Newgate BC 7 F2
New Germany NS 32 F6
New Germany NS 33 C4
New Glasglow QC 24 C5
New Glasgow QC 27 D1
New Glasgow NS 34 E3
New Glasgow ON 19 D4
New Glasgow ON 20 F3
New Glasgow PE 34 C2
New Glasgow PE 35 D4
New Grafton NS 33 D3
New Hamburg ON 19 B5
New Hamburg ON 20 D4
New Harbour NL 37 D7
New Harbour NS 34 F5-6
New Haven PE 34 C2
New Haven PE 35 E4
New Hazelton BC 3 F4
Newington ON 24 E3
Néwiska, lac QC 23 A3
New Liskeard ON 17 E7
New Liskeard ON 22 A1
New Liskeard ON 23 F1
New London PE 32 A8
New London PE 34 B1-2
New London PE 35 D3
New London Bay PE 35 D4
New Lowell ON 20 B6
New Lowell CA ON 20 B6
Newmans Beach ON 20 B8
Newmans Beach ON 21 E2
Newmans Cove NL 37 B7
Newmarket NB 32 C2
Newmarket ON 20 B7
Newmarket ON 21 E1
New Maryland NB 32 C2-3
New Melbourne NL 37 C7
New Mills NB 29 D7
New Mills NB 31 A5
New Minas NS 32 E7
New Minas NS 33 B4
New Norway AB 6 B6
New Norway AB 8 D7
New Osgoode SK 11 B6
New Osnaburgh (Mishkeegogamang) ON 13 D8
New Osnaburgh (Mishkeegogamang) ON 16 B4-5
New Perlican NL 37 D7
New Perth PE 34 C3
New Perth PE 35 E6
Newport PE 34 C3-4
Newport PE 35 E7
Newport QC 30 D2
Newport, pointe de QC 30 D2
Newport Corner NS 32 E8
Newport Corner NS 33 B5
Newport Corner NS 34 F1
Newport Station NS 32 E7
Newport Station NS 33 B-C5
New Richmond QC 29 D7
New Richmond QC 31 A6
New River NB 32 D-E3
New River Beach NB 32 E3
New River Beach PP NB 32 E3
New Ross NS 32 F7
New Ross NS 33 C4
New Ross Road NS 32 E7
New Ross Road NS 33 B4
New Russell NS 32 F7
New Russell NS 33 C4
Newry ON 19 A5
Newry ON 20 C4
New Salem NS 32 D6
New Salem NS 33 A3-4
New Sarepta AB 6 B6
New Sarepta AB 8 D7
New Sarum ON 19 D5
New Sarum ON 20 F4
Newstead MB 14 E1
Newstead NL 36 A7
Newstead NL 37 A5
Newstead NL 38 F6
New Tecumseth ON 20 B6
Newton MB 14 D4
Newton ON 19 A5
Newton ON 20 D4
Newton PE 35 D3
Newton Lake NL 36 C7
Newton Lake NL 37 C5
Newton Lake SK 11 F3
Newton Mills NS 33 A7
Newton Mills NS 34 F3
Newton-Robinson ON 20 B6
Newtonville ON 21 E3
New Town NS 33 A8
New Town NS 34 E4
Newtown NB 32 C5
Newtown NL 37 A7
Newtown NL 38 F8
New Tusket NS 33 D2
New Victoria NS 34 B-C8
Newville NL 38 E6
Newville Lake PP NS 32 D7
Newville Lake PP NS 33 A4-5
New Waterford NS 34 B-C8
New Westminster BC 2 E5
New-Wes-Valley NL 37 A6-7
New-Wes-Valley NL 38 F8
New World Island NL 38 E6
New Yarmouth NS 32 D6
New Yarmouth NS 33 A3-4
New Zealand PE 34 B4
New Zealand PE 35 D7
New Zion NB 31 F5-6
New Zion NB 32 B4
Ney Lake ON 15 F7
Neys ON 16 E7-8
Neys ON 17 C2
Neys PP ON 16 F7-8
Neys PP ON 17 C2
Nezan ON 16 D7
Nezan ON 17 B1
Nez Croche, lac NT 10 B-C8
Niagara Escarpment ON 20 A-B5 B-C6
Niagara Falls ON 20 E7-8
Niagara-on-the-Lake ON 20 E7-8
Niagara Peninsula ON 19 C8
Niagara Peninsula ON 20 E6-7
Niagara River ON 20 E8
Niapenco, Lake ON 19 C7
Niapenco, Lake ON 20 E6
Nibinamik (Summer Beaver) ON 16 A6
Nicabau, lac QC 26 C5
Nicabong QC 22 E5
Nicette, lac QC 29 A-B1
Nichcotéa, lac QC 22 B6
Nicholas-Denys NB 31 B6
Nichol Island NS 33 C7
Nicholson ON 17 D4
Nicholson Island NT 5 A6
Nicholson Island ON 21 E4-5
Nicholson Lake SK 9 A8
Nicholsville NL 36 A3
Nicholsville NL 37 A1
Nicholsville NL 38 F2-3
Nicholsville NS 32 E6
Nicholsville NS 33 B4
Nickel Plate PP BC 6 F1
Nicklin Lake MB 12 B8
Nicklin Lake MB 15 B2-3
Nicman QC 39 E2
Nicobi, lac QC 23 B8
Nicobi, lac QC 26 C2
Nicola BC 1 E8
Nicola BC 2 C8
Nicola BC 6 E1
Nicola Lake BC 2 C8
Nicola River BC 2 B7
Nicola River BC 6 E1-2
Nicolas, lac QC 22 B5-6
Nicolas-Denys NB 29 E7
Nicolet QC 24 B7-8
Nicolet QC 25 F7-8
Nicolet QC 27 C3
Nicolet, lac QC 27 D5
Nicolet, rivière QC 24 C8
Nicolet, rivière QC 27 C3 D4
Nicolet, RNOM de QC 24 B7-8
Nicolet, RNOM de QC 27 C2-3
Nicolet NMBS QC 24 B7-8
Nicolet NMBS QC 27 C2-3
Nicolet Sud-Ouest, rivière QC 27 C-D3 D4-5 E5
Nicolston ON 20 B6
Nicolum River PP BC 1 F8
Nicolum River PP BC 2 D7
Nicolum River PP BC 6 F1
Nictau NB 29 F5
Nictau NB 31 C4
Nictaux Falls NS 32 E6
Nictaux Falls NS 33 B3
Nictaux River NS 32 F6
Nictaux River NS 33 C3
Nictaux South NS 32 E6
Nictaux South NS 33 C3
Nigadoo NB 29 E8
Nigadoo NB 31 B6
Nigadoo River NB 29 E7
Niger Sound NU 39 C-D8
Night Hawk Lake ON 17 D6-7
Nightingale AB 7 B4-5
Nikip Lake ON 13 B7
Nile ON 19 A4
Nile ON 20 C3
Niles, lac NB 35 E1-2
Niles Lake NB 35 E1-2
Nilestown ON 19 C5
Nilestown ON 20 E4
Nilgaut, lac QC 22 C5
Nilkitkwa Lake PP BC 3 F5
Nilkitkwa River BC 3 E4-5
Nimpkish Lake BC 1 E4
Nimpkish Lake PP BC 1 E4
Nimpkish River BC 1 E4-5
Nimpo Lake BC 1 C5
Nimpo Lake BC 1 C5-6
Nine Mile Creek PE 34 C2
Nine Mile Creek PE 35 E4-5
Nine Mile River NS 32 E8
Nine Mile River NS 33 B6
Nine Mile River NS 34 F2
Nine Mile River ON 20 C3
Ninette MB 13 F2
Ninette MB 14 F2
Ninevah NS 32 F6
Ninevah NS 33 D3-4
Ninevah PP NS 32 F6
Ninevah PP NS 33 D3-4
Ninga MB 13 F2
Ninga MB 14 F1
Ningunsaw PP BC 3 D3
Niobe AB 8 F6
Nipawin SK 11 B6
Nipekamew-Sand Cliffs PPA SK 12 F5
Nipew Lake SK 12 E5
Nipi, lac QC 29 A2
Nipigon ON 16 E7
Nipigon ON 40 F1-2
Nipigon, lac ON 16 D6
Nipigon, lac ON 17 B1
Nipigon, lac ON 40 F1
Nipigon, Lake ON 16 D6
Nipigon, Lake ON 17 B1
Nipigon, Lake ON 40 F1
Nipigon Bay ON 16 E6-7
Nipigon Bay ON 17 C1
Nipigon River ON 16 E6
Nipigon River ON 17 B-C1
Nipin River SK 9 E8
Nipin River SK 12 E3
Nipishish Lake NL 39 B5
Nipisi River AB 9 E5
Nipissi QC 39 E2
Nipissing ON 17 F8
Nipissing ON 18 B8
Nipissing ON 22 D-E1
Nipissing, lac ON 17 F7-8
Nipissing, lac ON 18 A7-8
Nipissing, lac ON 22 D1
Nipissing, Lake ON 17 F7-8
Nipissing, Lake ON 18 A7-8
Nipissing, Lake ON 22 D1

Nipissing, réserve de chasse de la Couronne ON **17** E-F8
Nipissing, réserve de chasse de la Couronne ON **22** B-C1
Nipissing Beach ON **18** B8
Nipissing Crown Game Preserve ON **17** E-F8
Nipissing Crown Game Reserve ON **22** C1
Nipissing Ouest ON **17** F7
Nipissing Ouest ON **18** A7
Nipissing River ON **22** E2
Nipissis, lac QC **39** E2
Nipissis, rivière QC **39** D-E2
Nipisso, lac QC **39** E2-3
Nippers Harbour NL **38** F8
Nippers Harbour NL **39** F8
Nippon, lac du QC **26** D6
Nippon, lac du QC **28** B1
Nisbet AB **8** F6
Nisga'a Memorial Lava Bed PP BC **3** F3
Niska Lake SK **9** E8
Niska Lake SK **12** E3
Niskonlith Lake PP BC **6** D1-2
Nisku AB **6** B6
Nisku AB **8** B5
Nisling Range YT **4** C-D2
Nisling River YT **4** C2
Nistocaponano, rivière QC **26** C7
Nistowiak Lake SK **12** E6
Nisutlin Lake YT **4** D5
Nisutlin River YT **4** D5-6
Nisutlin River Delta NWA YT **4** E5
Nisutlin River Territorial Recreation Site YT **4** E5
Nitainlaii TP NT **5** D4
Nitchequon QC **40** E5-6
Nith River ON **19** B6
Nith River ON **20** D5
Nithburg ON **19** B5
Nithurg ON **20** D4
Nitinat BC **1** F6
Nitinat BC **2** E3
Nitinat Lake BC **2** E-F2
Nitinat River BC **2** E3
Nitinat River PP BC **2** E2-3
Niton AB **8** C4
Niton Junction AB **6** B4
Niton Junction AB **8** C3-4
Niverville MB **13** F4
Niverville MB **14** E5
Niverville, lac QC **25** B4-5
Niwelin Lake NT **5** C7
Niwene ON **18** C6
Nixon ON **19** C6
Nixon ON **20** F5
Nizard, lac QC **22** B6
Njau, lac des QC **26** B6
Nobel ON **18** D7
Nobleford AB **6** E6
Nobleford AB **7** E5
Nobleton ON **19** A7
Nobleton ON **20** C6
Noeick River BC **1** C5
Noel NS **34** F1
Noell Lake NT **5** B-C5
Noel Road NS **32** D8
Noel Road NS **33** B5
Noel Road NS **34** F1
Noel Shore NS **32** D8
Noel Shore NS **33** A6
Noel Shore NS **34** F1
Noels Pond NL **36** B-C2
Noëlville ON **17** F7
Noëlville ON **18** B6
Noganosh Lake ON **18** C6-7
Noggin Cove NL **36** A8
Noggin Cove NL **37** A6
Noggin Cove NL **38** F7
Nogies Creek ON **21** C-D3
Nogies Creek ON **21** D3
Noir, lac QC **28** A4
Noirclair, lac QC **39** E6
Noire, rivière QC **22** D3-6
Noire, rivière QC **24** D7
Noire, rivière QC **24** D8
Noire, rivière QC **25** E8
Noire, rivière QC **27** A-B4
Noire, rivière QC **27** E3
Noisy Creek YT **5** E-F4
Noisy River, RNP ON **20** B5-6
Noisy River PNR ON **20** B5-6
Nojack AB **6** B4
Nojack AB **8** B5
Nokomis SK **11** D5
Nokomis Lake SK **12** D6
Nokomis Lake SK **15** D1
Nolalu ON **16** F5
Nominingue QC **24** B3
Nominingue QC **25** F3
Nominingue, lac QC **24** B3
Nominingue, lac QC **25** F3
Nonacho NT **41** E2
Nonsuch MB **15** E5
Nootka Island BC **1** F4
Nopiming PP MB **13** E5
Nopiming PP MB **14** C7
Nopiming PP MB **16** C1
Nora NS **34** F1
Nora, lac QC **23** D2
Norah Willis Michener Game Preserve YT **4** E8
Norah Willis Michener Game Preserve YT **10** E1
Noral AB **8** A7-8

Noralee BC **1** A5
Norbertville QC **27** C4-5
Norboro PE **35** D3
Norbuck AB **8** D5
Norbury Lake PP BC **6** F4-5
Norbury Lake PP BC **7** E2
Norcan Lake ON **21** B6
Nord, rivière du QC **24** C-D5
Nordbye Lake SK **12** B6
Nordbye Lake SK **15** A1
Nordegg AB **6** C4
Nordegg AB **8** E3
Nordegg Eiver AB **8** E3-4
Nordegg River AB **6** B4
Nordenskiold River YT **4** C-D3
Nordenskiold Wetland Habitat Protection Area YT **4** D3
Nordique, ZEC QC **28** B-C8
Nordique, ZEC QC **29** B-C1
Nord-Ouest, pointe QC **30** B3
Norembega ON **17** C7
Norfolk CA ON **19** D6
Norfolk CA ON **20** F5
Norfolk County ON **19** C6
Norfolk County ON **20** F5
Norgate MB **13** D-E2
Norgate MB **14** C2
Norge SK **11** F4
Norham ON **21** E4
Norland ON **20** A8
Norland ON **21** C2
Norma AB **8** C8
Norman ON **19** D5
Norman ON **20** F4
Norman Bay NL **39** C8
Normand, lac QC **25** D6-7
Normand, lac QC **27** A2
Normand, lac QC **28** F2
Normandale ON **19** D6
Normandale ON **20** F5
Normandie, ZEC QC **25** C4
Normandin QC **26** D8
Normandin QC **28** B3
Normandin, lac QC **26** D5
Norman Range NT **5** F7-8
Norman's Cove NL **37** D6-7
Norman Wells NT **5** F7
Norman Wells NT **10** A1
Normétal QC **17** C8
Normétal QC **23** B2
Norquay ON **17** C7
Norquay SK **13** C1
Norris Arm NL **36** A6-7
Norris Arm NL **37** A4-5
Norris Arm NL **38** F5
Norris Arm NL **38** F6
Norris Arm North Side NL **36** A6
Norris Arm North Side NL **37** A4
Norris Arm North Side NL **38** F6
Norris Lake MB **11** A-B8
Norris Lake MB **13** B2
Norris Lake MB **14** C4-5
Norris Lake PP MB **13** E3
Norris Lake PP MB **14** C4
Norris Point NL **38** E-F2
Norris Point NL **39** F7
Norris Point NL **40** F8
Norsemans Pond NL **36** C8
Norsemans Pond NL **37** C6
Norstead PE **35** C2
North, Cape NS **34** A7
North, chenal ON **18** B2
Northam PE **35** C2
North Arm BC **1** B-C8
North Arm BC **6** B1
North Arm NL **36** A3
North Arm NL **37** A1
North Arm NL **38** F2
North Arm NT **10** D6-7
North Arm TP NT **10** C6
North Arm TP NT **41** D1
North Aspy River NS **34** A-B7
North Augusta ON **21** C8
North Augusta ON **24** F2
Northbank AB **8** B7
North Battleford SK **11** B3
North Bay NL **36** D5
North Bay NL **36** D2
North Bay NL **37** D3
North Bay ON **17** F8
North Bay ON **18** A8
North Bay ON **22** D1
North Bay SK **12** E5
North Bay Ingonish NS **34** A-B7
North Beach PP ON **21** E4-5
North Bedeque PE **35** D3
North Bend BC **1** E-F8
North Bend BC **6** C7
North Bentinck Arm BC **1** C4-5
North Big Salmon River YT **4** D4
North Branch ON **16** D1-2
North Branch ON **13** F6
North Branch ON **16** E2
North Branch Big Segovie River NB **29** F6-7
North Branch Big Segovie River NB **31** C5
North Branch Muskoka River ON **18** D-E8
North Branch Muskoka River ON **21** B1
North Branch Oromocto River NB **32** C2-3 D2
North Branch Renous River NB **31** D5

Northbrook ON **21** C5
North Brookfield NS **33** D3-4
North Bruce ON **18** F4
North Bruce ON **20** A3
North Buck Lake AB **8** A7
North Buck Lake AB **9** F6
North Buxton ON **19** E3
North Cape PE **34** A1
North Cape PE **35** A2
North Caribou Lake NT **5** C5
North Caribou Lake ON **13** B8
North Caribou Lake ON **16** A4
North Caribou Lake ON **40** E1
North Carleton PE **34** C1
North Carleton PE **35** E3
North Castor River ON **21** B8
North Castor River ON **24** E2
North Channel ON **18** B1-2
North Channel ON **21** E6
North Cobalt ON **17** E8
North Cobalt ON **22** A1
North Cobalt ON **23** F1
North Cooking Lake AB **8** C7
North Cowichan BC **2** E4
North Driftwood River PNR ON **17** B6
North Driftwood River, RNP ON **17** B6
North Duck River MB **11** C8
North Duck River MB **13** C1-2
North Earltown NS **34** E2-3
North East Arm NL **36** D4
North East Arm NL **37** D2
Northeast Cove NS **32** F7-8
Northeast Cove NS **33** C5
Northeastern Manitoulin and the Islands ON **18** B3-4
North East Harbour NS **33** F3
North East Margaree NS **34** C6
North East Margaree River NS **34** B6
North Enmore PE **35** C2
Northern Arm NL **36** A6
Northern Arm NL **37** A4
Northern Arm NL **38** F5
Northern Bathurst Island (Tuktusiuqvialuk) NP (proposed/projeté) NU **42** C3-4
Northern Bay NL **37** D7
Northern Bruce Peninsula ON **18** E4-5
Northern Indian Lake MB **15** C4
Northern Light Lake ON **16** F4-5
Northern Peninsula NL **38** C3-4
Northern Rocky Mountains PP BC **3** C6
North Etomami River MB **13** C4
Northfield NS **32** F5-6
Northfield NS **32** F6
Northfield NS **33** D3
Northgate SK **11** F7
North Gower ON **21** B8
North Gower ON **24** E2
North Grant NS **34** D4
North Gut St. Anns NS **34** C7
North Harbour NL **36** D8
North Harbour NL **37** D6
North Harbour NL **37** E7
North Harbour NS **34** A7
North Hatley QC **27** F4
North Head NB **32** F3
North Head NL **36** A2
North Head NL **38** F1-2
North Heart River NB **9** D-E4
North Kemptville NS **33** E2
North Kent Island NU **42** C4
North Kingston NS **32** E6
North Kingston NS **33** B3
North Klondike River YT **4** A2-3
North Knife Lake MB **15** C4
North Knife Lake MB **41** F1
North Knife River MB **15** B4-5
North Lake ON **32** C1
North Lake PE **34** B4
North Lake PE **35** D8
North Lake PE **35** D8
North Lake Provincial Campground ON **32** C1
North Lake, camping provincial NB **32** C1
North Lancaster ON **24** E4
Northleigh AB **8** C5
North Lochaber NS **34** E4-5
North McQuesten River YT **4** A-B3
North Milton PE **35** D-E4
Northminster SK **11** B2
North Monetville ON **17** F7
North Monetville ON **18** B6-7
North Moose Lake MB **11** A8

North Moose Lake MB **13** A2
North Mountain NS **32** E5-6
North Mountain NS **33** B3-4
North Mountain NS **34** D3
North Mountain NS **35** F6
North Nahanni River NT **10** D1-2
North Noel Road NS **32** D8
North Noel Road NS **33** B5
North Noel Road NS **34** F1
North Pelham ON **19** C8
North Pelham ON **20** E7
North Pender Island BC **2** F4
North Perth ON **19** A5
North Perth ON **20** C4
North Pine BC **3** E8
North Pine BC **9** D1
North Pond NL **36** B7
North Pond NL **37** B5
Northport NS **32** C8
Northport NS **34** D1
Northport NS **35** F2-3
North Port ON **21** E5
Northport PE **34** B1
Northport (Alberton South) PE **35** B2
North Portage ON **21** B2
North Portal SK **11** F7
Northport Beach PP NS **32** C8
Northport Beach PP NS **34** D1-2
Northport Beach PP NS **35** F3
North Preston NS **33** C6
North Ridge ON **19** E1-2
North River MB **15** B5
North River NL **37** D7
North River NL **39** B7
North River NL **39** B7
North River NS **32** F6
North River NS **33** C4
North River NS **34** C4
North River ON **18** A8
North River ON **18** F8
North River ON **20** A7
North River ON **21** C1
North River ON **21** D1-2
North River Bridge NS **34** B-C7
North River PP NS **34** B6-7
North Rustico PE **34** B2
North Rustico PE **35** D4
North Saanich BC **2** F4-5
North Saskatchewan River AB **6** A7
North Saskatchewan River AB **6** B-C5
North Saskatchewan River AB **8** C5-6 D6
North Saskatchewan River AB **8** D4
North Saskatchewan River AB **11** A1
North Saskatchewan River SK **11** B2-3
North Saugeen River ON **20** A-B4
North Seal River MB **12** A8 B7-8
North Seal River MB **15** B2-3
North Shoal Lake MB **14** C4
North Shore NS **34** B7
North Shore NS **34** D2
North Shore Waterway PP ON **17** F5
North Spirit Lake ON **13** C6
North Spirit Lake ON **13** C6
North Spirit Lake ON **16** A3
North Spirit Lake ON **16** A2-3
North Spruce Lake ON **22** C1
North Star AB **9** D3
North Sydney NS **34** C7
North Tea Lake ON **22** E2
North Tetagouche NB **29** E7
North Tetagouche NB **31** B6
North Thames River ON **19** B-D5
North Thames River ON **20** D-E4
North Thompson Islands PP BC **6** D1-2
North Thompson Oxbows East PP BC **6** C2
North Thompson Oxbows Manteau PP BC **6** C2
North Thompson Oxbows-Jensen Island PP BC **2** A8
North Thompson River BC **2** A8
North Thompson River BC **6** C-D2
North Thompson River PP BC **6** D1-2
North Tryon PE **35** E3
North Tweedsmuir PP BC **1** B5
North Twillingate Island NL **38** E6
North Twin Island NL **40** E3-4
North Twin Lake NL **36** A5
North Twin Lake NL **37** A3-4
North Twin Lake NL **38** F5
North Twin Lake NL **39** F8
Northumberland, détroit de NB **32** A-B7
Northumberland, détroit de NS **34** C-D4 C5

Northumberland, détroit de PE **35** D1-2 E7 F8
Northumberland PP PE **34** D3
Northumberland PP PE **35** F6
Northumberland Strait NB **32** A-B7
Northumberland Strait NS **34** D2-3
Northumberland Strait PE **35** D1-2 E7 F8
North Vancouver BC **1** F7
North Vancouver BC **2** D5
North Vermilion AB **9** B4
North Veuve River ON **18** A6
Northville ON **19** B3
Northville ON **20** C2
North Wabasca Lake AB **9** E5
Northwest Angle Inlet ON **14** E-F8
North West Arm NL **36** A8
North West Arm NL **37** A6
North West Arm NL **38** F7
Northwest Bay ON **13** F6
Northwest Bay ON **16** E2
North West Brook NL **36** C-D8
North West Brook NL **37** C-D6
Northwest Brook NL **37** F7
Northwest Gander River NL **36** B6-7 C6
Northwest Gander River NL **37** B4-5 C4
Northwest Miramichi River NB **29** F6-7
Northwest Miramichi River NB **31** C5-6
Northwest Passage TP NU **41** C4
Northwest Passage TP NU **42** F4
North West Point NT **10** D6
North West River NL **39** B5-6
North West River NL **40** E7-8
Northwest Upsalquitch River NB **29** E5-6
Northwest Upsalquitch River NB **31** B4
North Whirl ON **20** D4
North Wiltshire PE **34** C2
North Wiltshire PE **35** E4
North Wind Lake ON **16** D6-7
North Wind Lake ON **17** B1
North Woodslee ON **19** E1-2
North York ON **19** A8
North York ON **20** C7
North York ON **21** F1
Northwood ON **19** D3
Northwood ON **20** F2
Norton NB **32** C4
Norton NB **33** A2
Norton Lake MB **15** B6
Norton Shaw, Cape NU **42** C5
Norval ON **19** A7
Norval ON **20** C6
Norway NB **34** A1
Norway PE **35** A2
Norway Bay QC **21** A7
Norway Bay QC **22** F6-7
Norway Bay QC **24** D1
Norway House MB **13** A3
Norway House MB **21** B6
Norwegian Bay NU **42** C4
Norwich ON **19** C6
Norwich ON **20** E5
Norwood NS **33** E1
Norwood ON **21** D4
Nosbonsing ON **17** F8
Nosbonsing ON **18** B8
Nosbonsing ON **22** D1-2
Nosbonsing, Lake ON **22** D2
Notawassi, lac QC **22** B8
Notawassi, lac QC **25** D3
Notawassi, rivière QC **24** B-C8
Notawassi, rivière QC **25** D2
Notch Hill BC **6** D2
Notikewin AB **9** D3
Notikewin PP AB **9** C3-4 D3-4
Notikewin River AB **9** D3
Notre-Dame NB **31** F8
Notre-Dame NB **34** A1-2 B1
Notre-Dame, baie de NL **38** E5
Notre-Dame, lac QC **22** D7
Notre-Dame, lac QC **24** C1
Notre-Dame, monts QC **27** A-B7 C6
Notre-Dame, monts QC **28** B5
Notre-Dame, monts QC **29** B6-7
Notre-Dame, monts QC **31** A1-2 B1
Notre Dame Bay NL **38** E6
Notre-Dame-de-Ham QC **27** D4-5
Notre-Dame-de-la-Merci QC **24** B5
Notre-Dame-de-la-Merci NB **32** A-B7

Northumberland, détroit de PE **35** D1-2 E7 F8
Notre-Dame-de-la-Salette QC **22** F8
Notre-Dame-de-la-Salette QC **24** C2
Notre-Dame-de-Lorette QC **26** C8
Notre-Dame-de-Lorette QC **28** A3-4
Notre Dame de Lourdes MB **13** F3
Notre Dame de Lourdes MB **14** E3
Notre-Dame-de-Lourdes NB **29** F4
Notre-Dame-de-Lourdes NB **31** C2-3
Notre-Dame-de-Lourdes QC **24** B-C6
Notre-Dame-de-Lourdes QC **27** C4-5
Notre-Dame-de-Montauban-les-Mines QC **24** A8
Notre-Dame-de-Montauban-les-Mines QC **25** E8
Notre-Dame-de-Montauban-les-Mines QC **41** E8
Notre-Dame-de-Pierreville QC **27** B4
Notre-Dame-de-Pierreville QC **27** C3
Notre-Dame-de-Pontmain QC **22** D8
Notre-Dame-de-Pontmain QC **24** B2
Notre-Dame-de-Pontmain QC **25** F2
Notre-Dame-de-Portneuf QC **24** B4-5
Notre-Dame-des-Anges QC **24** A8
Notre-Dame-des-Anges QC **25** E8
Notre-Dame-des-Anges QC **27** A4
Notre-Dame-des-Bois QC **27** E6
Notre-Dame-des-Champs ON **21** A8
Notre-Dame-des-Champs ON **24** D2-3
Notre-Dame-des-Érables NB **29** E8
Notre-Dame-des-Érables NB **30** E1
Notre-Dame-des-Érables NB **31** B7
Notre-Dame-des-Monts QC **28** E7
Notre-Dame-des-Pins QC **27** C6-7
Notre-Dame-des-Prairies QC **24** C6
Notre-Dame-des-Prairies QC **27** D2
Notre-Dame-des-Sept-Douleurs QC **29** D1
Notre-Dame-de-Stanbridge QC **24** E7
Notre-Dame-de-Stanbridge QC **27** F2
Notre-Dame-du-Bon-Conseil QC **24** C8
Notre-Dame-du-Bon-Conseil QC **27** D4
Notre-Dame-du-Lac QC **29** E2
Notre-Dame-du-Laus QC **22** E8
Notre-Dame-du-Laus QC **24** B2
Notre-Dame-du-Mont-Carmel QC **24** B8
Notre-Dame-du-Mont-Carmel QC **24** E6
Notre-Dame-du-Mont-Carmel QC **25** E8
Notre-Dame-du-Mont-Carmel QC **27** F2
Notre-Dame-du-Nord QC **17** D-E8
Notre-Dame-du-Nord QC **22** A1-2
Notre-Dame-du-Portage QC **29** E1
Notre-Dame-du-Portage QC **29** E1
Notre-Dame-du-Rosaire QC **27** B7
Notre-Dame-du-Rosaire QC **28** B5
Notre Dame Junction NL **36** A7
Notre Dame Junction NL **37** A5
Notre Dame Junction NL **38** F6
Notre Dame PP NL **36** A7
Notre Dame PP NL **37** A5
Notre Dame PP NL **38** F6
Nottawa ON **18** F6-7
Nottawa ON **20** A5
Nottawaga Beach ON **18** F7
Nottawaga Beach ON **20** A6
Nottawasaga, baie ON **18** F6-7

Nottawasaga Bay ON **18** F6-7
Nottawasaga Bay ON **20** A5-6
Nottawasaga Lookout PNR ON **18** F6-7
Nottawasaga Lookout PNR ON **20** A5-6
Nottawasaga Lookout, RNP ON **18** F6-7
Nottawasaga Lookout, RNP ON **20** A5-6
Nottawasaga River ON **20** A-B6
Nottaway, rivière QC **40** F4
Nottingham Island NU **40** A4
Nottingham Island NU **41** D7
Nourse MB **14** D6
Nouveau, lac QC **39** B1
Nouveau-Québec (Chubb Crater), cratère du QC **40** B4-5
Nouveau-Québec (Chubb Crater), cratère du QC **41** D8
Nouvelle QC **29** D7
Nouvelle QC **31** A5
Nouvelle, rivière QC **29** C-D6
Nouvelle, rivière QC **31** A5
Nouvelle-France, cap de QC **40** B5
Nouvelle-France, cap de QC **41** D8
Novar ON **18** D8
Novar ON **21** A1
Novar ON **22** F1
Novellet, lac QC **23** B7
Novellet, lac QC **26** C1
Novra MB **11** C8
Novra MB **13** C1
Noyan QC **24** E6
Noyan QC **27** F2
Noyrot, lac QC **38** A-B1
Noyrot, lac QC **39** B6
Nuchatlitz Inlet BC **1** F4
Nuchatlitz PP BC **1** F4
Nueltin Lake NU **41** F1
Nukko Lake BC **1** A7
Nulki Lake BC **1** A6
Numabin Bay SK **12** D6
Numakoos Lake MB **15** C5
Numakoos PP MB **15** C4
Numogate ON **21** C7
Numogate ON **24** E1
Nunalla MB **15** A5
Nungesser Lake ON **13** D6
Nungesser Lake ON **16** B2
Nunim Lake SK **12** A6-7
Nunim Lake SK **15** A1
Nunn Lake SK **12** E-F6
Nunsti PP BC **1** C-D6
Nutak NU **40** C7
Nut Lake SK **11** C6
Nut Mountain SK **11** C6
Nuttby NS **33** A6
Nuttby NS **34** E2
Nuttlude Lake BC **3** C3
Nutzotin Mountains YT **4** C1
Nuvorak Point NT **5** A8
Nyanza NS **34** C6
Nyarling River NT **10** E-F7
Nymphes, ZEC des QC **24** A-B6
Nymphes, ZEC des QC **25** F6
Nymphes, ZEC des QC **27** B-C1

O

Oakbank MB **13** E4
Oakbank MB **14** D5
Oak Bay NB **2** F4-5
Oak Bay NB **32** E2
Oak Bay QC **29** D6
Oak Bay QC **31** A4-5
Oak Bay PP NB **32** E2
Oak Bluff MB **14** E5
Oak Brae MB **14** A1
Oakburn MB **11** E8
Oakburn MB **13** E1
Oakdale ON **19** D3
Oakdale ON **20** F2
Oakfield PP NS **32** E8
Oakfield PP NS **33** B6
Oak Hill NB **32** D2
Oak Island NS **34** D2
Oak Lake MB **11** E-F8
Oak Lake MB **13** F1
Oak Lake (IR) MB **13** F1
Oak Lake PP MB **13** E-F1
Oak Lake PP MB **11** E8
Oak Lake PP MB **13** E1
Oakland MB **13** E3
Oakland NS **32** F7
Oakland NS **33** D4-5
Oakland ON **19** C6

Oakland ON **20** E5
Oakner MB **11** E8
Oakner MB **13** E1
Oak Orchard ON **21** D3
Oak Park NS **33** F2
Oak Point MB **13** E3
Oak Point MB **14** C4
Oak Point MB **32** C4
Oak Point NB **33** A1
Oak Ridges ON **19** A8
Oak Ridges ON **20** C7
Oak Ridges ON **21** F1
Oak River MB **11** E8
Oak River MB **11** E8
Oak River MB **13** E1-2
Oak River MB **13** E2
Oak River MB **14** D1
Oakshela SK **11** E6-7
Oakview MB **14** B3
Oakville MB **13** E3
Oakville MB **14** D4
Oakville ON **19** B7-8
Oakville ON **20** D6-7
Oakwood ON **20** B8
Oakwood ON **21** D2
Oastler Lake PP ON **18** D7-8
Oatfield MB **14** B3
Oba ON **17** C4
Obabcata, lac QC **23** E7-8
Obabcata, lac QC **25** B2
Obabcata, lac QC **26** F2
Obabika Lake ON **17** E7
Obabika River PP ON **17** E7
Obakamiga Lake ON **17** C3
Oba Lake ON **17** C4
Obalski, lac QC **23** C4
Oban SK **11** C3
Obaska ON **23** E5
Obatagamau, lac QC **26** B4
Obatanga PP ON **17** B3
Obed AB **8** C2
Obedjiwan QC **26** C8
Obed Lake PP AB **6** A-B3
Obed Lake PP AB **8** C2
Oberon MB **14** C3
Oblats, lac des QC **22** D7
Oblats, lac des QC **24** B1-2
Oblats, lac des QC **25** F1-2
Obonga Lake ON **16** D5-6
Obonga-Ottertooth PP ON **16** D5
Obserguity lac ON **16** B5
Observatory Inlet BC **3** F3
Obustiga Lake ON **16** B5
Ocean Falls BC **1** C4
Ocean Man SK **11** E6-7
Ocean Pond NL **36** A8
Ocean Pond NL **37** A6
Ochak Lake SK **12** A6
Ochapowace SK **11** E7
Ochiichagwe'babigo'ining ON **16** D5
Ochre Beach MB **14** B1-2
Ochre River MB **11** D8
Ochre River MB **11** D8
Ochre River MB **13** D2
Ochre River MB **13** D2
Ochre River MB **14** B1
Ochre River NT **14** B1-2
Ochre River NT **10** C2-3
O'Connor Lake NT **10** D-E8
Octave, rivière QC **23** B4 C3
Octopus Islands PiMP BC **1** E5-6
Octopus Islands PMP BC **2** B2
Odanak QC **24** C7
Odanak QC **27** D3
O'Day MB **15** C5
Odei River MB **15** E3-4
Odell MB **31** D3
Odell River NB **31** D-E3
Odell River NB **32** A1-2
Odessa ON **21** D-E6
Odessa SK **11** E6
Odhill MB **15** F3
Odin, Mount NU **41** B8
Odin, Mount NU **42** F8
O'Donnel River BC **3** A1
O'Donnell River BC **4** F4-5
O'Donnell Point ON **18** E7
O'Donnell Point, RNP ON **18** E6-7
O'Donnell Point PNR ON **18** E6-7
O'Donnells NL **37** E7
Off Lake Corner ON **16** E2
Ogascanane, lac QC **22** B3
Ogden MB **34** E5
Ogden Mill NB **35** F1
Ogema SK **11** F5
Ogilvie MB **13** E2-3
Ogilvie MB **13** E1
Ogilvie Island NT **5** F7
Ogilvie Mountains YT **5** E-F2
Ogilvie River YT **5** F2
Ogoki ON **16** B8
Ogoki, Lake ON **16** C7
Ogoki, Lake ON **17** A2
Ogoki Reservoir ON **16** C7
Ogoki Reservoir ON **17** A1
Ogoki River ON **16** B8 C7
Ogoki River ON **16** C7
Ogoki River PP ON **16** C7
O'Hara, lac QC **22** E8
O'Hara, lac QC **24** C2
Ohaton AB **6** B6

Ohaton AB 8 D7
Ohaton AB 11 B1
Ohio NS 33 A8
Ohio NS 33 E2-3
Ohio NS 34 E4
Ohio-du-Barachois NB 35 E1
Ohsweken ON 19 C7
Ohsweken ON 20 E6
Oies, cap aux QC 28 E7
Oies, île aux QC 27 A7
Oies, île aux QC 28 F7
Oil City ON 19 C3
Oil City ON 20 F2
Oil Springs ON 19 C-D3
Oil Springs ON 20 F2
Oiseau Point ON 16 F8
Oiseau Point ON 17 D2
Oiseaux, rochers aux QC 30 E8
Ojibway PP ON 13 E7
Ojibway PP ON 16 D3
Oka QC 24 D5
Oka, parc d' QC 24 D5
Okanagan Centre BC 6 E2
Okanagan Falls BC 6 F2
Okanagan Falls PP BC 6 F2
Okanagan Lake BC 6 E-F2
Okanagan Lake PP BC 6 F2
Okanagan Mountain PP BC 6 E-F2
Okane, lac QC 25 C6
Okane, lac QC 28 F2
Okanese SK 11 D6
Okeover Arm BC 2 C3
Okeover Arm PP BC 2 C3
Okeover Arm PP BC 2 C3
Okipwatsikew Lake SK 12 E7
Okipwatsikew Lake SK 15 E1
Okla SK 11 C6
Okno MB 14 B5
Okotoks AB 6 D5-6 E5-6
Okotoks AB 7 C4
Olalla BC 6 F1
Old Barns NS 33 A6
Old Barns NS 34 E2
Old Bonaventure NL 37 C7
Old Canoe Lake NT 10 B7-8
Old Crow YT 5 C2
Old Crow Flats Special Management Area YT 5 C2-3
Old Crow Range YT 5 C1-2
Old Crow River YT 5 C2
Old Fort Bay AB 9 B7
Old Fort Bay AB 12 B2
Old Fort Nelson BC 3 B7
Old Fort Nelson BC 9 B1
Old Fort River AB 9 B7
Old Fort River AB 12 B2
Old Harry QC 30 E8
Old Horton Channel NT 5 A7
Old Man Lake NT 5 B5
Old Mans Pond NL 36 A3
Old Mans Pond NL 37 A1
Old Mans Pond NL 38 F2
Old Perlican NL 37 C7-8
Old Shoes Lake ON 16 A2
Old Shop NL 37 D-E7
Old Town Lunenburg NS 33 D5
Old Wives SK 11 E4-5
Old Wives Lake SK 11 E4-5
Old Wives Lake NMBS SK 11 E4-5
Oldcastle ON 19 E1
Oldenberg MB 14 D6-7
Oldham NS 33 B6
Oldman Dam AB 6 F5-6
Oldman Dam AB 7 E4
Oldman Dam PRA AB 6 F5-6
Oldman Dam PRA AB 7 E4
Oldman Lake SK 9 A8
Oldman River AB 6 E7
Oldman River AB 6 E-F5 F6
Oldman River AB 7 D7
Oldman River AB 7 E4
Oldman River MB 15 C4
Oldman River SK 9 A8
Oldman River SK 12 A4
Oldman River Reservoir AB 7 E4
Olds AB 6 C5
Olds AB 7 A4
O'Leary PE 32 A7
O'Leary PE 35 C1
Olga, lac QC 23 A6
Olha MB 11 D-E8
Olha MB 13 D-E1
Oliphant ON 18 E-F5
Oliphant ON 20 A3-4
Oliva, lac QC 26 C4
Oliver BC 6 F2
Oliver NS 34 E2
Oliver ON 19 E1
Oliver Cove PMP BC 1 C3-4
Oliver Lake SK 12 D6
Olomane, rivière QC 39 E5-6
Oman Lake SK 12 A4
Ombabika River ON 16 D6-7
Ombabika River ON 17 A1
Omemee ON 21 D2-3
Omerville QC 27 F4
Omineca Mountains BC 3 C4 D4-5
Omineca PP BC 3 E6
Omineca River BC 3 E5-6
Ominecetla Creek BC 3 E5-6

Ommanney Bay NU 41 A4
Ommanney Bay NU 42 D-E3
Omo, lac QC 26 A3
Ompah ON 21 C6
Onamakawash Lake ON 16 D5
Onaman Lake ON 16 D7
Onaman Lake ON 17 B1-2
Onanole MB 11 E8
Onanole MB 13 E2
Onanole MB 14 C1
Onaping ON 17 F6
Onaping Lake ON 17 E6
Onatchiway, lac QC 28 A6
Onatchiway, ZEC QC 28 A6
One Arrow SK 11 B5
One Island Lake PP BC 3 F8
One Island Lake PP BC 9 E1-2
Onefour AB 6 F8
Onefour AB 11 F1-2
100 Mile House BC 1 D8
100 Mile House BC 6 C1
108 Mile House BC 1 C-D8
108 Mile House BC 6 C1
150 Mile House BC 1 C7-8
Onésime, lac QC 28 D7
Onhda Lake NT 5 D6
Onion Lake SK 11 A2
Onondaga ON 19 C7
Onondaga ON 20 E6
Onoway AB 6 A-B5
Onoway AB 8 C5-6
Onslow NS 33 A6
Onslow NS 34 E2
Onslow Corners ON 21 A7
Onslow Corners QC 22 F7
Onslow Corners QC 24 D1
Ontaratue River NT 5 D6 E5
Ontario, lac ON 19 B8
Ontario, lac ON 20 D8
Ontario, lac ON 21 F4-6
Ontario, Lake ON 19 B8
Ontario, Lake ON 20 C-D8
Ontario, Lake ON 21 F4-5
Onyota'aka ON 19 C4
Onyota'aka ON 20 F3
Oona River BC 1 A2
Ootsa Lake BC 1 A-B5
Ootsa Lake BC 1 A5
Opachuanau Lake MB 12 D8
Opachuanau Lake MB 15 D2-3
Opakopa Lake ON 13 B7
Opal AB 6 A6
Opal AB 8 B6-7
Opapimiskan Lake ON 13 B8
Opasatica, lac QC 23 E2
Opasatika ON 17 B5
Opasatika Lake ON 17 C5
Opasatika River ON 17 A-B5
Opaskwayak MB 11 A7
Opaskwayak MB 13 A1
Opasquia MB 13 B6
Opasquia PP ON 13 B6
Opasquia PP ON 40 D1
Opawica, lac QC 23 A8
Opawica, lac QC 26 B2
Opawica, rivière QC 26 B3
Opawica, rivière QC 26 B2
Opeepeesway Lake ON 17 D-E5
Opémisca, lac QC 26 A4
Open Bay NL 39 B8
Open Hall NL 37 B7
Openshaw NL 37 F7
Opeongo Lake ON 21 A3
Opeongo Lake ON 22 F3
Opeongo River ON 21 A3-4
Opeongo River ON 22 F3-4
Opeongo River PP ON 21 A4
Opeongo River PP ON 22 F4
Opescal Lake NT 12 A5
Ophir ON 17 F4
Opikeigen Lake ON 16 B6-7
Opikinimika Lake ON 17 E6
Opinaca, rivière QC 40 E1
Opiscotéo, lac QC 39 C1
Opiscotéo, lac QC 40 E6
Opiscotiche, lac QC 39 C1-2
Opitsat BC 1 F5
Opitsat BC 2 D-E1
Opocopa, lac QC 39 C2
Optic Lake MB 12 F7
Opuntia Lake SK 11 C3
Opuntia Lake NMBS SK 11 C3
Opuskiamishes River MB 15 E6-7
Or, Cape d' NS 32 D6
Or, Cape d' NS 33 A4
Orange Bay NL 39 E8
Orangedale NS 34 D6
Orangeville ON 19 A6-7
Orangeville ON 20 C5-6
Orby Head PE 35 D4
Orcadia SK 11 D7
Orchre Pit Cove NL 37 D7-8
O'Regan's NL 36 D1
Oregon Jack PP BC 1 E8
Oregon Jack PP BC 2 B6-7
Orford, mont QC 27 F4
Orford River BC 2 A3 B2
Orient, rivière de l' QC 29 D3
Orient Bay ON 16 E6
Orient Bay ON 17 B1

Orignal, cap à l' QC 29 C2
Orignal, lac de l' QC 22 B4
Orillia ON 18 F8
Orillia ON 20 A7
Orillia ON 21 D1
Orion AB 6 F7
Orion AB 7 E8
Orion AB 11 F1
Orkney SK 11 F3
Orleans ON 21 A8
Orleans ON 24 D2-3
Orléans, île d' QC 27 A6
Ormes, rivière des QC 27 C4
Ormiston SK 11 F5
Ormsby ON 21 C4
Ormstown QC 24 E5
Ormstown QC 27 F1
Oro Beach ON 18 F8
Oro Beach ON 20 A7
Oro Beach ON 21 D1
Orok MB 11 A7-8
Orok MB 13 A1-2
Oro Lea Beach ON 18 F8
Oro Lea Beach ON 20 A7
Oro Lea Beach ON 21 D1
Oromocto NB 32 C3
Oromocto Lake NB 32 D2
Orono ON 20 C8
Orono ON 21 E2-3
Oro Station ON 18 F8
Oro Station ON 20 A7
Oro Station ON 21 D1
Orrville ON 18 D7-8
Orr Lake MB 15 E4
Orr Lake ON 18 F7
Orr Lake ON 18 F7
Orr Lake ON 21 C7
Orr Lake QC 24 F1
Orr Lake QC 24 C1
Orrell NB 31 D3
Orwell ON 19 D5
Orwell ON 20 F4
Orwell PE 34 C3
Orwell PE 35 E5-6
Orwell Bay PE 34 C3
Orwell Bay PE 35 E-F5
Orwell Corner Historic Village Provincial Heritage Site PE 35 E5
Orwell Cove PE 35 E5
Osaca ON 21 E3
Osage SK 11 E6
Osborne MB 13 F3-4
Osborne MB 14 E4-5
Osborne Corners ON 19 B6
Osborne Corners ON 20 E5
Osborne Lake MB 12 F8
Osborne Lake MB 15 F2
Osbourne Bay ON 13 F6-7
Oscar, lac QC 25 C6
Oscar, lac QC 28 E1-2
Oscar Peak BC 3 C4
Osceola ON 21 A5-6
Osceola ON 22 F5-6
Osgoode ON 21 B8
Osgoode ON 24 E2
Osgoode, rivière QC 27 C5
Oshawa ON 20 C8
Oshawa ON 21 F2
Osik Lake MB 12 E8
Osik Lake MB 15 E3
Osilinka River BC 3 E5-6
Oskélanéo QC 25 A3
Oskélanéo QC 26 B3
Oskélanéo, lac QC 25 A3
Oskélanéo, lac QC 26 B3
Osler SK 11 C4
Osnabruck Centre ON 24 E3
Osnaburgh Lake ON 13 D8
Osnaburgh Lake ON 16 B-C5
Osoyoos BC 6 F2
Osoyoos Lake BC 6 F2
Ospringe ON 19 A6-7
Ospringe ON 20 C5-6
Ospwagan Lake MB 15 D3
Ossokmanuan Lake NL 39 B-C3
Ossossane Beach ON 18 F7
Ossossane Beach ON 20 A6
Ostaboningue, lac QC 17 E8
Ostaboningue, lac QC 22 B2-3
Ostenfeld MB 13 E-F4
Ostenfeld MB 14 E6
Osterwick MB 13 F3
Osterwick MB 14 F4
Ostrander ON 19 C5-6
Ostrander ON 20 E4-5
Ostrea Lake NS 33 C7
Ostryhon Corners ON 19 C8
Ostryhon Corners ON 20 E7
O'Sullivan, lac QC 22 A7
O'Sullivan, lac QC 25 C2
O'Sullivan Lake ON 16 C7
O'Sullivan Lake ON 17 A2
Oswald MB 14 D4
Oswald Point BC 1 B3
Oswego Creek ON 19 C7-8
Oswego Creek ON 20 E6-7
Otasawian River ON 17 A4 B3-4
Otatakan Lake ON 16 C3-4
Otherside River SK 12 B4
Otis, lac QC 28 C6-7
Otish, monts QC 40 F5-6

Otonabee River ON 21 D3
Otoskwin-Attawapiskat PP ON 16 A6 B5
Otoskwin River ON 16 B4
Otoskwin River ON 16 B5-6
Otoskwin River ON 40 E1-2
Otosquen SK 11 B7
Otosquen SK 13 B1
Ottawa ON 21 A8
Ottawa ON 21 B8
Ottawa ON 24 D2
Ottawa, cap QC 30 A4
Ottawa, lac QC 25 D5
Ottawa, lac QC 28 F1
Ottawa Islands NU 40 C3
Ottawa Islands NU 41 E-F7
Ottawa River ON 17 E-F8
Ottawa River ON 21 A6
Ottawa River ON 22 C2
Ottawa River ON 22 E2-6
Ottawa River ON 24 D3-4
Ottawa River QC 23 F4-5
Ottawa River PP ON 21 A6
Ottawa River PP ON 22 F6
Otter Bay BC 2 E4
Otter Brook NS 33 A7
Otter Brook NS 34 F3
Otterburne MB 13 F4
Otterburne MB 14 E5
Otterburn Park QC 24 D7
Otterburn Park QC 27 E2
Otter Creek BC 1 E-F8
Otter Creek BC 2 C-D8
Otter Creek BC 6 E-F1
Otter Creek YT 4 C6
Otter Falls Territorial Recreation Site YT 4 D3
Otter Island ON 16 F8
Otter Island ON 17 D2
Otter Lake NL 39 B5
Otter Lake ON 18 D7
Otter Lake ON 18 D7
Otter Lake ON 21 C7
Otter Lake QC 24 F1
Otter Lake QC 24 C1
Otter Lake SK 12 E5
Otter Lake YT 4 C6
Otter Lake PP BC 1 F8
Otter Lake PP BC 2 D8
Otter Lake PP BC 6 E-F1
Otter Point NL 36 D3
Otter Point NL 36 D3
Otter Point NL 37 D1
Otter Point NL 37 D1
Ottertail Creek ON 22 C1
Otterville ON 19 C6
Otterville ON 20 E5
Otthon MB 11 D7
Otto MB 14 C4
Otto Fiord NU 42 A-B4
Otty Lake ON 21 C7
Otukamamoan Lake ON 13 F7
Otukamamoan Lake ON 16 E3
Ouapetec, rivière QC 39 D-E2
Ouareau, lac QC 24 B5
Ouareau, lac QC 25 F5
Ouareau, rivière QC 24 B5 C6
Ouareau, rivière QC 27 C-D1
Ouasiemsca, rivière QC 26 B-C7
Ouellette ON 17 F7
Ouellette ON 18 B6
Ouest, lac de l' QC 28 A4
Ouest, pointe de l' QC 30 D2
Ouest, pointe de l' QC 31 A7-8
Ouest, pointe de l' QC 40 F7
Ouimet Canyon, RNP ON 16 F5-6
Ouimet Canyon PNR ON 16 F5-6
Ouiscatis, lac QC 23 D8
Ouiscatis, lac QC 26 E2
Oungre SK 11 F6
Ours, lac à l' QC 39 E4
Ours, lac des QC 22 D8
Ours, lac des QC 24 B2
Ours, lac des QC 25 F2
Ours, rivière à l' QC 26 E8
Ours, rivière à l' QC 28 B-C3
Ours, rivière l' QC 24 D4
Ours Blanc, lac de l' QC 25 B4
Ours Blanc, lac de l' QC 26 F4
Ouse River ON 21 D4
Oustic ON 19 A6
Oustic ON 20 C5
Outaouais, lac des QC 23 F8
Outaouais, rivière des QC 22 A8
Outaouais, rivière des QC 22 C-D2 E5 F7
Outaouais, rivière des QC 23 E-F3

Outaouais, rivière des QC 23 F4-5
Outaouais, rivière des QC 24 D1
Outardes, baie aux QC 29 A3
Outardes, rivière aux QC 40 F6
Outer Cove NL 37 D8
Outer Duck Island ON 18 C2
Outer Island NS 33 F2
Outlet ON 21 D7
Outlet ON 21 F5
Outlook SK 11 D4
Outram NS 32 E5
Outram NS 33 B3
Outram SK 11 F6
Outram Lakes NT 5 D8
Overby Lake MB 15 B3
Overflow Bay MB 11 B8
Overflow Bay MB 13 B1-2
Overflowing River MB 11 B7
Overflowing River MB 13 B1
Overflowing River SK 11 B7
Overflowing River SK 13 B1
Overton MB 14 B3
Overton NS 33 E1
Owen, rivière QC 29 E3
Owen, ZEC QC 29 E3
Owen Channel ON 18 C4
Owendale AB 7 F5
Owen Sound ON 18 E-F5
Owen Sound ON 18 F5-6
Owen Sound ON 20 A4
Owen Sound ON 20 A4
Owikeno Lake BC 1 D4
Owl River AB 9 F6-7
Owl River AB 12 E-F1
Owl River MB 15 C6
Owl River MB 40 C1
Oxbow NB 31 D3
Oxbow SK 11 F7
Oxbow Creek ON 19 C4
Oxbow Creek ON 20 E3
Oxdrift ON 13 E6
Oxdrift ON 16 D3
Oxenden ON 18 E5
Oxford NS 32 C8
Oxford NS 34 D1
Oxford Centre ON 19 C6
Oxford Centre ON 20 E5
Oxford House MB 15 D5
Oxford Junction NS 32 C8
Oxford Junction NS 34 D-E1
Oxford Lake MB 15 F4-5
Oxford Mills ON 21 C8
Oxford Mills ON 24 E2
Oxley ON 19 F1
Oxtongue Lake ON 21 B2
Oxtongue Lake ON 21 B2
Oxtongue River ON 21 A2
Oxtongue River ON 22 F2
Oxtongue River-Ragged Falls PP ON 22 F2
Oyama BC 6 E2
Oyen AB 6 D8
Oyen AB 7 A8
Oyen AB 11 D2
Oyster Bed Bridge PE 34 C4
Oyster Bed Bridge PE 35 D4
Oyster Cove PE 35 D3
Oyster River BC 2 C2
Ozema MB 14 C1
Ozhiski Lake ON 16 B6

P

Paarens Beach PP BC 1 A6
Pabineau NB 29 E8
Pabineau NB 31 B6
Pabos QC 30 C-D2
Pabos Mills QC 30 D2
Pachena Bay BC 2 E1
Pachena Point BC 2 E1
Pacific Ocean BC 1 D2-3
Pacific Ocean BC 3 F1
Pacific Ranges BC 1 D5
Pacific Ranges BC 2 A-B3
Pacific Rim, RPN BC 1 F4-5
Pacific Rim (archipel Broken Group), RPN BC 2 E1-2
Pacific Rim (plage Long), RPN BC 2 E1
Pacific Rim (sentier de la Côte-Ouest), RPN BC 2 F2
Pacific Rim NPR BC 1 F4-5
Pacific Rim NPR (Broken Group Islands) BC 2 E1-2
Pacific Rim NPR (Long Beach) BC 2 E1
Pacific Rim NPR (West Coast Trail) BC 2 F2
Pacifique, océan BC 1 E2-3 F2-3
Pacifique, océan BC 3 F1
Packers, Mount YT 4 D3-4
Packington QC 29 E2
Packington QC 31 C1
Packs Harbour NL 39 B7-8
Pacquet NL 38 D5
Paddle Prairie AB 9 C3
Paddle River AB 6 A5
Paddle River AB 8 B5
Paddling Lake SK 11 B4
Paddockwood SK 11 B5
Padoue QC 29 C4

Pagato Lake SK 12 E6
Pagato Lake SK 15 D1
Pagato River SK 12 E7
Page, Mount YT 5 A2
Pageant AB 7 C5
Paguchi Lake ON 13 F7-8
Paguchi Lake ON 16 D8
Pagwachuan River ON 17 B3
Pagwachuan River ON 17 B3
Pagwa River ON 17 B3
Paimpont, lac QC 39 E5
Pain Court ON 19 E2
Painted Bluffs PP BC 2 A7-8
Paint Lake MB 15 E3
Paint Lake PP MB 15 E3
Paisley ON 20 B4
Paix, rivière de la AB 12 A1
Paix, rivière de la AB 41 F1
Paix, rivière de la BC 3 A8
Pakashkan Lake ON 13 F8
Pakashkan Lake ON 16 E5
Pakenham ON 21 B7
Pakenham ON 24 D1
Pakowki Lake AB 6 F7
Pakowki Lake AB 7 E-F8
Pakowki Lake AB 11 F1
Pakuashipi QC 38 B1
Pakuashini QC 39 E6-7
Pakukak Lake AB 15 A-B5
Pakwa Lake MB 15 F3
Pakwash Lake ON 16 C2
Pakwash Lake ON 16 C2
Pakwash PP ON 13 D6
Palgrave ON 20 C6
Palermo ON 19 B7
Palermo ON 20 D6
Palliser, lac QC 26 D5
Palliser, lac QC 28 B1
Palmarolle QC 17 C8
Palmarolle QC 23 C2
Palmer SK 11 E4
Palmer, rivière QC 27 C5-6
Palmer Rapids ON 21 B4-5
Palmer Road PE 35 B2
Palmerston ON 19 A5
Palmerston ON 20 C4
Palmerston, Cape NS 32 C8
Palmerston, Cape NS 34 D1
Palmerston/Canonto CA ON 21 B6
Palmerston Lake ON 21 B-C6
Palmyra ON 19 D3-4
Palo SK 11 C3
Palowki AB 7 E8
Palsen River MB 13 B5
Pambrun SK 11 E4
Pambrun, lac QC 26 C4
Pamialac Bay NL 39 A6
Pamialac Bay NL 39 A6
Pan Lake Fen, RNP ON 16 E8
Pan Lake Fen, RNP ON 17 C2
Pan Lake Fen PNR ON 16 E8
Pan Lake Fen PNR ON 17 C2
Panache, lac du QC 25 A8
Panache, lac du QC 26 F8
Panache, Lake ON 18 A-B5
Panache, rivière au QC 23 B-C8
Panache, rivière au QC 26 C2
Panache Lake ON 17 F6
Pancake Bay PP ON 17 C2
Panchia Lake NL 39 C3
Pandora Island NU 41 A4
Pandora Island NU 42 A4
Pangman SK 11 F5
Pangnirtuuq (Pangnirtung) NU 42 F7
Pangnirtung (Pangniqtuuq) NU 42 F7
Panmure ON 18 D7
Panmure Island PE 34 C4
Panmure Island PE 34 C3-4
Panmure Island PE 35 E7
Panmure Island PP PE 34 C4
Panmure Island PP PE 35 E7
Panny River AB 9 C-D5
Panorama BC 6 E4
Panorama BC 7 C1
Pansy MB 14 F6
Pantagruel Creek PP ON 16 D5
Panther Mountain AB 7 D4
Panther Mountain AB 7 D4
Papanakies Lake NT 10 B-C8
Papinachois QC 29 A3
Papinachois, rivière QC 29 A2-3
Papineau QC 24 D3
Papineau Creek ON 21 B4
Papineau Lake ON 21 B4
Papineau Lake ON 22 D3
Papineauville QC 24 D3
Papineau-Labelle, RF de QC 24 C3
Paquette Corners ON 19 E1
Paquetville NB 29 E8
Paquetville NB 31 B7
Paradis QC 23 E7
Paradis QC 25 A1
Paradis QC 26 F1
Paradise NL 37 B7

Paradise NL 37 D8
Paradise NS 32 C3
Paradise NS 33 C3
Paradise Hill SK 6 A-B8
Paradise Hill SK 11 B2
Paradise Lake NS 32 F5-6
Paradise Lake NS 33 C3
Paradise Point ON 18 E7
Paradise River NL 36 D7-8 E7
Paradise River NL 37 B3
Paradise River NL 37 D5-6
Paradise River NL 39 C7
Paradise River NL 39 B7
Paradise Sound NL 37 D6-7 E7
Paradise Sound NL 37 D5-6 E5
Paradise Valley AB 6 B7-8
Paradise Valley AB 11 B2
Paradise Valley BC 2 C2
Paragon AB 7 C7
Paramé, lac ON 13 F7-8
Paramé, lac QC 39 D7
Parent QC 25 B4
Parent QC 26 F4
Parent, lac QC 23 C6
Parent, lac QC 28 B1
Parham ON 21 C6
Paris ON 19 B-C6
Paris ON 20 D5
Parisville QC 27 B4
Park Court AB 8 C5
Park Head ON 18 F5
Park Head ON 20 A4
Park Lake PP AB 6 E6
Park Lake PP AB 7 E5
Parkbeg SK 11 E4
Parkbend AB 7 F4
Parkdale NS 32 F6
Parkdale NS 33 C4
Parker ON 19 A6
Parker ON 20 C5
Parker, Cape NU 42 D5
Parker, lac QC 25 A-B3
Parker, lac QC 26 B3
Parker Ridge NB 31 E5
Parker Ridge NB 32 A3
Parkers Cove NL 36 E7
Parkers Cove NL 37 E7
Parkers Cove NS 32 F5
Parkers Cove NS 33 C2
Parkerview SK 11 D6
Parkhill ON 19 C4
Parkhill ON 20 E3
Parkhill Creek ON 19 B3-4
Parkhill Creek ON 20 E2-3
Parkindale NB 32 C6
Parkins, Cape NB 1 E3-4
Parkinson ON 17 F5
Parkinson ON 18 A1
Parkland AB 6 E6
Parkland AB 7 D4
Parkman SK 11 F7
Parks Creek ON 21 D5
Parkside SK 11 B4
Parks Lake ON 16 E7
Parks Lake ON 17 B1
Parksville BC 1 F6
Parksville BC 2 D3
Parlee Beach PP NB 32 B7-8
Parlee Beach PP NB 35 E1
Parrsboro NS 32 D7
Parrsboro NS 33 A4-5
Parry SK 11 F5
Parry, Cape NT 5 A8
Parry, îles de NT 42 C2-3
Parry, îles de NU 42 C2-3
Parry Bay NU 41 B-C6
Parry Bay NU 42 F5-6
Parry Island (Wasauksing) ON 18 D6-7
Parry Islands NT 42 C2-3
Parry Islands NU 42 C2-3
Parry Passage BC 1 A1
Parry Peninsula NT 5 A8
Parry Sound ON 18 D7
Parry Sound ON 18 D7
Parsnip River BC 3 E7
Parson BC 6 D3-4
Parson's Pond NL 38 D2
Parson's Pond NL 39 F2
Parsons Lake NT 5 B4-5
Parsons Pond NL 38 D2
Parsons Pond NL 39 F7
Partridge Bay NL 39 C8
Partridge Breast Lake MB 15 C-D3
Partridge Creek ON 21 C5
Partridge Point NL 38 D4
Partridge Point NL 39 E8
Pasadena NL 36 A3
Pasadena NL 37 A1
Pascagama QC 29 A3
Pascagama, lac QC 26 D2
Pascagama, rivière QC 26 D3
Pascalis, lac QC 23 D-E5
Pas d'Eau Lake QC 15 E-F7
Pasfield Lake SK 12 B5
Pashkogan Lake ON 13 D8
Pashkokogan Lake ON 16 C5
Paskwachi Bay MB 12 C7
Paskwachi Bay MB 15 C-D1
Paspébiac QC 30 D1
Paspébiac QC 31 A7
Paspébiac, pointe de QC 31 A7
Pasqua SK 11 E5
Pasqua QC 11 E6
Pasqua Hills SK 11 B6

Pasquia River SK 11 A-B7
Pasquia River SK 13 A-B1
Passage Point NT 42 D2
Passamaquoddy Bay NB 32 E2
Pass Island NL 36 E5
Pass Island NL 37 E3
Pass Lake ON 16 F6
Pass Peak YT 4 D5
Patapédia, rivière QC 29 D4
Patapédia, rivière QC 31 A3
Patch Lake MB 15 E6
Paterson MB 11 A8
Paterson MB 12 F8
Paterson MB 13 A2
Paterson MB 15 F2
Path End NL 37 F7
Pathlow SK 11 B-C5
Patricia AB 6 D7
Patricia AB 7 C7
Patricia AB 11 D1
Patricia Beach PP MB 13 E4
Patricia Beach PP MB 14 C5-6
Patrick's Cove NL 36 F8
Patrick's Cove NL 37 E8
Patten River ON 23 A2 B1
Patterson NB 32 D3
Patterson, lac QC 26 D5
Patterson, lac QC 28 B1
Patterson, Mount YT 4 A4
Patterson Lake SK 9 C8
Patterson Lake SK 12 C3
Patterson Lake SK 15 A1
Patton ON 18 A1
Pattullo, Mount BC 3 E3
Patuanak SK 12 E3-4
Paudash ON 21 C4
Paudash Lake ON 21 C3-4
Paugh Lake ON 21 A4
Paugh Lake ON 22 F4
Pauingassi MB 13 C5
Pauingassi MB 16 A1
Paulatuk NT 5 B8
Paulatuk NT 41 A1
Pauline Cove YT 5 A3
Paul Lake PP BC 6 D1
Paull Lake SK 12 E5
Paull River SK 12 E5
Paulson BC 6 F2-3
Pavilion BC 1 D7-8
Pavilion BC 2 A6
Pawistik MB 15 E1-2
Pawistik MB 16 A1
Payne, lac QC 40 C5
Payne, lac QC 41 E1
Payne River ON 24 E3
Paynton SK 6 B8
Paynton SK 11 B3
Pays Plat ON 16 F7
Pays Plat ON 17 C1
Payuk MB 12 F7
Payuk MB 15 F1
Peace Arch Provincial/ State Park BC 2 E5
Peace Point AB 9 A-B6
Peace Point AB 12 A1
Peace Reach BC 3 E7
Peace River AB 9 B5
Peace River AB 9 B5
Peace River AB 12 A-B1
Peace River BC 3 E7-8
Peace River BC 9 D-E1
Peace River Corridor PP BC 3 E7-8
Peace River Corridor PP BC 9 D-E1
Peace River Valley AB 9 E3
Peace River Wildland PP AB 9 E3
Peachland BC 6 E2
Peacock AB 7 D5
Peacock Hills NU 41 C2
Peacock Point ON 19 D7
Peacock Point ON 20 F6
Peacock Point ON 20 F6
Peakes PE 34 C3
Peakes PE 35 E6
Pearce AB 7 E5
Pearson Lake AB 15 D-E4
Pearsonville NB 32 C4
Peary Channel NU 42 B3
Peavey AB 8 B6
Peavine AB 9 B4-5
Peawanuck ON 40 D2
Pebble Bay SK 11 B4
Pebble Beach BC 2 C2
Pebble Beach MB 14 B3
Pebonishewi Lake ON 17 D6
Pedley AB 8 C2
Peeagwon Lake ON 16 A6
Peebles SK 11 E6-7
Peel NB 31 D3
Peel NB 32 B1
Peel Channel NT 5 C4
Peel Point NT 41 A3
Peel Point NT 42 C2
Peel River YT 5 D4
Peel River Territorial Preserve NT 5 D-E5
Peel Sound NU 41 A4
Peel Sound NU 42 D-E4
Peerless SK 6 A8

Peerless SK **9** F8
Peerless SK **11** A3
Peerless SK **12** F3
Peerless Lake AB **9** D5
Peerless Lake AB **9** D5
Peers AB **6** B4
Peers AB **8** C3
Peesane SK **11** B6
Pefferlaw ON **20** B7
Pefferlaw ON **21** D1-2
Pefferlaw Brook ON **20** B7
Pefferlaw Brook ON **21** C1
Peggys Cove NS **32** F8
Peggys Cove NS **33** C-D5
Peguis MB **13** D3-4
Peguis MB **14** B4
Peguis MB **14** D5
Pekisko AB **7** C4
Pelé, cap NB **35** E1
Pelé Cape NB **35** E1
Pelee, Point ON **19** F2
Pelee, Point ON **19** F2
Pelee Island ON **19** F1-2
Pelee Island ON **19** F2
Pelee Passage ON **19** F2
Pèlerins, les QC **28** E8
Pèlerins, les QC **29** E1
Pelham ON **19** C8
Pelham ON **20** E7
Pelican AB **9** E6
Pelican AB **12** E1
Pelican Bay MB **11** B8
Pelican Bay MB **13** B2
Pelican Cove SK **11** C4
Pelican Islands PP Reserve
  MB **13** C3-4
Pelican Lake AB **9** E6
Pelican Lake MB **11** C8
Pelican Lake MB **13** A4
Pelican Lake MB **13** C2
Pelican Lake MB **13** F2
Pelican Lake MB **14** F2
Pelican Lake SK **11** A3
Pelican Lake SK **12** F6
Pelican Narrows SK **12** F6
Pelican Rapids MB **11** B8
Pelican Rapids MB **13** B2
Pelican River AB **9** E6
Pelican River AB **12** E1
Pellatt Lake NT **10** A8
Pelletier Lake MB **15** D4
Pelly SK **11** C7
Pelly SK **13** C1
Pelly Bay NU **41** B-C5
Pelly Bay NU **42** F4
Pelly Bay (Arviligjuat) NU
  41 C5
Pelly Bay (Arviligjuat) NU
  42 F4-5
Pelly Creek BC **3** D5
Pelly Crossing YT **4** C3
Pelly Island NT **5** A4
Pelly Mountains YT
  4 C4 D4-6
Pelly Point NU **41** B4
Pelly Point NU **42** F3
Pelly River YT **4** C3
Pelly River YT **4** C5 D5-6
Pemberton BC **1** E7
Pemberton BC **2** B5
Pemberton Icefield BC **1** E7
Pemberton Icefield BC **2** B4
Pemberton Meadows BC
  2 B5
Pembina AB **8** D4
Pembina River MB **13** F3
Pembina River MB **14** F3
Pembina River PP AB
  6 A-B5
Pembina River PP AB
  8 C4-5
Pembina Valley PP MB
  13 F3
Pembina Valley PP MB
  14 F3-4
Pembridge AB **8** B5
Pembroke NS **33** A7
Pembroke NS **33** B5
Pembroke ON **34** F3
Pembroke NS **34** F3
Pembroke ON **22** F5
Pembroke, Cape NU
  40 A-B3
Pembroke, Cape NU **41** D7
Pembroke Shore NS **33** E1
Pemichangan, lac QC **22** E8
Pemichangan, lac QC **24** C2
Pemichigamau Lake MB
  12 D8
Pemichigamau Lake MB
  15 D3
Pemmican Island PP
  Reserve MB **11** B8
Pemmican Island PP
  Reserve MB **13** B2
Pemmican Portage SK
  11 A7
Pemmican Portage SK
  13 A1
Pen Lake ON **21** A3
Pen Lake ON **22** F3
Pender Harbour BC **1** F6
Pender Harbour BC **2** D3
Pendleton ON **24** D3
Pendryl AB **8** D5
Penetanguishene ON **18** E7
Penguin Islands NL **36** E4
Penguin Islands NL **37** E2
Penhold ON **6** C5
Penhold AB **8** F6
Peninsula Lake ON **21** B2
Peninsular Lake ON **16** B6
Péninsule-Bruce, PN de la
  ON **18** D4-5
Penn SK **11** A4

Pennant NS **32** F8
Pennant NS **33** C6
Pennant SK **11** E3
Pennant Bay NS **33** D6
Pennant Point NS **33** D6
Pennfield NB **32** E2-3
Pennfield Ridge NB **32** E3
Penniac NB **32** B3
Penny BC **1** A8
Penny BC **6** A1
Pennycutaway River MB
  15 D-E6
Peno AB **8** B7
Penobsquis NB **32** C5
Penouille QC **30** B3
Penrhyn, Cape NU **41** C6
Penrhyn, Cape NU **42** F6
Penrose Island PMP BC
  1 D4
Pense SK **11** E5
Pentecôte, lac QC **39** F1
Pentecôte, rivière QC
  39 E-F1
Penticton BC **6** F2
Penzance SK **11** D5
Peoria AB **9** E3
Pepaw River SK **11** C7
Pepeekisis SK **11** D6
Perbeck AB **8** F7
Percé QC **30** C3
Percé, Cape NS **34** C8
Percé, rocher QC **30** C3
Perche, lac la QC **22** A5
Perch River SK **12** B5-6
Percival SK **11** E7
Percival River PE **35** C2
Percy Lake ON **16** C7-8
Percy Lake ON **17** A2
Percy Lake ON **21** B3
Percy Reach ON **21** E4
Perdix Blanche, lac de la
  QC **26** B8
Perdue SK **11** C4
Péré, lac QC **39** A1
Pereau NS **32** E7
Pereau NS **33** B4
Periard QC **30** B2
Péribonca, rivière QC **40** F5
Péribonka QC **28** B4
Péribonka, rivière QC
  28 A5
Périgny, lac QC **23** C7
Périgny, lac QC **26** C1-2
Perigord SK **11** C6
Perivale ON **18** C3
Perkins QC **22** F8
Perkins QC **24** D2
Perkinsfield ON **18** F7
Perkinsfield ON **20** A6
Perlson Lake NT **10** C7
Perras, lac QC **22** D-E8
Perras, lac QC **24** B3
Perras, lac QC **25** F3
Perrault Falls ON **13** E6
Perrault Falls ON
  16 C2-3 D2-3
Perrault Lake ON **16** D3
Perrier, lac QC **26** D4
Perrot, île QC **24** D5
Perrot, île QC **27** E1
Perry ON **17** D3-4
Perry Island NT **5** F7
Perry's Cove NL **37** D7-8
Perryvale AB **9** F6
Perryvale AB **8** A6-7
Perryvale AB **9** F6
Perseus, Mount BC **1** C8
Perseus, Mount BC **6** C1
Perth ON **21** C7
Perth ON **24** F1
Perth-Andover NB **31** E3
Perth-Andover NB **32** A1
Perth Road ON **21** D6-7
Peskawa Lake NS **33** D2-3
Peskawesk Lake NS **33** D3
Petabec Lakes SK **11** A7
Petabec Lakes SK **13** A1
Petaigan SK **11** B6
Petawaga, lac QC **22** B7-8
Petawaga, ZEC QC **25** D2
Petawaga, ZEC QC
  22 B7-8 C7-8
Petawaga, ZEC QC **25** D2
Petawanga Lake ON **16** B6
Petawawa ON **22** E5
Petawawa River ON **21** A2
Petawawa River ON
  22 E4-5
Petawawa Terrace PP ON
  22 E5
Peter, lac QC **25** A5
Peter Lake SK **12** C6
Peterbell ON **17** C4-5
Peterborough ON **21** D-E3
Peterborough, réserve de
  chasse de la Couronne
  ON **21** C-D4
Peterborough Crown Game
  Preserve ON **21** C-D4
Peter Lougheed PP AB
  6 E4-5
Peter Lougheed PP AB **7** C2

Peter Pond Lake SK **9** D-E8
Peter Pond Lake SK **12** E3
Petersburg ON **19** B6
Petersburg ON **20** D5
Peters Corners ON **19** B6-7
Peters Corners ON **20** D5-6
Petersfield MB **13** E4
Petersfield MB **14** D5
Peter's Forest, RNP ON
  21 E4
Peter's Forest PNR ON
  21 E4
Peter Snout NL **36** D3
Peter Snout NL **37** D1
Peterson SK **11** C5
Peters River NL **37** F7
Peters Road PE **35** F6
Peterville PE **35** A-B2
Pethei Peninsula NT **10** C8
Pethericks Corners ON
  21 A4
Petherton ON **19** A6
Petherton ON **20** C5
Petit-Cap NB **32** B7
Petit-Cap NB **34** C1
Petit-Cap QC **30** B3
Petit-Cap QC **32** C5
Petitcodiac NB **30** C2
Petitcodiac NB **32** C5
Petitcodiac River NB
  31 F7-8
Petitcodiac River NB
  32 B-C6
Petit-de-Grat Bridge NS
  34 E6-7
Petit-de-Grat Island NS
  34 E6-7
Petite-Aldouane NB **31** E8
Petite-Aldouane NB **32** A6
Petite-Anse QC **30** A2
Petite-Cascapédia, RF de la
  QC **29** C7
Petite-Cascapédia, RF de la
  QC **29** C8
Petite-Cascapédia, RF de la
  QC **31** A6
Petite Fourche, rivière QC
  30 B2
Petite-Lamèque NB **30** E2
Petite-Lamèque NB **31** B8
Petite Rivière NS **33** D4
Petite-Rivière-au-Renard
  QC **30** B3
Petite rivière aux Pékans QC
  39 C-D2
Petite rivière aux Rats QC
  26 B8
Petite rivière Cascapédia QC
  31 A6
Petite rivière Cascapédia Est
  QC **29** B7 C7-8
Petite rivière Cascapédia
  Ouest QC **29** B-C7
Petite rivière de la Baleine
  QC **40** D4-5
Petite rivière du Chêne QC
  27 C4
Petite rivière Manicouagan
  QC **39** C-D1
Petite rivière Matane QC
  29 B5
Petite rivière Nouvelle QC
  29 C6
Petite-Rivière-Ouest QC
  30 C2-3
Petite rivière Péribonka QC
  28 A-B4
Petite rivière Pikauba QC
  28 D5
Petite-Rivière-Saint-
  François QC-
  28 F7
Petites-Bergeronnes QC
  28 C8
Petites-Bergeronnes QC
  29 C1
Petites Plages, lac des QC
  26 A3
Petit Etang NS **34** B6
Petite-Vallée QC **30** A2
Petit Forte NL **36** E7
Petit Forte NL **37** E5
Petit Jardin NL **36** C1
Petit lac Batiscan QC **27** A4
Petit Lac des Esclaves AB
  9 E4-5
Petit lac du Caribou QC
  28 C3
Petit lac du Caribou QC
  25 A7
Petit lac du Cerf QC **22** E8
Petit lac du Cerf QC **24** B2
Petit lac du Cerf QC **25** F2
Petit lac du Nord QC **39** E2
Petit lac Hébert QC **26** C3
Petit lac Hertel QC **26** A7
Petit lac Jacques-Cartier QC
  28 F5
Petit lac Manicouagan QC
  39 D1
Petit lac Nominingue QC
  26 B8
Petit lac Nominingue QC
  24 B3
Petit lac Onatchiway QC
  28 A6
Petit lac Poigan QC **22** B7
Petit lac Poignan QC **25** D1

Petit lac Wayagamac QC
  25 C8
Petit lac Wayagamac QC
  27 C1
Petit-Matane QC **29** B4-5
Petit Mécatina, île QC
  40 E-F8
Petit Mécatina, île du QC
  39 E6
Petit Mécatina, rivière du
  NL **40** E7-8
Petit Mécatina, rivière du
  QC **39** D-E6
Petit Mécatina, rivière du
  QC **40** E7-8
Petitot River AB **9** A3
Petitot River AB **10** F5
Petitot River BC **3** A7
Petitot River BC **9** A1
Petitot River BC **10** F3
Petit Pabos QC **30** C2-3
Petit Pabos, rivière du QC
  30 C2
Petit-Rivière-de-l'Île NB
  30 E2
Petit-Rivière-de-l'Île NB
  31 A8
Petit-Rocher NB **29** F7-8
Petit-Rocher NB **31** B6
Petit-Rocher-Nord NB
  29 E7-8
Petit-Rocher-Nord NB
  31 B6
Petit-Rocher-Sud NB
  29 E7-8
Petit-Rocher-Sud NB **31** B6
Petit-Saguenay QC **28** C7-8
Petit-Shippagan QC **30** E2-3
Petit-Shippagan NB **31** A8
Petitsikapau Lake NL **39** A2
Petitsikapau Lake NL **40** D6
Petit-Témis, parc linéaire
  interprovincial
  29 D1 E1-2
Petit-Témis, parc linéaire
  interprovincial
  31 B-C1
Petit-Témis Interprovincial
  Linear Park QC **29** E2-3
Petit-Témis Interprovincial
  Linear Park QC **31** B-C1
Petley NL **36** D1
Petley NL **37** C6-7
Petownikip Lake ON **13** B7
Petpeswick Inlet NS **33** C7
Petre, Point ON **21** F5
Petrofka Bridge SK **11** B-C4
Petroglyphs PP ON **2** E3-4
Petroglyphs PP ON **21** D3-4
Petrolia ON **19** C3
Petrolia ON **20** E2
Pettapiece MB **14** D1
Petticoat Creek CA ON
  20 C8
Petticoat Creek CA ON
  21 F2
Petty Harbour NL **37** E8
Petty Harbour-Maddox
  Cove NL **37** D-E8
Pevensey NS **18** C8
Pevensey ON **21** A1
Pfister, lac QC **26** D4
Phare-de-l'Île-Bois Blanc,
  LHN du ON **19** E1
Phare-de-la-Pointe-Clark,
  LHN du ON **20** B2-3
Phare-de-Pointe-au-Père,
  LHN du QC **29** C2-3
Pheasant Rump SK **11** F7
Phelans NL **18** A4-5
Phelpston ON **18** F7
Phelpston ON **20** A6
Philip, River NS **32** C8
Philip, River NS **34** D-E1
Philip Edward Island ON
  18 B5
Philipsburg QC **24** D5
Philipsburg QC **27** E1
Philipsburg, RNOM de QC
  24 E7
Philipsburg, RNOM de QC
  27 F2-3
Philipsburg National
  Wilderness Area QC
  24 E7
Philipsburg National
  Wilderness Area QC
  27 F2-3
Philips Harbour NS **34** E6
Philipsville ON **21** C7
Philipsville ON **24** F1
Phillips Arm BC **1** E5
Phillips Arm BC **2** B2
Phillips Bay YT **5** B3
Phillipsburg ON **19** B5-6
Phillipsburg ON **20** D4
Phillips Grove SK **11** A4
Phillips Head NL **36** A6
Phillips Head NL **37** A4
Phillips Head NL **38** F5-6
Phillips Inlet NU **42** A7
Phillips River BC **2** A-B2
Phinney Cove NS **33** C2-3
Phippen SK **11** C3
Pian, lac QC **23** E2
Piapot SK **11** E2-3
Piapot SK **11** E5-6
Pibroch AB **6** A5
Pibroch AB **8** B6
Pic Chapman QC **27** E5
Pic Island ON **16** F8
Pic Island ON **17** C2
Pic Mobert South ON **16** F8

Pic Mobert South ON **17** C3
  20 F3-4
Pic River ON **16** E8
Pic River ON **16** F8
Pic River ON **17** C2
Pic River ON **17** C2
Picanoc, rivière QC **22** E6-7
Picanoc, rivière QC **24** C1
Picard, lac QC **25** D6
Picard, lac QC **28** F2
Piccadilly NL **36** B1
Pichogen River, RNP ON
  17 C4
Pichogen River PNR ON
  17 C4
Pickardville AB **6** A5
Pickardville AB **8** B6
Pickerel ON **18** B6
Pickerel Lake MB **13** C2
Pickerel Lake ON **16** F4
Pickerel Lake ON **18** C8
Pickerel Lake ON **18** C8
Pickerel Lake ON **21** A1
Pickerel Lake ON **21** A1
Pickerel Lake ON **22** E1
Pickerel Lake ON **22** F1
Pickerel Lake ON **22** F1
Pickerel River ON **18** B6
Pickerel River ON **18** B6-7
Pickerel River ON **17** C7-8
Pickerel River ON **18** B6
Pickering ON **20** C8
Pickering ON **21** F2
Pickering Village ON **20** C8
Pickering Village ON **21** F2
Pickhandle Lake Territorial
  Recreation Site YT
  4 C1-2
Pickle Lake ON **13** D8
Pickle Lake ON **13** D8
Pickle Lake ON **16** B5
Pickle Lake ON **40** E1
Picton ON **21** E5
Pictou NS **34** D3
Pictou Harbour NS **34** D3
Pictou Island NS **34** D3-4
Pictou Landing NS **34** D3
Picture Butte AB **6** E6
Picture Butte AB **7** E5-6
Pied-du-Lac QC **25** E9
Pied-du-Lac QC **31** C1
Piedmont QC **24** C5
Pie Island ON **16** F6
Pierce Lake ON **13** A6
Pierce Lake SK **9** F8
Pierce Lake SK **11** A2
Pierce Lake SK **12** F2
Pierceland SK **9** F8
Pierceland SK **11** A2-3
Pierre, lac QC **25** C4
Pierre, lac QC **28** C1
Pierrefonds QC **24** D5
Pierrefonds QC **27** E1
Pierrefonds, lac QC **23** B8
Pierrefonds, lac QC **26** C2-3
Pierreville QC **24** C7
Pierreville QC **27** D3
Pierriche, rivière QC **28** B7
Pierriche, rivière QC **28** D2
Pierson MB **11** F8
Pierson MB **14** A1
Pigeon Bay ON **19** F2
Pigeon Cove NS **34** B4
Pigeon Hill NB **30** E2-3
Pigeon Hill NB **31** A-B8
Pigeon Hill QC **24** D5
Pigeon Hill QC **27** F2-3
Pigeon Lake AB **6** B5
Pigeon Lake AB **8** D6
Pigeon Lake MB **14** D4
Pigeon Lake ON **21** D3
Pigeon Lake PP AB **6** B5
Pigeon River MB **13** C4-5
Pigeon River ON **16** F5
Pigeon River ON **20** B8
Pigeon River QC **21** E2
Pikangikum ON **13** C6
Pikangikum ON **16** B2
Pikangikum Lake ON
  13 C5-6
Pikangikum Lake ON **16** B2
Pikauba, lac QC **28** E6
Pikauba, rivière QC **28** D5
Pike, Mount YT **4** C6
Pike Bay ON **18** C5
Pike Lake ON **11** C4
Pike Lake PP SK **11** C4
Pike River QC **24** E7
Pike River QC **27** F2
Pikes Arm NL **38** E6-7
Pikianikijuan, lac QC **22** A7
Pikianikijuan, lac QC **25** B1
Pikitigushi River ON **16** C6
Pikogan QC **23** D4
Pikwitonei MB **15** E4
Pikwitonei Lake MB **15** E4
Pilger SK **11** C5
Pilley's Island NL **38** E-F5
Pilot Bay NL **38** E4
Pilot Butte SK **11** E5-6
Pilot Mound MB **13** F2-3
Pilot Mound MB **14** F3
Pilot Mount NL **4** D3
Pimodan, lac QC **24** B3
Pimodan, lac QC **25** F3
Pin, rivière du QC **28** B7
Pinafore Park NMBS ON
  19 D4-5
Pinafore Park NMBS ON
  20 F3-4

Pinafore Park, RNOM ON
  20 F3-4
Pinafore Park, RNOM ON
  19 D4-5
Pinantan Lake BC **6** D1-2
Pinawa MB **13** E4-5
Pinawa MB **14** D7
Pinawa Dam Provincial
  Heritage Park MB **13** E5
Pinawa Dam Provincial
  Heritage Park MB **14** D7
Pin Blanc, lac du QC **22** C4
Pincher AB **6** F5
Pincher AB **7** E4
Pincher Creek AB **6** F5-6
Pincher Creek AB **7** E4
Pinchi Lake BC **1** A6
Pinchi Lake BC **3** F6
Pincourt QC **24** D5
Pine, Cape NL **37** F7
Pine Bluff SK **11** A4
Pinecone Burke PP BC **1** F7
Pinecone Burke PP BC **2** D5
Pine Creek AB **9** F6
Pine Creek AB **12** F1
Pine Creek MB **11** C8
Pine Creek MB **13** C2
Pine Creek NT **10** C4
Pine Creek Station MB
  14 D2-3
Pine Dock ON **13** D4
Pine Dock MB **14** A5
Pine Falls MB **13** E4
Pine Falls MB **14** C6
Pine Grove ON **19** D6
Pine Grove ON **20** F5
Pine Hill QC **24** C4
Pinehouse Lake SK **12** E4
Pinehouse Lake SK **12** E4
Pinehurst NS **32** F6
Pinehurst NS **33** C-D4
Pinehurst Lake AB **9** F7
Pinehurst Lake AB **12** F1
Pinehurst Lake CA ON
  19 B6
Pinehurst Lake CA ON
  20 D-E5
Pineimuta Lake ON **16** A5-6
Pineimuta River ON **16** A5
Pineimuta River ON **16** A5-6
Pine Island SK **11** A7
Pine Island SK **13** A1
Pine Lake AB **6** C5-6
Pine Lake AB **8** F6-7
Pine Lake ON **21** C6
Pine LeMoray PP BC **3** F7
Pine Pass BC **3** F7
Pine Point NT **10** E6-7
Pine Point NT **41** E1
Pine Portage ON **16** E6
Pine Ridge MB **14** C6
Pine Ridge NB **31** E7-8
Pine Ridge NB **32** A6
Pine River BC **3** E7-8
Pine River NL **9** E1
Pine River MB **11** B7
Pine River MB **14** A1
Pine River ON **20** B6
Pine River ON **20** B3
Pine River SK **14** B4-5
Pine River Breaks PP BC
  3 E8
Pine River Breaks PP BC
  9 E1
Pine Tree Point ON **18** E4
Pinetree PE **35** C4
Pinette PE **35** F5
Pinette Harbour PE **35** F5
Pinette Point PE **35** F5
Pinette PP PE **34** C-D3
Pinette PP PE **35** F5
Pineville NB **31** D6
Pinewood ON **16** E1-2
Pinewood AB **8** B6
Piney MB **13** F4
Piney MB **14** F7
Pingualuit, parc QC **40** B5
Pingualuit, parc QC **41** E8
Pinkerton ON **20** B3-4
Pinkham AB **6** D8
Pinkham SK **11** D2
Pink Mountain BC **3** C4
Pink Mountain PP BC **3** C4
Pink River SK **12** D5-6
Pinnacles PP BC **1** B7
Pin Rouge, lac du QC **25** E8
Pins, rivière aux QC **29** B2
Pinsent's Arm NL **39** C8
Pintendre QC **27** B6
Pinware NL **38** A3-4
Pinware NL **39** D8
Pinware River NL **38** A3-4
Pinware River PP NL **38** A4
Pinware River PP NL
  39 D7-8
Pioneer AB **8** C3
Piopolis QC **27** E4
Pipestone AB **6** B5-6
Pipestone MB **11** F8
Pipestone MB **13** F1
Pipestone Bay ON **16** E6
Pipestone Lake MB **13** A4
Pipestone Lake ON **16** E6
Pipestone Lake SK **12** C4-5

Pipestone PP MB **11** F7-8
Pipestone PP MB **13** F1
Pipestone River ON **13** C7-8
Pipestone River ON **16** A4
Pipestone River ON **16** A5
Pipestone River ON **16** F3
Pipestone River ON **40** E1
Pipestone River SK
  12 B5 C4
Pipestone River PP ON
  13 C8
Pipestone River PP ON
  13 B-C8
Pipmuacan, réservoir QC
  40 F5-6
Pipun MB **15** F3
Pirates Cove PMP BC **1** F6-7
Pirates Cove PMP BC **2** E4
Piscatosine, lac QC **22** D7
Piscatosine, lac QC **25** E2-3
Pisew Falls PP MB **15** F3
Pisquid PE **35** D5-6
Pisquid River PE **35** D6
Piste-Chilkoot, LHN de la BC
  3 A1
Piste-Chilkoot, LHN de la BC
  4 F3-4
Pistolet Bay NL **38** A4-5
Pistolet Bay PP NL **38** A4-5
Pistolet Bay PP NL **39** D8
Pitaga NL **39** C2
Pitchimi Lake AB **9** B5
Pitman SK **11** E5
Pitman River BC **3** C4
Pitsitinu-Tugavik TP NU
  42 F8
Pitt Island BC **1** A-B3
Pitt Lake BC **2** D5
Pitt Meadows BC **1** F7
Pitt Meadows BC **2** D-E5
Pittock CA ON **19** C5
Pittock CA ON **20** E4
Pitts Ferry MB **14** A5
Pitts Lake NL **36** C8
Pitts Lake NL **37** C6
Pittston ON **24** E1
Pitukupi Lake ON **17** A4
Piusville PE **35** B1
Pivabiska, Lake ON **17** B4
Pivabiska River ON
  17 A5 B4
Placentia NL **36** E8
Placentia NL **37** E6
Placentia Bay NL **36** E7-8
Placentia Bay NL **37** E5-6
Plages, lac des QC **24** C3
Plainfield ON **21** D5
Plaisance QC **24** D3
Plaisance, baie de NL
  36 E7-8 F7-8
Plaisance, baie de NL
  37 E5-6 F5-6
Plaisance, rivière QC
  30 F7
Plamondon AB **9** F6
Plamondon AB **12** F1
Plantagenet ON **24** D3-4
Plantin, lac QC **22** C3
Plassez, lac QC **22** B-C4
Plaster PP NS **34** B7
Plaster Rock NB **31** D3-4
Plaster Rock-Renous PGA
  NB **31** D4
Plaster Rock-Renous, ZPF
  NB **31** D4
Plate Cove East NL **37** B7
Plate Cove West NL **37** B7
Plato SK **11** D3
Plattsville ON **19** B6
Plattsville ON **20** D5
Playgreen Lake MB **13** A3
Pleasant Bay NS **34** A6
Pleasant Camp BC **4** F3
Pleasant Corners ON **24** D3
Pleasant Creek YT **4** B5
Pleasant Grove PE **34** C2-3
Pleasant Grove PE **35** D4
Pleasant Hills AB **8** E6
Pleasant Lake NS **33** E1-2
Pleasant Point NS **33** C7
Pleasant Point ON **20** A8
Pleasant Ridge NB **31** E7
Pleasant Ridge NB **32** A5
Pledger Lake ON **17** A4
Pleasant River NS **33** D3-4
Pleasant Valley NS **33** B7
Pleasant Valley NS **34** F3
Pleasant Valley PE **35** D4
Pleasantview NL **36** A6
Pleasantview NL **37** A4
Pleasant View PE **35** B1-2
Pleasantville NS **33** D4
Pleasantville ON **21** E1
Plenty SK **11** C3
Plessisville QC **27** C4-5
Plétipi, lac QC **40** E-F6
Plevna ON **21** C5-6

Plongeon, lac au QC
  28 D7-8
Pluie, lac à la ON **13** F6
Pluie, lac à la ON **16** F3
Pluie, rivière à la ON **16** F2
Plumas MB **13** E2
Plumas MB **14** C-D3
Plum Coulee MB **13** F3
Plum Coulee MB **14** F4
Plum Lake MB **11** F8
Plum Lake MB **13** E-F1
Plummer ON **17** F4
Plumper Cove PMP BC **2** D4
Plunkett SK **11** C5
Plymouth NS **33** A7-8
Plymouth NS **33** E1
Plymouth NS **34** E3
Plympton NS **32** F4
Plympton NS **33** D1-2
Plympton Station NS **33** D2
Plympton-Wyoming ON
  19 C3
Plympton-Wyoming ON
  20 E2
Pocahontas AB **6** B3
Pocahontas AB **8** D1
Pocket Knife Lake NL
  39 A-B5
Pocknock, lac QC **22** D7
Pocknock, lac QC **24** B1
Pocknock, lac QC **25** F1
Pocologan NB **32** E3
Poe AB **6** B6
Poe AB **8** D8
Poe AB **11** B1
Pogamasing Lake ON **17** E6
Pohénégamook QC **29** E1-2
Pohénégamook, lac QC
  29 E1-2
Poigan, lac QC **22** B7
Poignan, lac QC **25** D1
Poincaré, lac QC **39** D7
Point Aconi NS **34** C8
Point Alexander ON **22** E4
Point Anne ON **21** E5
Point au Gaul NL **36** F6
Point au Gaul NL **37** F4
Point au Mal NL **36** B2
Point Clark ON **20** B2-3
Point Clark Lighthouse NHS
  ON **20** B2-3
Point Comfort QC **22** E8
Point Comfort QC **24** B-C2
Point Cross NS **34** B6
Point de Bute NB **32** C7
Point de Bute NB **35** F1
Pointe-à-Boisvert QC **29** D4
Pointe-à-la-Croix QC **29** D6
Pointe-à-la-Croix QC
  31 A4-5
Pointe-la-Frégate QC
  30 A2
Pointe-à-la-Garde QC
  29 D6
Pointe-à-la-Garde QC
  31 A5
Pointe au Baril ON **18** C6
Pointe au Baril Station ON
  18 C6-7
Pointe-au-Chêne QC **24** D4
Pointe-au-Père QC **29** C3
Pointe-au-Père, RNF de QC
  29 C2-3
Pointe-au-Père Lighthouse
  NHS QC **29** C2-3
Pointe-au-Père NWA QC
  29 C2-3
Pointe-aux-Anglais QC
  39 F1-2
Pointe-aux-Loups QC **30** F7
Pointe-aux-Outardes QC
  29 A3
Pointe-aux-Outardes, parc
  régional de QC **29** A3-4
Pointe aux Pins ON **17** C3-4
Pointe aux Pins ON **19** E3
Pointe-Calumet QC **24** D5
Pointe-Canot NB **31** B8
Pointe-Claire QC **24** D5
Pointe-Claire QC **27** E1
Pointe-de-l'Est, RNF de la
  QC **30** F8
Pointe de l'Est NWA QC
  30 F8
Pointe-des-Cascades QC
  24 D5
Pointe-des-Cascades QC
  27 E1
Pointe-des-Monts QC **39** E5
Pointe du Bois MB **13** E5
Pointe du Bois MB **14** D7
Pointe-du-Chêne NB
  32 B6-7
Pointe-du-Lac QC **25** F7-8
Pointe-du-Lac QC **27** C3
Point Edward ON **19** C2
Point Edward ON **20** E1
Pointe-Fortune QC **24** D4-5
Pointe-Jaune QC **30** B2-3
Pointe-Lebel QC **29** A3-4
Pointe-Parent QC **39** F5
Pointe-Pelée, PN de la QC
  19 F2
Pointe-Pelée, PN de la ON
  19 F2
Pointe-Prince-Edward, RNF
  de la ON **21** E-F6
Pointe-Saint-Pierre QC
  30 C3
Pointe-Sapin NB **31** D8
Point Escuminac NB **31** D8

Pointe-Taillon, parc de la QC 28 B4
Pointe-Verte NB 29 E7-8
Pointe-Verte NB 31 B6
Point Farms PP ON 19 A3-4
Point Farms PP ON 20 C2-3
Point La Haye NL 37 F7
Point La Nim NB 29 D6
Point La Nim NB 31 A5
Point Lake NT 10 A7
Point Lake NT 41 C2
Point Lance NL 36 F8
Point Lance NL 37 F6
Point Leamington NL 36 A6
Point Leamington NL 37 A4
Point Leamington NL 38 F5
Point May NL 36 F5-6
Point May NL 37 F3-4
Point Michaud NS 34 D7
Point Michaud Beach PP NS 34 D7
Point of Bay NL 36 A6
Point of Bay NL 37 A1
Point of Bay NL 38 F5-6
Point Pelee NP ON 19 F2
Point Pelee NP ON 19 F2-3
Point Prim PE 34 C2-3
Point Prim PE 35 F5
Point Rosie NL 36 F8
Point Rosie NL 37 F4
Points North Landing SK 12 B6
Point Tupper NS 34 D6
Point Verde NL 36 E8
Point Verde NL 37 E6
Point Wolfe River Gorge PCA NB 32 D5
Point Wolfe River Gorge PCA NB 33 A2-3
Point Wolfe River Gorge, ZCP NB 32 D5
Point Wolfe River Gorge, ZCP NB 33 A2-3
Poissant QC 24 A3
Poissant QC 25 E3
Poisson Blanc, lac QC 22 F8
Poisson Blanc, lac QC 24 C2-3
Poisson Blanc, lac (réservoir) QC 22 E8
Poisson Blanc, lac (réservoir) QC 24 C2
Poissons River NL 39 C3
Poivre, lac au QC 28 A6
Pokei Lake-White River Wetlands PNR ON 17 C-D3
Pokei Lake-White River Wetlands, RNP ON 17 C-D3
Pokemouche NB 30 E2
Pokemouche NB 31 B8
Pokemouche River NB 29 E8
Pokemouche River NB 30 F1
Pokemouche River NB 31 B7
Pokeshaw NB 29 E8
Pokeshaw NB 30 E1
Pokeshaw NB 31 B7
Pokesudie NB 31 B8
Pokesudie, île NB 30 E2
Pokesudie Island NB 30 E2
Pokiok NB 32 C2
Poland ON 21 B6
Polar Bear PP ON 40 D2-3
Police Island NT 5 F8
Police Island NT 10 A2
Police Outpost PP AB 6 F6
Police Outpost PP AB 7 F4-5
Pollards Point NL 38 E3-4
Pollards Point NL 39 F8
Pollett River NB 32 C5
Pollock Point NS 33 D4
Pollockville AB 6 D7
Pollockville AB 7 B7
Pollockville AB 11 D1
Polonais, lac des QC 25 D3
Polonia MB 13 E2
Polonia MB 14 C2
Poltimore QC 22 F8
Poltimore QC 24 C2
Polwarth SK 11 B4
Pomeroy MB 14 F4
Pomeroy NB 32 D2
Pommeroy, lac QC 22 B3
Pomquet NS 34 D4-5
Pomquet Beach PP NS 34 D4-5
Ponask Lake ON 13 A6-7
Ponask River ON 13 A6
Ponass Lake SK 11 C6
Pond Cove NL 38 B3
Pond Inlet NU 41 A6-7
Pond Inlet NU 42 B6
Pond Inlet (Mittimatalik) NU 41 A6
Pond Inlet (Mittimatalik) NU 42 D-E6
Pondosy Lake BC 1 B4-5
Ponds NS 34 D4
Pondville Beach PP NS 34 E6-7
Ponemah MB 13 E4
Ponemah MB 14 C5
Ponhook Lake NS 32 E7-8 F7
Ponhook Lake NS 33 C5
Ponoka AB 6 E6
Ponoka AB 8 E6
Pontbriand QC 27 C5

Pont-de-Béton-en-Arc-du-Lac-Canal, LHN du ON 20 A8
Pont-de-Béton-en-Arc-du-Lac-Canal, LHN du ON 21 D2
Ponteix SK 11 F3-4
Pontgrave NB 31 C8
Pontiac QC 21 A7
Pontiac QC 22 F7
Pontiac, ZEC QC 24 D1
Pontiac, ZEC QC 22 D6
Pont-Lafrance NB 30 F1-2
Pont-Lafrance NB 31 C7
Pont-Landry NB 30 F2
Pont-Landry NB 31 B7
Ponton MB 15 F3
Ponton River AB 9 B4
Pontrilas SK 11 B6
Pont-Rouge QC 27 B5
Pontypool ON 20 B8
Pontypool ON 21 E2-3
Poodiac NB 32 D5
Poodiac NB 33 A2
Poole ON 19 B5
Poole ON 20 C4
Poole's Cove NL 36 D6
Poole's Cove NL 37 D4
Poole's Corner PP PE 34 C3
Poole's Corner PP PE 35 E6
Pooley Island BC 1 B3-4
Pool's Cove NL 36 D6
Pool's Cove NL 37 D4
Pool's Island NL 37 A7
Pool's Island NL 38 F8
Poor Man SK 11 D5
Pope, lac QC 22 D8
Pope, lac QC 24 A2
Pope, lac QC 25 E2
Pope, RNF MB 11 E8
Pope, RNF MB 13 E1
Pope NWA MB 11 E8
Pope NWA MB 13 E1
Pope's Harbour NS 33 C8
Popham Bay NU 40 A6
Popkum BC 2 E6-7
Poplar ON 18 C2-3
Poplar Dale ON 17 F4
Poplarfield MB 13 D3-4
Poplarfield MB 14 B4
Poplar Grove PE 32 A7-8
Poplar Grove PE 34 B1
Poplar Grove PE 35 C2
Poplar Hill ON 13 C5-6
Poplar Hill ON 16 A2
Poplar Hill ON 19 C4
Poplar Hill ON 20 E3
Poplar/Nanowin Rivers PP Reserve MB 13 B4
Poplar Park MB 14 D5
Poplar Point MB 13 B3-4
Poplar Point MB 14 D4
Poplar Point PE 35 E7
Poplar Ridge AB 8 D4
Poplar River MB 13 B4 C5
Poplar River MB 14 B4
Poplar River NT 10 E3
Porc-Épic, lac au QC 28 D7
Porcher Island BC 1 A2
Porcupine ON 17 C6
Porcupine, Cape 39 B7
Porcupine Meadows PP BC 1 D8
Porcupine Meadows PP BC 2 A8
Porcupine Meadows PP BC 7 F4-5
Porcupine Plain SK 11 C6
Porcupine River NT 10 A2
Porcupine River SK 12 A5-6
Porcupine River YT 5 D3
Porphyry Island PP ON 16 F6-7
Porpoise Bay PP BC 2 D4
Porquis Junction ON 17 C6-7
Portage, lac du QC 27 D7
Portage, rivière du QC 30 C2-3
Portage Bay MB 13 D3
Portage Bay MB 14 A3
Portage Bay NT 10 B8
Portage-du-Fort QC 21 A6
Portage-du-Fort QC 22 F6
Portage Island NB 31 C7-8
Portage Island NWA NB 31 C7-8
Portage la Prairie MB 13 E3
Portage la Prairie MB 14 D3
Portage Lake MB 13 C3
Portage Lake NL 37 C1
Portage Vale NB 32 C5
Port Alberni BC 1 F6
Port Alberni BC 2 D-E2
Port Alberni Inlet BC 2 E2
Port Albert NL 38 E7
Port Albert ON 19 A3-4
Port Albert ON 20 C2-3
Port Albion BC 1 F5
Port Albion BC 2 E1-2
Port Alice BC 1 E4
Port Alma ON 19 E2-3
Port Anne NL 36 E7
Port Anne NL 37 E5
Port Anson NL 38 E5
Port Anson ON 18 C8
Port Anson ON 21 A1
Portapique NS 32 D8
Portapique NS 33 A5-6
Portapique NS 34 E1-2
Port au Choix, LHN de NL 38 C2-3
Port au Choix, LHN de NL 39 E7

Port au Choix NHS NL 38 C2-3
Port au Choix NHS NL 39 E7
Port-au-Persil QC 28 E8
Port au Port NL 36 C2
Port au Port NL 40 F8
Port au Port Bay NL 36 B2
Port au Port East NL 36 C2
Port au Port Peninsula NL 36 B-C1
Port au Port Peninsula NL 40 F8
Port au Port West-Aguathuna-Felix Cove NL 36 C1-2
Port-aux-Quilles QC 28 D8
Port Bickerton NS 34 F5
Port Blandford NL 36 C8
Port Blandford NL 37 C6
Port Bolster ON 20 B7
Port Bolster ON 21 D1-2
Port Bruce ON 19 D5
Port Bruce ON 20 F4
Port Bruce PP ON 19 D5
Port Bruce PP ON 20 F4
Port Burwell ON 19 D5
Port Burwell ON 20 F4
Port Burwell NL 40 A6
Port Burwell PP ON 19 D5
Port Burwell PP ON 20 F4
Port Carling ON 18 D-E8
Port Carling ON 21 B1
Port Carmen ON 18 C8
Port Carmen ON 21 A1
Port-Cartier QC 39 F1-2
Port-Cartier QC 40 F6
Port-Cartier-Sept-Îles, RF de QC 39 E1-2
Port Clements BC 1 A1
Port Clyde NS 33 F2
Port Clyde NS 33 F2
Port Colborne ON 19 C8
Port Colborne ON 20 F7
Port Coquitlam BC 1 F7
Port Coquitlam BC 2 D5
Port Credit ON 19 B8
Port Credit ON 20 D7
Port Cunnington ON 21 B2
Port Dalhousie ON 19 B-C8
Port Dalhousie ON 20 E7
Port-Daniel QC 30 D1
Port-Daniel QC 31 A7-8
Port-Daniel, baie de QC 30 D2
Port-Daniel, RF de QC 30 D2
Port-Daniel Nord, rivière QC 30 D1-2
Port Darlington ON 21 F2-3
Port de Grave NL 37 D7-8
Port Douglas BC 2 C6
Port Dover ON 19 D6-7
Port Dover ON 20 F5-6
Port Dufferin NS 33 C8
Port Dufferin NS 34 F4
Port Edward BC 1 A2
Port Elgin NB 32 B7
Port Elgin NB 34 D1
Port Elgin NB 35 E2
Port Elgin ON 18 F4
Port Elgin ON 20 A3
Port Eliza BC 1 E4
Port Elmsley ON 21 C7
Port Elmsley ON 24 F1
Porter Cove NB 31 E5
Porter Cove NB 32 A3
Porter Creek YT 4 E3-4
Porter Lake SK 12 D4
Porters Lake NS 33 C6
Porters Lake PP NS 33 C6
Porterville NL 36 A6-7
Porterville NL 37 A4-5
Porterville NL 38 F6
Port Essington BC 1 A3
Port Felix NS 34 E6
Port Franks ON 19 B3
Port Franks ON 20 D2
Port George NS 32 E6
Port George NS 33 B3
Port Glasgow ON 19 D4
Port Glasgow ON 20 F3
Port Greville NS 32 D7
Port Greville NS 33 A4
Port Hardy BC 1 E4
Port Hastings NS 34 D6
Port Hawkesbury NS 34 D6
Port Hébert, RNOM de NS 33 E4-5
Port Hill PE 32 A8
Port Hill PE 34 B1
Port Hill PE 35 C2-3
Port Hillford NS 34 F5
Port Hood NS 34 C5
Port Hood Station PP NS 34 C5
Port Hope ON 21 F3
Port Hope Simpson NL 39 C8
Port Howe NS 32 C8
Port Howe NS 34 D1
Portia MB 14 B2
Port Joli NS 33 F4
Port Joli Head NS 33 F4
Port Joli NMBS NS 33 E4-5
Port Joli, RNOM de NS 33 E4-5
Port Kirwan NL 37 D8
Port-la-Joye-Fort-Amherst, LHN de PE 34 C2-3

Port-la-Joye–Fort-Amherst, LHN de PE 35 E5
Port-la-Joye-Fort Amherst NHS PE 34 C2-3
Port-la-Joye-Fort Amherst NHS PE 35 E5
Port Lambton ON 19 D2
Port Lambton ON 20 F1
Portland NL 37 C6-7
Portland ON 21 C7
Portland ON 24 F1
Portland CA ON 21 D6
Portland Canal BC 3 F2-3
Portland Creek NL 39 E-F7
Portland Creek Pond NL 38 D3
Portland Creek Pond NL 39 E-F7
Portland Inlet BC 3 F2-3
Port la Tour NS 33 F2-3
Port Lewis QC 24 E5
Port l'Hébert NS 33 E5
Port l'Hébert NMBS NS 33 E3-5
Port Loring ON 18 B7
Port Lorne NS 32 E6
Port Lorne NS 33 B3
Port Louis BC 1 A1
Port Maitland NS 33 E1
Port Maitland ON 19 C8
Port Maitland ON 20 F7
Port Maitland Beach PP NS 33 E1
Port McNeill BC 1 E4
Port McNicoll ON 18 E-F7
Port McNicoll ON 20 A6
Port Malcolm NS 34 E6
Port Medway NS 33 D-E4
Port Mellon BC 1 F6
Port Mellon BC 2 D4
Port-Menier QC 39 F3
Port-Menier QC 40 F7
Port Moody BC 1 F7
Port Moody BC 2 D5
Port Morien NS 34 C8
Port Mouton NS 33 E4
Port Mouton NS 33 E3-4
Port Mouton Head NS 33 E4
Port Nelson MB 15 D6
Port Nelson MB 15 D6
Port Neville BC 1 E5
Port Neville BC 2 B1
Port Neville BC 2 B1
Portobello Creek, RNF NB 32 C3
Portobello Creek National Wilderness Area NB 32 C3
Port of Big Beaver SK 11 F5
Port of Carievale SK 11 F7-8
Port of Carievale SK 13 F1
Port of Climax SK 11 F3
Port of Coronach SK 11 F5
Port of Estevan Hwy. SK 11 F6
Port of Oungre SK 11 F6
Port of Torquay SK 11 F6
Port Perry ON 20 B8
Port Perry ON 21 E2
Port Philip NS 32 C8
Port Philip NS 34 D1
Portree NS 34 B6
Portreeve SK 11 D3
Port Renfrew BC 2 F3
Port Rexton NL 37 C7
Port Robinson ON 19 C8
Port Robinson ON 20 E7
Port Rowan ON 19 D6
Port Rowan ON 20 F5
Port Royal NS 32 F4-5
Port Royal NS 33 C2
Port Royal ON 19 D6
Port Royal ON 20 F5
Port-Royal, LHN de NS 32 F4-5
Port-Royal, LHN de NS 33 C1-2
Port-Royal NHS NS 32 F4-5
Port-Royal NHS NS 33 C1-2
Port Ryerse ON 19 D6-7
Port Ryerse ON 20 F5-6
Port-Saint-François QC 24 B7-8
Port-Saint-François QC 25 F7-8
Port-Saint-François QC 27 C3
Port San Juan BC 2 F3
Port Sandfield ON 18 D-E8
Port Saunders NL 38 C3
Port Saunders NL 39 E7-8
Port Saxon NS 33 F3
Port Severn ON 18 E8
Port Severn ON 21 C1
Port Shoreham NS 34 E5-6
Port Shoreham Beach PP NS 34 E6
Port Stanley ON 19 D4-5
Port Stanley ON 20 F4
Port Stanton ON 18 E8
Port Sydney ON 21 B1-2
Portugal Cove NL 37 D8
Portugal Cove South NL 37 F7-8

Portuguese Cove NS 33 C6
Port Union NL 37 B7-8
Port Wade NS 32 F4
Port Wade NS 33 C2
Port Williams NS 32 E7
Port Williams NS 33 B4
Postville NL 39 A6
Postville NL 40 D8
Poterie Inférieur, lac QC 25 D5
Poterie Inférieur, lac QC 27 A1
Poterie Inférieur, lac QC 28 E1
Poterie Supérieur, lac QC 25 D6
Poterie Supérieur, lac QC 27 A1
Poterie Supérieur, lac QC 28 F1
Potholes, RNP ON 17 D4
Potholes PNR ON 17 D4
Potomac River ON 18 A1-2
Pottageville ON 20 C6-7
Pottageville ON 21 E1
Pottawatomi CA ON 18 F5
Pottawatomi CA ON 20 A4
Potters Landing ON 18 E7
Pottles Bay NL 39 A7
Potts Lake AB 9 A7
Potts Lake AB 12 A2-3
Pouce Coupe BC 3 E8
Pouce Coupe BC 9 E2
Pouce Coupe River AB 9 E2
Pouch Cove NL 37 D8
Pougnet, lac QC 26 C3-4
Poularies QC 17 C8
Poularies QC 23 C2-3
Poulin-de-Courval, lac QC 28 B7
Poulter, lac QC 22 B6
Poutret, lac QC 39 B2
Poutrincourt, lac QC 26 C5
Poutrincourt, lac QC 28 A1
Poverty Cape PE 35 C3
Povungnituk, rivière de QC 40 B4-5
Povungnituk, rivière de QC 41 E8
Povungnituk Bay 40 C4
Povungnituk Bay QC 41 E7-8
Powassan ON 17 F8
Powassan ON 18 B8
Powassan ON 22 E1
Powell ON 11 B7
Powell MB 13 B1
Powell Lake BC 1 E6
Powell Lake BC 2 C3
Powell Peak BC 1 A-B4
Powell River BC 1 E-F6
Powell River BC 2 C3
Powells Point PP NS 34 D3-4
Powerscourt QC 24 E5
Powerview MB 13 E4
Powerview MB 14 C6
Pownal PE 35 E5
Pownal Bay PE 35 E5
Prairie River SK 11 B6
Prairie River Mouth, RNP ON 16 E7-8
Prairie River Mouth, RNP ON 17 C2
Prairie River Mouth PNR ON 16 E7-8
Prairie River Mouth PNR ON 17 C2
Prairies, Lake of the SK 11 D7
Prairies, Lake of the SK 13 D1
Prairies, PN des SK 11 F3-4
Prairies, rivière des QC 24 D6
Prairies, rivière des QC 27 E1
Prairie Siding ON 19 E2
Prairie View SK 11 E4
Prairie Creek NT 10 D-E2
Pratt MB 14 E3
Prawda MB 13 F5
Prawda MB 14 E7
Preacher Davis River NL 37 A6
Preacher Davis River NL 38 F5
Preble Island NT 10 D7
Precious Corners ON 21 E6
Pré-d'en-Haut NB 31 F8
Pré-d'en-Haut NB 32 B-C6
Preeceville SK 11 C7
Preissac QC 17 C8
Preissac QC 23 D3
Preissac, lac QC 23 D3
Prelate SK 6 D8
Prelate SK 11 D2
Prelude Lake NT 10 C7
Prelude Lake TP NT 10 C6-7
Premier Lake BC 12 A5
Premier Lake PP BC 6 E4
Premier Lake PP BC 3 D-E2
Premier-Puits-de-Pétrole-de-l'Ouest-Canadien, LHN du AB 7 F3-4
Premio QC 39 E2
Prentiss AB 8 E-F6
Prescott ON 21 C8

Prescott ON 24 F2
Prescott Island BC 1 A2
Prescott Island NU 41 A4
Prescott Island NU 42 D1
Presqu'île NS 34 B6
Prés du lac Grand, APP NB 32 C3-4
Prespatou BC 3 D8
Prespatou BC 9 D1
Presqu'île Point ON 21 E4
Presqu'île PP ON 21 E4
Press Lake ON 13 E7-8
Press Lake ON 16 A4
Preston NS 33 C6
Preston, lac QC 24 B3
Preston Lake ON 20 C7
Preston Lake ON 21 E1
Preston Lake SK 9 C8
Preston Lake SK 12 C3
Pretty River Valley PP ON 18 F6
Pretty River Valley PP ON 20 A5
Prévost QC 24 C5
Prévost, lac QC 25 D6
Prevost Island Unit (Gulf Islands NPR) BC 2 E4
Prevost River YT 4 E3
Price QC 29 B-C3
Price Island BC 1 C3
Prices Corner ON 18 F8
Prices Corner ON 20 A7
Prices Corner ON 21 D1
Price Settlement NB 30 F1
Price Settlement NB 31 C7
Priceville NB 31 E5
Priceville NB 32 A3
Priceville ON 20 B5
Priddis AB 6 D5
Priddis AB 7 C3
Priest Pond PE 34 B4
Priest Pond PE 35 C-D2
Prim, Point PE 34 C-D2
Prim, Point PE 35 F5
Primate SK 6 C8
Primate SK 11 C2
Primeau Lake SK 12 E4
Primeau Lake SK 12 E4
Primeau, lac QC 23 E-F8
Primeau, lac QC 25 B2
Primeau, lac QC 26 F2
Prim Point NS 34 F4
Prim Point NS 33 C2
Primrose ON 20 B6
Primrose PE 35 E6
Primrose Lake SK 9 F8
Primrose Lake SK 12 F2-3
Primrose Lake YT 4 E3
Primrose Lake Provincial Ecological Reserve SK 9 F8
Primrose Lake Provincial Ecological Reserve SK 12 F2-3
Primrose River YT 4 E3
Prince SK 11 B3
Prince Albert ON 20 B8
Prince Albert ON 21 E2
Prince Albert SK 11 B5
Prince Albert, PN de SK 11 A4
Prince Albert NP SK 11 A4
Prince Albert Peninsula NT 41 A2
Prince Albert Peninsula NT 42 D1-2 E1
Prince Albert Sound NT 41 A2
Prince Albert Sound NT 42 E1-2
Prince Charles Island NU 41 B-C7
Prince Charles Island NU 42 F6
Prince-de-Galles, cap du QC 40 B5
Prince-de-Galles, cap du QC 27 E1
Prince-de-Galles, île du NU 41 A4
Prince-de-Galles, île du NU 42 E3-4
Prince Edward Bay ON 21 E6
Prince Edward County ON 21 E5
Prince Edward Island NP PE 34 B2-3
Prince Edward Island NP PE 34 B3
Prince Edward Island NP PE 35 D4-5
Prince Edward Island NP PE 35 F3
Prince Edward Point NWA ON 21 E-F6
Prince George BC 1 B7
Prince George BC 3 F5-6

Prince of Wales Strait NT 41 A2
Prince of Wales Strait NT 42 D1
Prince Patrick Island NT 42 C2
Prince Regent Inlet NU 41 A5
Prince Regent Inlet NU 42 D-E4
Prince Rupert BC 1 A2-3
Prince Rupert BC 3 F3
Princess AB 6 D7
Princess Harbour MB 13 C4
Princess Louisa Inlet BC 2 B-C4
Princess Louisa PMP BC 1 E6
Princess Louisa PMP BC 2 B-C4
Princess Marie Bay NU 42 B5
Princess Park NB 32 C4
Princess Royal Channel BC 1 B3
Princess Royal Island BC 1 B3
Princeton BC 1 F8
Princeton BC 2 D8
Princeton BC 6 F1
Princeton NL 37 C7
Princeton ON 19 C6
Princeton ON 20 E5
Princeville QC 27 C4-5
Prince William NB 32 C2
Principal, lac du QC 26 D4
Principe Channel BC 1 B2-3
Pringle Lake ON 16 B2
Prinsta, lac QC 30 A7
Prinsta, rivière QC 30 A6-7
Prior Centennial Unit (Gulf Islands NPR) BC 2 E4-5
Pritchard BC 2 C2
Pritchard PP BC 6 D-E2
Procter BC 6 F3-4
Profits Point PE 32 A8
Profits Point PE 34 B1
Profitts Point PE 35 C3
Progress BC 3 E8
Progress BC 9 E1-2
Prongua SK 11 B3
Pronto East ON 17 F5
Pronto East ON 18 B2
Prophet River BC 3 C7
Prophet River BC 3 C7
Prophet River Hotsprings PP BC 3 C6
Prophet River Wayside PP BC 3 C7
Prosperous Lake NT 10 C6
Prosser Brook NB 32 C5
Proton Station ON 20 B5
Proulx, lac QC 28 A-B4
Proulx Lake ON 22 E-F3
Provancher, lac QC 23 E2
Providence, Lake NT 10 A7
Providence Bay ON 18 C3
Province House, LHN PE 34 C2
Province House, LHN PE 35 E5
Province House NHS PE 34 C2
Province House NHS PE 35 E5
Provost AB 6 C8
Provost AB 11 C2
Prud'homme SK 11 C5
Prudhomme Lake PP BC 1 A3
Ptarmigan Bay ON 14 E8
Ptarmigan Creek PP BC 1 B8
Ptarmigan Creek PP BC 6 B1
Pubnico NS 33 F2
Pubnico Harbour NS 33 F2
Puce ON 19 E1-2
Puff Island, RNP ON 16 F6-7
Puff Island, RNP ON 17 C1
Puff Island PNR ON 16 F6-7
Puff Island PNR ON 17 C1
Pugwash NS 32 C8
Pugwash NS 34 D1-2
Pugwash Harbour NS 35 F3
Pugwash Junction NS 32 C8
Pugwash Junction NS 34 D1-2
Pukaskwa, PN 16 F8
Pukaskwa, PN ON 17 D3
Pukaskwa NP 16 F8
Pukaskwa NP ON 17 D3
Pukaskwa River ON 16 F8
Pukaskwa River ON 17 D3
Pukaskwa River PP ON 17 D3
Pukatawagan MB 15 E1-2
Pukatawagan MB 15 E1
Pukatawagan (IR) MB 12 E7
Pukatawagan (IR) MB 15 E1
Pukatawagan Lake MB 12 E7
Pukatawagan Lake MB 15 E1

Pukeashun Mountain BC 6 D2
Pullen Island NT 5 A4-5
Pulp River MB 11 C8
Pulp River MB 13 C2
Pulp River MB 14 A1
Punnichy SK 11 D6
Puntchesakut Lake PP BC 1 B6-7
Puntledge River BC 1 E-F5
Puntzi Lake BC 1 C6
Purbeck's Cove NL 38 E4
Purcell Lake ON 16 B8
Purcell Mountains BC 6 E4
Purcell Mountains BC 7 C-D1
Purcell Wilderness Provincial Conservancy BC 6 E4
Purden Lake PP BC 1 A7
Purdon CA ON 21 C6
Purdy ON 21 B4
Pure Lake PP BC 1 A1
Purple Springs AB 6 E7
Purple Springs AB 7 E6-7
Purple Springs AB 11 E1
Purple Valley ON 18 E5
Purtuniq QC 40 B5
Purtuniq QC 41 D8
Purves MB 14 F3
Pushkin Hills, RNP ON 17 D7-8
Pushkin Hills, RNP ON 23 D1
Pushkin Hills PNR ON 17 D7-8
Pushkin Hills PNR ON 23 D1
Puskwakau River AB 9 E3
Puslinch ON 19 B7
Puslinch ON 20 D6
Pusticamica, lac QC 23 B7
Pusticamica, lac QC 26 C1-2
Putahow Lake MB 12 A7
Putahow Lake MB 15 A2
Putahow River MB 12 A7
Putahow River MB 15 A2
Putnam ON 19 C5
Putnam ON 20 E4
Puvirnituq QC 40 C4
Puvirnituq QC 41 E7-8
Puyjalon, lac QC 39 E4
Puzzle Lake PP ON 21 D6
Pynn's Brook NL 36 A3-4
Pynn's Brook NL 37 A1-2
Pynn's Brook NL 38 F3
Pyramid Falls PP BC 6 C2
Pyramid Mountain AB 6 B3
Pythonga, lac QC 22 D7
Pythonga, lac QC 24 B1
Pythonga, lac QC 25 F1

Q

Qamanit'uaq (Baker Lake) NU 41 D4-5
Qasuittuq (Resolute) NU 42 D4
Qaummaarviit, parc territorial NU 40 A5
Qaummaarviit TP NU 40 A5
Qaummaarviit TP NU 41 C-D8
Qikiqtarjuaq (Broughton Island) NU 42 E-F8
Qilaluqat TP NU 41 A6
Qilaluqat TP NU 42 D5-6
Qu'Appelle SK 11 E6
Qu'Appelle River SK 11 E6
Quaco Head NB 32 D5
Quaco Head NB 33 A2
Quadeville ON 21 B5
Quadra Island BC 1 E5
Quadra Island BC 2 B-C2
Qualicum BC 1 B5
Qualicum Beach BC 1 F6
Qualicum Beach BC 2 D2
Qualicum NWA BC 2 D2
Qualicum NWA BC 2 D3-4
Quantock SK 11 F4
Quaqtaq QC 40 B5
Quarry Bay ON 18 B1
Quarryville NB 31 D6
Quartz Lake ON 16 B8
Quathiaski Cove BC 1 E5-6
Quathiaski Cove BC 2 B2
Quatsino BC 1 E4
Quatsino PP BC 1 E4
Quatsino Sound BC 1 E3-4
Quatsino Subdivision BC 1 E4
Québec QC 27 B5
Québec QC 27 B6
Québec, arrondissement historique de QC 27 B5
Québec, Historic District of QC 27 B6
Queen Bess, Mount BC 1 D6
Queen Charlotte City BC 1 B1
Queen Charlotte Islands BC 1 B1
Queen Charlotte Sound BC 1 D-E4
Queen Charlotte Strait BC 1 D-E4

Queen Elizabeth Islands NT
    42 A3-4 B2-3
Queen Elizabeth Islands NU
    42 A3-4 B2-3
Queen Elizabeth PP AB
    9 D-E3
Queen Elizabeth TP AB
    10 F8
Queen Elizabeth TP NT
    41 E1-2
Queen Elizabeth II
  Wildlands PP ON 21 C2
Queen Mary, Mount YT 4 E1
Queen Maud Gulf
    41 C3-4
Queen Maud Gulf NU 42 F3
Queens NS 33 D3
Queensborough ON 21 D5
Queen's Cove NL 36 D8
Queen's Cove NL 37 D6
Queensland NS 32 F8
Queensland NS 33 C5
Queensland Beach PP NS
    32 F7-8
Queensland Beach PP NS
    33 C5
Queensport NS 34 E6
Queens Reach BC 2 C3-4
Queens Sound BC 1 C3
Queenston ON 20 E7-8
Queenston Heights NHS ON
    20 E8
Queenstown AB 6 E6
Queenstown BC 7 C5
Queenstown NB 32 C3-4
Queensville NS 34 D6
Queensville ON 20 B7
Queensville ON 21 E1
Quenouilles, lac aux QC
    24 B4
Quesnel BC 1 B7
Quesnel Lake BC 1 C8
Quesnel Lake BC 6 B-C1
Quesnel River BC 1 C7-8
Quetachou, baie QC
    39 E-F4
Quetico Lake ON 16 F3
Quetico PP ON 16 F4
Quetico PP ON 40 F1
Quévillon, lac QC 23 B6
Quibell ON 13 E6
Quibell ON 16 D2
Quidi Vidi NL 37 D8
Quiet Lake YT 4 D-E5
Quiet Lake Territorial
  Recreation Site YT
    4 D4-5
Quilchena BC 2 C8
Quilchena BC 6 E1
Quill Lake SK 11 C5-6
Quinan NS 33 E2
Quinn, lac QC 22 D8
Quinn, lac QC 24 A2
Quinn, lac QC 25 F2
Quinn Lake MB 15 B4
Quinn Lake NL 36 C4
Quinn Lake NL 37 C2
Quinsam BC 2 C2
Quinte, Bay of ON 21 E5
Quinte CA ON 21 E5
Quinte West ON 21 E4-5
Quinton SK 11 D5
Quinze, lac des QC 17 D8
Quinze, lac des QC 23 F2
Quirke Lake ON 17 F5
Quirke Lake ON 18 A3
Quirpon NL 38 A5
Quirpon Island NL 38 A5
Quisibis NB 29 F3-4
Quisibis NB 31 C2
Quisibis River NB 29 E-F3
Quispamsis NB 32 D4
Quispamsis NB 33 A1-2
Quoich River NU 41 D5
Quottoon Inlet BC 1 A3
Quttinirpaaq (Ellesmere
  Island) NP NU 42 A5
Quyon QC 21 A7
Quyon QC 22 F7
Quyon QC 24 D1

R

Raanes Peninsula NU
    42 B-C4
Rabbabou Bay SK 12 B6-7
Rabbabou Bay SK 15 B1
Rabbit, lac QC 22 A3
Rabbit, lac QC 23 F3
Rabbit Island NT 5 A7-8
Rabbit Island ON 18 C4
Rabbit Lake ON 17 E8
Rabbit Lake ON 22 B1
Rabbit Lake SK 11 B3
Rabbit Lake Mine SK
    12 B-C6
Rabbit Lake Mine SK 41 F3
Rabbit River BC 3 B4-5
Rabbit River BC 4 F7
Rabbitskin River NT 10 D4
Raccoon Lake NT 10 C5
Race Rocks, zone de
  protection marine de BC
    2 F3-4
Race Rocks Marine
  Protected Area BC 2 F3-4
Racine QC 24 D8
Racine QC 27 E4
Racine de Bouleau, rivière
  de la QC 39 C-D1
Racine Lake ON 17 D5

Racing Island NT 5 A8
Racing River BC 3 B5-6 C6
Rackham MB 11 E8
Rackham MB 13 E2
Rackham MB 14 C1
Rackla River YT 4 A4
Radiant Lake ON 22 E3
Radilih Koe (Fort
  Good Hope) NT 5 E7
Radisson QC 40 E4
Radisson SK 11 C4
Radisson Lake SK 11 C4
Radium Hot Springs BC
    6 E4
Radium Hot Springs BC
    7 C1
Radville SK 11 F6
Radway AB 6 A6
Radway AB 8 B7
Rae NT 10 C6
Rae NT 41 D1
Rae Creek YT 4 A3
Rae Lake NT 10 B5
Rae Lakes NT 10 B5
Rae Lakes NT 41 D2
Rae Strait NU 41 B4-5 C4
Rae Strait NU 42 F4
Rafferty Dam SK 11 F6
Rafferty Dam Reservoir SK
    11 F6
Rafter MB 12 E7
Rafter MB 15 E1-2
Ragged Falls/Oxtongue
  River PP ON 21 B2
Ragged Harbour NL 38 F7
Ragged Islands NL 39 A7
Ragged Lake ON 21 A2
Ragged Lake ON 22 D2
Raglan ON 20 B-C8
Raglan ON 21 E2
Ragueneau QC 29 A3
Ragueneau, rivière QC
    29 A3
Raimbault, lac QC 39 B-C1
Raimbault, rivière QC
    25 A7
Raimbault, rivière QC
    26 E-F7
Raimbault, rivière QC
    28 C2-3
Rain Lake ON 21 A2
Rainbow AB 7 A6
Rainbow Alley PP BC 3 F5
Rainbow Beach PP MB
    13 D2
Rainbow Beach PP MB
    14 B1-2
Rainbow Falls PP ON 16 E7
Rainbow Falls PP ON
    17 C1-2
Rainbow Haven
  Beach PP NS 33 C6-7
Rainbow Lake AB 3 B8
Rainbow Lake AB 9 B2
Rainbow River BC 3 C5
Rainham Centre ON 19 C7
Rainham Centre ON 20 F6
Rainier AB 6 E6-7
Rainier AB 7 D6
Rainy Lake ON 13 F6
Rainy Lake ON 16 E2
Rainy Lake ON 18 C8
Rainy Lake ON 21 A1
Rainy Lake ON 22 F1
Rainy River ON 16 E-F1
Rainy River ON 16 E1
Raisin River ON 24 E3-4
Raith ON 16 E5
Raleigh NL 38 A5
Raley AB 7 F5
Ralph SK 11 F6
Ralston AB 6 E7
Ralston AB 7 D7-8
Ralston AB 11 E1
Rama SK 11 C6
Rama Road ON 18 F8
Rama Road ON 20 A7
Rama Road ON 21 D1
Ramé, lac QC 23 C6
Ramea NL 36 E3-4
Ramea NL 37 E1-2
Ramea Islands NL 36 E4
Ramea Islands NL 37 E3
Rameau QC 30 C2-3
Ramore ON 17 D7
Rampart House Territorial
  Historic Site YT 5 C4
Ramparts River NT 5 E6
Ram River NT 10 D2
Ramsay Arm BC 2 B2
Ramsay Island BC 1 B2
Ramsey ON 17 E5
Ramsey Lake ON 17 E6
Ramusio, lac QC 39 A4
Ranch-Bar U,
  LHN du AB 6 E5-6
Ranch-Bar U,
  LHN du AB 7 C-D4
Rancheria YT 3 A3
Rancheria YT 4 E-F6
Rancheria Falls Territorial
  Recreation Site YT
    3 A2-3
Rancheria Falls Territorial
  Recreation Site YT 4 E6
Rancheria River YT 3 A3
Rancheria River YT 4 E-F6
Randboro QC 27 E-F5
Randolph MB 14 E5
Random Island NL 37 C7

Ranfurly AB 11 B1
Rang-des-Bosse NB 31 C2
Ranger SK 11 B3
Ranger Lake ON 17 E4
Ranger Lake ON 17 E-F4
Rangeton AB 8 C4-5
Rangifer, Mount YT 4 D4
Rang-Saint-Georges NB
    30 E1-2
Rang-Saint-Georges NB
    31 B7
Rankin ON 21 A5
Rankin ON 22 F5
Rankin Inlet NU 40 A2
Rankin Inlet NU 41 E5
Rankin Inlet (Kangiqiniq)
  NU 40 A1-2
Rankin Inlet (Kangiqiniq)
  NU 41 E5
Rannoch ON 19 B4-5
Rannoch ON 20 D3-4
Rantem NL 37 D6-7
Raoul-Blanchard, mont QC
    28 F6
Raper, Cape NU 41 A8
Rapid City MB 11 E8
Rapid City MB 13 E2
Rapid City MB 14 D1
Rapide-Blanc QC 25 B7
Rapide-Blanc-Station QC
    25 C7
Rapide-Blanc-Station QC
    28 E2-3
Rapide-Danseur QC
    17 C7-8
Rapide-Danseur QC 23 D2
Rapide-des-Cèdres QC
    23 C6
Rapide-Deux ON 17 D8
Rapide-Deux QC 23 E3
Rapides-des-Joachims, ZEC
  de QC 22 D4
Rapides-des-Joachims
  (Da Swisha) QC 22 D4
Rapide-Sept QC 23 F3
Rapid River BC 3 B4
Rapid View SK 6 A8
Rapid View SK 9 F8
Rapid View SK 11 A3
Rapson Bay ON 13 A7
Rapson Bay ON 15 F7
Rathbone Bay ON 13 B6
Rathtrevor Beach PP BC
    1 F6
Rathtrevor Beach PP BC
    2 D3
Rathwell MB 13 F3
Rathwell MB 14 E3
Rat Lake MB 12 D-E8
Rat Lake MB 15 B6
Rat Lake MB 15 F6
Rat Portage ON 13 F5
Rat Portage ON 14 E8
Rat River MB 12 D-E8
Rat River MB 13 F4
Rat River MB 14 F6
Rat River MB 15 E3
Rat River NT 5 C4
Rat River NT 10 E7-8
Rats, lac aux QC 25 A7-8
Rats, lac aux QC 25 C7
Rats, lac aux QC 26 C8
Rats, lac aux QC 26 E7-8
Rats, lac aux QC 28 A4
Rats, lac aux QC 28 C3
Rats, rivière aux QC
    25 C-D7
Rats, rivière aux QC 26 D8
Rats, rivière aux QC
    28 A-B4
Rats, rivière aux QC 28 F2
Rattling Brook NL 36 A6
Rattling Brook NL 37 A4
Rattling Brook NL 38 B5
Rattling Brook NL 38 F5-6
Ratz, Mount BC 3 C2
Raude, lac QC 39 A3-4
Raven AB 6 C5
Raven AB 8 F5
Raven Lake NT 5 C7
Raven Lake ON 18 B4
Raven Lake ON 21 B2
Raven Lake ON 22 D2
Ravenna ON 18 F6
Ravenna ON 20 A5
Ravenscliffe ON 18 D8
Ravenscliffe ON 21 B1
Ravenscliffe ON 22 F1
Ravenscrag SK 11 F2
Ravenshoe ON 20 B7
Ravenshoe ON 21 E1
Ravens Throat River NT
    4 B7-8
Ravensview ON 21 D-E7
Ravensworth ON 19 B3
Ravensworth ON 20 E2
Ravine ON 8 C4
Rawalpindi Lake ON 10 A6
Rawdon NS 32 E8
Rawdon NS 34 F1
Rawdon QC 24 C6
Rawdon QC 27 D1
Rawdon Gold Mines NS
    32 E8
Rawdon Gold Mines NS
    34 F1-2
Rawebb MB 11 A8
Rawebb MB 12 F8

Rawebb MB 13 A2
Rawhide Lake ON 17 F5
Raymond AB 6 F6
Raymond AB 7 E6
Raymond AB 18 D8
Raymond ON 21 B1
Raymore SK 11 D5-6
Reaboro ON 20 B8
Reaboro ON 21 D2-3
Reaburn MB 14 D4
Reader Lake MB 11 A7
Reader Lake MB 13 A1
Read Island PP BC 2 C2
Rearguard Falls PP BC
    6 B1-2
Rebecca Spit PMP BC 2 C2
Rebeca Lake NT 10 A5
Reboul, rivière QC 29 C8
Reboul, rivière QC 30 C1
Red Bank NB 31 D6
Red Bay NL 38 A4
Red Bay NL 39 D8
Red Bay ON 18 E5
Red Bay, LHN de NL 38 A4
Red Bay, LHN de NL 39 D8
Red Bay NHS NL 38 A4
Red Bay NHS NL 39 D8
Red Bluff PP BC 3 F5
Red Brook NL 36 C1
Red Cliff NL 36 B6
Red Cliff NL 37 B4
Red Cliff NL 37 D6
Red Cross Lake MB 15 F6
Red Deer AB 6 C5-6
Red Deer AB 8 F6
Red Deer, RNOM de AB
    6 C5-6
Red Deer, RNOM de AB
    8 F6
Red Deer Hill SK 11 B4-5
Red Deer Lake AB 8 E7
Red Deer Lake MB 11 B7
Red Deer Lake MB 11 B7
Red Deer Lake MB 13 B1
Red Deer Lake MB 13 B1
Red Deer Lake ON 18 A6
Red Deer NMBS AB 6 C5-6
Red Deer NMBS AB 8 F6
Red Deer Point MB 11 C8
Red Deer Point MB 13 C2
Red Deer River AB 6 C6
Red Deer River AB 6 D4-5
Red Deer River AB 6 D7-8
Red Deer River AB 7 A2
Red Deer River AB 7 B6
Red Deer River AB 7 B7-8
Red Deer River AB 8 F6-7
Red Deer River AB 11 D1
Red Deer River SK 11 B6-7
Red Earth SK 11 B6
Red Earth Creek AB 9 D4-5
Red Earth Creek AB 41 F1
Red Earth Lake SK 11 B6
Red Harbour NL 36 E7
Red Harbour NL 37 E5
Red Head NB 32 E7
Red Head NL 33 B1-2
Red Head NL 36 E7
Red Head Cove NL 37 D6
Red Indian Lake NL
    36 B4-5
Red Indian Lake NL
    37 B2-3
Red Island NL 36 E8
Red Island NL 36 E8
Red Island NL 37 E6
Red Island NL 37 E6
Red Islands NS 34 D6-7
Red Jacket SK 11 E7
Red Lake ON 13 D6
Red Lake ON 16 B-C2
Red Lake ON 16 C2
Red Lake Road ON 13 E6
Red Lake Road ON 16 D2
Red Lodge PP AB 6 C5
Red Lodge PP AB 8 F5-6
Red Mountain YT 4 A-B3
Red Pass BC 6 B2
Red Pheasant SK 11 C3
Red Pine Lake ON 21 B2
Red Point NS 34 D7
Red Point PE 35 C2-3
Red Point PE 35 D8
Red Point PP PE 34 C4
Red Point PP PE 35 D8
Red Rapids NB 31 D3
Red Rapids NB 32 A1
Red River AB 3 A4
Red River AB 4 F7
Red River MB 14 D5
Red River NL 14 F5
Red River NS 34 A6
Red River BC 1 A-B7
Red Rock ON 16 E6
Red Rock ON 17 C1
Red Rock Point NL 39 A7
Red Rocks NL 36 D1
Red Rose MB 13 D4
Red Rose MB 14 A4
Red Sucker Lake MB 13 A6
Red Sucker Lake MB 13 A6
Red Sucker Point, RNP ON
    16 E8
Red Sucker Point, RNP ON
    17 C2-3
Red Sucker Point PNR ON
    16 E8
Red Sucker Point PNR ON
    17 C2-3
Red Sucker River
    15 E-F6

Red Willow AB 6 C6
Red Willow AB 8 E7-8
Red Wine River NL 39 B4-5
Redberry Lake SK 11 B4
Redberry Lake NMBS SK
    11 C4
Redbridge ON 17 F8
Redbridge ON 18 A8
Redbridge ON 22 D2
Redcliff AB 6 E7
Redcliff AB 7 D8
Redcliff AB 11 E1
Redcliff Island NT 10 C-D8
Redcross Lake NL 36 C4
Redcross Lake NL 37 C2
Redditt ON 13 E5-6
Redditt ON 16 D1-2
Redfern-Keily Creek PP BC
    3 D6
Redfield SK 11 B3-4
Redfir Lake-Kapitagas
  Channel Ecological
  Reserve ON 39 C-D2
Redickville ON 20 B5
Redknife River NT 10 E-F4
Redland AB 7 B5
Rednersville ON 21 E5
Redonda Bay BC 2 B2
Redonda Bay BC 2 B2
Redonda Island BC 1 E6
Redrock Lake NT 10 B6
Redstone BC 1 C6
Redstone Flat BC 1 C6
Redstone Lake ON 21 B2-3
Redstone River NT 4 A-B8
Redstone River NT 10 B1-2
Redvers SK 11 F7
Redvers SK 13 F1
Redwater AB 6 A6
Redwater AB 8 B7
Redwood Meadows AB
    7 B3
Redwood Meadows AB
    6 D5
Reeces Corners ON 19 C3
Reeces Corners ON 20 D1
Reed Bay ON 21 E6-7
Reed Lake MB 12 F8
Reed Lake MB 15 F2
Reed Lake SK 11 E4
Reedy Creek MB 14 B3
Reef Icefield BC 6 D3
Reef's Harbour NL 38 B3
Reeve SK 11 C2
Reevecraig ON 21 B8
Refuge Cove BC 1 E6
Refuge Cove BC 2 C2-3
Refuge-du-Col-Abbot,
  LHN du AB 7 A-B1
Regent MB 11 F8
Regent MB 13 F1-2
Regent MB 14 F1
Regina SK 11 E5
Regina Beach SK 11 D-E5
Regina Beach PRS SK
    11 D5
Reginaville NL 37 F7
Regway SK 11 F5-6
Reid Lake SK 11 A7
Reid Lake NT 10 B8
Reid Lake SK 11 C3
Reid Lake TP NT 10 C-D7
Reid Lake TP NT 41 D2
Reidsville NL 36 A3-4
Reidville NL 37 A2
Reidville NL 38 B3
Reindeer Island MB 13 C3
Reindeer Lake NT 10 B7
Reindeer Lake SK 12 C7
Reindeer Lake SK 15 C1
Reindeer Mountain YT 4 B2
Reindeer River SK 12 E6
Reine-Charlotte,
  bassin de la BC 1 D3
Reine-Charlotte,
  îles de la BC 1 B1-2
Reine-Élisabeth, îles
  de la NT 42 A3-4 B2-3
Reine-Élisabeth, îles de la
  NU 42 A4 B2-3
Reine-Maud, golfe de la NU
    41 C3-4
Reine-Maud, golfe de la NU
    42 F3
Reinfeld MB 13 F3
Reinfeld MB 14 F3
Reinland MB 13 F3
Reinland MB 14 F3
Relais-Gabriel QC 39 D1
Reliance AB 7 E6
Reliance NT 41 D2
Relique, lac QC 23 A8
Relique, lac QC 26 B3
Remac BC 7 F2
Remblais-de-Southwold,
  LHN des ON 19 D4
Remblais-de-Southwold,
  LHN des ON 20 F4
Rembrandt MB 13 D3-4
Rembrandt MB 14 C5
Rémigny QC 17 D8
Rémigny QC 23 F2
Rémigny, lac QC 23 E-F2
Renard, lac du QC 30 A7
Renard, lac du QC 30 A7
Renards, pointe aux NL
    32 A6-7
Rencontre, lac de la QC
Rencontre East NL 36 D6-7
Rencontre East NL 37 E5
Rencontre West NL 36 D-E5
Rencontre West NL 37 D-E3

Rendezvous Island South
  PP BC 2 B2
Rene Brunelle PP ON
    17 B5-6
Reneault QC 17 C8
Reneault QC 23 D2
Renews NL 37 F8
Renews-Cappahayden NL
    37 F8
Renforth NB 32 D4
Renforth NB 33 A1
Renfrew NS 33 B6
Renfrew NS 34 F1-2
Renfrew ON 21 A8
Renfrew ON 24 C1
Renne, rivière le QC 24 D8
Renne, rivière le QC
    27 E3-4
Rennell Sound BC 1 A-B1
Rennie MB 13 E5
Rennie MB 14 E7
Renommée, pointe à la QC
    30 A-B2
Renous NB 31 D6
Renous River NB 31 D6
Renown MB 11 D5
Renown MB 13 C1-2
Renton ON 19 C6-7
Renton ON 20 F5-6
Renwer MB 11 C8
Renwer MB 13 C1-2
Repentigny QC 24 D6
Repentigny QC 27 D2
Repos, lac du QC 25 D5
Repulse Bay (Naujaat) NU
    41 C5-6
Reserve SK 11 C6-7
Réserve-Val-Marie,
  RNOM de SK 11 F3
Resolin, lac QC 22 D5
Resolute (Qasuittuq) NU
    42 D4
Resolution Bay NT 10 E7
Resolution Island NU
    40 B6
Resource SK 11 B5-6
Resthaven Icefield BC 6 B2
Resthaven Mountain AB
    6 B2
Restigouche River NB
    29 D-E5
Restigouche River NB
    29 E-F4
Restigo, ZEC QC 22 C3
Restigouche River NB
    31 B3-4
Restoule ON 17 F7-8
Restoule ON 18 B7-8
Restoule ON 22 E1
Restoule Lake ON 17 F7-8
Restoule Lake ON 18 B7
Restoule Lake ON 22 E1
Restoule PP ON 17 F7-8
Restoule PP ON 18 B7
Restoule PP ON 22 E1
Retty, lac QC 22 B6
Revelstoke BC 6 D3
Revelstoke, Lake BC
    6 D2-3
Revelstone, lac QC
    25 D-E4
Revenue SK 6 C8
Revenue SK 11 C3
Reward SK 6 C8
Reward SK 11 C2
Rexton NB 31 E8
Rexton NB 32 A6
Reykjavik MB 13 D3
Reykjavik MB 14 B2-3
Reynaud SK 11 C5
Rheault ON 18 A5
Rhéaume, lac QC 25 B6
Rhéaume, lac QC 28 D2
Rhein SK 11 D7
Rhineland MB 14 F3
Rhineland SK 11 D7
Ribstone AB 6 B7-8
Ribstone AB 8 D8
Ribstone Creek AB 6 B-C7
Ribstone Creek AB
    11 B2 C1
Rice Lake ON 17 D5
Rice Lake ON 21 E3
Rice Lake ON 21 E3
Rice Lake CA ON 21 E3
Rice Point PE 35 E4
Rice Point PE 35 E4
Rice River Provincial
  Reserve ON 11 B7
Rice River Provincial
  Reserve ON 13 A-B1
Riceton SK 11 E5-6
Riceville ON 24 D3
Rich, Cape ON 18 F6
Rich, Cape ON 20 A5
Richan ON 13 E6-7
Richan ON 16 D3
Richard SK 11 C4
Richard Collinson Inlet NT
    41 A3
Richard Collinson Inlet NT
    42 D2
Richard Point PE 35 E4
Richards Island NT 5 B4
Richards Landing ON 17 F4
Richardson SK 11 E5
Richardson Islands NU
    42 F2

Richardson Islands NU
    42 F2
Richardson Lake AB 9 B7
Richardson Lake AB 12 B2
Richardson Lake NMBS AB
    9 B7
Richardson Lake NMBS AB
    12 B2
Richardson Mountains YT
    5 C3 D3-4
Richardson River AB
    9 B-C7
Richardson River AB
    12 B-C2
Richardson River AB 41 F2
Richardson River Dunes
  Wildland PP AB 9 B-C7
Richardson River Dunes
  Wildland PP AB 12 B-C2
Richdale AB 6 D7
Richdale AB 7 A7
Richdale AB 11 C1
Riche, Point NL 38 C3
Riche, Point NL 40 E8
Richelieu QC 24 D6-7
Richelieu QC 27 E2
Richelieu, rivière QC
    24 C-D7 D6
Richelieu, rivière QC
    27 D-E2
Richer MB 13 F4
Richer MB 14 E6
Richfield NS 33 E1-2
Richibucto NB 31 E8
Richibucto NB 32 A6
Richibucto Cape NB
    31 D-E8
Richibucto Cape NB 32 A6
Richibucto River NB 31 E7
Richibucto River NB 32 A5
Richibucto-Village NB
    31 E8
Richibucto-Village NB
    32 A6
Richland BC 1 F6-7
Richmond BC 2 E4-5
Richmond ON 19 D5
Richmond ON 20 F4
Richmond ON 21 B8
Richmond ON 24 E2
Richmond PE 32 A7
Richmond PE 34 B1
Richmond PE 35 D2
Richmond ON 27 E4
Richmond Corner NB 31 B-C3
Richmond Corner NB 32 B1
Richmond Hill ON 19 A8
Richmond Hill ON 20 C7
Richmond Hill ON 21 F1
Richmond SK 11 E2
Richwood AB 6 E8
Richwood AB 8 E5
Rich Valley AB 6 A5
Rich Valley AB 8 B5
Rideau, canal ON 21 B-C8
Rideau, canal ON 21 D7
Rideau, canal ON 24 E2
Rideau, RNOM du ON 21 B-C8
Rideau, RNOM du ON 24 E2
Rideau Canal ON 21 B-C8
Rideau Canal ON 21 D7
Rideau Canal ON 24 E2
Rideau Ferry ON 21 C7
Rideau Ferry ON 24 F1
Rideau NMBS ON 24 E2
Rideau River ON 21 B-C8
Rideau River ON 24 E2
Rideau River PP ON 21 B8
Rideau River PP ON 24 E2
Ridge River ON 17 A4
Ridgetown ON 19 D3
Ridgeville MB 9 F3
Ridgeville MB 13 F4
Ridgeville ON 19 E8
Ridgeville ON 20 E7
Ridgeway MB 20 F8
Ridgewood AB 8 F6
Riding Mountain MB 11 D8
Riding Mountain MB
    13 D1-2
Riding Mountain MB 14 C1
Riding Mountain NP MB
    11 D8
Riding Mountain NP MB
    13 D1-2
Riding Mountain NP MB
    14 B-C1
Riel House NHS MB 14 E5
Rigaud QC 24 D4-5
Right Serpentine River NB
    29 F5
Right Serpentine River NB
    31 C4
Rigolet NL 40 D8
Rigolet (Kikiak) NL
    39 B6-7
Riley Brook NB 31 C3-4
Riley Lake ON 21 C2
Rimbey AB 6 C5
Rimbey AB 8 E5
Rimington AB 21 D5
Rimouski QC 29 F5
Rimouski, RF de QC 29 D3
Rimouski, RF de QC 31 A2

Rimouski, rivière QC 29 C3
Rimouski, rivière QC 29 D3
Rimouski, rivière QC
    31 A1-2
Rimouski-Est QC 29 C3
Ringwood ON 20 C7
Ringwood ON 21 E1
Riondel BC 6 F3-4
Riou Lake SK 12 B4-5
Ripault, lac QC 28 E6
Ripley ON 20 B3
Ripon QC 24 C3
Ripples NB 31 F5
Ripples NB 32 C3
Riske Creek BC 1 C7
Rissers Beach PP NS
    33 D4-5
Ristigouche, rivière QC
    31 B3
Ristigouche, rivière QC
    29 D5
Ristigouche, rivière QC
    31 A3-4 B3
Rita, lac QC 25 A6
Rita, lac QC 26 F6
Rita, lac QC 28 D2
Ritchie NB 32 C1-2
Ritchie NS 11 F5
Rivard, lac QC 26 B7
River aux Sables PP ON
    17 F5-6
River aux Sables PP ON
    18 A3-4
Riverbank NB 31 E3
Riverbank NB 32 A1
Rivercourse AB 6 B8
Rivercourse AB 11 B2
River Bourgeois NS 34 D6
River Brook NL 36 C2
River Charlo NB 29 D7
River Charlo NB 31 A5
Rivercrest MB 14 D5
Riverdale NS 33 D2
Riverdale PE 35 E4
River de Chute NB 31 E3
River de Chute NB 32 A1
River Denys NS 34 D6
River Drive Park ON 20 B7
River Drive Park ON 21 E1
Riverfield QC 24 E5-6
Riverfield QC 27 F1
River Glade NB 31 F7-8
River Glade NB 32 B5-6
Riverhead NL 37 D7
Riverhead NL 37 F7
River Hebert NS 32 D7
River Hills MB 13 E4-5
River Hills MB 14 D6-7
Riverhurst SK 11 D4
River John NS 34 D3
River of Ponds NL 39 E7-8
River of Ponds Lake NL
    38 C3
River of Ponds Lake NL
    39 E7
River Philip NS 32 C8
River Philip NS 34 E1
Riverport NS 33 D5
Rivers MB 11 E8
Rivers MB 13 E2
Rivers MB 14 D1
Rivers Inlet BC 1 D4
Rivers Inlet BC 1 C-D4
Rivers PP MB 11 E8
Rivers PP MB 13 E2
Rivers PP MB 14 D1
Rivers, Lake of the SK
    11 E-F5
Riverside NS 33 A7
Riverside NS 33 D4
Riverside ON 20 B3
Riverside MB 14 F4-5
Riverside-Albert NB 32 C6
Riverton MB 13 D4
Riverton MB 14 B5
Riverton NS 33 A7
Riverton PE 35 E6
River Valley ON 17 F7
River Valley ON 18 A7
Riverview NB 31 F8
Riverview NB 32 B6
Riverview ON 20 B5
Riverview Heights ON
    1 C8
Riverview Heights ON
    24 C2
Rivett Lake NT
    10 B7-8 C7-8
Rivière-à-Claude QC 29 B6
Rivière-à-Pierre QC 27 A4
Rivière-au-Portage NB
    31 D7-8
Rivière-au-Renard QC
    30 B5
Rivière-au-Tonnerre QC
    39 F2-3
Rivière-aux-Graines QC
    39 E-F3
Rivière-aux-Rats QC 25 D7
Rivière-aux-Rats QC 28 E2
Rivière-aux-Rats,
  ZEC de la QC 26 E5
Rivière-aux-Rats,
  ZEC de la QC 28 A4
Rivière-Beaudette QC
    24 E5
Rivière-Bersimis QC 29 B3
Rivière Black, APP NB
    31 D7

Rivière-Blanche, ZEC de la *QC* **28** F4
Rivière-Bleue *QC* **29** E2
Rivière-Bonaventure, ZEC de la *QC* **29** B-C8
Rivière-Bonaventure, ZEC de la *QC* **29** C8
Rivière-Bonaventure, ZEC de la *QC* **29** C-D8
Rivière-Bonaventure, ZEC de la *QC* **30** B-C1
Rivière-Bonaventure, ZEC de la *QC* **30** C-D1
Rivière-Cap-Chat, RF de la *QC* **29** B6
Rivière-Cap-Chat, ZEC de la *QC* **29** A-B6
Rivière-Cascapédia, RF de la *QC* **29** B-C7
Rivière-Cascapédia, RF de la *QC* **29** C7
Rivière-Cascapédia, RF de la *QC* **29** B6-7
Rivière-Dartmouth, ZEC de la *QC* **30** B2
Rivière-de-la-Chaloupe *QC* **30** A5-6
Rivière-de-Sable, RNOM de la *NS* **33** E3-5
Rivière des Escoumins, ZEC de la *QC* **28** C8
Rivière des Escoumins, ZEC de la *QC* **29** C8
Rivière-du-Loup *QC* **28** D8
Rivière-du-Loup *QC* **29** D1
Rivière-du-Portage *NB* **30** F2
Rivière-du-Portage *NB* **31** C8
Rivière-Éternité *QC* **28** C7
Rivière-Héva *QC* **17** D8
Rivière-la-Madeleine *QC* **29** A8
Rivière-la-Madeleine *QC* **30** A1
Rivière-Madeleine, ZEC de la *QC* **29** A8
Rivière-Madeleine, ZEC de la *QC* **30** A-B1
Rivière-Malbaie *QC* **28** E7
Rivière-Matane, RF de la *QC* **29** B4-5
Rivière-Matane, ZEC de la *QC* **29** B5
Rivière-Matapedia, RF de la *QC* **31** A4
Rivière-Matapédia, RF de la *QC* **29** C5
Rivière-Matapédia, RF de la *QC* **29** D5
Rivière-Matawin *QC* **25** E7
Rivière-Matawin *QC* **27** A3
Rivière-Moisie, RF de la *QC* **39** E-F2
Rivière-Nechako, RNOM de la *BC* **1** A6
Rivière-Ouelle *QC* **28** E7-8 F7-8
Rivière-Paspébiac *QC* **29** D8
Rivière-Paspébiac *QC* **30** D1
Rivière-Paspébiac *QC* **31** A7
Rivière-Patapédia, RF de la *QC* **29** D4-5
Rivière-Pentecôte *QC* **39** F1
Rivière-Petit-Saguenay, ZEC de la *QC* **28** D7
Rivière-Pigou *QC* **39** E-F2
Rivière Qui Barre *AB* **8** C6
Rivière-Ristigouche, RF de la *QC* **31** A4
Rivière-Sainte-Anne, RF de la *QC* **29** A-B6
Rivière-Sainte-Marguerite *QC* **28** C8
Rivière Sainte-Marguerite, ZEC de la *QC* **28** C7-8
Rivière-Saint-Jean, RF de la *QC* **39** E-F3
Rivière-Saint-Jean, RF de la *QC* **30** C2
Rivière-Saint-Jean, RF de la *QC* **30** B-C2
Rivière-Saint-Paul *QC* **38** A2
Rivière-Saint-Paul *QC* **39** D7
Rivière-Saint-Paul *QC* **40** E8
Rivière-Trois-Pistoles *QC* **29** D1-2
Rivière-Verte *NB* **29** F3-4
Rivière-Verte *NB* **31** C2
Rivière-York, ZEC de la *QC* **29** B8
Rivière-York, ZEC de la *QC* **30** B1
Rivière-York, ZEC de la *QC* **30** B2
Rivington *QC* **24** C4
Roachvale *NS* **34** E5
Roachville *NB* **32** C5
Road River *YT* **5** D4
Roaring River *MB* **11** C7-8
Roaring River *MB* **13** C1
Roaring River *ON* **16** D-E5
Robb *AB* **6** B3-4
Robb *AB* **8** D2
Robb Lake *NT* **10** A5-6
Robb Lake *NT* **39** E4
Roberge, lac *QC* **25** D8
Roberge, lac *QC* **27** A3

Robert, lac *QC* **26** C5
Robert-Bourassa, réservoir *QC* **40** E4
Robertine, lac *QC* **23** C8
Robertine, lac *QC* **26** D2
Robert's Arm *NL* **36** A6
Robert's Arm *NL* **37** A4
Robert's Arm *NL* **38** F5
Robert's Arm *NL* **39** F8
Roberts Creek PP *BC* **2** D4
Robertson, lac *QC* **23** C3
Robertson, lac *QC* **39** E6
Robertson Lake *NS* **33** A7
Robertson Lake *AB* **12** A1
Robertsonville *QC* **27** C5-6
Roberts Peak *BC* **1** B8
Roberts Peak *BC* **6** B1
Robertval *QC* **26** E8
Robertval *QC* **28** C4
Robeson, détroit de *NU* **42** A5
Robeson Channel *NU* **42** A5
Robichaud *NB* **32** B7
Robichaud *NB* **35** E1
Robin, rivière *QC* **23** C6
Robinhood *SK* **11** B3
Robinson *YT* **4** E4
Robinson River *NL* **36** C1-2
Robinsonville *NB* **29** E5-6
Robitaille *NB* **31** B4
Robitaille Lake *ON* **21** A4
Robitaille Lake *ON* **22** F4
Roblin *MB* **11** D7-8
Roblin *MB* **13** D1
Roblin *ON* **21** D6
Robsart *SK* **11** F2
Robson *BC* **6** F3
Robson, lac *QC* **23** F8
Robson, lac *QC* **25** B2
Robson, Mount *BC* **1** A8
Rocanville *SK* **11** E7
Rocanville *SK* **13** E1
Roc-Cousacouta, lac *QC* **25** D6
Roc-Cousacouta, lac *QC* **27** A1-2
Roc-Cousacouta, lac *QC* **28** F1-2
Roche, Pointe de *NT* **10** E6
Rochebaucourt *QC* **23** C5
Roche Lake PP *BC* **6** E1
Roche Percée *SK* **11** F7
Rocher, lac du *NT* **10** B8
Rocher River *NT* **10** D-E7
Rochers, rivière aux *QC* **39** E2
Rochers-aux-Oiseaux, RNOM des *QC* **30** E8
Roches Point *QC* **20** B7
Roches Point *ON* **21** E1
Rochester *AB* **6** A6
Rochester *AB* **8** A6
Rochester *AB* **9** F6
Rochester, lac *QC* **23** B7
Rochester, lac *QC* **26** C1
Rocheuses, contreforts des *AB* **6** B3 C3-4 D4
Rocheuses, contreforts des *AB* **7** C3 D3-4 E4
Rocheuses, contreforts des *AB* **8** D-E2 F3
Rocheuses, contreforts des *BC* **3** C6 D6-7
Rocheuses, montagnes *AB* **8** E1 F1-2
Rocheuses, montagnes *BC* **1** A-B8
Rocheuses, montagnes *BC* **3** D6 E6-7
Rocheuses, montagnes *BC* **6** D-E4 E-F5
Rocheuses, montagnes *BC* **7** C-E3
Rocheuses, montagnes *BC* **6** B2 C2-3
Rocheuses, montagnes *BC* **7** C1-2 D2
Rocheuses canadiennes, parcs des montagnes *AB* **6** C3-4
Rocheuses canadiennes, parcs des montagnes *AB* **7** A1-2
Rocheuses canadiennes, parcs des montagnes *BC* **7** B1-2
Rocheville *NB* **31** B7
Rochfort Bridge *AB* **6** A5
Rochfort Bridge *AB* **8** B4-5
Rochon Sands *AB* **6** C6
Rochon Sands *AB* **8** E7
Rochon Sands PP *AB* **8** E7
Rock Bay *BC* **1** E5
Rock Bay *BC* **2** B2
Rock Bay MPP *BC* **1** E5
Rock Bay MPP *BC* **2** B1-2
Rockburn *QC* **24** E5
Rock Creek *BC* **6** F2
Rockcliffe Park *ON* **21** A8
Rockcliffe Park *ON* **24** D2
Rockford *ON* **18** F5
Rockford *ON* **20** A4
Rock Glen CA *ON* **19** C3-4
Rock Glen CA *ON* **20** E2-3
Rockglen *SK* **11** F5
Rock Harbour *NL* **36** E7
Rock Harbour *NL* **37** E5
Rockhaven *SK* **11** B3

Rockingham *ON* **21** B5
Rock Island Lake *AB* **9** E6
Rock Lake *MB* **14** F2
Rock Lake *ON* **17** F4
Rock Lake *ON* **21** A3
Rock Lake *ON* **22** F3
Rock Lake-Solomon Creek Wildland PP *AB* **6** B2-3
Rock Lake-Solomon Creek Wildland PP *AB* **8** C1
Rockland *NS* **33** F3
Rockland *ON* **24** D3
Rocklin *NS* **33** A7
Rocklin *NS* **34** E3
Rocklyn *ON* **18** F6
Rocklyn *ON* **20** A5
Rock Point PP *ON* **19** D8
Rock Point PP *ON* **20** F7
Rockport *NB* **32** C7
Rockport *ON* **21** D7-8
Rock Ridge *MB* **13** C2
Rock River *YT* **4** E8
Rock River *YT* **5** D3
Rocksprings *ON* **21** C8
Rocksprings *ON* **24** F2
Rockton *ON* **19** B7
Rockton *ON* **20** D6
Rockwood *ON* **19** A7
Rockwood *ON* **20** D6
Rockwood, RNF de *MB* **14** D5
Rockwood CA *ON* **19** A7
Rockwood CA *ON* **20** D6
Rockwood NWA *MB* **14** D5
Rocky Bay *NL* **39** B8
Rocky Bay *ON* **16** E6
Rocky Bay *ON* **17** B1
Rocky Brook *NB* **31** D4 E5
Rocky Cove *NL* **39** A7
Rockyford *AB* **6** D6
Rockyford *AB* **7** B5
Rocky Harbour *NL* **38** E2
Rocky Harbour *NL* **39** F7
Rocky Island Lake *ON* **17** E5
Rocky Lake *MB* **11** A7
Rocky Lake *MB* **13** A1
Rocky Lane *AB* **9** B4
Rocky Mountain *NS* **33** A8
Rocky Mountain *NS* **34** E4
Rocky Mountain Foothills *AB* **6** B3 C4
Rocky Mountain Foothills *AB* **7** B2-3 C3
Rocky Mountain Foothills *AB* **8** D-E2 E-F3
Rocky Mountain Foothills *BC* **3** C6 D6-7
Rocky Mountain House *AB* **6** C4-5
Rocky Mountain House *AB* **8** E4-5
Rocky Mountain House, LHN *AB* **6** C4-5
Rocky Mountain House, LHN *AB* **8** F4-5
Rocky Mountain House NHS *AB* **6** C4-5
Rocky Mountain House NHS *AB* **8** F4-5
Rocky Mountains *AB* **8** E-F1
Rocky Mountains *BC* **1** A8
Rocky Mountains *BC* **3** B-C5
Rocky Mountains *BC* **6** B2 C2-3
Rocky Mountains *BC* **7** C1-2 D2
Rocky Point *BC* **2** F4
Rocky Rapids *AB* **6** B5
Rocky Rapids *AB* **8** D5
Rocky Ridge Pond *NL* **36** C-D1
Rocky Ridge Pond *NL* **37** C-D1
Rocky Saugeen River *ON*
Roddick, lac *QC* **22** E7-8
Roddick, lac *QC* **24** B2
Roddick, lac *QC* **25** F2
Roddickton *NL* **38** C4
Roddickton *NL* **39** E8
Rodef *AB* **8** B7
Roderick Haig-Brown PP *BC* **6** D2
Roderick Island *BC* **1** B-C3
Roderick Lake *ON* **13** D5
Roderick Lake *ON* **14** A8
Roderick Lake *ON* **16** B2
Rodgers Cove *NL* **36** A7
Rodgers Cove *NL* **37** A5
Rodgers Cove *NL* **38** F7
Rodin, lac *QC* **22** A6
Rodney *ON* **19** D4
Rodney *ON* **20** F3
Rodney Pond *NL* **36** B7
Rodney Pond *NL* **37** B5
Roe Lake *MB* **15** D3
Roe Lake *SK* **12** D3
Roebuck *ON* **21** C8
Roes Welcome Sound *NU* **40** A2-3
Roes Welcome Sound *NU* **41** B6
Roger, lac *QC* **17** D8
Roger, lac *QC* **23** F2-3
Rogers *MB* **14** C1
Rogers *SK* **11** E4
Rogerson Lake *NL* **36** C4-5
Rogerson Lake *NL* **37** C2-3

Rogers Pass *BC* **6** D3
Rogers Pass *BC* **6** D3
Rogersville *NB* **31** D7
Rogersville *NB* **32** A5
Rognons, lac aux *QC* **25** C8
Rognons, lac aux *QC* **28** E4
Rognons, lac aux *QC* **28** F4
Rohault, lac *QC* **26** C5
Rokeby *SK* **11** D7
Roland *MB* **13** F3
Roland *MB* **14** F4
Rolfe Lake *NT* **10** C7-8
Rolla *BC* **3** E8
Rolla *BC* **9** E2
Rollet *QC* **23** E2
Rolley Lake PP *BC* **1** F7
Rolley Lake PP *BC* **2** D5-6
Rollingdam *NB* **32** D2
Rolling Hills *AB* **6** E7
Rolling Hills *AB* **7** D7
Rolling Hills *AB* **11** E1
Rolling Hills Lake *AB* **7** D7
Rolling River *MB* **11** E8
Rolling River *MB* **13** E2
Rolling River *MB* **14** C1
Rollo Bay *PE* **34** C4
Rollo Bay *PE* **35** E7
Rollo Bay *PE* **35** D7
Rolly View *AB* **6** B6
Rolly View *AB* **8** D6
Rolphton *ON* **22** E4
Romaine, rivière *QC* **39** C3-4
Romaine, rivière *QC* **39** D-E4
Romaine, rivière *QC* **40** E-F7
Romaines *NL* **36** B-C2
Romance *SK* **11** C5
Roman Valley *NS* **34** E5
Roméo, lac *QC* **26** C3
Ronalane *AB* **7** D7
Rond, lac *QC* **26** C7
Rond, lac *QC* **28** A6
Rondeau Harbour *ON* **19** E3
Rondeau Park *ON* **19** E3
Rondelet, lac *QC* **22** A2
Rondelet, lac *QC* **24** E4
Rooms *NL* **38** E3-4
Roosville *BC* **6** F5
Roosville *BC* **7** F2-3
Root, Mount *BC* **4** F2
Root Lake *MB* **11** A7
Root Lake *MB* **13** A1
Root Lake *MB* **14** A1
Root River *NT* **10** C2
Rope Cove *NL* **36** B2
Roquemaure *QC* **17** C7-8
Roquemaure *QC* **23** C2
Rorey Lake *NT* **5** D6-7
Rorke Lake *NT* **15** F6-7
Rorketon *MB* **13** D2
Rorketon *MB* **14** A2
Rosa *MB* **13** F4
Rosa *MB* **14** F5
Rosaireville *NB* **31** D7
Rosalind *AB* **6** B6
Rosalind *AB* **8** E8
Rosalind *AB* **11** B1
Roscoe Bay PP *BC* **1** E6
Roscoe Bay PP *BC* **2** C2-3
Roscoe Inlet *BC* **1** C4
Rose *NS* **32** C8
Rose *NS* **33** A5
Rose *NS* **34** E1
Roseau River *MB* **13** F4
Roseau River *MB* **13** F4
Roseau River *MB* **14** F5-6
Rosebank *MB* **13** F3
Rosebank *MB* **14** F4
Rosebank *PE* **35** B2
Rosebary Lake *ON* **21** A2
Rose Bay *NS* **33** D5
Roseberry *PE* **35** F5-6
Roseberry River *ON* **13** B6-7 C7
Rosebery PP *BC* **6** E3
Rose Blanche *NL* **36** D2
Rose Blanche-Harbour Le Cou *NL* **36** D-E2
Rosebud *AB* **6** D6
Rosebud *AB* **7** B5
Rosebud Creek *YT* **4** B3
Rosebud River *AB* **7** B4-5
Rosedale *AB* **6** D6
Rosedale *AB* **7** A6
Rosedale *BC* **2** E6
Rosedale *ON* **20** A8
Rosedale *ON* **21** D2
Rosedale *ON* **21** D6
Rosehall *ON* **21** E5
Roseisle *MB* **13** F3
Roseisle *MB* **14** E3
Roseland *MB* **14** E1
Rose Lynn *AB* **7** A7
Rosemary *AB* **6** D7
Rosemary *AB* **7** C6
Rosemary *AB* **11** D1
Rosemere *QC* **24** D5-6
Rosemere *QC* **27** E1
Rosemont *ON* **20** B6
Roseneburg *MB* **14** F5
Roseneath *ON* **21** E3-4
Roseneath *PE* **34** C3
Roseneath *PE* **35** E6
Rosenfeld *MB* **13** F3-4
Rosenfeld *MB* **14** F4-5
Rosengart *MB* **13** F3

Rosengart *MB* **14** F4
Rosenhof *SK* **11** E3-4
Rosenort *MB* **13** F3
Rosenort *MB* **14** E4-5
Rose Point *BC* **1** A2
Rose Prairie *BC* **3** D8
Rose Prairie *BC* **9** D1-2
Roseray *SK* **11** E3
Rose River *YT* **4** D5
Rose Spit *BC* **1** A2
Rosetown *MB* **14** F4
Rosetown *SK* **11** D3
Roseville *ON* **19** B6
Roseville *ON* **20** D5
Roseville *PE* **35** B1
Rosewall Creek PP *BC* **1** F5-6
Rosewall Creek PP *BC* **2** D2
Roseway *NS* **33** F3
Roseway River *NS* **33** F3
Rosewood *MB* **14** E5-6
Rosiers, cap des *QC* **30** B3
Rosiers, rivière aux *QC* **29** A2-3
Rosiers, rivière des *QC* **27** D4
Roslin *BC* **21** D5
Roslyn Lake *ON* **16** E7
Roslyn Lake *ON* **17** B-C1
Ross *MB* **13** F4
Ross *MB* **14** E6
Ross, lac *QC* **25** A8
Ross, lac *QC* **26** E8
Ross, lac *QC* **28** C4
Ross Bay Junction *NL* **39** C2
Rossburn *MB* **11** E8
Rossburn *MB* **13** D-E1
Rosseau *ON* **18** D8
Rosseau *ON* **21** B1
Rosseau River *ON* **18** D8
Rosseau River *ON* **21** B1
Rosseau Road *ON* **18** D7
Rosseau, Lake *ON* **18** D8
Rosseau, Lake *ON* **21** B1
Rossendale *MB* **13** E3
Rossendale *MB* **14** E3
Ross Ferry *NS* **34** C7
Ross Ferry PP *NS* **34** C7
Rossfield *NS* **33** A8
Rossfield *NS* **34** E4
Rossignol, lac *NS* **33** D3
Rossignol, Lake *NS* **33** D3
Rossington *AB* **6** A5
Rossington *AB* **8** B6
Ross Island *MB* **13** A3
Ross Island *NB* **32** F3
Ross Lake *NT* **10** C7
Ross Lake PP *BC* **3** F4
Rossmore *ON* **21** E5
Ross Point *NU* **42** D2-3
Rossburn *BC* **6** F3
Rossmore *ON* **21** E5
Rosslyn Village *ON* **16** F5-6
Rossport *ON* **16** E7
Rossport, camping provincial *ON* **16** E6-7
Rossport, camping provincial *ON* **17** C1
Rossport Provincial Campground *ON* **16** E6-7
Rossport Provincial Campground *ON* **17** C1
Ross River *YT* **4** C6 C-D5
Ross River *YT* **4** C6 C-D5
Rossville *MB* **13** A3-4
Rossville *NB* **31** F3-4
Rossville *NB* **32** B2
Rossway *NS* **32** F4
Rossway *NS* **33** C1-2
Rosswood *BC* **3** F3-4
Rosthern *SK* **11** B4
Rostock *ON* **19** B5
Rostock *ON* **20** D4
Rothesay *NB* **32** D4
Rothesay *NB* **33** A1
Rothsay *ON* **19** A6
Rothsay *ON* **20** C5
Roti Bay *NL* **36** D5-6
Roti Bay *NL* **37** D4
Rottenfish River *ON* **13** A-B7
Rotterdam *NL* **37** A7
Rotterdam *NL* **36** B7
Rouge, Cape *NL* **38** B-C5
Rouge, Cape *NL* **39** A7
Rouge, lac *QC* **25** E4
Rouge, lac *QC* **26** C3
Rouge, parc *ON* **19** A8
Rouge, parc *ON* **20** C7
Rouge, parc *ON* **21** F1-2
Rouge, rivière *MB* **14** D5
Rouge, rivière *MB* **14** F5
Rouge, rivière *QC* **24** A4
Rouge, rivière *QC* **25** E5
Rouge Park *ON* **19** A8
Rouge Park *ON* **20** C7
Rouge Park *ON* **21** F1-2
Rouge River *ON* **19** A8
Rouge River *ON* **20** C7
Rouge River *ON* **21** F1
Rouge River *YT* **4** B5
Rouge-Matawin, RF de **24** A4
Rouge-Matawin, RF *QC* **25** E4
Rougemont *QC* **24** D7
Rougemont *QC* **27** E2-3
Rouging River PP *ON* **19** F6
Rougon *QC* **40** D1
Rouillard, lac *QC* **23** E4
Roulante Lake *NT* **10** A6-7
Rouleau *SK* **11** E5

Roulier *QC* **17** D8
Roulier *QC* **23** F2
Roundabout *NL* **36** F6
Roundabout *NL* **37** F4
Round Bay *NS* **33** F3
Round Harbour *NL* **36** D3
Round Harbour *NL* **37** D3
Round Harbour *NL* **38** F5
Round Hill *AB* **6** B6
Round Hill *AB* **8** D7
Round Hill *AB* **11** B1
Round Hill *NS* **32** F5
Round Hill *NS* **33** C3
Round Island *NS* **34** C8
Round Island Lake *ON* **22** F3
Round Lake *NS* **35** F2
Round Lake *ON* **18** A5
Round Lake *ON* **21** A5
Round Lake *ON* **21** A4
Round Lake *ON* **22** F1
Round Lake *ON* **22** F4-5
Round Lake Centre *ON* **21** A4
Round Lake Centre *ON* **22** F4
Round Lake PP *ON* **21** A5
Round Pond *NL* **36** C5-6
Round Pond *NL* **37** C3-4
Round Valley *AB* **8** C-D5
Rounthwaite *MB* **13** F2
Rounthwaite *MB* **14** E1
Routhierville *QC* **29** D5
Rouvray, lac *QC* **28** A6
Rouyn-Noranda *QC* **17** D8
Rouyn-Noranda *QC* **23** D2-3
Rouyn-Noranda *QC* **23** D2-3
Rowan Lake *ON* **13** F6
Rowan Lake *ON* **16** E2
Rowatt *SK* **11** E5
Rowdy Lake *ON* **13** E5
Rowdy Lake *ON* **14** C8
Rowena *NB* **31** D3
Rowley *AB* **6** C6
Rowley *AB* **7** A5
Rowley Island *NU* **41** B6
Rowley Island *NU* **41** B6
Rowley River *NU* **41** A-B7
Rowley River *NU* **42** E6
Roxbury *PE* **35** C2
Roxton Falls *QC* **24** D8
Roxton Falls *QC* **27** E3
Roxton Pond *QC* **24** D7-8
Roxton Pond *QC* **27** E3
Roxton-Sud *QC* **24** D8
Roxton-Sud *QC* **27** E3
Roy, lac *QC* **26** C4
Royal Geographical Society Islands *NU* **41** B-C4
Royal Geographical Society Islands *NU* **42** F2
Royal Park *AB* **8** C5
Royston *BC* **1** F6
Royston *BC* **2** D2
Ruaux, île aux *QC* **27** A6
Ruban, rivière *QC* **25** B5
Ruban, rivière *QC* **26** F5
Ruban, rivière *QC* **28** D1
Ruby Beach *SK* **11** B7
Ruby Lake PP *ON* **16** E6
Ruby Lake PP *ON* **17** C1
Ruby Range *YT* **4** D2
Rubyrock Lake PP *BC* **1** A5
Rubyrock Lake PP *BC* **3** F5
Ruckle PP *BC* **2** E4
Ruddell *SK* **11** C3
Ruddock *MB* **12** E7
Ruddock *MB* **15** E1
Rufus Lake *NT* **5** A6
Rugged Point PMP *BC* **1** A4
Ruis Lake *AB* **9** B5
Ruis Lake *AB* **12** B1
Ruisseau-à-Rebours *QC* **29** A7
Ruisseau-Big, RNF du *QC* **19** D6
Ruisseau-Big, RNF du *QC* **20** F5
Ruisseau-Castor *QC* **29** A6-7
Ruisseau-Vert *QC* **29** A3
Rumsey *AB* **6** C6
Rumsey *AB* **7** A5
Runciman *SK* **11** B6
Runnymede *SK* **11** D7
Runt Creek *YT* **5** D1-2
Rupert *QC* **22** F7
Rupert *QC* **24** C1-2
Rupert, lac *QC* **25** D4
Rupert, rivière *QC* **40** F4
Rupert Creek *MB* **15** C6
Rusagonis *NB* **32** C3
Ruscom River *ON* **19** E2
Ruscom Station *ON* **19** E2
Rushing River PP *ON* **13** F6
Rush Lake *ON* **14** D5
Rush Lake *SK* **11** E4
Rushoon *NL* **36** E7
Rushoon *NL* **37** E5
Rushton's Beach PP *NS* **34** D2
Rushy Pond Siding *NL* **36** A-B6
Rushy Pond Siding *NL* **37** A-B4
Russell Bay *NT* **10** A3

Russel Island *ON* **18** D4
Russell *MB* **11** D8
Russell *MB* **13** D1
Russell *ON* **24** E3
Russell, Cape *NT* **42** C2
Russell Inlet *NT* **5** A6
Russell Island *NL* **41** A4
Russell Island *NU* **42** D3
Russell Island Unit (Gulf Islands NPR) *BC* **2** F4
Russell Lake *MB* **12** D7
Russell Lake *MB* **15** D1
Russell Lake *NT* **10** C6
Russell Lake *SK* **12** C5
Russell Landing *ON* **21** B2
Russelldale *ON* **19** B4
Russelldale *ON* **20** D3
Russel Range *YT* **4** B4-5
Rustico Bay *PE* **35** D4
Rusty Lake *MB* **12** D8
Rusty Lake *MB* **15** D2-3
Rusty Mountain *YT* **4** A4-5
Rutan *SK* **11** C5
Rutherford *ON* **19** D3
Rutherford *ON* **20** F2
Rutherglen *ON* **17** F8
Rutherglen *ON* **22** D2
Ruthilda *SK* **11** C3
Ruth Lake *AB* **18** B8
Ruth Lake *AB* **22** E1
Ruth Lake PP *BC* **1** C8
Ruth Lake PP *BC* **6** C1
Ruthven *ON* **19** E2
Rutland *BC* **6** E2
Rutland *SK* **6** C8
Rutland *SK* **11** C2
Rutledge Lake *NT* **10** D8
Rutledge Lake *NT* **41** E2
Rutledge River *NT* **10** D8
Ruzé, lac *QC* **38** B1
Ruzé, lac *QC* **39** D-E6
Ryan *BC* **6** F1
Ryan Lake *MB* **12** B8
Ryan Lake *MB* **15** B3
Ryan PP *BC* **6** F4
Ryan PP *BC* **7** F1
Ryan Premises NHS *NL* **37** B7
Ryan River *BC* **2** B4-5
Rycroft *AB* **9** E3
Rydal Bank *ON* **17** F4
Ryerson *SK* **11** E7
Ryerson *SK* **13** E1
Ryersons Island *ON* **19** D6
Ryersons Island *ON* **20** F5
Rykerts *BC* **6** F4
Ryley *AB* **6** B6
Ryley *AB* **8** C-D8
Ryley *AB* **11** B1

Saddle River *AB* **9** E2-3
Sadene Lake *NT* **5** B7
Sadler Lake *SK* **12** E-F6
Sadowa *ON* **20** A7-8
Sadowa *ON* **21** C2
Sagamok *ON* **17** F5-6
Sagamok *ON* **18** B3
Saganaga Lake *ON* **16** F4
Saganaga Lake *ON* **16** F4
Saganaga Lake *ON* **40** F1
Saganagons Lake *ON* **16** F4
Saganash Lake *ON* **17** C5
Sagard *QC* **28** D7
Sagawitchewan Bay *MB* **13** A6
Sagehill *SK* **11** C4-5
Sagemace Bay *MB* **11** C8
Sagemace Bay *MB* **13** C2
Sagemace Bay *MB* **14** A1
Saglek Bay *NL* **40** C7
Saguenay *QC* **28** C5-6
Saguenay, parc du *QC* **28** C7 C-D8
Saguenay, rivière *QC* **28** C6
Saguenay River *QC* **28** C7
Saguenay-Saint-Laurent, parc marin du *QC* **28** C6-7 D8
Saguenay-Saint-Laurent, parc marin du *QC* **29** D1
Saguenay-St. Lawrence Marine Park *QC* **28** C7-8 D8
Saguenay-St. Lawrence Marine Park *QC* **29** D1
Sahanatien *ON* **18** E7-8
Sahdoanah Creek *BC* **3** A8
Sahdoanah Creek *BC* **9** A1-2
Sahtaneh River *BC* **3** B7
Sahtaneh River *BC* **9** A-B1
Saindon, lacs *QC* **40** D5
Saint-Adalbert *QC* **27** A8
Saint-Adelme *QC* **29** B5
Saint-Adelphe *QC* **24** E8
Saint-Adelphe *QC* **25** E8
Saint-Adelphe *QC* **27** B3-4
Saint-Adolphe *MB* **13** F4
Saint-Adolphe *QC* **24** C4
Saint-Adolphe-de-Dudswell *QC* **27** E5
Saint-Adolphe-d'Howard *QC* **24** C4
Saint-Adrien *QC* **27** E5
Saint-Adrien-d'Irlande *QC* **27** C-D5
Saint-Agapit *QC* **27** B4
St. Agatha *ON* **19** B6
St. Agatha *ON* **20** D5
Saint-Aimé-des-Lacs *QC* **28** E7
St. Alban's *NL* **36** D5-6
St. Alban's *NL* **37** D3-4
Saint-Alban *QC* **27** B4
St. Albert *AB* **6** A5-6 B5-6
St. Albert *AB* **8** C6
St. Albert *ON* **24** E3
Saint-Albert *QC* **27** D4
Saint-Alexandre *QC* **24** E7
Saint-Alexandre *QC* **27** E2
Saint-Alexandre *QC* **28** E8
Saint-Alexandre *QC* **29** E1
Saint-Alexandre-des-Lacs *QC* **29** C5
Saint-Alexis *QC* **24** C6
Saint-Alexis *QC* **27** D1
Saint-Alexis-de-Matapédia *QC* **29** D5
Saint-Alexis-de-Matapédia *QC* **31** A4
Saint-Alexis-des-Monts *QC* **24** B6-7
Saint-Alexis-des-Monts *QC* **25** F6-7
Saint-Alexis-des-Monts *QC* **27** B-C2
St. Almo *NB* **31** D3
Saint-Alphonse *MB* **14** E2-3
Saint-Alphonse *NS* **33** E1
Saint-Alphonse *QC* **24** E7
Saint-Alphonse *QC* **28** D8
Saint-Alphonse *QC* **30** D1
Saint-Alphonse *QC* **31** A6
Saint-Alphonse-Rodriguez *QC* **24** B5-6
Saint-Alphonse-Rodriguez *QC* **27** D1
Saint-Amable *QC* **24** D6
Saint-Amable *QC* **27** E2
Saint-Amand *NB* **31** C3
St. Ambroise *MB* **13** E3
St. Ambroise *MB* **14** D4
St. Ambroise Beach PP *MB* **13** E3
St. Ambroise Beach PP *MB* **14** D3-4
Saint-Ambroise-de-Kildare *QC* **24** C6
Saint-Ambroise-de-Kildare *QC* **27** D1
Saint-Amour, lac *QC* **22** B7
Saint-Amour, lac *QC* **25** C-D3
Saint-Anaclet *QC* **29** C3
Saint-André *NB* **29** F4
Saint-André *NB* **31** D3
Saint-André *QC* **28** E8
Saint-André *QC* **29** E1

Saint-André-Avellin QC **24** C-D3
Saint-André-de-Restigouche QC **29** D5-6
Saint-André-de-Restigouche QC **31** A4
Saint-André-de-Shediac NB **35** E1
Saint-André-du-Lac-Saint-Jean QC **28** C4
Saint-André-Est QC **24** D4-5
St. Andrew, Lake MB **13** D3-4
St. Andrew, Lake MB **14** A4-5
St. Andrews MB **14** D5
St. Andrews NB **32** E2
St. Andrew's NL **36** D1
St. Andrews NS **34** E5
St. Andrews ON **24** E4
St. Andrews PE **35** D5-6
St. Andrews Blockhouse NHS NB **32** E2
St. Andrews Channel NS **34** C7
Saint-Anicet QC **24** E4-5
St. Ann PE **35** D4
St.-Anne-du-Lac QC **27** D5-6
St. Ann Point NS **33** F2
St. Anns NS **34** C7
St. Anns ON **19** C8
St. Anns ON **20** E7
St. Anns Bay NS **34** B7
St. Anns PP NS **34** C7
Saint-Anselme NB **31** E8
Saint-Anselme NB **32** B6
Saint-Anselme QC **27** B6
St. Anthony NL **38** B5
St. Anthony NL **39** D8
St. Anthony Bight NL **38** B5
Saint-Antoine NB **31** E8
Saint-Antoine NB **32** A-B6
Saint-Antoine QC **24** C5
Saint-Antoine QC **27** D1
Saint-Antoine-Abbé QC **24** E5
Saint-Antoine-de-Tilly QC **27** B5
Saint-Antoine-sur-Richelieu QC **24** C6-7
Saint-Antoine-sur-Richelieu QC **27** D2
Saint-Antonin QC **29** E1
Saint-Apollinaire QC **27** B5
Saint-Armand NB **29** F4
Saint-Armand QC **24** E7
Saint-Armand QC **27** F2
Saint-Arsène QC **29** D1
Saint-Arthur NB **29** D-E6
Saint-Arthur NB **31** B4
Saint-Athanase QC **29** E1
Saint-Aubert QC **27** A7
Saint-Aubert QC **28** F7
Saint-Augustin QC **24** D5
Saint-Augustin QC **28** B4
Saint-Augustin QC **38** B1
Saint-Augustin QC **39** D6-7
Saint-Augustin QC **40** E8
Saint-Augustin, baie QC **39** E7
Saint-Augustin, rivière NL **40** E8
Saint-Augustin, rivière QC **38** A-B1
Saint-Augustin, rivière QC **39** D6-7
Saint-Augustin, RNOM de QC **38** B1-2
Saint-Augustin, RNOM de QC **39** E7
Saint-Augustin-de-Desmaures QC **27** B5
Saint-Augustin NMBS QC **38** B1-2
Saint-Augustin NMBS QC **39** E7
Saint-Augustin Nord-Ouest, rivière QC **39** D6
St. Barbe NL **38** B4
St. Barbe NL **39** D8
St. Barbe Bay NL **38** B3
Saint-Barnabé, île QC **29** C2-3
Saint-Barnabé-Nord QC **24** B7
Saint-Barnabé-Nord QC **25** F7
Saint-Barnabé-Nord QC **27** C2-3
Saint-Barnabé-Sud QC **24** D7
Saint-Barnabé-Sud QC **27** D2-3
Saint-Barthélemy QC **24** B6-7
Saint-Barthélemy QC **27** C2
Saint-Basile NB **29** F3
Saint-Basile NB **31** C2
Saint-Basile-de-Tableau QC **28** C7
Saint-Basile-le-Grand QC **24** D6-7
Saint-Basile-le-Grand QC **27** E2
Saint-Basile-Sud QC **27** B4
Saint-Benjamin QC **27** B7
St. Benedict SK **11** C5
Saint-Benoît QC **24** D5
Saint-Benoît-du-Lac QC **24** E8
Saint-Benoît-du-Lac QC **27** F4

Saint-Benoît-Labre QC **27** D6-7
St. Bernard NS **33** D1
Saint-Bernard QC **24** C7
Saint-Bernard QC **27** B5-6
Saint-Bernard QC **27** D2
Saint-Bernard, lac QC **24** A6
Saint-Bernard, lac QC **25** F6
Saint-Bernard, lac QC **27** B2
Saint-Bernard-de-Lacolle QC **24** E6
Saint-Bernard-de-Lacolle QC **27** F2
St. Bernardin ON **24** D3-4
St. Bernard's NL **36** E7
St. Bernard's NL **37** E5
St. Bernard's-Jacques Fontaine NL **36** E7
St. Bernard's-Jacques Fontaine NL **37** E5
Saint-Bernard-sur-Mer QC **28** E-F7
Saint-Blaise QC **24** E6
Saint-Blaise QC **27** F2
Saint-Bonaventure QC **24** C7
Saint-Bonaventure QC **27** D3
Saint-Boniface-de-Shawinigan QC **24** A-B7
Saint-Boniface-de-Shawinigan QC **25** F7
Saint-Boniface-de-Shawinigan QC **27** B2-3
St. Boswells SK **11** E4
St. Brendan's NL **37** B7
St. Bride, Mount AB **6** D4
St. Bride, Mount AB **7** A1
St. Bride's NL **36** F8
St. Bride's NL **37** F6
St. Brides AB **6** A7
St. Brides AB **11** A1
St. Brieux SK **11** C5
Saint-Bruno QC **28** C5
Saint-Bruno-de-Guigues QC **17** E8
Saint-Bruno-de-Guigues QC **22** A1
Saint-Bruno-de-Guigues QC **23** F1
Saint-Bruno-de-Kamouraska QC **28** E8
Saint-Bruno-de-Montarville QC **24** D6-7
Saint-Bruno-de-Montarville QC **27** E2
Saint-Cajetan QC **22** D7-8
Saint-Cajetan QC **24** A2
Saint-Calixte QC **24** C5
Saint-Calixte QC **27** D1
Saint-Camille QC **27** E5
Saint-Camille-de-Lellis QC **27** B7-8
Saint-Canut QC **24** D5
St. Carols NL **38** B5
Saint-Casimir QC **27** B4
Saint-Cassien-des-Caps QC **28** F6-7
St. Catherine's NL **37** E7
St. Catharines ON **19** B8
St. Catharines ON **20** E7
Saint-Célestin QC **24** B8
Saint-Célestin QC **27** C3
Saint-Césaire QC **24** D7
Saint-Césaire QC **27** E2-3
St. Chads NL **37** B6-7
Saint-Charles NB **31** E8
Saint-Charles NB **32** A6
St. Charles ON **17** F7
St. Charles ON **18** A6
St. Charles ON **18** A6
Saint-Charles, lac QC **27** A5
Saint-Charles-Borromée QC **24** C6
Saint-Charles-Borromée Q **27** D1-2
St. Charles Creek NT **10** A2
Saint-Charles-de-Bellechasse QC **27** B6
Saint-Charles-de-Bourget QC **28** C5
Saint-Charles-de-Drummond QC **24** C8
Saint-Charles-de-Drummond QC **27** D3-4
Saint-Charles-de-Mandeville QC **24** B6
Saint-Charles-de-Mandeville QC **25** F6
Saint-Charles-de-Mandeville QC **27** C2
Saint-Charles-Garnier QC **29** C4
Saint-Charles-sur-Richelieu QC **24** D7
Saint-Charles-sur-Richelieu QC **27** E2
Saint-Christophe-d'Arthabaska QC **27** D4-5
Saint-Chrysostome PE **32** A7
Saint-Chrysostome QC **24** E5-6
Saint-Chrysostome QC **27** F1
St. Clair, Lake ON **19** D2
St. Clair, Lake ON **20** F1
St. Clair Beach ON **19** E1

St. Clair NWA ON **19** E2
St. Clair River ON **19** C-D2
St. Clair River ON **20** E-F1
St. Clare Creek YT **4** D1
St. Claude MB **13** F3
St. Claude MB **14** E3
Saint-Claude QC **27** E4
Saint-Clément QC **29** D2
St. Clements ON **19** B5-6
St. Clements ON **20** D4-5
Saint-Cléophas QC **24** B6
Saint-Cléophas QC **25** F6
Saint-Cléophas QC **27** C2
Saint-Cléophas QC **29** C4
Saint-Clet QC **24** D-E5
Saint-Colomban QC **24** C5
St. Columban ON **19** B4-5
St. Columban ON **20** D3-4
Saint-Côme QC **24** B5-6
Saint-Côme QC **25** F5-6
Saint-Côme QC **27** C1
Saint-Côme-Linière QC **27** D7
Saint-Constant QC **24** D-E6
Saint-Constant QC **27** E1
St. Croix NB **32** D1-2
St. Croix Island International HS NB **32** E1-2
St. Croix PP NB **32** D1
Saint-Cuthbert QC **24** B6
Saint-Cuthbert QC **27** C2
Saint-Cyprien QC **27** C7
Saint-Cyprien QC **29** D2
Saint-Cyr QC **27** D4
Saint-Cyr, lac QC **23** C8
Saint-Cyr, lac QC **26** D2
St. Cyr, Mount YT **4** D5
Saint-Cyriac QC **28** C5
Saint-Cyrille-de-l'Islet QC **27** A7
Saint-Cyrille-de-Wendover QC **24** C8
Saint-Cyrille-de-Wendover QC **27** D3-4
St. Cyr Range YT **4** D5
Saint-Damase QC **24** D7
Saint-Damase QC **27** E2-3
Saint-Damase QC **29** B4
Saint-Damase-des-Aulnaies QC **28** F7-8
Saint-Damien QC **24** B6
Saint-Damien QC **25** F6
Saint-Damien QC **27** C2
Saint-Damien-de-Buckland QC **27** B6
Saint-Daniel QC **27** D5-6
Saint-David QC **24** C7
Saint-David QC **27** D2-3
St. David, Lake MB **13** C3
St. David, Lake MB **14** A4
Saint-David-de-Falardeau QC **28** B-C6
St. David's NL **36** C1-2
St. Davids ON **20** E7-8
Saint-Denis QC **24** C7
Saint-Denis QC **27** D2
Saint-Denis QC **28** E8
St. Denis SK **11** C4
St. Denis, RNF SK **11** C5
Saint-Denis-de-Brompton QC **27** E4
St. Denis National Wilderness Area SK **11** C5
Saint-Didace QC **24** B6-7
Saint-Didace QC **25** F6-7
Saint-Didace QC **27** C2
Saint-Dominique QC **24** D7
Saint-Dominique QC **27** E3
Saint-Dominique-du-Rosaire QC **23** C4
Saint-Donat QC **24** B5
Saint-Donat QC **25** F5
Saint-Donat QC **29** C3
Sainte-Adèle QC **24** C5
Sainte-Agathe MB **13** F3-4
Sainte-Agathe MB **14** E5
Sainte-Agathe QC **27** D4
Sainte-Agathe-des-Monts QC **24** C4-5
Sainte-Agnès QC **28** E7
Sainte-Agnès-de-Bellecombe QC **17** D8
Sainte-Agnès-de-Bellecombe QC **23** E2-3
Sainte-Agnès-de-Dundee QC **24** E4-5
Ste. Amelie MB **13** D2
Ste. Amelie MB **14** B2
Sainte-Anastasie QC **27** C5
Sainte-Angèle-de-Laval QC **24** B8
Sainte-Angèle-de-Laval QC **25** F8
Sainte-Angèle-de-Laval QC **27** C3
Sainte-Angèle-de-Mérici QC **29** C3-4
Sainte-Angèle-de-Monnoir QC **24** D6-7
Sainte-Angèle-de-Monnoir QC **27** E2
Sainte-Angèle-de-Prémont QC **24** B7
Sainte-Angèle-de-Prémont QC **25** F7
Sainte-Angèle-de-Prémont QC **27** C2
Ste. Anne MB **13** F4
Ste. Anne MB **14** E6
Sainte-Anne NB **31** B6
Ste. Anne, lac AB **6** A5
Ste. Anne, lac AB **8** C5

Ste. Anne, lac AB **8** C5
Sainte-Anne, lac QC **28** F8
Sainte-Anne, lac QC **39** E1
Sainte-Anne, mont QC **28** F6
Sainte-Anne, pointe QC **31** E8
Sainte-Anne, rivière QC **29** A6
Sainte-Anne, rivière QC **27** A5 B4
Sainte-Anne, rivière QC **28** F6
Sainte-Anne-de-Beaupré QC **27** A6
Sainte-Anne-de-Kent NB **31** E8
Sainte-Anne-de-Kent NB **32** A6
Sainte-Anne-de-la-Pérade QC **27** B4
Sainte-Anne-de-la-Rochelle QC **24** D8
Sainte-Anne-de-la-Rochelle QC **27** E3-4
Sainte-Anne-de-Madawaska NB **29** F4
Sainte-Anne-de-Madawaska NB **31** C2-3
Sainte-Anne-de-Portneuf QC **29** B2
Sainte-Anne-de-Prescott ON **24** D4
Sainte-Anne-des-Lacs QC **24** C5
Sainte-Anne-des-Monts QC **29** A6
Sainte-Anne-des-Monts QC **40** F6-7
Sainte-Anne-de-Sorel QC **24** C7
Sainte-Anne-de-Sorel QC **27** D2
Sainte-Anne-des-Plaines QC **24** C5-6
Sainte-Anne-des-Plaines QC **27** D1
Sainte-Anne-du-Lac QC **25** E3
Ste. Anne du Ruisseau NS **33** E2
Sainte-Apolline QC **27** B7
Sainte-Aurélie QC **27** C7
Sainte-Barbe QC **24** E5
Sainte-Béatrix QC **24** B6
Sainte-Béatrix QC **27** C1
Sainte-Blandine QC **29** C3
Sainte-Brigide-d'Iberville QC **24** E7
Sainte-Brigide-d'Iberville QC **27** E-F2
Sainte-Brigitte-de-Laval QC **27** A5-6
Sainte-Brigitte-des-Saults QC **24** C8
Sainte-Brigitte-des-Saults QC **27** D3
Sainte-Catherine QC **24** D6
Sainte-Catherine QC **27** E1
Sainte-Catherine-de-Hatley QC **27** F4
Sainte-Catherine-de-la-Jacques-Cartier QC **27** B4-5
Sainte-Cécile NB **29** D7
Sainte-Cécile NB **31** A-B8
Sainte-Cécile, mont QC **27** D-E6
Sainte-Cécile-de-Lévrard QC **27** B-C4
Sainte-Cécile-de-Masham QC **24** A7
Sainte-Cécile-de-Masham QC **27** F7
Sainte-Cécile-de-Masham QC **24** D1-2
Sainte-Cécile-de-Milton QC **24** D7
Sainte-Cécile-de-Milton QC **27** E3
Sainte-Cécile-de-Whitton QC **27** E6
Sainte-Christine QC **24** D8
Sainte-Christine QC **27** E3
Sainte-Christine-d'Auvergne QC **27** B4
Sainte-Claire QC **27** B6
Sainte-Claire, lac QC **19** D-E2
Sainte-Claire, lac ON **20** F1
Sainte-Claire, RNF ON **19** E2
Sainte-Clotilde-de-Beauce QC **27** C6
Sainte-Clotilde-de-Châteauguay QC **24** E6
Sainte-Clotilde-de-Châteauguay QC **27** F1
Sainte-Clotilde-de-Horton QC **24** C8
Sainte-Clotilde-de-Horton QC **27** D4
Sainte-Croix QC **27** B5
Sainte-Croix, lac NT **10** B5
Ste. Croix Cove NS **32** B2-3
Ste. Croix Cove NS **33** B2-3
Saint-Edgar QC **29** D7
Saint-Edgar QC **31** A6
Saint-Edmond QC **23** E4
Saint-Edmond QC **26** D8
Saint-Edmond QC **28** B3
Saint-Edmond-de-Grantham QC **24** C7
Saint-Edmond-de-Grantham QC **27** D3
Saint-Édouard QC **24** E6

Ste. Anne, lac AB **8** C5
Saint-Édouard QC **27** B4-5
Saint-Édouard QC **27** F1
Saint-Édouard-de-Frampton QC **27** B-C6
Saint-Édouard-de-Kent NB **31** E8
Saint-Édouard-de-Kent NB **32** A6
Saint-Édouard-de-Maskinongé QC **24** B7
Saint-Édouard-de-Maskinongé QC **25** F7
Saint-Édouard-de-Maskinongé QC **27** C2
St. Edward PE **35** B2
Sainte-Edwidge QC **27** F5
Sainte-Élisabeth QC **24** C6
Sainte-Élisabeth QC **27** D1
Sainte-Élisabeth-de-Proulx QC **28** B4
Sainte-Élisabeth-de-Warwick QC **27** D4
Sainte-Elizabeth MB **13** F3-4
Sainte-Elizabeth MB **13** F3-4
Sainte-Émélie-de-l'Énergie QC **24** B6
Sainte-Émélie-de-l'Énergie QC **25** F6
Sainte-Émélie-de-l'Énergie QC **27** C1
Sainte-Eulalie QC **24** C8
Sainte-Eulalie QC **27** C4
Sainte-Euphémie QC **27** B7
Sainte-Famille QC **27** A6
Sainte-Famille-d'Aumond QC **22** D7-8
Sainte-Famille-d'Aumond QC **24** A-B2
Sainte-Famille-d'Aumond QC **25** F2
Sainte-Félicité QC **27** A8
Sainte-Félicité QC **29** B5
Sainte-Flavie QC **29** B3
Sainte-Florence QC **29** C5
Sainte-Florence QC **31** A3-4
Sainte-Foy QC **27** B5
Sainte-Françoise QC **27** C4
Sainte-Françoise QC **29** D2
Ste. Genevieve MB **13** F4
Ste. Genevieve MB **14** E6
Sainte-Geneviève-de-Batiscan QC **24** A8
Sainte-Geneviève-de-Batiscan QC **25** F8
Sainte-Geneviève-de-Batiscan QC **27** B3-4
Sainte-Germaine-Boulé QC **17** C8
Sainte-Germaine-Boulé QC **23** C2
Sainte-Germaine-Station QC **27** C7
Sainte-Gertrude QC **24** B8
Sainte-Gertrude QC **25** F8
Sainte-Gertrude QC **27** C4
Sainte-Gertrude-Manneville QC **23** D4
Sainte-Hedwidge QC **25** A3-4
Sainte-Hedwidge QC **26** E8
Sainte-Hedwidge QC **28** C3-4
Sainte-Hélène QC **28** E8
Sainte-Hélène QC **29** E1
Sainte-Hélène-de-Bagot QC **24** D7-8
Sainte-Hélène-de-Bagot QC **27** D-E3
Sainte-Hélène-de-Chester QC **27** D5
Sainte-Hélène-de-Mancebourg QC **17** C7-8
Sainte-Hélène-de-Mancebourg QC **23** C2
Sainte-Hénédine QC **27** B6
Sainte-Jeanne-d'Arc QC **27** C4
Sainte-Jeanne-d'Arc QC **29** C4
Sainte-Julie QC **24** D6
Sainte-Julie QC **27** E2
Sainte-Julienne QC **24** C5-6
Sainte-Julienne QC **27** D1
Sainte-Justine QC **27** C7
Sainte-Justine-de-Newton QC **24** D4-5
Sainte-Clotilde-de-Beauce QC **27** C6
St. Eleanor's PE **32** A8
St. Eleanor's PE **34** C1
St. Eleanors PE **35** D3
Saint-Éleuthère QC **29** E1
St. Elias, Mount YT **4** E1
St. Elias Mountains YT **4** D1 E1-2
Saint-Élie QC **24** B7
Saint-Élie QC **27** B2
Saint-Élie-d'Orford QC **27** E4
Saint-Éloi QC **29** D1-2
Sainte-Louise QC **28** F7-8
Saint-Elphège QC **24** C7-8
Saint-Elphège QC **27** D3
Sainte-Luce QC **29** C3
Sainte-Lucie-de-Beauregard QC **27** B7-8
Sainte-Lucie-de-Doncaster QC **24** B5
Saint-Elzéar QC **27** C6
Saint-Elzéar QC **29** D8
Saint-Elzéar QC **29** E2
Saint-Elzéar QC **30** D1
Saint-Édouard QC **24** E6

Saint-Édouard QC **27** B4-5
Saint-Elzéar QC **31** A7
Sainte-Madeleine QC **24** D7
Sainte-Madeleine QC **27** E2
Sainte-Madeleine-de-la-Rivière-Madeleine QC **29** A8
Sainte-Madeleine-de-la-Rivière-Madeleine QC **30** A1
Sainte-Marcelline-de-Kildare QC **24** C5-6
Sainte-Marcelline-de-Kildare QC **27** C1
Sainte-Marguerite QC **24** C5
Sainte-Marguerite QC **27** B6
Sainte-Marguerite QC **29** C5
Sainte-Marguerite Nord-Est, rivière QC **28** B7 C7-8
Sainte-Marguerite, mont QC **27** C6
Sainte-Marguerite, rivière QC **28** C7-8
Sainte-Marguerite, rivière QC **39** E2
Sainte-Marguerite-de-Lingwick QC **27** E5-6
Sainte-Marguerite-Marie QC **28** B4
Sainte-Marguerite-Station QC **24** C5
Sainte-Marie, lac QC **27** C6
Sainte-Marie, lac QC **24** C2
Sainte-Marie, lac QC **22** E7-8
Sainte-Marie-de-Blandford QC **27** C4
Sainte-Marie-de-Kent QC **31** E8
Sainte-Marie-de-Kent NB **32** A6
Sainte-Marie-Saint-Raphaël NB **30** D2-3
Sainte-Marie-Saint-Raphaël NB **31** B8
Sainte-Marie-Salomé QC **24** C6
Sainte-Marie-Salomé QC **27** D1
Sainte-Marthe QC **24** D4-5
Sainte-Marthe-du-Cap QC **24** B8
Sainte-Marthe-du-Cap QC **25** F8
Sainte-Marthe-du-Cap QC **27** C3-4
Sainte-Marthe-sur-le-Lac QC **24** D5
Sainte-Martine QC **24** E5-6
Sainte-Véronique QC **24** A3
Sainte-Véronique QC **25** F3
Sainte-Victoire QC **24** C7
Sainte-Victoire QC **27** D2
Saint-Émile QC **27** A5
Saint-Émile-de-Suffolk QC **24** B8
Sainte-Mélanie QC **24** B6
Sainte-Mélanie QC **27** C1
Sainte-Monique QC **24** B8
Sainte-Monique QC **24** D5
Sainte-Monique QC **27** E1
Sainte-Monique QC **28** B4-5
Sainte-Odile QC **29** B5
Sainte-Paule QC **29** B4-5
Sainte-Perpétue QC **24** C8
Sainte-Perpétue QC **27** A8
Sainte-Perpétue QC **27** D3
Sainte-Pétronille QC **27** A6
Saint-Éphrem-de-Tring QC **27** C6
Saint-Épiphane QC **29** D1-2
Sainte-Praxède QC **27** D6
Ste. Rita MB **13** E4
Ste. Rita MB **14** E6
Sainte-Rita QC **29** D2
Sainte-Rita QC **31** A1
Sainte-Rosalie QC **24** D7
Sainte-Rosalie QC **27** E3
Sainte-Rose-de-Watford QC **27** C7
Sainte-Rose-de-Prescott ON **24** D4
Ste. Rose du Lac MB **13** D2
Ste. Rose du Lac MB **14** B2
Sainte-Rose-Gloucester NB **31** B8
Sainte-Sabine QC **24** E7
Sainte-Sabine QC **27** B7
Sainte-Sabine QC **27** F2
Sainte-Sabine-Station QC **27** B-C7
Sainte-Scholastique QC **24** D5
Sainte-Séraphine QC **27** D4
Sainte-Séraphine QC **27** D4
Saint-François QC **27** A6
Sainte-Sophie QC **24** C5-6
Sainte-Sophie QC **27** D1
Sainte-Sophie-de-Lévrard QC **24** C8
Sainte-Sophie-de-Lévrard QC **27** C4
Sainte-Sophie-de-Lévrard QC **25** F8
Sainte-Sophie-de-Lévrard QC **27** D3 E4
St. Esprit NS **34** D7
Saint-Esprit QC **24** C6
Saint-Esprit QC **27** D1

Saint-Esprit, rivière QC **24** C6
Saint-Esprit, rivière QC **27** D1
Sainte-Thècle QC **24** A8
Sainte-Thècle QC **25** E8
Sainte-Thècle QC **27** B3
Sainte-Thérèse QC **24** D5
Sainte-Thérèse QC **27** E1
Sainte-Thérèse, lac NT **10** A-B3
Sainte-Thérèse-de-Gaspé QC **30** C3
Sainte-Thérèse-de-la-Gatineau QC **22** D8
Sainte-Thérèse-de-la-Gatineau QC **24** B2
Sainte-Thérèse-de-la-Gatineau QC **25** F2
Saint-Étienne-de-Beauharnois QC **24** E5
Saint-Étienne-de-Beauharnois QC **27** F1
Saint-Étienne-de-Bolton QC **24** E8
Saint-Étienne-de-Bolton QC **27** F4
Saint-Étienne-de-Lauzon QC **27** B5-6
Saint-Étienne-des-Grès QC **24** B7
Saint-Étienne-des-Grès QC **25** F7
Saint-Étienne-des-Grès QC **27** C3
St. Eugene ON **24** D4
St. Eugene ON **24** D4
Saint-Eugène QC **24** C7
Saint-Eugène QC **26** D8
Saint-Eugène QC **27** D3
Saint-Eugène QC **28** A-B4
Saint-Eugène QC **28** F7
Saint-Eugène-de-Chazel QC **23** C3
Saint-Eugène-de-Guigues QC **22** A2
Saint-Eugène-de-Guigues QC **23** F2
Saint-Eugène-de-Ladrière QC **29** C2-3
Saint-Eugène-de-Ladrière QC **31** A1
Sainte-Ursule QC **24** B7
Sainte-Ursule QC **25** F7
Sainte-Ursule QC **27** C2
Saint-Eusèbe QC **29** E2
Saint-Eusèbe QC **31** B1
St. Eustache MB **13** E3
St. Eustache MB **14** D4
Saint-Eustache QC **24** D5
Saint-Évariste-de-Forsyth QC **27** D6
Saint-Fabien-de-Panet QC **27** B7-8
Saint-Fabien-sur-Mer QC **29** C2
Saint-Faustin-Lac-Carré QC **24** B4
Saint-Félicien QC **26** E8
Saint-Félicien QC **28** B3-4
St. Felix PE **35** B2
Saint-Félix-d'Otis QC **28** C6
Saint-Félix-de-Dalquier QC **23** C4
Saint-Félix-de-Kingsey QC **24** C8
Saint-Félix-de-Kingsey QC **27** D4
Saint-Félix-de-Valois QC **24** B6
Saint-Félix-de-Valois QC **27** C2
Saint-Ferréol-les-Neiges QC **27** A6
Saint-Ferréol-les-Neiges QC **28** F6
Saint-Fidèle-de-Mont-Murray QC **28** E8
Saint-Fidèle-de-Restigouche QC **29** D6
Saint-Fidèle-de-Restigouche QC **31** A4
Saintfield ON **20** B8
Saintfield ON **21** E2
St. Fintan's NL **36** C2
Saint-Flavien QC **27** B5
Saint-Fortunat QC **27** D5
St. Francis AB **6** B5
St. Francis AB **8** D5
St. Francis River NB **29** F2
St. Francis, Cape NL **36** C1
St. Francis Harbour NS **34** E6
St. Francois QC **27** A6-7
Saint-François, lac QC **24** E5
Saint-François, lac QC **27** D6
Saint-François, rivière QC **27** E5
Saint-François-d'Assise QC **29** D5

Saint-François-d'Assise QC **31** A3-4
Saint-François-de-Madawaska NB **29** F2
Saint-François-de-Madawaska NB **31** C1
Saint-François-de-Masham QC **21** A7
Saint-François-de-Masham QC **22** F7
Saint-François-de-Masham QC **24** D1
Saint-François-de-Sales QC **25** A8
Saint-François-de-Sales QC **26** E8
Saint-François-de-Sales QC **28** C4
Saint-François-du-Lac QC **24** C7
Saint-François-du-Lac QC **27** D2-3
St. François Xavier MB **13** E3-4
St. François Xavier MB **14** D4
Saint-François-Xavier-de-Brompton QC **27** E4
Saint-François-Xavier-de-Viger QC **29** D1-2
Saint-Frédéric QC **27** C6
St. Front SK **11** C5-6
Saint-Fulgence QC **28** C6
Saint-Gabriel QC **24** B6
Saint-Gabriel QC **25** F6
Saint-Gabriel QC **27** C2
Saint-Gabriel QC **29** C3-4
Saint-Gabriel-de-Kamouraska QC **28** F8
Saint-Gabriel-de-Valcartier QC **27** A5
Saint-Gédéon QC **27** D7
Saint-Gédéon QC **28** C4-5
St. George NB **32** E2
St. George ON **19** B6
St. George ON **20** D-E5
St. George, Cape NL **36** C1
St. George, Lake MB **13** C3-4
St. George, Lake MB **14** A4
Saint-Georges MB **13** E4-5
Saint-Georges MB **14** C6
St. George's NL **36** C2
St. Georges PE **34** C3-4
St. Georges PE **35** E7
St.-Georges QC **24** A8
St.-Georges QC **25** E8
St.-Georges QC **27** B3
St.-Georges QC **27** C7
St.-Georges, baie de NS **34** B1
St. George's Bay NL **36** C1
St. George's Bay NL **40** F8
St. Georges Bay NS **34** D5
St. Georges Channel NS **34** D6
Saint-Georges-de-Cacouna QC **29** D1
Saint-Georges-de-Clarenceville QC **24** E6-7
Saint-Georges-de-Clarenceville QC **27** F2
Saint-Georges-de-Malbaie QC **30** C3
Saint-Georges-de-Windsor QC **24** E4-5
St. George's Hill SK **9** E8
St. George's Hill SK **12** E3
Saint-Gérard QC **27** D5
Saint-Gérard-de-Berry QC **23** C4
Saint-Gérard-des-Laurentides QC **24** A7
Saint-Gérard-des-Laurentides QC **25** E-F7
Saint-Gérard-des-Laurentides QC **27** B3
Saint-Gérard-Majella QC **24** A6
Saint-Gérard-Majella QC **24** C7
Saint-Gérard-Majella QC **27** D1-2
Saint-Gérard-Majella QC **27** D3
Saint-Germain QC **28** E8
Saint-Germain, lac QC **22** F7-8
Saint-Germain, lac QC **24** C2
Saint-Germain, rivière QC **24** C-D8
Saint-Germain, rivière QC **27** D3-4 E4
Saint-Germain-de-Grantham QC **24** C8
Saint-Germain-de-Grantham QC **27** D3
Saint-Germains, lac QC **28** C6-7
Saint-Gervais QC **27** B6
Saint-Gilbert PE **32** A7
Saint-Gilbert PE **34** B1
Saint-Gilbert PE **35** D2
Saint-Gilbert QC **27** B4
Saint-Gilles QC **27** B5
Saint-Godefroi QC **30** D1-2
Saint-Godefroi QC **31** A7
Saint-Grégoire QC **24** B8
Saint-Grégoire QC **25** F8
Saint-Grégoire QC **27** E5
Saint-Grégoire-de-Greenlay QC **27** E4
St. Gregor SK **11** C5

St. Gregory, Cape NL 36 A2
St. Gregory, Cape NL 38 F1-2
St. Gregory, Mount NL 36 A2-3
St. Gregory, Mount NL 38 F2
Saint-Guillaume QC 24 C7
Saint-Guillaume QC 27 D3
Saint-Guillaume-de-Granada QC 17 D8
Saint-Guillaume-Nord QC 24 A5
Saint-Guillaume-Nord QC 25 E-F5
Saint-Guy QC 29 D2
Saint-Guy QC 31 A1
St . Helens ON 19 A4
St. Helens ON 20 C3
Saint-Henri QC 27 B6
Saint-Henri-de-Taillon QC 28 B4-5
Saint-Hermas QC 24 D5
Saint-Herménégilde QC 27 F5
Saint-Hilaire NB 29 F3
Saint-Hilaire NB 31 C1-2
Saint-Hilaire-de-Dorset QC 27 D6
Saint-Hilarion QC 28 E7
Saint-Hippolyte QC 24 C5
Saint-Hippolyte QC 27 D1
Saint-Honoré QC 27 D6
Saint-Honoré QC 28 C6
Saint-Honoré QC 29 E2
Saint-Hubert QC 24 D6
Saint-Hubert QC 27 E2
Saint-Hubert QC 29 D-E2
St. Hubert Mission SK 11 E7
Saint-Hugues QC 24 C7
Saint-Hugues QC 27 D2-3
Saint-Hyacinthe QC 24 D7
Saint-Hyacinthe QC 27 E3
Saint-Ignace NB 31 E7
Saint-Ignace NB 32 A5
Saint-Ignace-de-Loyola QC 24 C6-7
Saint-Ignace-de-Loyola QC 27 D2
Saint-Ignace-de-Stanbridge QC 24 E7
Saint-Ignace-de-Stanbridge QC 27 F2-3
Saint-Ignace-du-Lac QC 24 A5-6
Saint-Ignace-du-Lac QC 25 E5-6
Saint-Ignace-du-Lac QC 27 B1
St. Ignace Island ON 16 E6-7 F6-7
St. Ignace Island ON 17 C1
Saint-Irénée NB 30 F1-2
Saint-Irénée NB 31 B7
Saint-Irénée QC 28 E7
St. Isidore AB 9 E3-4
Saint-Isidore NB 30 F2
Saint-Isidore NB 31 B7
St. Isidore ON 24 D3-4
Saint-Isidore QC 24 E6
Saint-Isidore QC 27 B6
Saint-Isidore QC 27 F1
Saint-Isidore QC 30 C2-3
Saint-Isidore-d'Auckland QC 27 F5
Saint-Isidore-de-Bellevue SK 11 B4-5
St. Jacobs ON 19 B6
St. Jacobs ON 20 D5
Saint-Jacques NB 29 F3
Saint-Jacques NB 31 C1-2
St. Jacques NL 36 E6
St. Jacques NL 37 E4
Saint-Jacques QC 24 C6
Saint-Jacques QC 27 D1-2
St. Jacques-Coombs Cove NL 36 E6
St. Jacques-Coombs Cove NL 37 E4
Saint-Jacques-de-Leeds QC 27 C5-6
Saint-Jacques-le-Majeur-de-Wolfestown QC 27 D5
Saint-Jacques-le-Mineur QC 24 E6
Saint-Jacques-le-Mineur QC 27 F2
St. James, Cape BC 1 C2
Saint-Janvier QC 24 D5
Saint-Janvier QC 27 D1
Saint-Jean QC 27 B6
Saint-Jean, fleuve NB 29 F3
Saint-Jean, fleuve NB 31 D3
Saint-Jean, fleuve NB 32 C2
Saint-Jean, lac QC 26 E8
Saint-Jean, lac QC 28 B-C4
Saint-Jean, rivière QC 28 D7
Saint-Jean, rivière QC 30 B-C2 C1
Saint-Jean, rivière QC 39 E3
St. Jean Baptiste MB 13 F3-4
St. Jean Baptiste MB 14 F5
Saint-Jean-Baptiste QC 24 D7
Saint-Jean-Baptiste QC 27 E2
Saint-Jean-Baptiste-de-Restigouche NB 29 E5
Saint-Jean-Baptiste-de-Restigouche NB 31 B4

Saint-Jean-Chrysostome QC 27 B6
Saint-Jean-de-Brébeuf QC 27 C5
Saint-Jean-de-Cherbourg QC 29 B5
Saint-Jean-de-Dieu QC 29 D2
Saint-Jean-de-la-Lande QC 27 D6-7
Saint-Jean-de-la-Lande QC 29 E-F2
Saint-Jean-de-la-Lande QC 31 C1
Saint-Jean-de-Matapédia QC 29 D5
Saint-Jean-de-Matapédia QC 31 A3-4
Saint-Jean-de-Matha QC 24 B6
Saint-Jean-de-Matha QC 25 F6
Saint-Jean-de-Matha QC 27 C1
Saint-Jean-des-Piles QC 24 A7
Saint-Jean-des-Piles QC 25 E7
Saint-Jean-des-Piles QC 27 B3
Saint-Jean-Port-Joli QC 28 F7
Saint-Jean-sur-le-Lac QC 22 D8
Saint-Jean-sur-le-Lac QC 24 A2
Saint-Jean-sur-le-Lac QC 25 F2
Saint-Jean-sur-Richelieu QC 24 E6
Saint-Jean-sur-Richelieu QC 27 F2
Saint-Jérôme QC 24 C5
Saint-Jérôme QC 27 D1
St. Joachim ON 19 E2
Saint-Joachim QC 27 A6
Saint-Joachim-de-Courval QC 24 C8
Saint-Joachim-de-Courval QC 27 D3
Saint-Joachim-de-Shefford QC 24 D8
Saint-Joachim-de-Shefford QC 27 E3
Saint-Jogues QC 29 D8
Saint-Jogues QC 30 D1
Saint-Jogues QC 31 A7
St. John NB 32 D4
St. John NB 33 A-B1
St. John, Cape NL 38 D5
St. John, Lake NL 36 C7
St. John, Lake NL 37 C5
St. John Bay NL 38 C3
St. John Bay NL 39 E7
St. John Harbour NB 32 E4
St. John Harbour NB 33 B1
St. John Harbour NL 36 E4
St. John Harbour NL 37 E4
St. John Island NL 38 C3
St. John Island NL 39 E7
St. John River NB 31 C2
St. John River NB 31 F3
St. John River NB 32 B-C1
St. John's NL 37 D8
St. Jones Within NL 37 C7
St. Jones Without NL 37 D6-7
Saint-Joseph MB 13 F3-4
Saint-Joseph MB 14 F4-5
Saint-Joseph NB 31 F8
St. Joseph NS 32 C6
St. Joseph NS 33 A8
St. Joseph NS 33 D1-2
St. Joseph NS 34 E4
St. Joseph ON 19 B3-4
St. Joseph ON 20 D2-3
St. Joseph, lac QC 24 C4-5
St. Joseph, lac QC 24 A5
Saint-Joseph, lac QC 27 C5
St. Joseph, Lake ON 13 D8
St. Joseph, Lake ON 16 C4-5
St. Joseph, Lake ON 40 F1
Saint-Joseph-de-Beauce QC 27 C6
Saint-Joseph-de-Cléricy QC 17 D8
Saint-Joseph-de-Kamouraska QC 28 E8
Saint-Joseph-de-Kamouraska QC 29 E1
Saint-Joseph-de-la-Rive QC 28 E7
Saint-Joseph-de-Lepage QC 29 C3
Saint-Joseph-de-Madawaska NB 29 E3
Saint-Joseph-de-Madawaska NB 31 C2
Saint-Joseph-de-Mékinac QC 25 E7-8
Saint-Joseph-de-Mékinac QC 27 A3
Saint-Joseph-de-Sorel QC 24 C6-7
Saint-Joseph-de-Sorel QC 27 D2
St. Joseph du Moine NS 27 B-C4
St. Joseph Island ON 17 F4
St. Joseph's NL 36 E7
St. Joseph's NL 37 E5
St. Joseph's NL 37 E7

St. Joseph's Cove NL 36 D5-6
St. Joseph's Cove NL 37 D3-4
St. Joseph's Island, RNOM ON 17 F4
St. Joseph's Island NMBS ON 17 F4
Saint-Jovite QC 24 B4
Saint-Jude QC 24 C7
Saint-Jude QC 27 D2-3
St. Judes NL 36 A3-4
St. Judes NL 37 A1-2
St. Judes NL 38 F3
Saint-Jules QC 27 C6
Saint-Jules QC 29 D7
Saint-Jules QC 31 A6
Saint-Julien QC 27 D5
St. Julien's NL 38 C5
St. Julien's NL 39 D-E8
Saint-Just-de-Bretenières QC 27 B7-8
Saint-Juste-du-Lac QC 29 E2
Saint-Juste-du-Lac QC 31 B1
Saint-Justin QC 24 B7
Saint-Justin QC 25 F7
Saint-Justin QC 27 C2
St. Kyran's NL 36 E7-8
St. Kyran's NL 37 E5-6
St. Labre MB 13 F4-5
St. Labre MB 14 F6-7
Saint-Lambert QC 17 C7-8
Saint-Lambert QC 23 B2
Saint-Lambert QC 24 D6
Saint-Lambert QC 27 E2
Saint-Lambert-de-Lauzon QC 27 B5-6
St. Laurent MB 13 E3
St. Laurent MB 14 C4
Saint-Laurent QC 24 D5-6
Saint-Laurent QC 27 A6
Saint-Laurent QC 27 E1
Saint-Laurent, cap NB 35 F2-3
Saint-Laurent, Cape NB 35 F2-3
Saint-Laurent, fleuve ON 21 C-D8
Saint-Laurent, fleuve ON 24 B7 C6-7
Saint-Laurent, fleuve QC 24 E4-5
Saint-Laurent, fleuve QC 25 F8
Saint-Laurent, fleuve QC 27 C2-3 D2
Saint-Laurent, fleuve QC 28 E7-8 F7
Saint-Laurent, fleuve QC 29 B2-4
Saint-Laurent, fleuve QC 39 F2
Saint-Laurent, fleuve QC 40 F6-7
Saint-Laurent, golfe du NB 31 C8
Saint-Laurent, golfe du NL 36 A1-2
Saint-Laurent, golfe du NL 38 D1-2
Saint-Laurent, golfe du PE 32 A7
Saint-Laurent, golfe du PE 34 A3-5
Saint-Laurent, golfe du PE 35 B4-6
Saint-Laurent, golfe du QC 30 D5-6
Saint-Laurent, golfe du QC 39 F6
Saint-Laurent, golfe du QC 40 F7-8
St. Lawrence NL 36 F6-7
St. Lawrence NL 37 F4-5
St. Lawrence PE 35 B1-2
St. Lawrence, Bay NS 34 A7
St. Lawrence, Cape NS 34 A7
St. Lawrence, Gulf of NB 31 C8
St. Lawrence, Gulf of NL 36 A1-2
St. Lawrence, Gulf of NL 38 D1-2
St. Lawrence, Gulf of PE 32 A8
St. Lawrence, Gulf of PE 34 A3-5
St. Lawrence, Gulf of PE 35 B4-6
St. Lawrence, Gulf of QC 30 D5-6 E5-6
St. Lawrence, Gulf of QC 39 F5-6
St. Lawrence, Gulf of QC 40 F7-8
St. Lawrence Bay NS 34 A7
St. Lawrence Islands NP ON 21 D-E7 D8
St. Lawrence River ON 21 D8
St. Lawrence River ON 24 F2-3
St. Lawrence River ON 27 B-C4
St. Lawrence River QC 28 E8
St. Lawrence River QC 29 A4 B3-4

St. Lawrence River QC 39 F2
St. Lawrence River QC 40 F6-7
Saint-Lazare MB 11 E7-8
Saint-Lazare MB 13 E1
Saint-Lazare QC 24 D5
Saint-Lazare-de-Bellechasse QC 27 B6-7
Saint-Léandre QC 29 B4
Saint-Léolin NB 29 E8
Saint-Léolin NB 30 E1
Saint-Léolin NB 31 B7
St. Leon MB 13 F3
St. Leon MB 14 F3
Saint-Léon QC 28 B5
Saint-Léonard QC 29 F4
Saint-Léonard NB 31 C2-3
Saint-Léonard QC 24 D6
Saint-Léonard QC 27 A4
Saint-Léonard QC 27 E2
Saint-Léonard-d'Aston QC 24 C8
Saint-Léonard-d'Aston QC 27 C-D3 C4
Saint-Léonard-Parent QC 29 F4
Saint-Léonard-Parent NB 31 C2-3
St. Leonards NL 36 E8
St. Leonards NL 37 E6
Saint-Léon-de-Standon QC 27 B7
Saint-Léon-le-Grand QC 24 B7
Saint-Léon-le-Grand QC 25 F7
Saint-Léon-le-Grand QC 27 C3
Saint-Léon-le-Grand QC 29 C4-5
St. Lewis NL 39 C8
St. Lewis River NL 39 C7-8
Saint-Liboire QC 24 D7
Saint-Liboire QC 27 E3
Saint-Liguori QC 24 C6
Saint-Liguori QC 27 D1
Saint-Lin QC 24 C6
Saint-Lin QC 27 D1
St. Lina AB 6 A7
St. Lina AB 9 F7
St. Lina AB 11 A1
St. Lina AB 12 F1
St. Louis PE 35 B2
Saint-Louis QC 24 C7
Saint-Louis QC 27 D2
St. Louis SK 11 B5
Saint-Louis, lac QC 24 B4
Saint-Louis, lac QC 24 D5
Saint-Louis, lac QC 24 F4
Saint-Louis, lac QC 27 E1
Saint-Louis-de-Blandford QC 27 C4
Saint-Louis-de-France QC 24 B7-8
Saint-Louis-de-France QC 25 F7-8
Saint-Louis-de-France QC 27 C3
Saint-Louis-de-Gonzague QC 24 E5
Saint-Louis-de-Gonzague QC 27 C7
Saint-Louis-de-Gonzague QC 27 F1
Saint-Louis-de-Kent NB 31 D8
Saint-Louis-de-Kent NB 35 B4-6
Saint-Louis-de-l'Isle-aux-Coudres QC 28 F7
Saint-Louis-du-Ha! Ha! QC 29 E1-2
St. Louis Mission ON 18 F7-8
St. Louis Mission NHS ON 20 A6-7
Saint-Luc QC 24 D-E6
Saint-Luc QC 27 B7
Saint-Luc QC 27 E2
Saint-Luc QC 29 B5
Saint-Luc-de-Vincennes QC 24 A-B8
Saint-Luc-de-Vincennes QC 25 F8
Saint-Luc-de-Vincennes QC 27 B3-4
Saint-Lucien QC 24 C8
Saint-Lucien QC 27 D4
Saint-Ludger QC 27 D6-7
Saint-Ludger-de-Milot QC 28 B4
St. Lunaire NL 38 A5
St. Lunaire-Griquet NL 38 A5
St. Lupicin MB 14 E3
Saint-Magloire QC 27 B7
Saint-Majorique QC 30 B2
Saint-Majorique-de-Grantham QC 24 C7-8
Saint-Majorique-de-Grantham QC 27 D3
Saint-Malachie QC 27 B6-7
St. Malo MB 13 F4
St. Malo MB 14 F5
Saint-Malo QC 27 F5
St. Malo PP MB 13 F4
St. Malo PP MB 14 F5
Saint-Marc QC 23 D4
Saint-Marc-de-Latour QC 28 E8
Saint-Marc-de-Latour QC 29 B2
Saint-Marc-des-Carrières QC 27 B4
Saint-Marc-des-Carrières QC 29 A4 B3-4

Saint-Marc-du-Lac-Long QC 27 B1
Saint-Marc-du-Lac-Long QC 31 C1
Saint-Marcel QC 27 A8
Saint-Marcel-de-Richelieu QC 24 C7
Saint-Marcel-de-Richelieu QC 27 D2-3
Saint-Marcellin QC 29 C3
Saint-Marc-sur-Richelieu QC 24 D6-7
Saint-Marc-sur-Richelieu QC 27 E2
St. Margarets NB 31 D7
St. Margarets PE 34 B4
St. Margarets PE 35 D7
St. Margaret's Bay NL 38 B3
St. Margaret's Bay NS 32 F7-8
St. Margaret's Bay NS 33 C5
St. Margaret Village NS 34 A7
St. Martin MB 13 D3
St. Martin MB 14 A3
Saint-Martin QC 27 B5
St. Martin, Lake MB 13 D3
St. Martin, Lake MB 14 A3
Saint-Martin-de-Restigouche NB 29 E5
Saint-Martin-de-Restigouche NB 31 B3-4
St. Martin Junction MB 13 C-D3
St. Martin Junction MB 14 A3
St. Martins NB 32 D5
St. Martins NB 33 A2
St. Martins NS 33 D1
St. Mary, Cape NS 33 E1
St. Mary Reservoir AB 6 F6
St. Mary Reservoir AB 7 F5
St. Mary River BC 6 F4
St. Mary River BC 7 E1
St. Marys QC 32 C3
St. Mary's NL 37 F7
St. Mary's NS 33 A8
St. Mary's NS 34 F4
St. Marys ON 19 B5
St. Marys ON 20 D4
St. Mary's, Cape NL 36 B8
St. Mary's, Cape NL 37 F6
St. Marys, Cape NS 34 F5
St. Mary's Alpine PP BC 6 E-F4
St. Mary's Alpine PP BC 7 E1
St. Mary's Bay NL 37 F7
St. Mary's Bay NS 33 D1
St. Mary's Bay PE 35 E-F7
St. Mary's River NS 34 E-F4 F5
St. Mary's River NS 34 F4-5
St. Marys Road PE 34 C3
St. Marys Road PE 35 F6
Saint-Mathias-sur-Bonneterre QC 27 E5-6 F5
Saint-Mathias-sur-Richelieu QC 24 D6-7
Saint-Mathias-sur-Richelieu QC 27 E2
Saint-Mathieu QC 24 A7
Saint-Mathieu QC 24 E6
Saint-Mathieu QC 25 F7
Saint-Mathieu QC 27 B3
Saint-Mathieu QC 27 F1
Saint-Mathieu QC 29 D2
Saint-Mathieu QC 31 A1
Saint-Mathieu, lac QC 29 D2
Saint-Mathieu-d'Harricana QC 23 D4
Saint-Maure NB 31 A5
Saint-Maurice QC 24 B8
Saint-Maurice QC 25 F8
Saint-Maurice QC 27 C3
Saint-Maurice, RF du QC 25 D7
Saint-Maurice, RF du QC 27 A2
Saint-Maurice, rivière QC 25 A-B6
Saint-Maurice, rivière QC 25 D7
Saint-Maurice, rivière QC 27 B3-4
Saint-Maurice-de-Dalquier QC 23 C-D4
Saint-Maurice-de-l'Échouerie QC 30 B2-3
Saint-Maxime-du-Mont-Louis QC 29 A8
Saint-Médard QC 29 D2
Saint-Médard QC 31 A1
Saint-Méthode QC 26 D8
Saint-Méthode QC 28 B3
Saint-Méthode-de-Frontenac QC 27 D6
St. Michael AB 6 A6
St. Michael AB 8 B7
Saint-Michel QC 24 E6
Saint-Michel QC 27 F1
Saint-Michel, bras QC 27 B6
Saint-Michel-de-Bellechasse QC 27 A6
Saint-Michel-des-Saints QC 24 A5
Saint-Michel-des-Saints QC 25 E5

Saint-Michel-des-Saints QC 27 B1
Saint-Michel-de-Wentworth QC 24 C5
Saint-Modeste QC 29 D1
Saint-Moïse QC 29 C4
Saint-Narcisse QC 24 A8
Saint-Narcisse QC 25 F8
Saint-Narcisse QC 27 B3-4
Saint-Narcisse QC 27 B6 C5-6
Saint-Narcisse-de-Rimouski QC 29 C3
Saint-Nazaire QC 24 C-D8
Saint-Nazaire QC 27 D3
Saint-Nazaire QC 28 C5
Saint-Nazaire-de-Berry QC 23 C4
Saint-Nazaire-de-Buckland QC 27 B7
Saint-Nérée QC 27 B6-7
Saint-Nicéphore QC 24 C8
Saint-Nicéphore QC 27 D3-4
St. Nicholas PE 32 A7
St. Nicholas PE 34 C1
St. Nicholas PE 35 D2
Saint-Nicolas QC 27 B5
Saint-Noël QC 29 C4
St. Nora Lake ON 21 B2
Saint-Norbert MB 13 F4
Saint-Norbert QC 31 E8
Saint-Norbert NB 32 A6
Saint-Norbert QC 24 B6-7
Saint-Norbert QC 27 C2
Saint-Norbert-d'Arthabaska QC 27 D4-5
Saint-Norbert-du-Mont-Brun QC 17 C-D8
Saint-Norbert Provincial Heritage Park MB 14 E5
Saint-Octave-de-Métis QC 29 B-C4
Saint-Odilon QC 27 C6-7
St. Ola ON 21 C4
Saint-Omer QC 27 A8
Saint-Omer QC 28 F8
Saint-Omer QC 29 D7
Saint-Omer QC 31 A5
Saint-Onésime QC 28 F8
Saint-Ours QC 24 C7
Saint-Ours QC 27 D2
Saint-Ours, canal de QC 24 C7
Saint-Ours, canal de QC 27 D2
Saint-Ours Canal QC 24 C7
Saint-Ours Canal QC 27 D2
Saint-Pacôme QC 28 F8
Saint-Pamphile QC 27 A8
Saint-Pancrace, pointe QC 34 E-F4 F5
St. Pascal ON 24 D3
Saint-Pascal QC 28 E8
Saint-Patrice QC 28 D8
Saint-Patrice QC 29 D1
Saint-Patrice, lac QC 24 A3
Saint-Patrice, ZEC QC 22 D5
Saint-Patrice-de-Beaurivage QC 27 C5-6
St. Patrick's Channel NS 34 C6
St. Patricks Channel NS 34 E4-5
St. Patricks Channel NS 34 C6
St. Paul AB 6 A7
St. Paul AB 11 A1-2
Saint-Paul NB 31 F7
St. Paul NS 33 A7-8
St. Paul NS 34 E3
Saint-Paul QC 24 C6
Saint-Paul, lac QC 24 A3
Saint-Paul, lac QC 25 E3
Saint-Paul, lac QC 28 D4
Saint-Paul, rivière NL 40 E8
Saint-Paul, rivière QC 38 A2
Saint-Paul, rivière QC 39 D7
Saint-Paul-d'Abbotsford QC 24 D7
Saint-Paul-d'Abbotsford QC 27 E3
Saint-Paul-de-la-Croix QC 28 C-D1 E2
Saint-Paul-de-l'Île-aux-Noix QC 24 E6
Saint-Paul-de-l'Île-aux-Noix QC 27 F2
Saint-Paul-de-Montminy QC 27 B7
Saint-Paul-du-Nord QC 29 B-C2
Saint-Paulin QC 24 B7
Saint-Paulin QC 25 F7
Saint-Paulin QC 27 C2-3
St. Pauls NL 38 E2
St. Pauls NL 39 F7
St. Paul's Bay NL 40 E8
St. Pauls Inlet NL 38 E2
St. Pauls Inlet NL 39 F7
St. Pauls Station ON 19 B5
St. Pauls Station ON 20 D4
Saint-Père, lac QC 23 C7
Saint-Père, lac QC 26 D2
St. Peter's NL 34 B6
St. Peters PE 34 B3
St. Peters PE 35 D6
St. Peters Bay NS 34 B3
St. Peters Bay PE 35 D6

St. Peters Canal NHS NS 34 D6-7
St. Peters Harbour PE 35 D6
St. Peters Island PE 34 C2
St. Peters Island PE 35 E4-5
St. Peters Lake PE 35 D6
Saint-Philémon QC 27 B7
Saint-Philibert QC 27 C7
Saint-Philippe NB 31 F8
Saint-Philippe NB 32 B6
Saint-Philippe QC 24 D-E6
Saint-Philippe PE 35 D2
Saint-Philippe QC 27 E2
Saint-Philippe-d'Argenteuil QC 24 D4
Saint-Philippe-de-Néri QC 28 E8
Saint-Pie QC 24 D7
Saint-Pie QC 27 E3
Saint-Pie-de-Guire QC 24 C7-8
Saint-Pie-de-Guire QC 27 D3
Saint-Pierre QC 24 C6
Saint-Pierre QC 24 E5
Saint-Pierre QC 27 A6
Saint-Pierre QC 27 D1-2
Saint-Pierre QC 27 F1
Saint-Pierre, lac QC 24 B7
Saint-Pierre, lac QC 28 C3
Saint-Pierre-Baptiste QC 27 C5
Saint-Pierre-de-Broughton QC 27 C6
Saint-Pierre-de-Lamy QC 29 E2
Saint-Pierre-de-Sorel QC 24 C7
Saint-Pierre-de-Sorel QC 27 D2
Saint-Pierre-de-Wakefield QC 21 A8
Saint-Pierre-de-Wakefield QC 22 F8
Saint-Pierre-de-Wakefield QC 24 C-D2
Saint-Pierre-Jolys MB 13 F4
Saint-Pierre-Jolys MB 14 E5
Saint-Pierre-les-Becquets QC 24 B7
Saint-Pierre-Montmagny QC 27 A7
Saint-Placide QC 24 D5
Saint-Placide-de-Charlevoix QC 28 F6-7
Saint-Polycarpe QC 24 E4-5
Saint-Prime QC 26 E8
Saint-Prime QC 28 B-C4
Saint-Prosper PE 34 C3
Saint-Prosper QC 25 E8
Saint-Prosper QC 27 B4
Saint-Prosper QC 27 C7
Saint-Quentin NB 29 E4-5
Saint-Quentin NB 31 B-C3
Saint-Raphaël PE 35 D2
Saint-Raphaël QC 27 B6-7
St. Raphaels ON 24 E4
St. Raymond MB 14 E6
Saint-Raymond QC 27 A4
Saint-Rédempteur QC 24 D4-5
Saint-Rédempteur QC 27 B5
Saint-René QC 27 D7
Saint-René-de-Matane QC 29 B5
Saint-Robert QC 24 C7
Saint-Robert QC 27 D2
Saint-Robert-Bellarmin QC 27 D7
St. Roch PE 34 A1
St. Roch PE 35 D2
Saint-Roch QC 17 D8
Saint-Roch QC 24 B7
Saint-Roch-de-l'Achigan QC 24 C6
Saint-Roch-de-l'Achigan QC 27 D1-2
Saint-Roch-de-Mékinac QC 24 A7-8
Saint-Roch-de-Mékinac QC 25 E7-8
Saint-Roch-de-Mékinac QC 27 B3
Saint-Roch-des-Aulnaies QC 28 F7
Saint-Romain QC 27 D6
Saint-Romuald QC 27 B6
Saint-Rosaire QC 27 C4
Saint-Rosaire, rivière QC 27 C4
St. Rose NS 34 B4
Saint-Rose-du-Nord QC 28 C6-7
Saint-Samuel QC 24 C8
Saint-Samuel QC 27 D4
Saint-Samuel-Station QC 27 D6
Saints-Anges QC 27 C6
Saint-Sauveur NB 29 E-F8

Saint-Sauveur NB 30 F1
Saint-Sauveur NB 31 B7
Saint-Sauveur-des-Monts QC 24 C5
Saint-Sébastien QC 24 E7
Saint-Sébastien QC 27 D6
Saint-Sébastien QC 27 F2
Saint-Sébastien, mont QC 27 D6
Saint-Sévère QC 24 B7
Saint-Sévère QC 25 F7
Saint-Sévère QC 27 C2-3
Saint-Séverin QC 24 A8
Saint-Séverin QC 25 E8
Saint-Séverin QC 27 B3
Saint-Séverin QC 27 C6
St. Shotts NL 37 F7
Saint-Siméon QC 28 D8
Saint-Siméon QC 29 D8
Saint-Siméon QC 31 A6
Saint-Simon NB 30 E2
Saint-Simon NB 31 B8
Saint-Simon QC 24 C-D8
Saint-Simon QC 27 D3
Saint-Simon QC 29 C2
Saint-Simon-les-Mines QC 27 C7
Saint-Sixte QC 24 D3
Saints-Martyrs-Canadiens QC 27 D5
Saint-Stanislas QC 24 C8
Saint-Stanislas QC 25 E8
Saint-Stanislas QC 26 D8
Saint-Stanislas QC 27 B3
Saint-Stanislas QC 28 A4
Saint-Stanislas-de-Kostka QC 24 E5
St. Stephen NB 32 E2
St. Stephens NL 37 F7
Saint-Sulpice QC 24 C6
Saint-Sulpice QC 27 D2
Saint-Sylvère QC 24 B8
Saint-Sylvère QC 25 F8
Saint-Sylvère QC 27 C4
Saint-Sylvestre QC 27 C6
Saint-Télesphore QC 24 E4-5
St. Teresa PE 34 C3
St. Teresa PE 35 C2
Saint-Tharcisius QC 29 C5
Saint-Théodore-d'Acton QC 24 D7-8
Saint-Théodore-d'Acton QC 27 E3
Saint-Théophile QC 27 D7
St. Theresa Point MB 13 A5
St. Thomas NB 31 E8
St. Thomas NB 32 A6
St. Thomas NL 37 F7
St. Thomas ON 19 C-D5
St. Thomas ON 20 F4
St. Thomas, lac QC 24 C6
St. Thomas, lac QC 27 D2
St. Thomas, lac QC 25 D6-7
St. Thomas, lac QC 28 F2-3
St. Thomas, lac QC 28 F3
St. Thomas Bay BC 1 B4-5
Saint-Thomas-d'Aquin QC 24 D7
Saint-Thomas-d'Aquin QC 27 B5
Saint-Thomas-de-Caxton QC 24 B7
Saint-Thomas-de-Caxton QC 25 F7
Saint-Thomas-de-Caxton QC 27 C3
Saint-Thomas-Didyme QC 26 D7
Saint-Thomas-Didyme QC 28 B3
Saint-Thuribe QC 24 A8
Saint-Thuribe QC 25 E8
Saint-Thuribe QC 27 B4
Saint-Timothée PE 32 A7
Saint-Timothée PE 35 D2
Saint-Timothée QC 24 E5
Saint-Timothée QC 27 F1
Saint-Tite QC 24 A8
Saint-Tite QC 25 E8
Saint-Tite QC 27 B3
Saint-Tite-des-Caps QC 27 A6-7
Saint-Tite-des-Caps QC 28 F6
Saint-Ubalde QC 24 A8
Saint-Ubalde QC 27 B4
Saint-Ulric QC 29 B4
Saint-Urbain QC 28 E6-7 D7
Saint-Urbain-Premier QC 24 E6
Saint-Urbain-Premier QC 27 F1
Saint-Valentin QC 24 E6
Saint-Valentin QC 27 F2
Saint-Valère QC 27 C-D4
Saint-Valérien QC 24 D7-8
Saint-Valérien QC 27 E3
Saint-Valérien QC 29 C3
Saint-Vallier QC 27 A6
Saint-Venant-de-Paquette QC 27 F5
St. Veronica's NL 36 D5-6
St. Veronica's NL 37 D4
Saint-Vianney QC 29 C4-5
Saint-Viateur QC 24 B6-7
Saint-Viateur QC 27 C2

Saint-Victor QC **27** C6
St. Victor SK **11** F5
Saint-Victor, rivière QC
27 C-D6
St. Victor's Petroglyphs
Provincial Historic Park
SK **11** F4-5
St. Vincent AB **6** A7
St. Vincent AB **11** A1-2
St. Vincent's NL **37** F7
St. Vincent's-St. Stephen's-
Peter's River NL **37** F7
St. Walburg SK **6** A8
St. Walburg SK **11** A2-3
Saint-Wenceslas QC **24** B8
Saint-Wenceslas QC
27 C3-4
Saint-Wilfrid NB **31** C7
St. Williams ON **19** D6
St. Williams ON **20** F5
Saint-Yvon QC **30** A2
Saint-Zacharie QC **27** C7
Saint-Zénon QC **24** A5
Saint-Zénon QC **25** F5
Saint-Zénon QC **27** B1
Saint-Zéphirin-de-Courval
QC **24** C8
Saint-Zéphirin-de-Courval
QC **27** D3
Saint-Zotique QC **24** E5
Sainville River NT **5** E5
Sairs, lac QC **22** C3
Sakami QC **40** E4
Sakami, lac QC **40** E4
Sakami, rivière QC **40** E5
Sakimay SK **11** E6-7
Sakwatamau River AB **6** A4
Sakwatamau River AB
8 A3 B4
Sakwesew Lake SK **15** E6
Salaberry-de-Valleyfield
QC **24** E5
Salem NS **32** C7
Salem NS **34** D1
Salem ON **19** A6
Salem ON **20** C5
Salem ON **21** E4
Salem Road NS **34** D7
Salford ON **19** C5
Salford ON **20** E4
Salisbury NB **31** F7-8
Salisbury NB **32** B6
Salisbury Island NU **40** A4
Salisbury Island NU **41** D7
Salk, lac QC **26** C5
Salliq (Coral Harbour) NU
40 A4
Salliq (Coral Harbour) NU
41 D6-7
Salluit QC **40** B4
Salluit QC **41** D8
Sally's Beach PP PE **35** E7
Sally's Cove NL **38** E2
Sally's Cove NL **39** F7
Sallysout Creek BC **3** E4
Sally's Pond PP PE **34** C4
Salmo BC **6** F3
Salmon Arm BC **6** E2
Salmon Bay QC **38** A-B3
Salmon Beach NB **29** E4
Salmon Beach NB **30** E1
Salmon Beach NB **31** B6-7
Salmon Cove NL **36** B1
Salmon Cove NL **37** D7-8
Salmon Creek NB **31** F6
Salmon Creek NB **32** B4
Salmon Fork YT **5** D1-2
Salmon Inlet BC **1** F6
Salmon Inlet BC **2** D4
Salmon Lake ON **21** C3
Salmon Point ON **22** D1
Salmon River BC **1** A6-7
Salmon River BC **1** E5
Salmon River BC **2** B1
Salmon River NB **29** F4
Salmon River NB **31** C3
Salmon River NB **31** E7 F6
Salmon River NB **32** B4-5
Salmon River NL **36** D5
Salmon River NL **37** D3
Salmon River NL **38** R4
Salmon River NL **39** D-E8
Salmon River NS **33** A6-7
Salmon River NS **33** E1
Salmon River NS **34** E2
Salmon River ON **21** D5-6
Salmon River ON **21** D-E5
Salmon River Lake NS
34 E5
Salmon River Road NS
34 D7
Salmonier NL **36** F6-7
Salmonier NL **37** E7
Salmonier NL **37** F4-5
Salmonier Provincial Nature
Park NL **37** F7
Salon-de-Thé-des-Chutes-
Twin, LHN du QC **7** A1
Salone, lac QC **25** C5
Salone, lac QC **28** F1
Salsman PP NS **34** E5
Salt Harbour Island NL
38 E6-7
Salt Lake SK **11** F5
Salt Point MB **11** B8
Salt Point MB **13** B1
Salt Point MB **14** A1
Salt Pond NL **36** E6-7
Salt Pond NL **37** E4-5
Salt Prairie AB **9** E4
Salt River AB **9** A6
Salt River AB **10** F8
Salt River AB **12** A2

Salt River NT **9** A6
Salt River NT **10** F8
Salt River NT **41** E1-2
Salt Springs NS **33** A7
Salt Springs NS **34** E4-5
Salt Springs PP NS **33** A7
Salt Springs PP NS **34** E2-3
Saltair BC **2** E4
Saltcoats SK **11** D7
Salter SK **11** C3
Saltery Bay BC **1** F6
Saltery Bay BC **2** C3-4
Saltery Bay PP BC **1** E-F6
Saltery Bay PP BC **2** C3
Saltford ON **20** C3
Saltspring Island BC **2** E4
Salutation Cove PE
35 D2-3
Salvador SK **6** C8
Salvador SK **11** C2
Salvage NL **37** B7
Salvail, rivière QC **24** C-D7
Salvail, rivière QC **27** D2-3
Sam Lake **13** E7
Sambaa Deh Falls TP NT
10 E4
Sambo Creek YT **4** E6
Sambo Lake YT **4** E6
Sambro NS **32** F8
Sambro NS **33** C6
Sambro, Cape NS **33** C-D6
Samson, rivière QC
27 D-E7
Samuel de Champlain PP
ON **17** F8
Samuel de Champlain PP
ON **22** D2
Sanca BC **6** F4
San Clara MB **11** D7-8
San Clara MB **13** D1
Sanctuary SK **11** D3
Sandbanks PP NL **36** D3
Sandbanks PP NL **37** E1
Sandbanks PP ON **21** E-F5
Sandbar Lake PP ON
13 F7-8
Sandbar Lake PP ON
16 D-E4
Sand Beach NS **33** E1
Sandeau, lac QC **22** B4
Sandfield ON **18** C3-4
Sandfly Lake SK **12** E4-5
Sandford NS **33** E1
Sandford ON **20** B7
Sandford ON **21** E1-2
Sandford Island ON **18** B2
Sandford Lake ON **13** F7
Sandgren SK **11** D2
Sandhill ON **19** A7
Sandhill ON **20** C6
Sand Hill Cove NL **39** B8
Sand Hill River NL
39 B7-8 C7
Sand Hills Beach PP NS
33 F2
Sandhurst ON **21** E6
Sandilands MB **13** F4
Sandilands MB **14** F6
Sand Lake ON **13** E5
Sand Lake ON **16** D1
Sand Lake ON **18** B8
Sand Lake ON **18** C8
Sand Lake ON **21** A1
Sand Lake ON **21** C5
Sand Lake ON **21** D7
Sand Lake ON **22** D1
Sand Lake ON **24** F1
Sand Lakes PP MB **15** C3
Sand Lakes PP MB **41** F4
Sandon BC **6** E3
Sand Point NS **34** E6
Sandpoint Island PP ON
16 F2-3
Sand Pond NWA NS **33** E2
Sandridge MB **13** E3
Sandridge MB **14** C4-5
Sandringham NL **36** B8
Sandringham NL **37** B6
Sand River AB **9** F7
Sand River AB **12** E-F2
Sand River NS **32** D6
Sand River NS **33** A4
Sand River NS **33** A4
Sandville NL **34** D2
Sandwell PP BC **2** E4
Sandwich Bay NL **39** B7
Sandwich Bay NL **40** D8
Sandy, pointe NB **30** E2-3
Sandy Bar MB **13** C4
Sandy Bay MB **13** E3
Sandy Bay MB **14** C3
Sandy Bay SK **9** A7
Sandy Bay SK **11** A-B4
Sandy Bay SK **12** A3
Sandy Bay SK **12** E7
Sandy Bay SK **15** E1
Sandy Beach AB **6** A5
Sandy Beach AB **8** B-C6
Sandy Beach QC **30** D3
Sandybeach Lake ON **13** E7
Sandy Cove NL **37** B6-7
Sandy Cove NS **38** B4
Sandy Cove NS **38** E7
Sandy Cove NS **32** F6
Sandy Cove NS **33** D1
Sandy Cove ON **20** B7
Sandy Cove ON **21** D1

Sandy Harbour River NL
36 D7-8
Sandy Harbour River NL
37 D5-6
Sandy Hook MB **13** E4
Sandy Hook MB **14** C5
Sandy Island ON **18** D6-7
Sandy Island PMP BC
1 F5-6
Sandy Island PMP BC
2 D2-3
Sandy Islands, RNP ON
17 F3
Sandy Islands PNR ON
17 F3
Sandy Lake AB **6** A5
Sandy Lake AB **8** B6
Sandy Lake AB **9** E6
Sandy Lake AB **9** E5
Sandy Lake MB **11** E8
Sandy Lake MB **13** E2
Sandy Lake MB **14** C1
Sandy Lake NL **36** A4
Sandy Lake NL **37** A2
Sandy Lake NL **38** F3
Sandy Lake NL **39** F8
Sandy Lake NT **5** C5
Sandy Lake ON **13** B6
Sandy Lake ON **13** B6
Sandy Lake ON **40** E1
Sandy Lake ON **40** E1
Sandy Narrows SK **12** F6
Sandy Point NB **30** E2-3
Sandy Point NB **31** A8
Sandy Point NL **39** B6
Sandy Point NS **33** F3
Sandy Point ON **20** A8
Sandy Point ON **21** D2
Sanford MB **13** F3
Sanford MB **14** E4-5
Sangsues, lac aux QC
22 D4
Sanguedo AB **6** A5
Sangudo AB **8** B4-5
Sanikiluaq QC **40** D4
Sanirajak (Hall Beach) NU
41 B6
Sanirajak (Hall Beach) NU
42 F6
San Josef BC **1** E3-4
San Josef Bay BC **1** E3
San Juan River BC **2** F3
Sanmaur QC **25** B6
Sanmaur QC **28** D1
Sans Souci ON **18** D7
Santa Boca PP BC **1** F4
Santein, lac QC **39** E6
Santoy Lake ON **16** E7
Santoy Lake ON **17** C2
Saoyue and Ehdacho HS
(Saoyue Unit) NT
10 A3-4
Saoyue HS NT **41** C1
Sapawe ON **13** F7-8
Sapawe ON **16** E4
Saprae Creek Estates AB
9 D7
Saprae Creek Estates AB
12 D2
Sapton MB **14** D5-6
Sarah Island BC **1** B3
Sarah Lake NT **10** B5
Saratoga Beach BC **1** E5-6
Saratoga Beach BC **2** C5
Sarcee AB **7** A5
Sardis BC **1** F8
Sardis BC **2** E6
Sargeant Bay PP BC **2** D4
Sarita BC **2** E2
Sarnia ON **19** C2
Sarnia ON **20** E1
Sarsfield ON **24** D3
Sarto MB **13** F4
Sarto MB **14** E-F5
Sasaginnigak Lake MB
13 D5
Sasaginnigak Lake MB
14 A7
Sasaginnigak Lake MB
16 B1
Saseginaga, lac QC **22** B3
Saskatchewan, rivière SK
11 A6
Saskatchewan Beach SK
11 D5-6
Saskatchewan Landing PP
SK **11** E3
Saskatchewan Nord, rivière
AB **6** B4-5 C5
Saskatchewan Nord, rivière
AB **8** B8
Saskatchewan Nord, rivière
AB **8** E4
Saskatchewan Nord, rivière
AB **11** A1-2
Saskatchewan Nord, rivière
SK **6** B8
Saskatchewan Nord, rivière
SK **11** B-C4
Saskatchewan Point SK
13 C3
Saskatchewan River
Crossing AB **6** C3-4
Saskatchewan River
Crossing AB **8** F2
Saskatchewan Sud, rivière
AB **6** D-E8
Saskatchewan Sud, rivière
AB **7** C-D8
Saskatchewan Sud, rivière
AB **11** D2 E1-2

Saskatchewan Sud, rivière
AB **36** D7-8
Saskatoon SK **11** C-D4
Saskatoon Island PP SK
9 E-F2
Saskatoon Lake, RNOM de
AB **9** E-F2
Saskatoon Lake NMBS AB
9 E-F2
Saskeram Lake MB **11** A7
Saskeram Lake MB **13** A1
Sasquatch PP BC **1** F8
Sasquatch PP BC **2** D6-7
Sass River NT **10** F7
Satah Mountain BC **1** C6
Satah River YT **5** D4
Satellite Bay NT **42** C2
Saturna BC **2** E4-5
Saturna Island BC **2** E-F5
Saturna BC **2** E4-5
Schmitt, rivière QC **30** A6
Schoen Lake PP BC **1** E5
Schoenfeld SK **11** E3
Schoenwiese MB **14** E4
Schomberg ON **20** B-C6
Schoolhouse Lake PP BC
1 C8
Schoolhouse Lake PP BC
6 C1
Schreiber ON **16** E7
Schreiber ON **17** C1
Schreiber Channel, RNP ON
16 F7
Schreiber Channel, RNP ON
17 C1
Schreiber Channel PNR ON
16 F7
Schreiber Channel PNR ON
17 C1
Schuler AB **6** E8
Schuler AB **11** E2
Schultz Lake NU **41** D4
Schumacher ON **17** C6
Schutt ON **21** B4
Schwandt River SK **12** A6
Schwandt River SK **15** A1
Schwartz QC **22** F6-7
Schwartz QC **24** C1
Schyan, lac QC **22** D5
Schyan, rivière QC **22** E5
Scollard AB **8** F7
Scone ON **20** B4
Scot Bay ON **14** D8
Scotch Bay MB **14** C3
Scotch Bonnet Island NWA
ON **21** F4-5
Scotch Creek BC **6** D2
Scotchfort PE **35** D5
Scotch Line ON **21** C7
Scotch Line ON **24** F1
Scotch Ridge NB **32** D1
Scotch River SK **24** D3 E4
Scotch Settlement ON
18 F4-5
Scotch Settlement ON
20 A3
Scotchtown NB **32** C4
Scotchtown NS **34** C7-8
Scotch Village NS **32** E8
Scotch Village NS **33** B5
Scotch Village NS **34** F1
Scotfield AB **6** D7
Scotfield AB **7** A7
Scotfield AB **11** C-D1
Scotia ON **18** C8
Scotia ON **21** A1
Scotia ON **22** F1
Scotland ON **19** C6
Scotland ON **20** E5
Scots Bay NS **33** A4
Scots Bay PP NS **32** D7
Scots Bay PP NS **33** A4
Scotsburn NS **34** E3
Scotstown QC **27** E6
Scotsville NS **34** C6
Scott, Cape BC **1** D3
Scott, lac QC **26** A4
Scott SK **11** C3
Scott Islands BC **1** D3
Scott Islands PP BC **1** D-E3
Scott Lake SK **12** A4
Scotts Lake NL **36** C-D3
Scotts Lake NL **37** C-D1
Scottsville ON **20** E3
Scoudouc NB **31** F8
Scoudouc NB **32** B6-7
Scugog Lake SK **11** F4
Scugog ON **20** B8
Scugog, Lake ON **20** B8
Scugog, Lake ON **21** E2
Scugog River ON **20** B8
Scugog River ON **21** D-E2
Scudder ON **19** F2
Scurvy Creek YT **4** E6
Seagram Lake SK **12** A6
Sayabec QC **29** C4
Saymo-Aubindong-Gong
PP ON **17** E4
Sayward BC **1** E5
Sayward BC **2** B1
Scandia AB **6** E7
Scandia AB **7** D6
Scandinavia MB **14** C1-2
Scanlon Creek CA ON
20 B7
Scanlon Creek CA ON
21 E1
Scapa AB **6** E7
Scapa AB **8** F8
Scarborough ON **19** A8
Scarborough ON **20** C7
Scarborough ON **21** F1
Scarth MB **11** F8
Scarth MB **13** E-F1
Scaterie Island NS **34** C8

Scatter River Old Growth PP
BC **3** A6
Scent Grass, RNOM SK
11 B3
Scent Grass NMBS SK
11 B3
Sceptre SK **6** D8
Sceptre SK **11** D2
Schade Lake ON **13** B8
Schade River ON **13** B7-8
Schanzenfeld MB **13** F3
Schanzenfeld MB **14** F4
Schefferville NL **39** A2
Schefferville NL **39** A2
Schipa River BC **3** B-C5
Schist Lake MB **12** F7
Schist Lake MB **15** F1
Schmitt, rivière QC **30** A6
Schoen Lake PP BC **1** E5
Schoenfeld SK **11** E3
Schreiber ON **16** E7
Schreiber ON **17** C1
Searletown PE **32** B8
Searletown PE **34** C1
Searletown PE **35** E3
Searston NL **36** D1
Searston NL **37** B4
Searsville NB **32** C4
Sea Side NB **31** A5-6
Seattle, Mount YT **4** E2
Sea View PE **32** A8
Sea View PE **34** B1-2
Sea View PE **35** C3
Sea Wolf Island NWA NS
34 B5-6
Seba Beach AB **6** B5
Seba Beach AB **8** C5
Sebaskachu River NL
39 B5-6
Sébastien, lac QC **26** B4
Sebastopol ON **19** B5
Sebastopol ON **20** D4
Sebright ON **20** A7
Sebright ON **21** C1-2
Sebringville ON **19** B5
Sebringville ON **20** D4
Sechelt BC **1** F6
Sechelt BC **2** D4
Sechelt Inlet BC **2** D4
Sechelt Inlets MPP BC **1** F6
Sechelt Inlets MPP BC **2** D4
Sechelt Peninsula BC **1** F6
Sechelt Peninsula BC **2** D4
Sec Lake ON **22** E4
Second Burnt Pond NL **36** B7
2nd Burnt Pond NL **37** B5
Second Eel Lake NB **32** C1
Second Lac NB **29** E3
Second Peninsula ON **33** D5
Second Peninsula PP NS
33 D5
Secord ON **17** F7
Secord ON **18** A5-6
Secretan SK **11** E4
Sedalia AB **6** C7
Sedalia AB **7** A8
Sedalia AB **11** C1-2
Seddons Corner MB **13** E4
Seddons Corner MB **14** D6
Sedgewick AB **6** B7
Sedgewick AB **11** B1
Sedgman Lake PP ON
16 C7
Sedgman Lake PP ON
17 A1
Sedgwick, Mount YT **5** B2-3
Sedley SK **11** E6
Seebe AB **6** D5
Seebe AB **7** B3
Seeber Lake ON **13** A6
Seeber River ON **13** A6
Seech NS **34** C1
Seeley Lake PP BC **3** F4
Seeleys Bay ON **21** D7
Sefferville NL **36** B8
Seffernville NS **33** C4-5
Segise Lake ON **16** D2
Séguin, lac QC **22** B7
Séguin, lac QC **25** C-D2
Seguin Lake ON **22** F1
Seguin River ON **18** D7
Seibert Lake AB **9** F7
Seibert Lake AB **12** F1
Seignelay, rivière QC **39** C1
Seigs Corner MB **14** D6
Seine River ON **16** E-F3
Seine River Village ON
16 F3
Seize Îles, lac des QC **24** C4
Sekulmun Lake YT **4** D2
Selbaie QC **17** B8
Selby ON **21** E6
Seldom NL **38** F7
Selenite Point Provincial
Ecological Reserve SK
12 F4
Seline River ON **16** E4
Selkirk MB **13** E4
Selkirk MB **14** D5
Selkirk ON **19** C7
Selkirk ON **20** E6
Selkirk PE **35** D7
Selkirk Mountains BC **6** D3
Selkirk PP ON **19** D7

Selkirk PP ON **20** F6
Seller Lake MB **15** D5
Sellwood Bay NT **5** A7-8
Selma NS **32** E8
Selma NS **33** A6
Selma NS **34** E2
Selous, Mount YT **4** C5
Selwyn ON **21** D3
Selwyn YT **4** C2
Selwyn CA ON **21** D3
Selwyn Lake NT **41** E-F3
Selwyn Lake SK **12** A5
Selwyn Mountains YT
4 A5-6 C6-7
Semans SK **11** D5
Semenof Hills YT **4** D4
Semmens Lake MB **15** F5-6
Semple Lake MB **15** F5
Senate ON **6** F8
Senate SK **11** F2
Senlac SK **6** C8
Senlac SK **11** C3
Senneterre QC **23** D5-6
Senneville, lac QC **23** E4-5
Sentinel Peak BC **3** F7
Sentinel Peak BC **9** F1
Separation, Point NT **5** C4-5
Separation Lake ON **13** E6
Separation Lake ON **16** D2
Sept Frères, lac des QC
24 B3
Sept Frères, lac des QC
25 F3
Sept Îles, lac QC **18** F5
Sept Îles, lac QC **20** A4
Sept Îles, lac des QC **30** D2
Sept-Îles QC **39** F-E2
Sept-Îles QC **40** F6
Sept Milles, lac QC **22** C4-5
Sequart Lake NL **39** C-D4
Serath SK **11** D5
Sérigny, rivière QC **40** D6
Serpent, lac QC **22** E8
Serpent, lac QC **23** E7
Serpent, lac QC **24** B2-3
Serpent, lac QC **25** A2
Serpent, lac QC **26** E2
Serpent, rivière QC
23 D8 E7-8
Serpent, rivière QC **25** A2
Serpent, rivière QC **26** E2
Serpentine Lake NL **36** B2
Serpent River ON **13** D6-7
Serpent River ON **17** F5
Serpent River ON **18** A3
Serpent River ON **18** A3 B2
Serpent River ON **18** A2-3
Seseganaga Lake ON **16** D5
Seseganaga Lake ON **13** E8
Sesekinika ON **17** D7
Sesikinaga Lake ON **13** D7
Sesikinaga Lake ON
16 B-C3
Seton Lake BC **1** F6-7
Seton Lake BC **2** A-B6
Seton Portage BC **1** E7
Seton Portage BC **2** A6
Seton Portage Provincial
Historic Park BC **1** E7
Seton Portage Provincial
Historic Park BC **2** A-B6
Settee Lake MB **15** D4
Setting Lake MB **15** F3
Seul, lac QC **13** D7
Seul, lac QC **13** E7
Seul, lac ON **16** C-D3
Seul, lac ON **13** E7
Seven Persons AB **7** D-E8
Seven Persons AB **11** E1
Seven Persons Creek AB
6 F7-8
Seven Persons Creek AB
7 E8
Seven Persons Creek AB
11 E-F1
Seven Sisters Falls MB
13 E4-5
Seven Sisters Falls MB
14 D6-7
Seven Sisters PP BC **3** F4
70 Mile House BC **1** D8
70 Mile House BC **6** D1
Severn Bridge ON **18** D8
Severn Bridge ON **21** C1
Severn Falls ON **18** E8
Severn Falls ON **21** C1
Severn Lake ON **13** A7-8
Severn River ON **13** B7
Severn River ON **15** F8
Severn River ON **18** E8
Severn River ON **21** C1
Severn River ON **40** D1-2
Severn River PP ON **13** A8
Severn Sound ON **18** E7
Sevestre, lac QC **39** C1
Sevogle NB **31** D6

Shabo NL **39** B-C2
Shabogamo Lake NL **39** C2
Shabogamo Lake NL **40** E6
Shabumeni Lake ON **16** B3
Shabuskwia Lake ON
16 B6
Shackleton SK **11** D-E3
Shad Bay NS **32** F8
Shad Bay NS **33** C5-6
Shadd Lake SK **12** C5
Shadow Lake ON **20** A8
Shadow Lake ON **21** C2
Shaftesbury AB **9** E3
Shag, île QC **30** F7
Shag Harbour NS **33** F2
Shagwenaw Lake SK
12 E3-4
Shakespeare ON **19** B5
Shakespeare ON **20** D4
Shakespeare Island ON
16 D6
Shakespeare Island ON
17 B1
Shalalth BC **1** E7
Shalalth BC **2** A6
Shaler Mountains NT
41 A2-3
Shaler Mountains NT **42** E2
Shallop Cove NL **36** C2
Shalloway Cove NL **36** B8
Shalloway Cove NL **37** B6
Shallow Bay NT **5** B4
Shallow Lake ON **18** F5
Shallow Lake ON **20** A4
Shallow River, RNP ON
17 C7
Shallow River PNR ON
17 C7
Shamattawa MB **15** E7
Shamattawa MB **40** D1
Shamblers Cove NL
37 A6-7
Shamblers Cove NL **38** F8
Shamrock ON **21** B6
Shamrock PE **35** D3
Shamrock SK **11** E4
Shandro AB **8** B8
Shanklin NB **32** D4-5
Shanklin NB **33** A2
Shanly ON **21** C8
Shanly ON **24** F2
Shannon NB **32** C4
Shannon QC **27** A5
Shannon, lac QC **23** E4
Shannon, lac QC **25** B2
Shannon, lac QC **26** F2
Shannon Falls PP BC **1** F6-7
Shannon Falls PP BC **2** C5
Shannon Lake MB **12** F3-4
Shannon Lake MB **15** F1
Shannon Lake MB **15** B5
Shannonville ON **21** E5
Shanty Bay ON **18** F8
Shanty Bay ON **20** A7
Shanty Bay ON **21** D1
Shantz AB **7** A3
Shapio Lake NL **39** A5
Sharbot Lake ON **21** C6
Sharbot Lake ON **21** C6
Sharbot Lake PP ON **21** C6
Sharon ON **20** B7
Sharon ON **21** E1
Sharpe Lake MB **13** A6
Sharpe Lake MB **15** F6
Sharpes Creek ON **19** A4
Sharpes Creek ON **20** C3
Sharpewood AB **8** B3
Sharples Lake NT **10** B7
Sharp Mountain YT **5** D3
Sharun Lake NT **10** D5
Shaughnessy AB **6** E6
Shaughnessy AB **7** E5-6
Shaunavon SK **11** F3
Shaw, lac à QC **25** B8
Shaw, lac à QC **28** E3
Shawanaga Inlet ON
18 C6-7
Shawanaga Island ON
18 C6
Shawanaga Lake ON **18** C7
Shawinigan QC **24** A7-8
Shawinigan QC **25** F7-8
Shawinigan QC **27** B3
Shawinigan, lac QC **24** A7
Shawinigan, lac QC **25** E7
Shawinigan, lac QC **27** B2
Shawinigan-Sud QC
24 A7-8
Shawinigan-Sud QC
25 F7-8
Shawinigan-Sud QC **27** B3
Shaw Lake SK **12** D6
Shaw Lake SK **15** D1
Shawnigan-Sud **27** B3
Shaw Water BC **1** F8
Shawville QC **21** A6-7
Shawville QC **22** F6-7
Shawville QC **24** D1
Shearwater BC **1** C4
Sheaves Cove NL **36** B-C1
Shebandowan ON **16** F5
Shebandowan Lake ON
16 F5
Shebeshekong River ON
18 D6-7
Shedden ON **19** D4
Shedden ON **20** F3
Shedden Lake ON **18** A3
Shediac NB **31** F8
Shediac NB **32** B6-7

Shediac, baie de *NB*
　32 B6-7
Shediac, baie de *NB*
　35 D-E1
Shediac Bay *NB* **31** F8
Shediac Bay *NB* **32** B6-7
Shediac Bay *NB* **35** D-E1
Shediac Bridge *NB* **31** F8
Shediac Bridge *NB* **32** B6
Shediac Cape *NB* **31** F8
Shediac Cape *NB* **32** B6
Shediac Ridge *NB* **31** D7
Sheemahant River *BC*
　1 C-D5
Sheenboro *QC* **22** E5
Sheep River *AB* **6** E5
Sheep River *AB* **7** C3
Sheep River PP *AB* **6** E5
Sheep River PP *AB* **7** C3
Sheerness *AB* **6** D7
Sheerness *AB* **7** A7
Sheerness *AB* **11** D1
Sheet Harbour *NS* **33** B8
Sheet Harbour *NS* **33** B8
Sheet Harbour *NS* **34** F4
Sheet Harbour Road *NS*
　33 B7
Sheet Harbour Road *NS*
　34 F3
Sheffield *NB* **32** C3
Sheffield *ON* **19** B6-7
Sheffield *ON* **20** D5-6
Sheffield CA *ON* **21** D5
Sheffield Lake *NL* **36** A5
Sheffield Lake *NL* **37** A3
Sheffield Lake *NL* **38** F4
Sheffield Mills *NS* **32** E7
Sheffield Mills *NS* **33** B4
Sheguiandah *ON* **18** B4
Sheho *SK* **11** D6
Sheila *NB* **30** F2
Sheila *NB* **31** C7-8
Shekak River *ON*
　17 B4 C3-4
Shekatika *QC* **38** B1-2
Shekilie River *BC* **3** B8
Shekilie River *BC* **9** A-B2
Shelagyote Peak *BC* **3** E4
Shelburne *NS* **33** E3
Shelburne *NS* **33** E-F3
Shelburne *ON* **20** B5
Shelburne Harbour *NS*
　33 F3
Sheldon, Mount *YT* **4** C6
Sheldon Creek *ON* **20** B6
Sheldrake *QC* **39** E-F3
Shell Brook *SK* **11** B4
Shellbrook *SK* **11** B4
Shelley *BC* **1** A7
Shell Lake *SK* **11** B4
Shellmouth *MB* **11** D7
Shellmouth *MB* **13** D1
Shell River *MB* **11** D7-8
Shell River *MB* **13** D1
Shelter Bay *BC* **6** E3
Shelter Inlet *BC* **2** C2
Shelter Point *BC* **2** C2
Shemogue *NB* **32** B7
Shemogue *NB* **35** E1-2
Shemogue, havre de *NB*
　35 E2
Shemogue Harbour *NB*
　35 E2
Shemogue Head *NB* **35** E2
Shepard *AB* **7** B4
Shepherd Bay *NU* **41** E5
Shepherd Bay *NU* **42** F4
Shepherd Island NMBS *NL*
　38 C5-6
Shepody, RNF *NB* **32** C6
Shepody Bay *NB* **32** C6-7
Shepody NWA *NB* **32** C6
Sheppard Peak *BC* **3** C2
Sheppardville *NL* **36** A5
Sheppardville *NL* **37** A3
Sheppardville *NL* **38** F4
Sherard, Cape *NU* **42** D5
Sherborne Lake *ON* **21** B2
Sherbrooke *NS* **34** F4-5
Sherbrooke *PE* **35** D3
Sherbrooke *QC* **27** E4-5
Sherbrooke *QC* **27** F4-5
Sherbrooke Lake *NS* **32** F7
Sherbrooke Lake *NS* **33** C4
Sherbrooke PP *NS* **34** F4-5
Shergrove *MB* **13** D2-3
Shergrove *MB* **14** B2
Sheridan Lake *BC* **1** D8
Sheridan Lake *BC* **6** D1
Sherkston *ON* **19** C8
Sherkston *ON* **20** F7
Sherman Basin *NU* **41** C4
Sherridon *MB* **12** F7-8
Sherridon *MB* **15** F2
Sherrington *QC* **24** E6
Sherrington *QC* **27** F1
Sherritt Junction *MB* **15** F1-2
Sherritt Junction *MB*
Sherwood *MB* **14** C6
Sherwood *NS* **32** F7
Sherwood *NS* **33** C5
Sherwood Lake *ON* **13** E-F5
Sherwood Lake *ON* **14** E8
Sherwood Lake *ON* **16** D1
Sherwood Park *AB* **6** B6
Sherwood Park *AB* **8** C6-7
Sherwood River *ON* **21** A4
Sherwood River *ON* **22** F4
Sheshatshiu *NL* **39** B5-6

Shesheeb Bay PNR *ON*
　16 F6-7
Shesheeb Bay PNR *ON*
　17 C1
Sheshegwaning *ON* **18** B2
Sheslay River *BC* **3** C2
Shethanei Lake *MB* **15** B3-4
Shetland *ON* **19** D3
Shetland *ON* **20** F2
Shetland CA *ON* **19** D3
Shetland CA *ON* **20** F2
Shields *NS* **11** C4
Shigawake *QC* **30** D1-2
Shigawake *QC* **31** A7
Shikag Lake *ON* **13** E-F8
Shikag Lake *ON* **16** D4
Shillington *ON* **17** C6-7
Shilo *MB* **13** E2
Shilo *MB* **14** E2
Shingle Point *YT* **5** B3
Shinimicas Bridge *NS*
　32 C7-8
Shinimicas Bridge *NS* **34** D1
Shinimicas Bridge *NS* **35** F2
Shinimicas PP *NS* **32** C8
Shinimicas PP *NS* **34** D1
Shining Tree *ON* **17** E6
Ship Cove *NL* **36** C1
Ship Cove *NL* **38** E8
Ship Cove *NL* **37** E6
Ship Cove *NL* **38** A5
Ship Cove *NL* **39** D8
Ship Harbour *NL* **36** E8
Ship Harbour *NL* **37** E6
Ship Harbour *NS* **33** C7-8
Ship Harbour *NS* **33** B-C7
Ship Harbour East *NS* **33** C7
Shipiskan Lake *NL* **39** A4-5
Shipka *ON* **19** B4
Shipka *ON* **20** D3
Shipman *MB* **11** B5
Shippagan *NB* **30** E2
Shippagan *NB* **31** B8
Shipshaw *QC* **28** C5
Shipshaw, rivière *QC*
　28 B6 C5
Shipton *QC* **27** D4
Shirley Lake *ON* **21** A3
Shirley Lake *ON* **22** F3-4
Shoal Arm *NL* **38** E4-5
Shoal Bay *NL* **37** E8
Shoal Bay *NL* **38** E7
Shoal Brook *NL* **36** A4
Shoal Brook *NL* **37** A1
Shoal Brook *NL* **38** F2
Shoal Cove *NL* **38** B3
Shoal Harbour *NL* **36** C8
Shoal Harbour *NL* **37** C6
Shoal Lake *MB* **11** E8
Shoal Lake *MB* **13** E1-2
Shoal Lake *ON* **13** F5
Shoal Lake *ON* **14** E8
Shoal Lake *ON* **16** D1
Shoal Lake *SK* **11** B7
Shoal Lakes *MB* **13** D2
Shoal River (Sapotaweyak)
　MB **11** B8
Shoal River (Sapotaweyak)
　MB **13** B2
Shoalwater Bay *YT* **5** B3-4
Shoe Cove *NL* **37** D8
Shoe Cove *NL* **38** E5
Shoe Cove Point *NL* **37** A7
Shoe Cove Point *NL* **38** F8
Shonts *AB* **8** C7
Shoreacres *BC* **6** F3
Shores Cove *NL* **37** E8
Shorewood *MB* **14** C5
Shorncliffe *MB* **13** D4
Shorncliffe *MB* **14** B5
Short Hills PP *ON* **19** C8
Short Hills PP *ON* **20** F7
Shortdale *MB* **11** D8
Shortdale *MB* **13** D1
Shouldice *AB* **7** C5
Shrewsbury *ON* **19** E3
Shubenacadie *NS* **33** B6
Shubenacadie *NS* **34** F2
Shubenacadie Grand Lake
　NS **32** E8
Shubenacadie Grand Lake
　NS **33** B6
Shubenacadie Grand Lake
　NS **34** F2
Shubenacadie Provincial
　Wildlife Park *NS* **33** B6-7
Shubenacadie Provincial
　Wildlife Park *NS* **34** F2-3
Shubenacadie River *NS*
　33 A-B6
Shubenacadie River *NS*
　34 F2
Shulie *NS* **32** C6-7 D6-7
Shulie *NS* **33** A4
Shunacadie *NS* **34** C7
Shuswap *BC* **6** E4
Shuswap *BC* **7** C1
Shuswap Falls *BC* **6** E2
Shuswap Lake *BC* **6** D2
Shuswap Lake MPP *BC*
　6 D2
Shuswap Lake PP *BC* **6** D2
Shuswap River *BC* **6** D-E3
Shuswap River *BC* **6** E2
Sibbald *AB* **11** D2
Sibbald Point PP *ON* **20** B7
Sibbald Point PP *ON* **21** D1
Sibbeston Lake *NT* **10** D7
Sibley Peninsula *ON* **16** F6
Sibleys Cove *NL* **37** C7
Sicamous *BC* **6** D2
Sidewood *SK* **11** E3
Sidney *BC* **2** F4

Sidney *MB* **13** E2-3
Sidney *MB* **14** E2
Sidney Bay *ON* **18** B5
Sidney Spit Unit (Gulf
　Islands NPR) *BC* **2** F4-5
Siegas *NB* **31** D2
Siffleur Provincial
　Wilderness Area *AB* **8** F3
Siffleur Wilderness Area *AB*
　6 C4
Sifton *MB* **11** D8
Sifton *MB* **13** D2
Sifton *MB* **14** A1
Sifton Beach *MB* **14** A-B1
Sifton Range *YT* **4** D3
Siglunes *MB* **14** B3
Signal Hill, LHN de *NL*
　37 D8
Signal Hill NHS *NL* **37** D8
Sikanni Chief *BC* **3** D7
Sikanni Chief River *BC*
　3 C-D7
Sikanni Chief River *BC* **9** C1
Sikanni Old Growth PP *BC*
　3 C7
Sikanni Old Growth PP *BC*
　9 B1
Sik-e-dahk *BC* **3** F4
Silcox *MB* **15** C-D5
Silcox Creek *MB* **15** C-D6
Silent Lake PP *ON* **21** C4
Sillikers *NB* **31** D6
Siloam *ON* **20** B7
Siloam *ON* **21** E1
Silsby Lake *MB* **15** E5
Silton *SK* **11** D5
Silver *MB* **13** D4
Silver *MB* **14** C5
Silver Bay *MB* **14** B3
Silver Beach PP *BC* **6** D2
Silverberry River *NT*
　4 B8 C7
Silver Creek *BC* **2** D7
Silver Creek *BC* **6** E2
Silver Creek *ON* **21** A5
Silver Dollar *ON* **13** E8
Silver Dollar *ON* **16** D4
Silver Falls *MB* **13** E4-5
Silver Falls *MB* **14** C6
Silver Falls PP *ON* **16** E-F5
Silver Hill *ON* **19** D6
Silver Hill *ON* **20** F5
Silver Islet *ON* **16** F6
Silver Lake *ON* **13** E5-6
Silver Lake *ON* **16** D2
Silver Lake *ON* **18** B2
Silver Lake *ON* **20** A8
Silver Lake *ON* **20** A8
Silver Lake *ON* **21** C2
Silver Lake *ON* **21** D2-3
Silver Lake PP *BC* **1** F8
Silver Lake PP *BC* **2** D7
Silver Lake PP *ON* **21** C5-6
Silver Park *SK* **11** C5-6
Silver Plains *MB* **14** E5
Silver Ridge *MB* **13** D3
Silver Ridge *MB* **14** C3
Silver Star PP *BC* **6** E2
Silverthrone Glacier *BC*
　1 D5
Silverthrone Mountain *BC*
　1 D5
Silverton *BC* **6** E3
Silverton *MB* **11** D8
Silverton *MB* **13** D1
Silver Valley *AB* **9** E2
Silver Water *ON* **18** B2
Simard, lac *QC* **17** D8
Simard, lac *QC* **22** A3
Simard, lac *QC* **23** F3
Simcoe *ON* **19** D6
Simcoe *ON* **20** F5
Simcoe, Lake *ON* **18** F8
Simcoe, Lake *ON* **20** A7
Simcoe, Lake *ON* **21** D1
Simcoe Island *ON* **21** E6
Simcoeside *ON* **18** F8
Simcoeside *ON* **20** A7
Simcoeside *ON* **21** D1
Similkameen River *BC*
　6 F1-2
Simmie *SK* **11** E3
Simmons Peninsula *NU*
　42 C4
Simon, lac *QC* **24** B4
Simon, lac *QC* **24** C3
Simon, lac *QC* **26** A4-5
Simonds *NB* **31** F3
Simonds *NB* **32** B1
Simonette River *AB* **6** A3
Simonette River *AB* **8** A1
Simonhouse *MB* **11** A7
Simonhouse *MB* **12** F7
Simonhouse *MB* **13** A1
Simonhouse *MB* **15** F1
Simonhouse Lake *MB*
　11 A7-8
Simonhouse Lake *MB*
　12 F7-8
Simonhouse Lake *MB*
　15 F1-2
Simon Peak *BC* **6** C2-3
Simpson *SK* **11** D5
Simpson Island *ON* **16** E-F7
Simpson Island *ON* **17** C1
Simpson Islands *NT* **10** D7
Simpson Lake *NT* **5** D7
Simpson Lake *YT* **4** E7
Simpson Peak *BC* **3** A2-3
Simpson Peak *BC* **4** F5-6
Simpson Peninsula *NU*
　41 B6
Simpson Peninsula *NU*
　42 F4-5

Simpsons Corner *NS* **32** F6
Simpsons Corner *NS*
　33 C3-4
Simpson Strait *NU* **41** C4
Simpson Strait *NU* **42** F3-4
Sims Lake *NL* **39** B2
Simson PP *BC* **2** C3
Sincennes, lac *QC* **25** C5-6
Sincennes, lac *QC* **28** E1
Sinclair *MB* **11** F8
Sinclair *MB* **13** F1
Sinclair, lac *QC* **22** F7
Sinclair, lac *QC* **24** C1
Sinclair Mills *BC* **1** A7-8
Sinclair Mills *BC* **6** A1
Singhampton *ON* **20** B5
Sinnett *NS* **11** C5
Sintaluta *SK* **11** E6
Sioux Lake *MB* **12** B8
Sioux Lake *MB* **15** B3
Sioux Lookout *ON* **13** E7
Sioux Lookout *ON* **16** D3
Sioux Lookout *ON* **40** F1
Sioux Narrows *ON* **13** F6
Sioux Narrows *ON* **16** E2
Sioux Narrows PP *ON* **13** F6
Sioux Narrows PP *ON* **16** E2
Sioux Valley *MB* **11** E8
Sioux Valley *MB* **13** E1
Sipanok Channel *SK*
　11 A-B6
Sipiwesk *MB* **15** D4
Sipiwesk Lake *MB* **15** F4
Sir Alexander, Mount *BC*
　1 A8
Sir Alexander, Mount *BC*
　6 A1
Sir Alexander Mackenzie PP
　BC **1** C4
Sirdar *BC* **6** F4
Sirdar Mountain *AB* **6** E4-5
Sirdar Mountain *AB* **8** D1
Sir Douglas, Mount *BC*
　6 E4-5
Sir Douglas, Mount *BC* **7** C2
Sir James Macbrien, Mount
　NT **4** C7
Sir John Johnson House
　NHS *ON* **24** E4-5
Sirmilik NP *NU* **41** A6
Sirmilik NP *NU* **42** D5
Sir Richard Squires
　Memorial PP *NL* **36** A4
Sir Richard Squires
　Memorial PP *NL* **37** A2
Sir Richard Squires
　Memorial PP *NL* **38** F3
Sir Richard Squires PP *NL*
　39 F7-8
Sir Sandford, Mount *BC*
　6 C-D3
Sir Wilfrid Laurier, Mount
　BC **6** C2
Sir Wilfrid Laurier NHS *QC*
　24 C5-6
Sir Wilfrid Laurier NHS *QC*
　27 D1
Sir Winston Churchill PP *AB*
　9 F6
Sir Winston Churchill PP *AB*
　12 F1
Sir-Wilfrid, mont *QC* **22** C8
Sir-Wilfrid, mont *QC* **24** A2
Sir-Wilfrid, mont *QC* **25** E2
Sir-Wilfrid-Laurier, LHN de
　QC **24** C5-6
Sir-Wilfrid-Laurier, LHN de
　QC **27** D1
Sisib Lake *MB* **13** C2
Sisipuk Lake *MB* **12** E7
Sisipuk Lake *MB* **15** E1
Sissiboo Falls *NS* **32** E7
Sisson Branch Reservoir *NB*
　29 F5
Sisson Branch Reservoir *NB*
　31 C3
Sisson Ridge *NB* **31** D3
Sitdown Pond *NL* **36** C5
Sitdown Pond *NL* **37** C3
Sitidgi Lake *NT* **5** B-C5
Siwash Creek *YT* **3** A5
Siwash Creek *YT* **4** E8
Six Mile Lake PP *ON*
　18 E7-8
60th Parallel TP *NT* **9** A4
60th Parallel TP *NT* **10** F5-6
60th Parallel TP *NT* **41** E1
Sixty Mile River *YT*
　4 A1-2 B1-2
Skagit River *BC* **2** E7
Skagit Valley PP *BC* **1** F8
Skagit Valley PP *BC* **2** E7
Skagit Valley PP *BC* **6** F1
Skaha Lake *BC* **6** F2
Skaro *AB* **8** B7
Skead *ON* **17** F7
Skedans Point *BC* **1** B2
Skeena Mountains *BC*
　3 D3-4 E4
Skeena River *BC* **1** A3
Skeena River *BC* **3** E4
Skeetchestn *BC* **1** D8
Skeetchestn *BC* **2** A7
Skeetchestn *BC* **6** D1
Skeleton Lake *ON* **18** D8
Skeleton Lake *ON* **21** B1
Skelu Bay *BC* **1** A1
Skerryvore *ON* **18** C6
Skidegate *BC* **1** B1
Skidegate Inlet *BC* **1** B1-2
Skiff *AB* **6** F7
Skiff *AB* **7** E7
Skiff *MB* **11** F1
Skiff Lake *NB* **32** C1

Skiff Lake *NB* **32** C1
Skihist Mountain *BC* **1** E7-8
Skihist Mountain *BC* **2** B-C6
Skihist PP *BC* **1** E8
Skihist PP *BC* **2** B7
Skihist PP *BC* **6** E1
Skincuttle Inlet *BC* **1** C2
Skinner Lake *ON* **13** B8
Skinners Pond *PE* **35** A2
Skir Dhu *NS* **34** B7
Skoki Ski Lodge NHS *AB*
　6 D4-5
Skoki Ski Lodge NHS *AB*
　7 A1-2
Skookumchuck *BC* **2** C4
Skookumchuck *BC* **6** E4
Skookumchuck *BC* **7** E1-2
Skookumchuck Creek *BC*
　6 E4
Skookumchuck Creek *BC*
　7 D1
Skookumchuck Narrows PP
　BC **2** C4
Skootamatta Lake *ON*
　21 C5
Skootamatta River *ON*
　21 C-D5
Skownan *MB* **13** C2
Skukum, Mount *YT* **4** E3-4
Skylake *MB* **14** B4-5 C4-5
Sky Lake *ON* **18** E5
Slabtown *ON* **21** A5
Slate Falls *ON* **13** D7
Slate Island PP *ON* **16** F7
Slate Island PP *ON* **17** C2
Slate Islands *ON* **16** F7
Slate Islands *ON* **17** C2
Slate Lake *ON* **16** C3
Slave Bay *NT* **10** E6
Slave Lake *AB* **6** F5
Slave Point *NT* **10** E6
Slave River *AB* **9** A6
Slave River *AB* **10** F8
Slave River *AB* **12** A2
Slave River *NT* **10** E7
Slave River *NT* **41** E1 E2
Slavey Creek *AB* **9** A-B3
Sled Lake *SK* **11** A4
Sled Lake *SK* **11** C4
Sled Lake *SK* **12** F4
Sled River *SK* **12** F4
Sleeman *ON* **16** E1
Sleeper Islands *NU* **40** C3-4
Sleeper Islands *NU* **41** F7
Sleeping Giant PP *ON* **16** F6
Slemon Lake *NT* **10** C5-6
Slesse Park *BC* **2** E6-7
Sliammon *BC* **1** E6
Sliammon *BC* **2** C3
Slim Creek PP *BC* **1** A-B8
Slim Creek PP *BC* **6** A1
Slims River *YT* **4** D-E2
Slocan *BC* **6** E-F3
Slocan Park *BC* **6** F3
Slocan River *BC* **6** E-F3
Sloko River *BC* **3** A-B1
Sloko River *BC* **4** F4
Sloop Cove NHS *MB* **15** B5
Small Inlet PP *BC* **2** B2
Small Lake *MB* **15** C3
Small Point *NL* **37** D7-8
Small Point-Broad Cove-
　Blackhead-Adam's Cove
　NL **37** D7-8
Small River Caves PP *BC*
　6 B2
Smallwood Reservoir *NL*
　39 B3-4
Smallwood Reservoir *NL*
　40 D-E7
Smeaton *SK* **11** B5
Smelt Bay PP *BC* **1** E5-6
Smelt Bay PP *BC* **2** B2
Smelt Brook *NS* **34** A7
Smiley *NS* **6** C8
Smiley *SK* **11** C-D2
Smiley's PP *NS* **32** E8
Smiley's PP *NS* **33** B5
Smiley's PP *NS* **34** F1
Smith *AB* **9** F5
Smith, détroit de *NU* **42** B5
Smith, lac *QC* **24** C3
Smith Arm *NT* **41** B-C1
Smith Bay *NU* **42** C5
Smith Bay *ON* **18** C4
Smith Creek *ON* **19** A5
Smith Creek *ON* **20** C4
Smith Hill *MB* **14** F2
Smith Inlet *BC* **1** D4
Smith Island *NU* **40** B4
Smith Island *NU* **41** E7
Smith River *BC* **3** A5
Smith River *BC* **4** F8
Smith River *BC* **10** F1
Smiths Corner *NB* **31** E7
Smiths Corner *NB* **32** A5
Smiths Corner *NS* **33** B4-5
Smiths Corner *NS* **34** E1
Smiths Cove *NS* **32** F4-5
Smiths Cove *NS* **33** C5
Smiths Creek *NB* **32** C5
Smiths Falls *ON* **21** C7
Smiths Falls *ON* **24** F1-2
Smith's Harbour *NL* **38** E5
Smith Sound *BC* **1** D4

Smith Sound *NU* **42** B5
Smithville *NS* **33** F2-3
Smithville *ON* **19** C8
Smithville *ON* **20** E7
Smoke Lake *ON* **21** A2-3
Smoke Lake *ON* **22** F2
Smokey *NL* **38** A7
Smokey, Cape *NS* **34** B7
Smokey Head-White Bluff,
　RNP *ON* **18** D-E5
Smokey Head-White Bluff
　PNR *ON* **18** D-E5
Smoky Falls *ON* **17** B5-6
Smoky Lake *AB* **6** A6
Smoky Lake *AB* **8** B7-8
Smoky Lake *AB* **11** A1
Smoky Lake *AB* **12** F1
Smoky River *AB* **6** A2-3
Smoky River *AB* **8** A2
Smoky River *AB* **9** E3
Smoky River *ON* **18** A7-8
Smooth Rock Falls *ON*
　17 B-C6
Smooth Rock Falls *ON*
　40 F3
Smoothrock Lake *ON* **16** C5
Smoothstone Lake *SK*
　12 F4
Smoothstone River *SK*
　12 F4
Smuts *SK* **11** C4-5
Snag *YT* **4** C1
Snag Creek *YT* **4** C1
Snake Creek *ON*
　20 A3-4 B4
Snake Creek *ON* **21** B5
Snake Falls *ON* **13** D6
Snake Falls *ON* **16** B1
Snake Indian River *AB*
　6 B2-3
Snake River *AB* **22** F5
Snake River *YT* **5** F5
Snakes Bight *NL* **36** D1
Snare Lake *NT* **10** B6-7
Snare Lake *SK* **9** B8
Snare Lake *SK* **12** B8
Snare River *NT* **10** A-B7
Snare River *NT* **10** B6
Snare River *NT* **41** D1-2
Snare River *SK* **12** B-C4
Snegamook Lake *NL* **39** A5
Snelgrove *ON* **19** A7
Snelgrove *ON* **21** C6
Snelgrove Lake *NL* **39** A2-3
Sniatyn *AB* **8** B7-8
Snipe Lake *AB* **9** F4
Snipe Lake *SK* **11** D3
Snook Arm *NL* **38** E5
Snook's Harbour *NL* **36** C8
Snook's Harbour *NL* **37** C6
Snowball *ON* **20** C7
Snowball *ON* **21** E1
Snowbird Lake *NT* **41** E3
Snowden *SK* **11** B5
Snowdrift River *NT* **41** D-E2
Snowflake *MB* **13** F3
Snowflake *MB* **14** F3
Snow Lake *MB* **12** F8
Snow Lake *MB* **15** F2
Snow Road Station *ON*
　21 C6
Snowshoe Island *SK* **12** B6
Snowshoe Lake *ON* **16** C1
Snowshoe Peak *YT* **4** E2
Snug Harbour *ON* **39** D8
Snug Harbour *ON* **18** D6
Snug Harbour *ON* **18** E7
Snug Harbour *ON* **21** D2-3
Snyder *ON* **20** E8
Snyder Lake *MB* **12** A7
Snyder Lake *MB* **15** A1
Sober Island *NS* **33** B8
Sober Island *NS* **33** B8
Soda Creek *BC* **1** C7
Soda Creek (IR) *BC* **1** C7
Sointula *BC* **1** E4
Solace PP *ON* **17** E7
Soldiers Cove *NS* **34** D7
Solina *ON* **20** C8
Solina *ON* **21** E2
Solmesville *ON* **21** E5
Solsgirth *MB* **11** E8
Solsgirth *MB* **13** E1
Sombra *ON* **19** D2
Sombra *ON* **20** F1
Somerset *MB* **14** E-F3
Somerset *NS* **32** E6
Somerset *NS* **33** B4
Somerset, île *NU* **41** A4-5
Somerset, île *NU* **42** D4
Somerset, île *NU* **42** E4
Somerset Island *NU*
　41 A4-5
Somerset Island *NU* **42** D4
Somerville *NB* **32** B1
Somerville Island *BC* **3** F2-3
Somme *SK* **11** C6
Sommerfeld *MB* **13** F3-4
Sommerfeld *MB* **14** F4-5
Sonningdale *SK* **11** C3
Sonora *NS* **34** E5
Sonora Island *BC* **1** E5
Sonora Island *BC* **2** B2
Sonya *ON* **20** B8
Sonya *ON* **21** E2
Sooke *BC* **2** F4
Sooke Basin *BC* **2** F4

Sooke Potholes PP *BC* **2** F4
Soperton *ON* **21** D7
Soperton *ON* **24** F1
Sop's Arm *NL* **38** E3-4
Sop's Arm *NL* **39** F8
Sop's Island *NL* **38** E4
Sop's Island *NL* **39** F8
Sorcier, lac au *QC* **24** A6
Sorcier, lac au *QC* **25** E6
Sorcier, lac au *QC* **27** B2
Sorel *QC* **24** C7
Sorel *QC* **27** D2
Sorrento *BC* **6** D2
Soufflets River *NL* **38** C-D4
Soufflot, lac *QC* **22** A3
Soulier Lake *SK* **9** A8
Soulier Lake *SK* **12** A3
Soulis Pond *NL* **36** B8
Soulis Pond *NL* **37** B6
Soul Lake *MB* **13** B6
Sounding Creek *AB* **6** C-D7
Sounding Creek *AB* **7** A7
Sounding Creek *AB* **11** C1
Sounding Lake *AB* **6** C7-8
Sounding Lake *AB* **11** C2
Sound Island *NL* **36** D8
Sound Island *NL* **37** D6
Soup Harbour *ON* **21** F5
Sourd, lac du *QC* **24** B3
Souris *MB* **11** F8
Souris *MB* **13** F2
Souris *MB* **14** F1
Souris *PE* **34** C4
Souris *PE* **35** D7-8
Souris Beach PP *PE* **34** C4
Souris Beach PP *PE* **35** E7
Souris Line Road *PE*
　35 C-D3
Southey *SK* **11** D5-6
South Fork *SK* **11** F2-3
South Freetown *PE* **35** D3
Southgate River *BC* **1** D-E6
Southgate River *BC* **2** A3
South Gillies *ON* **16** F5
South Gloucester *ON* **21** B8
South Gloucester *ON* **24** E2
South Granville *PE* **35** D4
South Gut St. Anns *NS*
　34 C7
South Harbour *NS* **34** A7
South Hazelton *BC* **3** E4
South Head *NL* **37** B7-8
South Heart River *AB* **9** E4
South Henik Lake *NU* **40** A1
South Henik Lake *NU* **41** E4
South Huron *ON* **19** B4
South Huron *ON* **20** D3
South Indian Lake *MB*
　15 C3
South Junction *MB* **13** F6
South Junction *MB* **14** F7
South Kedgwick River *NB*
　29 D3-4
South Knife Lake *MB*
　15 B-C4
South Knife River *MB*
　15 B4-5
South La Cloche Mountains
　ON **18** B4-5
South Lake *PE* **35** D8
South Lake *PE* **35** D8
South Lake Ainslie *NS*
　34 C6
South Macmillan River *YT*
　4 C5-6
South Maitland *NS* **32** C8
South Maitland *NS* **33** A6
South Maitland *NS* **34** F2
South Maitland River *ON*
　19 A4
South Maitland River *ON*
　20 C3
South McQuesten River *YT*
　4 A3-4 B3
South Melville *PE* **35** E4
South Middleton *NS* **19** D6
South Middleton *ON* **20** F5
South Milford *NS* **32** F5
South Milford *NS* **33** C3
South Monaghan *ON* **21** E3
South Moose Lake *MB*
　11 A8
South Moose Lake *MB*
　13 A2
South Mountain *ON* **32** E6
South Mountain *NS* **33** B3-4
South Mountain *ON*
　21 B-C8
South Mountain *ON*
　24 E2-3
South Nahanni River *NT*
　4 C7 D8
South Nation River *ON*
　21 C8
South Nation River *ON*
　24 D2-3
South Ohio *NS* **33** E1
South Pender Island *BC*
　2 F5
South Pine River *ON* **20** B3
South Porcupine *ON*
　17 C-D6-7
Southport *MB* **13** E3
Southport *MB* **14** D3-4
Southport *NL* **37** C7
South Range *NS* **33** D2
South Rawdon *NS* **32** E8
South Rawdon *NS* **33** B5-6
South River *NL* **37** D7
South River *NS* **34** E5
South River *ON* **18** B8
South River *ON* **18** C8
South River *ON* **18** C8
South River *ON* **22** E1-2

South Bruce *ON* **20** B4
South Bruce Peninsula *ON*
　18 F5
South Bruce Peninsula *ON*
　20 A4
South Buxton *ON* **19** E3
South Castor River *ON*
　21 B8
South Castor River *ON*
　24 E2-3
South Cayuga *ON* **19** C7
South Cayuga *ON* **20** F6
South Charlo River *NB*
　29 D-E6
South Charlo River *NB*
　31 A-B5
South Cooking Lake *AB*
　6 B6
South Cooking Lake *AB*
　8 C6-7
South Dildo *NL* **37** D-E7
Southeast Bay *MB* **12** A8
Southeast Bay *MB* **15** A2-3
South East Bight *NL* **36** E7
South East Bight *NL*
　37 E5-6
Southeast Upsalquitch River
　NB **29** E6
Southeast Upsalquitch River
　NB **31** B5
Southend *SK* **12** D6
Southend (IR) *SK* **12** D6
Southern Bay *NL* **37** C6-7
Southern Harbour *NL* **36** D8
Southern Harbour *NL* **37** D6
Southern Indian Lake *MB*
　15 C-D3

South River ON 22 E1
South River Lake NS 34 E5
South Rushbrook Old Pine PP ON 17 F6
South Rustico PE 35 D4
South Saskatchewan River AB 6 E7
South Saskatchewan River AB 7 D7-8
South Saskatchewan River AB 11 E1
South Saskatchewan River SK 11 D3
South Saugeen River ON 20 B5 C4
South Seal River MB 15 B-C3
South Shoal Lake MB 14 D4
South Side NS 33 F2
Southside Antigonish Harbour NS 34 D5
South Slocan BC 6 F3
South Tetagouche NB 29 E7
South Tetagouche NB 31 B6
South Tweedsmuir PP BC 1 B-C5
South Twillingate Island NL 38 E6
South Twin Island NU 40 E3-4
South Twin Lake NL 36 B2
South Twin Lake NL 37 A4
South Twin Lake NL 38 F5
South Uniacke NS 32 E8
South Uniacke NS 33 B5-6 C5-6
South Victoria NS 32 C8
South Victoria NS 34 D1
Southview BC 2 C2-3
Southville NS 33 D2
South Wabasca Lake AB 9 E5
South Wellington BC 1 F6
South Wellington BC 2 E3-4
Southwest Gander River NL 36 B6-7
Southwest Gander River NL 37 B5
Southwest Head NB 32 F2
South West Lot 16 PE 35 D2
Southwest Margaree NS 34 C6
Southwest Margaree PP NS 34 C6
Southwest Miramichi River NB 31 D-E6
Southwest Miramichi River NB 31 E4-5
Southwest Miramichi River NB 32 A2-4
South West Port Mouton NS 33 E4
Southwest River NL 36 C8
Southwest River NL 37 C6
Southwest River PE 35 D3-4
Southwold ON 19 C-D4
Southwold ON 20 F3
Southwold Earthworks NHS ON 19 D4
Southwold Earthworks NHS ON 20 F3
South Wolf Island NL 39 B8
Southwood ON 18 E8
Southwood ON 21 C1
South Woodslee ON 19 E2
Sovereign SK 11 D3
Sovereign Lake SK 12 A5
Sowchea Bay PP BC 1 A6
Sowden Lake ON 13 F8
Sowden Lake ON 16 E4
Sowerby ON 17 F4-5
Sowerby ON 18 A1
Spaffordton ON 21 D6-7
Spalding SK 11 C5-6
Spallumcheen BC 6 E2
Spaniard's Bay NL 37 D7
Spanish ON 17 F5
Spanish ON 18 B3
Spanish River ON 17 E-F6
Spanish River ON 17 F6
Spanish River ON 18 A4
Spanish River ON 18 A4
Spanish River ON 18 A4 B3
Spanish River PP ON 17 E-F6
Spanish River PP ON 18 A4
Spanish Room NL 36 E7
Spanish Room NL 37 E5
Spanish Ship Bay NS 34 F4
Spar Island ON 16 F5
Sparkling Lake ON 13 E8
Sparrow Lake ON 18 E8
Sparrow Lake ON 21 C1
Sparta ON 19 D5
Sparta ON 20 F4
Sparwood BC 6 F5
Sparwood BC 7 E2-3
Spatsizi Headwaters PP BC 3 D3
Spatsizi Plateau PWP BC 3 C3-4
Spatsizi River BC 3 C-D4
Spear, cap NB 35 F3
Spear, Cape NB 35 F3
Spear Harbour NL 39 C8
Spearhill MB 13 D3
Spearhill MB 14 B3
Spear Lake SK 9 D8
Spectacle Lake PP 2 F4
Spector Lake MB 15 E7-8
Spectre Peak BC 3 C5

Spedden AB 6 A6-7
Spedden AB 8 B8
Spedden AB 11 A1
Spedden AB 12 F1
Spednic Lake NB 32 C-1
Spednic Lake, APP NB 32 C1
Spednic Lake PPA NB 32 C1
Spednic Lake Provincial Campground NB 32 C1-2
Speed River ON 19 A6
Speed River ON 20 C5
Speers SK 11 B-C4
Spence Lake MB 13 D2
Spence Lake MB 14 A2
Spence Lake AB 12 F1
Spencer, Cape NB 32 E4
Spencer, Cape NB 33 B2
Spencer Lake AB 9 F7
Spencer Lake AB 12 F1-2
Spencers Island NS 32 D6
Spencers Island NS 33 A4
Spencerville ON 21 C8
Spencerville ON 24 F2
Spences Bridge BC 1 E8
Spences Bridge BC 2 B7
Spence Settlement NB 32 B7-8
Spence Settlement NB 34 C1
Spence Settlement NB 35 E2
Sperling MB 13 F3
Sperling MB 14 E4
Speyside ON 19 A7
Speyside ON 20 D6
Spider Lake ON 18 D7
Spider Lake PP BC 2 D2-3
Spiers Lake NWA AB 6 C6-7
Spiers Lake NWA AB 8 C7
Spiers Lake NWA AB 11 C1
Spillars Cove NL 37 B7-8
Spiller Channel BC 1 C3-4
Spiller Inlet BC 1 C4
Spillimacheen BC 6 D4
Spillimacheen BC 7 B1
Spillimacheen River BC 6 D3-4
Spillway NL 36 A3-4
Spillway NL 37 A1-2
Spillway NL 38 F3
Spipiyus PP BC 2 D4
Spirit River AB 9 E2
Spirit Rock CA ON 18 E5
Spirit Wood Acres AB 11 D6-7
Spiritwood SK 11 B4
Spius Creek BC 1 E8
Spius Creek BC 2 C7
Spius Creek BC 6 E1
Split, Cape NS 32 D6-7
Split, Cape NS 33 A4
Split Lake MB 15 E4
Split Lake MB 15 E5
Split Lake (IR) MB 15 E4-5
Spoonbill Lake ON 16 B1
Spoon Lake ON 18 B5
Sporting Mountain NS 34 D6
Spotted Island NL 39 B8
Spragge ON 17 F5
Spragge ON 18 A2
Sprague MB 13 F5
Sprague MB 14 F7
Spray Lakes Reservoir AB 7 B-C2
Spray Valley PP AB 6 D5
Spray Valley PP AB 7 B-C2
Springbank, parc ON 20 E3
Springbank Park ON 20 E3
Spring Bay ON 18 C3
Spring Brook ON 21 D4-5
Springbrook PE 32 A8
Springbrook PE 34 B2
Springbrook PE 35 D4
Spring Coulee AB 6 F6
Spring Coulee AB 7 F5
Spring Creek ON 18 B7
Springdale AB 8 E5-6
Springdale NL 36 B3
Springdale NL 39 F8
Springfield NB 32 C2
Springfield NB 32 C4
Springfield NS 32 F6
Springfield NS 33 C3
Springfield ON 19 C5
Springfield ON 20 F4
Springfield PE 32 A8
Springfield PE 34 C2
Springfield PE 35 D3
Springfield West PE 35 C1
Springford ON 19 C5-6
Springford ON 20 E4-5
Springhill NS 32 F6
Springhill NS 34 E1
Springhill PE 32 A7
Springhill PE 34 B1
Springhill PE 35 C2
Springhill Junction NS 32 C7
Springhill Junction NS 34 E1
Springhouse BC 1 C7-8
Spring Lake ON 18 C7-8
Spring Lake ON 22 E1
Springmount ON 18 F5
Springmount ON 20 A4
Spring Point AB 7 E4
Springside BC 11 C6-7
Springstein MB 13 F3-4
Springstein MB 14 E4
Springtown ON 21 B6
Springvale ON 19 C6-7

Springvale ON 20 E5-6
Springvale PE 35 E4
Spring Valley ON 21 C8
Spring Valley ON 24 F2
Spring Valley PE 32 A8
Spring Valley PE 35 D3
Spring Valley SK 11 E5
Springville NS 33 A7-8
Springville NS 34 E3-4
Springville ON 21 E3
Springwater SK 11 C3
Springwater CA ON 19 D5
Springwater CA ON 20 F4
Springwater PP ON 18 F7
Springwater PP ON 20 A6
Spring Well MB 14 D6
Sproat Lake BC 1 F6
Sproat Lake BC 2 D2
Sproat Lake BC 2 D2
Sproat Lake PP BC 1 F6
Sproat Lake PP BC 2 D2
Sprott Lake MB 12 B8
Sproule, lac QC 25 D4
Sproule Peninsula NT 42 C2
Spruce Beetle Trail Territorial Recreation Site YT 4 E2
Spruce Brook NL 36 B2
Spruce Creek MB 14 B1
Sprucedale ON 18 D8
Sprucedale ON 21 A1
Sprucedale ON 22 F1
Sprucefield AB 8 B7
Spruce Grove AB 6 B5
Spruce Grove AB 8 C6
Spruce Home SK 11 B5
Spruce Island MB 11 B8
Spruce Island MB 13 B2
Spruce Island NT 5 E6
Spruce Islands, RNP ON 16 E1-2
Spruce Islands PNR ON 16 E1-2
Spruce Lake SK 6 A-B8
Spruce Lake SK 11 B3
Spruce Pond NL 36 C3
Spruce Pond NL 37 C1
Spruce River NL 16 A-B5
Spruce River SK 11 B4-5
Spruce View AB 6 C5
Spruce View AB 8 F5
Sprucewoods MB 13 E2
Sprucewoods MB 14 E2
Spruce Woods PP MB 13 E-F2
Spruce Woods PP MB 14 E2
Spry Bay NS 33 C8
Spry Bay NS 33 B8
Spry Bay PP NS 33 B7-8
Spry Cove PE 35 E7
Spry Harbour NS 33 B7-8
Spry Lake ON 18 E5
Spry Point PE 35 E7
Spudgels Cove NL 38 D2-3
Sputinow AB 6 A7-8
Sputinow AB 11 A2
Sputinow AB 12 F8
Spuzzum BC 2 D7
Squally Channel BC 1 B3
Squamish BC 1 F7
Squamish BC 2 D4-5
Squamish River BC 1 E7
Squamish River BC 2 B-C4
Squamish River BC 2 C4
Squanga Lake YT 4 E4
Square, lac NB 35 E1
Square Forks, rivière QC 29 C6-7
Square Islands NL 39 C8
Square Lake NB 35 E1
Squatec QC 29 D2
Squatec QC 31 B1
Squatec, rivière QC 29 D-E2
Squaw Island ON 18 C4
Squilax BC 6 D2
Squire ON 18 F5
Squire ON 20 A4
Squirrel Creek CA 21 E3
Squitty Bay PP BC 1 F6
Squitty Bay PP BC 2 D3
S.S. Klondike, LHN YT 4 E4
S.S. Klondike NHS YT 4 E4
Stackler, lac QC 26 C7
Stackler, lac QC 28 A2
Staffa ON 19 B4
Staffa ON 20 D3
Stafford River BC 1 D5
Stafford River BC 2 A2
Stag Harbour NL 38 E7
Stagleap PP BC 6 F3
Stait de Honguedo QC 30 A3-4 B4
Stallworthy, Cape NU 42 A4
Stalwart SK 11 D5
Stalwart, RNF de SK 11 D5
Stalwart National Wilderness Area SK 11 D5
Stamp River PP BC 1 F5-6
Stamp River PP BC 2 D2
Stampville ON 24 F2-3
Stanbridge East QC 24 E7
Stanbridge East QC 25 F2
Stanbridge Station QC 24 E7
Stanbridge Station QC 27 F2

Stanburn NS 32 F6
Stanburn NS 33 C4
Stanchel PE 35 D-E4
Standard AB 6 D6
Standard AB 7 B5
Standing Buffalo SK 11 D6
Stand Off AB 6 F5
Stand Off AB 7 E-F5
Stanger AB 8 C5
Stanhope NL 36 A6-7
Stanhope NL 37 A4-5
Stanhope NL 38 F6
Stanhope PE 34 C2-3
Stanhope PE 35 D5
Stanhope QC 27 F4-5
Stanley NB 31 F4
Stanley NB 32 B2-3
Stanley NS 32 E8
Stanley NS 33 B5
Stanley NS 34 F1
Stanley ON 16 F5
Stanley SK 12 E5
Stanley Bridge PE 32 A8
Stanley Bridge PE 34 B2
Stanley Bridge PE 35 D4
Stanley Cove NL 36 D5
Stanley Cove NL 37 D3
Stanley Mission SK 12 E5
Stanley Mission (Holy Trinity Anglican Church) PHS SK 12 E5-6
Stanley River PE 35 D4
Stanleyville ON 21 C7
Stanleyville ON 24 F1
Stanmore AB 7 A7
Stanstead QC 27 F4
Stanwell-Fletcher Lake NU 41 A4-5
Stanwell-Fletcher Lake NU 42 E4
Staples ON 19 E2
Star AB 6 A6
Star AB 8 B7
Star AB 11 A1
Star Blanket SK 11 D6
Star Lake NL 36 B4
Star Lake NL 37 B2
Starbuck MB 13 E-F3
Starbuck MB 14 E4
Starfish Lake NT 10 A7
Starkville ON 21 E3
Starratt-Olsen ON 13 D6
Starratt-Olsen ON 16 C2
Starvation Lake NT 10 A7
State Mount YT 4 D-E4
Stauffer AB 6 C5
Stauffer AB 8 F5
Stawamus Chief PP BC 1 F7
Stawamus Chief PP BC 2 C-D5
Stayner ON 18 F7
Stayner ON 20 A6
Stead MB 13 E4
Stead MB 14 C6
Steady Brook NL 36 B3
Steady Brook NL 37 B1
Steamboat Mountain YT 4 A4
Steel Lake ON 16 E7
Steel Lake ON 17 C2
Steel Lake ON 17 C2
Steel Narrows Provincial Historic Park SK 6 A8
Steel River ON 16 E7
Steel River ON 17 C2
Steel River PP ON 16 E7
Steel River PP ON 17 B-C2
Steele Narrows Provincial Historic Park SK 11 A2-3
Steelhead PP BC 1 D8
Steelhead PP BC 2 A8
Steelhead PP BC 6 D1
Steelman SK 11 F7
Steels Ferry MB 14 A3
Steels SK 11 B-C6
Steenburg Lake ON 21 C4
Steen River AB 9 A3
Steen River AB 9 A3
Steensby Inlet NU 41 B7
Steensby Inlet NU 42 E6
Steepbank River AB 9 D7
Steepbank River AB 12 D2
Steephill Lake SK 12 E6
Steep Rock MB 13 D2
Steep Rock MB 14 A3
Steep Rock Junction MB 14 A3
Steeprock Lake PP MB 11 B-C7
Steeprock Lake PP MB 13 B1
Steeves Mountain NB 31 F7-8
Steeves Mountain NB 32 B5-6
Stefansson Island NU 41 A3
Stefansson Island NU 42 D3
Steinbach MB 13 F4
Steinbach MB 14 E6
Stein Mountain BC 1 E7-8
Stein Mountain BC 2 B6
Stein River BC 1 E7-8
Stein River BC 2 B6
Stein Valley Nlaka'pamux PP BC 1 E7-8
Stein Valley Nlaka'pamux PP BC 2 B6

Stelcam SK 11 E5
Stella ON 21 E6
Stellako BC 1 A5-6
Stellako BC 1 A5-6
Stellarton NS 33 A7-8
Stellarton NS 34 E3
Stemwinder PP BC 2 D8
Stemwinder PP BC 6 F1
Stenen SK 11 C7
Stephansson House PHS AB 6 C4-5
Stephansson House PHS AB 8 F5-6
Stephenfield MB 14 E3-4
Stephenfield PP MB 14 E3
Stephens Island BC 1 B2
Stephens Lake MB 15 D5
Stephenville NL 36 B2
Stephenville Crossing NL 36 C2
Sterlet Lake NT 10 A8
Stettler AB 6 C6
Stettler AB 8 E-F7
Steven Creek ON 21 B8
Steven Creek ON 24 E2
Stevens ON 16 D-E8
Stevens ON 17 B2-3
Stevens Lake MB 12 B8
Stevens Lake MB 15 B2
Stevenson, Mount BC 1 B-C8
Stevenson, Mount AB 6 B1
Stevenson Lake MB 13 A4-5
Stevenson River MB 13 A5
Stevensville ON 20 E7-8
Steveville AB 7 C7
Stewardson Inlet BC 1 F5
Stewart BC 3 E3
Stewart YT 4 B2
Stewart, Mount YT 4 B2
Stewart Creek SK 11 A5
Stewart Crossing YT 4 B3
Stewartfield AB 8 B5
Stewart Hall ON 21 E3
Stewart Lake NT 4 A8
Stewart Lake NT 10 B1-2
Stewart Lake SK 11 A7
Stewart Property PNR ON 21 A-B6
Stewart Property, RNP ON 21 A-B6
Stewart River YT 4 A5
Stewart River YT 4 B2
Stewart River YT 4 B4
Stewarttown ON 19 A7
Stewartown ON 20 D6
Stewart Valley SK 11 E3
Stewiacke NS 33 B6
Stewiacke NS 34 F2
Stewiacke Cross Road NS 33 A7
Stewiacke Cross Road NS 34 E3
Stewiacke East NS 33 B6-7
Stewiacke East NS 34 F2
Stewiacke River NS 33 A-B7
Stewiacke River NS 34 E3 F2-3
Stickney NB 31 E3
Stickney NB 32 A-B1
Stikine BC 3 D2
Stikine Ranges BC 3 B-C4
Stikine River BC 3 C3-4
Stikine River BC 3 D2
Stikine River PP BC 3 C3
Still River ON 18 C6
Stillwater BC 1 F6
Stillwater BC 2 C3
Stillwater NS 34 F5
Stirling AB 6 F6
Stirling AB 7 E6
Stirling ON 21 D5
Stirling Creek ON 18 C8
Stirling Creek ON 21 A1
Stirling Creek ON 22 F1
Stirlingville AB 7 A4
Stirton ON 19 A5-6
Stirton ON 20 C4-5
Stittsville ON 21 B7-8
Stittsville ON 24 E2
Stivens, Point NT 5 A7
Stobart AB 7 B5
Stock Cove NL 37 B7
Stockdale ON 21 E4
Stockholm MB 8 F6
Stockholm SK 11 E7
Stockport PE 34 F5
Stockton MB 13 F2
Stockton MB 14 E2
Stoco ON 21 D5
Stoco Fen PP ON 21 D5
Stoke QC 27 E5
Stoke, rivière QC 27 E4-5
Stokes Bay ON 18 E4
Stokes Bay ON 18 E5
Stokes Point YT 5 A-B3
Stone BC 1 C7
Stonecliffe ON 22 D4
Stoneham NB 31 E6
Stonehaven NB 29 E8
Stonehaven NB 30 E1
Stonehaven NB 31 B7
Stonehouse NS 32 C8
Stonehouse NS 34 E7
Stonehurst West NS 33 D5
Stone Lake ON 17 A1
Stone Mountain PP BC 3 B5-6
Stoner BC 1 B7

Stone's Cove NL 36 D-E7
Stone's Cove NL 37 D-E5
Stone Valley NL 36 D5-6
Stone Valley NL 37 D3-4
Stoneville NL 38 F6-7
Stonewall MB 13 E4
Stonewall MB 14 D5
Stoney Creek ON 19 B7-8
Stoney Creek ON 20 E6-7
Stoney Point ON 19 E2
Stoney Point Beach MB 14 B1-2
Stony Creek NT 5 D4
Stony Hill MB 14 C4
Stony Island NL 39 C8
Stony Lake MB 12 B8
Stony Lake MB 15 B3
Stony Lake NL 36 B6
Stony Lake NL 37 B4
Stony Lake ON 21 D3-4
Stony Lake SK 12 A5
Stony Mountain MB 13 E4
Stony Mountain MB 14 D5
Stony Mountain Wildland PP AB 9 D7
Stony Mountain Wildland PP AB 12 D-E1
Stony Plain AB 6 B5
Stony Plain AB 8 C6
Stony Rapids SK 12 A5
Stony Rapids SK 41 F2-3
Stonyridge ON 21 D4
Stor Lake AB 12 B4
Storkerson Peninsula NU 41 A3
Storkerson Peninsula NU 42 E2-3
Storm, Cape NU 42 C4
Stormont NS 34 E-F5
Stormy Lake ON 13 F7
Stormy Lake ON 16 E3
Stormy Lake ON 18 B7-8
Stormy Lake ON 22 D-E1
Stornaway QC 27 D6
Stornoway SK 11 D7
Storthoaks SK 11 F7-8
Storthoaks SK 13 F1
Stouffville ON 20 C7
Stouffville ON 21 E1-2
Stoughton SK 11 F6-7
Stout Lake ON 13 C5
Stout Lake ON 16 A1-2
Strabane ON 19 B7
Strabane ON 20 D6
Straffordville ON 19 D5
Straffordville ON 20 F4
Straight River SK 12 A4-5
Straitsview NL 38 A5
Strand Bay NU 42 B4
Strangmuir AB 7 C5
Stranraer SK 11 C3
Strasbourg SK 11 D5
Stratford ON 19 B5
Stratford ON 20 D4
Stratford PE 34 C2-3
Stratford PE 35 E5
Stratford QC 27 D5
Strathallen SK 11 F4
Strathburn ON 19 D4
Strathburn ON 20 F3
Strathclair MB 11 E8
Strathclair MB 13 E2
Strathclair MB 14 C1
Strathcona ON 21 D6
Strathcona PE 35 E6
Strathcona PP BC 1 F5
Strathcona PP BC 2 C3
Strathcona-Westmin PP BC 1 F5
Strathcona-Westmin PP BC 2 D1
Strathgartney PP PE 34 C2
Strathgartney PP PE 35 E4
Strathlorne NS 34 C5-6
Strathmore AB 6 D6
Strathmore AB 7 B5
Strathnaver BC 1 B7
Strathroy ON 19 C4
Strathroy ON 20 E3
Stratton ON 16 E2
Strawberry Island ON 18 E5
Streak, Mount YT 4 E4
Streamstown AB 6 B7-8
Streamstown AB 11 B2
Streatfield Lake ON 16 A8
Streatfield River ON 16 A8
Streets Ridge NS 32 C8
Streets Ridge NS 34 E7
Streetsville ON 19 A7-8
Streetsville ON 20 D6-7
Strome AB 6 B6-7
Strome AB 8 D-E8
Strome AB 11 B1
Strongfield SK 11 D4
Stroud ON 20 B7
Stroud ON 21 D1
Struan SK 11 C4
Stry AB 8 B8
Stuartburn MB 13 F4
Stuartburn MB 14 F5-6
Stuart Channel BC 1 F7
Stuart Island BC 1 E5-6
Stuart Island BC 2 B2
Stuart Lake BC 1 A6
Stuart Lake PP BC 1 A6
Stuart River BC 1 A6
Stuart River PP BC 1 A6-7
Stubbs, lac QC 22 E6
Stuie BC 1 C5
Stukely-Sud QC 24 E8
Stukely-Sud QC 27 E3-4 F3-4
Stull Lake MB 15 F6

Stull Lake ON 15 F6
Stull River ON 15 F6
Stum Lake BC 1 C7
Stump Lake BC 2 B8
Stump Lake BC 2 B8
Stump Lake PP BC 6 E3
Stumplake Creek BC 2 B8
Stupart River MB 15 E5
Sturdies Bay BC 2 E4-5
Sturgeon PE 35 F7
Sturgeon Bay MB 13 C3
Sturgeon Bay ON 18 F7-8
Sturgeon Bay ON 20 A6
Sturgeon Bay PP ON 18 C6
Sturgeon Bay PP Reserve MB 13 C3
Sturgeon Bay PP Reserve MB 14 A4
Sturgeon Falls ON 17 F7-8
Sturgeon Falls ON 18 A8
Sturgeon Heights AB 9 F3
Sturgeon Lake AB 9 E3
Sturgeon Lake ON 13 E8
Sturgeon Lake ON 15 E7
Sturgeon Lake ON 16 D4
Sturgeon Lake ON 16 F4
Sturgeon Lake ON 18 F8
Sturgeon Lake ON 20 A8
Sturgeon Lake ON 21 D2
Sturgeon Lake SK 11 A7
Sturgeon Landing SK 11 A7
Sturgeon Landing SK 12 F7
Sturgeon Landing SK 15 F1
Sturgeon Point ON 20 A8
Sturgeon Point ON 21 D2
Sturgeon River ON 13 E5
Sturgeon River ON 14 C8
Sturgeon River ON 15 E7-8
Sturgeon River ON 17 F7
Sturgeon River ON 18 A7
Sturgeon River ON 18 F7-8
Sturgeon River ON 20 A6
Sturgeon River PP ON 17 E7
Sturgis SK 11 C7
Styx River ON 20 B4
Success SK 11 E3
Sucker Creek ON 18 B3
Sucker River SK 12 C6
Sud, pointe du QC 30 B6
Sud, rivière du QC 24 E7
Sud, rivière du QC 27 A6-7
Sud, rivière du QC 27 B6-7
Sud, rivière du QC 27 F2
Sudbury ON 17 F7
Sudbury ON 18 A5
Sudeten PP BC 3 E8
Sudeten PP BC 9 E1-2
Sud-Ouest, cap du QC 30 F7
Sud-Ouest, rivière du QC 29 C2
Suffield AB 6 E7
Suffield AB 7 D8
Suffield AB 11 E1
Suffolk PE 35 E5
Sugarbowl-Grizzly Den PP BC 1 A8
Sugarbowl-Grizzly Den PP BC 6 A1
Sugar Lake BC 6 E2-3
Sugarloaf PP NB 29 D6
Sugarloaf PP NB 31 A5
Suggi Lake SK 11 A6
Suggi Lake SK 12 F6
Sukunka Falls PP BC 3 F7
Sukunka Falls PP BC 9 E1
Sukunka River BC 3 E-F7
Sukunka River BC 9 E1
Suley Ann's Cove NL 38 E5
Sullivan QC 23 E4
Sullivan Bay BC 1 D4
Sullivan Lake AB 6 C6-7
Sullivan Lake AB 8 F8
Sullivan Lake AB 11 C1
Sully QC 29 C2
Sulphur Bay NT 10 E6
Sulphur Passage PP BC 1 F5
Sulphur Passage PP BC 2 D1
Sulphur Point NT 10 E6
Sultan ON 17 E5
Sulte, lac QC 25 A4
Sulte, lac QC 26 E-F4
Summer Beaver (Nibinamik) ON 16 A6
Summerberry MB 11 E6-7
Summerford NL 38 E6
Summerland BC 6 F1-2
Summers Corners ON 19 D5
Summers Corners ON 20 F4
Summerside NL 36 B3
Summerside NL 37 B1
Summerside PE 32 A8
Summerside PE 35 D2-3
Summerside Harbour PE 35 D2-3
Summerstown ON 24 E4
Summerstown Station ON 24 E4
Summerville NB 32 C2
Summerville NB 33 A1
Summerville NS 32 E7-8
Summerville NS 33 B5
Summerville NS 34 F1
Summerville PE 34 C3
Summerville PE 35 E6
Summerville Beach PP NS 33 E4

Summerville Centre NS 33 E3-4
Summit Lake BC 1 A7
Summit Lake BC 3 C4
Summit Lake BC 3 C7
Summit Lake PP BC 6 E3
Summit Pass BC 3 B6
Sunbury ON 21 D7
Sunbury Cove PE 35 D2
Sunbury Point PE 35 D2
Sundance MB 15 D5
Sundance MB 40 C1
Sundance Beach AB 6 B5
Sundance Beach AB 8 D5-6
Sundance PP AB 6 A3-4
Sundance PP AB 8 C2
Sunday Inlet BC 1 B1
Sunderland ON 20 B8
Sunderland ON 21 E2
Sunderland Channel BC 2 B1
Sundown MB 13 F4
Sundown MB 14 F6
Sundre AB 6 C5
Sundre AB 7 A3
Sundridge ON 18 C8
Sundridge ON 21 A1
Sundridge ON 22 E1
Sunnidale Corners ON 18 F7
Sunnidale Corners ON 20 A6
Sunny Brae NS 33 A8
Sunny Brae NS 34 E4
Sunnybrae PP BC 6 D2
Sunnybrook AB 8 D5-6
Sunny Corner NB 31 D6
Sunny Glen SK 11 C2-3
Sunny Lake NT 5 C5
Sunnynook AB 7 B7
Sunnynook AB 11 D1
Sunnyside NB 29 E7
Sunnyside NB 31 B5-6
Sunnyside NL 36 D7
Sunnyside NL 37 D6
Sunnyslope AB 7 A4
Sunnyville AB 8 D6-7
Sun-Oka Beach PP BC 6 F1-2
Sunset Beach AB 8 A6
Sunset Beach AB 9 F5
Sunset House AB 9 F3-4
Sunset Prairie BC 9 E1
Sunset Prairie BC 9 E1
Sunset, lac QC 25 E4
Sunville SK 11 E5
Sunville MB 14 C2
Suomi ON 16 F5
Superb SK 11 C2
Superb Mountain BC 1 D-E6
Superb Mountain BC 2 A3
Supérieur, lac ON 16 F7-8
Supérieur, lac ON 17 D-E1 E1-2
Supérieur, lac QC 29 C-D4
Supérieur, lac QC 31 A3
Superior, Lake ON 16 F7-8
Superior, Lake ON 17 D1-2 D-E2
Superior Junction ON 13 E7
Superior Junction ON 16 D4
Supreme SK 11 C4
Surbiton SK 11 C6
Surette Island NS 33 E-F2
Surf Inlet BC 1 B3
Surge Narrows BC 1 E5-6
Surge Narrows BC 2 B2
Surge Narrows PP BC 2 B2
Surprise Creek YT 5 C2
Surprise Lake BC 3 A1-2
Surprise Lake BC 4 F4-5
Surprise, lac QC 26 D5
Surprise, lac de la QC 26 C4
Surrey BC 1 F7
Surrey BC 2 E5
Surveyor Point PE 35 D8
Suskwa River BC 3 F4
Sussex NB 32 C5
Sussex Corner NB 32 C5
Sustut BC 3 E4-5
Sustut River BC 3 D-E5
Sutherland, RNOM SK 11 C4
Sutherland NMBS SK 11 C4
Sutherland River BC 1 A5-6
Sutherlands River NS 34 E4
Sutil, Cape BC 1 D3-4
Sutton ON 20 B7
Sutton ON 21 D1
Sutton QC 24 E8
Sutton QC 27 F3
Sutton PE 32 A8
Sutton, monts QC 24 E8
Sutton, monts QC 27 F3
Sutton Junction QC 24 E8
Sutton Junction QC 27 F3
Sutton Ridges ON 40 E2-3
Suwannee Lake MB 11 A8
Suwannee Lake MB 15 E2
Svendsen Peninsula NU 42 B3-4
Sverdrup, îles NU 42 B3-4
Sverdrup Islands NU 42 B3-4
Sverre, Cape NU 42 B4
Swale Island NL 37 B6-7
Swalwell AB 6 D6
Swalwell AB 7 A5
Swan Bay SK 12 C1
Swan Bay SK 15 C1
Swanger Cove NL 36 D5-6
Swanger Cove NL 37 D3-4

Swan Hills AB 8 A3
Swan Hills AB 8 A4
Swan Hills AB 9 F4
Swan Hills AB 9 F4
Swan Hills AB 9 F4
Swan Lake AB 9 D4
Swan Lake BC 3 E3
Swan Lake MB 11 C8
Swan Lake MB 11 C8
Swan Lake MB 13 B-C2
Swan Lake MB 13 C1-2
Swan Lake MB 13 F3
Swan Lake MB 14 F3
Swan Lake MB 14 E-F3
Swan Lake ON 13 A7
Swan Lake ON 14 D8
Swan Lake ON 15 F7
Swan Lake ON 16 D1
Swan Lake Kispiox River PP BC 3 E3-4
Swan Lake PP BC 9 E2
Swan Lakes SK 12 F4
Swannell Ranges BC 3 D4-5 E5
Swan Plain SK 11 C7
Swan Plain SK 13 C1
Swan River BC 3 E8
Swan River MB 11 C7-8
Swan River MB 13 C1
Swanson SK 11 C-D4
Swartz Bay BC 2 F4
Swastika ON 17 D7
Sweaburg ON 19 C5-6
Sweaburg ON 20 E4-5
Sweet Bay NL 37 B-C7
Sweet Grass SK 11 B3
Swift Current NL 36 D8
Swift Current NL 37 D7
Swift Current SK 11 E3-4
Swift Current Creek SK 11 E3
Swift Rapids ON 18 E8
Swift Rapids ON 21 C1
Swift River BC 1 B7-8
Swift River BC 3 A2
Swift River BC 4 F5
Swift River BC 3 A2
Swift River YT 3 A2
Swift River YT 4 F5-6
Swinburne, Cape NU 41 A4
Swinburne, Cape NU 42 E3
Swindle Island BC 1 C3
Swinton Park ON 20 B5
Swords ON 18 D7-8
Sybouts SK 11 F5-6
Sydenham ON 21 D6
Sydenham Lake ON 21 D6-7
Sydenham River ON 18 F5
Sydenham River ON 19 C3-4
Sydenham River ON 19 D2-3
Sydenham River ON 20 A4
Sydenham River ON 20 E2-3
Sydenham River ON 20 F1
Sydenham River ON 20 F1-2
Sydney NS 34 C8
Sydney Forks NS 34 C7
Sydney Harbour NS 34 B8
Sydney Inlet BC 1 F5
Sydney Inlet PP BC 1 F5
Sydney Lake ON 13 E5
Sydney Lake ON 14 C8
Sydney Lake ON 16 C1
Sydney Mines NS 34 C7-8
Sydney River NS 34 C7
Sydney River NS 34 C7
Sylvan MB 13 D4
Sylvan MB 14 B4-5
Sylvan ON 19 C3-4
Sylvan ON 20 E2-3
Sylvan Glen AB 8 A6
Sylvania SK 11 B6
Sylvan Lake AB 6 C5
Sylvan Lake AB 8 E5-6 F6
Sylvan Lake AB 8 F6
Sylvan Lake PP AB 6 C5
Sylvan Lake PP AB 8 F5-6
Sylvester NS 34 E3
Sylvia, Mount SK 3 C6
Sylvia Grinnell, parc territorial NU 40 A5-6
Sylvia Grinnell TP NU 40 A5-6
Symons Valley AB 7 B3-4
Syringa Creek BC 6 F3
Syringa PP BC 6 F3
Syroga, lac QC 25 B6
Syroga, lac QC 26 F6
Syroga, lac QC 28 D2

T

Taber AB 6 E-F7
Taber AB 7 E6
Table, cap de la QC 30 A7
Table Bay NL 39 B8
Table Head NL 39 D8
Table Lake NU 42 C4
Table Mountain 36 D1
Table Point Ecological Reserve NL 38 D2-3
Tabusintac NB 30 F2
Tabusintac NB 31 C7-8
Tabusintac Lagoon NB 30 F2
Tabusintac River NB 29 F8
Tabusintac River NB 29 F8
Tabusintac River NB 30 F1
Tabusintac River NB 31 C7

Taché, lac NT 10 B4
Tachick Lake BC 1 A6
Tachie BC 1 A6
Tachie BC 3 F6
Tachie River BC 3 F6
Tadek Lake NT 5 D7
Tadenac Lake ON 18 E7
Tadenet Lake NT 5 B7
Tadmore SK 11 C7
Tadoule Lake MB 12 B8
Tadoule Lake MB 15 B3
Tadoule Lake MB 15 B3
Tadoule Lake MB 41 F4
Tadoule Lake MB 41 F4
Tadoussac QC 28 D8
Tadoussac QC 29 D1
Taff Creek BC 3 D3
Taghum BC 6 F3
Tagish YT 4 E4
Tagish Bridge Territorial Recreation Site YT 4 E4
Tagish Lake YT 4 E-F4
Tahltan River BC 3 C2
Tahsis BC 1 E4
Tahsish-Kwois PP BC 1 E4
Tahsis Inlet BC 1 F4-5
Tahtsa Lake BC 1 A4
Tahtsa Reach BC 1 A4
Taizie NL 39 B2
Taka Lake NT 10 B5
Takhini YT 4 E3
Takhini Hotspring YT 4 E3-4
Takhini River YT 4 E3-4
Takijuq Lake NU 41 C2
Takipy MB 12 F7
Takipy MB 15 E1-2
Takla Lake BC 3 F5
Takla Lake BC 3 F5
Takla Landing BC 3 E-F5
Takla Lake PMP BC 3 F5
Taku River BC 3 B1
Takysie Lake BC 1 A5
Talbot Arm YT 4 D2
Talbot Creek YT 4 D2
Talbot Lake MB 11 A8
Talbot Lake MB 13 A2
Talbot Lake ON 20 A8
Talbot River ON 20 A7-8
Talbot River ON 21 D2
Talbotville Royal ON 19 C-D4
Talbotville Royal ON 20 F3
Tallman SK 11 B4
Tall Pines SK 11 C7
Tally Pond NL 36 B5
Tally Pond NL 37 B3
Talmage SK 11 F6
Talon, lac QC 27 B7
Talon, Lake ON 17 F8
Talon, Lake ON 22 D2
Taloyoak NU 41 B5
Taloyoak NU 42 F4
Taltapin Lake BC 1 A5
Taltson River NT 10 E7-8
Taltson River NT 41 E2
Talunkwan Island BC 1 B1-2
Tamarac, lac QC 23 E8
Tamarac, lac QC 25 A2-3
Tamarac, lac QC 25 A2-3
Tamarac, lac QC 26 F2-3
Tamarack ON 19 A7
Tamarack ON 20 C6
Tamarack Island BC 13 C4
Tamworth ON 21 D5-6
Tancook Island NS 32 F7-8
Tancook Island NS 33 D5
Tancredia QC 21 A6
Tancredia QC 22 F6
Tangamoose Lake ON 21 C6
Tangent AB 9 E3
Tangier NS 33 B-C7
Tangier Grand Lake NS 33 B7
Tangier Grand Lake NS 34 F3
Tangier River NS 33 B7
Tangier River NS 33 B7
Tanquary Fiord NU 42 A5
Tansleyville ON 16 D7
Tansleyville ON 17 B1
Tantallon NS 32 F8
Tantallon NS 33 C5
Tantallon SK 11 E7
Tantallon SK 13 E1
Tantalus PP BC 1 E-F7
Tantalus PP BC 2 C4
Tantramar, rivière NB 35 F1
Tantramar River NB 35 F1
Tanzilla River BC 3 B-C3
Tapani, lac QC 25 E3
Tappen BC 6 D2
Taquanyah CA ON 19 C7
Taquanyah CA ON 20 E6
Tara ON 18 F5
Tara ON 20 A4
Tarbotvale NS 34 B7
Tarnopol SK 11 B-C5
Tarpon Lake NT 10 A8
Tarzwell NS 34 D2
Tarzwell 17 D7
Tarzwell 23 E1
Taschereau QC 17 C8
Taschereau QC 23 C3
Taschereau, rivière QC 23 C-D5
Taseko Lakes BC 1 D6-7
Taseko River BC 1 C-D6
Taseko River BC 2 A4
Tasiujaq QC 40 C5
Tassé, lac QC 25 A4
Tassé, lac QC 26 F1

Tasseriuck Lake NT 5 B8
Tasso Lake ON 21 A2
Tasu Sound BC 1 B1
Tatachikapika River ON 17 D6
Ta Ta Creek BC 6 F4
Ta Ta Creek BC 7 E1
Tatagwa Lake SK 11 F6
Tatalrose BC 1 A5
Tatamagouche NS 34 D2
Tatamagouche Bay NS 34 D2
Tatamagouche PP NS 34 D2
Tatchu Point BC 1 E4
Tatchun Hills YT 4 C4
Tate SK 11 D5
Tate Lake NT 4 A8
Tate Lake NT 10 A-B2
Tathlina Lake NT 10 F5
Tathlina Lake NT 41 E1
Tatla Lake BC 1 C6
Tatla Lake BC 1 C6
Tatlatui Lake BC 3 D4
Tatlatui PP BC 3 D4
Tatlayoko Lake BC 1 D6
Tatlayoko Lake BC 1 D6
Tatlow, Mount BC 1 D6
Tatnam, Cape MB 15 C7
Tatnam, Cape MB 40 C1
Tatogga BC 3 C2
Tatonduk River YT 5 F2
Tatowaycho Lake MB 12 A7
Tatowaycho Lake MB 15 A2
Tatsamenie Lake BC 3 B2
Tatshenshini-Alsek PWP BC 4 F2-3
Tatshenshini River BC 4 E-F3
Tatuk Lake BC 1 B6
Taunton ON 20 C8
Taunton ON 21 E2
Taureau, réservoir QC 24 A6
Taureau, réservoir QC 25 E6
Taureau, réservoir QC 27 B1
Tavistock ON 19 B5
Tavistock ON 20 D4
Tawachiche, ZEC QC 25 D8
Tawachiche, ZEC QC 27 A3
Tawatinaw AB 6 A6
Tawatinaw AB 8 A6
Taweel PP BC 1 C-D8
Taweel PP BC 6 C1
Taxis River NB 31 E4-5
Taxis River NB 31 E5
Taxis River NB 32 A2-3
Taxis River NB 32 A3
Tay, canal ON 21 C7
Tay, canal ON 24 F1
Tay Canal ON 21 C7
Tay Canal ON 24 F1
Tay Creek NB 31 F4
Tay Creek NB 32 B2
Taye Lake YT 4 D-E3
Tay Lake YT 4 C5
Taylor BC 3 E8
Taylor, lac QC 9 D1-2
Taylor, lac QC 23 A7
Taylor, lac QC 26 B1-2
Taylor, Mount BC 3 F3
Taylor, Mount BC 7 E3
Taylor Arm PP BC 1 F5
Taylor Arm PP BC 2 D2
Taylor Corners ON 20 B8
Taylor Corners ON 21 D2
Taylor Head NS 33 C8
Taylor Lake ON 24 E1
Taylor Landing PP BC 3 E8
Taylor Landing PP BC 7 F8
Taylor River BC 3 D-E3
Taylor's Bay NL 36 F6
Taylor's Bay NL 37 F4
Taylors Head NS 33 C8
Taylorville AB 7 F5
Taymouth NB 31 F5
Taymouth NB 32 B3
Tay River YT 4 C4-5
Tazin Lake SK 9 A7-8
Tazin Lake SK 12 A3
Tazin Lake SK 41 E2
Tazin River SK 9 A7
Tazin River SK 12 A4
Tchaneta River NB 31 B5-6
Tchendferi Lake NT 5 D7
Tchentlo Lake BC 3 F5-6
Tchesinkut Lake BC 1 A5
Tchesinkut Lake BC 1 A5
Tchitogama, lac QC 28 B5
Tea Cove NL 36 B1-2
Tea Hill PE 35 E5
Tea Lake ON 21 A2
Tecumseh ON 19 E1
Tedji Lake NT 5 D7
Tee Lake QC 17 F8
Tee Lake QC 22 C2
Teepee Creek AB 9 E3
Teepee Mountain BC 6 E-F4
Tees AB 6 C6
Tees AB 8 E6-7
Teeswater ON 20 B3
Teeswater River ON 20 B3
Teeterville ON 19 C6
Teeterville ON 20 E5
Tehek Lake NU 41 D4
Tehkummah ON 18 C3
Telegraph Cove BC 1 E4-5
Telegraph Creek BC 3 C2
Telegraph Range BC 1 B6

Telford MB 14 E7
Telford MB 17 C-D7
Telfordville AB 6 B5
Telfordville AB 8 D5-6
Telkwa BC 3 F4
Telkwa River BC 3 F4
Tellier QC 39 E2
Temagami ON 17 E7
Temagami ON 22 B1
Temagami, Lake ON 17 E7
Temagami, Lake ON 22 B1
Temagami River ON 17 F7
Temagami River PP ON 17 F7
Témiscaming QC 17 F8
Témiscaming QC 22 C2
Témiscamingue QC 17 D8
Témiscamingue QC 17 E8
Témiscamingue, lac QC 17 F8
Témiscamingue, lac QC 22 A1-2
Témiscamingue, lac QC 22 A1-2
Témiscamingue, lac QC 23 F1-2
Témiscouata, lac QC 28 F8
Témiscouata, lac QC 31 B1
Temperance Vale NB 31 F3-4
Temperance Vale NB 32 B1-2
Tempest AB 7 E6
Temple NL 32 C1
Templeman ON 17 A6-7
Templeman NL 38 F8
Templeman, Mount BC 6 E3
Ten Mile Lake BC 36 A7
Ten Mile Lake NL 37 A5
Ten Mile Lake NL 38 B4
Ten Mile Lake NL 38 F6
Ten Mile Lake ON 18 A2
Ten Mile Lake PP BC 1 B7
Ten Mile Lake PP NL 33 D3-4
Ten Mile Pond NL 36 A8
Ten Mile Pond NL 37 A6
Ten Mile Pond NL 38 F7
Tenby MB 13 E2
Tenby MB 14 C2
Tenlen Lake NT 5 C6
Tennycape NS 32 D8
Tennycape NS 33 A5
Tennycape NS 34 F1
Tent Island NT 5 B4
Tepee Lake AB 9 D5
Terence MB 14 E1
Terence Bay NS 32 F8
Terence Bay NS 33 D6
Terrace BC 1 A3
Terrace Bay ON 16 E-F7
Terrace Bay ON 17 C2
Terrace Bay ON 40 F2
Terra Cotta ON 19 A7
Terra Cotta ON 20 C6
Terra Cotta CA ON 19 A7
Terra Cotta CA ON 20 C6
Terra Nova NL 36 B-C8
Terra Nova NL 37 B-C6
Terra Nova ON 20 B6
Terra-Nova, PN NL 36 B8
Terra-Nova, PN NL 37 B6
Terra Nova Lake NL 36 B8
Terra Nova Lake NL 37 B6
Terra Nova NP NL 36 B8
Terra Nova NP NL 37 B6-7
Terrasses, lac des QC 23 D8
Terrasses, lac des QC 26 D-E2
Terrebonne QC 24 D5-6
Terrebonne QC 27 D-E1
Terrenceville NL 36 D7
Terrenceville NL 37 D5
Terror Point BC 1 B2-3
Teslin YT 4 E5
Teslin Lake YT 4 E5
Teslin River BC 3 A-B2
Teslin River BC 4 D-E4
Teslin River YT 4 D-E4
Tessier SK 11 C4
Tessier, lac QC 25 A3
Tessier, lac QC 26 F3
Tessik Lake NU 40 A4
Tessik Lake NU 41 C7
Tetachuck Lake BC 1 B5
Tetagouche River NB 29 E7
Tetagouche River NB 31 B5-6
Tetcho Lake NT 10 F3-4
Tête, lac de la QC 23 D8
Tête, lac de la QC 26 E3
Tête d'Ours, lac NT 10 B8
Tete Jaune Cache BC 6 B2
Tetlit Gwinjik TP NT 5 D4
Tetlit Zheh (Fort McPherson) NT 5 D4
Tetrahedron PP BC 2 D4
Tetsa River PP BC 3 B6
Têtu QC 28 A6
Tetu Lake ON 16 D1
Teulon MB 14 C-D5
Teviotdale ON 19 A5-6
Teviotdale ON 20 C4-5
Tewkesbury QC 27 A5
Texada Island BC 2 D3
Texada Island BC 2 D3
Texas Creek BC 1 E7
Tezzeron Creek BC 3 F6
Tezzeron Lake BC 3 F6
Tha-anne River NU 40 A-B1
Tha-anne River NU 41 E4

Thackeray, RNP ON 17 C-D7
Thackeray, RNP ON 23 D1
Thackeray PNR ON 23 D1
Thackeray PNR ON 17 C-D7
Thaddeus Lake ON 16 D3
Thaïs, lac QC 26 D5-6
Thaïs, lac QC 28 B1
Thalberg MB 13 E4
Thalberg MB 14 C-D6
Thames River ON 19 B-C5
Thames River ON 19 D3
Thames River ON 20 E4
Thames River ON 20 F2
Thamesford ON 19 C5
Thamesford ON 20 E4
Thamesville ON 19 D3
Thamesville ON 20 F2
Thanet Lake ON 21 C4
Thaumur, lac QC 25 B3
The Anchorage PP ON 32 F2-3
The Arches PP NL 38 D2
The Arches PP NL 39 E-F7
The Backway NL 39 B7
The Banks Fishery—The Age of Sail NHS NS 33 D5-6
The Battlefords PP SK 11 B3
The Beaches NL 38 E4
The Beaches NL 39 F8
The Big Ridge NS 34 C6 D5
The Blue Mountains ON 18 F6
The Blue Mountains ON 20 A5
The Dalles ON 13 E5
The Dalles ON 14 D8
Thedford ON 19 B3
Thedford ON 20 E2
The Falls NS 34 E2
The Fundy Isles NB 32 E3
The Glades NS 32 C6
The Great Sand Hills SK 11 E2-3
The Gut CA ON 21 C4
The Hawk NS 33 F2
The Hiscock House PHS ON 37 C7-8
The Islands PP NS 33 E2-3
The Ledge NB 32 E2
The Massasauga PP ON 18 D7
The Middle Sand Hills ON 6 D8 E7
The Middle Sand Hills AB 7 C8
The Middle Sand Hills AB 11 D1-2 E1
The Mockbeggar Property & Cape Bonavista Lighthouse Provincial Historic Sites NL 37 B7-8
The Narrows ON 13 C3
The Narrows MB 14 A3
The Narrows MB 14 B3
The Old Settler BC 1 F8
The Old Settler BC 2 D7
The Pas MB 11 A7
The Pas MB 13 A1
The Pas SK 11 A-B7
The Pinery PP ON 19 B3
The Pinery PP ON 20 D2
The Porcupine Hills MB 11 C7
The Porcupine Hills MB 13 C1
The Porcupine Hills SK 11 C7
The Porcupine Hills SK 13 C1
The Queen Elizabeth The Queen Mother M'Nidoo M'Nissing PP ON 18 B2
The Ranee AB 6 B2-3
The Range NB 31 F6
The Range NB 32 B4
The Shoals PP ON 17 D4
The Shoals PP ON 33 B-C5
Thessalon ON 17 F4-5
Thessalon ON 18 A1
Thessalon Point ON 18 A1
The Tablelands NL 36 A5
The Tablelands NL 37 A1
The Tablelands NL 38 F2
Thetford Mines QC 27 C5
Thetis Bay QC 5 A3
Thetis Island BC 1 F6-7
Thetis Island BC 2 E4
Thetis Lake NT 10 B7
Thetlaandoa Creek BC 3 A7 B7-8
Thetlaandoa Creek BC 9 A1
Thévet, lac QC 39 D3
The Two Rivers SK 12 B6
The Willows NS 31 C7
The Wolves NB 32 E2
Thibaudeau MB 15 D5-6
Thibault MB 31 B3-4
Thicke Lake SK 12 A4
Thicket Portage MB 15 F3-4
Thickwood Hills SK 11 B4
Thinahtea Lake BC 9 A2
Thinahtea Lake BC 9 A2
Thinahtea PPA BC 3 A8
Thinahtea PPA BC 9 A2

Thinahtea PPA BC 9 A2
Third Lac MB 29 E3
Thirty Mile Lake YT 4 E5
Thirty Thousand Islands ON 18 C-D6
Thistlethwaite Lake NT 10 C7
Thlewiaza River NU 41 E4
Thluicho Lake SK 9 A7
Thluicho Lake SK 12 A3
Thoa River NT 41 E2
Thode SK 11 C4
Thoen, Mount BC 3 F4
Thomas Bay ON 18 C3-4
Thomasburg ON 21 D5
Thomas H. Raddall PP NS 33 E4
Thomas Hubbard, Cape NU 42 A4
Thomas Lake MB 15 A4
Thomas River NL 39 B5
Thomaston Corner NB 32 C2
Thomasville NS 33 D4
Thompson MB 15 F4
Thompson MB 16 F6
Thompson Island, RNP ON 16 F6
Thompson Island PNR ON 16 F6
Thompson Lake ON 22 D-E2
Thompson Landing NT 10 C8
Thompson Point MB 15 B6
Thompson River BC 1 D8
Thompson River BC 2 A-B7
Thompson River BC 6 D-E1
Thompson Sound BC 1 E5
Thompson Sound BC 1 D-E5
Thompson Sound BC 2 A1
Thompson Sound BC 2 A1
Thomsen River NT 42 D1
Thomson Lake SK 12 B4-5
Thomson Lake SK 13 E1
Thomson Station NS 32 C8
Thomson Station NS 34 E1
Thonokied Lake NT 10 D8
Thorah Island ON 18 F8
Thorah Island ON 20 A7
Thorah Island ON 21 B1
Thorburn NS 33 A8
Thorburn NS 34 E3-4
Thorburn Lake NL 36 C8
Thorburn Lake NL 37 C6
Thorhild AB 6 A6
Thorhild AB 8 B7
Thormanby Island BC 1 F6
Thornbury ON 18 F6
Thornbury ON 20 A5
Thornby QC 22 F6
Thorndale ON 19 C5
Thorndale ON 20 E4
Thorne ON 17 F8
Thorne ON 22 C2
Thorne River ON 15 F7
Thornhill MB 13 F3
Thornhill MB 14 F3
Thornhill ON 19 A8
Thornhill ON 20 C7
Thornhill ON 21 F1
Thornlea NL 37 D7
Thornloe ON 17 D7-8
Thornloe ON 23 F1
Thornton ON 20 B6
Thorold ON 19 C8
Thorold ON 20 E7
Thorold South ON 19 C8
Thorold South ON 20 E7
Thorsby AB 6 B5
Thorsby AB 8 D6
Thorsteinson Lake MB 15 C3-4 D3-4
Three Brooks NB 31 D3
Three Day Lake NT 10 C6
Three Day Lake NT 10 A1
Three Hills AB 6 C6
Three Hills AB 7 A6
Threehills Creek AB 6 D6
Threehills Creek AB 7 A6
Three Mile Bay ON 18 D8
Three Mile Lake ON 18 D8
Three Mile Lake ON 21 A1
Three Mile Lake ON 22 B1
Three Mile Lake ON 22 F1
Three Mile Plains NS 32 E7
Three Mile Plains NS 33 B-C5
Three Mile Rock NL 38 D8
Three Mile Rock NL 39 F7
Three Rock Cove NL 36 B1
Three Sisters Lakes PP BC 1 B7
Three Valley BC 6 D3
Throat River ON 13 C6
Throat River ON 16 A3
Throne AB 6 C7
Throne AB 11 C1
Thrums BC 6 F3
Thubun Lakes NT 10 D8
Thubun River NT 10 D8
Thultue Lake AB 9 A5
Thultue Lake AB 12 A1
Thunder Bay ON 16 F6
Thunder Beach ON 18 E7
Thunder Brook NL 36 B5-6
Thunder Brook NL 37 B3-4
Thunder Cape ON 16 F6
Thunderchild SK 11 B3
Thundercloud Creek NT 4 C7-8

Thundercloud Creek NT 10 C-D1
Thunder Creek SK 11 E4
Thunder Hill PP BC 6 E4
Thunder Hill PP BC 7 D1
Thunder Hills SK 12 F4
Thunder Lake PP AB 6 A5
Thunder Lake PP AB 8 B5
Thunder Mountain BC 1 C4-5
Thurston Bay PMP BC 1 E5
Thurston Bay PMP BC 2 B2
Thurston Lake AB 9 A3
Thurston Lake AB 10 F5
Thutade Lake BC 3 D4-5
Thuycoleein Lake MB 12 B7
Thuycoleein Lake MB 15 B2
Thwart Island NL 36 A6
Thwart Island NL 37 A4-5
Thwart Island NL 38 F6
Tian Head BC 1 A1
Tiblemont, lac QC 23 E5
Tice Lake MB 15 A2
Tichborne ON 21 C6
Tichfield SK 11 D4
Tickle Beach NL 36 D7
Tickle Beach NL 37 D5
Tickle Cove NL 37 D7
Tickle Harbour NL 36 D8
Tickle Harbour NL 37 D6
Tickle Harbour Point NL 37 D7
Ticouapé, rivière QC 26 D8
Ticouapé, rivière QC 28 B3
Tidal MB 15 B5
Tidd, Mount YT 4 C5-6
Tiddville NL 33 D1
Tide Head NB 29 D6
Tide Head NB 31 A4
Tide Lake AB 6 E7
Tide Lake AB 7 C7
Tide Lake AB 11 D-E1
Tide Lake ON 13 E6
Tide Lake, RNP ON 16 C1-2
Tide Lake PNR ON 13 E6
Tide Lake PNR ON 16 C1-2
Tides Point NL 36 E7
Tides Point NL 37 E5
Tidney River NL 33 E3
Tidnish NS 32 B7-8
Tidnish NS 34 D1
Tidnish NS 35 F2
Tidnish Bridge NS 32 B-C7
Tidnish Bridge NS 34 D1
Tidnish Bridge NS 35 F2
Tidnish Dock PP NS 32 B7-8
Tidnish Dock PP NS 34 D1
Tidnish Dock PP NS 35 F2
Tidnish Head NS 35 F2
Tiger Lake AB 8 B5
Tiger Lily AB 6 A5
Tiger Lily AB 8 B5
Tignish PE 34 A1
Tignish PE 35 A2
Tignish River PE 35 A-B2
Tignish Shore PE 34 A1
Tignish Shores PE 34 A1
Tika QC 39 E2
Tikamaganda Lake ON 17 E4
Tikirarjuaq (Whale Cove) NU 40 A-B2
Tikirarjuaq (Whale Cove) NU 41 E5
Tilbury ON 19 E2
Tilden Lake ON 17 F8
Tilden Lake ON 18 A8
Tilden Lake ON 18 A8
Tilden Lake ON 22 C1
Tilden Lake ON 22 C1
Tilley AB 6 E7
Tilley AB 7 C7
Tilley NB 31 D3
Tilley Road NB 31 E3
Tillicum Beach AB 8 D7
Tillsonburg ON 19 C5-6
Tillsonburg ON 20 F4-5
Tilney SK 11 E5
Tilston MB 11 F8
Tilston MB 13 F1
Tilt Cove NL 38 E5
Tilting NL 38 E7
Tilton NL 37 D7
Timber Bay SK 11 A5
Timber Island ON 21 E6
Timber Island PNR ON 21 E6
Timber Island, RNP ON 21 E6
Timberlea NS 32 F8
Timberlea NS 33 C6
Timber River NB 35 E-F2
Timeu AB 8 A5
Timiskaming, Lake ON 22 A1
Timiskaming, Lake ON 23 F1
Timiskaming, Lake QC 17 E8
Timmins ON 17 C-D6
Tim River ON 18 A8
Tim River ON 22 E-F2
Tincap ON 21 C-D8
Tincap ON 24 F2

Tinchebray AB 8 E8
Tincup Lake YT 4 C2-3
Tingwick QC 27 D4
Tintamarre, RNF NB 32 B-C7
Tintamarre, RNF NB 35 F1-2
Tintamarre NWA NB 32 B-C7
Tintamarre NWA NB 35 F1-2
Tin Town NB 29 E8
Tiny SK 11 C-D7
Tipahaskayo Lake MB 15 E6
Tip Top Mountain ON 16 F8
Tip Top Mountain ON 17 D2-3
Tisdale SK 11 B6
Titusville NB 32 D4-5
Titusville NB 33 A2
Tiverton ON 20 B3
Tizzard's Harbour NL 38 E6
Tlell BC 1 A1
Tlogotsho Range NT 10 E1-2
Toad River BC 3 B5-6
Toad River BC 3 B5
Toad River Hotsprings PP BC 3 B6
Toanche ON 18 E7
Tobacco Lake ON 18 B3
Toba Inlet BC 1 E6
Toba Inlet BC 2 B3
Toba River BC 1 E6
Toba River BC 2 B3-4
Tobeatic Lake NS 33 D3
Tobeatic Wilderness Area NS 33 D2
Tobeatic Wilderness Area NS 33 E2-3
Tobermory ON 18 D4
Tobin Lake SK 11 B6
Tobin Lake SK 11 B6
Tobique Narrows NB 31 D3
Tobique River NB 31 C4 D3
Toby, Mount BC 6 E4
Toby, Mount BC 7 D1
Toby Creek BC 7 D1
Tochcha Lake BC 3 F5
Tod, Mount BC 6 D1-2
Todagin South Slope PP BC 3 C3
Tod Creek BC 7 E3-4
Tod Lake MB 12 D7
Tod Lake MB 15 D1
Tofield AB 6 B6
Tofield AB 8 C7
Tofino BC 1 F5
Tofino BC 2 E1
Tofino Inlet BC 1 F5
Tofino Inlet BC 2 E1
Togo SK 11 D7
Togo SK 13 D1
Toker Point NT 5 A5
Toledo ON 21 C7
Toledo ON 24 F1-2
Tolman AB 7 A5
Tolmie Channel BC 1 B3
Tolstoi MB 14 F6
Tolstoi MB 14 F5
Tomahawk AB 6 B5
Tomahawk AB 8 C5
Tomasine, lac QC 22 C7
Tomasine, lac QC 24 A1
Tomasine, lac QC 25 E1
Tom Browne Lake ON 2 A-B1
Tombstone TP YT 4 A2
Tombstone TP YT 5 F2
Tomifobia QC 24 E8
Tomifobia QC 27 F4
Tomifobia, rivière QC 27 F4
Tomiko ON 18 A8
Tomiko Lake ON 17 F8
Tomiko Lake ON 18 A8
Tomiko Lake ON 22 C1
Tomiko River ON 18 A7-8
Tom Luscombe Brook NL 39 B6-7
Tomney, lac QC 23 D7
Tomney, lac QC 25 B2
Tomney, lac QC 26 F2
Tomogonops River NB 29 F7
Tomogonops River NB 31 B5-6
Tompkins NL 36 D1
Tompkins SK 11 E3
Tomslake BC 3 E8
Tomslake BC 9 E2
Tomstown ON 17 D7-8
Tomstown ON 23 E1
Toney River NS 34 D3
Tonkin NB 31 D7
Toodoggone River BC 3 D4-5
Toogood Arm NL 38 E6-7
Toosey BC 1 C7
Topaze Harbour BC 1 E5
Top of the World PP BC 6 E4-5
Top of the World PP BC 7 E2
Topley BC 1 A5
Topley Landing BC 3 F5
Topley PP BC 3 F5
Top Pond NL 36 D3
Top Pond NL 37 D1
Toquart Bay BC 2 E2
Torbay NL 37 D8
Tor Bay NS 34 E-F6

Tor Bay *NS* **34** F6
Tor Bay PP *NS* **34** E6
Torbrook *NS* **32** E6
Torbrook *NS* **33** B3
Torch River *SK* **11** A5 A-B6
Torch River *SK* **11** A-B6
Tormentine *NB* **35** E9
Tormentine, Cape *NB* **32** E6
Tormentine, Cape *NB* **34** C1
Tormentine, Cape *NB* **35** E3
Tornado Mountain *AB* **6** E5
Tornado Mountain *AB* **7** D3
Torngat Mountain NP
 (proposed/projeté) *NL*
 40 C6-7
Torngat Mountains *NL*
 40 C6-7
Toronto *ON* **19** A8
Toronto *ON* **20** C-D7
Toronto *ON* **21** F1-2
Toronto Islands *ON* **19** A8
Toronto Islands *ON* **20** D7
Toronto Islands *ON* **21** F1
Torquay *SK* **11** F6
Torrance *ON* **18** E8
Torrance *ON* **21** C1
Torrance Lake *MB* **15** D3
Torrington *AB* **6** C5-6
Torrington *AB* **7** A4
Tors Cove *NL* **37** E8
Tortue, lac à la *QC* **22** D7
Tortue, lac à la *QC* **24** A-B1
Tortue, lac à la *QC* **25** F1
Tortue, rivière à la *QC*
 24 E6
Tortue, rivière à la *QC* **27** F1
Tory Hill *ON* **21** C3
Toslow *SK* **36** E6
Toslow *NL* **37** E6
Totnes *SK* **11** E3
Totogan Lake *ON* **16** A5-6
Tottenham *ON* **20** B-C6
Touchwood Hills *SK*
 11 D5-6
Touchwood Hills Post
 Provincial Historic Park
 SK **11** D5-6
Touchwood Lake *AB* **9** D5
Touchwood Lake *AB* **12** F1
Touchwood Lake *MB* **13** A5
Touchwood Lake *MB* **15** F5
Touladi, lac *QC* **29** E2
Touladi, lac *QC* **31** B1
Toulnustouc, rivière *QC*
 39 E-F1
Tourbis, lac *QC* **25** D5
Tourelle *QC* **29** A6
Tourilli, lac *QC* **28** F5
Tour-Martello-de-Carleton,
 LHN de la *NB* **32** E4-5
Tour-Martello-de-Carleton,
 LHN de la *NB* **33** B1-2
Tourond *MB* **14** E5
Tourouvre, lac *QC* **25** B7
Tourouvre *QC* **28** E3
Tourville *QC* **27** A8
Tourville *QC* **28** F8
Toussaint, rivière *QC* **26** D4
Toutes Aides *MB* **13** D2
Toutes Aides *MB* **14** A2
Townsend *ON* **19** C6-7
Townsend *ON* **20** E5-6
Tracadie *NB* **30** F2
Tracadie *NB* **31** B8
Tracadie *NS* **34** D5
Tracadie Bay *PE* **34** C3
Tracadie Bay *PE* **35** D5
Tracadie Cross *PE* **34** C3
Tracadie Cross *PE* **35** D5
Tracadie-Sheila *NB* **30** F2
Tracadie-Sheila *NB* **31** B8
Tracadigache, pointe *QC*
 29 D7
Tracadigache, pointe *QC*
 31 A5-6
Tracey Mills *NB* **31** E3
Tracey Mills *NB* **32** A1
Tracy *NB* **32** C3
Tracy *QC* **24** C7
Tracy *QC* **27** D2
Trade Lake *SK* **12** E6
Trafalgar *NS* **33** A7-8
Trafalgar *NS* **34** E3
Trail *BC* **6** F3
Trail River *YT* **5** D-E4
Traîne, lac de la *QC* **25** D6
Traîne, lac de la *QC* **27** A2
Traîne, lac de la *QC* **28** F2
Train Lake *SK* **12** A4
Trainor Lake *NT* **10** F4
Tramore *ON* **21** A5
Tramore *ON* **22** F5
Tramping Lake *SK* **6** C8
Tramping Lake *SK* **6** C8
Tramping Lake *SK* **11** C3
Tramping Lake *SK* **11** C3
Tranquil Creek PP *BC* **1** F5
Tranquil Creek PP *BC* **2** D1
Tranquille *BC* **2** A8
Tranquille *BC* **6** D1
Trappist Monastery PP *MB*
 14 E5
Travaillant Lake *NT* **5** C5
Travaillant River *NT* **5** C-D5
Travellers Rest *PE* **35** D3
Travers *AB* **7** D6
Travers, lac de *QC* **27** A3-4
Travers, lac de *QC* **28** F3-4
Travers, Lake *ON* **22** E4
Traverse, lac de la *QC*
 26 A7
Traverse Bay *MB* **13** C6
Traverse Bay *MB* **14** C6

Traverse Bay *MB* **14** C6
Traverspine River *NL*
 39 C5-6
Troitsa Lake *BC* **1** B4
Trophy Mountain *BC* **6** C2
Trossachs *SK* **11** F5-6
Trousers Lake *NB* **31** D4
Trout Brook *NB* **31** C6
Trout Brook *NS* **34** C6
Trout Brook PP *NS* **34** C6
Trout Cove *NS* **32** F4
Trout Cove *NS* **33** C1
Trout Creek *ON* **18** B8
Trout Creek *ON* **22** E1
Troutly Lake *ON* **16** B6
Trout Lake *BC* **9** D5
Trout Lake *BC* **6** E3
Trout Lake *BC* **6** E3
Trout Lake *MB* **12** C8
Trout Lake *MB* **15** C3
Trout Lake *NT* **10** F3
Trout Lake *NT* **10** F3
Trout Lake *NT* **13** D6
Trout Lake *ON* **16** B2
Trout Lake *ON* **17** F7
Trout Lake *ON* **18** A4
Trout Lake *ON* **18** B6
Trout Lake *ON* **18** C7
Trout Lake *ON* **18** D7
Trout Lake *ON* **21** A2
Trout Lake *ON* **21** A4
Trout Lake *ON* **21** B5
Trout Lake *ON* **22** D1
Trout Lake *ON* **22** F2
Trout Lake *ON* **22** F4
Trout Lake PP *ON* **13** D6
Trout Lake PP *ON* **16** B-C2
Troutlake River *ON* **13** D6
Trout River *AB* **9** D5
Trout River *NL* **38** F2
Trout River *NL* **39** F7
Trout River *NT* **10** E3-4
Trout River Pond *NL* **37** A1
Trout River Pond *NL* **38** F2
Trouty *NL* **37** C7
Trowbridge *ON* **19** A5
Trowbridge *ON* **20** C4
Troy *NS* **34** D5
Troy *ON* **19** B6-7
Troy *ON* **20** E5-6
Troyes, lac *QC* **25** D5
Truax *SK* **11** E6
Trudeau Airport *QC* **24** D5-6
Trudeau Airport *QC* **27** E1
Trudel *NB* **31** B7
Truemanville *NS* **32** C7-8
Truemanville *NS* **33** A6
Truite, lac à la *QC* **22** B3-4
Truite, lac à la *QC* **26** B7
Truite, rivière à la *QC*
 29 B-C5-6
Truitt Peak *YT* **4** C4
Truman Dagnus Locheed PP
 BC **6** E2
Trunmore Bay *NL* **39** B7
Trunnel Island *ON* **18** B2
Turner's Bight *NL* **39** B7
Truro *NS* **33** A6
Truro *NS* **34** E2
Truro Heights *NS* **33** A6
Truro Heights *NS* **34** E2
Trutch *BC* **3** C7
Trutch Creek *BC* **3** C7
Trutch Creek *BC* **9** C1
Tryon *PE* **32** B8
Tryon *PE* **34** C2
Tryon *PE* **35** E3
Tryon Head *PE* **35** E3
Tsacha Lake *BC* **1** B6
Tsar Mountain *BC* **6** C3
Tsar Mountain *BC* **8** F1
Tsawwassen *BC* **2** E4-5
Tsea River *BC* **3** A7
Tsea River *BC* **9** A1
Tseepantee Lake *NT* **10** C3
Tseeteli Lake *MB* **12** B7-8
Tseeteli Lake *MB* **15** B2
Tsiigehtchic (Arctic Red
 River) *NT* **5** D4-5
Ts'il-os PP *BC* **2** A3-4
Tsinstikeptum *BC* **6** E2
Tsintsunko Lakes PP *BC*
 2 A8
Tsintsunko Lakes PP *BC*
 6 D1
Tsintu River *NT* **5** E7
Tsitsutl Peak *BC* **1** B-C5
Tsoko Lake *NT* **5** C7-8
Tsu Lake *NT* **10** E8
Tsulquate *BC* **1** D-E4
Tsuniah Lake *BC* **1** D6
Ts'yl-os PP *BC* **1** D6
Tuberose *SK* **11** D3
Tuchialic Bay *NL* **39** A7
Tuchitua *YT* **4** E6-7
Tuchitua River *YT* **4** D6
Tucho River *BC* **3** C4
Tuchodi River *BC* **3** C7
Tudor *AB* **7** B5
Tudyah Lake PP *BC* **3** F6-7
Tuffnell *SK* **11** D6
Tugaske *SK* **11** D4
Tugate *AB* **9** B3
Tuillé, lac *QC* **23** D7
Tuillé, lac *QC* **26** E1
Tuktoyaktuk *NT* **5** B5
Tuktusiujualuk (Northern
 Bathurst Island) NP
 (proposed/projeté) *NU*
 42 C3-4
Tuktut Nogait, PN *NT* **5** B7

Tuktut Nogait NP *NT* **5** B8
Tuktut Nogait NP *NT* **41** B1
Tulameen *ON* **1** F8
Tulameen *BC* **2** D8
Tulameen *BC* **6** F1
Tulita *NT* **10** A1-2
Tulita *NT* **10** A1-2
Tulliby Lake *AB* **6** A8
Tulliby Lake *AB* **11** A2
Tulsequah *BC* **3** B1
Tulugay River *YT* **5** B2-3
Tumbledown Dick Island *NL*
 39 B7
Tumbler Ridge *BC* **3** F8
Tumbler Ridge *BC* **9** E-F1
Tumeka Lake *BC* **3** D3
Tummel *MB* **11** D7
Tummel *MB* **13** D1
Tummel River *YT* **4** C4
Tunago Lake *NT* **5** E8
Tungsten *NT* **4** C-D7
Tunis *ON* **17** C6-7
Tunkwa PP *BC* **2** A7 B7-8
Tunkwa PP *BC* **6** D-E1
Tunnel Lake *ON* **17** F4-5
Tunstall *SK* **11** E2
Tunungayualok Island *NL*
 40 D7
Tununuk *NT* **5** B4
Tupper *BC* **3** E8
Tupper *BC* **9** E2
Tupper Lake *NS* **33** D3
Tupperville *NS* **32** F5
Tupperville *NS* **33** C3
Tupperville *ON* **19** D2
Tupperville *ON* **20** F1
Tuque, lac la *QC* **25** C7
Tuque, lac la *QC* **28** E2
Turbine *ON* **17** F6
Turbine *ON* **18** A4
Turcotte, lac *QC* **25** C7
Turcotte, lac *QC* **28** F3
Turgeon, lac *QC* **23** B3
Turgeon, rivière *QC* **17** B8
Turgeon, rivière *QC* **23** A2
Turin *QC* **6** E6
Turin *AB* **7** D6
Turkey Point *ON* **19** D6
Turkey Point *ON* **20** F5
Turkey Point *ON* **20** F5
Turkey Point PP *ON* **19** D6
Turkey Point PP *ON* **20** F5
Turks Cove *NL* **37** D7
Turmoil Lake *NT* **10** A5
Turnagain River *BC* **3** B4
Turnavik Island *NL* **39** A6
Turnberry *MB* **11** B7
Turnberry *MB* **13** B1
Turnbull *MB* **12** F8
Turnbull *MB* **13** A2-3
Turnbull, lac *QC* **25** C4
Turner Valley *AB* **6** E5
Turner Valley *AB* **7** C3
Turner Valley Gas Plant PHS
 AB **7** C3-4
Turnor Lake *SK* **9** D8
Turnor Lake *SK* **9** D8
Turnor Lake *SK* **12** D3
Turnor Lake *SK* **12** D3
Turnour Island *NL* **1** E5
Turtle Creek *NB* **31** F8
Turtle Creek *NB* **32** C6
Turtleford *SK* **6** B8
Turtleford *SK* **11** B3
Turtle Lake *NS* **34** C3
Turtle Lake *SK* **11** A3
Turtlelake River *SK* **6** B8
Turtlelake River *SK* **11** B3
Turtle Lake South Bay *SK*
 6 A-B8
Turtle Lake South Bay *SK*
 11 B3
Turtle Mountain PP *MB*
 11 F8
Turtle Mountain PP *MB*
 13 F2
Turtle Mountain PP *MB*
 14 F1
Turtle River *MB* **11** D8
Turtle River *MB* **14** B-C2
Turtle River *ON* **13** F7
Turtle River *ON* **16** E3
Turtle River PP *ON* **13** F7
Turtle River PP *ON* **16** E3
Turton Lake *NT* **5** E7
Tusket *NS* **33** E1-2
Tusket Island *NS* **33** F1
Tusket Lake *NS* **33** D2
Tusket River *NS* **33** D-E2
Tuxford *SK* **11** E5
Tuya Lake *BC* **3** B2
Tuya Mountains PP *BC*
 3 B2-3
Tuya River *BC* **3** B2-3
Tway *SK* **11** B5
Tway Lake, RNF *SK* **11** B5
Tway Lake National
 Wilderness Area *SK*
 11 B5
Tweed *ON* **21** B5
Tweed Island *NL* **36** A2
Tweed Island *NL* **38** F1

Tweed Lake *NT* **5** D8
Tweedside *NB* **32** C-D2
Tweedsmuir *SK* **11** D5
Tweedsmuir Glacier *BC* **1** B3
Tweedsmuir North PRA *BC*
 1 B4
Tweedsmuir South PRA *BC*
 1 B-C5
Twelve Mile Bay *ON* **18** E7
Twelve Mile Lake *SK* **11** F4
Twenty Mile Creek *ON*
 19 C7-8
Twenty Mile Creek *ON*
 20 E6-7
Twigge Lake *SK* **11** A6
Twigge Lake *SK* **12** F6
Twillingate *NL* **38** E6
Twin Butte *AB* **6** F5-6
Twin Butte *AB* **7** F4
Twin Falls *NL* **39** B3
Twin Falls Gorge TP *NT*
 10 F6
Twin Falls Gorge TP *NT*
 41 E1
Twin Falls Tea House NHS
 BC **7** A1
Twin Lakes *AB* **9** C3
Twin Lakes Beach *MB*
 14 D4
Twin Mount *YT* **4** D4-5
Twitya River *NT* **4** A6-7 B6
Twoforks River *SK* **12** F4-5
Two Hills *AB* **6** A6-7
Two Hills *AB* **8** C8
Two Hills *AB* **11** A1
Two River Lake *ON* **13** A7
Two Rivers, Lake of *ON*
 21 A3
Two Sheds *NS* **32** F6
Two Sheds *NS* **33** C3
Two Spring PPA *BC* **2** A6-7
Tyaughton Creek *BC* **2** A5
Tyee *BC* **1** C1-2
Tyendinaga *ON* **21** E5
Tyers River *YT* **4** D7
Tyhee Lake PP *BC* **3** F4-5
Tyndall *MB* **13** E4
Tyndall *MB* **14** D6
Tyner *SK* **11** D3
Tyne Valley *PE* **32** A7
Tyne Valley *PE* **34** B1
Tyne Valley *PE* **35** C2
Tyrconnell *ON* **19** D4
Tyrconnell *ON* **20** F3
Tyrone *ON* **20** C8
Tyrone *ON* **21** E2
Tyrrell *MB* **15** F3
Tyson Lake *ON* **18** B5
Tyvan *SK* **11** E6

U

Uchi Lake *ON* **16** C3
Ucluelet *BC* **1** F5
Ucluelet *BC* **2** E1
Udney *ON* **20** A7
Udney *ON* **21** D1-2
Udora *ON* **20** B7
Udora *ON* **21** E1
Uffington *ON* **21** C1-2
Ufford *ON* **18** D8
Ufford *ON* **21** C1
Ugjoktok Bay *NL* **39** A5-6
Uhlman Lake *MB* **15** D3
Uisage Bahn Falls PP *NS*
 34 C6-7
Uist, Lake *NT* **5** D7
Ukalta *AB* **8** B7
Ukkusiksalik (Wager Bay)
 NP *NU* **41** C-D5
Ukraina *MB* **11** D8
Ukraina *MB* **13** D2
Ukraina *MB* **14** A1
Ullswater *ON* **18** D8
Ullswater *ON* **21** B1
Ulverton *QC* **24** D8
Ulverton *QC* **27** D4
Umfreville *ON* **13** E7
Umfreville *ON* **16** D4
Umfreville Lake *ON* **13** E5
Umfreville Lake *ON* **14** D8
Umfreville Lake *ON* **16** C1
Umiujaq *QC* **40** D4
Umiujaq *QC* **41** F8
Uncas *AB* **8** C7
Uncha Mountains-Red Hills
 PP *BC* **1** A5
Underhill *MB* **11** F8
Underhill *MB* **13** F1-2
Underhill *MB* **14** E1
Underwood *ON* **20** B3
Ungava, péninsule d' *QC*
 40 C6
Ungava, péninsule d' *QC*
 40 B4-5 C4-5
Ungava, péninsule d' *QC*
 41 E8
Ungava Bay *QC* **40** B-C6
Union *NS* **34** A3
Union *NS* **34** E2-3
Union *ON* **19** D4-5
Union *ON* **20** F3-4
Union Bay *BC* **1** F5-6
Union Bay *BC* **2** D2
Union Corner *PE* **35** D2
Union Corner PP *PE*
 32 A-B7
Union Corner PP *PE* **34** C1
Union Corner PP *PE* **35** C1
Uniondale *ON* **19** B5
Uniondale *ON* **20** E4
Union Island *BC* **1** E4

Union Island *NT* **10** D8
Union Lake *NS* **6** A8
Union Passage PMP *BC*
 1 B3
Union Road *PE* **35** D-E5
Union Road *PE* **35** E6
Union Square *SK* **32** F6-7
Union Square *NS* **33** C4
Unionvale *PE* **35** C1-2
Unionville *ON* **19** A8
Unionville *ON* **20** C7
Unionville *ON* **21** F1
Unity *SK* **6** C8
Unity *SK* **11** C3
Unmingmaktok *NU* **41** B3-4
Unmingmaktok *NU* **42** F2
Unuk River *BC* **3** D-E2
Unwin *SK* **11** B2
Upham *SK* **32** D4-5
Upham *NB* **33** A2
Uphill *ON* **20** A8
Uphill *ON* **21** C2
Upper Adams River PP *BC*
 6 D2
Upper Arrow Lake *BC* **6** E3
Upper Beaver River *NT* **5** E5
Upper Beverley Lake *ON*
 21 C-D7
Upper Big Tracadie *NS*
 34 E5
Upper Blackville *NB* **31** E6
Upper Blackville *NB* **32** A4
Upper Branch *NS* **33** D4
Upper Brockway *NB* **32** D2
Upper Campbell Lake *BC*
 2 C1-2
Upper Cape *NB* **32** B8
Upper Cape *NB* **34** D1
Upper Cape *NB* **35** F2
Upper Carp Lake *NT*
 10 B6-7
Upper Clements *NS* **32** F5
Upper Clements *NS* **33** C2
Upper Clements PP *NS*
 32 F5
Upper Clements PP *NS*
 33 C2-3
Upper Clyde River *NS* **33** E3
Upper Cumins Lake *SK* **9** E8
Upper Cumins Lake *SK*
 12 E3
Upper Dorchester *NB* **32** C7
Upper Economy *NS* **32** D8
Upper Economy *NS* **33** A5
Upper Economy *NS* **34** E1
Upper Ferry *NL* **36** D1
Upper Foster Lake *SK* **12** D8
Upper Fraser *BC* **1** A7
Upper Fraser *BC* **6** A1
Upper Gagetown *NB* **32** C3
Upper Goose Lake *ON*
 13 C7
Upper Goose Lake *ON*
 16 B3
Upper Granville *NS* **32** F5
Upper Granville *NS* **33** C2-3
Upper Gulf Shore *NS* **33** A6
Upper Hainesville *NB* **31** F4
Upper Hainesville *NB* **32** B2
Upper Hammonds Plains *NS*
 32 F7-8
Upper Hammonds Plains *NS*
 33 C5
Upper Humber River *NL*
 38 F3
Upper Humber River *NL*
 39 F7-8
Upper Indian Pond *NL*
 36 A5
Upper Indian Pond *NL*
 37 A3
Upper Indian Pond *NL*
 38 F4
Upper Island Cove *NL*
 37 D7-8
Upper Kennetcook *NS*
 32 D-E8
Upper Kennetcook *NS*
 33 B6
Upper Kennetcook *NS*
 34 F2
Upper Kent *NB* **31** E3
Upper Kent *NB* **32** A1
Upper LaHave *NS* **33** D4
Upper Lakeville *NS* **33** C7
Upper Lawrencetown *NS*
 33 C6
Upper Liard *YT* **4** E7
Upper Liard *YT* **4** E-F7
Upper Lillooet PP *BC* **2** A-B4
Upper Lillooet PP *BC* **2** A-B4
Upper Linden *NS* **35** F2-3
Upper Madawaska River PP
 ON **21** A3-4
Upper Madawaska River PP
 ON **22** F3-4
Upper Manitou Lake *ON*
 13 F6-7
Upper Manitou Lake *ON*
 16 E3
Upper Margaree *NS* **34** C6
Upper Maugerville *NB*
 32 C3
Upper Middle River *NS*
 34 C6
Upper Mills *NB* **32** E1-2
Upper Musquodoboit *NS*
 32 A-B7
Upper Musquodoboit *NS*
 33 B6
Upper Nine Mile River *NS*
 32 E8
Upper Nine Mile River *NS*
 33 B6

Upper Nine Mile River *NS*
 34 F2
Upper Northfield *NS* **33** C4
Upper Northfield *NS* **33** D6
Upper Ohio *NS* **33** E3
Upper Point de Bute *NB*
 35 F1-2
Upper Rawdon *NS* **32** E8
Upper Rawdon *NS* **33** B6
Upper Rawdon *NS* **34** F1-2
Upper Rideau Lake *ON*
 24 F1
Upper River Denys *NS*
 34 D6
Upper Rockport *NB* **32** C7
Upper Roussay Lake NMBS
 SK **11** D6-7
Upper Sackville *NB* **35** F1
Upper Sackville *NS* **32** F8
Upper Sackville *NS* **33** C5-6
Upper Salmon Reservoir *NL*
 36 C3
Upper Salmon Reservoir *NL*
 37 C3
Upper Scotch Lake *ON*
 28 C6
Upper South River *NS* **34** E5
Upper South West Mabou
 NS **34** D5-6
Upper Stewiacke *NS*
 33 A-B7
Upper Stewiacke *NS* **34** F1
Upper Tantallon *NS* **32** F8
Upper Tantallon *NS* **33** C5-6
Upper Tracy *NB* **32** C3
Upper Twin Lake *ON* **16** D8
Upper Twin Lake *ON*
 17 A-B2
Upper Windigo Lake *ON*
 13 C7
Upper Windigo Lake *ON*
 16 A4
Upper Woods Harbour *NS*
 33 F2
Upper Woodstock *NB* **31** F3
Upper Woodstock *NB* **32** B1
Upsala *ON* **13** F8
Upsala *ON* **16** A4-5
Upsalquitch *NB* **29** E5-6
Upsalquitch *NB* **31** B4
Upsalquitch River *NB*
 29 E5-6
Upsalquitch River *NB* **31** B4
Upton *PE* **35** D-E6
Upton *QC* **24** D7-8
Upton *QC* **27** E3
Uqsuqtuuq (Gjoa Haven) *NU*
 41 B-C4
Uqsuqtuuq (Gjoa Haven) *NU*
 42 F3-4
Uranium City *SK* **9** A8
Uranium City *SK* **12** A3
Uranium City *SK* **41** F2
Urbainville *PE* **35** D2
Urbania *NS* **33** B6
Urbania *NS* **34** F2
Uren *SK* **11** E4
Urquhart Lake *NT* **5** B5
Ursula Channel *BC* **1** B3
Ursula Lake *NT* **10** A8
Usam Island *SK* **12** B6
Usam Island *SK* **15** B1
Usborne, lac *QC* **22** E6
Usherville *SK* **11** C6-7
Usk *BC* **3** F3-4
Usona *BC* **8** E6
Utica *ON* **20** B8
Utica *ON* **21** E2
Utikuma Lake *AB* **9** E4-5
Utikuma River *AB* **9** E4-5
Utopia *BC* **8** E6
Utopia, Lake *NB* **32** E2-3
Utopia PGA *AB* **32** E3
Utterson *ON* **18** D8
Utterson *ON* **21** B1
Uvajuq (Mount Pelly) TP *NU*
 41 B3
Uvajuq (Mount Pelly) TP *NU*
 42 F3
Uxbridge *ON* **20** B7
Uxbridge *ON* **21** E1-2

V

Vachon, rivière *QC* **27** C-D7
Vadzaih Van Tshik TP *NT*
 5 C5
Vail Point *ON* **18** E-F6
Vail Point *ON* **20** A4
Val-Alain *QC* **27** C5
Val-Barrette *QC* **24** A3
Val-Barrette *QC* **27** A-B5
Val-Belair *QC* **27** A-B5
Val-Brillant *QC* **29** C6
Val-Caron *QC* **17** F6-7
Val Côté *ON* **17** B4
Valcourt *QC* **24** D8
Valcourt *QC* **27** E3
Val-Comeau *NB* **30** F2
Val-Comeau *NB* **31** C8
Val-Comeau PP *NB* **30** F2
Val-Comeau PP *NB* **31** B-C8
Val-d'Amour *NB* **29** D6
Val-d'Amour *NB* **31** A4-5

Val-des-Lacs *QC* **24** B5
Val-des-Monts *QC* **21** A8
Val-des-Monts *QC* **22** F8
Val-des-Monts *QC* **24** C2
Val-d'Espoir *QC* **30** C3
Val-Doucet *NB* **29** E8
Val-Doucet *NB* **30** E-F1
Val-Doucet *NB* **31** B7
Valemount *BC* **6** B2
Valens CA *ON* **19** B6-7
Valens CA *ON* **20** D5-6
Valentia *ON* **20** B8
Valentia *ON* **20** B8
Valentia *ON* **21** E2
Vale Perkins *QC* **24** E8
Vale Perkins *QC* **27** F4
Valets, lac *QC* **23** D7
Valets, lac *QC* **26** E1
Valetta *ON* **19** E2
Val Gagné *ON* **17** B4
Valhalla Beach *MB* **14** C5
Valhalla Centre *AB* **9** E2
Valhalla PP *BC* **6** E3
Valin *QC* **28** C6
Valin *QC* **28** C4
Vallard *QC* **39** C1
Vallée-Jonction *QC* **27** C6
Vallentyne *ON* **20** B7-8
Vallentyne *ON* **21** E1
Valley *NS* **34** E2
Valley Bay *NL* **39** B6
Valley Centre *NS* **11** C3
Valley Mills *NS* **34** D6
Valleyfield *NL* **37** A6-7
Valleyfield *NL* **38** F8
Valleyfield *PE* **34** C3
Valleyfield *PE* **35** E6
Valley Pond *NL* **38** E6
Valley River
 (Tootinaowaziibeeng)
 MB **11** D8
Valley River
 (Tootinaowaziibeeng)
 MB **13** D1-2
Valleyview *AB* **9** F3-4
Valleyview PP *NS* **32** F8
Valleyview PP *NS* **33** C2-3
Vallican *BC* **6** F3
Val-Limoges *QC* **22** D8
Val-Limoges *QC* **24** A2
Val-Limoges *QC* **25** E2
Val-Limoges *QC* **24** A2
Val Marie *SK* **11** F3
Val Marie Reservoir NMBS
 SK **11** F3
Val-Morin *QC* **24** C5
Valmy, lac *QC* **23** D7
Valmy, lac *QC* **25** A1
Valmy, lac *QC* **26** E1
Val-Paradis *QC* **17** C8
Val-Paradis *QC* **23** B2
Valparaiso *SK* **11** B6
Val-Racine *QC* **27** E6
Val Rita *QC* **17** C8
Val-Saint-Gilles *QC* **17** C8
Val-Saint-Gilles *QC* **23** B2
Val-Senneville *QC* **23** E5
Val Therese *QC* **17** F6
Val-Viger *QC* **24** A3
Val-Viger *QC* **25** E3
Vananda *BC* **1** F6
Vananda *BC* **2** C3
Vanastra *ON* **19** A4
Vanastra *ON* **20** D3
Vance *SK* **11** C3
Vancouver *BC* **1** F6-7
Vancouver *BC* **2** D4-5
Vancouver, île de *BC* **1** F4
Vancouver, île de *BC*
 2 E3 F3-4
Vancouver Island *BC*
 1 F4-7
Vancouver Island *BC* **2** D1-2
Vancouver Island Ranges
 BC **1** E-F5
Vancouver Island Ranges
 BC **2** B-C1 D3 E2-3
Vanderhoof *BC* **1** A6
Vandekerckhove Lake *MB*
 12 D7
Vandekerckhove Lake *MB*
 15 D1-2
Vandorf *ON* **20** C7
Vandorf *ON* **21** E1
Vandry *QC* **25** B6
Vandry *QC* **28** D2-7
Vandura *SK* **11** E7
Vanel, lac *QC* **28** A6-7
Vanessa *ON* **19** C6
Vanessa *ON* **20** E5
Vanguard *SK* **11** E4
Vanier *ON* **21** A8
Vanier *ON* **24** C3
Vanier *QC* **27** B5
Vanier, île *NU* **42** C1
Vanier, lac *QC* **25** A5
Vanier, lac *QC* **26** E5
Vankleek Hill *ON* **24** D4
Van Koenig Point *NU* **41** B5
Van Koenig Point *NU* **42** F1
Vankoughnet *ON* **21** C2

Vanscoy SK 11 C4
Van Scoy Lake SK 11 C3-4
Vansittart Island NU 41 C6
Vantage SK 11 E4-5
Varennes QC 24 D6
Varennes QC 27 E2
Vargas Island BC 1 F5
Vargas Island BC 2 D1
Vargas Island PP BC 1 F4-5
Vargas Island PP BC 2 E1
Varna ON 19 B4
Varna ON 20 D3
Varney ON 20 B4
Vars ON 24 D3
Varty Lake ON 21 D6
Vaseux Lake, RNOM BC 6 F1-2
Vaseux Lake NMBS BC 6 F1-2
Vaseux Lake PP BC 6 F2
Vasey ON 20 A6
Vassan QC 23 E4
Vassar MB 13 F4-5
Vassar MB 14 F7
Vaudray, lac QC 23 E3
Vaudreuil-Dorion QC 24 D5
Vaudreuil-Dorion QC 27 E1
Vaudreuil-sur-le-Lac QC 24 D5
Vaughan NS 32 E7
Vaughan NS 33 C5
Vaughan ON 19 A8
Vaughan ON 20 C7
Vaughan ON 21 F1
Vauquelin, baie QC 23 E6
Vauxhall AB 6 E7
Vauxhall AB 7 D6
Vavenby BC 6 D2
Vawn SK 11 B3
Vega NS 6 A6
Vega AB 8 A5
Vega AB 9 F5
Vegreville AB 6 B6-7
Vegreville AB 8 C8
Vegreville AB 11 A1
Veillardville SK 11 B7
Vendée QC 24 C4
Venice AB 8 A8
Venise-en-Québec QC 24 E7
Venise-en-Québec QC 27 E2
Venison Islands NL 39 C8
Venlaw MB 11 D8
Venlaw MB 13 D1-2
Venn SK 11 D5
Vennachar ON 21 B5
Vennachar Junction ON 21 B5
Venosta QC 22 E-F7
Venosta QC 24 C1
Ventadour, lac QC 26 D5
Ventadour, lac QC 28 E4
Ventnor ON 21 C8
Vents River BC 3 A-B5
Vents River BC 4 F8
Vera SK 11 C2-3
Verchères QC 24 C6
Verchères QC 27 D2
Verdun QC 24 D6
Verdun QC 27 E1
Veregin SK 11 D7
Verlo SK 11 E3
Vermette Lake SK 9 E8
Vermette Lake SK 12 E3
Vermeulle, lac QC 39 A1
Vermilion AB 6 B7
Vermilion AB 11 B2
Vermilion Bay ON 13 E6
Vermilion Crossing BC 6 D4
Vermilion Crossing BC 7 B1
Vermilion Hills SK 11 E4
Vermilion Lake ON 16 D3
Vermilion Lake ON 18 A5
Vermilion Pass BC 6 D4
Vermilion Pass BC 7 B1
Vermilion PP AB 6 B7
Vermilion PP AB 11 B1-2
Vermilion River AB 6 B7
Vermilion River AB 11 A1 B1-2
Vermilion River MB 11 D8
Vermilion River MB 13 D2
Vermilion River MB 14 B1
Vermilion River ON 16 C4
Vermilion River ON 17 E6-7
Vermilion River ON 18 A4-5
Vermillion Bay ON 16 D2
Vermillon, lac QC 25 D6
Vermillon, lac QC 28 F2
Vermillon, rivière QC 25 C6
Vermillon, rivière QC 26 D6-7
Vermillon, rivière QC 28 A-B2
Vermillon, rivière QC 28 E-F2
Vermilyea Lake MB 13 A5
Vermilyea Lake MB 15 F5
Vermont, lac QC 28 B6
Verner ON 17 F7
Verner ON 18 A7
Verneuil, lac QC 26 B4
Vernon BC 6 E2
Vernon ON 21 B8
Vernon ON 24 E2-3
Vernon, Lake ON 18 D8
Vernon, Lake ON 21 B1
Vernon Bridge PE 35 E5-6
Vernon Lake BC 4-5
Vernon River PE 34 C3
Vernon River PE 35 E6
Vernonville ON 21 E4

Verona ON 21 D6
Verret NB 29 F3
Verret NB 31 C1-2
Verschoyle ON 19 C5
Verschoyle ON 20 E4
Versailles Lake SK 12 E6
Vert, lac QC 22 E8
Vert, lac QC 24 C2
Vert, lac SK 11 C5-6
Verte, baie NB 32 B7-8
Verte, baie NB 35 F2
Verte, baie NS 32 B7-8
Verte, baie NS 34 D1
Verte, île QC 29 D1
Verte, pointe QC 30 C3
Verte, rivière NB 29 E3
Verte, rivière NB 31 B2
Verte, rivière QC 29 D-E1
Vert Island ON 16 E6-7
Vert Island ON 17 C1
Verton, lac QC 39 D6
Vesuvius BC 1 F6-7
Vesuvius BC 2 E4
Veteran AB 6 C7
Veteran AB 11 C1
Vezina Lake NL 39 A3
Vianney QC 27 D5
Vibank SK 11 E6
Viceroy SK 11 F5
Vicomte-de-Melville, détroit du NT 42 D2-3
Vicomte-de-Melville, détroit du NU 42 D2-3
Victoire SK 11 B4
Victor AB 7 A6
Victor, lac QC 39 E5
Victoria AB 6 A6
Victoria AB 8 B7-8
Victoria AB 11 A1
Victoria BC 2 F4
Victoria NL 37 D7
Victoria ON 19 A7
Victoria ON 20 C6
Victoria PE 32 B8
Victoria PE 34 C2
Victoria PE 35 E3
Victoria, île NT 41 B2-3
Victoria, île NT 42 E2-3
Victoria, île NU 41 B2-3
Victoria, île NU 42 E2-3
Victoria Beach MB 13 B1
Victoria Beach MB 14 C5-6
Victoria Beach NS 32 F4
Victoria Beach NS 33 C2
Victoria Bridge NS 34 D7
Victoria Corner NB 31 F3
Victoria Corner NB 32 B1
Victoria Cove NL 37 A5-6
Victoria Cove NL 38 F7
Victoria Cross PE 34 C3
Victoria Cross PE 35 E6
Victoria Harbour NS 32 E6
Victoria Harbour NS 33 B3
Victoria Harbour ON 18 E7
Victoria Harbour ON 20 A6
Victoria Head NU 42 B5
Victoria Island NT 41 A2-3
Victoria Island NT 42 E2-3
Victoria Island NU 41 A2-3
Victoria Island NU 42 E1-3
Victoria Lake BC 1 E4
Victoria Lake NL 36 C3-4
Victoria Lake NL 37 C1-2
Victoria Lake ON 21 A3-4
Victoria Lake ON 22 F4
Victoria Lake ON 23 D1
Victoria Peak BC 1 E5
Victoria Peak BC 2 C1
Victoria PP PE 32 B8
Victoria PP PE 34 C2
Victoria PP PE 35 E3-4
Victoria River NL 36 B-C4
Victoria River NL 37 B-C2
Victoria Road ON 20 A8
Victoria Settlement PHS AB 6 A6
Victoria Settlement PHS AB 8 B8
Victoria Settlement PHS AB 11 A1
Victoria Strait NU 41 B4
Victoria Strait NU 42 F3
Victoria Vale NS 32 E6
Victoria Vale NS 33 B3
Victoriaville QC 27 D4-5
Victoria West PE 35 C2
Victor Lake PP BC 6 D3
Victory Lake NT 10 C7
Vidal Bay ON 18 B2
Vidir MB 14 B5
Vidora SK 11 F2
Vieillard, lac du QC 22 A4
Vieille, lac de la QC 22 C7
Vieille, lac de la QC 24 A1
Vieille, lac de la QC 25 E1
Vienna ON 19 D5
Vieux-Fort QC 38 A2
Vieux-Fort QC 39 D7
Vieux Fort, île du QC 38 A2
Vieux Fort, île du QC 39 D7
Vieux Lunenburg, le NS 33 D5
Vieux Poste, pointe du QC 39 F5
Viewfield SK 11 F6-7
Viau Royal BC 2 F4
Vignal, lac QC 39 B1
Vigneault, lac QC 39 E4-5

Vignerod, lac QC 25 D6
Vignerod, lac QC 28 F2
Viking AB 6 B6-7
Viking AB 8 D8
Viking AB 11 B1
Villa-Bellevue, LHN de la ON 21 E6-7
Village Bay BC 2 E4-5
Village Cove NL 38 E-F6
Village-des-Aulnaies QC 28 F7
Villa Marie NL 37 A6-7
Villa Nova ON 19 C6
Villa Nova ON 20 E5
Villebois QC 17 C8
Villebois QC 23 B2
Villebon, lac QC 23 C5
Villebon, lac QC 23 C7
Villejouin, lac QC 23 C7
Villejouin, lac QC 26 C1
Ville-Marie ON 17 E8
Ville-Marie QC 22 A1-2
Villemontel QC 23 C3
Villeneuve AB 6 A-B5
Villeneuve AB 8 C6
Villeroy QC 27 C4
Villier, lac QC 25 D5
Villier, lac QC 27 A1
Villier, lac QC 28 F1
Villiers ON 21 E4
Vilna AB 6 A6
Vilna AB 8 B8
Vilna AB 11 A1
Vimont, lac QC 22 B8
Vimont, lac QC 25 C2
Vimont, lac QC 26 B5
Vimy AB 6 A5-6
Vimy AB 8 B6
Vimy-Ridge QC 27 D5
Vin, Bay du AB 31 D7
Vincent Lake ON 13 D-E8
Vineland ON 19 C8
Vineland ON 20 E7
Vineland Station ON 19 C8
Vineland Station ON 20 E7
Vinemount ON 19 C7
Vinemount ON 20 E6
Vingt-Deuxième Mille, lac du QC 39 E4
Vinoy QC 24 C3
Vinsulla BC 2 A8
Vinton QC 22 F6
Violet ON 21 D-E6
Violet Grove AB 6 B5
Violet Grove AB 8 D4
Violet Hill ON 20 B6
Virago Sound BC 1 A1
Virden MB 11 E8
Virden MB 13 E1
Virgil ON 19 B8
Virgil ON 20 E7
Virgin Arm NL 38 E6
Virginia ON 20 B7
Virginia ON 21 D1
Virginia Falls NT 4 D8
Virginia Falls NT 10 D1
Virginiatown ON 17 D8
Virginiatown ON 23 E1
Virgin River SK 9 C8
Virgin River SK 12 C-D4
Viscount SK 11 C5
Viscount Island BC 2 A1
Viscount Melville Sound NT 42 D2-3
Viscount Melville Sound NU 42 D2-3
Vista MB 11 E8
Vista MB 13 D-E1
Vita MB 13 F4
Vita MB 14 F6
Vittoria ON 19 D6
Vittoria ON 20 F5
Vittrekwa River YT 5 D4
Vivian MB 13 E4
Vivian MB 14 D-E6
Vogar MB 13 D2
Vogar MB 14 B3
Voglers Cove NS 33 D4
Voisey's Bay NL 40 D7
Volant, rivière QC 29 A1-2
Volmer AB 8 C6
Voltigeurs, parc des QC 24 C7-8
Voltigeurs, parc des QC 27 D3
Vonda SK 11 C4-5
Voyageur PP ON 24 D4
Vulcan AB 6 E6
Vulcan AB 7 C-D5
Vuntut, PN YT 5 C2-3
Vuntut NP YT 5 B2

W

Waasagomach MB 13 A5
Waasis NB 32 C3
Waba ON 21 B6-7
Wababimiga Lake ON 16 D8
Wababimiga Lake ON 17 A2
Wabagishik Lake ON 18 A4
Wabakimi Lake ON 13 D8
Wabakimi Lake ON 16 C5
Wabakimi PP ON 16 C5
Wabakimi PP ON 40 F1
Wabamun AB 6 B5
Wabamun AB 8 C5
Wabamun Lake PP AB 6 B5
Wabamun Lake PP AB 8 C5
Wabana NL 37 D8
Wabano, rivière QC 26 E5
Wabano, rivière QC 28 B-C1
Wabasca-Desmarais AB 9 E5
Wabasca River AB 9 C4
Wabasca River AB 9 D-E5
Wabasca River AB 41 F1
Wabash ON 19 D3
Wabash ON 20 F2
Wabaskang Lake ON 13 E6
Wabaskang Lake ON 16 C2
Wabasseemoong (Whitedog) ON 16 D1
Wabassi River ON 16 B7 17 C4
Wabauskang ON 13 E6
Wabauskang ON 16 C2
Wabigoon ON 13 E-F7
Wabigoon ON 16 D3
Wabigoon Lake ON 13 E6 F7
Wabigoon Lake ON 13 F7
Wabigoon Lake ON 16 D3
Wabigoon River ON 13 E6
Wabimeig Lake ON 16 B8
Wabiskaw AB 9 E5
Wabos ON 17 F4
Wabu Creek ON 14 A8
Wabowden MB 15 F3
Wabuk Point ON 40 D2-3
Wabush NL 39 C2
Wabush NL 40 E6
Wabush Lake NL 39 C2
W.A.C. Bennett Dam BC 3 E7
Wachigabau, lac QC 23 A8
Wachigabau, lac QC 26 B2
Waco QC 39 D2-3
Waconichi, lac QC 26 A5
Wacouno, rivière QC 39 D2-3
Waddington, Mount BC 1 D5
Wade, Mount BC 4 F2
Wade Lake NL 39 B3
Wadena SK 11 C6
Wade's Landing ON 18 B8
Wadham Islands NL 38 E7
Wadin Lake AB 9 C4
Wadsworth Lake ON 21 A4
Wagarville ON 21 C-D6
Wager Bay NU 41 D5
Wager Bay (Ukkusiksalik) NP NU 41 C-D5
Waglisla-McLoughlin Bay BC 1 C3-4
Wagmatcook NS 34 C6
Wagosh Bay ON 18 B1
Wahgoshig (Abitibi) ON 17 C7
Wahleach Lake BC 2 E7
Wahnapitae ON 17 F7
Wahnapitae ON 18 A6
Wahnekewaning Beach ON 18 F7
Wahnekewaning Beach ON 20 A5
Wahpeton SK 11 B4-5
Wahwashkesh ON 18 C7
Wahwashkesh Lake ON 18 C7
Wainfleet ON 19 C8
Wainfleet ON 20 E7
Wainwright AB 6 B7
Wainwright AB 11 B2
Waiparous AB 6 D5
Waiparous AB 7 B3
Wakami Lake ON 17 E5
Wakami Lake PP ON 17 E5
Wakaw SK 11 C5
Wakaw Lake SK 11 C5
Wakefield QC 21 A7-8
Wakefield QC 22 F7-8
Wakefield QC 24 D2
Wakeham QC 30 B2
Wakeman River AB 9 C4
Wakeman Sound BC 1 D5
Wakes Cove PP BC 2 E4
Wakomata Lake ON 17 F5
Wakopa MB 13 F2
Wakopa MB 14 F1
Wakwayowkastic River ON 17 A6-7
Waldeck SK 11 E3
Waldemar ON 19 A6
Waldemar ON 20 C5
Walden NS 32 F7
Walden NS 33 D4
Waldersee MB 14 C2-3
Waldheim SK 11 C4
Waldo BC 6 F4-5
Waldron SK 11 D7
Waldron River NT 10 C4
Walesback Channel ON 18 B2-3
Wales Island BC 3 F2
Wales Island NU 40 B5
Wales Island NU 41 C5-6
Wales Island NU 41 D8
Wales Island NU 42 C5
Walford ON 17 F5-6
Walford ON 18 A-B3
Walhachin BC 2 A7
Walhachin BC 6 E1
Walhachin Oxbows PP BC 1 D-E8
Walhachin Oxbows PP BC 2 A7-8

Walhachin Oxbows PP BC 6 D1
Walker, lac QC 39 E1-2
Walker Creek SK 12 D5
Walker Lake MB 15 F4
Walker Lake SK 12 A6
Walkers Point ON 18 E8
Walkers Point ON 21 C1
Walkerton ON 20 B4
Walkerton ON 41 F1
Walkleyburg MB 14 D5-6
Wallace NS 34 D2
Wallace ON 19 A5
Wallace ON 20 C4
Wallace Bay, RNF de NS 34 D2
Wallaceburg ON 19 D2
Wallaceburg ON 20 F1
Wallace Island PMP BC 2 E4
Wallace Lake PP MB 13 D5
Wallace Lake PP MB 14 B7
Wallace Mountain AB 9 F4
Wallace Ridge NS 32 C8
Wallace Ridge NS 34 D2
Wallace River NS 34 E1-2 D2
Wallace Station NS 32 C8
Wallace Station NS 34 D2
Wallacetown ON 19 D4
Wallacetown ON 20 F2
Wallenstein ON 19 A5-6
Wallenstein ON 20 D4-5
Wall Island ON 18 C4
Walloper Lake PP BC 1 E8
Walloper Lake PP BC 2 B8
Walloper Lake PP BC 6 E1
Walpole SK 11 E7
Walpole SK 13 E1
Walpole Island ON 19 D2
Walpole Island ON 19 D2
Walpole Island ON 20 F1
Walpole Island ON 20 F1
Walsh AB 6 E8
Walsh AB 11 E2
Walsh ON 19 D6
Walsh ON 20 F5
Walsh Cove PP BC 2 E2
Walsh Creek YT 4 C-D4
Walsh Lake SK 12 B6
Walsh Lake SK 15 B1
Walsingham ON 19 D6
Walsingham ON 20 F5
Walter Cook Upland Caves PP Reserve MB 11 B8
Walter Cook Upland Caves PP Reserve MB 13 B2
Walters Falls ON 18 F6
Walters Falls ON 20 A5
Waltham QC 22 E6
Walton NS 32 D7-8
Walton NS 33 A-B5
Walton NS 34 F1
Walton ON 19 A4-5
Walton ON 20 C3-4
Walton River NS 32 D8
Walton River NS 33 B5
Walton River NS 34 F1
Wampum MB 13 F4-5
Wampum MB 14 F7
Wanapitei Lake ON 17 F7
Wanapitei PP ON 17 E-F7
Wanapitei River ON 17 E6-7
Wanapitei River ON 18 A4-5 B5
Wanasing Beach MB 14 C6
Wandering River AB 9 E6
Wandering River AB 9 F6
Wandering River AB 12 E1
Wandering River AB 12 E1
Waneta BC 6 F3
Wanham AB 9 E3
Wanipigow River MB 13 D4-5
Wanipigow River MB 14 B6-7
Wanless MB 11 A7
Wanless MB 13 A1
Wanup ON 17 F7
Wanup ON 18 A6
Wapah MB 14 B3
Wapaseese River ON 15 F7-8
Wapata Lake SK 12 B5
Wapawekka Hills SK 12 F5-6
Wapawekka Hills Provincial Ecological Reserve SK 12 F5-6
Wapeeminakoskak Lake MB 13 A5-6
Wapeeminakoskak Lake MB 15 F5-6
Wapekeka ON 13 A8
Wapella SK 11 E7
Wapella SK 13 E1
Wapesi Bay ON 13 C5
Wapesi Lake ON 13 C5
Wapesi River ON 13 D-E7
Wapikaimaski Lake ON 13 E8
Wapikaimaski Lake ON 16 D5
Wapikani River MB 15 E-F7
Wapiskau River SK 12 E6
Wapisu Lake MB 15 E2-3
Wapiti Lake PP AB 1 A8
Wapiti Lake PP AB 9 F1
Wapiti Mountain AB 6 C-D4
Wapiti Mountain AB 7 A2
Wapiti River BC 9 F2

Wapiyao Lake SK 12 A6
Wapiyao Lake SK 15 A1
Wapske NB 31 D3
Wapus, PN MB 15 C6
Wapus, PN MB 40 C1
Wapus, rivière QC 23 A8
Wapus, rivière QC 25 C-D2
Wapusk, PN MB 15 B-C6
Wapusk, PN MB 40 C1
Wapusk NP MB 15 B-C6
Wapusk NP MB 40 C1
Wapusk NP MB 41 F5
Wapus Lake SK 12 D7
Wapus Lake SK 15 D1
Wapuspimiskunow Lake MB 15 E6
Wapustagamau, lac QC 39 D6
Warburg AB 6 B5
Warburg AB 8 D5
Warburton ON 21 D7
Warburton Bay NT 10 B7
Ward NB 35 F1
Ward, baie QC 27 D5
Ward, lac QC 22 D6
Warden AB 8 F7
Warden QC 24 D8
Warden QC 27 E3
Wardlow AB 6 D7
Wardlow AB 7 B7
Wardlow AB 11 D1
Wardner BC 6 F4
Wardner BC 7 F2
Wardner PP BC 6 F4
Wardner PP BC 7 F2
Wardsville ON 19 D3-4
Wardsville ON 20 F2-3
Wareham NL 36 A8
Wareham NL 37 A6
Wareham NL 38 F7-8
Warfield BC 6 F3
Warina ON 24 E3
Warkworth ON 21 E4
Warkworth Lake MB 15 B5-6
Warman SK 11 C4
Warminster ON 20 A7
Warminster ON 21 D1
Warneford River BC 3 C5-6
Warner AB 6 F6-7
Warner AB 7 F6
Warpath River MB 13 C3
Warren AB 8 E3
Warren MB 14 D4
Warren ON 17 F7
Warren ON 18 A6
Warren Grove PE 35 E4
Warren Landing MB 13 A3
Warren Point NT 5 A5
Warsaw ON 21 D3
Warsaw Caves CA ON 21 D3-4
Warspite AB 6 A6
Warspite AB 8 B7
Warspite AB 11 A1
Warspite AB 12 F1
Wartburg ON 19 B5
Wartburg ON 20 D4
Wartime SK 11 D3
Wart Lake ON 17 E4
Warwick AB 6 A6-7
Warwick AB 8 C8
Warwick AB 11 A1
Warwick ON 19 C3
Warwick ON 20 E2
Warwick QC 27 D4
Warwick CA ON 19 C3
Warwick CA ON 20 E2
Wasa BC 6 F4
Wasa BC 7 E2
Wasaga Beach ON 18 F7
Wasaga Beach ON 20 A6
Wasaga Beach PP ON 18 F7
Wasaga Beach PP ON 20 A6
Wasagaming MB 13 E2
Wasagaming MB 14 C1
Wasa Lake PP BC 6 F4
Wasa Lake PP BC 7 E1
Waseca SK 11 B2
Wasekamio Lake SK 9 D8
Wasekamio Lake SK 12 D3
Wasel AB 8 B8
Washademoak Lake NB 32 C4
Washago ON 18 E-F8
Washago ON 20 A7
Washago ON 21 C1
Washburn Lake NU 42 E-F2
Washi Lake ON 16 B7
Washington ON 19 B6
Washington ON 20 D5
Washow Bay MB 13 D4
Washow Bay MB 14 B5
Wasi Lake ON 18 C7
Wasi River ON 22 D-E2
Waskada MB 11 F8
Waskada MB 13 F1
Waskaganish QC 40 F4
Waskaiowaka Lake MB 15 D4
Waskatenau AB 8 B7
Waskatenau AB 11 A1
Waskesiu Lake SK 11 A4
Waskesiu Lake SK 11 A4-5
Waskwei Island SK 12 D3
Waspison Lake SK 12 B6-7

Waspison Lake SK 15 B1
Wastina AB 7 A7-8
Waswanipi QC 23 A8
Waswanipi QC 23 A7
Waswanipi QC 26 B1-2
Waswanipi, lac QC 23 A7
Waswanipi, lac QC 26 B1
Waswanipi, rivière QC 23 A7
Waswanipi, rivière QC 26 B1-2
Watabeag Lake ON 17 D7
Watagheistic, île QC 39 E6
Watapi Lake SK 9 E8
Watapi Lake SK 12 E2
Watchorn PP MB 13 D3
Watchorn PP MB 14 B3
Water's Edge AB 8
Waterborough NB 32 C4
Waterbury Lake SK 12 C5
Waterdown ON 19 B7
Waterdown ON 20 D6
Waterford NB 32 C5
Waterford NS 32 F4
Waterford NS 33 C1
Waterford ON 19 C6
Waterford ON 20 E5
Waterford PE 35 A1-2
Waterford CA ON 19 C6
Waterford CA ON 20 E5
Waterfound River SK 12 B5-6
Waterhen MB 11 C8
Waterhen MB 13 C2
Waterhen (IR) MB 13 C2
Waterhen Lake MB 13 C2
Waterhen Lake SK 9 F8
Waterhen Lake SK 11 A3
Waterhen Marsh SK 11 B5
Waterhen River SK 9 F8
Waterhen River SK 11 A3
Waterhen River SK 12 F3
Waterloo NS 33 D4
Waterloo ON 19 B6
Waterloo ON 20 D5
Waterloo ON 24 E8
Waterloo ON 27 E3
Waterloo, lac QC 25 D4
Waterloo Lake SK 9 A7-8
Waterloo Lake SK 12 A3
Waternish NS 34 F4
Waterside NB 32 C6
Waterside NB 33 A3-4
Waterside Beach PP NS 34 A3
Waterton ON 21 D7
Waterton-Glacier AB 7 F4
Waterton-Glacier, parc international de la paix AB 7 F4-5
Waterton-Glacier International Peace Park AB 7 F4
Waterton Lakes NP AB 6 F5
Waterton Lakes NP AB 7 F4
Waterton Park AB 6 F5
Waterton Park AB 7 F4
Waterton Reservoir AB 7 F4
Waterton River AB 6 F6
Waterton River AB 7 F4
Watervale NS 34 E3
Watervale PE 35 E5
Water Valley AB 6 D5
Water Valley AB 7 A3
Waterville NS 32 E6
Waterville NS 33 B1
Waterville NL 37 C6-7
Waterville NS 32 E-F7
Waterville NS 33 B4
Waterville NS 33 C5
Waterville QC 27 F4-5
Watford ON 19 C3
Watford ON 20 E2
Wathaman Lake SK 12 D6
Wathaman River SK 12 D5-6
Watino AB 9 E3
Watopeka, rivière QC 27 E4-5
Watrous SK 11 D5
Watshishou, lac QC 39 E4
Watshishou, rivière QC 39 F4
Watshishou, RNOM QC 39 F4-5
Watshishou NMBS QC 39 F4-5
Watson SK 11 C5-6
Watson, lac QC 22 B3
Watson Lake YT 3 A4
Watson Lake YT 4 E7
Watson Lake YT 3 A4
Watson Lake YT 4 E7
Watson Point MB 15 B6
Watson River YT 4 E3-4
Wattla Lake NT 10 D7
Watt Section NS 33 B8
Watt Section NS 34 F4
Watts Point Calcareous Ecological Reserve NL 38 A4
Waubamik ON 18 D7
Waubaushene ON 20 A6-7
Waubaushene ON 21 C1

Waubaushene Beaches, RNP ON 18 E-F8
Waubaushene Beaches, RNP ON 21 C1
Waubaushene Beaches PNR ON 18 E-F8
Waubaushene Beaches PNR ON 21 C1
Waubuno Channel ON 18 B3-4
Waubuno Creek ON 19 B3
Waubuno Creek ON 20 E4
Wauchope SK 11 F7
Wauchope SK 13 F1
Waugh AB 8 B6
Waugh MB 13 F5
Waugh MB 14 E7
Waugh MB 8 B6
Waugh Creek YT 5 F3
Waupoos ON 21 E6
Waupoos Island ON 21 E6
Wauquimakog Lake ON 18 B-C7
Wauzhushk Onigum 16 D1
Wavell ON 17 D7
Waverley NS 32 F8
Waverley NS 33 C6
Waverley ON 18 F7
Waverley ON 20 A6
Waverley Game Sanctuary NS 33 C6
Wavy Lake ON 18 A5
Wawa ON 17 D3-4
Wawagosic, rivière QC 17 B8
Wawagosic, rivière QC 23 A-B3
Wawanesa MB 13 F2
Wawanesa MB 14 E2
Wawang Lake ON 13 F8
Wawang Lake ON 16 E4
Wawanosh CA ON 19 A4
Wawanosh CA ON 20 C3
Waweig NB 32 E-F2
Wawota SK 11 E7
Wayagamac, lac QC 25 D8
Wayagamac, lac QC 28 F3
Way Bay NL 39 B3
Wayerton NB 29 F7
Wayerton NB 31 C6
Wayne AB 6 D6
Wayne AB 7 A5-6
Wayow Lake SK 12 A6
Ways Mills QC 27 F4
Weagamow Lake ON 13 B7
Weagamow Lake ON 13 B7
Weagamow Lake ON 40 E1
Weald AB 8 C2
Weasel Creek AB 8 E7
Weatherall Bay NU 42 C3
Weaver Lake AB 13 A4
Weaver Settlement NS 33 D1-2
Weaver Siding NB 31 E6
Weaver Siding NB 32 A4
Webb SK 11 E3
Webber Cove PE 35 D3
Webber Lake MB 13 A6
Webber Lake MB 15 F6
Webbwood ON 17 F6
Webbwood ON 18 A3-4
Webb Island NL 39 A7
Weber, Mount BC 3 F3-4
Wecho Lake NT 10 B6-7
Wecho River NT 10 B-C6
Wedding, rivière QC 23 B6
Wedge Mountain BC 1 E7
Wedge Mountain BC 2 C5
Wedgeport NS 33 F2
Weedon QC 27 E5
Weedon Centre QC 27 D-E5
Weekes SK 11 C6-7
Weequed Lake ON 18 A4
Weeskinisht Peaks BC 3 F4
Weeteeam Bay NL 1 C3
Weiden AB 14 A1-2
Weir QC 24 C4
Weirdale SK 11 B5
Weir River MB 15 D5-6
Weir River MB 15 D5-6
Weirs Pond NL 36 A8
Weirs Pond NL 37 A6
Weirs Pond NL 38 F7
Weitzel Lake SK 12 C4
Wekusko MB 13 A2
Wekusko MB 15 F2
Wekusko MB 15 F7
Wekusko Falls PP MB 12 F8
Wekusko Falls PP MB 15 F7
Wekusko Lake MB 15 F7
Wekweti NT 10 B6
Wekweti NT 41 D6
Welcome ON 21 E3
Weldon NB 32 C6
Weldon NS 34 F4
Welland ON 19 C8
Welland ON 20 E7
Welland Canal ON 19 C8
Welland Canal ON 20 E7
Wellandport ON 19 C8
Wellandport ON 20 E7
Welland River ON 19 C7-8
Welland River ON 20 E6-7
Wellbore Channel BC 2 B1
Wellburn ON 19 B5
Wellburn ON 20 E4
Wellers Bay National Wilderness Area ON 21 E4-5

Wellesley ON 19 B5-6
Wellesley ON 20 D4-5
Wellesley Lake YT 4 C2
Welling AB 6 F6
Welling AB 7 E5
Wellington BC 2 D3-4
Wellington NS 32 E8
Wellington NS 33 B6
Wellington ON 21 E5
Wellington PE 32 A7
Wellington PE 34 C1
Wellington PE 35 D2
Wellington Bay NU 41 B3
Wellington Bay NU 42 F2
Wellington Channel NU 42 D1-2
Wellman ON 21 D4
Wells BC 1 B8
Wells BC 6 B1
Wells, lac QC 22 E8
Wells, lac QC 24 C2
Wells Cove BC 1 C1
Wells Gray PP BC 6 C1-2
Wells Lake MB 12 C7
Wells Lake MB 15 C2
Wellwood MB 13 E2
Wellwood MB 14 D2
Welsford NB 32 D3
Welsford NB 33 A1
Welsford NB 34 D3
Welshpool NB 32 E2
Welshtown NS 33 E2-3
Welwyn SK 11 E7
Welwyn SK 13 E1
Wembley AB 9 F2
Wemindji QC 40 E4
Wemyss ON 21 C6-7
Wemyss ON 24 F1
Wenasaga River ON 13 D7
Wendake Beach ON 18 F7
Wendake Beach ON 20 A6
Wendigo Lake ON 23 E1
Wendover ON 24 D3
Wenebegon Lake ON 17 E5
Wenebegon River ON 17 E5
Wenebegon River PP ON 17 E4-5
Wenesaga Lake ON 16 C3
Wenesaga River ON 16 C3
Wentworth NS 32 C8
Wentworth NS 34 E1-2
Wentworth Centre NS 32 C8
Wentworth Centre NS 34 D-E2
Wentworth PP NS 32 C8
Wentworth PP NS 34 E2
Wentworth Station NS 32 C8
Wentworth Station NS 33 A6
Wentworth Station NS 34 E2
Wentzel Lake AB 9 B5
Wepusko Bay SK 12 D6
Wepusko Bay SK 15 C-D1
Werner Lake ON 13 E5
Werner Lake ON 14 C8
Werner Lake ON 16 C1
Wernham Lake MB 15 D4
Weslemkoon ON 21 C5
Weslemkoon Lake ON 21 B-C5
Wesley Corner ON 20 B-C7
Wesley Corners ON 21 E1
Wesleyville NL 37 A4
Wesleyville NL 38 F8
Wessonneau, ZEC QC 25 D7
Wessonneau, ZEC QC 28 F2
West Alba NL 34 D6
West Arichat NS 34 E6
West Arm NL 36 A6
West Arm NL 37 A4
West Arm NL 38 F5
West Arm ON 17 F7
West Arm ON 18 A6
West Arm PP BC 6 F3
West Baccaro NS 33 F2-3
Westbank BC 6 E2
West Baptiste AB 9 F5
West Bay NL 39 B7
West Bay NS 34 D6
West Bay NS 34 D6
West Bay ON 18 B3
West Bay ON 18 B4
West Bay ON 18 B7
West Bay ON 18 C3
West Bay, RNP ON 16 D6
West Bay Centre NL 36 B1-2
West Bay PNR ON 16 D6
West Bay Road NS 34 D6
West Bend SK 11 D6
West Berlin NS 33 E4
Westbourne MB 13 E3
Westbourne NB 34 C7-8
West Branch Musquash River NB 32 D3
West Branch River John NS 34 E3
Westbridge BC 6 F2
West Brome QC 24 E8
West Brome QC 27 F3
West Brook NS 32 D7
West Brook NS 33 A4
Westbrook ON 21 D-E6
West Brook Ecological Reserve NL 36 A5
West Brook Ecological Reserve NL 37 A3
West Brook Ecological Reserve NL 38 F4
West Caledonia NS 33 D3

West Cape PE 32 A6-7
West Cape PE 35 C1
West Channel NT 5 B-C4
Westchester Station NS 32 C8
Westchester Station NS 34 E1-2
West Chezzetcook NS 33 C6-7
West Coal River YT 4 D-E7
Westcock NB 35 F1
Westcott AB 7 A3
West Cove AB 8 C5
West Dalhousie NS 32 F5-6
West Dalhousie NS 33 C3
West Devon PE 35 C2
West Ditton QC 27 E5
West Dover NS 32 F8
West Dover NS 33 C-D5
West Elgin ON 19 D4
West Elgin ON 20 F3
West End SK 11 E7
West English River PP ON 16 D4
Westerdale AB 7 A3
Western Arm NL 38 E4
Western Bay NL 37 D7
Western Blue Pond NL 38 D3
Western Brook Pond NL 38 E2
Western Channel ON 18 B5
Western Duck Island ON 18 C1-2
Western Head NL 38 E2
Western Head NS 33 E4
Western Head NS 33 F3
Western Head NS 34 E4
Western Island NL 38 D5
Western Peninsula ON 13 F5
Western Peninsula ON 14 E8
Western Shore NS 32 F7
Western Shore NS 33 C4
Westerose AB 6 B5
Westerose AB 8 D6
Westfield NB 32 D3
Westfield NB 33 A1
Westfield NS 34 D3
West Fiord NT 42 C4
West Flamborough ON 19 B7
West Flamborough ON 20 D-E6
Westgate MB 11 B7
Westgate MB 13 B1
West Gore NS 32 E8
West Gore NS 33 B5
West Gore NS 34 F1
West Grant Island ON 18 B1
West Gravenhurst ON 18 E8
West Gravenhurst ON 21 C1
West Green Harbour NS 33 F3
West Guilford ON 21 B3
West Hants NS 32 E7-8
West Hants NS 33 B5
West Hart River YT 4 F4
West Hart River YT 5 F3
West Hawk Lake MB 13 F5
West Hawk Lake MB 14 E7
West Hawk Lake MB 14 E7
West Hawk Lake MB 16 D1
West Jeddore NS 33 C7
West Kettle River BC 6 E2
West LaHave NS 33 D4
West Lake NL 36 B5
West Lake NL 37 B3
West Lake PP BC 1 A7
West Leicester NS 32 C7
West Leicester NS 34 D1
West Linden NS 35 F2
Westlock AB 6 A5
Westlock AB 8 B6
West Lorne ON 19 D4
West Lorne ON 20 F3
West Mabou Harbour NS 34 C5
West McGillivray ON 19 B4
West McGillivray ON 20 E3
Westmeath ON 22 F6
Westmeath PP ON 22 F6
West Micmac Lake NS 39 A6
West Montreal River ON 17 D6-7
West Montrose ON 19 A6
West Montrose ON 20 D5
Westmoreland PE 32 B8
Westmoreland PE 35 E3-4
Westmount NS 34 C7-8
West New Annan NS 34 E2
West Nipissing ON 17 F7
West Nipissing ON 18 A7
West Northfield NS 32 F7
West Northfield NS 33 D4
Westover ON 19 B7
West Point PE 32 A7
West Point PE 35 C1
West Point PE 35 C1
West Pompey Island NL 39 A-B7
West Pond NL 36 A8
West Pond NL 37 A3

West Pond NL 38 F4
West Poplar River SK 11 F4
Westport NL 38 E4
Westport NL 39 F8
Westport NS 33 D1
Westport ON 21 C7
Westport ON 24 F1
West Prairie River AB 9 F4
West Pubnico NS 33 F2
West Quaco NB 32 D5
West Quaco NB 33 A2
Westray MB 11 B7
Westray MB 13 A1
West Redonda Island BC 2 B2
Westree ON 17 E6
West River NL 38 B4
West River NT 5 B7
West River PE 35 E4
West River Sheet Harbour NS 33 B7-8
West River Sheet Harbour NS 34 F3
West River Station NS 33 A7
West River Station NS 34 E3
West Road (Blackwater) BC 1 B6
West Sackville NB 35 F1
West St. Andrews NS 33 B6-7
West St. Andrews NS 34 F2
West St. Modeste NL 39 D8
West St. Peters PE 34 C3
West St. Peters PE 35 D6
West Sandy Island, RNP ON 17 F7
West Sandy Island, RNP ON 18 A-B7
West Sandy Island PNR ON 17 F7
West Sandy Island PNR ON 18 A-B7
West Shawnigan Lake PP BC 2 F4
West Sheet Harbour NS 33 B7-8
West Sheet Harbour NS 34 F3-4
West Shoal Lake MB 14 D4
West Texada Island PP BC 2 D3
West Thurlow Island BC 1 E5
West Thurlow Island BC 2 B1
West Twin PP BC 1 B8
West Twin PP BC 6 B1
West Vancouver BC 1 F7
West Vancouver BC 2 D5
Westview SK 11 D6
Westville NS 33 A7
Westville NS 34 E3
Westward Ho AB 6 C5
Westward Ho AB 7 A3
Westwold BC 6 E1-2
Wetaskiwin AB 6 B6
Wetaskiwin AB 8 D6
Wetetnagami, lac QC 23 C7
Wetetnagami, lac QC 26 D1-2
Wetetnagami, rivière QC 23 B-C7
Wetetnagami, rivière QC 26 C2-D1
Weyakwin SK 11 A5
Weyakwin SK 12 F5
Weyakwin Lake SK 11 A4
Weyakwin Lake SK 12 F4
Weybridge NL 36 C8
Weybridge NL 37 C6
Weyburn SK 11 F6
Weyburn Lake NT 10 C5
Weymer Creek PP BC 1 E4-5
Weymontachie QC 25 B6
Weymontachie QC 28 D1
Weymouth NS 33 D1-2
Weymouth Falls NS 33 D2
Weymouth North NS 33 D1-2
Whaleboat Island PMP NS 2 E4
Whale Channel BC 1 B3
Whale Cove (Tikirarjuaq) NU 40 A-B2
Whale Cove (Tikirarjuaq) NU 41 E5
Whalen Corners ON 19 B4
Whalen Corners ON 20 D3
Whapmagoostui QC 40 D4
Wharncliffe ON 17 F4-5
Wharncliffe ON 18 A1
Wha Ti NT 10 C5
Whatshan Lake BC 6 E3
Wheatland MB 11 E8
Wheatland MB 13 D1
Wheatland MB 14 D1
Wheatley ON 19 E2
Wheatley PP ON 19 E2
Wheatley River PE 35 D4
Wheeler, rivière QC 40 D6
Wheeler Lake NT 10 C6
Wheeler River SK 12 C5
Whelan SK 6 A8
Whelan SK 11 A2-3
Whim Road PE 35 E-F6
Whipple Point NS 33 D1
Whippoorwill Bay ON 18 E5
Whiskers Point PP BC 3 F7

Whiskey Jack Lake MB 12 B7
Whiskey Jack Lake MB 15 B1
Whisky Lake ON 18 A3
Whitbourne NL 37 E7
Whitby ON 20 C8
Whitby ON 21 F2
Whitchurch-Stouffville ON 20 B-C7
Whitchurch-Stouffville ON 21 E1
Whitchurch ON 20 C7
Whitecap Lake MB 15 D5
Whitechurch ON 20 D3
White City SK 11 E6
Whiteclay Lake ON 16 C6
White Cloud Island ON 18 A-B7
Whitecourt AB 6 A4
Whitecourt AB 8 B3-4
Whitedog ON 14 D8
Whitedog (Wabasseemoong) ON 16 D1
Whitefish ON 17 F6
Whitefish ON 18 A5
Whitefish Bay ON 13 F6
Whitefish Bay ON 13 F6
Whitefish Bay ON 16 D1
Whitefish Falls ON 17 F6
Whitefish Falls ON 18 A4
Whitefish Lake AB 6 A6
Whitefish Lake AB 8 A8
Whitefish Lake AB 9 F6
Whitefish Lake AB 11 A1
Whitefish Lake AB 12 F1
Whitefish Lake MB 15 E6
Whitefish Lake ON 16 F5
Whitefish Lake ON 17 D4
Whitefish Lake ON 17 F6
Whitefish Lake ON 18 D7-8
Whitefish Lake ON 18 A5
Whitefish Lake ON 21 A3
Whitefish Lake PP MB 11 C7
Whitefish Lake PP MB 13 C1
Whitefish River NT 5 E8
Whitefish River YT 10 E-F1
White Fox SK 11 B6
White Goat Provincial Wilderness Area AB 8 F2
White Goat Wilderness Area AB 6 C3-4
White Gull AB 9 F6
White Gull Creek SK 11 A5
White Gull Lake SK 11 A5
Whitehaven Harbour NS 34 E-F6
White Head NL 32 F3
Whitehead NS 34 E6
White Head Island NB 32 F3
Whitehead Pond NL 36 C-D8
Whitehead Pond NL 37 C-D6
White Hill NS 34 A7
Whitehorn Mountain BC 1 B8
Whitehorn Mountain BC 6 B1
Whitehorse YT 4 E4
Whitehorse Wildland PP AB 6 B3
Whitehorse Wildland PP AB 8 D1-2
White Island NU 41 C6
White Lake ON 16 E8
White Lake ON 17 C3
White Lake ON 21 B6
White Lake ON 24 D1
White Lake Peatlands, RNP ON 17 C3
White Lake Peatlands PNR ON 17 C3
White Lake PP BC 6 D2
White Lake PP ON 17 C3
Whitelaw AB 9 F2
White Mud Falls MB 13 E4
White Mud Falls MB 14 C6
Whitemud Falls Wildland PP AB 9 D7

Whitemud Falls Wildland PP AB 12 D2
Whitemud Lake MB 15 F5
Whitemud Lake MB 16 C3
Whitemud River AB 9 D3
White Oak Lake ON 18 A5
White Otter Lake ON 13 F7
White Otter Lake ON 16 E3
White Partridge Lake ON 22 E3-4
White Pass BC 3 A1
White Pass BC 4 F4
White Pelican (Stum Lake) PP BC 1 C7
White Point BC 1 A1
White Point NS 33 B4
White Point NS 34 A7
White Rapids NB 31 D6
White Ridge PP BC 1 F5
White Ridge PP BC 2 C1
White River BC 2 B-C1
White River BC 6 E4
White River BC 7 D2
White River ON 16 F8
White River ON 17 C2-3
White River ON 17 C3
White River YT 4 C1
White River PP BC 1 E5
White River PP BC 2 B-C1
White Rock BC 1 F7
White Rock BC 2 E5
White Rock NS 32 E7
White Rock NS 33 B5
Whitesail Lake BC 1 B4
Whitesail Reach BC 1 A-B5
Whitesand Bay MB 12 D1
Whitesand Bay MB 15 D1
Whitesand PP ON 16 E6
Whitesand River AB 9 A4
Whitesand River AB 11 C-D6 D7
White Sands NS 34 D3-4
White Sands PE 35 F6-7
Whiteshell PP MB 13 E5
Whiteshell PP MB 14 D-E7
Whiteshell PP MB 16 C-D1
Whiteshell River MB 14 D-E7
Whiteshell River MB 16 C-D1
Whiteshore Lake SK 11 C3
Whites Brook NB 29 E5
Whites Brook NB 31 A4
Whites Cove NB 32 C4
Whitestone NS 33 D4
Whitestone Lake ON 17 D7
Whitestone Lake ON 18 A4
Whitestone Lake ON 18 B4
Whitestone River YT 5 E2-3
Whiteswan Lake PP BC 6 E4-5
Whiteswan Lake PP BC 7 D2
White Swan Lakes SK 11 A5
Whitevale ON 20 C7-8
Whitevale ON 21 F2
Whitewater AB 14 F1
Whitewater Lake MB 11 F1-2
Whitewater Lake MB 14 C1
Whitewater Lake MB 14 F1
Whitewater Lake ON 16 C6
Whitewater Lake ON 18 A5
White Water Lily Lake NT 5 E8
Whiteway NL 37 D7
Whitewolf Lake NT 10 A6-7
Whitewood SK 11 E7
Whitford AB 8 B8
Whitford Lake AB 8 B8
Whithorn MB 11 B7
Whithorn MB 13 A1
Whiting River BC 3 C1
Whitkow SK 11 B3-4
Whitla AB 7 D-E8
Whitmire Lake MB 12 A-B7
Whitmore Lake MB 15 B2
Whitney NB 31 D6
Whitney ON 21 A3
Whitney ON 22 F3
Whitney Lakes PP AB 6 A7
Whitney Lakes PP AB 11 A2
Whitson Lake ON 18 A5
Whitworth QC 28 E3
Wholdaia Lake NT 41 E3
Whonnock BC 2 E6
Whycocomagh NS 34 C6
Whycocomagh (IR) NS 34 C6
Whycocomagh PP NS 34 C-D6
Whytewold MB 13 E4
Whytewold MB 14 C4
Wiah Point BC 1 A1
Wiarton ON 18 E5
Wiarton ON 20 A4
Wiashgamic, lac QC 23 D6-7
Wiashgamic, lac QC 26 E1
Wiau Lake AB 9 E7
Wiau Lake AB 12 E2
Wickaninnish Bay BC 2 E1
Wicked Point MB 13 B3

Wickham NB 32 C4
Wickham NB 33 A1
Wickham QC 24 C8
Wickham QC 27 D3
Wicklow NB 31 E3
Wicklow NB 32 A1
Wicklow ON 21 E4
Wicksteed Lake ON 17 E-F8
Wicksteed Lake ON 22 C1
Widdifield Forest PP ON 17 F8
Widdifield Forest PP ON 18 A8
Widewater AB 9 E5
Widigo, lac QC 25 E2
Wiggly Lake BC 6 F8
Wigwam River BC 6 F5
Wigwam River BC 7 F3
Wigwascence Lake ON 13 C8
Wigwascence Lake ON 16 A5-6
Wikwemikong ON 18 C4
Wikwemikonsing ON 18 C4
Wilberforce ON 21 C3
Wilber Lake ON 21 A5
Wilber Lake ON 22 F5
Wilbert SK 6 B8
Wilbert SK 11 B2-3
Wilcox BC 1 E5
Wilcox Lake ON 20 C7
Wilcox Lake ON 21 E1
Wild Bight NL 38 A5
Wild Bight NL 39 F8
Wildcat NS 33 D3-4
Wildcat Hill PWP SK 11 B6-7
Wildfield ON 19 A7
Wildfield ON 20 C6
Wild Goose ON 16 F6
Wildhay River AB 6 A-B3
Wildhay River AB 8 C1
Wild Horse AB 6 F8
Wild Horse AB 11 F2
Wilding Lake NL 36 C4
Wilding Lake NL 37 C2
Wildnest Lake SK 11 D7
Wildnest Lake SK 15 F1
Wildwood AB 6 B4-5
Wildwood AB 8 C5
Wildwood CA ON 19 B5
Wildwood CA ON 20 D4
Wildwood Lake ON 19 B5
Wildwood Lake ON 20 D4
Wilfrid ON 20 B7-8
Wilfrid ON 21 D2
Wilkesport ON 19 D2
Wilkesport ON 20 F1
Wilkie SK 11 C3
Wilkins Lake ON 22 A5
Wilkinson Range YT 4 C4
Wilkins Strait NU 42 B2 C3
Willard Lake ON 13 E6
Willard Lake ON 16 D2
Willbeach NS 33 D4
Willen MB 11 E8
Willen MB 13 E1
Willesden Green AB 8 C5
William, lac QC 27 E5
William A. Switzer PP AB 6 B3
William A. Switzer PP AB 8 C1
William E. de Garthe Memorial PP NS 33 C5 D5-6
William Lake MB 13 A2-3
William Lake PP MB 13 F2
William Lake PP MB 14 F1
William Point SK 9 A7-8
William Point SK 12 A3
William River SK 9 B8
William River SK 12 A3
Williams Beach BC 2 C2
Williamsburg NB 31 F4
Williamsburg NB 32 B2
Williamsburg ON 24 E3
Williamsdale NS 32 C7
Williamsdale NS 34 E1
Williamsford ON 20 A4
Williamstown NB 32 B1
Williamstown ON 24 E4
Williamsport NL 38 C4
Williamsport NL 39 E8
Williams Harbour NL 39 C8
Williams Island, RNP ON 17 A6
Williams Island PNR ON 17 A6
Williams Lake BC 1 C7
Williams Lake BC 6 D3
Williamson PP AB 9 A8
Willingdon AB 6 A6-7
Willingdon AB 8 B8
Willingdon, Mount AB 6 C4
Willingdon, Mount AB 7 A1
Willis Island NL 37 B6-7
Williston Lake BC 3 E6

Willisville ON 17 F6
Willisville ON 18 B4
Willmar SK 11 F7
Willmore Wilderness PP AB 6 A-B2
Willow Beach ON 19 E1
Willowbrook SK 11 D6-7
Willow Bunch SK 11 F5
Willow Bunch Lake SK 11 F5
Willow Creek AB 6 E5-6
Willow Creek AB 7 D-E4
Willow Creek ON 20 B3
Willow Creek SK 11 F2
Willow Creek PP AB 6 E5-6
Willow Creek PP AB 7 D4
Willowdale AB 11 B4
Willowdale NS 34 E4
Willow Grove ON 19 C7
Willow Lake NT 10 D4
Willowlake River NT 10 C3-4
Willow River BC 1 B7
Willow River BC 1 A7
Willow Run AB 8 F6
Willows SK 11 F5
Willowview MB 14 C4
Wilmer BC 6 E4
Wilmer BC 7 C1
Wilmot NS 32 E6
Wilmot NS 33 B3
Wilmot PE 32 A8
Wilmot PE 35 C1
Wilmot River PE 35 D3
Wilmot Valley PE 35 D3
Wilno ON 21 A4
Wilno ON 22 F4-5
Wilson, lac QC 23 B7
Wilson, lac QC 26 C1
Wilson, Mount BC 4 C6
Wilson Creek BC 2 D4
Wilson Island NT 10 D7
Wilson Island ON 16 F7
Wilson Island ON 17 C1
Wilson Lake NL 39 B-C4
Wilson Lake ON 18 C6-7
Wilson Landing BC 6 E2
Wilson River MB 11 D8
Wilson River MB 13 D2
Wilsons Beach NB 32 E2
Wilsonville ON 19 C6
Wilsonville ON 20 E5
Wilstead ON 21 D7
Wilton ON 21 D6
Wiltondale NL 36 A3
Wiltondale NL 37 A1
Wiltondale NL 38 F2-3
Wiltondale NL 39 F7
Wimapedi Lake MB 12 E-F8
Wimapedi Lake MB 14 A1-2
Wimapedi River MB 12 E8
Wimapedi River MB 15 E2-3
Wimborne AB 6 C6
Wimmer SK 11 C5-6
Winagami Lake AB 9 E4
Winagami Lake, Lake AB 9 E4
Winagami Wildland PP AB 9 E4
Winawiash, lac QC 22 A4
Winchell, lac QC 22 A8
Winchell, lac QC 25 C2
Winchester ON 24 E3
Winchester Cape PE 35 C2-3
Winchester Springs ON 24 E3
Windermere BC 6 E4
Windermere BC 7 C1
Windermere ON 18 D8
Windermere ON 21 B1
Windermere Lake BC 7 C1
Windermere Lake ON 17 D4
Windermere Lake PP BC 6 E4
Windermere Lake PP BC 7 C1
Windfall Lake ON 18 C5
Windflower Lake NT 10 C4
Windham Centre ON 19 C6
Windham Centre ON 20 E5
Windigo QC 25 B6
Windigo QC 28 E2
Windigo, lac QC 22 C8
Windigo, lac QC 25 B6
Windigo, rivière QC 25 B6
Windigo, rivière QC 28 D2
Windigo Bay, RNP ON 17 A6
Windigo Bay PNR ON 17 A6
Windigo Lake ON 13 B-C7
Windigo Lake ON 16 A4
Windigo Point PP ON 13 E6-7
Windigo Point PP ON 16 D6
Windigo River ON 13 B7
Windsor NB 32 A1-2 B1-2
Windsor NS 32 E7
Windsor NS 33 B5
Windsor ON 19 E1
Windsor QC 27 E4
Windsor Forks NS 32 E7
Windsor Forks NS 33 B5
Windsor Lake ON 16 A7
Windthorst SK 11 E6-7
Windward Sands ON 21 D3
Windy Arm YT 4 E4

Windygates MB 13 F3
Windygates MB 14 F3-4
Windy Lake MB 15 F5
Windy Lake ON 18 A4
Windy Lake SK 11 A7
Windy Lake SK 12 F7
Windy Lake SK 13 A1
Windy Lake SK 15 F1
Windy Lake PP ON 17 F6
Winefred Lake AB 9 E7
Winefred Lake AB 15 F1 (?)
Winefred River AB 9 E7
Wine Harbour NS 34 F5
Winfield AB 6 B5
Winfield AB 8 D5
Winfield BC 6 E2
Wingard SK 11 B4
Wingdam BC 1 B7-8
Winger ON 19 C8
Winger ON 20 E7
Wingham ON 19 A4-5
Wingham ON 20 C3-4
Wing Pond NL 36 A8
Wing Pond NL 37 A6
Wings Point NL 36 A7
Wings Point NL 37 A5
Wings Point NL 38 F6-7
Winisk Lake ON 40 E2
Winisk River ON 40 E2
Winkler MB 13 F3
Winkler MB 14 F4
Winlaw BC 6 F3
Winnange Lake PP ON 13 F5
Winneway QC 17 D8
Winneway QC 22 A3
Winneway, lac QC 22 A3
Winnifred AB 7 D7
Winnipeg MB 13 E3-4
Winnipeg MB 14 C5
Winnipeg, lac MB 13 B-C3
Winnipeg, lac MB 14 B5-6 C5-6
Winnipeg, Lake MB 13 B3
Winnipeg, Lake MB 14 B5-6
Winnipeg, rivière MB 13 E5
Winnipeg, rivière MB 14 C-D7
Winnipeg, rivière MB 16 C1
Winnipeg Beach MB 13 E4
Winnipeg Beach MB 14 C5
Winnipeg Beach PP MB 13 E4
Winnipeg Beach PP MB 14 C5
Winnipegosis MB 11 C5
Winnipegosis MB 13 C-D2
Winnipegosis MB 14 A1-2
Winnipegosis, lac MB 11 B-C8
Winnipegosis, lac MB 13 B-C2
Winnipegosis, lac MB 14 A1-2
Winnipegosis, Lake MB 11 B8
Winnipegosis, Lake MB 13 B2
Winnipegosis, Lake MB 14 A1-2
Winnipeg River MB 13 E5
Winnipeg River MB 14 C-D7
Winnipeg River MB 16 C1
Winokapau Lake NL 39 C4
Winona ON 19 C7-8
Winona ON 20 E6-7
Winslow North PE 35 D4-5
Wintego Lake SK 12 C6
Wintego Lake SK 15 E1
Winter SK 11 B-C2
Winterbourne ON 19 A-B6
Winterbourne ON 20 D5
Winter Cove Unit (Gulf Islands NPR) BC 2 E3-4
Winterhouse NL 36 B1
Winter Harbour BC 1 E3-4
Winterhouse Brook NL 36 A3
Winter House Brook NL 37 A1
Winter House Brook NL 38 F2
Wintering Lake MB 15 E3-4
Wintering Lake ON 13 E7
Wintering Lake ON 16 E1
Wintering Lake ON 17 F6
Winter Island NU 41 C6
Winter Lake NT 10 A7
Winterland NL 36 E6
Winterland NL 37 E4
Winter River NT 10 A6-7
Winterton NL 37 D7
Winthrop ON 19 A4
Winthrop ON 20 D3
Winton Creek ON 21 D-E6
Wire Cache PP BC 6 C-D2
Wirral NB 32 D3
Wirral NB 33 A1
Wiseman's Corners ON 18 C8
Wiseman's Corners ON 21 A1
Wiseton SK 11 D3
Wishart SK 11 D6
Wishart Point NB 30 F2
Wishart Point NB 31 C7-2
Wisla MB 14 C1
Wistaria BC 1 A5
Witchai Lake MB 15 E4
Witchekan Lake SK 11 B3-4

Witchekan Lake SK 11 B4
Witch Lake BC 3 F6
Witegoo River ON 15 F8
Withers Lake ON 13 A7
Withers Lake ON 15 F7
Withrow AB 8 E5
Witless Bay NL 37 E8
Witless Bay Ecological
Reserve NL 37 E8
Wittenburg NS 33 B6-7
Wittenburg NS 34 F2
Wivenhoe MB 15 E5
Wiwa Hill SK 11 E4
W.J.B. Greenwood PP ON
17 E7
W.J.B. Greenwood PP ON
22 A1
Woburn QC 27 E6
Woking AB 9 E2
Wolf Bay QC 39 E-F6
Wolf Creek AB 6 B4
Wolf Creek AB 8 C3
Wolf Creek AB 8 C3
Wolfe SK 11 C3
Wolf Island ON 21 E7
Wolfe Lake ON 21 C6
Wolf Island PP ON 21 D3
Wolf Lake AB 9 F7
Wolf Lake AB 12 F2
Wolf Lake NL 36 C4-5
Wolf Lake NL 37 C2-3
Wolf Lake ON 21 C3
Wolf Lake YT 4 E5
Wolf River AB 9 F7
Wolf River AB 12 F2
Wolf River BC 2 C1
Wolf River ON 18 B7
Wolf River ON 4 E5
Wolf River YT 4 E5
Wolfville NS 32 E7
Wolfville NS 33 B4-5
Wôlinak QC 24 B8
Wôlinak QC 25 F8
Wôlinak QC 27 C3-4
Wollaston, Cape NT 41 A2
Wollaston, Cape NT 42 E1
Wollaston, lac SK 12 C6
Wollaston, lac SK 15 C1
Wollaston, lac SK 41 F3
Wollaston Lake ON 21 C4
Wollaston Lake SK 12 B6
Wollaston Lake SK 12 C6
Wollaston Lake SK 15 B1
Wollaston Lake SK
15 B-C1
Wollaston Lake SK 41 F3
Wollaston Lake SK 41 F3
Wollaston Lake Landing SK
12 C5-6

Wollaston Peninsula NU
41 B2
Wollaston Peninsula NU
42 E1
Wolseley SK 11 E6
Wolseley Bay ON 17 F7
Wolseley Bay ON 18 B7
Wolseley Bay ON 18 B7
Wolseley River ON 18 B6
Wolsey, Lake ON 18 B-C2
Wolverine Lake NT 10 C7-8
Wolverine Lake YT 4 D6
Wolverine River AB 9 C4
Wolverine River MB
15 A-B4
Wolverine River NT 5 C6
Wolverton ON 19 B6
Wolverton ON 20 D5
Woman Lake ON 13 D6
Woman Lake ON 16 B3
Woman River Forest PP ON
17 E5
Wonowon BC 3 D7
Wonowon BC 9 D1
Wood, Mount YT 4 D1
Wood Bay NT 5 A6
Woodbridge ON 19 A7-8
Woodbridge ON 20 C6-7
Wood Buffalo AB 9 A6
Wood Buffalo AB 9 B6
Wood Buffalo AB 9 C-D7
Wood Buffalo AB 10 F7
Wood Buffalo AB 12 A1
Wood Buffalo AB 12 C2
Wood Buffalo, PN AB
9 B-C6
Wood Buffalo, PN AB 10 F7
Wood Buffalo, PN AB 12 B1
Wood Buffalo NP AB 9 A5-6
Wood Buffalo NP AB 12 A1
Wood Buffalo NP AB 41 E1
Wood Buffalo NP NT 10 F7
Wood Buffalo NP Fort
Chipewyan Visitors
Centre AB 9 B6-7
Wood Buffalo NP Fort
Chipewyan Visitors
Centre AB 12 B2
Wood Buffalo NP Fort Smith
Visitors Centre AB
9 A6-7
Wood Buffalo NP Fort Smith
Visitors Centre AB
12 A2
Woodburn ON 19 C7
Woodburn ON 20 E6
Woodfibre BC 1 F6-7
Woodfibre BC 2 D4

Woodford ON 18 F5-6
Woodford ON 20 A4
Woodham ON 19 B4
Woodham ON 20 D3
Woodhouse AB 7 D4-5
Woodington ON 18 D8
Woodington ON 21 B1
Wood Islands PE 34 D3
Wood Islands PE 35 F6
Wood Islands PE 35 F6
Wood Islands PP PE 34 D3
Wood Islands PP PE 35 F6
Wood Lake MB 15 C3
Wood Lake ON 21 B-C2
Wood Lake SK 12 E-F6
Woodland Beach ON 18 F7
Woodland Beach ON
20 A6
Woodland Caribou PP ON
13 D5
Woodland Caribou PP ON
14 B8
Woodland Caribou PP ON
16 B-C1
Woodlands MB 13 E3
Woodlands MB 14 D4
Woodley SK 11 F6-7
Woodmore MB 14 F5
Wood Mountain SK 11 F4
Wood Mountain (IR) SK
11 F4
Wood Mountain Post
Provincial Historic Park
SK 11 F4
Woodnorth MB 11 E8
Woodnorth MB 13 E1
Wood Point NB 32 C7
Woodridge MB 13 F4-5
Woodridge MB 14 F6
Wood River SK 11 E-F4
Woodrow SK 11 F4
Woods, Lake of the ON
13 F5-6
Woods, Lake of the ON
14 E-F8
Woods, Lake of the ON
16 D1
Woods, Point of the MB
15 B5
Woods Bay ON 18 D7
Woods Harbour NS 33 F2
Woodside MB 13 E2-3
Woodside MB 14 D3
Woodside NB 35 E2
Woodside, LHN ON 19 B6
Woodside, LHN ON 20 D5
Woodside NHS ON 19 B6
Woodside NHS ON 20 D5

Woodside River YT 4 C6
Woods Island NL 36 A2-3
Woods Island NL 38 F2
Woods Lake NL 39 A3
Woodstock NB 31 F3
Woodstock NB 32 B1
Woodstock NL 38 E5
Woodstock ON 19 C5
Woodstock ON 20 E4
Woodstock (IR) NB 31 F3
Woodstock (IR) NB 32 B1
Woodview ON 21 D3-4
Woodville ON 20 A8
Woodville ON 21 D2
Woodville Mills PE 35 E7
Woody Island NL 36 D8
Woody Island NL 37 D6
Woody Lake SK 12 E-F6
Woody Point NL 38 F2
Woody Point NL 39 F7
Woody River SK 11 C7
Woody River SK 13 C1
Wooler ON 21 E4
Woolford PP AB 6 F6
Woolford PP AB 7 F5
Woosey Lake MB 12 F8
Woosey Lake MB 15 F2
Wopmay Lake NT 10 A5
Wopmay River NT 10 A5
Worby MB 14 E2
Worcester SK 11 E-F6
Wordsworth SK 11 F7
Work Channel BC 3 F2-3
Worsley AB 9 D2
Worthington ON 17 F6
Worthington ON 18 A4
Woss BC 1 E4
Woss Lake BC 1 E4
Woss Lake PP BC 1 E4-5
Wostok AB 8 B7-8
Wostok AB 11 A1
Wotton QC 27 D5
Wottonville QC 27 D5
Woyenne BC 1 A5
Wreck Cove NL 36 D1
Wreck Cove NL 36 E6
Wreck Cove NL 37 E4
Wreck Cove NS 34 B7
Wreck Cove Lakes NS
34 B7
Wrentham AB 6 F7
Wrentham AB 7 E6
Wright QC 22 E7
Wright QC 24 B1-2
Wright Bay NT 5 A7-8
Wright Lake ON 16 B4

Wrigley NT 10 C2
Wrigley Creek NT 4 C-D8
Wrigley Creek NT 10 D1
Wrigley Lake NT 4 A-B8
Wrigley Lake NT 10 B1
Wrigley River NT 10 C2
Writing-On-Stone PP AB
6 F7
Writing-On-Stone PP AB
7 F7
Writing-On-Stone PP AB
11 F1
Wrong Lake MB 13 C4
Wroxeter ON 19 A5
Wroxeter ON 20 C4
Wroxton SK 11 D7
Wroxton SK 13 D1
Wunnummin Lake ON
40 E1
Wuskwatim Lake MB
15 E3
Wycliffe BC 6 F4
Wycliffe BC 7 E1
Wyebridge ON 18 F7
Wyebridge ON 20 A6
Wyecliff AB 8 C7
Wyevale ON 18 F7
Wyevale ON 20 A6
Wylie Lake AB 9 A7
Wylie Lake AB 12 A2
Wymark SK 11 E3
Wymbolwood Beach ON
18 F7
Wymbolwood Beach ON
20 A6
Wyndham-Carseland PP AB
6 D6
Wyndham-Carseland PP AB
7 C5
Wynndel BC 6 F4
Wynniatt Bay NT 41 A3
Wynniatt Bay NT 42 D-E2
Wynyard SK 11 C-D6
Wyoming ON 19 C3
Wyoming ON 20 E2
Wyse Corner NS 33 B6
Wyse Corner NS 34 F2
Wyvern NS 33 A5
Wyvern NS 34 E1

X Y____

Xhlawit (Alder Peak) BC
3 F3
Yahk BC 6 F4

Yahk BC 7 F1
Yahk Mountain BC 6 F4
Yahk Mountain BC 7 F2
Yahk PP BC 6 F4
Yahk PP BC 7 F1
Yale BC 1 F8
Yale BC 2 D7
Yamachiche QC 24 B7
Yamachiche QC 25 F7
Yamachiche QC 27 C2-3
Yamaska QC 24 C7
Yamaska QC 27 D2-3
Yamaska, parc de la QC
24 D8
Yamaska, parc de la QC
27 E3
Yamaska, rivière QC
24 C-D7
Yamaska, rivière QC
27 D2-3
Yamaska-Est QC 24 C7
Yamaska-Est QC 27 D2-3
Yamba Lake NT 10 A7-8
Yankeetown NS 32 F8
Yankeetown NS 33 C5
Yarbo SK 11 E7
Yarbo SK 13 D1
Yard Creek PP BC 6 D2
Yarker ON 21 D6
Yarmouth NS 33 E1-2
Yarmouth NS 33 E1
Yarrow BC 2 E6
Yates AB 8 C3
Yates River AB 9 A4
Yathkyed Lake NU 40 A1
Yathkyed Lake NU 41 E4
Yatton ON 19 A6
Yatton ON 20 C-D5
Yehiniko Creek BC 3 C2
Yellow Creek SK 11 B5
Yellow Grass SK 11 F5-6
Yellowhead BC 6 B2-3
Yellowhead Pass AB 6 B3
Yellowhead Pass NHS AB
6 B2-3
Yellowknife NT 10 D6
Yellowknife NT 41 D1
Yellowknife Bay NT
10 D6
Yellowknife River NT
10 B-C6
Yellowknife River NT
41 D2
Yellowquill SK 11 C6
Yeltea Lake NT 5 D6
Yelverton Bay NU 42 A4
Yeoford AB 8 D5

Yeo Island BC 1 C4
Yeo Island ON 18 D4
Ymir BC 6 F3
Yoho, PN BC 6 D4
Yoho, PN BC 7 B1
Yoho NP BC 6 D4
Yoho NP BC 7 A1
Yolakom River BC 1 D7
Yolakom River BC 2 A5-6
Yonge Mills ON 21 D8
Yonker SK 11 B2
York ON 19 A8
York ON 19 C7
York ON 20 C7
York ON 20 E6
York ON 21 F1
York PE 34 C2
York PE 35 E5
York, lac QC 29 B8
York, lac QC 30 B1
York, rivière QC 30 B1-2
York-Baillargeon, ZEC QC
30 B2
York Centre QC 30 B2
York Factory MB 15 D6
York Factory, LHN MB
15 D6-7
York Factory, LHN MB
40 C1
York Factory NHS MB
15 D6-7
York Factory NHS MB
40 C1
York Harbour NL 36 A2
York Harbour NL 38 F1
York Landing MB 15 E4
York Mills NB 32 C2
York Point PE 35 E4-5
York River ON 21 B4
Yorkton SK 11 D7
Youbou BC 2 E3
Youghall NB 29 E8
Youghall NB 31 B6
Young SK 11 C5
Young Lake SK 12 A5
Youngs Cove NB 32 C4
Young's Cove NS 32 E5
Young's Cove NS 33 C2
Youngs Cove Road NB
32 C4
Youngs Point ON 21 D3
Young's Point PP ON 9 F3
Youngstown AB 6 D7
Youngstown AB 7 A7-8
Youngstown AB 11 C-D1
Youngstown ON 21 D3

Yser, lac QC 23 F6
Yukon, fleuve YT 5 F1-2
Yukon/Northwest
Territories Border TP YT
5 D3-4
Yukon River YT 4 B-C2
Yukon River YT 4 C-D4
Yuquot BC 1 F4
Yusezyu River YT 4 C-D6

Z____

Zakwaski Mountain BC
1 E8
Zakwaski Mountain BC
2 C7
Zakwaski Mountain BC
6 E1
Zama City AB 9 A-B3
Zama Lake AB 9 B2-3
Zangeza Bay SK 12 C7
Zangeza Bay SK 15 B-C1
Zawale AB 8 B7-8
Zayas Island BC 3 F2
Zbaraz MB 13 D3-4
Zbaraz MB 14 B4-5
Zealand NB 31 F4
Zealand NB 32 B2
Zealandia SK 11 D3
Zeballos BC 1 E4
Zebulon Lake NT 10 A5
Zebulon River NT 10 A5
Zed Lake MB 12 D7
Zed Lake MB 15 D1-2
Zed Lake PP MB 12 D7
Zed Lake PP MB 15 D1-2
Zehner SK 11 E5
Zelana MB 14 A1
Zelma SK 11 C5
Zeneta SK 11 D-E7
Zengle Lake SK 12 C6-7
Zengle Lake SK 15 C1
Zenon Park SK 11 B6
Zephyr ON 20 B7
Zephyr ON 21 E1
Zeta Lake NU 41 B3
Zeta Lake NU 42 E2-2
Zhoda MB 13 F4
Zhoda MB 14 F6
Zinto Lake NT 10 B5
Zionz Lake ON 13 D7
Zionz Lake ON 16 B4
Zoria MB 14 A1
Zurich ON 19 B4
Zurich ON 20 D3

CREDITS AND ACKNOWLEDGMENTS

PHOTOS AND ILLUSTRATIONS

2–3 Gunter Marx/Corbis/Magmaphoto.com; **4** Corbis/Magmaphoto.com; **6–7** Dale Wilson; **8–9** Pat Morrow/Firstlight.ca; **9** WorldSat; **10** Dimension DPR; Mike Grandmaison; Barrett & MacKay; **10–11** WorldSat; **11** S. Fick/Canadian Geographic Magazine; Corbis/Magmaphoto.com; RD; Roy Tanami/Ursus; RD (2); Mike Grandmaison; Bill Ivy/Ivy Images; Mike Grandmaison; **12** Brian Milne; Dimension DPR (*centre*); S. Fick/Canadian Geographic Magazine (*left*); PhotoDisc (*background*); Digital Vision/Picture Quest; Firstlight.ca (2); Mike Grandmaison; Firstlight.ca; **12–13** John de Visser; **13** WorldSat; Geovisuals; Dimension DPR; Yves Lachance; B.C. Hydro; **14** Dimension DPR; S. Fick/Canadian Geographic Magazine (2); Québec Amérique International; Jacques Perrault (*left*); **14–15** Vision Quest; **15** S. Fick/Canadian Geographic Magazine (2); WorldSat; Calgary Transit; Luc Vidal/Ponopresse; Skyart; Gerald J. Blackmore; **16** Robert McCaw; Christian Bucher/Parks Canada; Gunter Marx; Bob Herger; Mike Grandmaison; Canadian Geographic Magazine (*left*); Rob Curtis/Vireo; Brian Milne; Ivy Images; Barrett & MacKay; **16–17** Dimension DPR; **17** Barrett & MacKay; The Slide Farm; Ivy Images; Barrett & MacKay; **18** W. Lynch/Ivy Images; Dimension DPR (*centre*); W. Lynch/Ivy Images; RD (*bottom*); **18–19** Doug Plummer/Superstock; **19** Dimension DPR; Jacques Perrault; Exxon Mobil Canada Ltd.; **20** RD (3, *left*); Dimension DPR (*map*); WorldSat (*globe*); Québec Amérique International; Ursus Photography; **20–21** Gunter Marx; **21** Québec Amérique International; RD (5); Gunter Marx (*centre*); Les Bazso; **22** RD (3); Dimension DPR (*maps*); Jacques Perrault; Potash Corp. of Saskatchewan; **22–23** T. Kitchen/Firstlight.ca; **23** Todd Korol; Jacques Perrault (*charts*); RD (2); Dimension DPR (*map*); Todd Korol; **24** RD (4); WorldSat (*globe*); Dimension DPR (*maps*); Natural Resources Canada; Jacques Perrault; **24–25** Mark Burnham/Firstlight.ca; **25** Québec Amérique International; Geovisuals; Jacques Perrault; O. Samson-Arcand–I. Clement/Alpha Presse; **26** Robert McCaw; Dimension DPR; Canadian Geographic Magazine (*centre*); Jacques Perrault; **26–27** WorldSat (*map*); Paul G. Adam Publiphoto; **27** Jacques Perrault; Lou Wise/Aerographic; WorldSat; Jacques Perrault; **28** RD (5); Dimension DPR (*maps*); Jacques Perrault (*chart*); D. Daigle/Ivy Images; **28–29** Barrett & MacKay; **29** Barrett & MacKay; WorldSat; Jamie Roach/Sunnyside Media; HMDC (2); Corbis/Magmaphoto.com; **30** Dimension DPR; RD; Royal BC Museum PN158 & PN296; J. Mathan Blair/Corbis/Magmaphoto.com; RD; Harry Foster/Canadian Museum of Civilization RbJu-1:767; **30–31** Dimension DPR; **31** Ingalik Indian Spirit Mask, photograph courtesy of the Smithsonian National Museum of Natural History; Glenbow Museum A3372a; Museum of the American Indian, Heye Foundation; Canadian Museum of Civilization PgHq1:1 S2004-653; **32** Dimension DPR; Manitoba Archives; Glenbow NA-2798-6; National Archives C-036-146; musée des Arts décoratifs, Paris; Manitoba Archives; **33** Barrett & MacKay; Jacques Perrault; Provincial Archives of Nova Scotia; Immigration Canada; Royal BC Museum; Royal Ontario Museum 955.217.15; **34–35** Dimension DPR (*maps*); PhotoDisc; Symbols of Canada; **36** NASA; **37** Yves Lachance; Dimension DPR; Jacques Perrault; Comstock; Publiphoto; Getty Images; Rocky Mountaineer Railtours; **38** Firstlight.ca; Stockbyte; Index Stock/Maxx Images; Firstlight.ca (2); **39** Jacques Perrault; Corbis/Firstlight.ca; WorldSat; Stockbyte; Jacques Perrault (*charts*); Comstock; **40** WorldSat; Corbis/Magmaphoto.com (3); Barrett & MacKay (*bottom*); **41** Publiphoto; **42** Anne-Marie Weber/Taxi/Getty Images; General Motors; Getty Images; **43** AP; Arthur Erickson; Barry Blackmore/Taxi/Getty Images; André de Chastenet/Le Figaro Magazine; Variel Alt/Corbis/Magmaphoto.com (*eye*); Imtek Imagineering Inc./Corbis/Magmaphoto.com; **44–45** Bob Herger; **134–135** Corbis/Firstlight.ca; **150–151** © First Light/Corbis/Magmaphoto.com.

CHARTS AND MAPS

Statistics Canada is the source of information reproduced/adapted in these pages: **23** Oil and natural gas production; **24** Mining output; **29** Atlantic crops; **33** Ten top countries of birth; **36–37** Canada online, Container cargo by commodity, Canada's busiest airports; **38–39** Natural increase, Languages, Religions, Levels of educational attainment, Top fields of study, Marriage, Divorce, Household size, Where we work, Rates of disease, Life expectancy, Age pyramids; **40–41** Urban and rural population.

Statistics Canada information is used with the permission of the Ministry of Industry, as Minister responsible for Statistics Canada. Information on the availability of the wide range of data from Statistics Canada can be obtained from Statistics Canada's Regional Offices, its World Wide Web site at http://www.statcan.ca, and its toll-free access number 1-800-263-1136.

Thematics base map data: Canada Base Map Series, Natural Resources Canada.

"NATURAL BALANCE" SIDEBARS

The Publisher also wishes to thank the following individuals for their assistance in the preparation of the "Natural Balance" sidebars: Richard Bailey, Fisheries and Oceans Canada, Kamloops, British Columbia; Marc Choma, Canadian Wireless Telecommunications Association, Ottawa; Peter Kingsmill, Redberry Lake Pelican Reserve, Saskatchewan; Tina McCaffrey, Land Reclamation Program, City of Sudbury; Cory McPhee, Inco Limited, Sudbury, Ontario; Joshua Matlaw, Earthroots, Toronto; Stephanie Meakin, Inuit Circumpolar Conference, Ottawa; Mike Peters, Inco Limited, Sudbury, Ontario; Maureen Reed, University of Saskatchewan, Saskatoon, Saskatchewan; Barb Scott, Fundy Model Forest, Sussex, New Brunswick; Jim Shevchuk, Principal, Hafford Central School, Hafford, Saskatchewan; Leslie Welsh, Sustainable Energy Section, Environment Canada, Ottawa.

BOOKS AND PAPERS

The editors acknowledge their indebtedness to the following books, papers, and other materials that were consulted as reference and illustration sources:

Across This Land: A Regional Geography of the United States and Canada, John C. Hudson, John Hopkins University Press, 2002; *Canada: A People's History,* Vol. I, Don Gillmor & Pierre Turgeon, McClelland & Stewart Ltd., 2000; *Canada: A People's History,* Vol. II, Don Gillmor, Achille Menaud & Pierre Turgeon, McClelland & Stewart Ltd., 2001; *Canada: A Regional Geography,* John Warkentin, Prentice-Hall Canada, 1997; *Canada: Land of Diversity,* Bruce Clark, John K. Wallace, Prentice-Hall Canada, 1995; *Canada's Cold Environments,* Hugh M. French and Olav Slaymaker (editors), McGill-Queen's University Press, 1993; *The Canada We Want: Competing Visions for the New Millenium,* John F. Godfrey with Rob McLean, Stoddart, 1999; *The Canadian Encyclopedia,* McClelland & Stewart Ltd., 2000; *Canadian Geographic Quiz Book,* Douglas MacLean, Fitzhenry and Whiteside, 1999; *Canadian Global Almanac,* John Wiley and Sons, 2003; *Canadian Oxford World Atlas* (4th edition), Quentin H. Stanford (editor), Oxford University Press, 1998; *The Climates of Canada,* David Phillips, Environment Canada, 1990; "Competing on Creativity: Placing Ontario's Cities in North American Context—A report for the Ontario Ministry of Enterprise, Opportunity and Innovation and the Institute for Competitiveness and Prosperity," Meric S. Gertler, Gary Gates, Richard Florida, Tara Vinodrai, 2002; *Comprendre la terre,* Éditions Québec Amérique inc., 2001; *Comprendre le climat et l'environnement,* Éditions Québec Amérique inc.; "Economic Benefits of British Columbia's Provincial Parks," British Columbia Ministry of Water, Land and Air Protection, 2001; *Facts at Your Fingertips,* The Reader's Digest Association (Canada) Ltd., 2003; *The First Canadians,* Fraser Symington, Natural Science of Canada, 1970; *The Fitzhenry and Whiteside Book of Canadian Facts and Dates,* Jan Myers (revised and updated by Larry Hoffman, Fraser Sutherland), Fitzhenry & Whiteside, 1991; *Historical Atlas of Canada: Canada's History Illustrated with Original Maps,* Derek Hayes, Douglas & Mcintyre, 2002; *Historical Atlas of Canada: From the Beginning to 1800,* Vol. I, R. Cole Harris (editor), Geoffrey J. Matthews (cartographer/designer), University of Toronto Press, 1987; *Historical Atlas of Canada: The Land Transformed, 1800–1891,* Vol. II, R. Louis Gentilcore (editor), Geoffrey J. Matthews (cartographer/designer), University of Toronto Press, 1993; *Historical Atlas of Canada: Addressing the Twentieth Century,* Vol. III, Donald Kerr, Deryck W. Holdsworth, Susan L. Laskin (editor), Geoffrey J. Matthews (cartographer/designer), University of Toronto Press, 1990; "Human Activity and the Environment," Statistics Canada, 2003; *The Illustrated Natural History of Canada* series, Natural Science of Canada Ltd., 1970; *Life in 2030: Exploring a Sustainable Future for Canada,* John B. Robinson et al., University of British Columbia Press, 1996; "Metropoles and Peripheries: The Evolution of City-Regions in Contemporary Canada—Research findings from a study supported by the Canadian Centre for Community Renewal," Paul B. Reed; *Native Trees of Canada,* R. C. Hosie, Fitzhenry & Whiteside Ltd., 1979; *The Natural Landscapes of Canada: A Study in Regional Earth Science,* J. Brian Bird, J. Wiley and Sons Canada, 1981; "Quality of Life in Canada: A Citizens' Report Card," Joseph H. Michalski, Canadian Policy Research Networks, 2002; *Reaching North,* Jamie Bastedo, Red Deer College Press, 1998; *The Regional Geography of Canada,* Robert M. Bone, Oxford University Press, 2002; *Rethinking the Future,* Patricia Elliott (editor), Fifth House, 1993; *Shield Country,* Jamie Bastedo, Arctic Institute of North America of the University of Calgary, 1994; *The State of Canada's Environment—1996,* Minister of Public Works and Government Services; *Sustainable Development of Canada: National and International Perspectives,* O. P. Dwivedi, Broadview Press, 2001; *Through Indian Eyes,* The Reader's Digest Association (Canada) Ltd., 1996; *2000 Fact Book,* British Columbia Ministry of Forests; "Urban Sustainability—Research Perspectives," Don Drummond, Strategies for Urban Sustainability Conference, Edmonton, Alberta, Sept. 9, 2003; *The Weather,* John Lynch, Firefly, 2002.

Endsheets
The spectacular northern lights, or aurora borealis, *occur in night skies over Canada's Arctic and subarctic regions.* (Photo: Mike Grandmaison.)

Printed and bound in Canada by Transcontinental/Imprimerie Interglobe Inc.